Clinical Handbook of Air Pollution-Related Diseases

Fabio Capello • Antonio Vittorino Gaddi
Editors

Clinical Handbook of Air Pollution-Related Diseases

Springer

Editors
Fabio Capello
Alder Hey Children's Hospital
Liverpool
United Kingdom

Antonio Vittorino Gaddi
Scientific Director of Health Lab
of GTechnology Foundation
Modena
Italy

ISBN 978-3-030-09680-9 ISBN 978-3-319-62731-1 (eBook)
https://doi.org/10.1007/978-3-319-62731-1

Softcover re-print of the Hardcover 1st edition 2018

Printed on acid-free paper

This Springer imprint is published by Springer Nature
The registered company is Springer International Publishing AG
The registered company address is: Gewerbestrasse 11, 6330 Cham, Switzerland

Foreword 1

Only Knowledge Can Defeat Pollution

We are certainly living in a world that is rapidly changing and, in fact, I am writing this article closed up in my home in the north-east of Italy because of the incredibly high temperatures, consistently more than 40°C, that are burning our land in this summer of 2017.

From a certain point of view, "global warming" is another big challenge/battle that humankind has to face, as pollution can be correctly considered "global pollution." But as we very well know, not everybody believes that global warming is a real issue and, similarly, not everybody believes that global pollution is a real challenge. Only correctly informed people can adequately face the challenges of life; consequently, only consistently informed people can face the challenges of pollution. Only knowledge, which means education, can fight pollution's threat.

I was very pleased to be asked to write this short introduction because this book will help people to understand and evaluate the challenges that pollution and, more specifically, air pollution represent for humankind. The challenge is growing every day, not just in a linear way but exponentially, so it is necessary to start fighting pollution immediately. What does it mean to fight pollution and how can this be managed? These are the questions that I focus on in my short introduction.

First of all, pollution has to be considered a "global issue." Thus, only a global and collective approach can hope to be effective. A global and collective approach means that nations should cooperate, which requires a strong and global political commitment. Global, continental, national, and local Institutions need to cooperate to assure that everyone can manage their own and specific part in the full battle.

The most important global organizations (United Nations, World Health Organization, etc.) and the biggest worldwide companies (Google, Facebook, General Electrics, IBM, etc.) should cooperate to address, together, the problem of pollution. Then, extending a previously expressed concept, the global political commitment also becomes the global social responsibility of the biggest institutions and private companies and also the ethical responsibility of every single person: man or woman, white or black, American or African, European or Asian, rich or poor. However, the only way to leverage an ethical global response from individuals is to adequately inform and educate all humankind: old and young, professors and students, owners and employees.

Obviously, research and innovative technologies will play an important role in effectively fighting pollution, as readers of this book will understand very well, but only a global approach that involves each and every person can have a chance of success. A global approach also means that everyone must be committed to take responsibility and manage their own part of the whole process. However, to be really committed a person needs to understand the problem in depth and the full value of the possible solutions. Because it is so important that everyone understands the problem without having to be an expert in pollution, I started looking for available information on pollution. I began by searching for information that everyone could find using a TV and/or computer (games, apps, *Youtube*, *Instagram* with 331,937 public posts with #pollution, MOOC, etc.).

Obviously, different strategies and different methodologies can be used but a global educative program should be launched. To be successful, we need to be able to get in touch with every single inhabitant of the world to inform and educate them.

Coming back to some of the web searches that I launched looking for available information, I discovered that entering "pollution solutions" immediately introduced several other related searches (air pollution definition, air pollution causes, air pollution effects, air pollution map, water pollution, water pollution causes and solutions, land pollution, soil pollution, etc.), each with enormous numbers of references. Similarly, by entering "pollution how to defeat," I discovered some very important content, such as the following examples:

- "How can we defeat pollution as individuals?" (*Physics Forums*)
  - "Hi, I will tell what I am doing for my part in pollution control. (1) I switch off my vehicle when i am in signal. (2) I separate my plastic waste from garbage before disposing to garbage person. (3) I avoid getting plastic bags from grocery shop. instead i am using cloth bags. (4) I have started switching off my electrical equipments when ever it is not in use (even if it is 1 min). (5) I am started avoiding print outs of my materials. i have started reading soft copies." https://www.physicsforums.com/threads/how-can-we-defeat-pollution-as-individuals.373691/
- "Top 5 ways to fight pollution" by Jacob Shwartz-Lucas (*Progress.org*)
- "Air pollution 'kills 7 million people a year'" (*The Guardian*)
- "China vows to defeat pollution with energy 'revolution'" (*The Telegraph*)
- "Lets defeat pollution" (*Youtube*); https://www.youtube.com/watch?v=hd1RJtdFdgw

Using the web, everyone can find plenty of useful information and start understanding the problem. There are also many practical suggestions regarding what individuals can do to fight global pollution. Looking for information using the web, students and researchers can also find many well-organized sources of technical information. Searching for "pollution MOOC" (massive open online courses) leads to several excellent online courses and university level courses, including the following:

- *Coursera.org*; Environmental science and sustainability courses
  - The Age of Sustainable Development (Columbia University)
  - Introduction to Sustainability (University of Illinois at Urbana-Champaign)
  - Municipal Solid Waste Management in Developing Countries (École Polytechnique Fédérale de Lausanne)
  - Greening the Economy: Sustainable Cities (Lund Universities)

- Introduction to Environmental Law and Policy (The University of North Carolina at Chapel Hill)
- Our Energy Future (University of California, San Diego)
- Global Environmental Management (DTU Technical University of Denmark)

• *EDX.org*

- Sustainability Science – A Key Concept for Future Design (created by UTokyoX)
- Energy Within Environmental Constraints (created by HarvardX)
- Reclaiming Broken Places: Introduction to Civic Ecology (created by CornellX)
- Energy — The Technology You Must Know in the 21st Century (created by OECx)

In addition, there are millions of videos on *Youtube* focusing on pollution that everyone could discover. Examples include "The Devastating Effects of Pollution in China" and cartoon movies such as "Air Pollution for Kids," "Water Pollution for Kids," Environmental Pollution Animation," "Pollution Video for Kids – Pollution: Meaning and Definition," and "Pollution: Biggest World Problem!" (an awesome animation video).

These simple searches show that there is already plenty of free documentation available for use and that the conditions now exist for launching a worldwide campaign to encourage people to know, study, and start applying measures that everyone can use every day to fight pollution! To launch a worldwide campaign, it would be useful to find at least one big sponsor to act, for example, as Rotary International did in the "End Polio Now" campaign. I hope that I have help to promote understanding of how important it is to implement such a process.

To finish, I want to emphasize again that pollution is a problem of SUSTAINABILITY for the Earth. LONG-TERM SUSTAINABILITY needs to be achieved SOON because the overall situation is becoming worse and worse every day and we need, together, to achieve and maintain SUSTAINABLE DEVELOPMENT!!!

Stephen Hawking, the famous scientist and cosmologist, claims that the human race should look for new planets to which we can migrate. I personally do not want to accept this idea and, on the contrary, want to save and defend our great and beautiful planet Earth. I like and accept a GLOBAL approach and want to live the global dimension of our gorgeous planet, but I am not ready to accept a UNIVERSAL approach/strategy/overview.

With these few words, I hope I have expressed my support for the "Global View and Approach" that the pollution issue/problem deserves and requires, and is the only way to try to manage and overcome pollution!

Giampaolo Amadori
CEO, Fondazione Alma Mater of the
University of Bologna, Bologna, Italy

Foreword 2

Air pollution has gained attention in recent decades in studies of global climate change, raising a heated (and not always innocent) debate about the role of natural versus anthropic drivers. However, its influence on human health has not experienced a similar systematic approach, despite an abundance of studies on the topic and ample empirical evidence from medical practice. This might be ascribed to the fact that the relationship between air pollution and human health is more complex than the study of climate change because it involves another highly nonlinear system—the "homo patient."

In this respect, air pollution can be viewed as a compound of airborne chemicals, particulates of different sizes, organic molecules, and biological matter (active or inert). Each component can lead to consequences alone or in synergy with the others. It can penetrate the human body through inhalation, ingestion, or skin absorption. It can interfere with normal physiological or biochemical pathways. Furthermore, it can affect humans not only as a function of the duration and intensity of the exposure, but also as a function of the recipient's condition. Thus, it is a paradigmatic nonlinear system, whose effects can manifest immediately, or days, weeks, months, and even years later.

Despite a humble disclaimer about its comprehensiveness, this book provides a very integrated view of a complex topic. The discussion begins with its physical and chemical properties and continues with detailed analyses of the health effects on various systems of the human body. Finally, legal and sociological aspects are examined. This text strongly champions the need to develop a new branch of medicine that holistically addresses diseases relataed to air pollution. Furthermore, it does not shy away from suggestions for guidelines to attack the problem.

In today's society, risk analysis should be always a primary tool of policymakers and even citizens at large. In the classical definition, risk equals hazard × exposure × vulnerability. This book provides the tools to discuss each term of this equation. Thus, it is a unique publication, as well as a good read for both students and scholars. As with every respected scientific book, its conclusions are open to discussion—except for the urgency to increase awareness about the extent and severity of the threat, which has the potential to alter society as we know it. The effects of air pollution are now compounded by other factors, such as population growth, urbanization, and developing countries—all of which tend to increase the problem in exponential ways. However, our ecosystem is finite and cannot withstand this exponential exploitation. Considering that humans have existed for less than 0.1% of the age of Earth, we should always remember this: Nature does not need people, but people need Nature.

Sergio Bertolucci
Alma Mater Studiorum, University of Bologna,
Bologna, Italy

Foreword 3

The complex phenomenon of air pollution is creating an urgency for a courageous cultural revolution. The world's economies, businesses, and enterprises cannot take a neutral stance. They can influence the current situation, with the ability to sit simultaneously on the prosecution's bench and in the defendant's chair.

No one wants the world to regress. However, it is essential to look at the situation in a new way. We need to create positive and sustainable changes while restoring the values and the intentions destroyed by uncontrolled development. Businesses have been instrumental in helping to advance humankind in past centuries. Today, true "enterprises" can again play a fundamental role in our progress.

The complexity of the problems examined in this book means that the solutions cannot be delegated solely to institutions. The solutions require the action of the business sector, which should make its resources—managerial, financial, and technical/scientific—available to the community for facing this challenge.

The challenge in question is triple in nature. The first is a cultural challenge, which concerns everyone on the planet. It requires a change of mindset in the way that the natural environment is considered. The second challenge is management. There is a need to reshape business processes, redesign products, and restructure organizational models to preserve economic efficiency while safeguarding the environment. The role of businesses in society should also be preserved, as they are called to produce wealth and occupations. The third challenge is one of innovation. Resources and technical-scientific expertise should be shared to create innovative solutions to problems such as emissions and air pollution (e.g., sustainable mobility, energy production).

The attitude and attention of companies toward society and the environment must be sustained and improved within clear lines of economic policy and a defined system of rules, which identify the most appropriate instruments to achieve sustainability objectives. At the same time, the absence of the institutions or their weaknesses in effective planning cannot be a justification for not taking action in the business world. Businesses must be a "credible witness" of necessary and not deferable transformation.

This volume, which is set up as a handbook, precisely defines the priorities and the logic according to which individuals, companies, and governments can define objectives and prioritize actions to combat the causes and consequences of pollution at the individual level—namely, the initiatives aimed at defending each one of us.

Cristiano Benassati
President, GTechnology Foundation
Modena, Italy

Contents

1 Pollution and Air Pollution–Related Diseases: An Overview

Antonio Vittorino Gaddi and Fabio Capello

1.1 Introduction

This book come from the realization that we are witnessing nowadays a radical change in human health, in terms of human pathology—namely, the alteration of that state of well-being that everyone should rightfully aspire to—and of life and death cycles; those are the basic determinants of phylogenetics. This change is due to modification of environmental factors strictly related to human activities. Most of those, in fact, can be classified as polluting activities, which are able to determine different effects according to their physical, biological, and chemical characteristics.

Air pollutants coming from anthropogenic sources are relevant when compared with the fraction that comes from natural sources (e.g., methane, radon, volcanic emissions, etc.); that share in several countries is rising regardless of the attempts and the proclaimed acts to contain harmful emissions, although with regional differences that suggest differentiated strategies for the area [1, 2]. Local policies may cause different trends, suggesting the implementation of different and more effective measures in selected areas.

Theoretically, in effect, the share of human-related emissions is controllable and can be reduced. However, the short-, medium-, and long-term effects on human health are increasing and will maintain the same trend in coming decades, even in the event of full and effective completion of all the antipollution plans coming from individual governments and institutions. This is due to the fact that people and children exposed today will more likely show the outcomes of their exposure to air

A.V. Gaddi (✉)
Scientific Director of Health Lab of GTechnology Foundation,
Modena, Italy
e-mail: profgaddi@gmail.com

F. Capello
Alder Hey Children's Hospital, Liverpool L14 5AB, UK

F. Capello, A.V. Gaddi (eds.), *Clinical Handbook of Air Pollution-Related Diseases*,
https://doi.org/10.1007/978-3-319-62731-1_1

pollutants after many years, even if the overall amount of pollution is meanwhile reduced.

On the other hand, other elements—such as genetic transmission of mutated alleles to offspring secondary to selection or to direct action of pollutants—may emerge; at the moment, there are no guarantees that the phenomenon will be contained in the future. For that reason, it is very difficult to produce forecasts and reliable epidemiological predictions.

Nevertheless, we do not think that catastrophic positions or positions of cautious optimism—however desirable—are appropriate, at least from a scientific point of view. A scientific approach is indeed needed so that accurate measures and data can be obtained in order to deduce proper conclusions. This is a well-known process for researchers, scientists, and medical doctors. The same doctors, in fact, have been able, over the centuries, to face every new hazard coming from natural or artificial sources. Medicine has found useful answers even when those threats were completely unknown or out of control, and when the quality of the estimates of the outcomes was poor.

So, since we are already facing a global health emergency, since there is already a clear drift of human pathology and of its determinants, reflected also in whole ecosystems, and since several species, among which is the human race, are involved, we believe that action from all health professional bodies is of paramount importance.

All the areas need to be covered: clinical practice, epidemiological studies, clinical and experimental research, and medical and health education.

The promotion of medical literacy in this term is crucial; the sensitivity of people and doctors to this subject is limited and often based on misconceptions. There are many examples; one is the idea that air without smog is always healthy. At the same time, there is a need for constant awareness of the problem, avoiding cyclically activated alarmism—often triggered by media—alternating with long periods of complete silence.

In recent years, several tens of thousands of experimental or theoretical studies have been published on the effects of air pollutants on health; almost all those publications have focused only on a few selected fields of research—namely, lung diseases, cardiovascular diseases, and a few others. The ratio between the number of articles and the relevance of health problems is not always obvious and is likely to create a new "GAP 10/90" [3] also in air pollution–related diseases; this is particularly true when the critical problem of indoor pollution in less-developed areas of the planet is considered [4].

This division into sectors of research is largely attributable to the division of medicine into subject areas, causing, at the same time, a vanishing of clinical vision; individuals are not interesting anymore. What seems to matter in modern medicine is common denominators that allow health professionals to create those virtual entities that many doctors call diseases—virtual because, as good practice should suggest, there are no diseases but sick persons [5], who may or may not be clustered into groups of people with similar medical conditions, which we arbitrarily call diagnoses (see Chap. 8 and 14).

Obviously, also, shareable methodological choices, which in some cases are necessary, have to face the consequences of this approach (like, for example, in clinical epidemiology, the need to study univocally defined events).

For these reasons, in defining the chapters of this book, we have tried to give the same emphasis to both the emerging disciplinary issues and the traditional ones; we have also tried to underline the need for a more patient-centered vision.

Thus, promoting an original approach, we have also decided to analyze the effects of air pollution on the different categories that compose society (the elderly, future mothers, children); nowadays, the alternative would be a mere assay of cardiopulmonary diseases, in which the action of air pollution is well known, dismissing the complexity of the human body as a whole, as well as the emerging knowledge of the damage of air pollutants in other tissues, organs, or systems, and their interactions.

The other reasoned choice has been the involvement of the greatest possible number of experts from different cultural backgrounds and from different—and not necessarily medical—fields; this was done in order to give the most exhaustive picture of the problem, and also in an attempt to create an antipollution task force. The goal was to start a roundtable among all the actors who work in the field of pollution, even if they come from apparently differing areas; for this reason, in a clinical handbook, physicians, chemists, sociologists, engineers, experts, and management scientists have their say. This comes in an attempt to build a text where all facets of the pollution problem can be presented and analyzed, at the same time trying to find the answers to those questions that the real society is facing today.

This is why also in a clinical book—aimed at giving practical and usable information for medical doctors, public health officers, and health workers—we confront ourselves with the political implications of air pollution. From these same considerations came the decision—taken with the humility of those who dedicate their lives to research and science—to try to suggest to governments and decision makers what we consider the best possible solutions available at the moment.

We are also pleased that other authors who treat the theme of air pollution—such as Vallero in *Fundamentals of Air Pollution* [6], an essential reference book on this subject—feel the need to propose some challenges for the future, taking the share of responsibility that falls on those who have thoroughly studied the problems.

On the other hand, we are aware of the fact that any attempt to be completely exhaustive on the subject of air pollution—and, generally speaking, in science—is pretentious and unavoidably defective. This is the same nature of science to avoid dogmas and to eschew the promotion of mainstream's theories. Those remain in fact the biggest limitations to the achievement of innovative ideas, original research, and, ultimately, new discoveries.

This innovative approach, nonetheless, should be fostered also because professionals from different fields may not be aware of how their expertise can positively affect also other areas of research, even when links are not so evident.

For the same reason, every topic treated in this book could be exposed and exploded in many chapters or in monographic books. Our attempt in this sense has been to define the most complete clinical vision (which does not necessarily mean a medical vision), to the best of our abilities, in order to join everyone's effort to reach the final goal: the improvement of people's lives.

1.2 The Scientific Point of View

As we have seen, such a complex subject in which all the different fields of science and research come together cannot be treated according to the traditional formal and schematic subdivisions of medicine (e.g., cardiology, nephrology, neurology, etc.), even when multidisciplinary and integrated visions are considered.

If we could, we would have proposed, on the one hand, analysis of clinical cases, starting from their description "in order to say what we have seen"[1] and not what we read in a paper or an essay; on the other hand, we might have suggested using the vision of systems medicine (that, however, is still struggling to define its methodological canons). It means analysis of complete sets of data coming from single individuals—the measure of everything, taking advantage of the opportunities offered to us. This is possible because the pollutants are researchable and measurable, albeit with difficulty in the individual as in populations. There are no other fields of research, in truth, where the study of a single case is so close to the epidemiological facts.

We did not do this, for several reasons—also related to our capacity—but mainly for one motive: we did not have enough data for a solid scientific foundation.

In any case, the topic of complexity is so central in this matter that we wanted to deal with it from the first chapters of this book.

A system is complex when its behavior is not attributable to that of its individual components but depends also on their interactions; even more, a system is complex because the internal relations to the system are nonlinear and for them we may not have interpretative models, also from a mathematical standpoint.

The truth is that we believe that no one today can define models that could account for the complexity when the scale is the human being; even more, we doubt that this can be achieved even in simpler systems such as a prokaryotic cell.

Therefore, we are aware of the limitation of this study, but, at the same time, we are convinced that those restraints would be exacerbated and would become insuperable if we ignored the fundamental aspects of complexity, using simple—if not simplistic—models, as happens sometimes in medicine.

So, this challenge is welcomed for more valid reasons. The first is that the specific field of air pollution–related diseases is the most appropriate sector in which to foster this approach, precisely because of the convergence of disciplinary interactions and different skills, and because the prevalence and severity of this problem is affected—not least—by economic interests of a size never seen before in health.

The second reason is that the field is almost virgin in terms of disciplinary contamination. It was said above that research is generally discipline oriented, but this

[1] Augusto Murri on the clinical skill of Antonio Cardarelli, an eminent clinician, who was reportedly able to "read the patient as an open book."

did not result in theoretical constructs to be demolished. It could be a spontaneous scientific revolution or, rather, the sum of many small revolutions that flow into one. Through measurement and collection of data, a new world is somehow rising.

The third reason is that newly available methods and technological solutions allow study of living beings in a totally innovative way; these are appropriate to address the issues of complexity and to produce sufficient data to understand more. And, for the purposes of medical research, it is essential to have ideas and gather knowledge about the "etiologies," which, in this case, are known or knowable (namely, the pollutants).

And if, on the one hand, we will always have to face the complications that have come from existence and from the sum of idiopathic factors, on the other hand, there will always be one or more other possible starting points.

Yet other reasons exist, secondary to the discovery of new biological codes [7] and the existence of multiple forms of language. This is the key to interpreting the complexity at unimaginable levels and opening the way to epochal discoveries able to create the turning point that could take research to the next level.

References

1. Liu M, Huang Y, Ma Z, Jin Z, Liu X, Wang H, et al. Spatial and temporal trends in the mortality burden of air pollution in China: 2004–2012. Environ Int. 2017;98:75–81.
2. Silva RA, Adelman Z, Fry MM, West JJ. The impact of individual anthropogenic emissions sectors on the global burden of human mortality due to ambient air pollution. Environ Health Perspect. 2016;124(11):1776–84.
3. World Health Organization. Global Forum for Health Research: the 10/90 report on health research 2001–2002. Geneva; 2002.
4. Martin WJ 2nd, Glass RI, Araj H, Balbus J, Collins FS, Curtis S, et al. Household air pollution in low- and middle-income countries: health risks and research priorities. PLoS Med. 2013;10(6):e1001455.
5. Sadegh-Zadeh K. Handbook of analytic philosophy of medicine. Dordrecht: Springer; 2015.
6. Vallero D. Fundamentals of air pollution. Cambridge, MA: Elsevier Science; 2014.
7. Barbieri M. Code biology—a new science of life: Springer International Publishing: Cham; 2015. p. 234.

2 Physical and Chemical Properties of Airborne Particulate Matter

Laura Tositti

2.1 Introduction

Airborne particulate matter (from now on abbreviated in PM) is one of the main atmospheric pollutants also known as a criteria pollutant as well as an environmental indicator. It affects the radiative balance of the planet and its hydrological cycle, it may play a role in visibility [1], but above all, it has been historically associated with adverse effects on human health (see, e.g. [2]), and therefore it soon started to be included in monitoring systems and networks for the safeguard of the population and of the environment.

In Europe PM is currently subjected to environmental regulation which introduced as reference metrics PM10 since 1995 (EU Directive) and more recently PM2.5 [3]. Though both terms are internationally used in atmospheric research as well as in environmental protection, they still appear to be widely misunderstood or misused by the public. To some extent even current terminology anticipates the complexity of this pervasive and ubiquitous pollutant.

Presently in all the most developed countries as well as in rapidly developing ones, research and regulations concerning PM are implemented, enforced and continuously updated with the aim of both protecting the population and the environment and expanding the understanding of this topic. This is of foremost importance since the life and sanitary costs associated with air pollution are estimated to be very high. A recent study by WHO (World Health Organization) has established that "Air pollution costs European economies US$1.6 trillion a year in diseases and deaths" (http://www.euro.who.int/en/media-centre/sections/press-releases/2015/04/air-pollution-costs-european-economies-us$-1.6-trillion-a-year-in-diseases-and-deaths,-new-who-study-says) [4].

L. Tositti
Department of Chemistry "G. Ciamician", University of Bologna, Bologna, Italy
e-mail: laura.tositti@unibo.it

F. Capello, A.V. Gaddi (eds.), *Clinical Handbook of Air Pollution-Related Diseases*,
https://doi.org/10.1007/978-3-319-62731-1_2

In spite of a general consensus on PM hazard, its link with human health is not straightforward, since the knowledge of the processes and mechanisms in which PM is involved, albeit wide and in exponential development in the last decade[1] [5, 6], is not yet definitive for a series of reasons which will be illustrated in the present chapter.

In order to establish a relationship between PM and health effects, it is fundamental to focus on the physico-chemical characteristics of this pollutant, as the key factor between PM and its adverse effects on health is/are the biochemical interaction/s with human cells and organs.

This presumes (a) a contact preceded by several potential ways of access (translocation) into the human body [7], and (b) once the contact has occurred, chemical interactions between PM and living substrates capable to trigger adverse effects.

None of the passages outlined, albeit long recognized and deeply investigated, is presently fully understood.

Evidence of the tight connections between PM and health derives mainly from epidemiological studies [see, e.g. [4, 8, 9] and references therein]. This means that, although the association is statistically substantiated, the inference is indirect and does not provide any mechanistic connection. However, even in this latter field, significant progress has continuously been achieved, providing a solid decisional platform on which air quality standards are set up and implemented in air pollution management/legislation. Though recent work in this field starts to face the problem directly, i.e. integrating epidemiological results with PM properties with the aim of optimizing the association between health effects and diseases (see, e.g. [9, 10]), the major gap between the comprehension of the links between PM and its detrimental effects on health remains typically mechanistic, i.e. in spite of the huge and ever-increasing data from in vitro and in vivo experiments, the way and/or the unambiguously and mostly responsible chemical species are not well known.

2.2 Background Information

PM, or more correctly “atmospheric aerosol”, is defined as a suspension of particles and droplets in a mixture of atmospheric gases. Both aerosols and, to some extent, gases in the air are not conservative, i.e. they are unstable (subjected to more or less rapid transformations and removal processes). There is a complex interplay between gases and particles in the atmosphere, since the mixture of gases contains not only abundant but fairly inert gases such as nitrogen (N_2) and oxygen (O_2) but also a series of trace gases which are relevant to climate and above all to chemical reactivity. Table 2.1 reports a detailed list of atmospheric components relevant to the present discussion.

[1] According to [5], the publication rate of scientific papers on aerosol science has grown from units article/year in the 1980s to the present 1500–2000 articles/year in the present time, based on ISI Web of Science.

Table 2.1 Gaseous chemical composition of the atmosphere (adapted from [11])

Constituent	Chemical formula	Mole fraction in dry air	Major sources
Nitrogen	N_2	78.084%	Biological
Oxygen	O_2	20.948%	Biological
Argon	Ar	0.934%	Inert
Carbon dioxide	CO_2	400 ppm	Combustion, ocean, biosphere
Neon	Ne	18.18 ppm	Inert
Helium	He	5.24 ppm	Inert
Methane	CH_4	1.8 ppm	Biogenic, anthropogenic
Hydrogen	H_2	0.55 ppm	Biogenic, anthropogenic, photochemical
Nitrous oxide	N_2O	0.33 ppm	Biogenic, anthropogenic
Carbon monoxide	CO	50–200 ppb	Photochemical, anthropogenic
Ozone (troposphere)	O_3	10–500 ppb	Photochemical
Ozone (stratosphere)	O_3	0.5–10 ppm	Photochemical
NMHC	$CxHy$	5–20 ppb	Biogenic, anthropogenic
Chlorofluorocarbon 12	CF_2Cl_2	540 ppt	Anthropogenic
Chlorofluorocarbon 11	$CFCl_3$	265 ppt	Anthropogenic
Methyl chloroform	CH_3CCl_3	65 ppt	Anthropogenic
Carbon tetrachloride	CCl_4	98 ppt	Anthropogenic
Nitrogen oxides	NOx	10 ppt–1 ppm	Soils, lightning, anthropogenic
Ammonia	NH_3	10 ppt–1 ppb	Biogenic
Hydroxyl radical	OH	0.05 ppt	Photochemical
Hydroperoxyl radical	HO_2	2 ppt	Photochemical
Hydrogen peroxide	H_2O_2	0.1–10 ppb	Photochemical
Formaldehyde	CH_2O	0.1–1 ppb	Photochemical
Sulphur dioxide	SO_2	10 ppt–1 ppb	Photochemical, volcanic, anthropogenic
Dimethyl sulphide	CH_3SCH_3	10–100 ppt	Biogenic
Carbon disulphide	CS_2	1–300 ppt	Biogenic, anthropogenic
Carbonyl sulphide	OCS	500 ppt	Biogenic, volcanic, anthropogenic
Hydrogen sulphide	H_2S	5–500 ppt	Biogenic, volcanic

1 ppm =1 molecule over 1 million or 10^6; 1 ppb = 1 molecule over 1 billion or 10^9; 1 ppt = 1 molecule over 1 million or 10^{12}

The chemical behaviour of trace gases is largely driven by photochemistry, i.e. chemical reactivity induced by incoming solar radiation in the UV-vis field, capable to transfer its energy to a series of molecules triggering pervasive oxidative reactivity. This characteristic behaviour of present-day Earth's atmosphere is largely physiological, i.e. natural, and as such, it is usually not perceived as significant. Instead this means that gases which make up our atmosphere, and that we inhale, are capable of interacting each other and also with all the materials, both abiotic and biotic, exposed to them.[2] Among the most noticeably active gases, there is ozone, a secondary pollutant which is well known for its relevant oxidant properties and for this

[2]The overall oxidative behaviour of the planetary atmosphere is defined as *oxidant capacity* of the atmosphere; details of these fundamental properties of the Earth's atmosphere are beyond the scope of this contribution and can be found in [11].

reason subjected to restrictions and monitoring, VOCs (volatile organic compounds) and finally NOx which are largely released as by-products of any combustion and participate to atmospheric chemistry during daytime but are responsible of oxidation capacity of the atmosphere at night-time. The complex chemistry which is triggered by sunlight within this complex gaseous medium is based on (relatively) fast and effective radical mechanisms, which are mostly typical of the natural atmosphere, but which may be remarkably enhanced in areas strongly affected by anthropogenic emissions and has, anyhow, sensibly increased from pre-industrial time up to date [12].

Why introducing gas chemistry in a text concerning PM? Primarily because PM is transported by atmospheric gases many of which are reactive, therefore its carrier has potential to interact on its own, and secondly because many of the oxidative processes occurring among gaseous species lead to the production of a large fraction of particles as described further on. As a third, non-negligible, issue, suspended particles are capable of adsorbing and desorbing gases, often providing active surfaces on which VOCs and other trace gases are initially bound, though release may often occur after their chemical transformation [13]. Both the latter cases represent general examples of the so-called multiphase chemistry as recently outlined in [14].

This brief overview is therefore meant to provide a focus on the extremely dynamical chemical framework in which PM is found and investigated, but of which, with the exception of aerosol specialists, there is limited awareness.

2.2.1 Exposure and Dosimetry

In order to assess the risk represented by the contact between a potentially hazardous pollutant, such as PM, it is necessary to introduce a suitable definition of exposure and set up a convenient quantitative relationship accounting for the causal effects on health between the pollutant considered and the damages observed.

In this brief overview, we adopt the approach provided by [15] based on the fundamental expression proposed by the US Academy of Science in terms of inhalation exposure [16], though the adoption of the exclusive inhalation pathway may be reductive, as research in the last decade has revealed how, especially for ultrafine particles (see further on in this chapter) may include other biokinetic pathways such as through the olfactive nerve and/or the cute as reported in [7]. According to [16], the following exposure equation holds:

$$E = \sum C_{ijk} t_{jk},$$

where E is the exposure produced by PM inhalation, C_{ijk} is the concentration of a given aerosol species (single or composite) i, assumed to be constant in time t_{jk} in each of the so-called microenvironments[3] j where an individual k spends time, and t_{jk} is the time spent by subject k in microenvironment j.

[3] The "microenvironment" as compared to the environment as a whole indicates a portion of space in which PM is well characterized and which is able to provide an inherent interaction/outcome in individuals.

In the case of an average individual, Moschandreas et al. [15] explicit the equation above as follows:

$$E_{ijk} = \sum_{ik}^{\text{outdoors}} C_{ik} t_{ik} + \sum_{ik}^{\text{indoors}} C_{ik} t_{ik} + \sum_{ik}^{\text{occupational}} C_{ik} t_{ik} + \sum_{ik}^{\text{in-transit}} C_{ik} t_{ik}.$$

The rationale for the equation above is that PM produces distinct but cumulative exposures as a function of the single microenvironment characterized by their own emission sources of PM, with distinctive composition and bioactive properties, as well as of the time a single individual spends in each (the so-called personal cloud) [17] and references included. In particular the equation roughly points out to the main microenvironments such as *outdoor air*, which refers to the basic environmental compartment in which air pollution including PM is formed and emitted; *indoor air* which presently constitutes an increasing matter of concern[4] covering both the field of residential environments together with hospitals and other working environment, i.e. offices, not classifiable among the occupational environments in a strict sense; *occupational environments* in which an individual is exposed to specific PM sources at levels usually well above the average environmental or even natural background concentration, though this matter is typically treated by occupational health and safety regulations; and *in-transit* term which refers to the particles characterizing the transportation condition (vehicles, bus, off-road vehicles, bike, etc.) for the considered individual as each of the possible situations is affected by the emission sources active, for example, during commuting or similar.

Like in all the exposure/dose approaches, the duration (time) of contact with a given agent is a fundamental parameter in the risk assessment procedure. However, though the theoretical formulation of the exposure problem is apparently simple, none of the quantities mentioned is straightforward. The exposure time is modulated by the habits of individuals requiring a wide spectrum of possibilities; the concentration C presumes not only the knowledge and therefore a detailed characterization of PM composition in each microenvironment but, above all, implies the known and quantitative association between a given bioactive species and its effect on human health upon inhalation.

In this respect, even after several years from the publication of the mentioned paper by [15] and in spite of an ever-increasing availability of results and progress in the understanding of PM toxicity, the application of this exposure model is an extremely challenging task.

The reason for this difficulty will be explained in the following paragraphs.

The exposure to PM has been found to be associated to excess mortality; a relevant update for Europe has recently estimated that PM2.5 is responsible for 467,000 premature deaths across 41 EU countries (http://www.eea.europa.eu/highlights/stronger-measures-needed/table-10-1-premature-deaths, visited on 15 December 2016).

[4] In this respect, it is worth to consult guidelines and reports by the World Health Organization available at http://www.who.int/indoorair/publications/en/ (page visited on 12th October 2016).

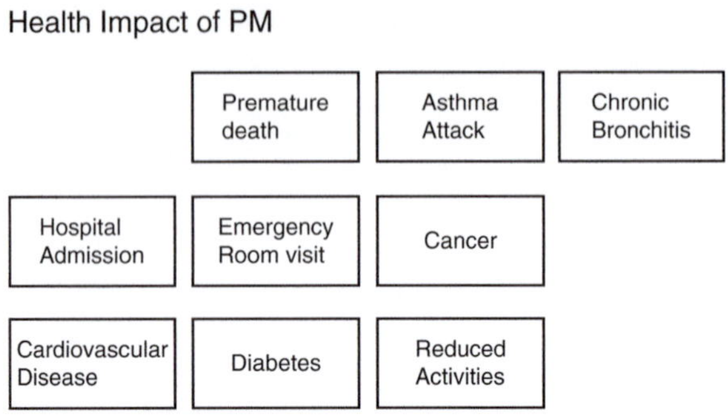

Fig. 2.1 Health effects that have been associated to PM

The principal health outcomes, including casualties, from PM exposure, are well documented and have been recently summarized by [18] and are reported in the following scheme (Fig. 2.1).

2.3 Aerosol Fundamentals

PM behaviour cannot be understood without introducing its basic properties on the basis of a series of classification modes which help to disclose and arrange PM complexity in a rational way.

The first fundamental concept is the distinction between *primary* and *secondary* aerosol. Primary aerosol is constituted by particles which are emitted into the atmosphere as soon as they are formed. Secondary aerosols are those produced following the emission of their gaseous precursors subjected to irreversible oxidation processes whose polar products are contextually condensed. The term "secondary" contains therefore a temporal implication, i.e. a time lag between an emission phase involving gas (primary substances) undergoing chemical reactions accompanied by a physical state modification, i.e. heterogeneous processes usually defined as "gas-to-particle conversions". It is found that ambient aerosols are made up of 40–90% in mass of secondary components.

PM has both *natural and anthropogenic sources*, which are mixed in variable ratios with the relative distance and intensity of each active source. In both cases, aerosol can be either primary or secondary. Table 2.2 reports aerosol contributions from natural and anthropogenic sources on the global scale expressed as average annual fluxes [19].

As it will be described in detail further on in this chapter, data reported in the table above clearly shows how natural and anthropogenic emissions are comparable and compositionally extremely complex, even though the focus is on sources and on the major chemical components of PM, as related to their specific formation process. Another non-negligible issue is that, beyond biogeochemical processes on which PM is tightly related, it is dynamically recirculated by atmospheric circulation throughout the atmosphere. As a result, particles are also subjected to continuous mixing, so that PM is always a complex mixture of both natural and anthropogenic PM fractions, no matter the location and the time of sampling.

The second basic characteristic of PM is its wide *particle size* range. Aerosol particles have a mean size variable from the nanometre ($1/10^9$ m) up to hundred

Table 2.2 Aerosol contributions from natural and anthropogenic sources on the global scale expressed as average annual fluxes

Source	Present flux		
	Low	High	Best
Natural			
Primary			
Soil dust (mineral aerosol)	1000	3000	1500
Sea-salt	1	10,000	1.3
Volcanic dust	4	10,000	33
Biological debris	26	80	50
Secondary			
Sulphates from biogenic sources	60	110	90
Sulphates from volcanic SO_2	4	45	12
Organics from biogenic NMCH[a]	40	200	55
Anthropogenic			
Primary			
Industrial dust etc.	40	130	100
Black carbon (soot and charcoal)	10	30	20
Secondary			
Sulphates from SO_2	120	180	140
Biomass burning (w/o blackcarbon)	50	140	80
Nitrates from NO_x	20	50	36
Organics from anthropogenic NMCH[a]	5	25	10
Total	2390	24,000	3450

[a]*NMHC* non-methane hydrocarbons

micrometres (where micrometre corresponds to $1/10^6$ m). However, aerosol particle size is tightly bound to the source process responsible of particle formation. As a result, PM is characterized by a multimodal behaviour with frequency distribution peaks around well-defined particle sizes. We can now introduce the four PM modes experimentally detectable with present-day instrumentation. PM distribution modes are as follows:

- Condensation nuclei peaking at 10–12 nm
- Aitken nuclei with a maximum at 70–80 nm
- Accumulation mode covering the range 200–800 nm
- Coarse particles >2.5 μm

Figure 2.2 depicts the four aerosol particle modes as a function of size, together with the principal mechanisms of formations. This scheme shows how the condensation nucleation mode, i.e. particles in the smallest size, derives by the chemical transformation of gaseous precursors through gas-to-particle reaction, initially in the free molecule state (average molecular size ~0.4 nm), which soon after initiates aggregation and clustering in an intermediate size range presently not yet detectable, until the nanometric size is reached.

Once particles in this fraction are formed, they continuously undergo accretion processes which encompass vapour condensation upon their surface and coagulation of other particles until the Aitken nuclei mode is reached [20]. Ageing of aerosol and further coalescence/condensation processes promote the

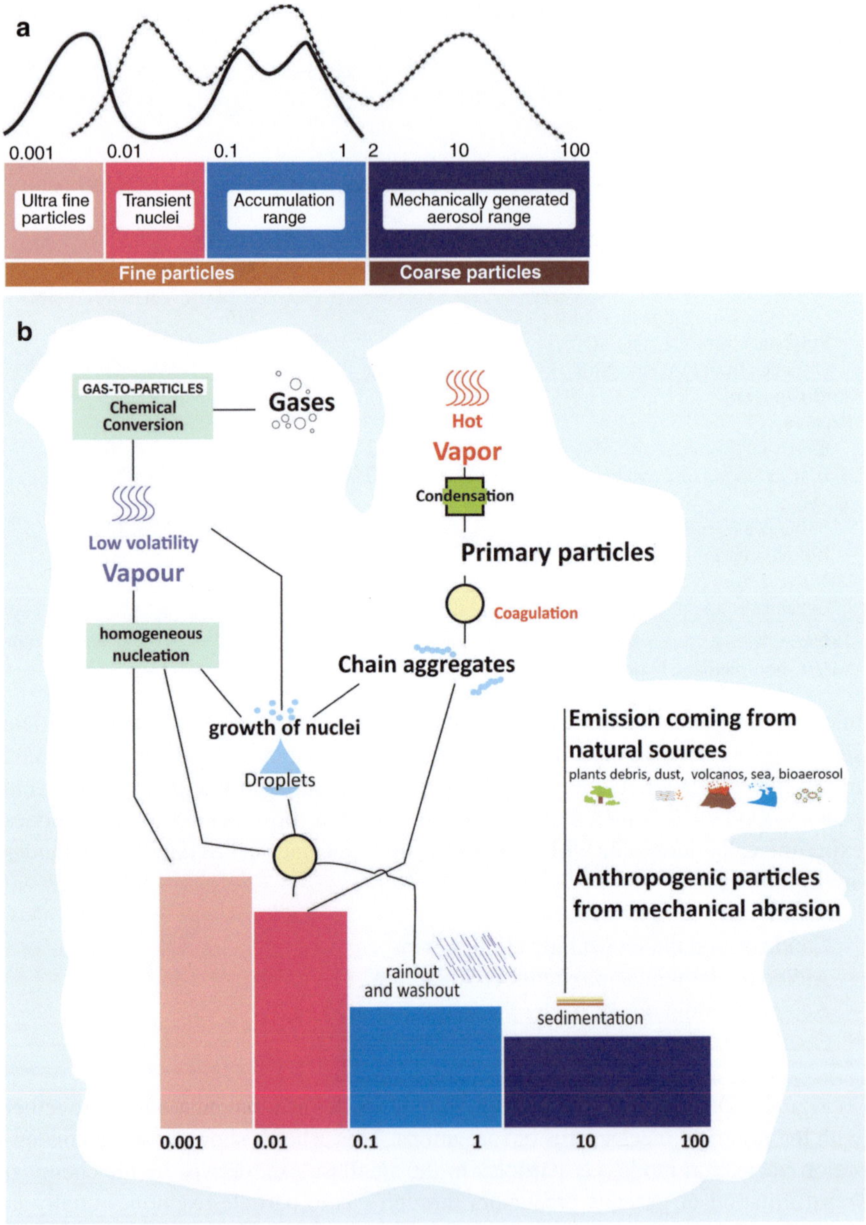

Fig. 2.2 The aerosol particle modes as a function of size and the principal mechanisms of formations

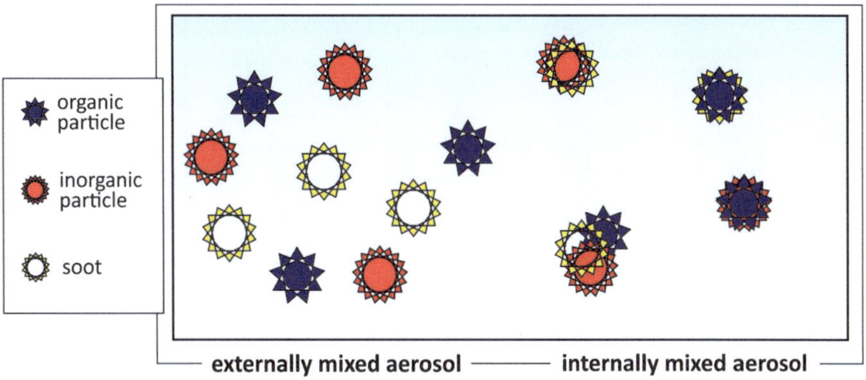

Fig. 2.3 A simplified scheme of externally and internally mixed aerosols

growth of aerosol size into the so-called accumulation mode. In the latter modes, particles are more both externally and internally mixed, i.e. they do not only include many different chemical species forming a mixture of different particles (externally mixed aerosol), but each single particle is constituted by distinct chemical species (Fig. 2.3) [21]. This circumstance allows to anticipate two consequent and extremely relevant properties of particles which are not or are very limitedly spherical and have complex composition as a result of the multiple sources on which they depend.

Finally, coarse particles usually derive from mechanical processes such as solid abrasion (both natural and anthropogenic), wind erosion of rocks and soil, organic debris, volcanic emission and sea-salt aerosol.

Another interesting feature of aerosol is the occurrence of biological particles whose diversity covers the whole size range of ambient particles from nanometric viruses through bacteria up to pollen and spore in the supermicron range. This component is somehow less treated than abiotic components; indeed, they represent a very important contribution to every aerosol population, either outdoor or indoor, including the associated health risks.

Figure 2.4 depicts a comprehensive scheme of the various types of ambient particles as a function of size.

Another simpler, widely used classification of particles as a function of size, independent of frequency distribution of particles, is based on defined threshold size. Given the broad size range and the presence of two more or less evident saddles between the particle modes, we can distinguish among *coarse* particles with a diameter $D > 2.5$ μm,[5] *fine* particles with 0.1 μm $< D < 1.0$ μm which covers the accumulation mode and *ultrafine* particles with $D < 0.1$ μm. Ultrafine particles (from now on UFP) include the so-called nanoparticles, which are defined as those

[5] This threshold value is conventional rather than physically based as the transition between the fine and the coarse mode is not so sharply defined; sometimes the threshold is fixed at 1 μm.

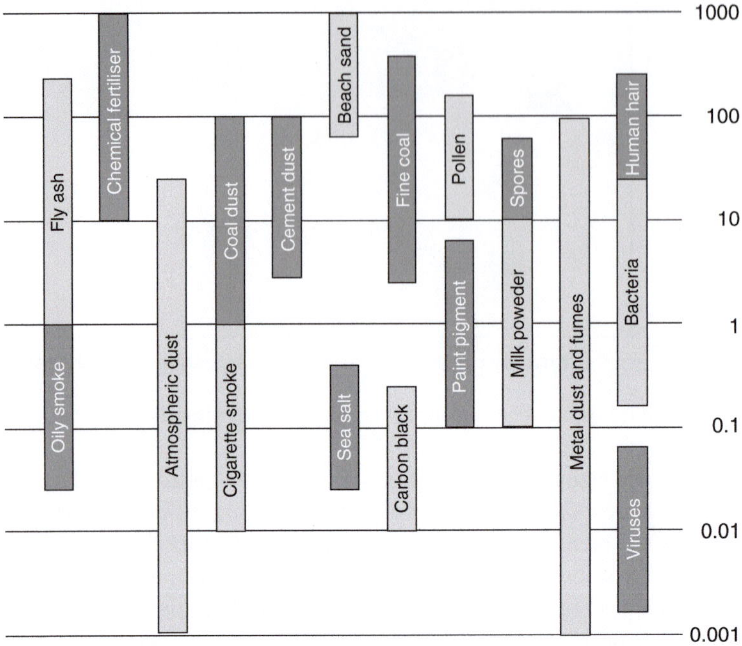

Fig. 2.4 Synthetic overview of aerosol types as a function of size

particles in which at least one of the three dimensions is smaller than 100 nm. Since particles in this latter size range are highly penetrating in the respiratory system, they represent a significant health concern. Attention is presently given to ambient nanoparticles which derive from gas-to-particle reactions in the atmosphere or are produced unintentionally in high-temperature processes such as the pervasive application of combustion both on the individual and on the industrial scale and many others such as industrial manufacturing, etc. [2]. Modern technologies however have brought to light and to the general attention of both research and policymakers the importance of engineered nanoparticles whose production leads to health risk of professionally exposed personnel as well as to threats to the environment [22, 23]. Considering that nanoparticles are both natural and anthropogenic (which include both unintentional and engineered ones) and since their size is connected to an enhanced reactivity [22], there is a need of clarifying both their potentiality in terms of both positive technological advance (new performing materials as well as drugs) and of adverse effects on health and environment. Since the latter have clearly emerged, nanoparticles are exponentially increasingly investigated, and as concerns nanotechnology in Europe, they are presently covered in REACH (EC 1907/2006) and CLP (Regulation 1272/2008 on classification, labelling and packaging) regulations (see, e.g. https://echa.europa.eu/regulations/nanomaterials).

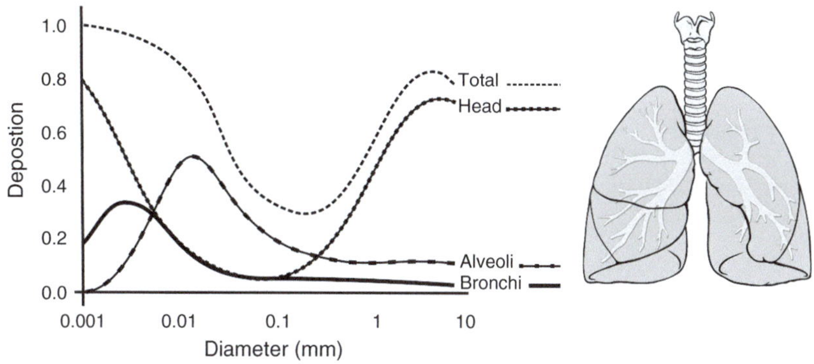

Fig. 2.5 Particle deposition in a normal adult mouth-breathing (1.2 m^3/h) male human subject at rest, as a function of particle size

The particle size has two fundamental implications:

1. Accessibility into the body by breathing system and/or by translocation as a function of size
2. Residence time of particles in the atmospheric medium

While accessibility as a function of size is fairly well known, at least when considering the respiratory system (Fig. 2.5), atmospheric residence time controls human exposure to particles. This means that exposure is not uniform over the size range spanning between the nanometre and several micrometre sizes and depends on particle size in a non-obvious way. Exposure and related dosimetry will sensibly differ as a function of several parameters.

Therefore, while the finer the particle, the deeper the depth reached in the lungs down to the alveoli, it is relevant to evaluate the availability of particles in each size mode, a process which is affected by the removal mechanisms of particles and by their resulting lifetime.

The *atmospheric residence time* τ of an aerosol is defined as the average time it spends in the atmosphere as a mutual result of the processes of formation/production against those of removal. Not only aerosol size is affected by production processes, but even PM removal from the atmosphere is strongly size dependent. It is generally accepted that atmospheric particles are substantially removed through two main pathways, i.e. respectively wet and dry deposition. Wet deposition includes rain, snow, hail and fog; it is largely dominant over dry deposition relying on very efficient mechanisms largely mediated by particle composition and resulting hygroscopicity, though it may depend on precipitation regimes and therefore on local climatology. The processes are complex and still matter of investigation though it is ascertained that most of the submicron particles are removed through cloud processing and wet deposited by hydrometeors (see, e.g. [24] for details). Dry deposition,

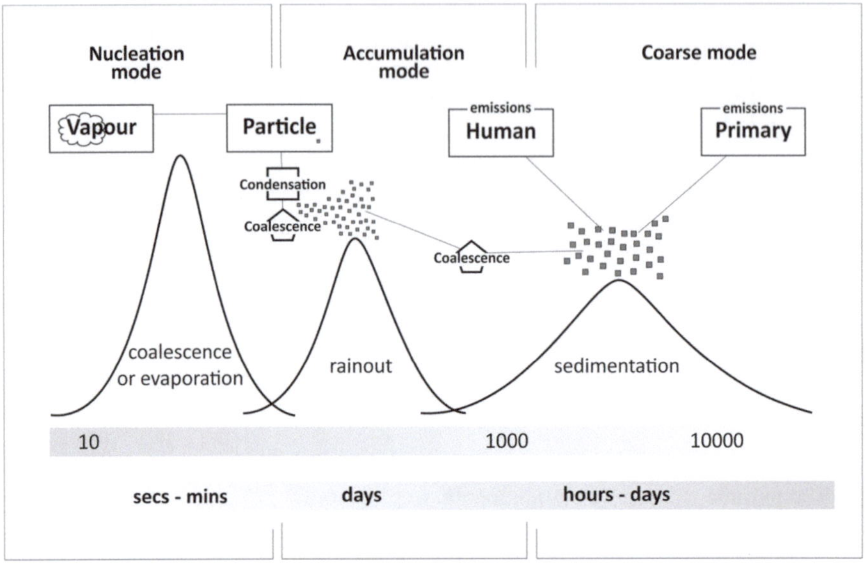

Fig. 2.6 Size distribution of aerosol, formation and removal processes and resulting lifetimes (Data available from available at https://allaboutaerosol.com/aerosol-in-the-atmosphere/). For the sake of simplicity, the nucleation modes are represented as a single mode

i.e. PM removal in the absence of precipitation, differs on account of PM size; it may occur by gravitational settling when particles are coarse (i.e. >1 μm) or by impaction when particles are <1 μm. In the UFP range, dry removal is controlled by particle-particle impact followed by coalescence and coagulation, which determine the disappearance of smallest particles in favour of the accumulation mode; these processes are statistically promoted by the highest number density of UFP. Finally, particles may move from the smallest size up to the larger one by growth, as both gases and water vapour may condense and deposit on the particle surface contributing to their accretion. At this stage, it is apparent the close relationships among gaseous species, UFP and accumulation mode particles, whose interplay is controlled by complex thermodynamic conditions and by aerosol sources, properties and intensities.

While residence time of particles removed by wet deposition is substantially associated with atmospheric water cycle and may reach up to 8–9 days on average (i.e. the mean τ of atmospheric water), in the case of dry deposition, the residence time is highly variable and dependent on size (see Fig. 2.6).

It can be noticed that the shortest-lived particles are both the coarse and the ultrafine particles, which include nanoparticles. The accumulation mode particles are small enough to stay in suspension in the atmosphere until atmospheric motion will end up in impacts on surfaces or will be affected by wet deposition.

Moreover, aerosol lifetime is dependent on location and height in the atmosphere as reported in Fig. 2.7.

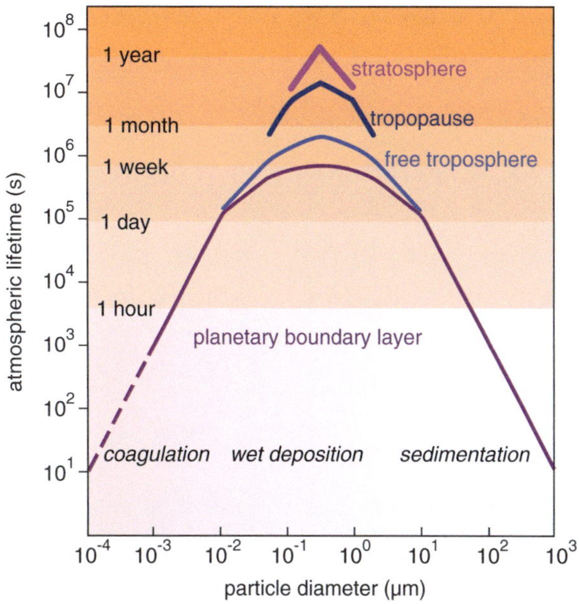

Fig. 2.7 Atmospheric lifetime of aerosol as a function of size and of atmospheric height (Data available from http://elte.prompt.hu/sites/default/files/tananyagok/AtmosphericChemistry/ch09s02.html)

2.4 Number of Particles, Mass and Surface vs. Size

Additional PM properties varying with particle size are volume and mass, mutually proportional, surface and number density, i.e. the number of particle per unit volume of air. Typically, volume and mass are largest in the coarse mode and decrease with decreasing size, while surface and number density increase more than proportionally as size decreases. Figure 2.8 shows the variation of these parameters in a well-known diagram obtained from the fundamental text [1].

It can be observed that volume (and consequently mass) is maximum in the coarse mode, whose population instead is very low, while UFPs present a negligible overall volume and mass but an overwhelmingly high number density. This is remarkably relevant to health issues, as, due to their tiny size, UFPs are most likely to reach the deepest section of the respiratory system and access the alveoli; this means that size and population may substantially increase the potential for biochemical interactions in the body when particle chemistry is favourable.

Interestingly the surface term, i.e. the specific surface area[6] of particles, is maximum in the fine mode defined as *accumulation mode*. This circumstance is relevant

[6] Specific surface area of solids is a parameter defined as the total area of a grain-sized material per unit mass and is measured in m^2/mass (g or kg) of material. Owing to a basically, though not

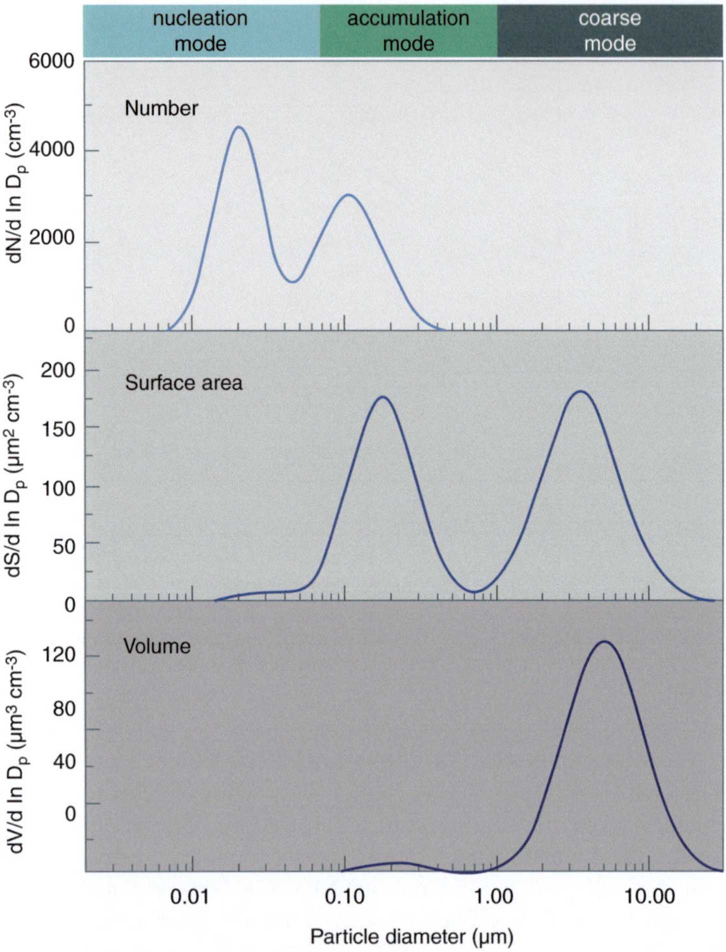

Fig. 2.8 Behaviour of PM particle number, surface and volume as a function of particle size

to PM behaviour in air, as a high specific surface area promotes interactions (adsorption and desorption) with reactive gases in which it is suspended. The longer the lifetime, the higher the potential not only for PM ageing but also for complex physico-chemical interactions and consequent influence on the overall atmospheric chemistry, including effects on health, materials and environment. In the case of health-related issues, recent research has actually identified specific surface area as an extremely efficient parameter in the assessment of the dosimetry of acute toxicity from fine and ultrafine particles [25].

Since the number of particles (or better their number density) is remarkably huge for both fine and ultrafine particles, this parameter is complementary to mass and

exclusively, geometrical reason, overall surface of an unconsolidated material increases exponentially with decreasing size on a constant mass basis. We omit visual examples for the sake of brevity and suggest the following reference for a deeper insight [25].

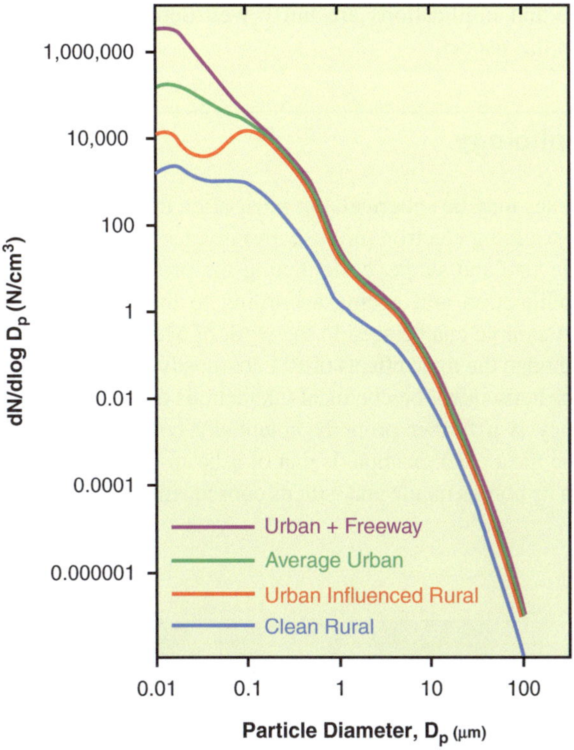

Fig. 2.9 Particle number density vs. size distribution in several ambient conditions ranging from clean background to man-impacted situations

vcomposition of PM. All of these properties are highly variable in time and space and produce distinct distribution pattern resulting from the source location and intensity, atmospheric mixing processes and biogeochemistry, all the factors being equally relevant.

Figure 2.9, reproduced from data collected by [1], visualizes PM distribution pattern, which we introduce.

This diagram offers several opportunities to understand PM behaviour. Firstly, it can be appreciated how clean air even in remote areas is not depleted in particles, i.e. particles pertain also to the natural atmosphere. The main difference with an anthropized airshed is their concentration or more correctly their number density. Both axes are logarithmic; therefore the increase in particle number in urban conditions is far more sensible than it appears at a first sight; the worst situation is found at a kerbside, i.e. an area highly urbanized and affected by vehicles, pointing to the importance of traffic in heavily impacting the PBL (planetary boundary layer) and consequently affecting human exposure. In practice, regardless of the previously described size bins, the total particle number of PM in the lower troposphere ranges between $10^2 \div 10^4$ cm^{-3}, equivalent to a mass range of $1 \div 100$ μg m^{-3} between remote and anthropized sites, as reported by [26].

Finally, this diagram does not cover UFPs which usually include even smaller and above all more abundant particles, as previously described; therefore even if the overall picture and implications are fairly well depicted, it gives an incomplete information on this aspect.

2.5 Morphology

Airborne particles may be spherical, but most often they are not. Figure 2.10 representing SEM (scanning electron microscopy) images provides wide information on aerosol particle size and shape, enlightening the processes responsible of particle formation, modification and ageing according to the wide range of sources and mechanisms. A sample can be seen in the work of Megido et al. [27].

Moreover, though the toxic effects of PM are mostly related to its chemical composition and to the consequent biochemical interactions occurring on living tissues, particle morphology is a further property potentially conducive of adverse effects on health [27]. The most clear example is that of asbestos, whose harmfulness is largely associated with its fibrous nature and with its consequent accessibility to the lungs [28].

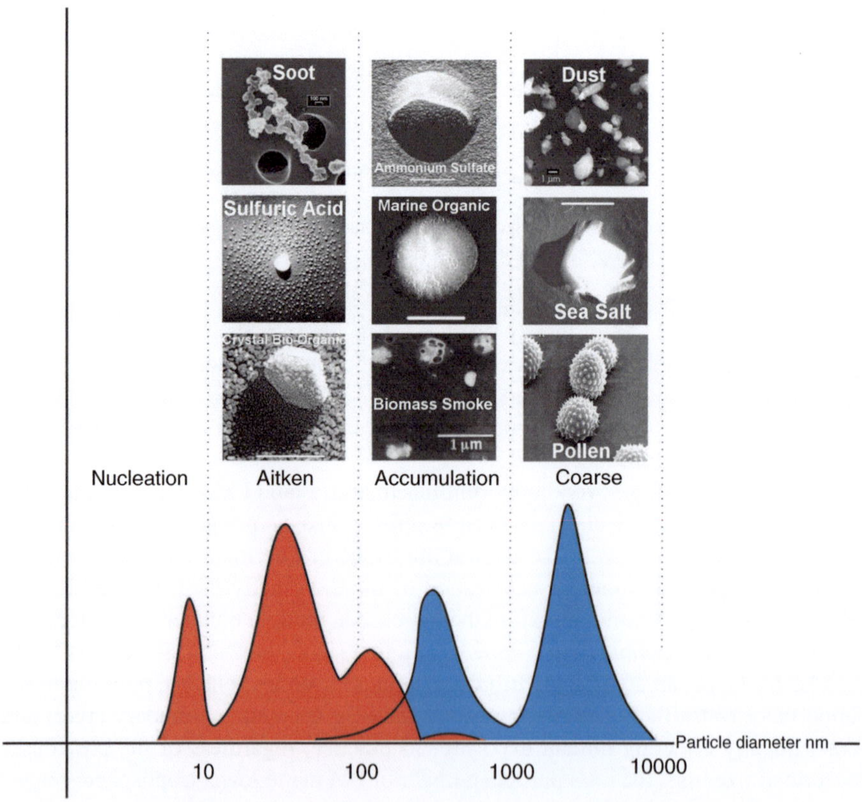

Fig. 2.10 Types of particles and their composition as a function of size (available at http://capita.wustl.edu/aerosolintegration/specialTopics/Integration/Capter4Drafts/CHAPTER4000708.htm)

Thanks to the energy dispersive X-ray probes usually coupled to electronic microscopy devices, they can provide semi-quantitative[7] elemental information on single particle whose chemical composition may be deduced, revealing not only the actual sources of ambient PM including primary and secondary ones but shed light also on the occurrence of internally mixed particles as well as on the effects of hygroscopicity on particle swelling by water absorption and deliquescence process.

In the following a series of TEM (transmission electron microscopy) images from a historical and popular paper by [29] are reported. The figure depicts many of the concepts so far described in terms of dynamics of aerosol formation and ambient production/modification during its lifetime in atmospheric suspension and resulting shape.

Though SEM and TEM analyses are widely used especially on a single particle basis, quantitative analysis by microscopic techniques may be used for quantitative analysis, provided the approach is based upon the analysis of hundreds to thousands particle per single aerosol sample to account for the huge statistical variability in terms of size, composition and shape.

In addition to the structural properties so far described, another fundamental morphological characteristic of PM typical of aged particles is the occurrence of stratifications often with different chemical species on the surface of or inside the particles. As a result, chemical species on the surface are potentially more prone to interaction with the surrounding chemical environment. This means that while insoluble components will be more active in adsorption/desorption mechanisms, the soluble ones could be released upon contact with solvents or solutions, while occurrence of metals on the surface might trigger redox reactions, with implications on atmospheric chemistry through catalytic processes or on oxidative stress through ROS (reactive oxygen species) formation when in contact with living substrates.

2.6 Aerosol Metrics and Sampling

The scope of this paragraph is to introduce the reader to a fundamental process in PM characterization and understanding: its sampling. As previously illustrated, air is a heterogeneous system so that particles, usually solid and/or liquid, as a function of atmospheric relative humidity, can be separated and accumulated until a sufficient amount has been collected for subsequent analysis. The required PM amount is dictated by the analytical approach applied, since the family of analytes and atmospheric concentration is typically very low, even in very polluted conditions.

[7] For the sake of completeness, it must be remarked under particular conditions which requires the analysis of hundreds to thousands particles per sample: also SEM analysis can be employed even for quantitative elemental analysis though this is far less convenient than other instrumental/analytical techniques, as the analysis of large numbers of samples is usually needed in order to collect sufficient information to cover the two main objectives of PM characterization, namely, source profiling and identification of health-related species.

The experimental techniques for aerosol investigations typically require highly specialized and skilled experts owing to the extreme complexity of the topic and to the non-conservative nature of the aerosol material. Therefore, since technical details are beyond the scope of this overview, we suggest other information sources for a deeper insight of this, for example, [30], while in the following of this paragraph, we provide the basic knowledge necessary to correctly understand PM fundamentals in the framework of health-related problems.

Moreover, the present contribution is not intended to be exhaustive of the topic, as the current technology continuously offers new solutions and improvements. Therefore, this section will be limited to the basic tools used in PM studies by the largest part of the aerosol community with the aim of determining aerosol composition and its links to health and environmental hazards.

Instruments for PM investigation are substantially subdivided into offline and online devices. Basically, the most widely used methods to collect aerosol particles are based on air filtration using porous membranes and/or by impaction, both pertaining to the offline class. In both cases a pump draws an air flux through a mechanical system allowing for particle collection and concentration on a support suitable for the following weighing and analysis. Flow rates of sampler range from a few L/min in low volume samplers up to several tens of m^3/min for high volume. The choice of a given flowrate is usually dictated by the standard metrics required by the environmental laws/regulations in each country or by the average concentration of the chemical species investigated. One of the basic requirements for current sampling protocols is the need for a constant flowrate during each sampling to allow for the correct size selection of particles during the sampling itself; consequently, samplers are provided with a compensation system to overcome the head loss produced by filter loading. The overall complexity in the sampling management is such that continuous verification and recalibration of samplers are strictly required.

Another fundamental issue in sampling is the choice of the filtering material as it must guarantee the minimal influence on the analytical results; therefore specific filters/supports for air pollution monitoring are available; in this framework the choice for the most suitable material will also depend on the class of analytes of interest and among the chosen ones those with the minimum blank compatibly with detection limits, type of environment investigated (background, rural, urban, kerbside) and obviously costs.

For practical and rigorous details on sampling and chemical speciation of PM, we suggest to consult, for example, US-EPA guidelines by J.C. Chow and J.G. Watson available at https://www3.epa.gov/ttn/amtic/files/ambient/pm25/spec/drispec.pdf (page visited last on 14th October 2016).

In the last decades, aerosol collection has been integrated by other devices which allow to access other fundamental properties, such as particle counting, or even to identify single aerosol particle allowing for the simultaneous determination of its mass and composition. They are usually online devices, i.e. particles are continuously measured during the operating time; data collected are usually integrated over adjustable time intervals with a time resolution suitable to describe the order of processes and variability. These instruments will be briefly described in the final part of this paragraph.

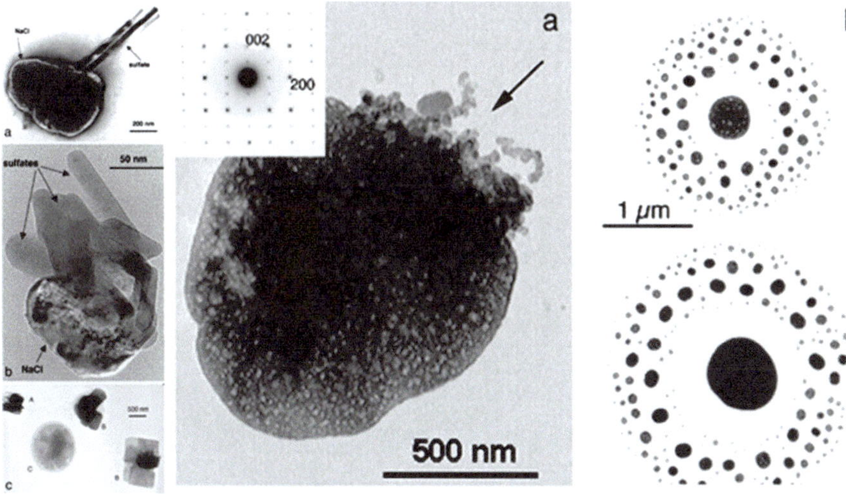

Fig. 2.11 Examples of SEM images of aerosol particles showing their morphology and on the right, modifications due to hygroscopicity of some components (deliquescence)

2.7 Particle Collection

The older sampling systems consisted basically of pumps which filtrated airborne particles on a porous medium allowing for their separation from the gaseous phase, for which just the flowrate was known. Owing to air and aerosol properties, which were soon discovered to be extremely variable, filtration efficiency is scarcely reproducible being affected by filter porosity and material and pump flux but also by PM size distribution, atmospheric concentration and composition, as well as by the thermodynamic conditions of the atmosphere. As a consequence, there was a need to leave the older standard known as TSP (total suspended particulate) per unit volume of filtered ambient air to a more reproducible and consistent metric allowing for data homogeneity and subsequent evaluation.

The current PM metrics, which are substantially adopted worldwide, are based on samplers which:

a. Filtrate aerosols based on a size cut-off,[8] i.e. prevent the collection of particles larger than a conventionally chosen size threshold but collect quantitatively PM below this level

[8]According to IUPAC (International Union of Pure and Applied Chemistry), cut-off is defined as "The size of particles at which the retention efficiency of an instrument device drops below a specified value under defined conditions" (http://goldbook.iupac.org/C01481.html, visited September 12th, 2016).

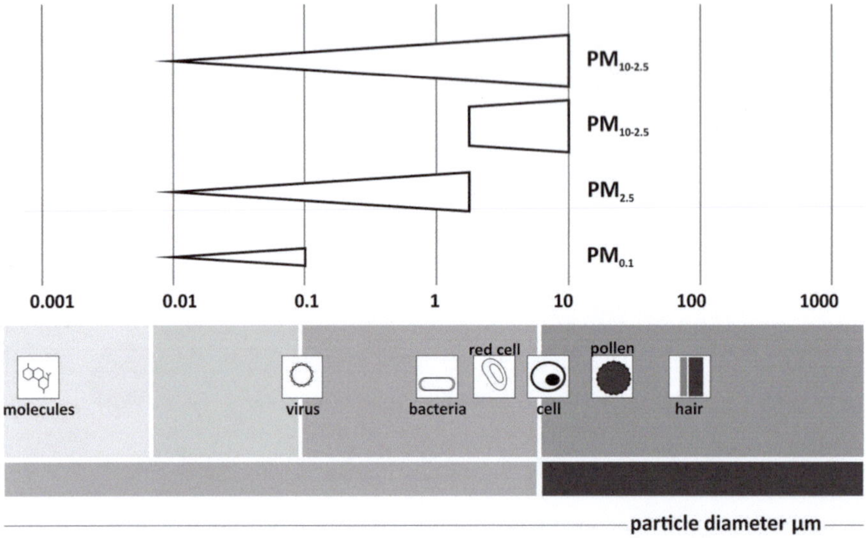

Fig. 2.12 Aerosol sampling metrics based on PM10 and PM2.5 including ultrafine and coarse fractions [31]. The length of the arrows indicates the size range of collected PM

b. Have pumps constantly working at a given flux, i.e. are provided with an automatic system which corrects the pressure drop when the particle load on the membrane is too high

This means that sampling PM is per se a very accurate and metrologically controlled operation, carried out with certified instruments, i.e. not simple mechanical devices, whose use requires systematic calibration and maintenance protocols. The rationale for excluding larger particles from sampling is that coarse particles are far less numerous as seen before, but above all they are unable to access the deeper part of the respiratory system; therefore they do not represent a substantial health hazard.

Omitting herein the technical details by which a reliable aerosol sampling is achieved, as well as other instrumental methods used more often in more specialized sectors of atmospheric research, we can now define the current metrics adopted in atmospheric research as well as in governmental air quality network.

PM sampling is substantially based on PM_{10} and $PM_{2.5}$ metrics, where PM_{10} is defined as particulate matter with a mean aerodynamic[9] diameter equal to or lower than 10 µm, while $PM_{2.5}$ covers particulate matter with a mean aerodynamic diameter equal to or lower than 2.5 µm. Consequently, both standards physically exclude particles above the size indicated from being retained, while $PM_{2.5}$ is actually included in PM_{10}. The following diagram depicts the particle size range covered by the described PM references (Fig. 2.12).

[9]Aerodynamic diameter does not necessarily coincide with the geometrical diameter of aerosol particles but defines a virtual diameter of particles with their own density and shape but terminal velocity equivalent to that of a spherical particle of unitary diameter and density.

UFPs are therefore collected either in PM_{10} and in $PM_{2.5}$, while $PM_{2.5}$ excludes the coarse fraction, further focusing onto the finest particles more prone to access the respiratory system. Though apparently trivial, this clarification is felt as due, as, even after many years of application, the PM_x concept is still frequently incorrectly reported, generating confusion in the public.

Beside filtration systems, PM can be collected by impaction on suitable supports, always based on pumps coupled to collectors. Most frequently these devices can be designed to size-segregated aerosol, i.e. to collect finite amounts of PM in subsequent ranges of diameters. Fractionation is very important not only in connection with the wide particle size range of ambient aerosols but above all because this sampling approach allows to characterize the distinct chemistry of PM as a function of size. As previously described, the mechanisms of formation of PM are multiple and complex, highly distinct in the various modes. The knowledge of the mentioned processes together with local characterization as PM is strongly affected by local sources of emission and requires the accessibility of samples representing this huge variability. Therefore, once PM has been collected on filters or supports, they may be weighted to produce the primary parameter used in PM metrics, i.e. its mass load per unit volume of filtered air.[10] Clearly, in most cases, PM collection on a support is preliminary to offline analysis, including subsequent processing for chemical speciation, covering substantially the whole analytical range of methodological and instrumental solutions available and in continuous development.

Online methods are usually preferred when it is necessary to focus on the time modulation of atmospheric chemistry and on the inherent modifications of PM, all factors highly affecting potential health effects. In these cases, it is possible to perform semi-continuous analyses coupling size-selective sampling heads like PM_{10} or $PM_{2.5}$ either to denuders (devices used to sample the vapours supposed in equilibrium with some of PM components) and/or to a filtering station both connected to a solubilization unit and an ion chromatograph like in MARGA (monitoring of aerosols and gases in ambient air) (Metrohm) or in AIM (ambient ion monitor) (Dionex Thermofisher). These devices allow sampling and analysis of finite amounts of PM according to accurately planned time sequences at the sub-daily time scale providing invaluable detailed information on aerosol evolution as a result of meteorology and photochemistry. Another interesting solution for semi-continuous analysis is the VACES (versatile

[10] As a general rule, gravimetry of aerosol samples is the most accurate way to achieve the determination of PM mass loads, even though the method is lengthy and demanding, requiring standardized protocols to overcome potential bias in PM sample handling. Moreover, the weighing operations are characterized by an extremely accurate performance compared to many other measuring operations; this characteristic is of basic importance as PM samples can be very limited and therefore difficult to characterize accurately. Recently PM_x samplers may be equipped with tools allowing for an automatized determination of aerosol loading based, for example, on the attenuation of a weak beta-beam or others. In this way air mass load can be monitored also at the sub-daily scale (PM_x is usually collected on a 24-h basis) as the measurement can be carried out during the sampling without any interruption, increasing the operation information and efficiency and reducing costs of the operation; nevertheless, a systematic and frequent check against the standard method is highly recommended due to drifts and bias especially important at low PM_x concentrations.

aerosol concentration enrichment system) [32] which allows to sample ambient particulates with standardized sampling heads thereafter collecting PM into a water slurry; again this sampling scheme is extremely useful not only because it allows to fractionate PM samplings in time intervals adequate to the scientific objectives but also because it is very flexible, allowing to split the samples injectable in whatever are the analytical tools planned. This solution is particularly appealing in both speciation and toxicity test as it prevents the potential interference from membrane materials.

Together with sampling strategies for the chemical characterization of PM according to its fundamental properties, we mention herein further instrumental techniques aimed at integrating aerosol analysis and size classification by online techniques operating on a semi-continuous basis.

PM can be analysed for aerosol number density, a parameter which is directly related to the population of aerosol particles in the atmospheric unit volume, using physical properties allowing to count and to classify by size each aerosol particle detected.

Though there are several technical solutions available to this scope, particles cannot be detected and counted with a unique instrument covering the whole size range. The most widely used are the so-called optical particle counters (OPCs), based on light scattering of a laser beam by each single aerosol particle, which allow to measure PM population from 200 nm upwards. These instruments therefore, though very useful and widely available, do not cover the particles of smaller size, associated with nucleation processes and to the initial steps of aerosol growth whose relevance has been explained before in this chapter. In the case of particles smaller than 200 nm, other instruments are required based on other physical techniques, such as particle ionization and subsequent acceleration in an electric field or on condensation of cold vapours on their surface, in order to let the particles grow in size and achieve a size detectable by the OPC.

Whatever the instrumentation available, the availability of particle size spectra even with an OPC which cannot access ultrafine particles provides a fundamental integration in PM characterization since it produces direct and instant information on particle size with emphasis on submicron fractions prone to access the respiratory system to depth, at high time resolution when PM mass load data only rarely reach the sub-daily frequency. At the same time, this approach is very efficient in providing prompt information on the influence of pollution sources (particle size and population, distance and intensity of the sources) and degree of ageing of PM as related to lifetime as a function of size and to its "in-transit" transformations.

Moreover, number density data can be used to obtain gravimetric data such as PM_{10} or $PM_{2.5}$, on the basis of suitable inversion processes based on the material density of particles. This step presumes the knowledge of PM average chemical composition, information not always promptly available and above all subjected to time/space fluctuations at each single site. This means that OPCs allow the coupling of PM size distribution to a more or less approximated mass estimate at a time resolution higher than traditional metrics, a highly desirable approach as it provides a high degree of detailed information on the behaviour of the airshed investigated and to its influence on the population and on the environment.

In order to emphasize the relevance of collecting particle size data, it is worth to know how in last years the scientific community is strongly debating about the possibility of modifying the current PM metrics based on PM_{10} and $PM_{2.5}$ shifting towards the number density concentration one. Though this parameter is very relevant to health issues and to all the other environmental and climatic problems associated with PM, in our opinion a holistic approach including both the traditional and the proposed metrics is highly preferable as they do contribute synergistically to a better understanding of PM multidimensional phenomenology and to its potential or actual health outcome.

As a last example of aerosol instrumental techniques presently available, we would like to mention aerosol mass spectrometry (AMS), a relatively recent family of powerful instruments based on single particle counting coupled to mass spectrometry analysis for particle identification; this means instruments are capable of:

1. Sampling atmospheric particles using a pump station interfaced with size-selective inlets
2. Detecting and identifying the composition of each particle by mass spectroscopy analysis

Some details on the methods can be found, for example, at http://cires1.colorado.edu/jimenez/ams.html and http://fy.chalmers.se/OLDUSERS/molnar/lectures/Measurement%20Methods%20II-filer/ATOMFS-3800.pdf (visited 14 September 2016), showing the outstanding features of AMS in terms of aerosol characterization potential; at the same time, there are three major difficulties in its routine applicability. First of all, the costs are very high, but another non-negligible aspect is the difficulty of associating mass to composition, given that we are not dealing with molecules but with particles which in most cases are random combinations of different chemical species (see the concept of internally mixed aerosols), and even in the case of identified particles, the environmental representativeness of the collected data may be very limited. In fact, even in the case of highly qualified investigations, the information achievable covers the identification of hundreds of thousands to millions of particles, which corresponds to huge efforts, but needs to be compared to real conditions. Ambient air in fact usually contains 10^5–10^8 particles/cm^3, and in EU, a daily $PM_{2.5}$ or PM_{10} sample is obtained filtering about 56 m^3/d. AMS is therefore an extremely powerful technique to investigate the complex physico-chemical processes behind aerosol nucleation and ageing processes, rather than in long-term surveys/investigations.

2.8 Aerosol Composition

2.8.1 A Dynamic Framework

As previously described, PM size distribution is very tightly associated with the formation processes and with the ageing of particles. A straight consequence of this complex phenomenology is not only that aerosol composition is extremely differentiated but that it varies with size, as well as in time and space, i.e. it has an extremely transient character. A smart and effective metaphor for ambient aerosol composition is that it resembles an "exploded pharmacy" as from [26]. In fact, each aerosol sample is a mixture of widely different chemical species on account of the physical mixing of the emissions of a series of sources as well as of their transformation during their transit in the atmosphere. Gases and particles are efficiently stirred and transported in the air including both natural and anthropogenic materials as a result of a restless atmospheric circulation. At each site in the troposphere[11] and all over the planet and at each time, PM composition has a unique composition which may be affected by local, distant and even remote sources whose relative importance stochastically depends on variable source emissions and intensities, transport conditions and degree of chemical transformations.

The efficiency of atmospheric transport clearly emerges when analysing data from catastrophic events, such as the deposition of radioactive fallout. A typical example is the redistribution pattern of radioactive plumes emitted in nuclear accidents such as Chernobyl in 1986 and Fukushima in 2011, when trace amounts of aerosol-borne radionuclides emitted from these point sources spread in the troposphere and were detected all over the whole Northern Hemisphere [33–35]. In the case of Fukushima accident after the tsunami impact in March 2011, the radioactive plume took about 3 weeks to reach Italy after the main emission event [36].

This highly dynamic situation constitutes the basis for the introduction of the Convention on Long-Range Transport Transboundary Air Pollution (CLRTAP),[12] which is related to the aim of quantitatively detecting the amount of atmospheric pollutants produced in a country with respect to aliquots produced by neighbour countries, adding to local sources through atmospheric transport. Moreover, meteorological conditions significantly affect absolute PM concentrations not only in determining which sources are entrained in air masses during transit but also in terms of wet scavenging which removes particles from the atmosphere, as well as of photochemistry related to solar periodicities, which largely modulate chemical reactions in the troposphere and products. All the listed factors participate to the aerosol budget in a given dynamical atmospheric volume through complex chains of input/output mechanisms.

In order to capture the overall properties of ambient aerosols at a given location (the so-called receptor site), it is necessary to plan a sufficiently high number of PM samplings followed by demanding and complex sequences of chemical analyses, the latter necessary to cover the extremely huge range of chemical species which constitute ambient aerosol mass. The need for such complex and costly experimental activity has several objectives:

- Understanding the composition of aerosols and reconstructing their mass balance
- Identifying the main aerosol sources through chemical fingerprinting
- Addressing the chemical species responsible for health adverse effects

[11] The troposphere is the innermost layer of the Earth's atmosphere, in contact with the planetary surface and containing 75–80% of the atmospheric mass.

[12] United Nations Economic Commission for Europe UNECE: http://www.unece.org/env/lrtap/welcome.html and http://emep.int/index.html (both sites visited on 26th September 2016)

In practice, owing to the extremely variable nature of PM, the experimental approach required for PM characterization is typically based on the analysis of hundreds of samples for a single station, encompassing several types of different analytical techniques followed by comparatively complex statistical and chemometric techniques by which it is possible to reach the objectives outlined. While some details on this approach will be briefly described later on in this chapter, information on aerosol components are provided in Chap. 3.

References

1. Seinfeld J, Pandis S. Atmospheric chemistry and physics: from air pollution to climate change. 2nd ed. Hoboken, NJ: Wiley; 2006.
2. Claxton L. The history, genotoxicity and carcinogenicity of carbon-based fuels and their emissions: 1. Principles and background. Mutat Res. 2014;762:76–107.
3. Directive 2008/50/EC on ambient air quality and cleaner air for Europe.
4. WHO Regional Office for Europe. Economic cost of the health impact of air pollution. Copenhagen: OECD; 2015. Retrieved from 13Aug 2016: http://www.euro.who.int/__data/assets/pdf_file/0004/276772/Economic-cost-health-impact-air-pollution-en.pdf
5. Fuzzi S, Baltensperger U, Carslaw K, Decesari S, Denier van der Gon H, Facchini M, et al. Particulate matter, air quality and climate: lessons learned and future needs. Atmos Chem Phys. 2015;15:8217–99.
6. West J, Cohen A, Dentener F, Brunekref B, Zhu T, Armstrong B, et al. What we breathe impacts our health: improving understanding of the link between air pollution and health. Environ Sci Technol. 2016;50:4895–904.
7. Oberdöster G, Oberdöster E, Oberdöster J. Nanotoxicology: an emerging discipline evolving from studies. Environ Health Perspect. 2005;113(7):823–39.
8. Pope C III, Dockery D. Health effects of fine particulate air - critical review. J Air Waste Manage Assoc. 2006;56:709–42.
9. Thurston G, et al. Ischemic heart disease mortality and long-term exposure to source-related components of U.S. fine particle air pollution. Environ Health Perspect. 2016;124(6):785–94.
10. Zanobetti A, Dominici F, Wang Y, Schwartz J. A national case-crossover analysis of the short-term effect of PM2.5 on hospitalizations and mortality in subjects with diabetes and neurological disorders. Environ Health. 2014;13(1):11. https://doi.org/10.1186/1476-069X-13-89.
11. Prinn R. 4.01 ozone, hydroxyl radical, and oxidative capacity. Treat Geochem. 2003;4:1–19.
12. Monks P. Gas-phase radical chemistry in the troposphere. Chem Soc Rev. 2005;34:376–95.
13. George C, Ammann M, D'Anna B, Donaldson DJ. Heterogeneous photochemistry in the atmosphere. Chem Rev. 2015;115:4218–58.
14. Pöschl U, Shiraiwa M. Multiphase chemistry at the atmosphere–biosphere interface influencing climate and public health in the anthropocene. Chem Rev. 2015;115:4440–75.
15. Moschandreas D, Watson J, D'Abreton P, Scire J, Zhu T, Klein W, Saksena S. Chapter three: methodology of exposure modeling. Chemosphere. 2002;49:923–46.
16. NAS. Human exposure assessment for airborne pollutants: advances and opportunities. Washington DC: National Academy Press; 1991.
17. Claxton L. The history, genotoxicity, and carcinogenicity of carbon-based fuels and their emissions: part 5. Summary, comparisons, and conclusions. Mutat Res. 2015;763:103–47.
18. Kim K-H, Kabir E, Kabir S. A review on the human health impact of airborne particulate matter. Environ Int. 2015;74:136–43.
19. Andreae MO. Climatic effects of changing atmospheric aerosol levels. World Survey Climatol. 1995;16:347–98.
20. Finlayson-Pitts B, Pitts J. Chemistry of the upper and lower atmosphere—theory, experiments, and applications. 2nd ed. San Diego, California: Academic Press; 2000.

21. Li WJ. A review of single aerosol particle studies in the atmosphere of East Asia: morphology, mixing state, source, and heterogeneous reactions. J Clean Prod. 2016;112:1330–49.
22. Biswas P, Wu C-Y. Nanoparticles and the environment. J Air Waste Manage Assoc. 2005;55:708–46.
23. Smita S, Gupta S, Bartonova A, Dusinska M, Gutleb A, Rahman Q. Nanoparticles in the environment: assessment using the causal diagram approach. Environ Health. 2012;11(S13):1–11.
24. Ohata S, Moteki N, Mori T, Koike M, Kondo Y. A key process controlling the wet removal of aerosols: new observational evidence. Nature Sci Rep. 2016;6:34113. https://doi.org/10.1038/srep34113.
25. Donaldson K, Stone V, Clouter A, Renwick L, MacNee W. Ultrafine particles. Occup Environ Med. 2001;58:211–6.
26. Pöschl U. Atmospheric aerosols: composition, transformation, climate and health effects. Angew Chem. 2005;44(46):7520–40. https://doi.org/10.1002/anie.200501122.
27. Ruzer L, Harley N. Aerosols handbook: measurement, dosimetry, and health effects. Boca Raton, FL: CRC Press; 2005.
28. Abdul-Jabbar S, Martini L. A review: are all inhaled fibres, such as asbestos, toxic? Int Arch Clin Pharmacol. 2015;1(1):1–5.
29. Buseck P, Pòsfai M. Airborne minerals and related aerosol particles: effects on climate and the environment. Proc Natl Acad Sci. 1999;96(7):3372–9.
30. Kulkarni P, Baron PA. Aerosol measurement: principles, techniques, and applications. 3rd ed. Hoboken, NJ: Wiley-Blackwell; 2011.
31. Kim I-S, Jang J-Y, Kim T-H, Park J, Shim J, Kim J-B, et al. Guidelines for the prevention and management of cardiovascular disease associated with fine dust/Asian dust exposure. J Korean Med Assoc. 2015;58(11):1044–59.
32. Kim S, Jaques PA, Chang M, Froines JR, Sioutas C. Versatile aerosol concentration enrichment system (VACES) for simultaneous in vivo and in vitro evaluation of toxic e&ects of ultra'ne, 'ne and coarse ambient particles part I: development and laboratory characterization. Aerosol Sci. 2001;32:1281–97.
33. Battiston G, Degetto S, Gerbasi R, Sbrignadello G, Tositti L. Fallout distribution in Padua and Northeast Italy after the Chernobyl nuclear reactor accident. J Environ Radioact. 1988;8:183–91.
34. Masson O, Baeza A, Bieringer J, Brudecki K, Bucci S, Cappai M, et al. Tracking of airborne radionuclides from the damaged fukushima dai-ichi nuclear reactors by european networks. Environ Sci Technol. 2011;45:7670–7.
35. Tositti L, Brattich E, Cinelli G, Previti A, Mostacci D. Comparison of radioactivity data measured in PM10 aerosol samples at two elevated stations in northern Italy during the Fukushima event. J Environ Radioact. 2012;114:105–12.
36. Schmid O, Stoeger T. Surface area is the biologically most effective dose metric for acute nanoparticle toxicity in the lung. J Aerosol Sci. 2016;99:133–43.

The Relationship Between Health Effects and Airborne Particulate Constituents 3

Laura Tositti

3.1 Major and Minor Components

PM is mainly constituted by solid abiotic particles including mineral insoluble phases; biological particles such as pollen grains, spores, mould, bacteria and viruses; fractal agglomerates based on elemental carbon assemblies originated in combustions; and larger amounts of physically less defined, fairly polar material composed of inorganic salts and organic matter, more or less deliquescent as a function of relative humidity in the air masses.

Independently from the natural or the anthropogenic origin, the average composition of aerosols includes major and minor components virtually covering the whole periodic table of elements in combinations of minerals, salt oxides, etc., while the carbonaceous fraction, which is often very abundant, is reputed to account up to several hundreds of different compounds.

As explained further on in this paragraph, major components make up the mass of PM and, to a first approximation, do not include particularly noxious substances but comprehend species not exclusively related to natural/anthropogenic sources. Minor and, above all, trace species play the most important role in this field of knowledge. In fact, in spite of their negligible mass over PM composition, *trace species are the most informative* as they do not only provide the basis for efficient chemical fingerprinting used in the identification of PM emission sources, but they also include the most relevant components from the biological point of view.

In general, the most abundant components of particulate matter are both natural and anthropogenic and may have both a primary and a secondary origin as reported in Table 2.2.

L. Tositti
Department of Chemistry "G. Ciamician", University of Bologna, Bologna, Italy
e-mail: laura.tositti@unibo.it

F. Capello, A.V. Gaddi (eds.), *Clinical Handbook of Air Pollution-Related Diseases*, https://doi.org/10.1007/978-3-319-62731-1_3

Among primary species, PM contains *mineral debris* deriving from soil and road abrasion, *biological debris and microorganisms,*[1] *sea salt* and *organic and elemental carbon and trace elements*, while secondary components, which are usually the most abundant species, include *SIA* (secondary inorganic aerosol) and *SOA* (secondary organic aerosol) [1].

In order to understand the average compositional of PM, we refer to the recent overview by [2] depicting the partitioning of the major components of PM_{10} and $PM_{2.5}$ across Europe.

3.2 Crustal Material

This component which is usually found in the coarse fraction mainly derives from the mechanical disruption (abrasion) and resuspension of geological materials such as rocks and soils; it is typically of mineral origin and therefore fairly insoluble in most cases. Owing to its (relatively) large size and short residence time in the atmosphere, it is mainly locally originated.

Most noticeable is the so-called mineral dust, which can be advected from the local up to the synoptic scale as a result of transport mainly from the huge desert areas in Africa and Asia, often significantly affecting local PM levels [3]. Though essentially natural, mineral dust transport is carefully monitored and investigated, as many climatological data suggest that frequency and intensity of these phenomena are increasing as a result of climate change operated by humans (see, e.g. https://www.ipcc.ch/publications_and_data/ar4/wg1/en/ch7s7-5-1-1.html, visited 25th November 2016). In addition to the pure meteo-climatic effects, there is more than a basic evidence suggesting that mineral dust plays a key role in atmospheric processes through active heterogeneous chemistry mediated by its surface area [4], eventually exacerbating atmospheric pollution [5] and finally inducing health outcomes [6].

Where the most widely considered crustal fraction of PM is mainly coarse in size as a result of its mechanical origin, the occurrence of crystalline silica (classified in Group I by IARC—International Agency for Research on Cancer—in particular in the form of quartz and cristobalite [7]) in each of the sources listed below cannot be

[1] This carbonaceous fraction spans from bacteria, viruses, pollen and spores, i.e. living organisms which are more than bioactive species and which are always present in PM, down to biological debris from the decomposition of animals and plants and from the human body. As such, the living components may cause infections, diseases and allergies mostly on a seasonal basis, with moulds being most active in indoor environments. Though widely investigated in the field of pathology, they are far less examined than all the other chemical compounds of PM especially in aerosol science even if a holistic approach should correctly suggest major efforts for including also the biological component in the overall evaluation of the real environmental and health effects of PM. For this reason, though the authors are aware of the importance of bioaerosol, they will not provide details on it in the present chapter, while they indicate how this topic is of increasing interest especially as far as indoor environments.

ruled out, leading to potential occupational and nonoccupational exposure. Therefore, the evaluation of the crustal component should not be underscored.

Mineral dust [8, 9] includes also submicron components leading to likely inhalation exposure. In this framework, also volcanic emission must be included among the mineral sources deserving special attention as they include also highly aggressive PM components accompanied by abundant mineral fractions as reported by literature inspired by the recent volcanic eruption of Eyjafjallajökull (Iceland) in 2010 [10] and more in general to integrate the well-established knowledge of PM [11, 12].

Indeed, mineral components are historically related to adverse effects on human health as in the case of asbestosis and silicosis, for which the distinction between natural and anthropogenic origin has very labile borders. What is undoubted is that silicates are well known as noxious agent with very clearly identified health outcomes, though their mechanistic influence is not yet thoroughly elucidated [13].

In addition to natural sources, there are other mineral sources related to human activities which may sensibly affect PM composition. The distinction between them is not always straightforward. In fact, crustal materials are remobilized and resuspended in mixtures which are often difficult to solve owing to similarity and more or less direct connections. In urban environments, the mineral components do not only include what is generally identified as soil resuspension (typically larger in the warm season owing to drier conditions) but also road debris produced by traffic. This component, extremely heterogeneous, may be partly similar to crustal components from asphalt as well as ceramic fragments from vehicle brakes and catalysers, though it is typically accompanied by other peculiar components which reveal their origin such as copper and iron (from brakes) and zinc (from tyres).

Other work activities involved in the emission of crustal materials with large potential to confound with natural emissions, range from agriculture, cave and mining work to tunnel boring (as well as other underground construction activities), the building and demolition sector and the building material industry covering the production of cement, bricks and tiles. The latter ones, besides grinding and mixing rocks and minerals in the coarse size emissions, include also high-temperature operations contributing to the submicron component. Finally, an interesting and, to some extent, emerging emission source owing to the huge increase in the maritime traffic is the shipyard sector including emission of crustal particles as a result of sandblasting over metal surfaces as found by the authors within a current investigation on the source apportionment of particulate matter in a coastal industrial location in northern Italy (unpublished data).

A detailed overview of the component herein described can be found in [7].

3.3 Sea Salt

The sea-salt component is typical of coastal regions and consists in crystals consistent with the average composition of seawater (see Table 3.1), as usually accepted, given the conservativeness of chemical compounds in salinity.

Table 3.1 Average composition of seawater

Ion	% of total salinity	Concentration (g/L)
Chloride	55.03	17.6096
Sodium	30.59	9.7888
Sulphate	7.68	2.4576
Magnesium	3.68	1.1776
Calcium	1.18	0.3776
Potassium	1.11	0.3552
Bicarbonate	0.41	0.1312
Bromide	0.19	0.0608
Borate	0.08	0.0256
Strontium	0.04	0.0128
Fluoride	0.003	0.00096
Other	0.001	0.00032

The formation of this aerosol fraction is mechanical and is mainly caused by the so-called whitecappings, an interaction among wave breakings, rising gas bubbles and wave breakings and bursting at the sea surface, followed by the production and release of seawater droplets which dehydrate once suspended in air; after water evaporation, the suspension contains salt crystals which mix with all the other aerosol particles. Sea-salt aerosol is typically produced in the coarse fraction which prevents its transfer farther inland, unless under the influence of very strong wind events. Reflecting the composition of seawater, according to data reported in the table above, its composition is not expected to produce adverse effects on the human health, though it may interact with other aerosol components in areas affected by atmospheric pollution and ageing, such as the strong secondary acidic components (i.e. HNO_3) which may release HCl upon exchange reaction with NaCl and consequent formation of $NaNO_3$ in the aerosol phase [14].

3.4 Secondary Inorganic Aerosol

Secondary inorganic aerosol (SIA) is composed of ammonium, sulphates and nitrate, where the anions are produced by oxidation, respectively, of SO_2 and NO_x, two of the so-called "classical" gaseous pollutants.

SO_2 is mainly emitted by the combustion of condensed fossil fuels, and in particular coal and oil, but has also substantial natural sources such as volcanoes and some volatile biogenic organic sulphides emitted by marine phytoplankton which, following seawater/air diffusion, is oxidized in the troposphere to SO_2 (secondary contribution), while NO_x are almost entirely a by-product of all the combustion systems, deriving from the combination of atmospheric N_2 and O_2 within the combustion environments which are characterized by extremely exothermic conditions and high temperature, promoting the reaction between the elements in the gaseous state.

SO_2 and NO_x[2] are oxidized (mainly by OH• radical, therefore within photochemically driven processes), respectively, to H_2SO_4 and HNO_3 both highly hygroscopic and polar, leading to a phase variation from the gaseous to the condensed phase upon reaction, so that these reactions are essentially the prototype examples of gas-to-particle conversion reactions, i.e. some of the most relevant formation mechanisms of PM in the troposphere.

These irreversible chemical reactions controlled by radical mechanisms in the atmosphere lead to the formation of strong acidic species with kinetically complex mechanisms and slow gas-to-particle conversion rate of about 5–6%/h [15], the latter being an average value depending on daily and seasonal conditions, as the reactions involved are temperature dependent and photochemically driven, modulated by meteorology, transport, photochemistry and latitude. More details on the chemical reactions involved can be found in [16] or in [17].

The strong secondary acids produced are largely neutralized by atmospheric gaseous ammonia (NH_3), the most abundant alkaline substance in the troposphere, mainly emitted by the natural nitrogen cycle, in turn more or less affected by the agricultural industry and to a lesser extent by the catalytic converters of gasoline cars[3] and finally by industry. This reaction leads to the formation of ammonium sulphates and nitrate salts which are among the most abundant component of PM; in Europe, it accounts for about 34% by weight of PM_{10} [18]. The neutralization process may reach completeness according to stoichiometry as a function of the relative availability of the various emitting sources and of their lifetimes; this means that in case of an unbalance between acids and bases, aerosols may display an acidic character potentially expressed upon inhalation when in contact with lung fluids, similarly to what is observed in environmental conditions for hydrometeors (rain, snow, fogs and hail).[4]

The adverse consequences of acid rains, one of the major environmental emergencies of the recent past, peaked around the 1970s–1980s among the most developed countries, nowadays are fairly well known and have been largely mitigated to a great extent in Europe and the Western USA [19, 20] but are still a current problem in developing countries whose rapid industrialization still rely on older technologies and in the Eastern USA [21]. It is to note that according to [1] and references therein included, non-neutralized acidity of particle is thought to be noxious or at least capable to initiate subsequent inflammatory processes in the respiratory system.

[2] It must be pointed out that NO_x in the air is subjected to competitive reactions leading not only to SIA and therefore to PM but also to ozone, largely contributing to the overall oxidation chemistry and to the oxidation capacity of the atmosphere.

[3] Ammonia is produced by excess reduction of NO_x in the three-way vehicle catalysers, and its occurrence/relevance in an urban environment is a function of the traffic level.

[4] The hydrological cycle, which includes the return of water evaporated from the oceans to the Earth's surface through precipitation events meteorologically driven, necessarily relies on the availability of aerosol particles; this means, in brief, that the occurrence of fine and ultrafine PM is firstly a natural driver of cloud formation but also that when PM is affected by atmospheric pollution, cloud processing will consequently affect the physico-chemical properties of hydrometeors.

3.5 Carbonaceous Aerosols

The carbonaceous fraction is among the dominant aerosol components and is estimated as being "at least one third of PM composition" according to [22].

The characterization of the carbonaceous fraction is still one of the most challenging aspects of aerosol science, owing to the complexity of the mixtures including several hundreds of different compounds. In fact, the carbonaceous component includes an extremely wide range of molecular species, with highly variable molecular weights and polarity in relation to functional groups and heteroatoms and finally with more or less pronounced volatility and chemical lability; in addition, not only single species but even each class of compounds is present in very small amounts requiring extreme analytical skills for their identification and quantification. The marked reactivity of the atmospheric environment, with a strong oxidizing character, further contributes to the instability of the carbonaceous fraction which, once emitted, may undergo further "in transit" reactions. As previously described, atmospheric reactivity is not only responsible of the production of secondary aerosol from gaseous precursors, but, once formed, PM may further be modified through atmospheric oxidants, and it may actively participate to atmospheric chemistry acting as a mediator through adsorption and desorption of vapour substances in a complex interplay of multiphase reactions at the surface of the particles themselves.

The overall carbonaceous fraction is an abundant component of PM, quantitatively comparable to SIA, and is subdivided in:

- *Elemental carbon*
- *Organic carbon* which includes both primary and secondary derivatives
- *Carbonates* mostly inorganic, i.e. of geological origin

Carbonates are usually treated separately from the former carbonaceous fractions; in most cases, they are excluded from the total carbon calculation (usually defined as the sum of elemental and organic carbon) but are included in the closure of PM mass balance, though they have been analysed in detail for source apportionment purposes in very specific local environments as, for example, in [23].

Elemental carbon (from now on EC for the sake of simplicity, see below for more details) is a typical by-product of all combustions and is related to pyrolysis mechanisms which occur in all burners and flames under localized oxygen-deficit conditions. EC sources are therefore associated with a range of activities, from energy production, manufactures and transformation to biomass burning, transportation, heating, cooking and smoking from the large-scale industry down to individual habits and lifestyles.

EC particles are not chemically modified once formed; they are characterized by the typical black colour and a very fine grain size, in the ultrafine range (see figure where soot particles can be clearly identified in TEM imaging), which rapidly grow to larger composite structures characterized by huge specific surface area and a fractal geometry [24]. EC mass fraction in PM may range from a few unit percent up to around 10% as typically reported in Europe, where this parameter may even reach higher values as reported in [25]. According to [26] who include in their review the latest evaluation on the relationship between black carbon[5] and health outcomes and to the recent report by the World Health Organization [28], EC can be assumed as a "universal carrier" of biochemically active species. In fact, this allotropic form of carbon is known for being a good electronic conductor; as such, this property is associated, among others, with the agglomeration of the smallest units, suggesting further potential interactions with cells. It is interesting to note how such properties are the basis of nanotechnology and of the success of this range of materials, where it is not only involved in the development of smart products, but it is already investigated in connection with exposure and health risks in professionals [29].

While the direct action of EC on health has not been definitely outlined yet, there is more than an evidence that this component is highly biochemically active through surface processes mediated by adsorption/desorption mechanisms on adsorbed VOCs resulting in efficient multiphase chemistry and in the production of secondary organic bioactive species [30].

Strictly speaking, *organic carbon* (OC) refers to the elemental carbon fraction occurring in complex molecular moieties of PM defined as "organic matter" accounting for the heteroatoms (oxygen and nitrogen, beyond hydrogen) and the multiplicity of organic compounds in which carbon is structurally combined. OC occurs both in the sub- and super-micron fractions of PM. Coarse OC includes both biological particles (pollen, spores, bacteria, etc.) and biological debris from plant and animal degradation processes; as such, they may have seasonal modulations in the outdoor environment. Fine and ultrafine organic aerosol may be classified in primary (POA) and secondary (SOA) organic aerosol. While POA is mainly associated with combustion and therefore to EC, SOA is produced by chemical reactions and by gas-to-particle conversion of VOCs in a complex way [31]. As previously mentioned, its characterization is still one of the most challenging aspects in PM investigations owing to the elevated number of components, highly variable in mass, structure, polarity, reactivity and volatility [32]; the principal components are summarized in Table 3.2 from [31].

Owing to the relative low amounts of each single class and a fortiori of individual compounds, only part (though ever increasing) of the aerosol organic compounds have been identified so far [32], and among the known ones, only a few are regulated and monitored as hazardous compounds. The latter typically include PAHs (polycyclic aromatic hydrocarbons) and dioxins, both originating from combustion and whose toxicity, mutagenicity and carcinogenicity have long been recognized. The complexity in OC characterization and impact assessment reflects not only the analytical difficulties as previously described but also the range of sources as well as the physico-chemical conditions occurring in the atmosphere.

[5] Strictly speaking, elemental carbon and black carbon are not synonyms and are complemented by even further terms, i.e. soot and refractory carbon and more. This is due to the dependency between the carbon-related parameter measured, the experimental/instrumental approach adopted and its structural properties, a series of conditions which have prevented so far from a general consensus on terminology and definitions. For an exhaustive update on the topic we address to [27].

Table 3.2 Principal organic particulate components [31]

Substance classes	Proportions[a]	Sources
Aliphatic hydrocarbons	10^{-2}	Biomass, fossil-fuel combustion
Aliphatic alcohols and carbonyls	10^{-2}	Biomass, SOA/ageing
Levoglucosan	10^{-1}	Biomass burning
Fatty acids and other alkanoic acids	10^{-1}	Biomass, SOA/ageing
Aliphatic dicarboxylic acids	10^{-1}	SOA/ageing
Aromatic (poly-)carboxylic acids	10^{-1}	SOA/ageing, soil/dust
Multifunctional aliphatic and aromatic compounds (OH, CO, COOH)	10^{-1}	SOA/ageing, soil/dust
Polycyclic aromatic hydrocarbons (PAHs)	10^{-3}	Fossil-fuel combustion, biomass burning
Nitro- and oxy- PAHs	10^{-3}	Fossil -fuel combustion, biomass burning, SOA/ageing
Proteins and other amino compounds	10^{-1}	Biomass
Cellulose and other carbohydrates	10^{-2}	Biomass
Secondary organic oligomers/polymers and humic-like substances	10^{-1}	SOA/ageing, soil/dust

[a]Characteristic magnitudes of the mass proportions in fine OPM

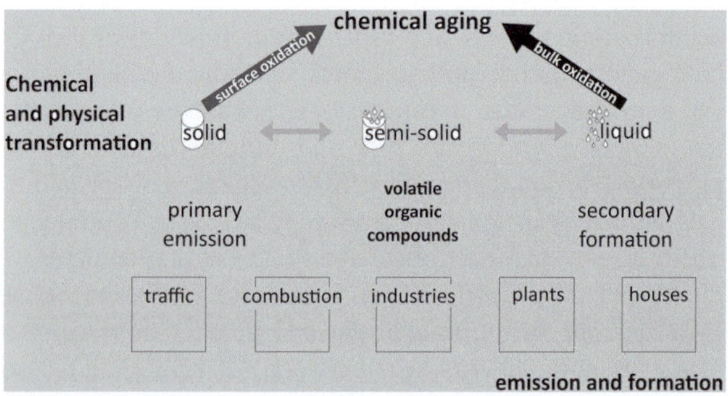

Fig. 3.1 Scheme depicting the cycle of organic aerosol

In fact, organic substances, besides deriving from countless emission sources, are sensitive to photochemical influence by solar radiation, both directly (photodissociation) and indirectly (through oxidation mechanisms triggered by OH• radical and others and/or cross-reactions with chemical species in the gaseous phase, such as, e.g. ozone) as well as to NO_3 radical, yielding a series of other organic derivatives of secondary origin. As a result, atmospheric chemistry significantly and remarkably affects both the stability of primary/secondary organics and the formation of secondary species (SOA) from anthropogenic and biogenic VOCs but also from pre-existing POA. The ensemble of reactions to which aerosol and in particular OC undergo during its permanence in the air is usually accounted for as "ageing" of PM [30]. The scheme in Fig. 3.1 depicts the complex of different chemical interactions involving PM once it has formed.

While SOA is estimated as ranging between 50 and 85% of OC mass [33], OC and, above all, SOA speciation may be affected by the complexity in the oxidation mechanisms involved, characterized by a number of steps affecting the build-up not only of the final, most stable species but also of intermediates, both in the radical and in the molecular state.

Following oxidation reactions which are subjected to day/night modulation, derivatives result enriched in oxygen and nitrate such as, respectively, carbonyls and carboxylic species and in nitro-derivatives. These products are therefore progressively more polar than their precursor, which facilitates aggregation into/onto particles; most interestingly some derivatives may even be more hazardous than their precursors as recently observed for nitro-PAHs and quinones, both known for their adverse effects when inhaled [34].

Interestingly, the single-chemical organic species, which make up OA, are usually occurring at concentration levels of around the per-mil level or less, as compared to the overall bulk aerosol carrier in which they are transported. As an example, if we take the EU metrics for airborne particulate matter in the air quality EU networks (EU Directive), as a reference, we find that the PM_{10} limits are 40 and 50 $\mu g/m^3$, according to the time interval over which the parameter is assessed (1 year and 24 h, respectively), while for $PM_{2.5}$, included in PM_{10}, the threshold is 25 $\mu g/m^3$; these values must be compared with PAHs and in particular with benzoapyrene (BaP) threshold in PM_{10}, which has been set to 1 ng/m^3. This class of micropollutants which has been historically associated with mutagenicity and carcinogenicity is mostly found below this concentration threshold. However, we would like to draw the reader's attention towards the comparison among all the threshold values mentioned; it can be easily noticed that BaP limit in air, even considering that, according to the EU standards, is a weighted mean of several PAHs, is of four orders of magnitude smaller than the massic carrier, be it PM_{10} or $PM_{2.5}$. This simple evaluation points out to one of the fundamental issues in aerosol toxicology: a reactive species in PM is able to interact adversely with human cells at extremely low concentration levels compared to other fields of toxicology.

3.6 Trace Elements

Another class of substances of major concern in PM characterization is the one represented by *trace elements*, including metals and metalloids. Usually this component does not exceed altogether more than a few unit percent by mass of PM_{10}, virtually covering a largest part of the periodic table of elements, radionuclides included, and therefore substantially contributing together with OC to what we could define "*chemodiversity*",[6] previously described. Similarly to organic substances, therefore, the concentrations of each single element are several orders of magnitude smaller than the PM mass loading [35].

[6] N.B. the term "chemodiversity" is a neologism by Laura Tositti, aimed at including all the different chemical species occurring at environmental level and covering elements with their compounds (i.e. chemical speciation in a literal sense by definition contributing to the mass balance of the single element) resulting from natural and anthropic sources and interactions and including also isotopic fractionation. This term should also include the speciation of the carbonaceous fraction, which is typically abundant and highly differentiated.

With the exception of the major inorganic elements such as SIA, sea salt in coastal areas and some mineral components associated with geological materials, whose major constituents are, respectively, sulphur, nitrogen, sodium chloride and the other salinity compounds, silicon, aluminium and calcium, the other elements occur in PM in much lower and highly differentiated concentration levels, which reflect both the natural abundance in natural sources and the anthropogenic influence, with human activities often sensibly altering the natural concentration ratios of elements. Among the anthropogenic sources, high-temperature processes are largely, even if not exclusively, responsible of trace element release and emission; among them, large- and small-scale combustion are the most important ones (see, e.g. [36, 37]). Combustion is pervasive in our society as it is at the basis for energy production, transformation industry, waste management and transport, not to mention myriad of individual activities such as driving, cooking, heating, smoking and even lighting candle or incense, i.e. activities apparently negligible on the individual scale but which assume a quantitatively remarkable relevance when treated collectively [36].

The occurrence of elements in aerosols depends on several factors. Firstly, all the condensed fuels contain inorganic elements, in large part metals, which occur in living matter as essential elements with a very specific bioactive role in structural, functional and/or enzymatic chemistry. Both fossil and "fresh" fuels (e.g. biomass fuels) present the same characteristic, though the processes of fossilization determine an enrichment in the inorganic component of the carbonaceous substrates as a function of age and of the associated degree of mineralization, maximum for certain coals. According to their use, fossil fuels may also be mixed with metal additives usually to optimize the combustion process and to reduce polluting by-products as well as for lubrication purposes.

The rationale in connecting trace inorganic species to combustion, which is mostly associated with carbonaceous fuels and their derivatives/by-products, is that the heat released during combustion promotes the release of trace elements in vapour phase (as elements, oxides, chlorides, etc.) plus ultrafine and fine aerosol phases, which thereafter undergo redistribution and transport through the atmosphere at local and distant scales as a function of the particle size.[7] Their fate, which is very similar to the volcanic emissions made of gases, vapours and particles as function of the high temperature, affects air quality at various space and time scales determining exposure to humans and to the environment which is fairly challenging to assess.

Since trace elements are numerous in each single aerosol source, be it natural or anthropogenic, and often arranged in characteristic and identifiable ensembles, they are extremely useful fingerprints to assess emission sources and the extent of their impact on a receptor site, that is, the location at which the exposure is to be estimated (see further on in the source apportionment section). For example, we report the following Table 3.3 from [38] and/or the most recent by Calvo (Table 3.4) [39],

[7] It is to note that the emission of pollutants can be highly minimized when modern mitigation technologies are adopted, especially when compared to old industrial facilities and/or vehicles, etc. Nevertheless, minimization and up-to-date technologies do not mean zero emission; as a result regulations and monitoring are required and must be continuously updated and implemented.

Table 3.3 Elemental fingerprints of typical emission sources [38]

Characteristic elements emitted from various combustion sources	
Emission source	Characteristic elements emitted
Oil-fired power plants	V, Ni
Motor vehicle emissions	Br, Pb, Ba
Refuse incineration	Zn, Sb, Cu, Cd, Hg
Coal combustion	Se, As, Cr, Co, Cu, Al
Refineries	V
Non-ferrous metal smelters	As, In (Ni smelting), Cu
Use of pesticides	As
Iron and steel mills	Mn
Plant producing Mn metal and Mn chemicals	Mn
Copper refinery	Cu

Table 3.4 Up-to-date fingerprints based on trace elements [39]

Technogenic tracers	Steel industry	Cr, Ni and Mo
	Copper metallurgy	Cu and As
	Ceramic industries	Ce, Zr and Pb
	Heavy industry (refinery, coal mine, power stations)	Ti, V, Cr, Co, Ni, Zn, As and Sb
	Petrochemical industry	Ni and V
	Oil burning	V, Ni, Mn, Fe, Cr, As, S and SO_4^{2-}
	Coal burning	Al, Sc, Se, Co, As, Ti, Th, S, Pb and Sb
	Iron and steel industries	Mn, Cr, Fe, Zn, W, and Rb
	Non-ferrous metal industries	Zn, Cu, As, Sb, Pb and Al
	Cement industry	Ca
	Refuse incineration	K, Zn, Pb and Sb
	Biomass burning	K and Br
	Firework combustion	K, Pb, Ba, Sb and Sr
	Vehicle tailpipe	Platinum group elements, Ce, Mo and Zn
	Automobile gasoline	Ce, La, Pt, SO_4^{2-} and NO_3^-
	Automobile diesel	S, SO_4^{2-} and NO_3^-
	Mechanical abrasion of tyres	Zn
	Mechanical abrasion of brakes	Ba, Cu and Sb

reporting a series of the most frequently studied emission sources with their peculiar elemental tracers.

Trace element associations (emitted as “co-pollutants”) and source profiling can be achieved on the basis of multivariate methods applied to extensive datasets of analytical data carried out on series of PM samples. Clearly the resolution of such complex aerosol mixtures is hardly a straightforward process; nevertheless, this is a key stage in aerosol characterization not only in the source identification problems but also in the individuation of biochemically significant species.

The partitioning of trace elements as a function of particle size is another fundamental characteristic of PM as they are not uniquely attributable to mineral phases typical of the coarse fraction. Instead, using size-segregating samplers, it was possible to determine how inorganic species may be differently distributed over the wide size range of aerosol particles, often showing bimodal distributions when a single element is generated by two different significant processes. Some studies even pointed out how the same element occurring in different size modes usually shows different chemical behaviours according to distinct chemical speciation, in this case taking the stricter definition of this term, another parameter relevant both to atmospheric chemistry and to health. This point is particularly important as it allows us to introduce one of the fundamental issues in aerosol toxicology, i.e. the importance of the chemical form in triggering a biochemical interaction [40].

It is to note how the listed sources together with their so-called chemical fingerprints are largely traceable to combustion and/or high-temperature processes; overall this means that the conditions under which the mentioned processes are conducted are very likely to release complex volatilized phases, which may be emitted to the airshed as a function of the size of the process and of the mitigation steps applied (not quantitatively efficient even in the most modern configurations), as well as of the relative volatility of the chemical species involved. Condensation in the cooler environment leads thereafter to condensation of trace elements and to their transfer from the vapour phase into the particulate one.

As a result of high-temperature processes, trace elements populate the whole size range of PM, including the fine and ultrafine ones, indicating accessibility or more correctly easier and likely translocation inside the human body. Moreover, elements in the submicron fractions have been found to be more soluble and therefore potentially more bioactive and harmful.

Trace elements in the coarse fraction are considered less prone to affect human health for at least two reasons: (a) coarse particles are less numerous and access only the upper part of the respiratory system, and (b) they are often in a hardly soluble mineral phase which makes them less bioavailable, if not chemically inert, as a result of their immobilization in the crystal lattice of mineral phases mostly based on highly insoluble silicates. However, the outlined properties do not allow to neglect a priori a role in toxicity from coarse particles, which sometimes have shown an adverse effect in toxicity tests.

Owing to decades of research in the field of atmospheric pollution, we have been long aware of the toxicity of numerous trace elements which are currently subjected to regulation in the European countries.[8] We refer in particular to lead, arsenic, cadmium, nickel and mercury whose concentration level in ambient PM_{10} is presently regulated. With the exception of mercury (Hg), which is toxic and usually treated separately from the others even within the same EU Directive including these five elements in PM_{10}, all the other elements are classified by IARC as carcinogenic to humans (Group I) including As, Cd and Ni while Pb is included among the probably carcinogenic to human (Group II). European air quality standards set atmospheric

[8] European Directives 1999/30/EC and 2004/107/EC relating to lead, arsenic, cadmium and nickel; in Italy D. L. n. 155/2010, which includes also mercury (HG) and BaP previously discussed.

thresholds at very low ambient concentration levels for these elements, with values several orders of magnitude lower than the PM_{10} which acts as a carrier but is also a criteria pollutant in itself. The threshold concentrations for the listed elements, not to be overcome on an annual basis, are, respectively, 500 ng/m^3 for lead, 6 ng/m^3 for As, 5 ng/m^3 for Cd and 20 ng/m^3 for Ni (see list of European air quality standards at http://ec.europa.eu/environment/air/quality/standards.htm, visited on November 18th 2016). A large part of the knowledge concerning the hazard from specific trace elements derives from professional exposure outcomes, where the exposure to hazardous chemicals may be substantially higher than at environmental levels [41]. Moreover, the United States Environmental Protection Agency (US-EPA) has published a list of hazardous substances (HAPs, i.e. hazardous air pollutants; https://www.epa.gov/haps) which includes not only all the above-mentioned elements but also Cl, Cr (Group I in IARC classification), Co, Mn, Sb and Se. Each of these inorganic species is classified as potentially toxic as detailed in specific datasheet at the URL specified.

From a toxicological point of view, many trace elements are often defined as "heavy metals", a questionable, though still widespread, term improperly used to define their harmfulness [42]; for many of them, e.g. Hg (mercury) and Pb (lead), adverse effects on humans are related to their strongly "soft" character, a property which indicates a high affinity towards the sulphur-containing biomolecules (proteins, enzymes), which lead to the perturbation of homeostasis and to the inhibition of fundamental metabolic activity [43].

In aerosol science, what is mostly relevant in terms of trace element chemical role is the redox chemistry they can express, especially when transition elements are concerned, a behaviour which is extremely important both for the atmospheric chemistry implications as well as for the health-related issues. Trace elements, in fact, may interact both with abiotic and biotic substrates through complex radical mechanisms often including Fenton-like steps such as the in-cloud oxidation of SO_2 catalysed by iron—a well-known atmospheric process—[17] or, in the case of biological systems, the induction of ROS on cells and biological tissues by several elements such as Fe, Ni, Cu, Cr, Mn, Zn and V [44].

It is important to remind that in most cases some trace elements (as well as most of the PM components) do not exclusively belong to one single emission source; in this case the attribution requires a cautious evaluation of the data in order to solve adequately any ambiguity or confounding.

The overall aerosol chemistry so far described can be summarized by the following diagram (Fig. 3.2) depicting the multiphase interplay between the various components of PM as previously described.

The following scheme instead provides a synthesis of the other dynamic framework which characterizes ambient aerosol cycling in the troposphere (Fig. 3.3).

All the information provided so far on PM and illustrated in the short pictorial forms here above constitute the basic knowledge on which any health-related study is based.

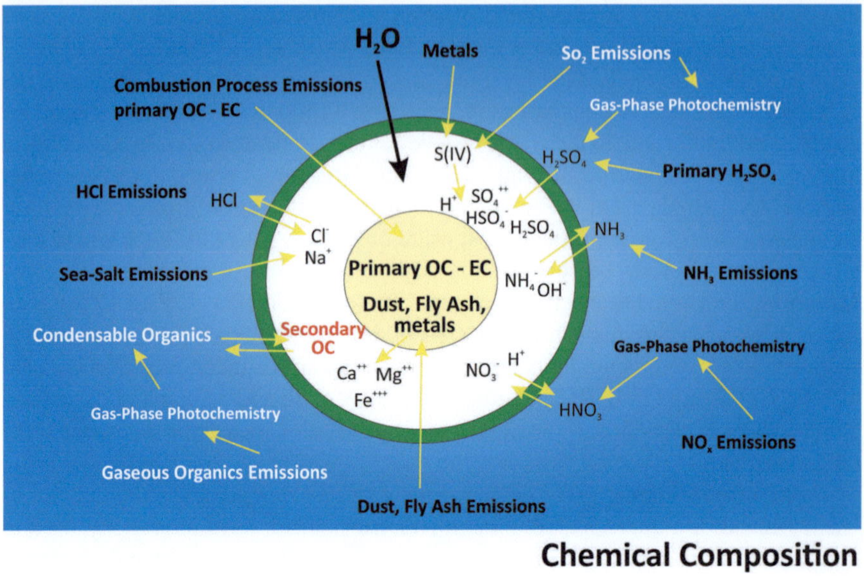

Fig. 3.2 Simplified scheme of aerosol particle structure, physico-chemical behaviour, reactivity and origin

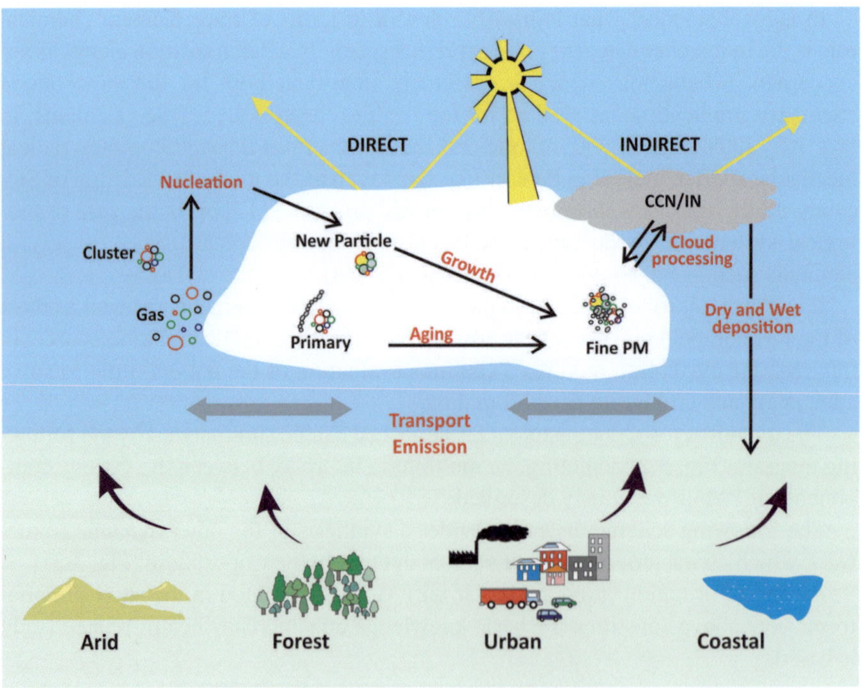

Fig. 3.3 The cycle of fine ambient particulate matter in an average urban framework

3.6.1 Source Apportionment

For the sake of completeness, it is important to introduce a non-negligible part of the aerosol research, i.e. the association of the complex and dynamic PM chemistry summarized in the previous paragraph with the respective emission sources.

The identification and quantification of PM emission sources, defined as source apportionment, are an ever-increasing field of application of atmospheric science, providing extremely efficient tools not only for research purposes but also in real-world applications such as environmental management and forensic investigations. Since this topic enables the reliable identification and quantification of the ensemble of emission sources active over a territory, this information can be successfully cross-linked to epidemiological data in order to enlighten the association between the chemistry of specific sources with adverse health effects.

The methods used for source apportionment date back to the late 1960s of the past century [45] but are continuously updated and optimized in agreement with the exponentially growing of scientific understanding in the field of atmospheric science and consequently in the field of aerosol science.[9]

Source apportionment is based on a series of modelling tools overall known as "receptor models" [46]. The name receptor models or receptor-oriented models arises from the fact that these methods are focused on the behaviour of the ambient environment at the point of impact as opposed to the source-oriented dispersion models that focus on the emissions, transport, dilution and transformations that occur beginning at the source and following the pollutants to the sampling or receptor sites. While source-oriented models are predictive models that may be applied both to PM and gaseous species and calculate ambient concentrations starting from source emissions and atmospheric dispersion models, receptor models are diagnostic models, which identify the sources and calculate their contribution starting from measured concentrations of PM mass and chemical compounds in the sampling site (receptor).

While extensive work on this ever-growing field of investigation is based on the experimental characterization of the aerosol plumes emitted by all the classified sources in order to estimate their input to troposphere both qualitatively and quantitatively, major efforts and on the opinion of the authors, with a higher degree of reliability, are carried out on post-emission studies.

The former approach, in fact, substantially allows to collect information on the primary pollutants emitted by a given source, providing invaluable qualitative information for the identification of the chemical fingerprints of each emission source based, for example, on trace elements or on more conservative chemical species. In most cases, though, a large fraction of the primary emissions are subjected to substantial post-emission chemistry owing to the extremely active, photochemically modulated atmospheric processing. These processes lead in fact to the most

[9] Fuzzi et al. [2] estimated from the ISI Web of Science database that the average number of ambient aerosol papers published in refereed journals increased from a few tens per year in the 1980s to the present day 1500–2000 papers per year.

important atmospheric secondary pollutants such as ozone and a conspicuous fraction of PM, as previously outlined. Moreover, for each distinct source, whether it is natural or technogenic, a large degree of variability is possible, not only in terms of intensity/flux but also in terms of composition, contributing to limited comparability and to the impossibility of applying solely a scaling factor.

The basic idea of source apportionment is that aerosol particles retain elemental-chemical composition characteristic of their origin: the PM composition at the receptor site is a combination of the compositions of the aerosols emitted by the different sources. The main hypothesis is then that mass conservation can be assumed and a mass balance analysis can be used to identify and apportion sources of airborne particulate matter in the atmosphere. The common approach is to obtain a dataset constituted by a large number of chemical constituents such as elemental concentrations in a number of samples. The assumption is that the measured concentrations derive from the summation of the mass contributions of a number of independent sources or source types [47]. The mathematical expression of the mass balance equation that accounts for all m chemical species in the n samples as contributions from p independent sources is

$$x_{ij} = \sum_{k=1}^{p} g_{ik} f_{kj} + e_{ij},$$

where $i = 1,\ldots, n$ samples; $j = 1, \ldots, m$ species; and $k = 1, \ldots, p$ sources.

Receptor models based on the mass balance equation require the validity of the following assumptions [45, 48, 49]:

1. All the sources that give a significant contribution have been identified (otherwise the mass reconstruction is poor).
2. Source chemical profiles are constant in time (or at least during the sampling period).
3. Source chemical profiles are constant in space, i.e. they do not change during transport from the source to the receptor site.
4. Source profiles are each other linearly independent (not collinear), in order to correctly disentangle the corresponding source contributions.

The mass balance equation above can be solved in a number of different ways, depending on the available information. The problem can be divided into two classes: the case in which the source profiles are known and the case in which the source profiles are unknown.

The solution can be achieved by different numerical approaches mostly based on multivariate techniques, whose description is well beyond the scope of this chapter. Details on the methods, which rely on costly and demanding chemistry as well as a high degree of numerical skill together with a continuous attention to technological developments and the changing emission panorama, can be found in [45, 46, 48, 49].

The availability of the above tools is of utmost importance as they allow to proceed to a series of potential actions ranging from the correct evaluation of environmental impacts, support within forensic investigations and consequent actions, addressing the improvement of an emission scenario in terms of environmental policy and decision-making.

The results of receptor modelling may be crucial also in the identification of emission sources whose chemical fingerprint is found to statistically correlate with health outcomes. Nevertheless, as seen, for example, in the paragraphs on carbonaceous aerosols and on trace elements, emission sources are typically characterized by a spectrum of different chemicals; therefore the cause-effect link between specific component (and or arrays of components) and the insurgence of health effects has not been resolved so far in a definitive way as illustrated in the next paragraph.

3.6.2 Aerosol Components and Bioactive Mechanisms

While epidemiological results provide a substantial and highly reliable body of evidence of the close connections between them, there are many difficulties in establishing the biochemical mechanisms between PM and its health impacts, a necessary step either to support epidemiological evidence or to plan mitigation actions, when needed.

Firstly, individual exposure and susceptibility provide another highly variable source of uncertainty [50]. Moreover also the health outcomes are not unique, as seen in the scheme previously reported in figure, showing that there are several possible paths through which PM can express its toxicological potential but bearing in mind that this synopsis mainly accounts for exposure by inhalation, by far the most important exposure path though not the exclusive one, implicitly including translocation mechanisms to organs and tissues else than the respiratory tract [51].

Secondly, the extreme complexity of PM mixtures and their variability as previously outlined require huge analytical efforts to characterize aerosol in detail and covering as much as environmental variance as possible.

Finally, testing the toxicity of PM encompasses countless and extremely differentiated methodologies, including cell-free, cell-based and in vivo experiments, which certainly provide a lot of valuable information but cannot rely on data comparability.

In this kaleidoscopic framework, the available and ever-increasing information suggest that PM toxicity is mainly due to the induction of oxidative stress, as deduced from the detection of ROS and/or of specific biomarkers like cytokines upon inhalation. This implies that biochemically relevant behaviour of PM is basically a redox chemistry with a marked oxidant character. According to [52] “Oxidative stress results when the generation of reactive oxygen species (ROS), or free radicals, exceeds the available antioxidant defences”. In brief it is clear that PM may exceed a “safety threshold” in the physiological response as a function of its properties and composition which depend on one or more emission sources contributing to the inhaled PM, on the basis of a given dose and exposure.

The need for a metric of PM toxicity based on its chemical behaviour has inspired the introduction of the concept of *oxidative potential* [53], which has been defined as “a measure of the capacity of PM to oxidize target molecules” [52]. At the time of the paper by [53], the introduction of this parameter was expected to substitute PM_{10} metrics to assess the health risk from PM, circumstance not accomplished so far. In fact, the oxidative potential can be determined, for instance, according to one of the many methods reviewed by [54], who provided also a rough but useful classification between cell-free and cell-based tests. Since then, no major standardization effort has been carried out, leaving a high number of degrees of freedom in the methodological choice. Among the most relevant papers the authors arbitrarily acknowledge, it is worth to cite the following examples, respectively, used for the cell-based testing such as [44] and for the cell-free testing, i.e. [52], both apparently reporting highly performing results.

In spite of the huge amount of heterogeneous toxicity data currently available, there is an emerging and agreed paradigm according to which PM toxicity can be formulated.

The principal, though still considered potential, mechanistic association between PM and adverse health effects by inhalation is considered *inflammation*, a cellular response activated by chemicals present in PM and mediated by ROS, in order to counteract the injuries they caused [55].

The PM properties and parameters which have been found to give the highest correlation with health effects either from epidemiological studies or by experiments above are number of ultrafine particles, surface area, transition metals (in particular Fe and Cu), soot (especially when produced by diesel engines) and organic compounds like PAHs, which can undergo ageing during PM permanence in the troposphere, leading to the formation of even more toxic oxygenated and nitro-derivatives such as quinones, phenols, epoxides, lactones, etc. [34].

All the mentioned chemical species share a marked redox behaviour via electron transfer processes, which has been found to trigger the production of ROS[10] [34, 55] and RNS (reactive nitrogen species) [30]. Both ROS and RNS may be produced in situ in the lungs or may occur in the atmosphere owing to the marked oxidant behaviour. In both cases, they will cause oxidative stress, whose biology is extensively reviewed in [56].

A pictorial scheme of the possible pathways of ROS to the lung is reported in Fig. 3.4. It is to note that the right column in Fig. 3.4, concerning the in situ ROS formation by PM, includes a list of all the gaseous and particulate species described in this section but also includes asbestos, which is historically associated to professional, accidental and environmental inhalation exposure producing fatal diseases, and fly ash deriving from coal burning in energy production but also from home cooking in Asia, the latter case being of major concern at present time, suggesting common mechanistic patterns, at least in the initial phase. In their dense paper, [34] devote a relevant part of their discussion to an extremely suggestive issue in which the ROS-inflammation paradigm is extended from the more classical PM component so far described to the proteins associated to pollen and spores responsible for allergies. The authors, in fact, on observing the relevant increase in recent decades

[10] ROS includes singlet oxygen ($^1O_2$), superoxide radical ($O_2^{\,-}$), hydrogen peroxide (H_2O_2), hydroperoxyl (HO_2) and hydroxyl radical ($\cdot OH$).

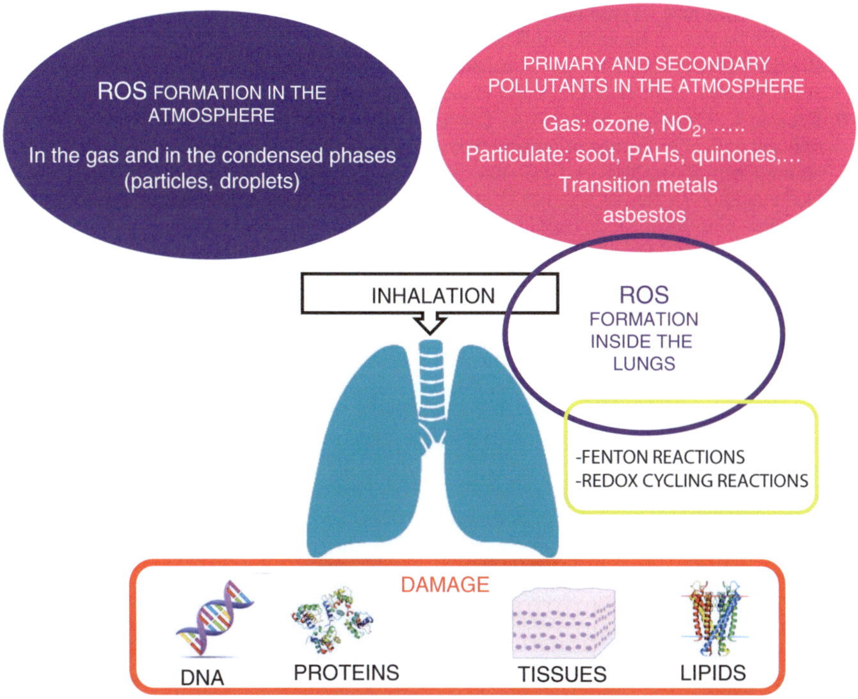

Fig. 3.4 The scheme depicts the two ways by which the lung may be accessed by ROS, inducing adverse effects on cells

of this v of diseases in urban population more than in rural one, hypothesize a multiphase interaction between gaseous tropospheric oxidants such as ozone and NO_2 leading to the nitration of the protein molecules and to enhanced allergenicity.

References

1. Harrison R, Yin J. Particulate matter in the atmosphere: which particle properties are important for its effects on health? Sci Total Environ. 2000;249:85–101.
2. Fuzzi S, Baltensperger U, Carslaw K, Decesari S, Denier van der Gon H, Facchini M, Gilardoni S. Particulate matter, air quality and climate: lessons learned and future needs. Atmos Chem Phys. 2015;15:8217–99.
3. Stafoggia M, al. Desert dust outbreaks in southern Europe: contribution to daily pm10 concentrations and short-term associations with mortality and hospital admissions. Environ Health Perspect. 2016;124(4):413–9.
4. Zender H, Charles S. Mineral dust and global tropospheric chemistry: relative roles of photolysis and heterogeneous uptake. J Geophys Res. 2003;108(D21):4672. https://doi.org/10.1029/2002JD003143.
5. He H, Wang Y, Ma Q, Ma J, Ji D, Tang G, Hao J. Mineral dust and NOx promote the conversion of SO2 to sulfate in heavy pollution days. Nature. 2014;4:4172. https://doi.org/10.1038/srep04172.
6. Sajani SZ, Miglio R, Bonasoni P, Cristofanelli P, Marinoni A, Sartini C, et al. Saharan dust and daily mortality in Emilia-Romagna (Italy). Occup Environ Med. 2011;68:446–51.

7. IARC. Silica dust, crystalline, in the form of quartz or cristobalite. International agency for research on cancer. Lyon: World Health Organization; 2012.
8. Brattich E, Riccio A, Tositti L, Cristofanelli P, Bonasoni P. An outstanding Saharan dust event at Mt. Cimone (2165 m a.S.L., Italy) in march 2004. Atmos Environ. 2015;113:223–35.
9. Denjean C, Cassola F, Mazzino A, Triquet S, Chevaillier S, Grand N, et al. Size distribution and optical properties of mineral dust aerosols transported in the western Mediterranean. Atmos Chem Phys. 2016;16:1081–104.
10. Gudmundsson G. Respiratory health effects of volcanic ash with special referenceto Iceland. A review. Clin Respir J. 2011;5(1):2–9.
11. Tam E, Miike R, Labrenz S, Sutton A, Elias T, Davis J, et al. Volcanic air pollution over the island of Hawai'i: emissions, dispersal, and composition. Association with respiratory symptoms and lung function in Hawaii Island school children. Environ Int. 2016;92-93:543–52.
12. Vignelles D, Roberts T, Carboni E, Ilyinskaya E, Pfeffer M, Dagsson Waldhauserova P, et al. Balloon-borne measurement of the aerosol size distribution from an Icelandic flood basalt eruption. Earth Planet Lett. 2016;453:252–9.
13. Fubini B, Fenoglio I. Toxic potential of mineral dusts. Elements. 2007;3:407–14.
14. Finlayson-Pitts B, Pitts J. Chemistry of the upper and lower atmosphere—theory, experiments, and applications. 2nd ed. San Diego, CA: Academic; 2000.
15. Gaffney G, Marley N. The impacts of combustion emissions on air quality and climate–from coal to biofuels and beyond. Atmos Environ. 2009;43:23–36.
16. Seinfeld J, Pandis S. Atmospheric chemistry and physics: from air pollution to climate change. 2nd ed. Hoboken, NJ: Wiley J; 2006.
17. Hewitt C. The atmospheric chemistry of sulphur and nitrogen in power station plumes (millennial review). Atmos Environ. 2001;35:1155–70.
18. Aksoyoglu S, Ciarelli G, El-Haddad I, Baltensperger U, Prévôt A. Secondary inorganic aerosols in Europe: sources and the significant influence of biogenic VOC emissions especially on ammonium nitrate. Atmos Chem Phys Discuss. 2016;17:7757–73. https://doi.org/10.5194/acp-2016-739.
19. Cazzuli O, Lanzani G, Giudici G, Tebaldi G. La qualità dell'aria in Lombardia: l'analisi delle serie storiche e l'evoluzione degli indicatori di pressione. TECAIR 2005, (p. 10). Rimini; 2005.
20. Jones A, Harrison R. Temporal trends in sulphate concentrations at European sites and relationships to sulphur dioxide. Atmos Environ. 2011;45(4):873–82.
21. Vet R, Artz R, Carou S, Shaw M, Ro C-U, Aas W, et al. A global assessment of precipitation chemistry and deposition of sulfur, nitrogen, sea salt, base cations, organic acids, acidity and pH, and phosphorus. Atmos Environ. 2014;93:3–100.
22. Hammonds M, Heal M, Mark D. Insights into the composition and sources of rural, urban and roadside carbonaceous PM10. Environ Sci Technol. 2014;48:8995–9003. https://doi.org/10.1021/es500871k.
23. Cuccia E, Piazzalunga A, Bernardoni V, Brambilla L, Fermo P, Massabò D, et al. Carbonate measurements in PM10 near the marble quarries of Carrara (Italy)by infrared spectroscopy (FT-IR) and source apportionment by positivematrix factorization (PMF). Atmos Environ. 2011;45(35):6481–7.
24. Long C, Nascarella M, Valberg P. Carbon black vs. black carbon and other airborne materials containing elemental carbon: physical and chemical distinctions. Environ Pollut. 2013;181:271–86.
25. Putaud J-P, Raes F, Van Dingenen R, Brüggemann E, Facchini M, Decesari S, et al. A European aerosol phenomenology—2: chemical characteristics of particulate matter at kerbside, urban, rural and background sites in Europe. Atmos Environ. 2004;38(16):2579–95.
26. Cassee R, Héroux M-E, Gerlofs-Nijland M, Kelly F. Particulate matter beyond mass: recent health evidence on the role of fractions, chemical constituents and sources of emission. Inhal Toxicol. 2013;25(14):802–12.
27. Chakrabarty RK, Baumgardner AD, Lack D, Moosmüller H, McMeeking G, Chakrabarty R, Baumgardner D. Characterizing elemental, equivalent black, and refractory black carbon aerosol particles: a review of techniques, their limitations and uncertainties. Anal Bioanal Chem. 2014;406(1):99–122.

28. Janssen N, Gerlofs-Nijland M, Lanki T, Salonen R, Cassee F, Hoek G, et al. Health effects of black carbon. Copenhagen: WHO Regional Office for Europe; 2012. Retrieved from: http://www.euro.who.int/__data/assets/pdf_file/0004/162535/e96541.pdf
29. Biswas P, Wu C-Y. Nanoparticles and the environment–critical review. J Air Waste Manage Assoc. 2005;55:708–46.
30. Pöschl U, Shiraiwa M. Multiphase chemistry at the atmosphere–biosphere interface influencing climate and public health in the anthropocene. Chem Rev. 2015;115:4440–75.
31. Pöschl U. Atmospheric aerosols: composition, transformation, climate and health effects. Angew Chem. 2005;44(46):7520–40. https://doi.org/10.1002/anie.200501122.
32. Nozière B, Kalberer M, Claeys M, Allann J, D'Anna B, Decesari S, et al. The molecular identification of organic compounds in the atmosphere: state of the art and challenges. Chem Rev. 2015;115(10):3919–83.
33. Glasius M, Goldstein A. Recent discoveries and future challenges in atmospheric organic chemistry. Environ Sci Technol. 2016;50:2754–64.
34. Shiraiwa M, Selzle K, Pöschl U. Hazardous components and health effects of atmospheric aerosol particles: reactive oxygen species, soot, polycyclic aromatic compounds, and allergenic proteins. Free Radic Res. 2012;46(8):927–39.
35. Tositti L, Sandrini S *Variabilità Spazio-temporale dei Livelli di Aerosol Ambientale (Frazione PM10) in Provincia di Ferrara.* PM10 and trace element data analysis and interpretation, ARPA (Regional Environmental Protection Agency) Emilia- Romagna, Sezione di Ferrara; 2008.
36. Claxton L. The history, genotoxity and carcinogenity of carbon-based fuels and their emissions: 1. Principles and background. Mutat Res. 2014;762:76–107.
37. Lighty J, Veranth J, Sarofim A. Combustion aerosols: factors governing their size and composition and implications to human health. J Air Waste Manage Assoc. 2000;50(9):1565–618.
38. Zhang L, Morawska L. Combustion sources of particles. 1. Health relevance and source signatues. Chemosphere. 2002;49(9):1045–58.
39. Calvo A, Alves C, Castro A, Pont V, Vicente A, Fraile R. Research on aerosol sources and chemical composition: past, current and emerging issues. Atmos Res. 2013;120-121:1–28.
40. Tchounwou P, Yedjou C, Patlolla A, Sutton D. Heavy metal toxicity and the environment. In: Molecular, clinical and environmental toxicology, Volume 101 of the series experientia supplementum, vol. 101. Basel: Springer; 2012. p. 133–64. doi:10.1007/978-3-7643-8340-4_6.
41. Hopke P, Rossner A. Exposure to airborne particulate matter in the ambient, indoor, and occupational environments. Clin Occup Environ Med. 2006;5:747–71.
42. Duffus J. Heavy metals—a meaningless term. Pure Appl Chem. 2002;74(5):793–807.
43. Merian E, Anke M, Ihnat M, Stoeppler M. Elements and their compounds in the environment: occurrence, analysis and biological relevance, three-volume set, 2nd, completely revised and enlarged edition, vol. 1. Weinheim: Wiley VCH; 2004.
44. Saffari A, Daher N, Shafer M, Shauer J, Sioutas C. Global perspective on the oxidative potential of airborne particulate matter: a synthesis of research findings. Environ Sci Technol. 2014;48:7576–83.
45. Hopke P. Review of receptor modeling methods for source apportionment. J Air Waste Manage Assoc. 2016;66(3):237–59.
46. Brattich E. Origin and variability of PM10 and atmospheric radiotracers at the WMO-GAW station of Mt. Cimone (1998–2011) and in the central Po Valley. PhD Thesis University of Bologna; 2014.
47. Shirmohammadi F, al. Oxidative potential of coarse particulate matter (PM10–2.5) and its relation to water solubility and sources of trace elements and metals in the Los Angeles Basin. Environ Sci. 2015;17:2110–21.
48. Belis C, Larsen B, Amato F, El Haddad I, Favez O, Harrison R, et al. European guide on air pollution source apportionment with receptor models. Luxembourg: Publications Office of the European Union: European Commission JRC Institute for Environment and Sustainability; 2014. doi: 10.2788/9307
49. Watson J, Chow J. Receptor models and measurements for identifying and quantifying air pollution sources. In: Murphy B, Morrison R, editors. Introduction to environmental forensics. 3rd ed. Cambridge, MA: Academic; 2015. p. 677–706.

50. Sacks J, Stanek L, Luben T, Johns S, Buckley B, Brown J, Ross M. Particulate matter–induced health effects: who is susceptible? Environ Health Perspect. 2011;119(4):449–54.
51. Elder A, Oberdörster G. Translocation and effects of ultrafine particles outside the lung. Clin Occup Environ Med. 2006;5(4):785–96.
52. Janssen N, Aileen Y, Strak M, Steenhof M, Hellack B, Gerlofs-Nijland M, et al. Oxidative potential of particulate matter collected at sites with different source characteristics. Sci Total Environ. 2014;472:572–81.
53. Borm P, Kelly F, Künzli N, Schins R, Donaldson K. Oxidant generation by particulate matter: from biologically effective dose to a promising, novel metric. Occup Environ Med. 2007;64(2):73–4.
54. Landreman A, Shafer M, Hemming J, Hannigan M, Schauer J. A macrophage-based method for the assessment of the reactive oxygen species (ROS) activity of atmospheric particulate matter (PM) and application to routine (daily-24 h) aerosol monitoring studies. Aerosol Sci Technol. 2008;42(11):946–57.
55. Frampton M. Inflammation and airborne particles. Clin Occup Environ Med. 2006;5(4):796–815.
56. Xia T, Kovochich M, Nel A. The role of reactive oxygen species and oxidative stress in mediating particulate matter injury. Clin Occup Environ Med. 2006;5(4):817–36.

Environment/Climate Air Pollution

4

Carlo Cacciamani, Vanes Poluzzi, and Marco Deserti

4.1 Basic Concepts

The Earth's atmosphere is conventionally considered as formed by nitrogen (78%), oxygen (21%), and other trace gases (1%). The variation of the natural composition and amount of trace gases can substantially change the climate and the life on the planet. Trace gases include pollutants and greenhouse gases.

The WHO defines *air pollution* as a contamination of the indoor or outdoor environment by any chemical, physical, or biological agent that modifies the natural characteristics of the atmosphere. The same definition is adopted by the European Law about air quality, like the Directive 2008/50/EC of the European Parliament and of the Council of 21 May 2008 on ambient air quality and cleaner air for Europe, where "ambient air" shall mean outdoor air in the troposphere.

Greenhouse gases (sometimes abbreviated GHG) are (see IPCC: https://www.ipcc.ch/ipccreports/tar/wg1/518.htm) "those gaseous constituents of the atmosphere, both natural and anthropogenic, that absorb and emit radiation at specific wavelengths within the spectrum of infrared radiation emitted by the Earth's surface, the atmosphere, and clouds. Water vapor (H_2O), carbon dioxide (CO_2), nitrous oxide (N_2O), methane (CH_4), and ozone (O_3) are the primary greenhouse gases in the Earth's atmosphere. Moreover there are a number of entirely human-made greenhouse gases in the atmosphere, such as the halocarbons and other chlorine- and bromine-containing substances, dealt with under the Montreal Protocol. Beside CO_2, N_2O, and CH_4, the Kyoto Protocol deals with the greenhouse gases sulfur hexafluoride (SF_6), hydrofluorocarbons (HFCs), and perfluorocarbons (PFCs). This process is the fundamental cause of the greenhouse effect.

C. Cacciamani (✉)
National Department of Civil Protection, Rome, Italy
e-mail: carlo.cacciamani@protezionecivile.it

V. Poluzzi • M. Deserti
Arpae Emilia Romagna, Bologna, Italy

F. Capello, A.V. Gaddi (eds.), *Clinical Handbook of Air Pollution-Related Diseases*,
https://doi.org/10.1007/978-3-319-62731-1_4

Particulate matter is a category of pollutants: it is formed by a mixture of solid particles and liquid droplets suspended in the air. Some particles, such as dust, dirt, soot, or smoke, are large or dark enough to be seen with the naked eye. Others are so small they can only be detected using an electron microscope. These particles come in many sizes and shapes and can be made up of hundreds of different chemicals. Some are emitted directly from a source, such as construction sites, unpaved roads, fields, smokestacks, or fires. Most particles form in the atmosphere as a result of complex reactions of chemicals such as sulfur dioxide and nitrogen oxides, which are pollutants emitted from power plants, industries, and automobiles. (https://www.epa.gov/pm-pollution/particulate-matter-pm-basics#PM).

Atmospheric concentration is the amount of gas or particulate matter present in air referred to a defined quantity of the mixture or its components. Trace gases concentration is usually expressed as mass concentration (micrograms per cubic meter or milligrams per cubic meter), or as parts per billion (ppb) or parts per million (ppm), depending from the substances. Concentration of particulate matter or aerosol is sometimes expressed as number of particles per cubic meter. International organizations and states adopt limit values for atmospheric pollutants. Standard is usually expressed as the maximum mass concentration.

As a consequence that most pollutants are emitted at the Earth's surface, most atmospheric processes relevant for air quality occurs in the *boundary layer*, the part of the troposphere that is directly influenced by the presence of the Earth's surface and responds to surface forcings with a timescale of about 1 h or less [1]. Nevertheless some persistent pollutants and GHG emitted by human activities can modify the atmosphere over time scales ranging between 1 year and 100 year (see 4.1.3). The thickness of the boundary layer is called boundary layer height or mixing height. *Mixing height* is defined as the height of the stratum contiguous to the surface of the Earth in which substances emitted will be mixed or, in other words, the layer in which the potential temperature (the temperature that a particle would acquire if brought adiabatically—namely, with no exchange of heat—to a pressure of 1000 millibars) is nearly constant. As a general rule, we can assume that the concentration of pollutants emitted at the surface decreases when the mixing height increases. The thickness of the boundary layer can vary from less than 30 m, in conditions of strong static stability, up to more than 3 km, in the presence of intense convective motions [2].

A main feature of the boundary layer is *turbulence*. There are two causes of turbulent transport:

- The viscosity, which requires the vanishing of the wind speed at the surface, with the result that a weak wind also induces the formation of vortices
- The radiative heating of the surface, which causes convective motions

The radiative heating of the surface, and as consequence the mixing of pollutants in the boundary layer, is regulated by the surface energy budget. The *surface energy budget (SEB)* describes the balance between radiation, conduction, and convective heat flows toward the surface of the ground and backward to the atmosphere and, with the surface temperatures and heat fluxes, determine the stability conditions in the lower atmosphere [3].

Atmospheric concentration of pollutants and mixing height as well, depends from *turbulent flows* and *winds* at different scales, determining the movement of air masses or advection. *Advection* can be defined as movement of air masses. Movement of air in mixed contaminants is one of the more important things that have to be considered when speaking of atmospheric pollution. Different are the scales that need to be considered to understand how is the bulk motion and, hence, how and where the pollutants can be transported.

The large-scale or macroscale winds. Influenced by the differences of pressure, they can be considered for the transport of pollutants at long distances. In these categories of winds, the distances covered vary from a few hundred to a few thousands kilometers; the horizontal motion is prevailing than the vertical one, and they are subject to the Coriolis force.

The local-scale winds and breezes. The scale of this kind of winds is typically from ten to hundreds of kilometers. They originate from the differences of the temperatures between neighboring regions and then are often convective. In this category, it can be included the sea and land breezes as well as the mount-valley winds. They can be very important, especially in absence of other type of winds, for the transport of the pollutants between nearby zones.

4.1.1 Time-Space Scales

From the complexity and variety of atmospheric processes, follow that pollutants concentration at surface is characterized by space-time variability at all scales characteristic of the atmosphere. Fig 4.1 summarizes the characteristic time-space scales of most pollutants and their lifetime. Atmospheric diffusion theory is briefly discussed in 4.2.3.

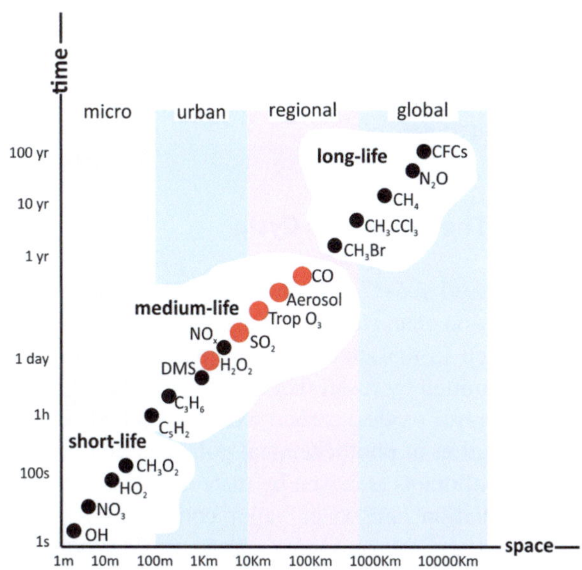

Fig. 4.1 Time-space scales

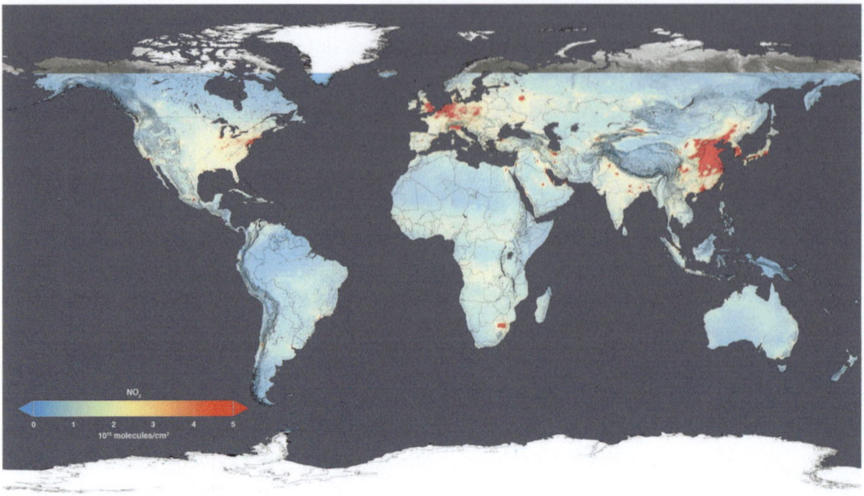

Fig. 4.2 This global map shows the concentration of nitrogen dioxide in the troposphere as detected by the Ozone Monitoring Instrument aboard the Aura satellite, averaged over 2014 (http://svs.gsfc.nasa.gov/12094); *blue* and *green* colors denoting lower concentrations, and *orange* and *red* areas indicating higher concentrations

During the last decade, the global patterns of pollutants and GHG substances have been well monitored by space observations. An example of the global distribution of nitrous diioxide (NO_2) is shown in the Fig 4.2, a NO_2 is an important tracer of main pollution sources because it is rapidly formed by the oxidation of nitrogen monoxide (NO), coming from any combustion process. The global distribution of NO_2 put into evidence major hot spots that are related to the densely populated and heavily industrialized areas of the planet.

The more extended hot spots at global level are located mainly in Northern Hemisphere: East China, Europe, and East Coast of the United States. Less extended hot spots appear in Indian Peninsula, Central Asia, South Africa, and West Coast of the United States.

4.1.2 The Pollutants Cycle

As discussed above, pollutants concentration is related to emission intensity and to dispersion characteristics of the atmosphere. Some pollutants as ozone and PM are formed in the atmosphere by transformation of primary pollutants that are mainly emitted by fossil fuel combustion. Some primary pollutants, like nitrogen oxides, sulfur oxides, carbon oxides, and hydrocarbons, are main precursors for the formation of photochemical pollution and particulate. The formation of secondary pollutants is driven by meteorological conditions such as the intensity of solar radiation and water vapor content of the atmosphere. Finally, all these pollutants are removed from the atmosphere by several sinks as wet deposition by acid rains and dry deposition on Earth's surface. This chain of emission sources – dispersion by winds and atmospheric turbulence, chemical transformation, and deposition on the surface – is the pollution cycle.

4.1.2.1 The Nitrogen Cycle

Being present for 78% of the total of the molecules, the atmosphere is surely the more important source of this nitrogen for the processes involving the living organism. However, to be used for these scopes, the nitrogen must be fixed, or converted, in reduced species like ammonia (NH_3) or ammonium ion (NH_4^+) and in oxidized species like nitrogen oxides (NO_x). The first process brought to the reduced species is realized by bacteria, while the second one, the oxidation process, is acted by the ionization of nitrogen, due to lightning or cosmic radiation, to bring—after reactions involving the oxygen—to nitrogen oxides which are deposited on the ground and so used as source of nitrogen. The atmospheric pollution originated by the different anthropogenic activities provides abnormal amount of nitrogen that can have the property to change the natural cycle sintetically described above.

4.1.2.2 The Sulfur Cycle

The sulfur cycle is quite complex to be described in details; this contributes to the levels of sulfur compounds in the atmosphere derived principally from the volcanic eruption, seas, lake, rivers, and land biota. Also the human activities bring in the atmosphere important amount of compounds of sulfur like SO_2. A lot of chemical reactions involving molecules in different phases (gas and liquid) as well as sulfur in many oxidation states and photolysis in different layers of atmosphere are necessary to explain what happens to sulfur compounds in air and its deposition on the surface of the Earth.

4.1.2.3 The Carbon Cycle

As it is known, the molecules based on carbon are fundamental for life in the Earth, and CO_2 is one of the gases, coming also from human activities that partially influence the greenhouse effect. The main forms of carbon in the atmosphere are carbon dioxide and methane. Both of these gases contribute partially to the greenhouse effect, but the higher concentration of CO_2 and its larger lifetime than methane (despite its lower greenhouse effect per volume) make carbon dioxide more important for the consequences of the impact on climate.

The continuous exchange between terrestrial ecosystem, oceans, sediments, and atmosphere of carbon is important to maintain an equilibrium between sources and sinks and then keep the climate system of the Earth sufficiently stable. Human activities with important emission of carbon-based chemicals and other substances have strongly disrupted this equilibrium since the beginning of industrial era.

Furthermore, carbon chemistry in troposphere influences a lot of reactions which products are often of a great impact again for the climate system and for human health.

These cycles, together with the cycle of oxygen and water, if disturbed by the emissions of anthropogenic activities can bring different concentrations of compounds, particles, and substances in the air that all the people breath, having an impact on the human health.

4.1.3 The Global Climate System

At the global scale, the global climate is regulated by the amount of energy the Earth receives from the Sun. However, the global climate is also affected by the other energy flows taking place within the climate system itself.

From the physical, chemical, and biological point of view, the climate is a dynamic system ("climate system") extraordinarily active and complex, which includes several distinct components (ocean, atmosphere, cryosphere, biosphere, lithosphere) interacting with each other over space and time scales also significantly different. The status of the climate is determined both by external causes (called forcing), as the solar irradiation, and by internal causes, due to the inherent variability of the climate system itself.

The atmosphere plays a crucial role in the climate system. Despite their relative scarcity, greenhouse gases, including carbon dioxide and methane, have a dramatic effect on the amount of stored energy within the atmosphere and consequently on the climate of the Earth. These greenhouse gases trap heat in the lower atmosphere that is trying to escape into space and, in so doing, make the surface of the Earth warmer. This trapping heat is called natural greenhouse effect and keeps the Earth warm 33°C than it would be otherwise. Over the past 200 years, anthropogenic emissions of greenhouse gases have increased the natural greenhouse effect, which could be the cause of global warming.

Atmosphere, however, is not an isolated system. The energy flows take place between the atmosphere and the other parts of the climate system, most significantly to the world's oceans. For example, ocean currents move heat from hot equatorial latitudes to colder polar latitudes. A large part of heat is also transferred through the moist processes. Water evaporates from the surface of oceans, and the heat released is stored within the clouds in the atmosphere and then is released during the condensation phase and the production of rain.

The ice cover, over land and sea, collectively known as the cryosphere, has a significant impact on Earth's climate. The cryosphere includes the Antarctic, the Arctic Ocean, Greenland, Northern Canada, Northern Siberia, and most of the high mountain ranges of the world, where sub-zero temperatures persist throughout the year. Snow and ice reflect solar radiation. Without the cryosphere, much more energy would be absorbed on the surface of the Earth, and consequently the temperature of the atmosphere would be much higher.

Finally, vegetation makes food by photosynthesis of carbon dioxide and water in the presence of sunlight. Through this use of carbon dioxide in the atmosphere, plants have the ability to regulate the global climate. In the oceans, the microscopic plankton uses carbon dioxide dissolved in water for photosynthesis and the building of carbonate shells. When the plankton dies, these carbonate shells are deposited on the sea bottom and "block" so to say the carbon dioxide from the atmosphere. This "biological pump" reduces the concentration of carbon dioxide in the atmosphere, weakening the natural greenhouse effect and thus helping to reduce the Earth's surface temperature.

4.1.4 Radiative Transfer

The interaction between the incident solar radiation with the atmosphere and the Earth's surface determines the amount of solar energy absorbed by the system and the heating distribution of the atmosphere and soil. The atmosphere is almost transparent to solar radiation, while about half of solar energy is absorbed by the oceans and land. In order to achieve the energy balance, the heat absorbed from the solar radiation must be returned to space through emission of the Earth. In this process, it is fundamental the transmission of energy in the form of IR radiation through the atmosphere and the vertical transport of heat from the atmospheric motions as well. The transmission properties of the atmosphere are given by its gaseous composition, the amount and chemical composition of the aerosols, and the presence, the type, and characteristics of clouds. As already mentioned the composition is such as to enable an efficient absorption and emission of IR radiation. This feature, together with the relative transparency to solar radiation, as has already been said makes the Earth's surface warmer than it would be without atmosphere.

The absorption of infrared radiation in the atmosphere is caused by a minority of gas molecules, which constitute a very small fraction of the total mass. The climate system is very sensitive to the abundance of these elements and also to the changes of their composition in the atmosphere, regardless of their natural or anthropogenic origin. Even small changes in the composition can greatly influence the flow of energy through the climate system causing severe climate changes.

To understand how the climate depends on the composition of the atmosphere, it is necessary to analyze the physical processes by which the electromagnetic radiation interacts with gases and particles in the atmosphere. The radiative transfer equations represent the mathematical basis for describing these physical processes that determine the radiation flux in the atmosphere.

The interactions between radiation and matter are described in terms of interaction between photons and molecules and include:

a. Scattering, where the photon changes phase or direction
b. Absorption, where the photon transfers its energy to the substance that has absorbed it and this process produces an increase of internal energy of matter

Absorption: In the processes of absorption, energy of the molecules can only be stored if the energy of the photon is the difference between the energy of two permitted quantum states of the molecule. The permitted transitions between the energy levels of the molecules within the atmosphere then define the frequencies of radiation that will be effectively absorbed and/or emitted. If there are not permitted transitions corresponding to the photon energy, then the radiation has high probability of crossing the atmosphere. In this process, the absorption predominates selectively by the permanent gas constituents of the atmosphere such as O_2, and water vapor and, to a lesser extent, by minority contaminants.

Scattering: The scattering processes are of three types: the Rayleigh scattering, Mie scattering, and those nonselective. Rayleigh scattering interests gas having a size smaller than the wavelength of the incident radiation. In these cases, the radiation spreads in all directions and has its maximum effectiveness for the shortwave radiation (blue, violet). This is the reason for which we can see the blue light come from all regions of the sky. The Rayleigh scattering is also responsible for the red sky at sunset when the Sun's rays pass through a greater thickness of Earth's atmosphere and therefore have a greater number of diffusing centers, so that not only "blue" photons but also the yellow ones are widespread: the result is that sunlight is deprived of all spectrum components except the red. The Mie scattering interests aerosols larger than the gas molecules. It has the same effect throughout the visible spectrum. Finally, the nonselective scattering processes interests particle whose sizes are much greater than the wavelength of the incident radiation (e.g., cloud of drops) and act on all wavelengths, generating white appearance (the clouds appear white).

Because of these extinction processes, the solar radiation undergoes an attenuation that follows an exponential curve described by the known law of Bouguer–Lambert–Beer (hereafter B-L-B). This law expresses the variation of intensity of the solar radiation through the atmosphere due to absorption. Not considering the atmospheric emission and scattering, but only the absorption, the law of B-L-B says that the absorption is proportional to the intensity of radiation and the amount of absorbent along the route path.

The heating rate determined by the absorption of direct radiation to the Earth is proportional to the flow itself which is maximum at the outer atmosphere. This is what happens in the case of absorption of UV radiation by O_2 and N_2, in the higher atmosphere, which produces the maximum rate of heating at high altitude and is responsible for the rapid growth of T in thermosphere. The same applies to the absorption of UV radiation by ozone in stratosphere, for which the maximum heating occurs at the top of the stratosphere around 50 km, despite the maximum concentration of the gas which is between 25 and 30 km.

4.1.5 Global Effects

As already anticipated, solar radiation is the main source of energy of the climate system. The surface of our planet is warming because of the energy received from the Sun in the form of light radiation. Not all the incident energy is however absorbed: a part of it is reflected both by the Earth's surface and from the atmosphere. The fraction of incident energy that is reflected is measured by the "albedo." Without the atmosphere and oceans, the equilibrium temperature of the surface of our planet would be −18°C, much less than that actually observed and would be due to the adjustment by the distance of the Earth from Sun and from Earth's albedo.

The presence of the thin (10–15 Km height) layer of atmosphere around the Earth introduces a significant redistribution of heat between the Earth's surface, the lower and higher atmosphere. The incident solar rays pass through the atmosphere (which is almost transparent to visible solar radiation) and are only partly reflected and absorbed by the surface Earth and the ocean.

Surface Earth and ocean heat and reemit radiation, in turn, part of the absorbed energy in the form of infrared radiation. A fraction of this is absorbed by some gases in the atmosphere: the "greenhouse gases" (GHGs).

The energy absorbed by greenhouse gases is reemitted in all directions, and that part going back to the Earth surface contributes to an additional heating. In consequence of this, the surface temperature reaches +15°C on average (33°C higher than −18°C of the radiative equilibrium temperature in absence of atmosphere!). This phenomenon, quite naturally, is called the "greenhouse effect," and it is greater if larger is the amount of greenhouse gases present in the atmosphere.

In recent years, there has been an overall increase in emissions of "greenhouse gases," which today have reached levels never touched in the last 800 thousand years. *Since 1970, CO_2 emissions have increased by about 90%, with emissions from fossil fuel combustion and industrial processes contributing about 78% of the total greenhouse gas emissions increase from 1970 to 2011. Agriculture, deforestation, and other land-use changes have been the second-largest contributors* (From: EPA—https://www.epa.gov/ghgemissions/global-greenhouse-gas-emissions-data#Trends). Greenhouse gas emission has produced an overall increase in the temperature of the planet, estimated to be about one degree Celsius since the beginning of industrial era (the mid-late 1800s) to date. Furthermore, the intense climate anomalies that occurred over the past decades have led the scientific community to admit the existence of a change in the observed climate, due in large part to human activity. During the last century, man has caused a profound change in the Earth's atmosphere, as regards chemical species that contribute to the determination of radiative equilibrium of our planet. Even small variations in the concentrations of "greenhouse gases," able to reissue the absorbed radiation from the Earth, can change the energy force of the planet and the balance of the Earth's climate system both globally and at regional and local scale. From reading the last fifth report of the IPCC published in 2013 (see: http://www.ipcc.ch/report/ar5/wg1/), some important conclusions can be underlined as follows:

- The growing set of available observations presents the image of a world in the process of general warming.
- Greenhouse gas emissions and aerosols due to human activities continue to grow and alter the atmosphere and affect the climate.
- New and more stringent tests confirm that the observed warming can be attributed to human activities and in particular to the increased emission of greenhouse gases over the last 50 years, which are the main cause of that warming.
- All scenarios lead to a future state of the world characterized by a global rise in temperatures and sea levels and also to profound changes of extreme events.

- The climate change will persist for many centuries and may be more or less intense depending on the greenhouse of mainly anthropogenic gas emission scenarios and then in relation to the different economic development scenarios that the world will know perhaps.
- The climate changes defined above can produce large impacts and alter the very conditions of "risk" for humanity.
- In order to decrease the damages of climate change, it is necessary to act both on the causes, through:
  a. Mitigation policies (drastic reduction of greenhouse gas emissions to try to keep the warming below a safety threshold determined by 1.5–2° of increase temperature, as stated in the recent meeting of the COP21 in Paris in December 2015).
  b. Including the implementation of relevant adaptation policies, in order to reduce the effects of climate change that, in any case, the humanity is already experiencing in various modes and entities and will continue to persist in the coming decades.

4.2 Regional and Local Effects on Climate and Air Quality of pollutants

4.2.1 Effects on Climate

The climate changes are and will be increasingly even at the local scale. In addition to the general overall increase of the temperature signal, evident even at a local scale for example in Italy are more and more frequent prolonged periods of thermal anomaly, such as the exceptional hot summer of year 2003 which was second only to that exceptional 2003 (http://www.metoffice.gov.uk/learning/learn-about-the-weather/weather-phenomena/case-studies/heatwave) and then of 2012 (http://science.nasa.gov/science-news/science-at-nasa/2012/03aug_summer2012/). By now (2017), it has become almost "normal" summers considered extremely hot 30–40 years ago, with prolonged periods of heat waves, which unfortunately cause heavy impacts to health of the people and also procure prolonged periods of drought, leading to water scarcity impacting heavily, for example, in agriculture but also for hydro-potable uses.

Concerning rainfall occurrence, more and more often very intense and rapid events are observed, causing "flash" floods and therefore damage and often, human life losses. In this regard, we can mention, for example, in Italy, the flood of Genoa in October of 2014 where it rained more than 500 mm of precipitation in a few hours. These quantities are more typical of tropical areas of the planet than the territories in the middle latitudes. The list of the floods in recent years is absolutely alarming and unfortunately set to grow rapidly in the coming years.

The special report on extreme events, published in 2012 by the IPCC Working Group 2 (http://www.ipcc.ch/report/ar5/wg2/), has shown how exposure and vulnerability to weather events has increased a lot over the past 20–30 years, greatly amplifying the risk conditions already measurable in terms of damage. According to the

World Meteorological Organization, only in the last decade 2001–2010, there were 370,000 deaths well attributable to the impacts of extreme weather events (WMO, "The Global Climate 2001-2010, a decade of climate extremes." Summary report, report n. 1119, http://www.unep.org/pdf/wmo_report.pdf), including all phenomena classified as "extreme": heat waves, long periods of intense cold, and prolonged droughts, which have caused major disturbances in turn floods and landslides. The increase in loss of life is of the order of 20% over the period 1991–2000.

These changes will continue to have impacts on humans and the environment in lives both directly and indirectly, by interacting with the entire social and economic system. Vulnerabilities associated with many systems susceptible to climate change concern as already mentioned the water resource but also the natural ecosystems, coastal areas, the industry, the production capacity, agriculture, and health.

Regarding the Mediterranean area, the most significant impacts that climate chaange will determine can be summarized as it follows:

- The greater frequency of intense precipitation will have a very high impact by increasing the hydrogeological-hydraulic risk in already very exposed areas. In parallel, the occurrence of more frequent events of intense precipitation alternated with long dry periods will alter the hydrological cycle.
- The majority of episodes of drought frequency will have negative repercussions on the availability of water resources (water scarcity) resulting in increased "competition" for water supply between different sectors of society.
- The rise in sea level and increased marine invasion events of the most low-lying coastal areas will accelerate coastal erosion, growing salinity in the estuaries, and deltas because of the entry of the salt wedge.
- The increase in average and extreme temperatures can rise to increased frequency and duration of heat waves with great inconvenience to the health of populations, remarkable progress also in other human activities (e.g., tourism), and increased occurrence of fires.
- The further depletion of surface and underground waters and therefore the most withering of the territory will have negative consequences on agriculture, reductions in yield and quality of agricultural production.

4.2.2 Effects on Air Quality

Many pollution hot spots are observed in large urban areas, that are the sites of most anthropogenic pollutant emissions and where the vast majority of citizens live. Urbanized surfaces influence atmospheric dispersion, hence the pollutant concentration, in several ways:

- The building and ground covering materials alter radiative transfers and the energy budget. As a consequence of the alteration of the surface energy balance, an increase of air temperature inside the urban canopy is often observed (this is known as the *urban heat island*).

- Urban surface reaction to precipitation and evaporation is different from natural surface. As a consequence, the horizontal water flows within the ground are different from their counterparts in natural grounds. In densely built districts, the vegetation is scarce, thus reducing the evapotranspiration.
- The high roughness of the urban area close to the ground influences the turbulent flow and modifies the structure of the boundary layer (*urban boundary layer*).

The above mentioned features determine the hydrostatic stability conditions in the lower atmosphere, which regulate the mixing of pollutants, whereas the mixing height parameter determines the available volume for pollutant mixing. For that reason, the surface energy balance (SEB) is the key component of any model aiming to simulate dynamical and thermodynamical patterns above the surface. Following Fisher et al., for a city, the surface radiation budget and SEB are written [4]

$$Q^{*} = K\downarrow - K - + L\downarrow - L - = \mathrm{QH} + \mathrm{QE} + \mathrm{QG} + \mathrm{QF}\left[\mathrm{W\ m^{-2}}\right]$$

where Q^* is the net all-wave radiation; $K\downarrow$ the incoming shortwave radiation; $K\uparrow = \alpha 0K\downarrow$ the outgoing, reflected shortwave radiation where α0 is surface albedo; $L\downarrow$ the incoming longwave radiation from the sky and surrounding environment seen from the observation point; $L\uparrow = \varepsilon_0\sigma \mathrm{Ts}4 + (1 - \varepsilon_0)L\downarrow$ the outgoing longwave including both that emitted from the surface consistent with its emissivity ε0 and absolute surface temperature Ts and the reflected incoming longwave; QH the turbulent sensible heat flux; QE the turbulent latent heat flux; and QG the soil heat flux. QF represents additional sources of energy due to combustion of fuels (see the cost 715 WG2, report for a review of experimental results showing heat island in North America, Basel, Marseille, Bologna, Birmingham).

The mixing height depends on the vertical variation of temperature in the atmosphere. A special type of situation is the *temperature inversion*, i.e., the temperature of air in the atmosphere is cooler the higher up in altitude you go. Temperature inversions can result in peak pollution episodes. Inversions can be caused by several atmospheric mechanisms, such as subsidence, fronts, radiation, and advection. Episodes of bad air quality are most frequently connected to radiational and advective inversions. These situations are common during winter period in the valley, such as the *Po valley*, northern Italy.

As mentioned above, the *flow structure* is modified by the urban surface due to the increased roughness. The flow modification from wind observation can be summarized as follows:

- The urban wind speed is usually smaller than the rural reference.
- For no wind conditions at the rural site, measurable wind speeds are observed at the urban site. This is attributed to the (Fig. 4.3) thermally driven flows generated by the urban heat island [5].

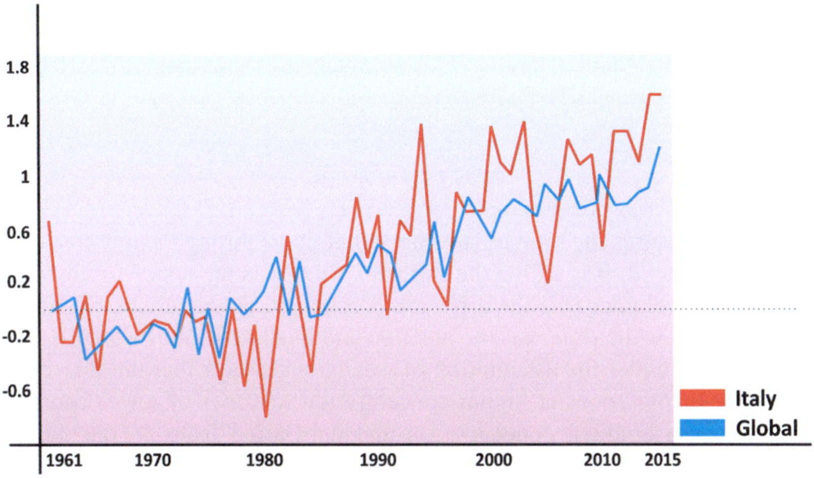

Fig. 4.3 Time series of temperature anomaly (averaged over land, Italy), with respect to Climate average (1961–1990). *Red line*: Italy (Font: ISPRA, Italy) and *Blue line* (global average, http://www.scia.isprambiente.it/Documentazione/RAPPORTOCLIMA2015.pdf)

4.2.3 Atmospheric Chemical and Physical Processes Governing Pollutant's Concentration

This chapter summarizes the main chemical and physical factors influencing the pollutants concentration in the lower atmosphere. Basic concept and definitions were introduced in 4.1. It is assumed that theoretical basic principles are assimilated and the discussion will go further, neglecting the mathematical aspects, on the following main topics:

- Atmospheric diffusion
- Physical-chemical reactions and transformations in the troposphere

4.2.4 Atmospheric Diffusion

How the pollutants mix with the other gases of the atmosphere is the question that the commonly called "atmospheric diffusion theory" [6] tries to answer to. Different aspects must be taken into account to comprehend the physical basis of the diffusion (or dispersion) process of pollutants or, more in general, of particles, atoms, molecules, substances, or compounds. Below are presented the principal factors, or aspects, which can influence—in different ways—the atmospheric diffusion.

It is now necessary to comprehend the main scientific fundamentals to understand how the factors discussed above influence the atmospheric diffusion.

1. The molecular diffusion that is well mathematically described with the two laws of Fick (4.1 and 4.2) which formulas (for one dimension) are

$$J = -D\partial\varphi / \partial x \tag{4.1}$$

$$\partial\varphi / \partial t = D\partial^2\varphi / \partial x^2 \tag{4.2}$$

where J measures the total of substance that flows through a unit area during a unit time interval, it is called the diffusion flux; φ is the concentration of the substance per volume; D is the diffusion coefficient, characteristic of each single substance. t is the time; x is the position and its dimension is a length. Fick's first law (4.1), under the assumption of steady state, says that the flux of substances goes from zones at higher concentration to zones of lower concentration, through a gradient concentration and depending from D, the diffusion coefficient. The second one (4.2) gives the variation of the concentration of substances with the time.

2. The movement of air masses due to advection or winds and the turbulent diffusion that is normally faster than the molecular diffusion due to chaotic motions.

To simulate what happens at a pollutant that mixes in the air motion, considering the advection, the molecular diffusion, and the turbulent flows, two approaches have been developed:

a. A Eulerian approach that centers the analysis on an infinitesimal volume at a specific reference frame fixed in space and time.
b. A Lagrangian approach that follows the trajectories and the change of position of the species as they move.

From these models, it is possible to calculate the mean concentration of substances coming from continuous sources in air, under the assumption of steady state, homogeneous turbulence, and no presence of chemical reactions and transformations. It can be demonstrated that solutions of both the models—Eulerian and Lagrangian—bring to a Gaussian equation of the following type:

$$C(x,y,z) = q / 2\pi\bar{u}\sigma_Y\sigma_Z \; \exp\left[-\left(y^2 / 2\sigma_Y^2 + z^2 / 2\sigma_Z^2\right)\right] \tag{4.3}$$

where C is the concentration of the substance considered; $\sigma_y^2 = 2K_{yy}\, x/\bar{u}$ is the variance in y direction; $\sigma_z^2 = 2K_{zz}\, x/\bar{u}$ is the variance in z direction; K is the eddy diffusivity; q is the emission rate of the substance considered; u is the wind speed.

These approaches, under known environmental conditions as wind velocity and emission rate of the substance considered, can be used to calculate the concentrations of species at a specific point from an emission source and hence can be useful to understand the levels of contaminants in air pollution.

4.2.5 Physical-Chemical Reactions and Transformations in the Troposphere

As it is known, the atmosphere is the mix of gases surrounding our planet. To understand its fragility, it is interesting to note that the ratio of its thickness with respect to the diameter of the Earth is similar to the ratio of the peel of an apple and its diameter.

The objective of this chapter is to understand the main physical-chemical aspects that rule the concentration of pollutant or, more in general, of the substances in the lower troposphere where all the forms of life live.

4.2.5.1 Atmospheric Chemistry

It is now important to analyze what are the main types of chemical reaction that can happen in the atmosphere. Surely one of the class of reactions particularly important and that must be considered to comprehend the product and the new compounds that can be found in atmosphere are those in which a reactant or a catalyst is a photon; in other words, the entire set of reaction that goes under the definition of atmospheric photochemistry.

Most of the energy flux reaching the Earth is an electromagnetic radiation emitted by the Sun; for this reason, the importance of the photochemistry in atmosphere is high. A lot of chemical species are able, if the energy $h\nu$ of the photon colliding is appropriated, to go to an excited state becoming more unstable and more reactive. After this step, the excited species A^*can go to the following types of reaction:

Emission of radiation $A^* \rightarrow A + h\nu$
Dissociation $A^* \rightarrow B + C$
Ionization $A^* \rightarrow A^+ + e$
Direct reaction with other species $A^* + B \rightarrow C + D$

Other mechanisms involving the species can be the isomerization and the collisional deactivation that can bring to a physical quenching, an energy transfer, or an electron transfer.

One of the principal sets of known photochemical reactions is the ones that bring the important processes of pollution called photochemical smog, which is considered the mixture of new compounds originating from reactions between nitrogen oxides and volatile organic compounds reacting at the presence of solar radiation. The set of these reactions produce a haze composed by an important number of secondary pollutants that have the property to decrease the visibility and are often dangerous for the human health. A lot of oxidized and/or nitro-organic molecules can be produced during these processes, and one of the principal substances that can reach very high level of concentration (in some cases more than a hundred ppb) is surely the ozone.

The main reactions that happen during photochemical smog episodes involving nitric oxides and volatile organic compounds (VOC) are the following:

1. $NO + O_2 \rightarrow NO_2 + hv \rightarrow O + NO$
2. $O + O_2 + M \rightarrow O_3 + M$
3. $O_3 + NO \rightarrow O_2 + NO_2$
4. $VOC + NO + O_2 + hv \rightarrow O_3 + HNO_3 + SVOCs$

The reaction no. 4 indicates as a product the SVOCs—secondary volatile organic compounds—which are often oxidized organics like peroxy or peroxyacyl radicals. These compounds can finally react with NO_x to give peroxy nitrates or peroxyacyl nitrates (PANs).

Other important reactions—for example, when a molecule of excited oxygen, originated by the decomposition of ozone, reacts with water in gas phase or when the photolysis of nitrous acid HONO occurs—are those that bring to the formation of hydroxyl radical (OH·). This species, because of its involvement in many reactions that lead to product such as highly reactive free radicals or new products, is surely one of the more important to understand what happens in the tropospheric chemistry.

Further physical-chemical reactions and transformation that can happen in atmosphere are those that bring a formation of new particles (gas to particle conversion) or the ones that involve species in different phases. Because of the strong impact that can have the different kind of aerosol—or particulate matter—on human health and on the climate of the Earth, the next paragraph will be dedicated to the analysis of the atmospheric aerosol behavior.

4.2.5.2 Aerosol Size Distribution

The atmospheric aerosol is made of particles in suspension in air whose chemical and physical nature is variable and it depends on its origin.

The current European legislations require only the systematic monitoring of the mass concentration of the fractions PM10 and PM2.5 aerosol, whose measurement methodology is based on the aerodynamic diameter of the particles that compose them.

However, for the potential impacts on the environment and health, it is also important to study the numerical concentration of the fractions of aerosols, particularly those of less than 1 μm. In the lower troposphere in effect, in urban and in rural areas, measures of numerical concentrations of aerosol particles showed values ranging from a few hundred up to 10^7–10^8 particles per cubic centimeter (#/cm^3). Given this wide size ranges, it is central to better understand the conditions in which certain levels of concentrations occur and what are the distributions of the different fractions that compose the atmospheric aerosol.

It is known that the submicronic particles (diameter <1 μm), expressed as the number in a volume of sampled air, can be numerically abundant in the ambient air, but their contribution in terms of mass concentration is negligible compared to the size of particulate above the micrometer.

It is necessary—in order to rule out possible misunderstanding—to define which fractions of aerosol one refers when talking about mass concentration or number concentration. For the mass concentration, the particulate with aerodynamic diameter up to 2.5 m (PM2.5) is defined as "fine"; "coarse" particles are those with a diameter greater than this value.

Instead, the classification in terms of number concentration defines:

- "Ultrafine particles" (UFP) aerosol size less than 100 nanometers in diameter
- "Non-ultrafine" (NoUFP) particles with a diameter greater than 100 nm

The term "nanoparticle" is used to identify, by some authors, the particles that have dimensions less than 50 nm; instead the "coarse" particulate is represented by all the material larger than 1 μm.

Generally, atmospheric aerosols could be divided into three size ranges (trimodal distribution), defined as follows:

- Nucleation interval (~3–25 nm)
- Aitken nuclei interval (~25–90 nm)
- Accumulation interval (~90–1000 nm)

In the size distribution of the aerosol measurements, in urban and suburban atmosphere with the presence of direct emissions from traffic, it has been frequently observed a classification of the particles in more than three intervals. As already mentioned, the diameters of the aerosol particles are distributed in a rather wide range, and their numerical concentrations vary considerably depending on the size. For this reason, in order to improve the representation of the distribution, it is necessary to characterize the dimensional range of different width.

The best representation of the size distribution of the aerosol in the atmosphere is defined by a log-normal distribution in which the three parameters that characterize the single mode ,*i*th are the particle number, the geometrical deviation and the geometric mean diameter.

The numerical concentrations and the relative size distributions of the aerosol particles may have different values in different environmental conditions, result of the set of chemical and physical parameters of the analyzed atmosphere portion. The weather period, the time zone of the day, and the possible presence of local sources of emission are parameters that greatly influence the aerosol size distribution.

The nucleation interval is generally defined in a dimensional range between values close to 1 nm and 25 nm, with the upper limit which can vary from 20 to 30 nm depending on the reference bibliography. The "nucleation interval" is so called because, within this restricted size range, there is the formation of new particles through different processes. The particles in the range of nucleation derive from direct emission or formation in the atmosphere through the conversion processes due to:

- The rapid cooling and dilution of the gases and/or vapors produced by emissions
- Different chemical nature reactions involving precursors in the atmosphere

Generally, high concentrations of particles of this size range are found near the emission sources, with values that decrease rapidly in time and in space. In fact, particles in the interval of nucleation, once formed and suspended in the atmosphere, can easily get in contact with each other since they are subject to the interaction of different hydrodynamic forces, electrical and gravitational (Brownian motions). Collisions between the individual nanometer-sized particles, as well as with those of larger dimensions, generate aggregation processes that increase the particles of the nucleation to the range of accumulation.

The interval of the Aitken nuclei is generally defined in a size range between about 25 nm and 90 nm, with the extreme values that, also in this case, vary from a few nanometers. The aerosol belonging to this dimensional interval grows by coagulation and condensation of the particles of nucleation range but can also be produced and emitted directly into the atmosphere from combustion sources.

The particles in the interval of accumulation (~90–1000 nm) are generally constituted by carbonaceous compounds, such as soot or dust. These can be derived from engines burning fuel and by lubricating oils for diesel or gasoline vehicles, either by coagulation processes of nucleation particles.

Unlike particle nucleation, the aerosol of "accumulation" has a lifetime in the atmosphere rather long and can also be transported at a long distance. Such particles in fact cannot be removed from the collision phenomena; these particles coagulate very slowly.

As already affirmed, the size fractions closer to 1 nm particles are named in the so-called nucleation mode. The term nucleation means the physical/chemical process that leads to the formation of new particles from precursors in vapor or gas phase and from transformation processes. Causes that lead to formation of new particles are complex and depend on different factors, such as emission of precursors, preexisting aerosols, meteorological parameters (in particular solar radiation, temperature, and relative humidity), and photochemical processes.

The formation of atmospheric aerosol is closely related to chemistry; in particular, the species present in the atmosphere that seem to have a key role in this process are sulfuric acid, ammonia, and organic compounds in the gas phase or vapor. Processes involving other substances are currently studied.

The most part of particles of nanometric dimensions ($d \leq 20$ nm) in the atmosphere seems to result mainly from gas-particle transformation processes, with nucleation phenomena of homogeneous and heterogeneous type.

The homogeneous nucleation is nucleation of a substance in vapor phase on embryo molecules of the same type, in the absence of foreign substances. The heterogeneous nucleation is nucleation of a substance in vapor phase to another in the different phase as an ion or a solid particle.

The particle size distribution of aerosols suspended in a fluid, however, can also change as a result of the forces due to Brownian, hydrodynamic, electric, and gravitational action. After these movements, it can occur coagulation phenomena in which the particles come together and form a single agglomeration growing in size.

The aerosol particles have the property of having a high surface/mass ratio, and then on the surface of the aerosols many interactions may occur between vapor molecules and particles. The reactions, which are the basis of such interactions and that can cause a change of composition and density, generate a small dimensional change and do not necessarily lead to their growth.

The motivations for having attention to the particles distribution in the atmosphere are found in the impact that its size fractions have on human health (a), on the climate (b), and on global and regional hydrological cycles (c).

(a) Numerous studies have shown that fraction PM2.5 is associated with the increase of diseases and deaths related to cardiovascular and respiratory disorders.

The sub-micrometer particles (<1 μm) and the ultrafine particles (<100 nm), represent an area of research that still needs further epidemiological and toxicological studies and insights.

The particles belonging to the smaller size fractions (<100 nm), when inhaled through the respiratory tract can settle and accumulate in deeper tissues, have immediate and sustained impact on epithelial and immune cells, penetrate into the circulatory system, and accumulate in body organs human.

The fraction of nanometer size (<50 nm) has high mobility in the atmosphere and can easily be inhaled by humans through the respiratory tract. While PM2.5 particulate is deposited along the respiratory tract, a very high number of particles of nanometric dimensions can reach the pulmonary alveoli by reducing the capacity of defense of alveolar macrophages. This condition leads to an increase of the time of contact between the particles and the epithelial cells of the lungs.

(b) The atmospheric aerosol influences the Earth's climate directly because particles that compose it reflect and adsorb the solar radiation and indirectly because particles act as condensation nuclei for the formation of clouds (cloud condensation nuclei) that reflect the solar radiation in space.

When a light ray hits a particle, the electric charges present in it are excited by generating an oscillating movement. The electrical charges excited reradiate the energy in all directions (scattering) and can be that part of the incident radiation which is converted into thermal energy (absorption).

The radiative forcing of dust and the impact on the radiation balance in the atmosphere are complex because it is attributable to scatter or to absorb solar radiation and emit in the infrared frequencies. All quantitative and qualitative changes in atmospheric aerosol cause a change in the radiation balance and in the Earth's surface temperature. The amplitude of "heating" and "cooling" induced depends on the capacity of particles to reflect the solar radiation toward the space. The knowledge of the aerosol chemical composition, its size distribution, and optical properties (absorption coefficient, "single scattering" albedo, and optical depth) is critical to get an estimate of aerosol radiative "forcing" on a regional scale.

With the many aerosol components that contribute to the radiative "forcing", the aerosol containing carbon ("carbon black" in particular) plays a dominant role because it has a strong ability to absorb solar radiation and contributes significantly to the warming of atmosphere at global and regional levels.

Conversely, aerosols containing compounds with atoms of sulfur or elemental carbon reflect the Sun's rays favoring the cooling of the atmosphere.

The indirect effect of the aerosol on the climate takes place when the increase of numerical concentration of particles, produced from anthropogenic sources, contributes to an increase of the concentration of so-called condensation nuclei. These are particles that come in the processes of formation of clouds in the atmosphere.

A part of solar radiation is reflected from the tops of the clouds to outer space, and a part is absorbed within the cloud. An increase in the number of "droplets" brings to an intensification of the cloud ability to reflect light, to an increase of the cloud extension and persistence in time.

(c) Another effect connected directly to the composition the clouds is rainfall events. The precipitation formation within the cloud depends on the size distribution and on the drops that constitute the cloud: for specific water content, the presence of a little amount of drops, but characterized by large dimensions, produces rainfall events. A characteristic growth of the droplets in clouds that happen effectively in turbulent motions and at temperature above 0°C is the coalescence, in which large drops are pushed updraft and smaller droplets encountering increase its size. In the presence of drops of about 200 μm size, these currents become ineffective to the resuspension, and then the increased drops begin to fall increasing its size.

The decrease of rainfall, due to the reduction of the size of cloud droplets, gives an increase of the cloud lifetime.

References

1. Stull RB. An introduction to boundary layer meteorology. Cambridge, MA: Kluwer Academic Publishers; 1988. p. 2.
2. Holton JR. An introduction to dynamic meteorology. Cambridge, MA: Academic; 1992.
3. Piringer M, Joffre S. The urban surface energy budget and mixing height in European cities: data, models and challenges for urban meteorology and air quality. Final Rep WG2 Cost Action. 2005;715:194. http://www2.dmu.dk/AtmosphericEnvironment/cost/docs/WG2_final.pdf
4. Fisher B, Kukkonen J, Piringer M, Rotach MW, Schatzmann M. Meteorology applied to urban air pollution problems: concepts from COST 715. Atmos Chem Phys. 2006;6(2):555–64. http://www2.dmu.dk/AtmosphericEnvironment/cost/docs/Cost715-final.pdf
5. Baklanov A, Burzynski J, Christen A, Deserti M, De Ridder K, Emeis S, Joffre S, Karppinen A, Mestayer P, Middleton D, Piringer M, Tombrou M. COST715 WG2: The urban surface energy budget and mixing height in European cities: data, models and challenges for urban meteorology and air quality. In: Piringer M, Joffre S, editors. The urban surface energy. Budget and mixing height in European cities: data, models and challenges for urban meteorology and air quality, Final Report of Working Group 2 of COST-715 Action: Demetra Ltd. isbn:954-9526-29-1. http://www 2.dmu.dk/AtmosphericEnvironment/cost/docs/WG2_final.pdf.
6. Seinfeld JH, Pandis SN. Atmospheric chemistry and physics–from air pollution to climate change. 2nd ed. Hoboken, NJ: Wiley; 2006.

Spatial Distribution of Air Pollutants in Cities

5

Silvana Di Sabatino, Riccardo Buccolieri, and Prashant Kumar

5.1 Introduction

A growing amount of world's population lives in cities. This trend is expected to continue posing the problem of air pollution, particularly in megacities of developed and developing countries. Exposure to air pollutants is a problem due to the variety of pollutants. Both the outdoor and the indoor exposure are of concern. Driving forces of air pollution in cities are economic development, urbanisation, energy consumption, transportation/motorisation and rapid population growth. Transportation, industry and power generation are the main sources contributing to outdoor air pollution [1] and in particular traffic-induced emissions. Pollutants such as respirable particulate matter, nitrogen dioxide, carbon monoxide and hydrocarbons are directly emitted by vehicles, and the most affected group is the urban inhabitants, especially the population residing in close vicinity of the urban roadways and streets as well as the pedestrians [2].

S. Di Sabatino (✉)
Department of Physics and Astronomy, Alma Mater Studiorum, University of Bologna, Viale Berti Pichat 6/2, Bologna 40127, Italy
e-mail: silvana.disabatino@unibo.it

R. Buccolieri
Dipartimento di Scienze e Tecnologie Biologiche ed Ambientali, University of Salento, S.P. 6 Lecce-Monteroni, Lecce 73100, Italy
e-mail: riccardo.buccolieri@unisalento.it

P. Kumar
Department of Civil and Environmental Engineering, Faculty of Engineering and Physical Sciences (FEPS), University of Surrey, Guildford GU2 7XH, UK

Environmental Flow (EnFlo) Research Centre, FEPS, University of Surrey, Guildford GU2 7XH, UK
e-mail: p.kumar@surrey.ac.uk

F. Capello, A.V. Gaddi (eds.), *Clinical Handbook of Air Pollution-Related Diseases*, https://doi.org/10.1007/978-3-319-62731-1_5

The study of pollutant dispersion in the built environment involves a multi-scale (as well as a multidisciplinary) approach. Pollutants emitted in the atmosphere are dispersed over a wide range of horizontal length scales, from macro-scale to meso-scale and to micro-scale. Pollutants which are transported to the city or which are locally emitted are dispersed at different horizontal local scales: street (of order 10–100 m), neighbourhood (100 m–1 km) and city (10–20 km) [3]. Their final spatial distribution is determined by several factors, such as the meteorology and the morphological characteristics of the city, as well as the population density and the type, nature and spatial location of sources. Important parameters for dispersion around buildings are in particular building geometry and morphology (in terms of building arrangements and packing density), wind speed, wind direction, turbulence, atmospheric stability, temperature, humidity and solar radiation [4], together with the presence of obstacles such as trees, low barriers and parked cars [5]. As a consequence, local-induced wind fields consist of complex flow features such as recirculation zones and stagnation points which in turn strongly govern the dispersion of pollutants [6].

This chapter, after briefly introducing the process of urbanisation and the problem of outdoor air pollution, focuses on some of the above factors, with attention to the influence of city morphology and wind direction on the final outdoor spatial distribution of pollutants. The effect of urban obstacles is also discussed. The recent research towards the problem of ultrafine particles, which are considered to show even higher health impacts compared with fine particles, is also briefly addressed.

5.2 The Process of Urbanisation

The process of urbanisation refers to the relative growth in urban population followed by an increase in economic, political and cultural development of cities with respect to rural areas. There is a trend towards urbanisation as a natural consequence of economic development based on industrialisation and post-industrialisation [7].

Since the beginning of the industrial revolution, people have increasingly settled in urban areas. In 1990 43% (2.3 billion) of the world's population lived in urban areas; by 2015, this had grown to 54% (4 billion), and 60% of the global population (4.9 billion people) is expected to live in cities by 2030. Rapid urbanisation is mainly associated with the expansion of built-up areas, converting agricultural land into residential and industrial areas. The increase in urban population has not been evenly spread throughout the world. Asia has the highest number of people living in urban areas, followed by Europe, Africa and Latin America (Fig. 5.1). In particular, urban growth rates have been much faster in some regions than others. The highest growth rate between 1995 and 2015 was in the least developed regions with Africa being the most rapidly urbanising showing a rate almost 11 times more rapid than that in Europe. As the urban population increases, the land area occupied by cities has increased at an even higher rate [9].

The process of urbanisation is particularly leading to an increase of large cities (between 5 and 10 million inhabitants) and megacities (10 million or more

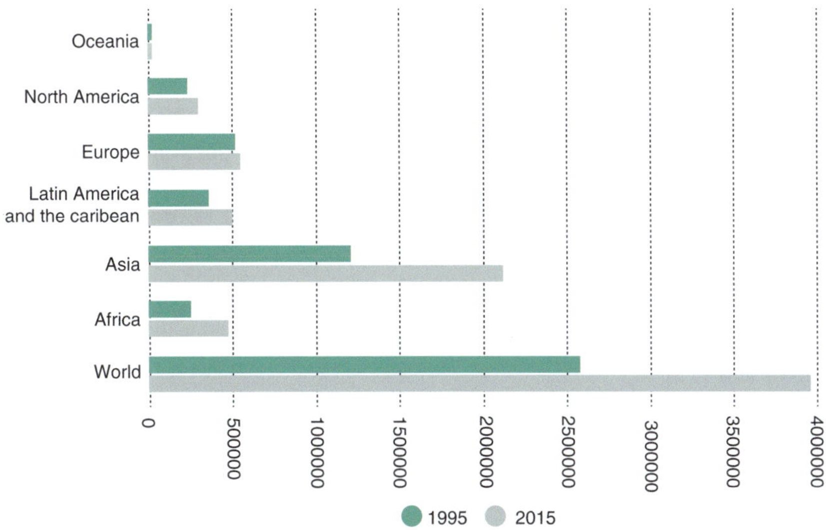

Fig. 5.1 Urban population at mid-year (1995–2015), source: United Nations [8] (adapted from Ref. [9] © UN-Habitat, reprinted with permission of UN-Habitat)

inhabitants), particularly in the low- and middle-income regions of the world. In 1995 there were 22 large cities and 14 megacities; by 2015, both categories of cities had doubled, with 44 large cities and 29 megacities. Most megacities are located in developing countries, and this trend will continue as several large cities in Asia, Latin America and Africa are projected to become megacities by 2030 [9].

In spite of economic growth and improvement of basic facilities, cities, especially in developing countries, suffer from negative consequences of urbanisation, such as environmental and air quality degradation, as well as respiratory problems. People living in large urban areas, especially in developing countries, where the health risks of air pollution may be underappreciated and pollution controls lacking, are routinely exposed to concentrations of airborne pollutants that have been shown to cause negative health effects in both the short and long term [10].

Much of air pollution caused by traffic and industries is concentrated in urban areas, and visible air pollution, known as smog, is present in such areas. Outdoor air pollution and photochemical smog can occur in any environment where there are large and continuous emissions of primary air pollutants. However, specifics of climate and geography play an important role in the persistence and severity of the pollution. As urbanisation, global population and economic development increase, not only short-range but also long-range transport of pollution becomes an increasingly important factor for cities since airflow transports pollutants from upwind areas contributing to local air quality together with locally emitted pollutants [11]. Effective measures should thus be taken as soon as possible to build a comfortable and liveable environment to live in and enjoy the advantages of urbanisation.

5.3 The Urban Environment

Pollutants emitted into the atmosphere are dispersed over a wide range of horizontal scales. Macro-scale or synoptic scale is the scale of atmospheric motions with a typical range of many hundreds of kilometres, including phenomena as cyclones and tropical cyclones. Mesoscale is the scale of atmospheric phenomena having horizontal scales ranging from a few to several hundred kilometres, including thunderstorms, squall lines, fronts, precipitation bands in tropical and extratropical cyclones and topographically generated weather systems such as mountain waves and sea and land breezes. Micro-scale is the scale of atmospheric motions for which the effects of planetary rotation are unimportant or spatial scales are of 2 km or less [12].

Before discussing the spatial distribution of air pollutants in cities, it is useful to briefly introduce the atmospheric boundary layer (ABL) and in particular the urban boundary layer (UBL). The ABL is the region of the atmosphere directly influenced by the earth's surface. In this region physical quantities (flow velocity, temperature, moisture, etc.) display rapid fluctuations and the vertical mixing is strong. The ABL is turbulent, particularly in the urban environment where the main disturbing features are the buildings of different heights and shapes, and the interaction between wind and buildings generates complex local flows. The vertical structure of the atmospheric boundary layer, i.e. the UBL, is composed of a roughness sub-layer (RSL) near the ground and an inertial sub-layer (ISL) above it. Both the RSL and ISL are encompassed within the surface layer (SL) and above which the urban outer layer extends to a height where the wind is unaffected by the earth's surface and the UBL adopts a classical ABL structure [6]. Specifically, the RSL can be defined of depth between the ground and 2-5H, where H is the mean building height. Within this layer, roughness elements exert a significant drag on the flow. Flow is highly spatially dependent; turbulence can dominate the mean flow; and turbulence has different characteristics from the flow in the ISL above, where turbulence is homogeneous and fluxes vary little with height. Near the ground surface, the buildings form the so-called urban canopy layer (UCL), defined as the layer up to mean roof height and where local complex flow regimes occur [13] (Fig. 5.2).

The UBL structure is determined not only by urban surface characteristics but also by mesoscale thermal circulations. As a consequence, pollutant dispersion in the built environment is intrinsically a multi-scale problem. For example, the pollutant transport process in an urban area could be described such that the pollutants are released into the ambient air in somewhere of the upwind direction and diffuse with the air flows that are affected by many factors, pass through building groups in the neighbourhood scale, arrive at a building located in a street canyon and finally enter into a room through the open windows and affect the health of residents [14].

Britter and Hanna [3] proposed a simple approach to describe urban scales. Length scales are street (of order 10–100 m), neighbourhood (100–1 km) and city (10–20 km). At these scales, the urban morphology becomes homogeneous (i.e. a single house or street, a collection of buildings of similar height and shape in a neighbourhood, a town or city which is rougher than the surrounding rural area)

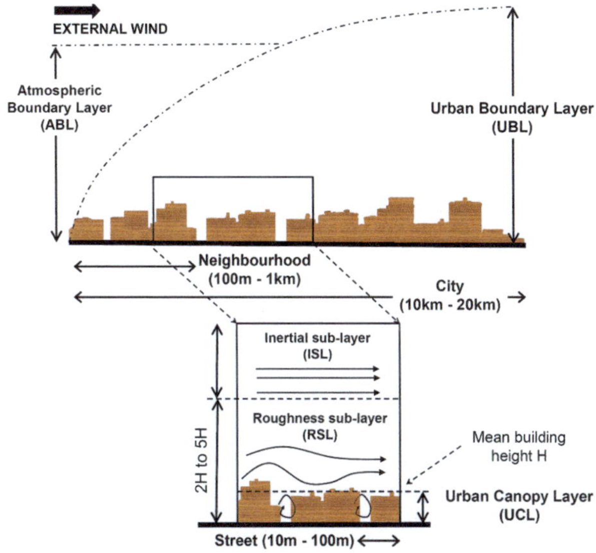

Fig. 5.2 Schematic diagram of the urban boundary layer (UBL) with indication of the characteristic horizontal scales (*top*) and schematic diagram of roughness sub-layer (RSL) and inertial sub-layer (ISL) with *arrows* indicating streamlines (*bottom*). Please note that the vertical scale is exaggerated (adapted from [13] © The Author. Published by Elsevier)

(Fig. 5.2). The smallest length scale is that of the vehicle wake where the mixing and dilution of pollutants occur faster than at any other scale. Knowledge of both the flow and mixing at various urban scales is essential for dispersion modelling of pollutants. A summary of the key flow and mixing features at these urban scales is shown in Fig. 5.3 [15], and the influence on dispersion of pollutants is described in the following section.

5.4 Distribution of Air Pollutants in Cities

As already discussed in Sect. 5.3, urban outdoor pollutant dispersion (i.e. advection by mean wind and diffusion by turbulence) can be classified as micro-scale dispersion and refers to small-scale meteorological phenomena that affect small areas (micro-scales are likely to be of the order of metres) compared to large-scale meteorological phenomena (macro-scale and mesoscale). The processes controlling the pollution phenomenon act at a range of spatial and temporal scales spanning the depth of the whole UBL. Progress in simulating air quality and city ventilation depends on accurate observations and modelling of UBL processes (see the recent review by [13]).

At the micro-scale, two different dispersion regimes are distinguished in the literature: the near-field dispersion that concerns the near vicinity of the pollutant source and for which the relevant turbulence time and length scales controlling

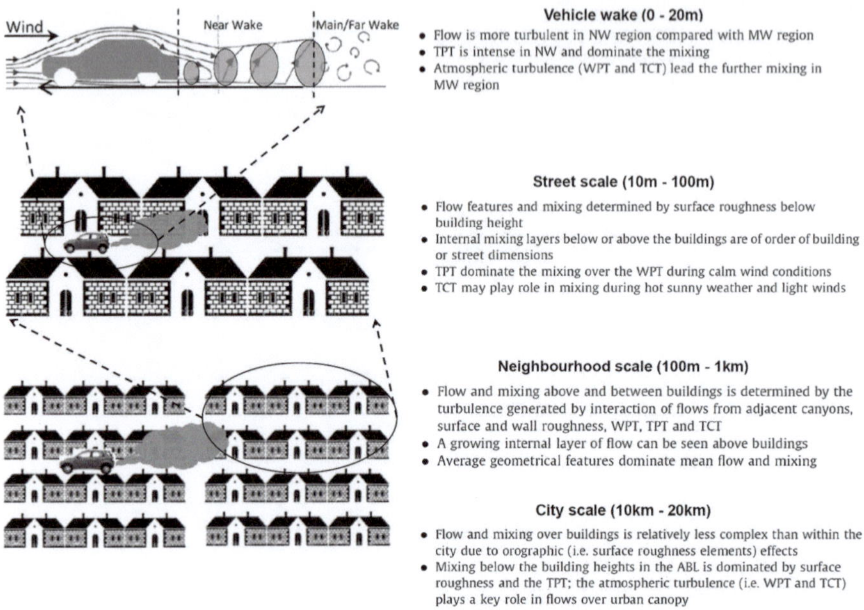

Fig. 5.3 Description of key flow and mixing features at urban scales. *NW* near wake, *MW* main/far wake, *WPT* wind-produced turbulence, *TPT* traffic-produced turbulence, *TCT* thermally produced turbulence (buoyancy) by temperature differences between air and urban surfaces (adapted from [15] © Elsevier, reprinted with permission of Elsevier)

dispersion are related to the mean building height and to spacing between buildings and the far-field dispersion of interest in plumes with a flow structure and a vertical dimension larger than the urban canopy height for which the dispersion is governed by the atmospheric boundary layer scales. In the near-field dispersion case, which is our interest here, pollutants released from various sources inside the UCL are mixed and dispersed over and around buildings because of the interactions between many physical processes that contribute to its evolution, including the dynamics of flow over urban topography and the building configurations. This leads to different exposure of pedestrians as they move through the city. Specifically, the wind flows in the atmospheric boundary layer over buildings are complex and exhibit a wide range of physical phenomena including large low-speed areas, strong pressure gradients, unsteady flow regions, three-dimensional effects and the confluence of boundary layers and wakes [6].

5.4.1 Monitoring, Experimental and Modelling Approaches

Several approaches are available for monitoring, assessing and predicting the distribution of air pollutants in cities.

For the purpose of monitoring air quality, *monitoring stations* monitor gaseous pollutants and particulate matter using several techniques depending on the specific application. They are located in places where air pollution hotspots are expected, but also must be reasonable with respect to population exposure over the averaging times associated with the legislative values. The use of such stations in cities is appropriate for monitoring the temporal trend of air pollutants, but requires high costs of maintenance and can be only used in a limited number of sites due to practical constraints which do not allow a detailed mapping of exposure in large areas.

For the study of pollutant dispersion and distribution, the approaches commonly employed are full-scale field measurements, physical modelling, operational modelling and computational fluid dynamics (CFD) simulations (see the recent review by [6]) which are briefly discussed below.

In *full-scale field measurements* (or real field test), flow and concentration tests are performed under real atmospheric conditions. They have the advantage that the real situation is studied, and the full complexity of the problem is taken into account. However, they are limited by their low spatial resolution, uncontrollable meteorological conditions and complex building configurations.

In comparison to field measurements, *physical modelling* (or reduced-scale experiments), which is usually performed in the wind tunnel or water channel, has the advantage that boundary conditions can be chosen based on the problem and the stationary flow conditions can be maintained throughout the test. However, physical modelling also suffers from the limited set of spatial points. In addition to the high costs (i.e. personnel, instrumentation and others), the setup can be time-consuming and requires adherence to similarity criteria that can be a problem for many applications such as buoyant flows.

Dispersion models, which are widely used in both industrial and research community, constitute a relatively economic approach to urban pollution study. With high spatio-temporal resolutions, they provide comprehensive information on flow and pollutant transport and thus are widely employed for assessing urban air quality as well as temporal and spatial variations. They give insights into the physical and chemical processes that govern the dispersion and transformation of atmospheric pollutants. Depending on the application (air quality and traffic management, urban planning, interpretation of monitoring data, people's exposure, etc.), a specific dispersion model to be employed can be used. They can be classified according to their physical or mathematical principles (e.g. reduced scale, box, Gaussian, CFD) and their level of complexity (e.g. screening, semiempirical, numerical) [16].

For regulatory purposes, those preferred are *operational models* which are mostly based on the Gaussian dispersion model, often referred to as 'fast response models'. These models can be modified for various purposes and are commonly used for many comprehensive formal evaluations because they are designed to enable many different cases. They may include many complicated dispersion processes such as atmospheric stratification, buoyancy, chemistry, deposition, concentration fluctuations, etc. However, they cannot explicitly treat the detailed plume behaviour affected by building obstacles [17].

At the micro-scale, *numerical simulation with CFD*, even though computational expensive, is the preferred way of investigation and the most suitable for studies of various physical flow and dispersion processes in complex geometries such as cities. CFD offers some advantages compared to other methods since it provides results of flow features at every point in space simultaneously and does not suffer from potentially incompatible similarity because simulations can be conducted at full scale. Due to the rapid development in computer hardware, CFD is increasingly used for such studies [6].

5.4.2 Flow and Dispersion in Street Canyons

Many fundamental studies have been conducted providing fundamental knowledge on flow dynamics and pollutant dispersion in the presence of obstacles (see experimental and modelling reviews by Ahmad et al. [2], Li et al. [18], Di Sabatino et al. [19], Blocken [20], Yazid et al. [21] and Lateb et al. [6]).

Streets (commonly named street canyons) are the basic architectural structures in urban environments. They constitute the basic geometric unit of cities and represent highly polluted zones around buildings. They show a distinct climate where micro-scale meteorological processes dominate. The most important features of street canyon microclimate are the wind-induced flow patterns, such as air recirculation. These micro-scale meteorological processes not only affect the local air quality but also the comfort of the city's populations. High pollution levels have been observed in street canyons where pedestrians, cyclists, drivers and residents may be exposed to pollutant concentrations exceeding air quality legislative standards.

Most studies focussed on flow and dispersion in the idealised case of a street canyon consisting of two or more building blocks with an external wind blowing perpendicularly to the street axis. In such case, the flow field is composed of two main vortex structures, the canyon vortex and the corner vortices. The first leads to an accumulation of traffic emissions typically at one side of the street (at the leeward wall) and a reduction at the wind ward wall, while the second leads to reduced pollutant concentrations at canyon ends (Fig. 5.4). A 3D street canyon has three determining factors of flow regime, namely, the relative *height* (H), *width* (W) and *length* (L) of the canyon, in contrast to a 2D canyon with two factors, H and W. Therefore, the flow field formed in a 3D street canyon is highly complex in comparison to the 2D case [17].

The case of a *wind blowing perpendicularly to the street axis* is commonly assumed as the worst scenario leading to high pollution levels within the street. Based on field measurements and modelling results, Oke [24] identified three flow regimes for wind direction perpendicular to the street axis in neutral stratification, i.e. with no buoyancy effects due to temperature differences between air and urban surfaces (Fig. 5.5). For widely spaced buildings (aspect ratio $H/W < 0.3$), the flow fields associated with the buildings do not interact (isolated roughness flow regime). At closer spacing ($0.3 < H/W < 0.7$), the wake behind the upwind building is disturbed by the recirculation created in front of the windward building (wake interference flow regime). Further reducing spacing ($H/W > 0.7$) results in the skimming flow regime, where a stable recirculation is developed inside the canyon and the ambient flow is decoupled from the street flow. In this latter case, traffic-produced

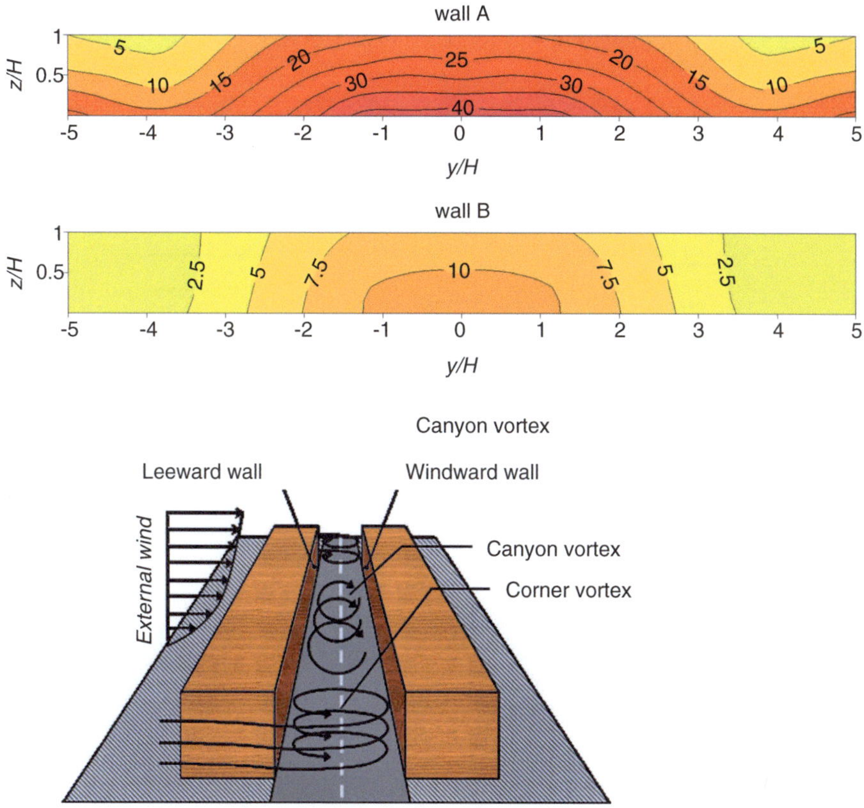

Fig. 5.4 3D sketch of canyon and corner vortices (*top*) (*top* figure adapted from [22] © Elsevier, reprinted with permission of Elsevier). Normalised concentration contours at wall A (leeward) and wall B (windward) (*bottom*) within an idealised 3D street canyon model with perpendicular wind (H is the building height and $y/H = 0$ denotes the middle of the canyon wall; bottom figure from [23] © CODASC, KIT)

pollutants could not be easily ventilated resulting in high pollutant concentration and poor air quality. The SF regime, which is widely studied in the literature, has different numbers of recirculation depending on the aspect ratio H/W. The aspect ratio L/H also affects the flow regime transitions [18].

The strength of the wind-induced recirculation inside street canyons also depends on other factors, such as the roof-level wind velocity and direction, the roof shape, the buoyancy (thermal) effects, the mechanical turbulence induced by moving vehicles, etc. For example, the flow induced in a street by an *external wind nearly parallel to the street axis* produces a strong channelling of flow and pollutant plume within the street. For the 45° inclined case, the canyon and corner vortices are not as pronounced as for the perpendicular case, and the concentration pattern at both walls of the street canyon is due to ventilation caused by the superposition of the canyon vortex and the corner eddy which lead to a complex 3D helicoids

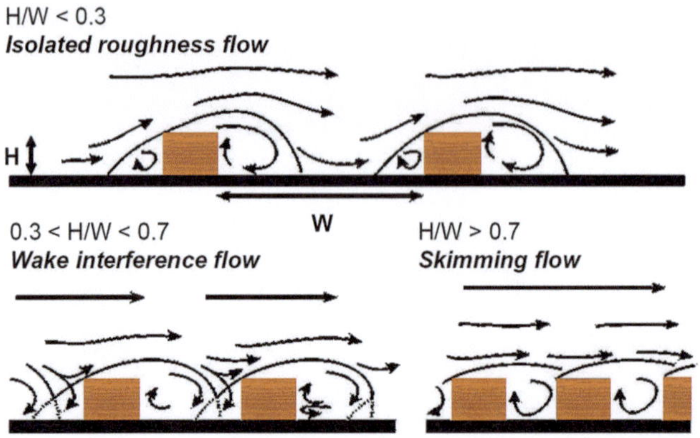

Fig. 5.5 Flow regimes associated with different building-height-to-street-width ratios H/W for perpendicular flow (adapted from [24] © Elsevier, reprinted with permission of Elsevier)

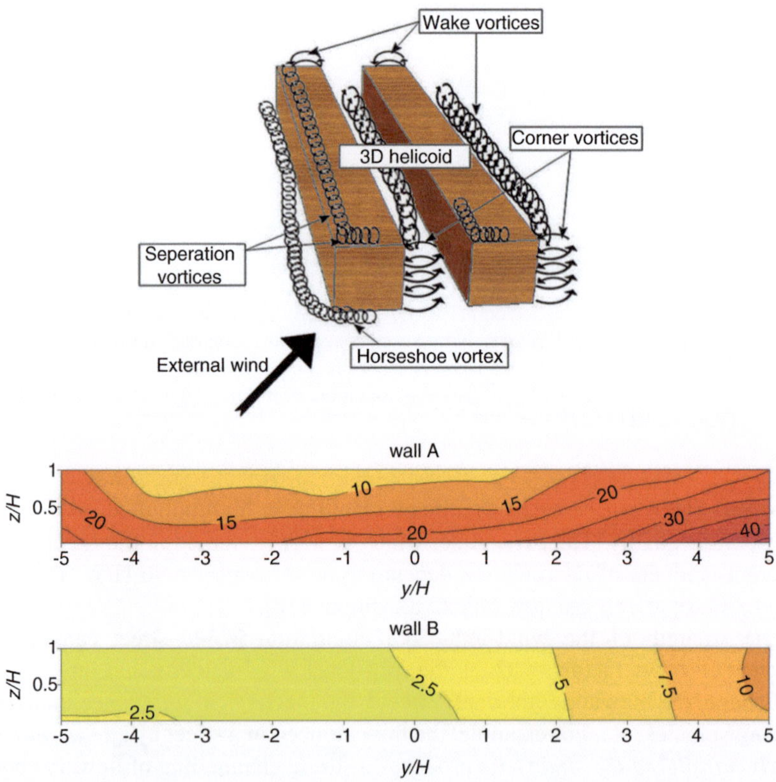

Fig. 5.6 Flow structure inside a long 3D idealised street canyon with 45° oblique wind (*top*) (*top* figure adapted from [21] @SAGE). Normalised concentration contours at leeward wall A and windward wall B (*bottom*) (*H* is the building height and $y/H = 0$ denotes the middle of the canyon wall; *bottom* figure from [23] © CODASC, KIT)

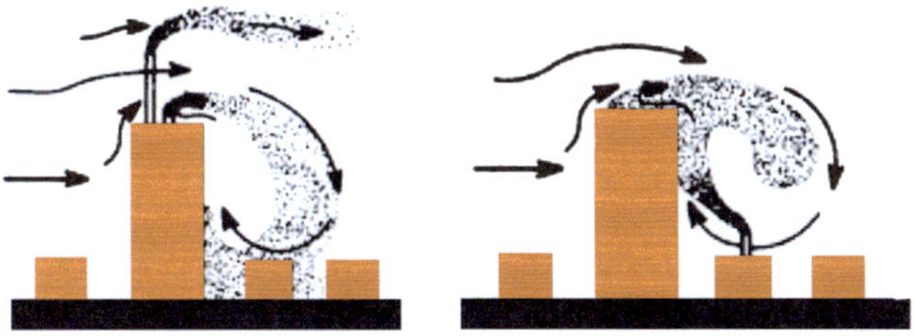

Fig. 5.7 Example sketch of the impact of asymmetric buildings on the dispersion of pollutants emitted from sources located at different positions (adapted from [25] © Routledge, reprinted with permission of Taylor & Francis)

(Fig. 5.6, top). If the street is sufficiently long, advection along the street axis may cause an accumulation of pollutants resulting in concentrations at the downwind end of the street that can be equal or exceed those produced with a perpendicular wind (Fig. 5.6, bottom). The increasing pollutant concentration towards the downstream end of the street indicates that the flow along the street axis becomes the dominant transport mechanism. Further, the higher the wind speed was, the more effectively the pollutants tended to be diluted.

Due to the irregularities of building configurations in real life, a number of past studies have also looked into the effects of different street canyon configurations and roof shapes. It was found that interesting airflow variations and different pollutant distributions were induced by asymmetric building configurations. As an example Fig. 5.7 shows the impact of the presence of a taller building on plume pattern.

The presence of street intersections significantly enhances the complexity of flow and dispersion patterns (see the reviews by [2, 26]). The complicated flow pattern resulted in an important influence on dispersion and mixing within the intersection, making the pollutant concentrations vary significantly around the intersection depending on the wind direction (Fig. 5.8).

Flow and pollutant patterns in street canyons also depend on solar radiation which induces *heat fluxes* (buoyancy, thermal-produced turbulence) produced by the temperature difference between the canyon walls and the air within the canyon. For a single wall heated or multiple walls heated, the leeward and/or ground heating intensifies the flow within the street canyon and promotes better pollutant removal. If the windward wall is involved in multiple wall heating, the vortex intensity is slightly weakened and reduced pollutant removal (see the review by [21]).

5.4.3 Flow and Dispersion in More Complex Geometries

Urban planning has evolved leading to a great variety of urban forms, from urban sprawl to compact (high density) cities. While the first are usually associated with negative environmental, social and economic effects, the latter has been proposed as the main 'sustainable' alternative. The compact city is generally attributed to high-density urban development having increased socio-economic diversity providing

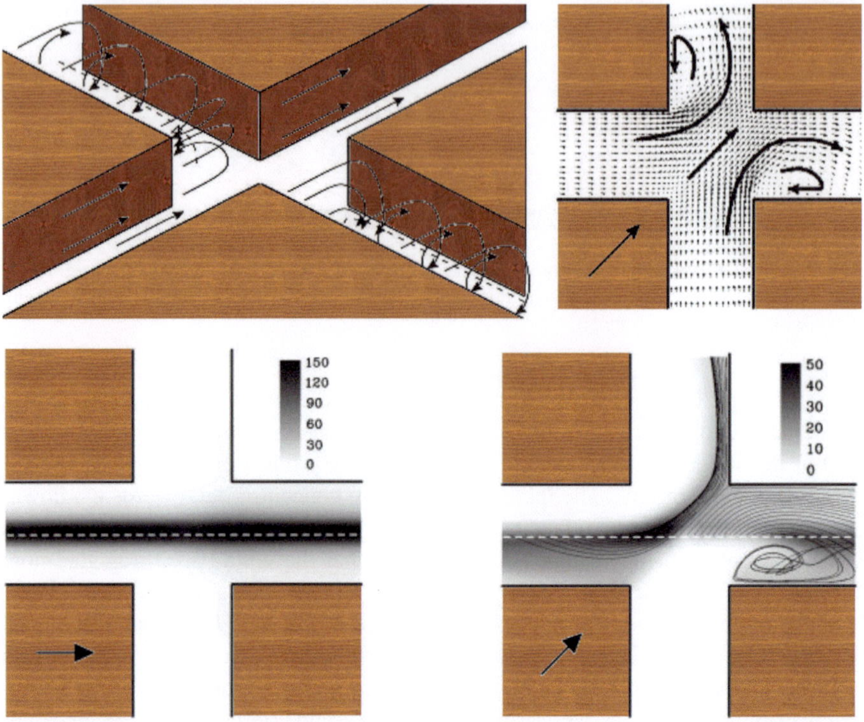

Fig. 5.8 Flow (*top*) and pollutant normalised concentration (*bottom*) in a street intersection with parallel (*left*) and oblique wind (*right*) (adapted from [27] © Elsevier, reprinted with permission of Elsevier)

opportunities for social interactions and exchanges and minimising environmental degradation. But, at the same time, the high density can lead to poor air quality conditions due to decreased ventilation conditions with respect to sparse cities.

Similarly to what was observed at street canyon scale, pollutant concentrations within more complex geometries (such as a neighbourhood or a city) are sensitive to dynamical and morphological factors (wind direction and velocity, street canyon aspect ratios, street intersections and upstream buildings, roof shape, packing density). The influence of these factors has been studied in the literature, and recent reviews have highlighted trends and directions for future research (e.g. see [4, 6, 17, 19]).

The *packing density* of a neighbourhood or a city is usually expressed through the morphometric parameters λ_p (planar area index, i.e. the ratio between the planar area of buildings and the lot area) and λ_f (frontal area index, i.e. the ratio between the frontal area of buildings and the lot area) [28]. Most of the studies focussed on pollutant dispersion in and around *regular arrays of obstacles* (mostly cubes or parallelepipeds) with low packing density; see, for example, the well-known Mock Urban Setting Test (MUST) [29] which helped to understand the processes affecting the plume structure, e.g. channelling, lateral dispersion, detrainment, secondary-source dispersion and plume skewing. The geometry of such low-density obstacle arrays is similar to that of North American or European suburban neighbourhoods [28], but very different from central neighbourhoods of most European cities, where buildings are regularly and densely packed and characterised by high traffic levels and a high density of population. Tests in real geometries were performed within the Joint Urban 2003 project [30] conducted in Oklahoma City to study atmospheric transport and diffusion in an urban environment and the London experiment (DAPPLE project, [31]) which involved field measurements, wind-tunnel experiments and computational modelling.

An approach to determine street level pollutant dilution has recently been offered by the implementation of ventilation concepts to study the so-called city breathability, suggesting a way to classify city neighbourhoods into three groups, *sparse*, *compact* and *very compact cities*, depending on its packing density [32]. Figure 5.9 shows an example of vectors of normalised velocity magnitude and normalised mean age of air (which gives information on ventilation efficiency) for a sparse canopy (left) and a very compact canopy (right). It is visible that the mean age of air, that is, the time taken by a parcel of air to reach a given place after it enters the urban canopy layer, is larger in the compact city, which corresponds to poorer ventilation.

In a recent paper by Buccolieri et al. [33], who studied the ventilation of regular arrays of parallelepipeds representing compact cities, a high variability of ventilation conditions was found depending on the incident wind angle and slight modifications of the array geometry. For low wind angles, a clear effect of channelling of pollutants along the street parallel to the wind direction was observed. This effect, combined with relative low exchanges at street intersections and with the overlying atmosphere, resulted in poor ventilation conditions. Conversely, large wind angles were shown to enhance transversal mean transfers at street intersections and vertical exchange with the overlying atmosphere. For increasing incident angles, the vertical

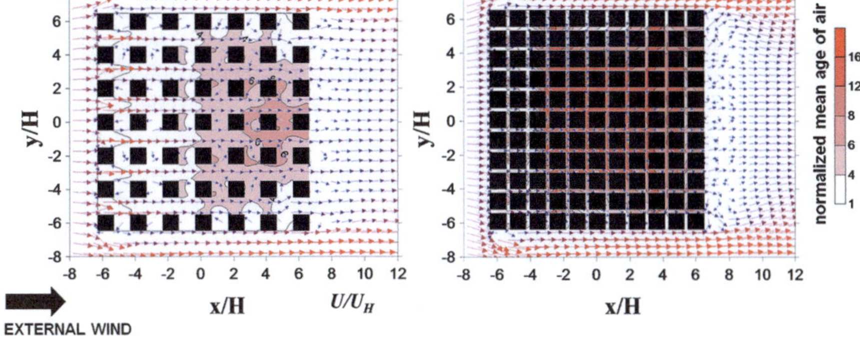

Fig. 5.9 Flow pattern and ventilation efficiency for arrays of cubes of different packing density: spars canopy (*left*) and very compact canopy (*right*) (adapted from [32] © Elsevier, reprinted with permission of Elsevier)

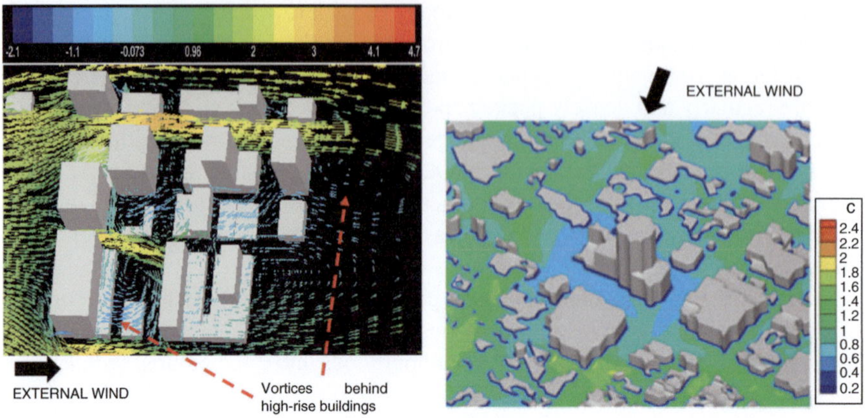

Fig. 5.10 Flow pattern coloured by the velocity component along the external wind direction (*left*) and mean age of air (*right*) in real geometries (*right* figure adapted from [36] © Springer, reprinted with permission of Springer)

transfer increased due to the enhancement of the mean counterpart of the total flux. This effect was associated with a stronger interaction between the mean flow developing above the canopy and that within it.

Recently awareness about the creation of low wind speed areas and poor outdoor ventilation around *high-rise buildings*, where the approaching wind is quickly weakened due to the strong drag force, has also increased (e.g. see recent papers by [34–36]). Their results revealed the importance of the wind incident direction for wind circulation within the re-entrant corners of closely constructed high-rise buildings. As an example Fig. 5.10 (left) shows the complicated flow patterns which develop around a real geometry, and strong reverse flow behind high-rise buildings may occur leading to an accumulation of pollutants. This is strictly dependent on the overall morphology of the city, and values of the mean age of air may be high in areas crowded with low-rise buildings and relatively low (thus corresponding to better ventilation) in areas around high-rise buildings because of the strong winds from higher altitudes (Fig. 5.10, right).

These studies have highlighted that interaction between the atmospheric approaching flow and a city results in complicated flow patterns between buildings, along streets, stagnant zones and wake regions. Air mass approaching a city either can enter the streets or flows above the buildings or around it. Exploring ventilation conditions in a city neighbourhood is thus crucial for the assessment of air quality scenarios and their impacts on people's health.

5.4.4 The Effects of the Presence of Urban Obstacles

A number of methods have been demonstrated to influence flow patterns in the built environment under varying urban geometrical and meteorological conditions, such as trees and vegetation (porous) as well as noise barriers, low boundary walls and parked cars (solid). Advantages and disadvantages of those methods have been recently reviewed by Gallagher et al. [5] and are briefly summarised below.

Trees and vegetation, which have been commonly employed as an aesthetical element, can affect pollutant dispersion in the built environment. The recent reviews by Janhall [37] and Abhijith et al. [38] provides a detailed overview of the impact of several green infrastructures on dispersion. Tree parameters (crown height, leaf density tree height and spacing) have been found to impact air quality, and, despite their porous nature, trees have been reported to act more like a solid barrier. In general, when trees are too close with interfering crowns, polluted air in the street canyon is not able to mix with cleaner air of the underlying atmosphere because of the closed tree canopy which results in the tunnel effect. By using less sprawling trees or placing trees at a distance from each other, the mixing with cleaner air can be strengthened (Fig. 5.11). Comprehensive studies are however still missing, and thus final guidelines are not still available to urban planners.

Noise barriers can be found alongside busy arterials and high-speed, high-traffic highways in most cities as solid high walls. Those barriers may lead to a

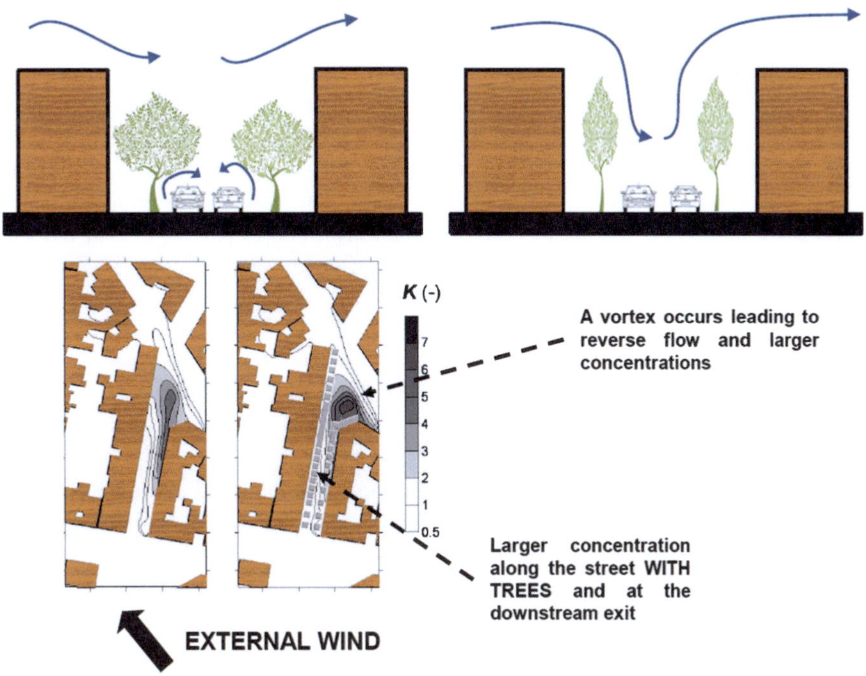

Fig. 5.11 Sketch of the flow in an idealised street canyon with the perpendicular wind in the presence of trees of different crown size (*top*) (top figure adapted from http://www.urbangreenbluegrids.com/air © atelier GROENBLAUW, reprinted with permission of atelier GROENBLAUW—www.atelierGROENBLAUW.nl). Top view of the pattern of normalised concentration K in a real street canyon (*bottom*) (*bottom* figure adapted from [39] © Inderscience Publishers, reprinted with permission of Inderscience Publishers)

reduction in pollutant concentrations downwind of the barriers. However, similar to the vegetation studies, an increase in upwind concentrations has been found due to the recirculation of pollutants in the zone in front of the structure. The height and layout of the noise barrier (i.e. continuous barrier or the combination of the barrier and vegetation) were found to have the greatest influence on the dispersion of pollutants along the highway, as well as wind conditions. Overall, studies suggest that noise barriers are a feasible passive method of pollution reduction and present less variable factors in their effectiveness than porous barriers. However, the relationship between vegetation and noise barriers is more complex and still lacks of comprehensive studies preventing the developing of design guidelines.

Low boundary walls can improve urban air quality by enhancing pollutant dispersion in street canyons. They act as a baffle at street level and increase the distance between the pollutant source and human receptor in most wind conditions. However, the majority of the research focussed on idealised street canyon models, typically overestimating the improvements in air quality compared to real world measurements. The development of guidelines to provide practical instructions for implementing low boundary walls in a street canyon environment still requires further research as current findings are limited.

Parked cars have a different impact on the pollutant dispersion depending on the parking bay layout and an occupancy rate of the parking bays. Due to the spacing between vehicles, parked cars impact air flow differently to the other barrier types. For example, parallel parked cars can provide improvements in air quality in all wind conditions. Varying street geometry and the noncontinuous nature of the parking bay may prevent the occurrence of strong vortices which enhances dispersion at street level. Guidelines to promote the retrofitting of parking bay designs in street canyons presents the greatest potential to ensure parked cars can promote pollutant dispersion.

5.5 The Air Pollution and the Problem of the Ultrafine Particles

Air pollution is composed of a variety of gaseous pollutants and particulate matter. These pollutants arise from a variety of sources such as road vehicles, industrial emissions and wood-burning [40]. The majority of cities worldwide are facing exceeding levels of air pollutants over the regulatory limits [41]. Air pollutants can be classified into primary and secondary. Primary pollutants are those directly emitted into the atmosphere by human activities (vehicle engines, industrial production, etc.) and by natural processes (windblown dust, volcanic activity, etc.). Secondary pollutants are formed within the atmosphere when primary pollutants react with sunlight, oxygen, water and other chemical compounds present in the air. As for public health, major pollutants are particulate matter (PM), tropospheric ozone (O_3), nitrogen dioxide (NO_2) and sulphur dioxide (SO_2). NO_2 and SO_2 are produced by the combustion of fossil fuels and contribute to photochemical smog, as well as to acid rain. O_3 is a major component of photochemical smog, an air pollution phenomenon that forms when primary pollutants like NO_2 and carbon monoxide (CO) react with sunlight to form a variety of secondary pollutants. PM is mainly attributed to the combustion of fossil fuels, especially coal and diesel fuel, and is composed of tiny particles of solids and liquids including ash, carbon soot, mineral salts and oxides, heavy metals such as lead, and other organic compounds. PM includes particles from a few nanometre size to 10s of a micron. For example, PM_{10} refers to particles with an aerodynamic diameter $\leq$ 10 μm while $PM_{2.5}$ to particles with an aerodynamic diameter $\leq$ 2.5 μm. In general, the smaller the size of the particles, the larger their effects are on human health. For example, the smaller particles show much larger health effects compared with their larger-sized particles [42].

Together with conventional air pollutants, which are currently under regulation, there is an increase interest towards unregulated pollutants, such as airborne ultrafine particles (UFPs; diameter less than 100 nm). They differ from larger counterparts in their dynamic nature which contribute to substantial temporal and spatial variability which increases with decreasing particle size. UFPs are considered to show even higher health impacts compared with fine particles [43]. These particles are so small in size that they contribute almost a negligible mass, unlike fine or coarse particles, but contain significant numbers and therefore are represented by particle number concentrations (PNC). In typical urban environments, ultrafine particles contribute up to 80% of the total PNCs [44]. The dominant source of ultrafine particles in the urban environment is road vehicles. The fresh emissions cool and dilute rapidly just after their emissions to produce nucleation mode particles (<30 nm), while the larger size particles are part of exhaust emissions [45]. The emissions from these vehicles undergo intense transformation processes such as nucleation, condensation, coagulation and dry deposition to change their concentrations between the source and receptor [15, 46].

A study by Kumar et al. [47] reviewed the concentrations of particles in various cities worldwide. They reported that average PNCs in the European cities are much lower compared with those in Asian cities. This is expected due to a much lower sulphur content in the fuel. Recent Euro 5 and 6 emission standards included, for the first time, the limits of particle number emissions for the heavy-duty vehicles. However, there are yet no standards to control their ambient concentrations in urban areas. The European Committee for Standardisation (CEN) at the European level is currently taking an initiative to standardise the measurement methods for ultrafine particles, which is likely to result in regulatory guidelines in the near future in Europe.

A recent review [48] reported on average PNCs in different places of urban areas. The place with higher concentrations, usually referred to as pollution hotspots, of ultrafine particles is found close to the intense sources (industrial plants, motorways, harbours) or in microenvironments with low air exchange rates such as street canyons. The other most prominent pollution hotspots of ultrafine particles, which are important from an exposure perspective, are traffic intersections. At the traffic intersections, the vehicles go under variable driving conditions such as acceleration, deceleration and stop and go, resulting in much higher emissions compared with traffic under free-flow traffic conditions [49].

There is still a need for further studies, developing the exposure profiles in varying city environments worldwide to assess the exposure of the urban commuters in complex urban environments. Moreover, there is a clear need for dispersion models, which could treat the ongoing transformation process that is unique to ultrafine particles. Moreover, there is still a lack of standardised methods to measure them highlighting a need to develop robust instrumentation that could produce reproducible results and are affordable. Unlike other pollutants such as coarse or fine particles, the progress on miniaturisation of ultrafine particles is still behind compared with those for coarse or fine particles or the gaseous pollutants. Such a progress in the case of ultrafine particles could also allow the high spatial mapping of ultrafine particles in complex urban environments to design appropriate mitigation measures.

Conclusions

Outdoor air pollution is a major environmental health problem affecting people in developed and developing countries. The 'WHO Air quality guidelines' provide an assessment of health effects of air pollution and thresholds for health-harmful pollution levels. In 2014, 92% of the world population was living in places where the WHO air quality guidelines levels were not met. A 2013 assessment by WHO's International Agency for Research on Cancer (IARC) concluded that outdoor air pollution is carcinogenic to humans, with PM most closely associated with increased cancer incidence [50].

With the rate of urbanisation expected to increase in the next years, countries are forced to face serious challenges in addressing the problem of air pollution. The prevention and control of air pollution in cities imply the study and the knowledge of those processes and mechanisms which govern its spatial distribution. As shortly discussed in this chapter, the dispersion of pollutants is strongly affected by several factors, such as variations in street canyon geometry, wind speed and direction, thermally stability conditions, tree planting and moving vehicles. While some of these aspects have been extensively studied in the literature, there are some aspects of research which still lack of attention, and future works are expected to focus more extensively on thermal investigation, turbulence produced from moving vehicles, deposition and aerodynamics effects of trees and study the effect of combining various physical factors [21].

Such studies are crucial for providing suggestions on better designing a city for improved wind and thermal conditions as well as people's health. It is important to develop design guidelines which should include building dimensions, width of streets, allowable emissions from stationary sources (e.g. industry), nonstationary sources (e.g. vehicles) and location of industries within the city. Cool roofs and tree planting, such as other barriers, should also be investigated in detail to provide a cooler and safer environment [51].

References

1. Hilber C, Palmer C. Urban development and air pollution: evidence from a global panel of cities. Grantham Research Institute on Climate Change and the Environment, Working Paper No. 175. http://www.lse.ac.uk/GranthamInstitute/wp-content/uploads/2014/12/Working-Paper-175-Hilber-Palmer-2014.pdf; 2014. Accessed 30 Dec 2016.
2. Ahmad K, Khare M, Chaudhry KK. Wind tunnel simulation studies on dispersion at urban street canyons and intersections – a review. J Wind Eng Ind Aerodyn. 2005;93:697–717. https://doi.org/10.1016/j.jweia.2005.04.002.
3. Britter RE, Hanna SR. Flow and dispersion in urban areas. Annu Rev. Fluid Mech. 2003;35:469–96. https://doi.org/10.1146/annurev.fluid.35.101101.161147.
4. Blocken B, Tominaga Y, Stathopoulos T. CFD simulation of micro-scale pollutant dispersion in the built environment. Build Environ. 2013;64:225–30. https://doi.org/10.1016/j.buildenv.2013.01.001.
5. Gallagher J, Baldauf R, Fuller CH, Kumar P, Gill LW, McNabola A. Passive methods for improving air quality in the built environment: a review of porous and solid barriers. Atmos Environ. 2015;120:61–70. https://doi.org/10.1016/j.atmosenv.2015.08.075.
6. Lateb M, Meroney RN, Yataghene M, Fellouah H, Saleh F, Boufadel MC. On the use of numerical modelling for near-field pollutant dispersion in urban environments. A review. Environ Pollut. 2016;208:271–83. https://doi.org/10.1016/j.envpol.2015.07.039.
7. Soubbotina TP. Beyond economic growth. An introduction to sustainable development. 2nd ed. Washington, DC: World Bank; 2004. https://openknowledge.worldbank.org/handle/10986/14865 Accessed 30 Dec 2016
8. United Nations. World urbanization prospects: the 2014 revision (CD-ROM edition). New York: Department of Economic and Social Affairs, United Nations; 2014. https://esa.un.org/unpd/wup/Publications/Files/WUP2014-Report.pdf Accessed 30 Dec 2016
9. UN-Habitat. Urbanization and development: emerging futures. World cities report 2016. Nairobi: United Nations Human Settlements Programme (UN-Habitat); 2016. http://wcr.unhabitat.org/main-report Accessed 30 Dec 2016
10. Ericson B, Hanrahan D, Kong V (2008) The World's worst pollution problems: the top ten of toxic twenty. Project report Blacksmith Institute, New York. Available from: http://www.worstpolluted.org. Accessed 30 Dec 2016.
11. Parrish DD, Stockwell WR. Urbanization and air pollution: then and now. EOS Earth & Space Science News. https://eos.org/features/urbanization-air-pollution-now; 2015. Accessed 30 Dec 2016.
12. AMS. Meteorological glossary. American Meteorological Society. http://glossary.ametsoc.org/wiki/Main_Page; 2015. Accessed 30 Dec 2016.
13. Barlow JF. Progress in observing and modelling the urban boundary layer. Urban Clim. 2014;10:216–40. https://doi.org/10.1016/j.uclim.2014.03.011.
14. Cui P-Y, Li A, Tao W-Q. Buoyancy flows and pollutant dispersion through different scale urban areas: CFD simulations and wind-tunnel measurements. Build Environ. 2016;104:76–91. https://doi.org/10.1016/j.buildenv.2016.04.028.
15. Kumar P, Ketzel M, Vardoulakis S, Pirjola L, Britter R. Dynamics and dispersion modelling of nanoparticles from road traffic in the urban atmospheric environment - a review. J Aerosol Sci. 2011;42:580–603. https://doi.org/10.1016/j.jaerosci.2011.06.001.
16. Vardoulakis S, Fisher BEA, Pericleous K, Gonzalez-Flesca N. Modelling air quality in street canyons: a review. Atmos Environ. 2003;37:155–82. https://doi.org/10.1016/S1352-2310(02)00857-9.
17. Tominaga Y, Stathopoulos T. CFD simulation of near-field pollutant dispersion in the urban environment: a review of current modeling techniques. Atmos Environ. 2013;79:716–30. https://doi.org/10.1016/j.atmosenv.2013.07.028.

18. Li X-X, Liu C-H, Leung DYC, Lam KM. Recent progress in CFD modelling of wind field and pollutant transport in street canyons. Atmos Environ. 2006;40:5640–58. https://doi.org/10.1016/j.atmosenv.2006.04.055.
19. Di Sabatino S, Buccolieri R, Salizzoni P. Recent advancements in numerical modelling of flow and dispersion in urban areas: a short review. Int J Environ Pollut. 2013;52:172–91. https://doi.org/10.1504/IJEP.2013.058454.
20. Blocken B. 50 years of computational wind engineering: past, present and future. J Wind Eng Ind Aerodyn. 2014;129:69–102. https://doi.org/10.1016/j.jweia.2014.03.008.
21. Yazid AWM, Sidik NAC, Salim SM, Saqr KM. A review on the flow structure and pollutant dispersion in urban street canyons for urban planning strategies. Simulation. 2014;90(8):892–916. https://doi.org/10.1177/0037549714528046.
22. Gromke C, Ruck B. Influence of trees on the dispersion of pollutants in an urban street canyon - experimental investigation of the flow and concentration field. Atmos Environ. 2007;41:3287–302. https://doi.org/10.1016/j.atmosenv.2006.12.043.
23. CODASC. Concentration data of street canyon database. Laboratory of Building- and Environmental Aerodynamics, IfH. Karlsruhe Institute of Technology. http://www.windforschung.de/CODASC.htm; 2008. Accessed 23 Nov 2016.
24. Oke TR. Street design and urban canopy layer climate. Energ Buildings. 1988;11:103–13. https://doi.org/10.1016/0378-7788(88)90026-6.
25. Oke TR. Boundary layer climates. 2nd Edition. Abingdon: Routledge; 1987.
26. Tiwary A, Robins A, Namdeo A, Bell M. Air flow and concentration fields at urban road intersections for improved understanding of personal exposure. Environ Int. 2011;37:1005–18. https://doi.org/10.1016/j.envint.2011.02.006.
27. Soulhac L, Garbero V, Salizzoni P, Mejean P, Perkins RJ. Flow and dispersion in street intersections. Atmos Environ. 2009;43:2981–96. https://doi.org/10.1016/j.atmosenv.2009.02.061.
28. Di Sabatino S, Leo LS, Cataldo R, Ratti C, Britter RE. Construction of digital elevation models for a southern European city and a comparative morphological analysis with respect to northern European and north American cities. J Appl Meteorol Climatol. 2010;49:1377–96. https://doi.org/10.1175/2010JAMC2117.1.
29. Yee E, Biltoft CA. Concentration fluctuation measurements in a plume dispersing through a regular array of obstacles. Bound-Layer Meteorol. 2004;111:363–415. https://doi.org/10.1023/B:BOUN.0000016496.83909.ee.
30. Allwine KJ. Overview of JOINT URBAN 2003 an atmospheric dispersion study in Oklahoma City. Proceedings of the AMS symposium on planning, nowcasting, and forecasting in the urban zone, Seattle, Washington, USA, January 11-15; 2004.
31. Wood CR, Barlow JF, Belcher SE, Dobre A, Arnold SJ, Balogun AA, Lingard JJN, Smalley RJ, Tate JE, Tomlin AS, Britter RE, Cheng H, Martin D, Petersson FK, Shallcross DE, White IR, Neophytou MK, Robins AG. Dispersion experiments in central London: the 2007 DAPPLE project. Bull Am Meteorol Soc. 2009;90:955–69. https://doi.org/10.1175/2009BAMS2638.1.
32. Buccolieri R, Sandberg M, Di Sabatino S. City breathability and its link to pollutant concentration distribution within urban-like geometries. Atmos Environ. 2010;44:1894–903. https://doi.org/10.1016/j.atmosenv.2010.02.022.
33. Buccolieri R, Salizzoni P, Soulhac L, Garbero V, Di Sabatino S. The breathability of compact cities. Urban Clim. 2015;13:73–93. https://doi.org/10.1016/j.uclim.2015.06.002.
34. Iqbal QMZ, Chan ALS. Pedestrian level wind environment assessment around group of high-rise cross-shaped buildings: effect of building shape, separation and orientation. Build Environ. 2016;101:45–63. https://doi.org/10.1016/j.buildenv.2016.02.015.
35. Mao J, Yang W, Gao N. The transport of gaseous pollutants due to stack and wind effect in high-rise residential buildings. Build Environ. 2015;94:543–57. https://doi.org/10.1016/j.buildenv.2015.10.012.
36. Tominaga Y. Visualization of city breathability based on CFD technique: case study for urban blocks in Niigata City. J Vis. 2012;15:269–76. https://doi.org/10.1007/s12650-012-0128-z.
37. Janhall S. Review on urban vegetation and particle air pollution – deposition and dispersion. Atmos Environ. 2015;105:130–7. https://doi.org/10.1016/j.atmosenv.2015.01.052.

38. Abhijith KV, Kumar P, Gallagher J, McNabola A, Baldauf R, Pilla F, Broderick B, Di Sabatino S, Pulvirenti P. Air pollution abatement performances of green infrastructure in open road and built-up street canyon environments – A review. Atmos Environ. 2017;162:71–86.
39. Di Sabatino S, Buccolieri R, Pappaccogli G, Leo LS. The effects of trees on micrometeorology in a real street canyon: consequences for local air quality. Int J Environ Pollut. 2015;58:100–11. https://doi.org/10.1504/IJEP.2015.076587.
40. Kumar P, Wiedensohler A, Birmili W, Quincey P, Hallquist M. Ultrafine particles pollution and measurements. Compr Anal Chem. 2016a;73:369–90.
41. Kumar P, Khare M, Harrison RM, Bloss WJ, Lewis AC, Coe H, Morawska L. New directions: air pollution challenges for developing megacities like Delhi. Atmos Environ. 2015;122:657–61. https://doi.org/10.1016/j.atmosenv.2015.10.032.
42. Heal MR, Kumar P, Harrison RM. Particles, air quality, policy and health. Chem Soc Rev. 2012;41:6606–30. https://doi.org/10.1039/c2cs35076a.
43. HEI. HEI review panel on ultrafine particles. Understanding the health effects of ambient ultrafine particles HEI perspectives 3. Boston, MA: Health Effects Institute; 2013. https://www.healtheffects.org/publication/understanding-health-effects-ambient-ultrafine-particles Accessed 30 Dec 2016
44. Kumar P, Robins A, Vardoulakis S, Britter R. A review of the characteristics of nanoparticles in the urban atmosphere and the prospects for developing regulatory controls. Atmos Environ. 2010;44:5035–52. https://doi.org/10.1016/j.atmosenv.2010.08.016.
45. Kittelson DB, Watts WF, Johnson JP. Nanoparticle emissions on Minnesota highways. Atmos Environ. 2004;38:9–19. https://doi.org/10.1016/j.atmosenv.2003.09.037.
46. Buonanno G, Morawska L, Stabile L, Wang L, Giovinco G. A comparison of submicrometer particle dose between Australian and Italian people. Environ Pollut. 2012;169:183–9. https://doi.org/10.1016/j.envpol.2012.03.002.
47. Kumar P, Morawska L, Birmili W, Paasonen P, Hu M, Kulmala M, Harrison RM, Norford L, Britter R. Ultrafine particles in cities. Environ Int. 2014;66:1–10. https://doi.org/10.1016/j.envint.2014.01.013.
48. Kumar P, de Fatima AM, Ynoue RY, Fornaro A, de Freitas ED, Martins J, Martins LD, Albuquerque T, Zhang Y, Morawska L. New directions: from biofuels to wood stoves: the modern and ancient air quality challenges in the megacity of São Paulo. Atmos Environ. 2016b;140:364–9. https://doi.org/10.1016/j.atmosenv.2016.05.059.
49. Kumar P, Goel A. Concentration dynamics of coarse and fine particulate matter at and around the signalised traffic intersections. Environ Sci. 2016;18:1220–35. https://doi.org/10.1039/C1036EM00215C.
50. WHO. Ambient (outdoor) air quality and health. http://www.who.int/mediacentre/factsheets/fs313/en; 2016. Accessed 30 Dec 2016.
51. Hajra B. A review of some recent studies on buoyancy driven flows in an urban environment. Int J Atmos Sci. 2014;2014:362182. https://doi.org/10.1155/2014/362182.

6 Outdoor Air Pollution: Effects, Measure and Perception

Giulio Lolli

Outdoor air pollution, or environmental pollution, is often associated to modern life and industrialization. We believe it is an actual problem, or at most we date it back to the industrial revolution in the nineteenth century. We would think differently if we consider what Seneca wrote about Rome in 61 AD, so almost 2000 years ago in a preindustrial era: "As soon as I had escaped the heavy air of Rome and the stench of its smoky chimneys, which when stirred poured forth whatever pestilent vapours and soot they held enclosed, I felt a change in my disposition".

Air pollution was even then perceived as serious and with severe health effects. What relates Seneca's Rome with the modern times is the size of the city. The imperial capital was more than 1 million inhabitants, an incredible concentration of people for the time; today where most the pollution is concentrated is also in the urban areas, where several millions of people live. It is more a matter of how much more than what is emitted (i.e. how dangerous it is) compared to the ability of the environment to balance it [1]. Moreover not only pollution is often related to human activities, this is an objective fact, but what matters most is how it is perceived subjectively. Measured by satellites, the highest world concentration and amount of particulate matter (PM), one of the most discussed pollutants, is located in the Sahara region. It is generated naturally by the abrasion of the sand particles under the wind shear and can reach several g/m^3. Nobody talks about it. In comparison, several orders of magnitude less PM (μg/m^3) in one of our metropolis have much more severe consequences and generate huge media coverage. In the following pages, we will try to discuss not only what the air pollution effectively and objectively is but also how it is perceived by the people in contact with it and what is then its impact on our life.

G. Lolli
Centro Studi Internazionale SIT, Bologna, Italy
e-mail: g.lolli@giuliololli.com

F. Capello, A.V. Gaddi (eds.), *Clinical Handbook of Air Pollution-Related Diseases*,
https://doi.org/10.1007/978-3-319-62731-1_6

6.1 Perceptive Distinction

The distinction between objective (instrument measurable) and subjective (human perceived) pollution is becoming the focus of the research in the recent years. From one side the public authorities (regulators) have to define objective criteria for environmental and health protection; at the same time, they have to address the public opinion, driven by subjective perception. A small concentration of a non-toxic but very odorous substance could be sometimes more critical than an undetectable one. For this reason the environmental protection industry is also developing assessment methods which are a combination of objective concentration and perceivable concentration. Examples of that are sort of "electronic noses", whose measured value is crossed with panels of "real noses", so people should correlate the objective measure with the subjective perception. This is the actual frontier of research [2–6].

Another important distinction is the target group of people affected by the pollution. Children and older people are often affected by air pollution more in comparison to healthy adults. Air pollution is related with pathologies of the respiratory system, which are accentuated if associated to other pathologies, like cardiovascular disorders, typical of the ageing population. In small children the airways have a smaller section and thus are prone to blocking or restrictions, which are associated to respiratory diseases like asthma. Localized pollution near a hospital, kindergarten or hospice should then be minimized.

Thus pollution is defined as multidimensional problem where substance, concentration, timing (when), location (where) and population should be considered at the same time.

Based on this multidimensional approach, the type of pollution which affects our lives at most is the one associated with transportation, heating or more in general combustion processes. These processes have a strong local impact: the big metropolis where the traffic and heating density are at most is also where the majority of the people live; thus the effects are also greater and perceived at most. These processes can also have a long-lasting and diffuse effect. The emitted CO_2 is also responsible of global warming, which is a planetary effect. If pollutants like nitrogen or sulphur oxides are emitted, they can generate acid rains thousands of kilometres away from where the (diffuse) emission took place. We will see what these pollutants are, how they can be mitigated with modern environmental technologies and what their effects on human beings are.

6.2 Particulate Matter (PM)

Particulate matter is a complex mixture of particles, either solid or liquid (aerosol) suspended in air. They vary very much in size, composition and origin. Primary particles originate mainly from combustion processes and consist of a carbonaceous core with different chemicals (sulphates, metals or polycyclic aromatics) adsorbed onto the surface. The core is formed by an incomplete combustion, where locally the oxygen amount was not high enough. High temperatures in the flame and locally low concentration of oxygen lead to thermal pyrolysis of the combustible with formation of solid carbonaceous particles, which are then less reactive and are often expelled as PM. For this reason, for example, diesel engines which inject locally the fuel into the combustion chamber and use a self-induced detonation are more prone to generate PM compared to gasoline engine where a premixed air-fuel vapour mixture is taken into the combustion chamber in the aspiration phase and ignited by a spark. For the same reason, for example, heating systems based on methane or natural gas are intrinsically cleaner in terms of PM compared to systems running on oil, coke or wood itself. The natural gas mixes readily with air, and its combustion is practically complete, leading no particulate residue. Oil has to be vaporized first with air, leading to a comparable risk as in combustion engines. Coke or wood is solid. Their combustion is more complicated, and the chance to have locally oxygen deficiency is much higher. They are also large producers of PM, even if wood is publicized as environmentally friendly, as is renewable; from an emission prospective, it is not so clean.

Besides the primary particles generated during the (incomplete) combustion process, secondary particles are generated by chemical reactions between the primary particles and gases present in the atmosphere like nitric oxides, sulphur oxide or ozone, which are strong oxidants, generating aerosols of metal sulphates or nitrates. Thus the specific composition of PM varies strongly by region, time of the year and of the day and weather conditions and is not only determined by the emitting source. For this reason PM is often classified based on diameter more than composition. Particles with a diameter less than 10 μm are named PM10; particles with a diameter less than 2.5 μm are called PM2.5 (or fine particulates), and particles with a diameter less than 0.1 μm are called PM0.1 (or ultrafine particulates).

Particle size is the most important factor in determining where particles are deposited in the lung. Compared with large particles, fine particles can remain suspended in the atmosphere for longer periods and be transported over longer distances. Some studies [7, 8] suggest that fine particles have stronger respiratory effects in children than large particles. Particles greater than 10 μm rarely make it past the upper airways, whereas fine particles smaller than 2 μm can make it as far as the alveoli. Many studies have noted an association between particulate air pollution and mortality among people of all ages. Some data suggest that exposure to particulate matter may be associated with decreased birth weights. There are data from Brazil, Central and Eastern Europe and China to support this association [9–12].

6.3 Ozone

Ozone is often associated with the ozone depletion in the upper atmosphere and not to pollution. This "good" ozone is a naturally occurring form of oxygen that provides a protective layer shielding the Earth from harmful ultraviolet radiation. "Bad" ozone occurs in the lower atmosphere. Ozone is the major component of urban smog and a potent respiratory irritant that can also synergistically enhance a child's reaction to other air pollutants and pollen. Ozone is a secondary air pollutant formed in the atmosphere from a chemical reaction between hydrocarbons and nitrogen oxides in the presence of heat and sunlight. One important fact about ozone is that it requires sunlight for its formation, so it tends to peak on hot summer afternoons from 3 to 5 pm. As it is formed by hydrocarbons released into the atmosphere, it is strongly associated to the hydrocarbon combustion processes. It is an important pollutant in many parts of the industrialized world. It is rarely measured in developing countries.

Patients may better understand the effects of acute ozone exposure if clinicians describe it in terms with which they are familiar, e.g. it is "like sunburn of the lungs". After exposure to ozone, people with asthma have increased bronchial reactivity to subsequent allergens. Most of the acute respiratory effects such as cough and shortness of breath are thought to be reversible. Recent studies show that long-term exposure to ozone is associated with decrements in lung function that persist into the second decade [13].

Chronic exposure to ozone pollution has been associated with de novo development of chronic lung disease, mild pulmonary fibrosis and modest increases in small airway obstruction. Lifetime exposure to ambient ozone is negatively associated with lung function measures that reflect small airway physiology. There is an association of exposure to ozone with chronic phlegm, wheeze (apart from colds) and higher composite respiratory index. These are considered early indicators for pathological changes that might progress to chronic obstructive pulmonary disease. Some evidence has linked ozone to chronic lung scarring, especially at the bronchi alveolar junction [14].

6.4 NO_x

Air is composed primarily of Nitrogen (N_2 ca. 80%) and Oxygen (O_2 ca. 20%). Nitrogen is an inert part, while Oxygen is more reactive and is necessary for life and combustion. Only at very high temperatures, like the ones present in a flame or combustion engine, radical processes occur, and these two gases combine forming nitrogen oxide (NO). NO in contact with O_2 can oxidize further to nitrogen dioxide (NO_2). Being derived and interchangeable molecules, in environmental chemistry they are often referred as NO_x. In engine optimization, this process has to be balanced with formation of PM. Increasing the engine temperature, improve the combustion and reduce PM; however increase NO_x formation. Finding the right optimum between PM and NO_x is the tricky part, especially in diesel engines.

NO_x are very reactive. Especially NO_2 can photochemical react with O_2 to form Ozone and NO. If this process is wanted in the troposphere, regenerating the protecting ozone layer, it is very dangerous at ground level. NO_2 can also react with water forming HNO_3 or nitric acid. This contributes also to acid rains. It is also clear that when NO_x gets in contact with the water present in our lungs, an acidification of the pulmonary tissues occurs. NO_x has also an effect on blood vessel tonicity. Exposure to NO_x, especially in children, has been scientifically associated to increase respiratory disorders and increase morbidity [15, 16].

6.5 CO

Carbon monoxide (CO) poisoning is probably better known for the cases of suffocation which happens inside households with old combustion systems. CO is formed by an incomplete combustion of hydrocarbons or fuels in general. This incomplete combustion is often associated to a malfunctioning system. The main issue arises from the fact that CO is completely odourless, so it cannot be perceived and leads to death by fixing itself to the iron present in the haemoglobin, leading to tissue suffocation and death. Infants and children have an increased susceptibility to CO toxicity because of their higher metabolic rates. Children with existing pulmonary or haematological illness (such as anaemia) or other conditions that compromise oxygen delivery are also more susceptible to adverse effects of exposure to CO at lower levels than are healthy children. Chronic exposure to low levels of CO causes headaches [17].

As odourless gas, CO is not perceived as a problematic pollutant. Large amounts of CO are, for example, contained in cigarette smoke. Combustion engines produce modest amounts of CO due to the fast and incomplete combustion, but this is normally completely abated in the post combustion treatment.

6.6 SO_2

Sulphur dioxide (SO_2) is formed by the combustion of sulphur. A previously important source of SO_2 was in the last 30 years the combustion of hydrocarbons. S occurs naturally in oil. The amount varied depending on the field where it is extracted. The heavier fractions like diesel or naphtha produced from sour oilfields could contain several % of S. The petrochemical industry developed the so-called hydrodesulphurization processes, and modern regulated fuels are considered sulphur-free, as they contain less than 5 parts per million (ppm) of S. The actual main source of atmospheric SO_2 are today volcanic activities or specific industrial activities, especially related to the mineral industry. These activities are generally located far away from urban centres, so these are not perceived as critical.

6.7 CO_2

Carbon dioxide (CO_2) is not properly a pollutant. It is naturally contained in the atmosphere at around 400 parts per million (ppm) or 0.04% and is breathed out by every living organism. However its level in the atmosphere has increased in the last century generating the so-called greenhouse effect or global warming. The reason is that CO_2 absorbs very strongly in the infrared spectrum. The sunlight that reaches the ground (energy) is then absorbed and re-emitted as infrared radiation (heat). It can then escape into the space or be absorbed by CO_2 or other molecules increasing the atmosphere temperature. Maintaining a low and acceptable CO_2 concentration would prevent more serious effects on the long term.

Managing CO_2 emissions also collide slightly with the reduction of the other pollutants, especially for transportation. In the last year, due to a manipulation scandal, diesel engines are now under a critical eye. As we said before, due to the higher pressures and temperature they can reach, and for the injection method they use, they are prone to higher emissions of pollutants, especially NO_x and PM compared to gasoline engine. Moreover in the last 40 years, the gasoline engines had developed the so-called three-way catalytic converter [18]. This is an exhaust posttreatment based on a mixture of precious metals which can reduce to extremely low levels the NO_x and CO emissions. It works by monitoring the oxygen level in the exhaust gases, to maintain a relatively fat mixture, so that the amount of CO or partially burned hydrocarbon and NO_x is balanced. The reaction which the catalyst promotes (in a very simplified way) is

$$2\,CO + 2\,NO \rightarrow 2\,CO_2 + N_2$$

Thus it converts the pollutant CO and NO into N_2 which is inert and CO_2 which is unavoidable.

In diesel engines the direct motor control is not possible as the engine must operate always in lean mode (so with excess of oxygen). Special NO_x treatments have been developed also for diesel engines; however they require a postinjection of a reducing agent. The most widely used is a urea solution in many countries known as "AdBlue". This must be tanked separately and filled up every 2–3 refuelling. This technology is widely spread for heavy-duty trucks but is not well accepted by the normal city driver.

An alternative technology includes an NO absorbent, which is regenerated periodically by post-burning of fuels for a short period of time. This is often associated to filters for the particulate matter which are also regenerated in a similar way. These technologies are however relatively complex and costly, resulting in general in diesel cars emitting more NO_x and PM than gasoline ones.

The advantage of the diesel engine is that, due to the higher pressures and temperatures, it is more efficient. The mileage is about 20% higher compared to a gasoline counterpart. CO_2 emissions, so greenhouse emissions, for a modern diesel car, are around 100 g CO_2/km while for gasoline around 120 g CO_2/km. If we bring CO_2 in the pollution emission spectrum, it is then difficult to compare diesel and gasoline engines as we have to compare different emission levels of different pollutants. Meeting the Kyoto objectives without the CO_2 efficiency of diesel engines would be difficult.

Another important CO_2 source besides combustion engines, so transportation, is energy production. The quote of renewable energy produced worldwide is increasing, and the largest economies are fortunately leading. China is the main producer and user of photovoltaic panels and wind turbines. Germany is also very close to meet its 2020 targets. Even if the quote of "green" renewable and low-CO_2-impact energy increase, a complete replacement is still far away, as on-demand and flexible energy sources are needed to guarantee supply and stabilize the network. This base supply will be guaranteed by traditional power plants. The first large-scale experiments of carbon capture are also available. In 2012 the

German power giant RWE had constructed in its lignite power plant in Niederaußen a CO_2 separation and washing unit which delivers high purity CO_2 for the beverage and chemical industry [19]. Part of this CO_2 could then used by the nearby polyurethane producer Covestro in its "Dream Production" process. This process produces one of the two components required to generate polyurethane foams. Soft foams are used in upholstery (i.e. car seats, mattresses or sofas), while hard foams in building insulation. The process bound the CO_2 into the polymer backbone and replaces up to 30% of fossil raw materials with it. Not only is it a carbon sequestration and a saving of raw materials, but when used as building insulator, such a hard foam can save in its life cycle up to seven times the CO_2 required to fabricate it [20].

6.8 Natural Pollutants

Besides the anthropogenic pollutants, there is a large variety of naturally occurring pollutants. For example, huge quantities of methane are generated naturally by anaerobic fermentation processes. Methane is a greenhouse gas with a global warming potential (GWP) 25 times higher than CO_2. Every year 600 million tons of methane is released into the atmosphere (equivalent to 15 billion tons of CO_2). Thirty percent of these are related to fermentation under seas or lakes, like in Siberia during the summer period, when the ice cap melts. 39% are generated by the fermentation of grass in the kettle stomach and 17% in the cultivation of rice under water. These processes are only partially related to human activities, even if intensive farming and breeding practices can sharpen it.

6.9 Pollens

Another very important natural source of pollution is given by plants. Each one of us, not only the people affected by allergies, is aware of the pollen season. Pollens are emitted by plants during their reproduction cycle. Pollen is the male reproductive structure of flowering plants. They are intrinsically active biological species, and this can trigger biological activities, like immune responses. Pollen exposure has long been recognized as a stimulant for symptoms of allergic disease, especially for allergic rhinitis (hay fever). Pollen grains range in size from about 10 to 100 μm; the most common types are in the range of 15–30 μm. However, pollen allergens have been documented in air on much smaller particles. In this they are comparable to PM. Pollen is produced seasonally. In general, tree pollens are released early in the year, grasses during late spring and early summer and weed pollens in the late summer and autumn.

There is an association between grass pollen counts and admissions of patients with asthma in Mexico City in both dry and wet seasons [21]. In England, thunderstorms following periods of high pollen counts are more likely to lead to asthma epidemics [22].

6.10 Mould

Moulds are an important pollutant of the outdoor air. Exposure to moulds can cause severe asthma morbidity and mortality. Sixty species of moulds have spores that are proven allergenic. Many patients with respiratory allergies are particularly sensitive to moulds. Odds of death from asthma are twice as high on days with outdoor mould spore counts above 1000 spores/m^3 [23]. Daily increases in mould spore counts are associated with daily increases in hospital admissions for asthma [24]. There may also be a synergistic effect between ozone and some mould spores. That is, the combined effects of exposure to ozone and mould spores are greater than the effects of either exposure alone [25, 26].

Conclusions

As we have seen, the sources of air pollution are much diversified. Natural pollution can be as severe as man-made. The perception of the pollution is also very much critical. The most perceived and with human health effect pollution is the one which happens in the cities, as there is where most people live. This is clearly mainly man-made. Many pollutants are also linked one with the other. For example, a reduction of PM could be associated with an increase in NO_x and vice versa. The effects on the health are often synergistic. Ozone, as irritant of the airways, increases the effect on the health of many other pollutants like pollens, mould or PM. Under this light it is clear that the topic of air pollution is very complex and not only driven by measurable parameters. The most important aspect is to find the right balance between our life, our health and our interaction with the environment.

References

1. G. T. Miller and S. Spoolman, Living in the environment: principles, connections, and solutions, Brooks / Cole, Belmont (CA), 2011. ISBN 0-495-55671-8.
2. Lucernoni F, Tapparo F, Capelli L, Sironi S. Evaluation of an odour emission factor (OEF) to estimate odour emissions from landfill surfaces. Atmos Environ. 2016;144:87–99.
3. Capelli L, Dentoni L, Sironi S, Del Rosso R. The need for electronic noses for environmental odour exposure assessment. Water Sci Technol. 2014;69(1):135–41.
4. Capelli L, Sironi S, Del Rosso R. Electronic noses for environmental monitoring applications. Sensors. 2014;14(11):19979–20007.
5. Capelli L, Sironi S, Del Rosso R. Odour emission factors: fundamental tools for air quality management. Chem Eng Trans. 2014;40:193–8.
6. Sironi S, Capelli L, Dentoni L, Del Rosso R. Odour regulation and policies. In: Odour impact assessment handbook. Hoboken, NJ: Wiley; 2012. p. 175–86.
7. World Health Organization. Air quality guidelines. Geneve: WHO Regional Office for Europe; 2002.
8. Hajat S, Armstrong B, Wilkinson P, Busby A, Dolk H. Outdoor air pollution and infant mortality: analysis of daily time-series data in 10 English cities. J Epidemiol Community Health. 2007;61(8):719–22.
9. Bobak M, Leon D. The effect of air pollution on infant mortality appears specific for respiratory causes in the postneonatal period. Epidemiology. 1999;10(6):666–70.

10. Dickey J. Part VII. Air pollution: overview of sources and health effects. Dis Mon. 2000;46(9):566–89.
11. Ha E, Lee J, Kim H, Hong Y, Lee B, Park H, Christiani D. Infant susceptibility of mortality to air pollution in Seoul, South Korea. Pediatrics. 2003;111(2):284–90.
12. Kaiser R, Romieu I, Medina S, Schwartz J, Krzyzanowski M, Künzli N. Air pollution attributable postneonatal infant mortality in U.S. metropolitan areas: a risk assessment study. Environ Health. 2004;3:4.
13. Tager IB. Air pollution and lung function growth. Am J Respir Crit Care Med. 1999;160(2):387–9.
14. Gauderman W, Gilliland G, Vora H, Avol E, Stram D, McConnell R, Thomas D, Lurmann F, Margolis H, Rappaport E, Berhane K, Peters J. Association between air pollution and lung function growth in southern California children: results from a second cohort. Am J Respir Crit Care Med. 2002;166(1):76–84.
15. O'Connor G, Neas L, Vaughn B, Kattan M, Mitchell H, Crain EF, Evans R 3rd, Gruchalla R, Morgan W, Stout J, Adams G, Lippmann M. Acute respiratory health effects of air pollution on children with asthma in US inner cities. J Allergy Clin Immunol. 2008;121(5):1133–9.
16. Chen G, Song G, Jiang L, Zhang Y, Zhao N, Chen B, Kan H. Short-term effects of ambient gaseous pollutants and particulate matter on daily mortality in shanghai, China. J Occup Health. 2008;50(1):41–7.
17. Kind T. Carbon monoxide. Pediatr Rev. 2005;26(4):150–1.
18. Votsmeier M, Kreuzer T, Gieshoff J, Lepperhoff G. Automobile exhaust control. In: Ullmann's encyclopedia of industrial chemistry. Hoboken, NJ: John Wiley & Sons; 2009.
19. RWE AG. Chemcial conversion of CO2, RWE AG, [Online]. Available from: http://www.rwe.com/web/cms/en/542048/rwe-power-ag/innovations/coal-innovation-centre/chemical-conversion-of-co2/. [Accessed 23 09 2016].
20. Covestro AG. CO2 Project, Covestro AG, 20 07 2016. [Online]. Available from: http://www.covestro.de/de/projects-and-cooperations/co2-project. [Accessed 23 09 2016].
21. Rosas I, McCartney H, Payne R, Calderón C, Lacey J, Chapela R, Ruiz-Velazco S. Analysis of the relationships between environmental factors (aeroallergens, air pollution, and weather) and asthma emergency admissions to a hospital in Mexico City. Allergy. 1998;53(4):394–401.
22. Newson R, Strachan D, Archibald E, Emberlin J, Hardaker P, Collier CG. Acute asthma epidemics, weather and pollen in England, 1987-1994. Eur Respir J. 1998;11(3):694–701.
23. O'Hollaren M, Yunginger J, Offord K, Somers M, O'Connell E, Ballard D, Sachs M. Exposure to an aeroallergen as a possible precipitating factor in respiratory arrest in young patients with asthma. N Engl J Med. 1991;324(6):359–63.
24. Dales R, Cakmak S, Judek S, Dann T, Coates F, Brook J, Burnett R. Influence of outdoor aeroallergens on hospitalization for asthma in Canada. J Allergy Clin Immunol. 2004;113(2):303–6.
25. Jenkins HS, Devalia JL, Mister RL, Bevan AM, Rusznak C, Davies RJ. The effect of exposure to ozone and nitrogen dioxide on the airway response of atopic asthmatics to inhaled allergen. Am J Respir Crit Care Med. 1999;160:33–9.
26. Molfino N, Wright S, Katz I, Tarlo S, Silverman F, McClean P, Szalai J, Raizenne M, Slutsky A, Zamel N. Effect of low concentrations of ozone on inhaled allergen responses in asthmatic subjects. Lancet. 1991;338(8761):199–203.

Indoor Air Pollution

7

Kati Huttunen

Abbreviations

CO	Carbon monoxide
CO_2	Carbon dioxide
COPD	Chronic obstructive pulmonary disease
ETS	Environmental tobacco smoke
IAQ	Indoor air quality
IHD	Ischaemic heart disease
LPS	Lipopolysaccharide
PAH	Polycyclic aromatic hydrocarbons
PM	Particulate matter
UFP	Ultrafine particle
VOCs	Volatile organic compounds

7.1 History of Indoor Air Pollution

The importance of indoor air was recognized already in ancient Rome, where oils were burned to provide light and homes were heated with wood, vegetation and animal dung [1]. Studies on skeletons buried by volcanic eruptions and lungs of ancient human mummies have revealed signs of inflammation and anthracosis typically caused by long-term exposure to coal and soot [2]. Biological contaminants in indoor air are not new either; the adverse effects of moisture-damaged indoor environment have been reported for at least 2500 years, as the problem with "plague of

K. Huttunen
Department of Environmental and Biological Sciences, University of Eastern Finland, Kuopio, Finland
e-mail: kati.huttunen@uef.fi

F. Capello, A.V. Gaddi (eds.), *Clinical Handbook of Air Pollution-Related Diseases*,
https://doi.org/10.1007/978-3-319-62731-1_7

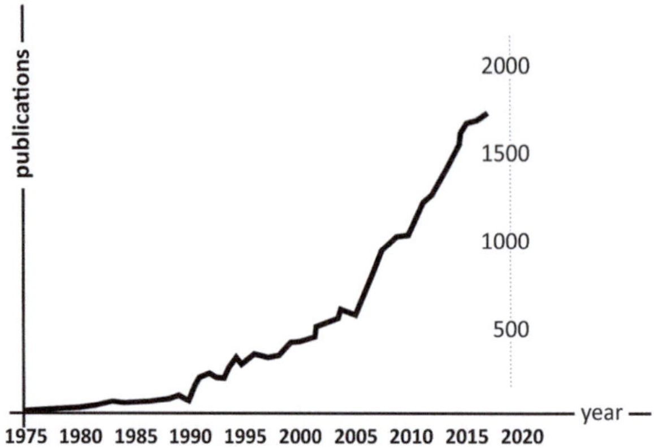

Fig. 7.1 Number of indoor air-related papers published in 1975–2016. Bibliographic search of Web of Knowledge http://apps.webofknowledge.com, topic "indoor air", accessed 9.1.2017. (update of bibliographic search by [4])

the house" is mentioned already in the Old Testament (Third book of Moses, Leviticus 14, 34–57).

The first published investigations on impact of poor ventilation on health appeared in the mid-nineteenth century, showing how the levels of CO_2, microbes and total organic matter are linked with ill health [3]. The number of scientific studies on indoor air pollution has increased rapidly since the 1960s and even doubled during the last decade (Fig. 7.1). The focus of interest has slightly shifted from decade to decade along the changes in buildings and behaviour of people: In the 1960s radon and tobacco smoke were the original (and still existing) hot topics, followed by formaldehyde and asbestos in the 1970s, house dust mites and sick building syndrome in the 1980s, asthma and allergies in the 1990s and finally combustion particles and moisture-damaged buildings around the new millennium. The recent trends include a strong research interest on the protective exposures and synergistic effects of pollutants, showing a gradient towards studying "exposomes" instead of single exposures. Fortunately the exposure to many of the indoor pollutants with known health risks (e.g. asbestos, formaldehyde, naphthalene) has been reduced due to regulatory efforts or discontinued use, but the exposure to chemicals such as plasticizers, surfactants and flame retardants has been on the rise, and many of these chemicals are suspected endocrine disruptors [5].

7.2 Importance of Indoor Air Quality (IAQ)

The quality of indoor air is not a trivial thing. A comparative risk assessment of burden of disease caused by different risk factors ranked household air pollution from solid fuels as third on the list of top ten global health risks. In addition, another indoor-related risk cluster, tobacco smoking (including second-hand smoking), was ranked second on the same list, right after high blood pressure [6]. The numbers are staggering no matter which way you look at the problem: in 2012 household air pollution was estimated to cause 4.3 million premature deaths annually [7] in addition to the increased medication use, sick leaves and lowered performance in learning and working. The leading cause of death related to indoor air pollution is stroke, followed by ischaemic heart disease (IHD), chronic obstructive pulmonary disease (COPD), lower respiratory infections and lung cancers [7]. The adverse health effects are not distributed evenly among social economic groups as household air pollution affects particularly women and children [8].

Reducing exposure to air pollutants indoors would have a major effect on public health as people spend approximately 90% of their time in different indoor environments [9]. The close proximity to pollutant source makes the indoor emissions even more critical; it is thousand-fold more likely for pollutants emitted indoors to be inhaled compared to outdoor emissions [10]. The indoor exposure to smoke from cooking and heating as well as pollutants originating from ambient air could be controlled with relatively simple actions, yet about half of the world's households are still affected by indoor smoke from solid fuels [11, 12]. Unfortunately adoption of new techniques is not always straightforward or enough to reduce pollutant concentrations to recommended levels [13]. Sometimes the intervention efforts may even have unintended adverse consequences, e.g. tighter constructions aiming to improve thermal comfort may lead to accumulation of indoor contaminants or condensation of water in building structures.

7.3 Indoor Air Pollutants

Altogether there are more than 60 risk factors associated with increased burden of disease by household air pollution [8]. Rated by their impact on burden of disease, the top five indoor exposure agents are particulate matter (particularly combustion particles), bioaerosols, radon, carbon monoxide and volatile organic compounds [12]. Also asbestos, nitrogen and sulphur oxides, ozone and carbon dioxide are highlighted in the following chapter for their role as significant indoor contaminants. Many of the risk factors are linked with more than one source, and most of the sources are linked with several pollutants (Table 7.1).

In addition to the ambient air, the other main sources of indoor exposure include combustion appliances (cooking and heating), smoking, water systems (leaks, condensation leading to moisture damage), building site (source for radon from soil or groundwater) as well as materials and chemicals used indoors [12]. However, also environmental conditions such as relative humidity and temperature affect the

Table 7.1 The key exposure agents and their main emission sources in indoor environments

Pollutant	Source
Particulate matter	Ambient air, combustion sources (stoves, fireplaces, furnaces, candles), tobacco smoke, cooking, resuspension of dust
Bioaerosols	Moisture-damaged building materials (microbial growth), ambient air (pollen), building occupants, pets (dander), pests (insects, rodents)
Radon	Uranium-bearing bedrock and groundwater, construction materials
Carbon monoxide	Combustion sources, gas appliances, tobacco smoke, ambient air
Volatile organic compounds	Household chemicals (cleaning agents, perfumes, deodorants, pesticides, solvents), combustion sources, tobacco smoke

indoor air quality or, at least, how the indoor air quality is perceived by the occupants. For example, odorous compounds such as formaldehyde influence the perceived indoor air quality already at very low concentrations, which may result even in acute effects of the exposed individuals even though the health-based guidelines are met [14].

Ventilation is one of the most important factors affecting indoor air quality, diluting the exposure agents originating from indoors. A comparison of the alternative control strategies showed that, in addition to the adequate ventilation, also filtration of the incoming air and controlling the indoor sources are necessary to reduce the indoor exposures to an acceptable level [15] although, in some cases, ventilation may even be source of contaminants if not designed or maintained properly [16].

7.3.1 Particulate Matter

Over half of the total annual burden of disease associated with indoor exposures is related to exposure to outdoor air fine particulate matter indoors; thus particulate matter is by far the most significant exposure agent impacting public health [12, 17]. Particularly harmful particulates originate from combustion appliances (cooking, heating), carrying a mixture of chemicals such as polycyclic aromatic hydrocarbons (PAH) along them. Other important indoor sources of PM are candles, printers and vacuum cleaners [18]. As far as health effects are concerned, the most relevant particles are within the size fraction of inhalable particles (aerodynamic diameter $\leq$ 100 μm), which can be further classified into thoracic (aerodynamic diameter $\leq$ 10 μm) and respirable fractions (aerodynamic diameter $\leq$ 4 μm) according to their ability to penetrate further into the lungs. However, particle penetration into the lungs is dependent also on the activity levels and breathing route, which have been estimated to significantly decrease the actual 50% cut sizes for thoracic and respiratory fractions [19]. In addition to mass concentration, size distribution and composition of the airborne particles, also the shape of the particles affects the adverse health effects associated with the exposure. For example, mineral fibres like asbestos are notorious for their ability to cause mesothelioma, lung cancer and asbestosis, and for this reason the use of asbestos is banned in most of Europe. However, due to heavy asbestos use during earlier decades and the long latency period of the onset of these diseases, Europe is still carrying most of the global asbestos-related disease burden, and correspondingly the countries that still have not banned asbestos are likely to have a substantial burden of asbestos-related disease in the future [20].

7.3.2 Gaseous Pollutants

The most dangerous gaseous component of indoor air pollution is considered to be radon, a radioactive, lung carcinogenic compound with a linear dose-response curve, no apparent threshold and increased risk of adverse health effects already at concentrations below the action levels regulated by health authorities. The exposure to radon is heavily dependent on the location of the building, as the main source of radon is the soil beneath the building containing uranium and its decay products. The elevated indoor air concentrations could be lowered with relatively easy interventions, which emphasizes the importance of educating the general public about the possible issues with radon [21, 22].

Combustion products (e.g. carbon monoxide, nitrogen and sulphur dioxide) originating from traffic (outdoor air), cooking or heating are another important group of gaseous exposure agents partly responsible for the adverse health effects of household fuel combustion. Surprisingly, indoor exposure to carbon monoxide is among the highest causes of acute lethal intoxication in Europe and most likely even underestimated as a public health risk [12]. Carbon monoxide is a colourless, non-irritant, odourless and tasteless gas, which explains why emissions from faulty, poorly maintained or poorly ventilated combustion appliances may lead to toxic or even lethal exposures. In addition to emissions from incomplete combustion and exhaust from vehicles, tobacco smoke can be a significant source for CO exposure [21].

Volatile organic compounds (VOCs) are a group of irritating volatile compounds emitted from materials and household chemicals, cooking and heating systems and also infiltrated indoors with ambient air. Epidemiological data is available only for some of the most common VOCs such as benzene, naphthalene and formaldehyde, which are all either known or suspected carcinogens and respiratory toxicants [21]. There are indications that mixtures of VOCs are more potent than individual compounds, leading to probable underestimation of the public health role of VOCs [12].

Exposure to ozone is linked with location of the building, as the main source of ozone is ambient air, where ozone is formed in photochemical reactions. However, also office appliances such as printers and photocopiers can produce significant amounts of ozone indoors, and increased levels of ozone may be intentionally produced in the process of disinfecting or removing unwanted odours in indoor spaces [23].

Concentration of carbon dioxide (CO_2) in occupied indoor environment is higher than outdoors due to CO_2 produced by occupants themselves, although CO_2 is emitted also from combustion processes. The levels causing acute toxicity are clearly higher than the typical indoor concentrations, but peak CO_2 levels have been used as an indicator of poor air quality or inadequate ventilation. In addition, there are indications that already low levels of CO_2 impair cognitive functions such as decision-making [24] and accumulating evidence on the effect of low ventilation rates on learning and school performance [25–28].

7.3.3 Bioaerosols

Biological material is a part of both particulate and gaseous indoor air pollutants, but merits a chapter on its own as bioaerosols are in the top three exposure agents associated with burden of disease attributed to IAQ, either due to exposure to bioaerosols from outdoor air or increased microbial exposure due to building dampness. Bioaerosols are a heterogeneous group of compounds originating from microbes (viruses, protozoa, bacteria, fungal spores, fragments and metabolites), plants (pollen, plant debris), animals (fragments from pets and pests) and humans themselves. To some extent it is normal to be exposed to biological materials in indoor air, considering that shedding of the skin and commensal bacteria of the occupants themselves are major contributors to the biological fraction [29]. There are indications that exposure to diverse microbial components such as lipopolysaccharide (LPS), $\beta(1 \rightarrow 3)$glucans, extracellular polysaccharides and muramic acid may even be necessary or beneficial for the development of immune system and protective for atopy and asthma [30], posing a major challenge for the research on the bipolar health effects of environmental exposure to microbes.

One of the major sources for microbial exposure indoors is excess moisture in building structures, which enables the growth of fungi and bacteria present in building materials and dust. However, the adverse health effects associated with dampness in buildings are not necessarily related only to bioaerosols, as the moisture damage may increase also e.g. emissions of VOCs from materials. Exposure to moisture-damaged indoor environment is the most important health risk determinant after exposure to combustion particles, particularly for respiratory symptoms and asthma [12]. In the absence of reliable bioindicators of exposure or diagnostic tools to recognize mould-exposed individuals, it is recommended to monitor and fix leaks in water systems as well as replace moisture-damaged or mouldy materials proactively.

Exposure to pollen and other bioaerosols originating from outdoor air is highly dependent on season, location of the building (surrounding vegetation) and filtering capacity of the ventilation system. Exposure to bioaerosols from outdoor air is an important cause of allergic rhinitis and asthma and considered to be as important contributor to burden of disease e.g. exposure to radon indoors. Worryingly, climate change is predicted to increase pollen production and even allergenicity of some plant proteins [12, 31].

Conclusions

People spend the majority of their lives in indoor environments such as homes, offices, health care facilities and public buildings, so the poor quality of indoor air is a major public health issue. However, indoor air exposures are problematic both for research and regulatory systems because of the wide range of pollutants, sources and causes. World Health Organization has established guideline values and numerical standards for indoor air pollutants to assist e.g. building managers, architects and engineers, to achieve good air quality [21]. Similarly, WHO has published a specific guideline aiming to control mould growth indoors [32] and indoor combustion of solid fuels [33]. Unfortunately the mechanism of the adverse health effects and "safe" exposure levels are not known for many of the indoor air pollutants. Reduction of harmful exposure would be possible with relatively simple actions, but the interventions may fail due to the lack of knowledge or motivation to change the behaviour of the occupants and traditionally used appliances in the households.

References

1. Makra L, Brimblecombe P. Selections from the history of environmental pollution, with special attention to air pollution. Int J Environ Pollut. 2004;22(6):641–6.
2. Capasso L. Indoor pollution and respiratory diseases in ancient Rome. Lancet. 2000; 356(9243):1774.
3. Carnelly D, Haldane JS, Anderson AM. The carbonic acid, organic matter and microorganisms in air, more especially in dwellings and schools. Philos Trans R Soc Trop Med Hyg. 1887;178:61–111.
4. Colbeck I, Nasir ZA. Indoor air pollution. In: Lazaridis M, Colbeck I, editors. Human exposure to pollutants via dermal absorption and inhalation, Environmental pollution, vol. 17: Springer; 2010. https://doi.org/10.1007/978-90-481-8663-1.
5. Weschler CJ. Changes in indoor pollutants since the 1950s. Atmos Environ. 2009;43(1):153–69.
6. Lim SS, Vos T, Flaxman AD, Danaei G, Shibuya K, Adair-Rohani H, et al. A comparative risk assessment of burden of disease and injury attributable to 67 risk factors and risk factor clusters in 21 regions, 1990–2010: a systematic analysis for the global burden of disease study 2010. Lancet. 2012;380(9859):2224–60.
7. World Health Organization (WHO). Deaths from household air pollution, 2012. Global health observatory data. http://www.who.int/gho/phe/indoor_air_pollution/burden/en/; 2012. Accessed 10 Jan 2017.
8. Smith KR, Bruce N, Balakrishnan K, Adair-Rohani H, Balmes J, Chafe Z, Dherani M, Hosgood HD, Mehta S, Pope D, Rehfuess E, Others in the HAP CRA Risk Expert Group. Millions dead: how do we know and what does it mean? Methods used in the comparative risk assessment of household air pollution. Annu Rev. Public Health. 2014;35:185–206. https://doi.org/10.1146/annurev-publhealth-032013-182356.
9. Klepeis NE, Nelson WC, Ott WR, Robinson JP, Tsang AM, Switzer P, Behar JV, Hern SC, Engelmann WH. The National Human Activity Pattern Survey (NHAPS): a resource for assessing exposure to environmental pollutants. J Expo Anal Environ Epidemiol. 2001;11(3):231–52.
10. Lai AC, Thatcher TL, Nazaroff WW. Inhalation transfer factors for air pollution health risk assessment. J Air Waste Manage Assoc. 2000;50(9):1688–99.
11. Asikainen A, Carrer P, Kephalopoulos S, de Oliveira FE, Wargocki P, Hänninen O. Reducing burden of disease from residential indoor air exposures in Europe (HEALTHVENT project). Environ Health. 2016;15(Suppl 1):S35. https://doi.org/10.1186/s12940-016-0101-8.
12. Jantunen M, Oliveira Fernandes E, Carrer P, Kephalopoulos S. Promoting actions for healthy indoor air (IAIAQ). Luxembourg: European Comission Directorate General for Health and Consumers; 2011.
13. Thomas E, Wickramasinghe K, Mendis S, Roberts N, Foster C. Improved stove interventions to reduce household air pollution in low and middle income countries: a descriptive systematic review. BMC Public Health. 2015;15:650. https://doi.org/10.1186/s12889-015-2024-7.
14. Wolkoff P. Indoor air pollutants in office environments: assessment of comfort, health, and performance. Int J Hyg Environ Health. 2013;216(4):371–94. https://doi.org/10.1016/j.ijheh.2012.08.001.
15. Hänninen O, Asikainen A. Efficient reduction of indoor exposures. Health benefits from optimizing ventilation, filtration and indoor source controls. THL Report 2/2013. National Institute for Health and Welfare; 2013.
16. Zuraimi MS. Is ventilation duct cleaning useful? A review of the scientific evidence. Indoor Air. 2010;20(6):445–57. https://doi.org/10.1111/j.1600-0668.2010.00672.x.
17. Ji W, Zhao B. Estimating mortality derived from indoor exposure to particles of outdoor origin. PLoS One. 2015;10(4):e0124238. https://doi.org/10.1371/journal.pone.0124238.
18. National Academies of Sciences, Engineering, and Medicine. In: Butler DA, Madhavan G, Alper J, editors. Health risks of indoor exposure to particulate matter: workshop summary. Washington DC: The National Academies Press; 2016.

19. Brown JS, Gordon T, Price O, Asgharian B. Thoracic and respirable particle definitions for human health risk assessment. Part Fibre Toxicol. 2013;10:12. https://doi.org/10.1186/1743-8977-10-12.
20. Kameda T, Takahashi K, Kim R, Jiang Y, Movahed M, Park E-K, Rantanen J. Asbestos: use, bans and disease burden in Europe. Bull World Health Organ. 2014;92:790–7. https://doi.org/10.2471/BLT.13.132118.
21. World Health Organization (WHO). WHO guidelines for indoor air quality: selected pollutants. Copenhagen: WHO Regional Office for Europe; 2010. ISBN: 9789289002134
22. Vogiannis EG, Nikolopoulos D. Radon sources and associated risk in terms of exposure and dose. Front Public Health. 2015;2:207. https://doi.org/10.3389/fpubh.2014.00207.
23. Tuomi T, Engström B, Niemelä R, Svinhufvud J, Reijula K. Emission of ozone and organic volatiles from a selection of laser printers and photocopiers. Appl Occup Environ Hyg. 2000;15(8):629–34.
24. Satish U, Mendell MJ, Shekhar K, Hotchi T, Sullivan D, Streufert S, Fisk WJ. Is CO2 an indoor pollutant? Direct effects of low-to-moderate CO2 concentrations on human decision-making performance. Environ Health Perspect. 2012;120(12):1671–7. https://doi.org/10.1289/ehp.1104789.
25. Haverinen-Shaughnessy U, Shaughnessy RJ. Effects of classroom ventilation rate and temperature on students' test scores. PLoS One. 2015;10(8):e0136165. https://doi.org/10.1371/journal.pone.0136165.
26. Maddalena R, Mendell MJ, Eliseeva K, Chan WR, Sullivan DP, Russell M, Satish U, Fisk WJ. Effects of ventilation rate per person and per floor area on perceived air quality, sick building syndrome symptoms, and decision-making. Indoor Air. 2015;25(4):362–70. https://doi.org/10.1111/ina.12149.
27. Mendell MJ, Eliseeva EA, Davies MM, Lobscheid A. Do classroom ventilation rates in California elementary schools influence standardized test scores? Results from a prospective study. Indoor Air. 2016;26(4):546–57. https://doi.org/10.1111/ina.12241.
28. Petersen S, Jensen KL, Pedersen ALS, Rasmussen HS. The effect of increased classroom ventilation rate indicated by reduced CO2 concentration on the performance of schoolwork by children. Indoor Air. 2016;26:366–79. https://doi.org/10.1111/ina.12210.
29. Rintala H, Pitkäranta M, Täubel M. Microbial communities associated with house dust. Adv Appl Microbiol. 2012;78:75–120. https://doi.org/10.1016/B978-0-12-394805-2.00004-X.
30. Heederik D, von Mutius E. 2012. Does diversity of environmental microbial exposure matter for the occurrence of allergy and asthma? J Allergy Clin Immunol. 2012;130(1):44–50. https://doi.org/10.1016/j.jaci.2012.01.067.
31. Peden D, Reed CE. Environmental and occupational allergies. J Allergy Clin Immunol. 2010;125(2 Suppl 2):S150–60. https://doi.org/10.1016/j.jaci.2009.10.073.
32. WHO, World Health Organization. WHO guidelines for indoor air quality: dampness and mould. Copenhagen: WHO Regional Office for Europe; 2009. ISBN: 798289041683
33. WHO, World Health Organization. WHO guidelines for indoor air quality: household fuel combustion. Copenhagen: WHO Regional Office for Europe; 2014. ISBN: 9789241548878

8 Air Pollution and the Health Preservation: The Point of View of a Clinician and of a Cardiologist

Antonio Vittorino Gaddi and Michela Dimilta

"It is a challenge preventing and taking care of diseases caused by pollution, rather it is the challenge of this century and maybe of those upcoming".

8.1 Introductions and Definitions

Air pollution remains a major issue in terms of health promotion. Although the implication for the patient may be clear, the way through clinicians approach to those diseases or health issues related to air pollution is complex and not necessarily easy to define.

Some clarifications may be useful to understand the discussion on air pollution and clinical practice better. Therefore some definitions are needed:

a. With the term "diseases", we do not refer to taxonomies which are proper of medical specialities but rather to any change or alteration in state of health that may undermine someone's wellness.
b. "Polluting substances" refer to molecules, particles or compounds (air, water, ground, food) which come from different sources, such as chemical substances, toxins, biological active or inactive components or radiation; however, the most common air pollutants are particularly considered as most of the available studies mainly refer to a selected number of toxins.
c. With the expression "caused by polluting substances", we intend any alteration of the state of health caused or triggered by pollution. It includes all the conditions and outcomes (symptoms, signs, fatal or nonfatal events) linked to one or more pollutants (causing addictive or synergic effects).

A.V. Gaddi (✉)
Scientific Director of Health Lab of GTechnology Foundation, Modena, Italy
e-mail: profgaddi@gmail.com

M. Dimilta
Pharmacy, Unibo, Caravelli Lab e Techobios EuroGenLab, Bologna, Italy

F. Capello, A.V Gaddi (eds.), *Clinical Handbook of Air Pollution-Related Diseases*,
https://doi.org/10.1007/978-3-319-62731-1_8

A further consideration is needed: comorbidity, which includes also the idea of multifactoriality of some physiopathological processes, plays a crucial role in several medical conditions and in different stages of life, from maternity to childhood up to old age.

Some aspects of the complexity which derives from all these interactions are augmented exponentially by the number of pollutants and by environmental and genetic factors that take part and modulate physiopathological processes. This problem will be discussed in deep in a further chapter (see Chap. 14). Clinical considerations and methodological implications will be developed, instead, in the chapter on old age and air pollution (see Chap. 10).

In addition, since we are just starting to understand the real effect of air pollution on human health. A new clinical discipline is needed. This discipline has to be created from the scratches in order to give answers to real needs of people exposed to pollutants, and this is an urgent need.

Thus, we will try to analyse the most relevant situations from a clinical point of view; those are the topics that ask for prompt answers and actions, asked from governments and exposed people.

Diseases caused by pollution represent a big challenge for the following reasons:

a. Considering the practical results that can be realistically achieved, this is a field of medicine where theoretically final and relevant results can be reached. This has practical implication in terms of public health and prevention but also for the management of single cases affecting individual patients. In fact, the cause of disease (viz. one or more pollutants) is known, measurable and removable. If we consider air pollutants as risk factors (considering the definition given by McGill) for the development of a disease, the logical consequence is that removing such factors we should be able to avoid the effects and consequently the disease; it also means that within certain temporal and exposure limits, the effect of pollutants on human health may be reversible.
b. In order to achieve a result to avoid the development of diseases caused by pollution, a complete action is required. It should combine prevention (pre-primary and primary prevention) and therapeutic management of the disease that in the particular case of acute or chronic exposure to pollution results in the so-called secondary and tertiary prevention (depending on the timing of the intervention which can be premature or late) [1] as summarized in Table 8.1. Air pollution in fact can be considered a relatively new event, when compared to the natural evolution of the human race, and so are its effects on health? Thus, strategies of disease control and prevention can be set up ex novo, allowing each country to properly design public health policies aimed to prevention; furthermore, diseases caused by pollution share common physiopathological processes; this allows the implementation of an action aimed to an e*arly, integrated, efficient prevention, w*here prevention is intended as a continuous process that embraces the pre-primary, primary, secondary and tertiary one. This is particularly true when the harm caused by pollutants is part of a long-lasting almost obsessive—and for this reason predictable—process. It must be necessarily an advantage for the practical medicine.

Table 8.1 Some relevant aspects of clinical and preventive medicine integrated

Preventive and clinical medicine integrated believes:
Integrations between preventive lines destined to different conditions/diseases, in particular if they are based on risk and/or common aetiopathogenetic factors, in particular for diseases at high prevalence
Integrate information (sanitary and health education) related to prevention: citizen, patient and families must receive comprehensible and specially coherent signals which must also be coherent with whom those given by all medical staff
Integrate strategies of primary (and pre-primary) and secondary (tertiary) prevention
Integrate curative stages with those preventive when it is possible
Integrate the concept of prevention both with whom of conservation or protection of health and with that of wellness
Integrate each other actors of prevention: citizens, patient, doctor, personnel of sanity area, employees of clinic, districts, teacher, etc. (defining precisely roles, teaching them how to practice preventive actions with perseverance and coherence with each other, practically avoiding negative interferences!)
Integrate normative documents and simplify virtuous interactions between different administration and between them and private individuals (and mass media); that means predicting and removing major obstacles which could prevent the first realization and the maintenance of preventive campaign
Integrate at the best facilities for prevention and care, proposing—when it is possible—solutions which are not based only on hospitals or districts or on specific kind of persons. In particular, it should be avoided that different areas (physical or functional) have different projects for what concerns objective and means
Integrate actions aimed at modifying behaviour and/or identifying risk conditions, etc. with those that are formative (of any level) or even just informative
Remember that formation (intended as education) is the base of any possible kind of prevention

c. The challenge can be won only if we are able to measure the effects, with a step-by-step process. The development of strategies based on strict and controlled procedures however based on logical, methodological and steady physiopathological principles is not enough. Measuring the final outcomes is a key, also to modulate the action. Moreover, the best available knowledge cannot be sufficient to create proper responses even when driven by the two fundamental ethical principles of maximum advantage and of non-harm. Because of the long-lasting effects of air pollution, which can show up even decades after the exposure, any planned intervention on the population takes time and involves lots of professionals for a very long time (years or decades) together with a lot of resources. Hence, markers are needed to dynamically evaluate the outcomes of the intervention and at the same time the state of health of that same population (meaning monitoring the effective implementation of the strategies of prevention and whether those strategies are actually effective in improving the quality of life of the individuals and of the population as a whole). However, the reader must not be amazed if some standards which are part of preventive and curative medicine do not exist or only few are known. Often, in fact, the definition and the certification of an outcome are tricky and may be disputable as well as its interpretation. Some of the standards defined for the evaluation of acute stages, for instance (e.g. the result of a chirurgical operation or the duration of a recovery), although used worldwide, provide only short misrepresented visions (standards of result, process, pertinence, cost, replaced or actual, predictive and perspective or historical, etc.). The group of diseases caused by pollution represents on the contrary one of those fields of medicine where a system of bank marking of rational phenomenon can be proposed, and the expected outcomes can theoretically be reproduced in chronic and on entire populations. Considering this from a technological point of view, the problem is not easy to identify, but this approach appears possible at least.

d. Another element in favour of this kind of vision is that this new class of diseases caused by pollutant:
   - From an aetiological point of view, can be equalled to infective diseases (bacterial, viral and maybe even those subviral and linked to new kind of agents) which spread out through air and food.
   - Seem to depend on damages at cellular level; the mechanisms of damage seem to strengthen those of other factors of genetic diseases.

This approach resembles a more traditional one in medicine, that is, synthetically, the idea that removing environmental risk factors, common to numerous pathologies, is an augmented probability of intervening favourably on more than one disease [2]; in lots of cases, the efficacy has been demonstrated experimentally. This is also in strict analogy with what happens in the fight against cigarette smoke (prevention of atherosclerosis in every district, myocardial infarction, cerebral vascular disease, any ischemic cerebral vascular disease, claudicatio intermittens, claudicatio penis, COPD, lots of tumour, not only those of the lung, etc.).

Furthermore, tobacco smoke is in each aspect an “individual and local anthropogenic pollutant” that increases the effect of those coming from environment and industries.

8.2 Diseases Caused by Pollution: The Clinical Vision

The clinical vision starts from the individual; this human being (man, woman, child) is a mixture of good and bad *elements* (which can be considered “internal”) and is exposed during his life to good and bad *elements* (or “external”); according to a different vision, these *elements* are risk and protective factors, although this can be considered a narrow interpretation; in a wider scheme, those are biochemical process, micro- and macromolecules that generate complex micro- and macro-interactions which in themselves are not even good or bad. Yet, those create effects that could trigger further effects that may change the state of tissues, organs, apparatuses and the whole organism; these results can be good or bad, improving or worsening the state and/or the perception of health of an individual.

The same idea of disease is a convention that, indeed, does not exist from a strict ontological point of view. Diagnoses are just convenient definitions created by us, generally based on the “least common denominator” of aggregations that are not numerically true. In fact, the single patient is the only thing that exists. The unique individual never corresponds with a single definition of disease, not even in the most difficult and extreme cases.

8.2.1 Should We Classify Chemical Substances and Diseases or Should We Standardise Individuals?

Is classification really useful in the medical practice? Or neither of these proposed choices should be followed? In clinical practice the last and irrevocable purpose is the cure and/or conservation of the state of wellness (prevention intended in all its possible meaning). This is also necessary and sufficient to start individual or mass actions which could exclude the need to create classifications and diagnostic cadres.

In other words, occidental traditional medicine is based on the following sequence:

1. Diagnosis (trying to understand the mechanisms of diseases and to predict its evolution, symptoms, outcomes, etc.)
2. Prognosis (*quo* ad vitam *vel quo ad valetudinem*)
3. Therapy (to improve prognosis)

Any efficient medical action that improves the prognosis (in spite of how the result has been achieved and provided that the risk–benefit ratio is acceptable) should be considered justifiable and possible. This is particularly true in complex and serious scenario such as the "pandemic" caused by pollution.

In our opinion the debate on "health and pollution" should not be based on disciplinary divisions; the analysis cannot follow traditional (cardiology, pneumology, etc.) or interdisciplinary (as neuro-psycho-endocrine-immunology that is the current trend nowadays) approach. It should rather be principally based on individual cases and should be developed according the system medicine's viewpoint.

On the other hand, the majority of scientific publications in air pollution derives from studies coming from single disciplinary frameworks, with a clear predominance of pneumology. There are lots of studies where sufficient cognitive elements are missing; so, the clinical vision based on scientific evidences is limited, and this is the reason why we have only information limited to single organ and specific diseases. Few studies, instead, have tried to measure different parameters that take also into account the state of health and of general wellness of single individuals. For these reasons, the systematic analysis of literature cannot be unmerged in chapters organized in organs or disciplines.

Nevertheless, we have tried to keep an open mind, trying to approach the problem according to a new clinical vision; thus, some chapters of this textbook refer to specific categories of people particularly exposed and sensitive to the effect of air pollution (elderly, children, pregnant women), underlying when possible the problem of comorbidity and analysing problems from an interdisciplinary perspective.

Therefore, a not interdisciplinary clinical description is hereby offered, based on the few available studies; a more relevant overview of the available literature will be presented in other chapters (viz. those on cardiovascular and cardiopulmonary disease).

However, a multimedia and interdisciplinary approach remains the first step towards a more clinician vision. This has also been demonstrated by studies about the complexity of asthmatic syndromes, where a holistic approach suggests the existence of "variants underlying particular endotypes gene" which interact in turn with detrimental environmental stimuli (e.g. smoking, viral infection and air pollution) that determines the final clinical view, in this case the diagnosis of asthma versus COPD or other pulmonary conditions (c.d. Dutch hypothesis) [3]. In addition, recent extensive studies, using the method of wide-genome study, are giving further data about specific genes which are responsible of the effects of some pollutants [4].

This is an interesting vision, as it is based on a genetic and environmental classification identifying progressively restricted groups of people whose prognosis is better defined.

Paradoxically, some epigenetic studies—that analyse the relationship between exposition to pollutants and preterm birth—suggested a new system to identify women at risk of preterm birth, allowing "the development of new preterm birth prevention measures" [5]. There is a lot of interest about studies of epigenetic markers related to pollution, and this kind of research could lead to some results of clinical interest in a few years [6]. Nevertheless, today they still suggest that the control of the space-time distribution of air pollutants is an appropriate policy for birth defence and protection [7].

There are few case reports in literature about people exposed to pollution, despite entity of the phenomenon. Some of these concern the induction of supraventricular arrhythmias caused by voluntary exposition to fine particulates [8] or ventricular arrhythmias [9] or *cor pulmonale* (caused directly and not by cigarette smoke) [10] or other pulmonary diseases, also linked to wood smoke or other kind of indoor pollutants [11]. It should not be underestimated those single cases concerning exposition to low concentration of "occult" carbon monoxide [12]. Other clinical reports are omitted as they are not exhaustive and anecdotal and however based on the description of each specialized aspect.

Therefore, today, doctors are confused and are put in a difficult position when dealing with a patient exposed to pollutants, also relying only on the few suggestions that come from literature and the available few guidelines for prevention and treatment [13–14]. The majority of those guidelines focused on single specialities and fields of application; this is not useful as this is a problem that needs the cooperation of experts coming from different areas and of everyone involved (people, chemical or climatology expert, engineer, nurse and doctor, public health officers, policymakers and so on).

This approach, however, is simply missing. There are some specific guidelines for single pathologies or specific pollutants (such as those of WHO about formaldehyde) [15]; and we can find several standards used to define normal, optimal or acceptable values of the concentration of single chemicals in the air [16–21]. On the Internet, recommendation to reduce polluting emissions and to measure pollutants or measures for individual protection are freely available, like those of US Environment Protection Agency (EPA), the United Nations Environment

Programme, from the Chinese government—that are deeply modifying the industrial framework in order to reduce pollutants emission—and the European Environment Agency. Some of them are mentioned somewhere else in this book though the clinical information provided in these documents are modest and in any case not easily accessible for both healthcare workers and patients.

Therefore, today, preventive and curative campaigns (refer to mainstream specified in the chapter "Main actions and possible claims against pollution") allow and justify the adoption of all kind of behaviours suggested by the practical common sense of doctors and healthcare workers, of those guided by the knowledge of the human physiopathology and of those based on the advices coming from any study which get successful results, the only restrain being the feasibility of the suggested actions when applied to specific patients and/or specific communities.

According to a physiopathological approach, it is essential to reduce individual exposition to pollutants by using any kind of means. This can be obtained by reducing the exposition time, by completely removing or reducing drastically contacts with pollutants and/or finally by reducing the pollutant doses. Practical recommendations are exposed in detail in the chapter "Basic Principles for treatments of air pollution related diseases". The basic concept is however the use of masks or intranasal spirals or other personal protective equipment in order to avoid the contact with pollutants in the first place (e.g. inhalation, contact with eyes and mucosae, contact with open wound on the skin or even contact on intact skin for specific pollutants) and also to avoid particulate and volatile organic compounds, when it is possible, or gas through the use of air depuration systems, changing air when external pollution is inferior (or less dangerous) than the one indoor. Those are relatively simple rules. In addition, if people could measure pollutant substances, they can easily and promptly detect the danger, increasing their defences at the same time, as authoritatively suggested by EPA.

Somehow, we are dealing with rules dictated by the common sense, partially already substantiated by the favourable results coming from literature. Unfortunately, the technology is still inadequate in this field. Most of the products available on the market (as masks, air purifying, antipollution devices) are ineffective, underutilised or badly employed.

Some years ago, in an article provocatively titled "From Good Intentions to Proven Interventions: effectiveness of Action to reduce the Health Impacts of Air Pollution", Luisa V Giles [22] identified two principal strategies:

a. Reduce individual baseline cardiovascular risk factors (and we also suggest the extension to non-cardiovascular risk factors).
b. Incorporate everything about health problems depending on pollution "into land-use decisions".

That approach seems to be even today totally acceptable: in particular, the first suggestion falls entirely within the holistic vision of medicine. The favourable or not favourable positions coming from literature to each preventive and curative intervention are mentioned in the following chapters.

References

1. Gaddi AV, Capello F, Gaddi P, Cevenini M, Carnevali L, Benedetto D. Integrated prevention of cardiovascular disease: current and future perspectives. Intern Emerg Med. 2012 (7 supp.):57–62.
2. Descovich, G. C. (1988, October). From the Brisighella Study to the Brisighella Project. In Epidemiology of Atherosclerosis (Satellite Meeting, 8th International Symposium on Atherosclerosis, Porto Cervo, 1988).-Roma: CIC Edizioni Internazionali (pp. 71-76).
3. Hizawa N. Clinical approaches toward asthma and COPD based on the heterogeneity of disease pathogenesis. Clin Exp Allergy. 2016;46(5):678–87.
4. Gref A, Kebede Merid S, Gruzieva O, Ballereau S, Becker A, Bellander T, et al. Genome-wide interaction analysis of air pollution exposure and childhood asthma with functional follow-up. Am J Respir Crit Care Med. 2016;195(10):1373–83.
5. Lin VW, Baccarelli AA, Burris HH. Epigenetics—a potential mediator between air pollution and preterm birth. Environmental Epigenetics, Volume 2, Issue 1, 1 January 2016, dvv008, https://doi.org/10.1093/eep/dvv008.
6. Tan C, Jin YT. Current progress in research on epigenetic effects of air pollution. J Zhejiang Univ (Med Sci). 2011;40(4):451–7.
7. Tu J, Tu W, Tedders SH. Spatial variations in the associations of term birth weight with ambient air pollution in Georgia, USA. Environ Int. 2016;92-93:146–56.
8. Ghio AJ, Bassett M, Montilla T, Chung EH, Smith CB, Cascio WE, et al. Case report: supraventricular arrhythmia after exposure to concentrated ambient air pollution particles. Environ Health Perspect. 2012;120(2):275–7.
9. Emilsson K. Suspected association of ventricular arrhythmia with air pollution in a motorbike rider: a case report. J Med Case Rep. 2008;2:192.
10. Opotowsky AR, Vedanthan R, Mamlin JJ. A case report of cor pulmonale in a woman without exposure to tobacco smoke: an example of the risks of indoor wood burning. Medscape J Med. 2008;10(1):22.
11. Diaz JV, Koff J, Gotway MB, Nishimura S, Balmes JR. Case report: a case of wood-smoke-related pulmonary disease. Environ Health Perspect. 2006;114(5):759–62.
12. Candura SM, Fonte R, Finozzi E, Guarnone F, Taglione L, Manzo L, et al. Indoor pollution: a report of 2 clinical cases of occult carbon monoxide poisoning. G Ital Med Lav. 1993;15(1–4):67–70.
13. Wang R, Yang Y, Chen R, Kan H, Wu J, Wang K, et al. Knowledge, attitudes, and practices (KAP) of the relationship between air pollution and children's respiratory health in shanghai, China. Int J Environ Res Public Health. 2015;12(2):1834–48.
14. Cai DP, He YM. Daily lifestyles in the fog and haze weather. J Thorac Disease. 2016;8(1):E75–7.
15. Nielsen GD, Larsen ST, Wolkoff P. Re-evaluation of the WHO (2010) formaldehyde indoor air quality guideline for cancer risk assessment. Arch Toxicol. 2016;91(1):35–61.
16. Kawamoto T, Pham TT, Matsuda T, Oyama T, Tanaka M, Yu HS, et al. Historical review on development of environmental quality standards and guideline values for air pollutants in Japan. Int J Hyg Environ Health. 2011;214(4):296–304.
17. Zhang H, Zhou ZQ, Zhao HY, Xiong ZQ. Pollution level and health risk assessment of heavy metals in atmospheric PM(2).(5) in Nanjing before and after the youth olympic games. Huan Jing Ke Xue. 2016;37(1):28–34.
18. Zhao H, Wang S, Wang W, Liu R, Zhou B. Investigation of ground-level ozone and high-pollution episodes in a megacity of eastern China. PLoS One. 2015;10(6):e0131878.
19. Yin HL, Li SP, Ye ZX, Yang YC, Liang JF, You JJ. Pollution level and sources of organic phosphorus esters in airborne PM2.5 in Chengdu City. Huan Jing Ke Xue. 2015;36(10):3566–72.
20. Xia Z, Duan X, Tao S, Qiu W, Liu D, Wang Y, et al. Pollution level, inhalation exposure and lung cancer risk of ambient atmospheric polycyclic aromatic hydrocarbons (PAHs) in Taiyuan, China. Environ Pollut. 2013;173:150–6.
21. Wang N, Yi L, Shi L, Kong D, Cai D, Wang D, et al. Pollution level and human health risk assessment of some pesticides and polychlorinated biphenyls in Nantong of Southeast China. J Environ Sci. 2012;24(10):1854–60.
22. Giles LV, Barn P, Kunzli N, Romieu I, Mittleman MA, van Eeden S, et al. From good intentions to proven interventions: effectiveness of actions to reduce the health impacts of air pollution. Environ Health Perspect. 2011;119(1):29–36.

Legal and Ethical Implications of Air Pollution

9

Chiara Rabbito

9.1 The Birth and the Development of Environmental Ethics and the Need of Legal Obligations

The study of ethics as applied to the environment began to develop in the 1970s. However many of the reflections and in-depth environmental ethics are rooted in philosophy, literature, and science, which, on several occasions and in several variants, stressed the important link between man and nature and the close living relationship between all bodies and the Earth.

A decisive contribution was provided by Charles Darwin's studies (*The Origin of Species*, 1859, and *The Descent of Man*, 1871) [1, 2] as its evolutionary theory. It provided a vision of such breadth and complexity to contradict of the simplistic concept of the centrality of man in the natural world and its hierarchical superiority over all other species.

Another critical step in environmental ethics development was represented by the birth of a new science: ecology that studies the relationship between the environment and living organisms and, therefore, between the environment and man.

Do not forget another branch of science that has seen the birth in the last century: ethology. It studied the biological, social, and mental similarity between animals and humans. Because animals demonstrate emotions, sensitivity, and consciousness, they are given the status of sentient beings, with characteristics of cognition and sociality in some cases very similar to those of humans.

The environmental ethics father was Aldo Leopold, who, in his collection of essays entitled *A Sand County Almanac* [3], spoke for the first time of a "land ethic" in which there is a Copernican Revolution of the ethical principles: nonethic that

C. Rabbito
"Security and Privacy" Research Group of SIT, Italian Society for Digital Health and Telemedicine, Rome, Italy

Rabbito Law Firm, Via Piave 44, 40068 San Lazzaro, BO, Italy
e-mail: crabbito@avvocatorabbito.it

F. Capello, A.V. Gaddi (eds.), *Clinical Handbook of Air Pollution-Related Diseases*,
https://doi.org/10.1007/978-3-319-62731-1_9

comes from man but an ethics which is born from the Earth, as a balanced system of reciprocal interactions between symbiotic organisms with the same importance. The main innovation, of great influence to the ethical environment, is that the view proposed by Leopold is not individualistic or humanistic but holistic: you do not get to the environment starting from man but is regarded as a single whole community the human, the animal, and the natural one.

Another influential article was *The Shallow and the Deep, Long-Range Ecology Movement: A Summary*, written by the Norwegian philosopher Arne Næss [4] and published by the journal *Inquiry*, in which the philosopher stresses the need to rethink the man-environment relationship, to protect the life and the nature.

In general, environmental ethics will reflect on three fundamental propositions:

1. The Earth and its creatures have a moral status; they are therefore worthy of ethical concern.
2. The Earth and its creatures have an intrinsic value; therefore they have a moral value as such, not an instrumental value resulting from the fact that responds to human needs.
3. Men should take into consideration the fact of living in an ecosystem, which, in addition to men, includes other forms of life and the environment.

Not only, then, can one of our actions have an effect far from where the action took place, but we must also consider that our isolated action could occur within an existing biological system in which a small effect, sometimes insignificant, it may interact with other systems, causing changes in other forms of interdependent life.

Moreover, since environmental issues almost always involve actions that may have an effect on living creatures including men, every action that affects the environment must be assessed in the light of the common good. Therefore, environmental ethics is also seen as part of social ethics.

In the contemporary ethical debate, there are many voices that seek to recognize the nature and the beings that have intrinsic value, a value in itself, by the very fact of their existence within the terrestrial habitat. This recognition of an intrinsic value is opposed to the vision, often dominant in the past, of the instrumental value of the Earth and its creatures, and evaluated the utility report that man could draw [5, 6].

Given the seriousness and urgency of the current situation, and in order to protect the environment and natural beings that inhabit it, it (who is "it"?) has gradually asserted the thought that, next to the recognition of their moral dignity, it is also necessary to impose certain virtuoso behaviors through binding regulations, as emanation of the power of states, and predict, in case of noncompliance with those behaviors, specific legal responsibilities.

Since the end of the twentieth century, most Member States' legal systems have begun to standardize environmental issues.

We describe below the main legislative measures and international agreements aimed at protecting the environment, with particular reference to those pertaining to the protection of our atmosphere and air pollution limitation.

9.2 The Kyoto Protocol

One of the most important global agreements regarding environmental pollution is the Kyoto Protocol.

Kyoto Protocol is an international agreement concerning the so-called "global warming."

It started in the United Nations Framework Convention on Climate Change (UNFCCC) in 1992 during the UN Conference on Environment and Development which was held in Rio de Janeiro (Earth Summit).

In 1995 the UNFCCC participants met in Berlin to define the main objectives regarding emissions. The meeting continued on December 11, 1997, in the Japanese city of Kyoto where the participants agreed in general terms on emissions targets and signed the Kyoto Protocol.

The Kyoto treaty, however, became effective only in 2004, since for the treaty to become effective, it required that the treaty was ratified by no less than 55 signatory nations and nations that had ratified would produce at least 55% of polluting emissions; the latter condition has been reached only in November 2004, when Russia ratified the Protocol on February 16, 2005.

The treaty mandates a reduction in the emissions of pollution elements (carbon dioxide and five other greenhouse gases, such as methane, nitrous oxide, hydrofluorocarbons, perfluorocarbons, and sulfur hexafluoride) by an amount not less than 8.65% of emissions recorded in 1985—regarded as the base year—in 2008–2012.

Assuming that the Earth's atmosphere contains 3 million megatons (Mt) of CO_2, the Protocol requires industrialized countries to reduce their emissions by 5% of these gases. Human activities enter 6000 Mt of CO_2 per year into the atmosphere, of which 3000 by the industrialized countries and 3000 by the developing countries; so, with the Kyoto Protocol, they should enter 5850 annually instead of 6000.

To date, 175 countries and a regional economic integration organization (EEC) have ratified the Protocol or have started the procedures for ratification. These countries account for 61.6% of global emissions of greenhouse gases.

The Kyoto Protocol also provides for the acceding countries the possibility of using a system of flexible mechanisms for the purchase of emissions credits.

The aim of the flexible mechanisms is to maximize the reachable reductions with the same investment.

The mechanisms are:

- Clean Development Mechanism (CDM) allows industrialized countries and "economies in transition" to implement projects in developing countries that produce environmental benefits in terms of reducing greenhouse gas emissions and economic and social development of the host countries and at the same time generate emission credits (CER) for countries that promote the activity.
- Joint implementation (JI) enables developed countries and "economies in transition" to implement projects to reduce greenhouse gas emissions in another country of the same group and to use credits, in conjunction with the host country.

- Emissions trading (ET) enables the exchange of emission credits between industrialized countries and economies in transition; a country that has achieved a reduction of their greenhouse gas emissions exceeding its target can thus release (using ET) such "credits" in a country that has not been able to meet its commitments to reduce greenhouse gas emissions.

9.3 The Doha Conference and the CPO21

A major limitation of the Kyoto Protocol has been represented by the non-participation of the United States, alone responsible for 36.2% of total emissions, which signed the protocol without ratifying it, and Canada, whose emissions have increased by 18.2% in the period 1990–2012, who withdrew without signing the protocol.

After the Kyoto Protocol, there were other negotiations culminated in 2009 in the Copenhagen Conference which was attended by delegations from 192 countries, all members of the UNFCCC. The conference did not have the desired result as they failed to reach an effective agreement between the industrialized countries, historically responsible for global warming, and emerging countries, which did not intend to limit their growth. The result was a "nonbinding" political declaration by the participating states that has limited to a maximum of 2° temperature increase. This result is considered by many experts to not be enough, as they estimate the temperature increase caused by the greenhouse equal to 3°C.

Major successes can be attributed to the Doha Conference in 2012, which initiated the second phase of the Kyoto Protocol beginning January 1, 2013. The conference adopted an amendment that defines the rules governing the second phase of the Protocol.

The second phase will run for 8 years, in order to ensure that there is no gap between its conclusion and entry into force of the new global agreement in 2020.

The participating states also agreed on a timetable to adopt a universal climate agreement by 2015 and settled ways and means to gradually increase funding and technological support to developing countries in order to mobilize $100 billion, both for the adaptation and mitigation, within 2020.

The most recent and relevant event aimed to limit global emissions of gases responsible for the greenhouse effect was represented at the Paris Conference on climate change (COP 21), held from November 30 to December 12, 2015.

This was the 21st annual session of the conference of the UN Framework Convention on Climate Change (UNFCCC) of 1992 and the 11th session of the parties of the Kyoto Protocol of 1997.

The objective of the conference was to reach, for the first time in over 20 years of mediation by the United Nations, a binding and universal climate agreement, accepted by all nations.

The result was the so-called Paris Agreement, a global agreement on reducing climate change, the text of which met with the consensus of the representatives of the 196 participating parties. The agreement, however, as the Kyoto Protocol, will become legally binding if ratified by at least 55 countries which together represent at least 55% of global emissions of greenhouse gases.

According to the organizing committee, the key outcome was to provide for an agreement to secure the objective of limiting the increase in global warming to less than 2°C compared to preindustrial levels.

In the 12-page document, members agreed to reduce their carbon monoxide production "as soon as possible" and to do their best to keep global warming between 1.5°C and 2°C. The agreement provides for human emission of greenhouse gases to be zero, to be achieved during the second half of the twenty-first century.

Each country that ratifies the agreement will be required to set a goal of reducing emissions, but the amount will be voluntary. There will be a mechanism to force a country to set a target by a specific date but no retribution if the target set is not satisfied: there will only be a "name and shame" system or a list of noncompliant countries, with the aim to encourage them to implement the plan on climate.

The differences with the Kyoto Protocol thus consist in:

- The absence of a "punishment mechanism" as judged ineffective and responsible for the exit of some countries from the Protocol
- Application of the principle of "transparency" to the industrialized countries as well as to developing countries

9.4 The Fight Against Air Pollution by the European Union

The European Union from its very beginning has shown sensitivity to the problems caused by air pollution and tried to limit it in order to protect health and the environment, because according to EU figures, it is the main cause of environment-related premature death.

The legal basis of the actions taken by the EU against air pollution can be found in Articles 191 to 193 of the Treaty on the Functioning of the EU (TFEU)[1] [7].

---

[1] Article 191
(ex Article 174 TEC)

1. Union policy on the environment shall contribute to pursuit of the following objectives:
   - Preserving, protecting and improving the quality of the environment,
   - Protecting human health,
   - Prudent and rational utilization of natural resources,
   - Promoting measures at international level to deal with regional or worldwide environmental problems, and in particular combating climate change.
2. Union policy on the environment shall aim at a high level of protection taking into account the diversity of situations in the various regions of the Union. It shall be based on the precautionary principle and on the principles that preventive action should be taken, that environmental damage should as a priority be rectified at source and that the polluter should pay.

   In this context, harmonization measures answering environmental protection requirements shall include, where appropriate, a safeguard clause allowing Member States to take provisional measures, for non-economic environmental reasons, subject to a procedure of inspection by the Union.
3. In preparing its policy on the environment, the Union shall take account of:
   - Available scientific and technical data,
   - Environmental conditions in the various regions of the Union,
   - The potential benefits and costs of action or lack of action,

- The economic and social development of the Union as a whole and the balanced development of its regions.

4. Within their respective spheres of competence, the Union and the Member States shall cooperate with third countries and with the competent international organizations. The arrangements for Union cooperation may be the subject of agreements between the Union and the third parties concerned.

   The previous subparagraph shall be without prejudice to Member States' competence to negotiate in international bodies and to conclude international agreements.

Article 192
(ex Article 175 TEC)

1. The European Parliament and the Council, acting in accordance with the ordinary legislative procedure and after consulting the Economic and Social Committee and the Committee of the Regions, shall decide what action is to be taken by the Union in order to achieve the objectives referred to in Article 191.
2. By way of derogation from the decision-making procedure provided for in paragraph 1 and without prejudice to Article 114, the Council acting unanimously in accordance with a special legislative procedure and after consulting the European Parliament, the Economic and Social Committee and the Committee of the Regions, shall adopt:

   a. Provisions primarily of a fiscal nature;
   b. Measures affecting:

   - Town and country planning,
   - Quantitative management of water resources or affecting, directly or indirectly, the availability of those resources,
   - Land use, with the exception of waste management;

   c. Measures significantly affecting a Member State's choice between different energy sources and the general structure of its energy supply.

      The Council, acting unanimously on a proposal from the Commission and after consulting the European Parliament, the Economic and Social Committee and the Committee of the Regions, may make the ordinary legislative procedure applicable to the matters referred to in the first subparagraph.

3. General action programs setting out priority objectives to be attained shall be adopted by the European Parliament and the Council, acting in accordance with the ordinary legislative procedure and after consulting the Economic and Social Committee and the Committee of the Regions.

   The measures necessary for the implementation of these programs shall be adopted under the terms of paragraph 1 or 2, as the case may be.
4. Without prejudice to certain measures adopted by the Union, the Member States shall finance and implement the environment policy.
5. Without prejudice to the principle that the polluter should pay, if a measure based on the provisions of paragraph 1 involves costs deemed disproportionate for the public authorities of a Member State, such measure shall lay down appropriate provisions in the form of:

   - Temporary derogations, and/or
   - Financial support from the Cohesion Fund set up pursuant to Article 177.

Article 193
(ex Article 176 TEC)

The protective measures adopted pursuant to Article 192 shall not prevent any Member State from maintaining or introducing more stringent protective measures. Such measures must be compatible with the Treaties. They shall be notified to the Commission.

Although in recent decades the air pollution in Europe has decreased in general terms, the Union's objective is to achieve "levels of air quality that do not give rise to significant negative impacts on human health and the environment" and has not yet been reached. Especially in urban areas ("hot spots"), where most of the Europeans live, the air quality standards are often violated. The most problematic contaminants are currently fine particles and ground-level ozone.

That's why the latest EU laws on air pollution aim to achieve full compliance with the existing legislation on air quality by 2020 and establish new long-term plan for 2030.

Among the results achieved are substances such as sulfur dioxide (SO_2), carbon monoxide (CO), benzene (C_6H_6), and lead (Pb) which decreased considerably in the European Union from the 1970s to today.

There are three legal mechanisms available to the EU in the field of air pollution: the definition of general air quality standards designed to limit the concentration of air pollutants in the environment; setting limits (national) for total emissions of pollutants; and the elaboration of specific legislation aimed at pollution sources, for example, aimed at limiting industrial emissions or to establish rules on vehicle emissions, energy efficiency, and fuel quality.

The main measures taken by the European Union: first it must be mentioned Directive 2008/50/EC [8] on ambient air quality that incorporates much of the legislation. It aims to reduce air pollution to levels which minimize harmful effects on human health or the environment.

According to the strategy on air pollution of 2005 (reduce the concentration of fine particles PM2.5 by 75%, that of ground-level ozone (O_3) by 60% and the acidification and eutrophication by 55%), the main atmospheric pollutants, which it limits the dispersion in the environment, are sulfur dioxide, nitrogen dioxide, nitrogen oxides, particulate matter, lead, benzene, carbon monoxide, and ozone.

Member States are required to define zones to assess and manage the air quality, monitor long-term trends, and ensure that the information on air quality is made available to the public. The measures are also designed to maintain ambient air quality where it is good, but, if they exceed the limit values, it is necessary to take measures.

The directive introduces for the first time an air quality objective environment for fine particulate matter (PM 2.5).

The Directive (2004/107/CE) [9] fixes target values (less stringent than the limit) for arsenic, cadmium, nickel, and polycyclic aromatic hydrocarbons.

Directive 2001/81/EC [10] defines national emission limits for certain atmospheric pollutants and establishes national emission ceilings for four atmospheric pollutants (SO_2, NO_x, VOC, and ammonia (NH_3)), the main chemicals responsible for acidification, ozone at ground level, and soil eutrophication, in order to reduce the harmful effects, setting as terms of reference years 2010 and 2020. It requires that states shall annually communicate information concerning emissions and projections for all pollutants concerned and draw up programs for the progressive reduction of national emissions to conform to the individual national emission ceilings.

Member States had to comply with the limits set by 2010; however, at least one limit has not been respected by several states, sometimes over many years.

At the end of 2013, the commission presented a package of measures entitled "Clean Air for Europe," which had two main objectives: compliance with applicable legislation and the reduction of the impacts of air pollution on long-term health and on the environment. It proposes the revision of the National Emission Ceilings Directive, updating the national limits for 2020 and 2030 for the four pollutants currently regulated (SO_2, NO_x, VOC, and ammonia (NH_3)), as well as other two: fine particulate and methane (CH_4). The package also includes a proposal for a new directive on the limitation of emissions of certain pollutants originating from combustion plants (in addition to large combustion plants which are already covered) and a proposal for ratification of the amended Gothenburg Protocol to the Economic Commission for Europe Convention (UNECE) on air pollution to abate acidification, eutrophication, and ozone at ground level.

9.5 EU Measures Relating to Transport and Industrial Emissions

In view of their polluting potential, a series of directives were issued to limit pollution caused by road transports. Directives set emission limits for different categories of vehicles such as cars, light commercial vehicles, trucks, buses, and motorcycles; they also govern the quality of fuel and its sulfur and lead content.

As of September 2014, the Euro 6 emission standards for passenger cars and light vans have been extended to all new vehicle models. In 2015 the Euro 6 came into force for the registration and sale of all new cars and new light vans (light commercial vehicles and cars for special needs have the terms extended for 1 year). It sets emission limits for a number of air pollutants, especially for nitrogen oxides (NO_x) and particulate matter (PM).

Member States are obliged to deny the approval, registration, sale, or introduction of vehicles (and their replacement pollution control devices) that do not meet the limits in question.

Regulation (EC) No. 595/2009 [11] was issued in order to limit emissions from heavy-duty vehicles (buses and trucks), and from January 2013, the limits are those set by Euro 6.

Regulation (EC) No. 715/2007 [12] lays down rules for in-service conformity, durability of pollution control devices, on-board diagnostic (OBD) systems, and the measurement of fuel consumption. It also regulates the accessibility of information for the repair and maintenance of the vehicle to independent operators.

In addition, in order to further reduce pollution from car emissions, the EU introduced a ban on the sales of leaded petrol and the obligation to make sulfur-free fuels available within the union.

Directive 2009/33/EC [13], on the promotion of clean and energy-efficient road transport, requires public authorities to take account of energy and environmental impacts on the purchase of vehicles, in order to promote and stimulate the market of clean vehicles and energy-efficient.

Moving on to the naval transport, in order to reduce air pollution caused by ships, Directive 2012/33/EU [14] restricts the sulfur values of marine fuels in European seas. The general sulfur limit will be reduced from 3.5% to 0.5% by 2020, in accordance with the limits approved by the IMO. In certain sulfur emission control areas (SECA), such as the Baltic Sea, the English Channel, and the North Sea, the directive from 2015 will apply a standard even more stringent: 0.1%.

Additional emission limits have been provided for non-road mobile machinery, such as excavators, bulldozers, and chain saws, for agricultural and forestry tractors and recreational craft.

9.6 The Ozone Problem and the Montreal Protocol

The Montreal Protocol aimed to protect the air, in particular with regard to the ozone layer. The ozone layer is threatened by a number of substances that have caused its reduction, with very dangerous consequences for human health and the environment.

The first official study on the subject was commissioned by the World Meteorological Organization (WMO), which in 1975 made public the report entitled: "The ozone layer changes as a result of human activities," which identified in chlorine floro carbides (CFC) and in halon gas primarily responsible for the depletion of the stratospheric ozone layer (which is between 10 and 50 km altitude, a crucial filter of the UV rays of the sun). In the same year, the UNEP (United Nations Environment Program) convened a meeting of experts designated by governments and intergovernmental organizations in order to verify and analyze the collected data.

The work of the experts confirmed the correlation between depletion of the ozone layer in the atmosphere and the increase in the presence of gases mentioned above and foresaw, if they are not taken appropriate action, destruction of the ozone layer by 2050.

This realization by the states led to the Convention (Vienna Convention, 1985) which was followed by the signing and ratification of the "Montreal Protocol" signed September 16, 1987, entered into force on January 1, 1989, and subject to the reviews of 1990 (London), 1992 (Copenhagen), 1995 (Vienna), 1997 (Montreal), and 1999 (Beijing). More than 192 states have ratified it.

The Montreal Protocol sets targets and timelines for reducing the production and use of substances harmful to the stratospheric ozone layer, in order to eliminate them completely. Substances identified as harmful are chlorofluorocarbons, carbon tetrachloride, 1,1,1 trichloroethane, halons, HCFCs, and methyl bromide.

In almost all existing uses, CFCs have been replaced by hydrochlorofluorocarbons (HCFCs). HCFCs are of molecules with similar properties, and therefore good substitutes for CFCs, but with a lower ozone depletion potential (ODP). However, they are not totally harmless for the ozone layer. For this reason, even HCFCs have been incorporated in a phased program of reduction of use leading to the suspension of their production by the end of 2019.

The Meeting of the Parties of the Montreal Protocol (MOP) meets every year in order to assess the validity and effectiveness of the Protocol control measures imposed, update the implementing rules and, where necessary, make changes to the treaty.

The developing countries have, compared to the developed countries, a lengthening of the time, called "grace period" of 10 years.

In 1990, the second conference of the parties (London, 1990), the Multilateral Fund (MLF) for ozone was established. The Fund is the financial mechanism of the Protocol that allows developing countries to achieve the objectives of the Protocol through the implementation of projects:

- The technological conversion
- Technical assistance
- Institutional strengthening
- Training and information activities

The Fund is financed by developed countries every 3 years. The Executive Committee of the Multilateral Fund (Executive Committee) meets three times a year which has functions of:

- Monitoring the financial situation, the allocation of donations, and funding of projects
- Evaluation of the implementation of the agencies' activities (UNEP, UNDP, UNIDO, World Bank)
- Evaluating and approving projects
- Issuance of guidelines and procedures
- Definition of intervention strategies

The European Union made the Montreal Protocol operational in 1994 with the EC Regulation 3093/94, which aims to lay down uniform rules in all Member States for the gradual elimination of harmful substances. The 1994 regulation was repealed and replaced by EC Regulation 2037/00. This regulation anticipates the total banning of production and use of CFCs by 2015.

Beginning January 1, 2010, the new Regulation (EC) 1005/2009 [15] on substances that deplete the ozone layer and Regulation (EU) No. 744/2010 [16] in regard to the critical uses of halons which replace the previous Regulation (EC) 2037/2000 [17] were in effect.

With regard to the results achieved, it must be said that they are reasonably positive as the entry into force of the Montreal Protocol, the atmospheric concentrations of CFCs, and related hydrocarbons have stabilized or declined in much of the world but in China and India have increased.

9.7 The Regulatory Bodies and Environmental Pollution Control Bodies

9.7.1 IARC: International Agency for Research on Cancer

It is an intergovernmental agency affiliated to the World Health Organization (WHO) of the United Nations. It is headquartered in Lyon and is involved in conducting scientific research concerning the causes of cancer and mechanisms of carcinogenesis. One of their tasks is to conduct a systematic and comprehensive review of all the scientific literature published in journals subject to peer review related to carcinogenicity evaluation of a given agent.

The experts and researchers of the IARC are also instructed to classify the carcinogenic potential of each agent identified by approved scientific literature.

The classification is made with regard to the following categories:

- Sufficient: experts believe that it is detectable that there is a causal relationship between exposure to the agent of interest and cancer in humans.
- Limited: a positive relationship was observed between exposure and cancer; the causal relationship is possible but cannot exclude errors or the role of chance.
- Inadequate: the available studies are insufficient to prove a causal relationship, or no data is available on cancer in humans.
- Lack of carcinogenicity: There are several studies on the basis of which it can be assumed that there is no association between exposure to the agent and cancer (although the possibility of a very small risk cannot be excluded).

Based on the above classification agents, mixtures and exposures are divided into five groups:

- Group 1—carcinogenic to humans: This category is used when there is sufficient evidence of carcinogenicity in humans. In June 2016, 118 agents have been classified as carcinogenic to humans.
- Group 2A—probably carcinogenic to humans: This category is used when there is limited evidence of carcinogenicity in humans and sufficient evidence in animal experiments. In some cases an agent may be classified in this category when there is inadequate evidence in humans, sufficient evidence in experimental animals, and strong evidence that the mechanism of carcinogenicity observed in animals also applies to humans. Exceptionally, an agent may be classified in this category solely on the basis of limited evidence of carcinogenicity in humans. The evidence of carcinogenicity in humans of a "probable carcinogen" is greater than those of a "possible carcinogen." In June 2016, 80 agents were classified as probably carcinogenic to humans.
- Group 2B—may be a human carcinogen: This category is used for agents for which there is limited evidence of carcinogenicity in humans and less than sufficient evidence of carcinogenicity in experimental animals. In June 2016, 289 agents have been classified as possible human carcinogens.
- Group 3—not classifiable: This category is used for agents for which the evidence of carcinogenicity is inadequate in humans and inadequate or limited in experimental animals. Exceptionally may be placed in this agent group for which the evidence in humans is inadequate but the evidence is sufficient in animals, while there is strong evidence that the carcinogenicity mechanisms in animals are not operative in humans. Also classified in this group are agents that do not fall under any other category. In June 2016, 502 agents are not classified according to their carcinogenicity to humans.
- Group 4—probably not carcinogenic to humans: This category is used for agents for which there is no evidence of carcinogenicity in humans and in experimental animals. In June 2016 one agent was classified as probably not carcinogenic to humans.

9.7.2 European Environment Agency (EEA)

The EEA is an EU body which aims to monitor, through a network system, the environmental conditions of the territory of European States. It is based in Copenhagen.

It is formed by representatives of the Governments of Member States, a representative of the European Commission, and two scientists appointed by the European Parliament; there is also advisory contribution of a committee of scientists.

The Agency was established in Directive 1210/1990 of the EEC and amended by EEC Directive 933/1999; it became operational in 1994.

All EU Member States are automatically participants in the Agency; other European countries may join as a result of special agreements with the EU.

The Agency is responsible for compliance with EU legislation on air emissions and air quality; it gives its opinion on the union's policies on air pollution and air quality, along with long-term strategies aimed at improving air quality in Europe.

In addition, the EEA collects and makes available the data on air pollution in Europe, analyzes and evaluates the trends related to air pollution and containment measures, and studies and reports about the interactions between air pollution and phenomena such as climate change. It also assesses how transport, industrial activity, and energy consumption impact pollution.

9.7.3 The US Environmental Protection Agency (EPA)

The US Environmental Protection Agency (EPA) is the US Agency chartered to protect the environment and is responsible for determining the acceptable levels of exposure to pollutants and to protect the environment and human health through compliance with the laws passed by the US Congress.

The Agency was created based on the proposal of President Nixon and became operational on December 2, 1970. It is headed by a director, who is appointed by the US President and confirmed by a vote of the Congress. The administrator has the rank of a cabinet member of the US government.

9.8 A Case Worthy of Analysis: The Chinese Situation and the Current Legal Measures

As it is known, in China the rapid and broad industrial development has caused, in recent decades, the progressive worsening of the situation of environmental pollution, including air pollution [18].

A particularly serious situation, in the case of air pollution, occurred in Beijing in 2012–2013 and 2013–2014 winters.

The 2012–2013 winter recorded very high levels of concentration of capital finer particulate matter (PM2.5), with peaks of 886 mg/m^3, higher by more than 35 times the limit of 25 mg/m^3 considered acceptable within 24 hours by the "WHO" [19].

During winter 2013–2014, the situation repeated itself: the levels of PM2.5 have reached 500 mg/m^3 and were maintained at high-risk levels for over a week.

In 2014 the report of the Ministry of Environmental Protection on air pollution monitoring of Chinese cities determined that in 2013 only 3 out of 74 cities recorded a defined air quality "healthy."

To counter air pollution, the Chinese government has repeatedly taken measures to curb the phenomenon, including the following among others:

- In 2012 an air monitoring system has been established in 74 Chinese cities, with real-time publication of data on the concentrations of ozone and PM2.5, and this has contributed greatly to raising awareness of the citizens with respect to air pollution.
- In September 2014, the "National Action Plan, Pollution Prevention and Control of the Air" was approved for the years 2013–2017, with an investment by the government of 277 billion dollars, and the goal by 2017 is to bring the levels of PM2.5 at 60 mg/m^3 in Beijing [20, 21].
- Legislation was approved that requires the reduction of the sulfur content in fuels by the oil industry, conforming to EU standards. The new standard requires a reduction of emissions by the most polluting vehicles currently circulating in the country and the adoption of advanced technologies in the control of vehicle emissions.
- A new goal was established for renewable energies.
- An encouraging result was achieved in the reduction of SO_2 emissions, the main precursor of fine particles, and forming of acid rain: the aggressive policy introduced in 2006 with the 11th Five-Year Plan has produced an important result, even surpassing the target set of 10% reduction, mainly thanks to the installation of desulfurization systems in many of the country's electrical power plants.

In terms of legislation, the first framework to reduce pollution and protect the environment was passed in 1987. It is the "Environmental Protection Law," a law which provided guidelines on environmental protection. Since then, two revisions were carried out, respectively, in 1995 and 2000.

However, in the first decade of this century, that legislation was beginning to be inadequate in the face of the rapid and extensive industrialization process undergone in the country and the increase in Chinese water, air, and soil pollution.

With the obsolescence of the existing legislation, on April 24, 2014, the National People's Congress approved the reform of the Environmental Protection Law (EPL). The law as amended came into effect January 1, 2015.

The current EPL contains more than 20 additional articles as compared to the previous version and introduces tough new measures and sanctions on environmental protection.

On the penalties, the amendment introduced covers the criteria with which they are imposed: in front of the ineffectiveness of one-time fines as a deterrent paid which the polluters felt free to continue in the unlawful conduct, the new amendment introduces a system of penalties which provides a continuum of fines, imposed at regular intervals and gradually increased.

In the previous system, it was much less expensive for a company to pay the fine and continue its polluting behavior, rather than to adapt or replace the equipment in accordance with the standards for environmental protection. The current system, based on the principle of eliminating any gains from illegal activities, attempts to correct this distorted outcome.

The law also provides information and reporting requirements for the transparency purposes: business and local government authorities need to disclose specific information regarding the quality of the environment, the control of air, and any environmental accidents.

It must also be made known the fines imposed and information about the collection and the methods of use of the penalties paid by companies in order to discharge polluting substances (Arts. 53, 54 EPL).

Companies have also the obligation to declare the names of the pollutants used and the method of treatment of the waste, in addition to information about the implementation and operation of appropriate system for pollution control (Art. 55 EPL).

Noted the growing awareness of the Chinese public on the matter, the legislature formalizes in the section entitled "Information disclosure and public participation," the right to the involvement of citizens, directly and severely affected by the problem. To the point that is legitimized, the role of so-called *whistle-blower* (literally "spy") means any natural or legal person has the right to report, remaining anonymous, the environmental damage caused by any institution or individual or any violation of the law by the bodies responsible for environmental control and of the implementation regulations.

Before starting a business in a number of industrial categories, companies will have to make an environmental impact assessment (EIA) and make it public in order to allow even private citizens affected by the operation to deliver its opinion on the compatibility of the project with respect to the environment and possibly to report alleged wrongdoing to the competent authorities (Art. 19 EPL). Investments in real estate, on the other hand, need to first perform an ECA (energy conservation assessments), the lack of which can lead to delays in construction.

In the opposite perspective, there are reward measures for companies that adhere diligently to standards, such as a tax incentive (Arts. 21:22 EPL).

In terms of legal protection, in accordance with Art. 58 EPL, have locus standi in front of the competent Courts all the Chinese NGOs who are registered with the Department of Civil Affairs (engaged in the field of environmental protection) and that they are operating for at least 5 years without having committed any offense.

This law helps to complete the Chinese regulatory framework on pollution, for which, taken as a whole, cannot be attributed to serious shortcomings or specific gaps [22].

The issue of the fight against pollution in China is therefore now moving on the side of the effective application of the rules rather than on the lack of them. So it is no longer a matter of verifying and discussing the existence of the rules but instead on enforcing their application and the subsequent effectiveness.

9.9 Conclusions in Perspective *"de jure condendo"*

The phenomenon of environmental pollution, and the damage it causes, doesn't know territorial limits. National borders have, in relation to these problems, no meaning, neither as a deterrent or barrier. Pollution is now a global issue, and as such it must be understood by the law.

The above is especially true for air pollution.

It follows that the protection of the environment, and air quality as far as we are concerned, should be done through regulations and measures as global as possible or at least international. Lawmakers have already taken note of this peculiarity of the subject matter of their discipline.

A very similar situation occurred with the spread of information-communication technologies and the consequent widespread worldwide use of the Web.

As it has become clear to the legislators of many states, regulation of everything that is related to the use of the web can no longer be characterized by a local footprint, regional or even national, risking its complete ineffectiveness.

The web phenomenon is worldwide and the same should be the laws, in every respect: the prosecution of computer criminals and the protection of privacy in data processing but also in regard to e-commerce and provision of ICT services, including their taxation.

A similar perspective must accompany the legislation of environment protection. Surpassed then the nationalistic perspective, there are now some decades of joint effort by states, in particular of the leading manufacturers of polluting substances in the sense of legislating in a concordant manner and choose commitments and common goals.

There are international agreements achieved: the conventions and the working groups that have been described in the preceding pages. Unfortunately, the instrument in international law which has been used so far is not the best way to get relevant results quickly and territorially uniform, because it is not in most cases rules and direct application but the legal instruments that require national ratification and more detailed rules issued at the state level.

The times are then long (assuming that the rules are endorsed) and the outcomes are uncertain.

The above is with the exception of the European Union, in case of regulations, since in this case it is directly applicable within each Member State. Unfortunately, the principle of national sovereignty does not allow doing the same in the case of protocols, agreements, and international conventions. Therefore, these are problems that cannot be solved with the address rules.

Instead an effort is needed by all countries of the world in the sense of adoption of stringent rules, to apply them immediately and punish those who do not apply the rules. A real commitment as territorially uniform as possible is required. The law can be the driver but does not constitute the decisive factor; this will rather be sought in the global awareness of the need for intervention. Rather, the pulse will be ethical, cultural, or political.

The severity of the phenomenon of environmental pollution is significantly proven to the jurist by a progressive increase both in Europe and globally of recourse to criminal law. The polluting behavior was recognized in many cases as a criminal offense and was therefore provided by criminal sanctions.

Can the use of criminal law be an effective environmental protection?

The issue is debated, but the use of criminal sanctions is certainly testimony of the importance of the environment as a protected asset and the severity of the attacks it suffers, as in the case of the Chinese legal framework that we have analyzed.

Unfortunately, moving environmental protection under criminal law is affected by the well-known problems that this protection can provide, as often stressed by environmental groups. First of all, the length of trials may make it impossible or very difficult to obtain a limiting action or reduction of the damage caused to the environment, while it would be much more important to be timely before more damage is done.

Another aspect is related to the necessity of reporting of pollutants behavior while it is produced: it can be difficult to identify those committing a violation when they are doing it and where in the country, especially if it is very broad.

As we have seen in China's new criminal law to respond to this problem, the formation of the "*whistle-blowers*" is statute to reward informants.

You may agree or disagree with this type of solution, but it is unquestionably the need for monitoring as most comprehensive and frequently as possible throughout the country to block, if possible outset, each pollutant behavior rated as unlawful. This is not a simple task, nor inexpensive.

From what has been said, it is clear how the legislation to combat pollution, including air pollution, requires joint efforts and consistent enforcement by all states, timeliness of action, efficacy, and measures actually disincentive of dangerous or forbidden behavior, regardless of the laws used and the sanctions proposed.

References

1. Darwin C. The origin of species by means of natural selection, or the preservation of favoured races in the struggle for life. London; 1859.
2. Darwin C. The descent of man, and selection in relation to sex. London; 1871.
3. Leopold A. Sand County Almanac. Milano; 1997.
4. Næss A. The shallow and the deep, long range ecology movement: a summary, inquiry. Oslo; 1973.
5. Alfons A. Etica dell'ambiente. Brescia; 1988
6. Edwards A. The sustainability revolution: portrait of a paradigm shift. New Society Publishers: Philadelphia; 2005.
7. Treaty on the Functioning of the EU (TFEU), the link: http://eur-lex.europa.eu/legal-content/IT/TXT/?uri=celex%3A12012E%2FTXT.
8. Directive 2008/50/EC http://eur-lex.europa.eu/LexUriServ/LexUriServ.do?uri=OJ:L:2008:152:0001:0044:IT:PDF.
9. Directive 2004/107/EC, at the link http://eur-lex.europa.eu/LexUriServ/LexUriServ.do?uri=OJ:L:2005:023:0003:0016:it:PDF.
10. Directive 2001/81/EC http://eur-lex.europa.eu/LexUriServ/LexUriServ.do?uri=OJ:L:2001:309:0022:0030:it:PDF.

11. Regulation (EC) No. 595/2009, at the link http://eur-lex.europa.eu/legal-content/IT/TXT/?uri=celex%3A32009R0595.
12. Regulation (EC) No. 715/2007, at the link http://eur-lex.europa.eu/legal-content/it/ALL/?uri=CELEX%3A32007R0715.
13. Directive 2009/33/EC, at the link http://eur-lex.europa.eu/LexUriServ/LexUriServ.do?uri=OJ:L:2009:120:0005:0012:IT:PDF.
14. Directive 2012/33/EU, to the link http://eur-lex.europa.eu/LexUriServ/LexUriServ.do?uri=OJ:L:2012:327:0001:0013:it:PDF.
15. Regulation (EC) 1005/2009, to the link http://eur-lex.europa.eu/LexUriServ/LexUriServ.do?uri=OJ:L:2009:286:0001:0030:IT:PDF.
16. Regulation (EU) No. 744/2010 at http://eur-lex.europa.eu/legal-content/IT/TXT/?uri=CELEX%3A32010R0744.
17. Regulation (EC) 2037/2000 to the link http://eur-lex.europa.eu/legal-content/IT/ALL/?uri=CELEX%3A32009R1005.
18. Delang CO. China's pollution problemss. Routledge; 2016.
19. Sinisgalli E. La tosse pechinese. L'inquinamento atmosferico in Cina, http://www.saluteinternazionale.info/2014/10/la-tosse-pechinese-linquinamento-atmosferico-in-cina/#biblio.
20. Editorial. (Barely) living in smoke: China and air pollution. Lancet. 2014;383–845.
21. Du Y, Xu N. Stalled bid to cut car fumes, 2012. Chinadialogue.net.
22. Lovisetti G. China: into force the new Law on Environmental Protection, 2015. http://www.diritto24.ilsole24ore.com/art/avvocatoAffari/mercatiImpresa/2015-02-10/cina-vigore-nuova-legge-protection and environment-1.

10 Air Pollution in Infancy, Childhood and Young Adults

Fabio Capello and Giuseppe Pili

10.1 Background

Air pollution remains a major concern for the health of younger people: their different physiology and lifestyles can affect them deeply and in different terms when compared to the adults. Studies on the effect of air pollution on children can be dated as early as the eighteenth century: not surprisingly one of the first epidemiological studies in oncology, which is linked to the onset of cancer [1, 2] with the exposure to pollutants of children hired as chimney sweeper. Today we are far from the level of pollution of the industrial revolution, and children's labour is not welcome anymore as in the past. Yet, in urban area children are constantly exposed to air pollution in their everyday activities. Those that live in less polluted or in greener areas commute to schools exposed to pollution. Extra-scholastic activities take children around the city; most of those are outdoor activities that take place when the concentration of pollutants in the air is higher. In poor-income areas, where the number of children in the population is higher, air pollution is often out of control and exacerbated by routine activities. In those same settings, there is no awareness of the risk to air pollutions, and there are basic or no procedures to reduce exposure to pollutants.

The risk on children health is somehow of more concern compared to the ones in adults: children can have lifelong consequences that can also deeply affect their wellbeing and their choices and consequently the same quality of life. Early exposures, besides, mean an increased risk for chronic outcomes for those pollutants that have cumulative effects.

F. Capello (✉)
Alder Hey Children's Hospital, Liverpool L14 5AB, UK
e-mail: info@fabiocapello.net

G. Pili
Department of Child and Adolescent Psychiatry and Neurological Disorders,
ASL 1 Imperiese, Sanremo 18038, Italy

F. Capello, A.V. Gaddi (eds.), *Clinical Handbook of Air Pollution-Related Diseases*,
https://doi.org/10.1007/978-3-319-62731-1_10

The effect of air pollution on respiratory system in children is nowadays well-known. Recent studies show how other systems and organs are affected [3] and how exposure to pollutant can become a risk factor for developing chronic non-communicable disease in the adulthood [4]. But as we will see, every aspect of the developing process can be affected by air pollution. This is not an unpredictable observation: a child's organism grows rapidly and undergoes to a number of changes over the months or the years. This process can be easily altered by external influences. These involve behavioural and affective stimuli, microorganisms, physical and chemical interactions as well as the combinations of those.

For these reasons, lately scientists are becoming more aware of the implications of air pollution in childhood. They promoted a new set of studies that helped to better understand how air pollution can compromise the natural growing of a child and how it can affect both his or her everyday life and his or her adult life.

10.1.1 Why Children Differ from Adults When Exposed to Air Pollution

As we have briefly seen, children are normally more vulnerable than adults to the effect of air pollution. This is not an idea that can be given for granted, as children can otherwise be less affected than adults to other pathological processes: children do not normally have chronic and multiple conditions, can have more healing resources when exposed to acute injuries and infections, do not normally take medicines for long-term or lifelong treatments or do not have consolidated habits that can be risk factors for the development of diseases.

But from a different point of view, children are more fragile than adults. They are more vulnerable because their body is developing and changing rapidly over the months or the years. And children comprise a wide category of individuals that range from few days of life to the teenage.

When it comes to air pollution, besides, environmental, behavioural and physiological factors deeply affect the way children are exposed to pollutants compared to adults [3, 5–12]. The main characteristics are synthetized in Table 10.1.

Some of these factors are peculiar only for children. Others may be in common with adults but can result in more harmful or long-term effect in children. Others affect children as well as adults, but children can proactively change the circumstances that cause them to be exposed to pollutants. Passive smoking, for instance, is an environmental factor but also a behavioural factor. While adult can decide to quit smoking or avoid smoking at home, children normally cannot take similar decisions. Thus, smoking is a relatively non modifiable circumstance that could affect children, while it is a potentially easily modifiable factor when individual adults are considered.

Other behavioural factors are more related on children's activities and have to be considered: children spend more time outside. Children engaged in physical outdoor activities are exposed to outdoor pollution which in urban areas is strongly associated with traffic. Because of the increase in the respiratory rate and to bronchodilation secondary to the exercise, the amount of pollutant inhaled has increased.

Table 10.1 Susceptibility of children to air pollution when compared to adult exposure to environmental factors

Behavioural	Environmental	Physiological
• More time spent outside • More physical activities done outside • Ingestion of dust and soil (pica behaviour) • Time spent commuting (walking, biking, car, public transport) • Less awareness and proactive preventive behaviours • Children labour (in selected settings) • Malnutrition and selective micronutrient deficiency (in selected settings) • Mouth breathing • More time spent outside in summer time when the level of pollution is higher • Different timing when compared to adults in terms of daily or seasonal exposure to pollutants • Time spent at school (are often closer to busy roads when compared to residential areas) normally at daily traffic pollution peak • Different behaviours across the life stages • Active smoking during adolescence, as part of normal challenging behaviour of the teenage • Pica behaviour (ingestion of dust on object contaminated by air pollution) • Poor control of cough and expectoration in smaller children • More engagement in increased physical activities as children are normally more active than adults, that results in increased oral breathing and respiratory rate	• Indoor pollution • Household air pollution and cooking fumes • Time spent in school (the quality of indoor and outdoor air cannot be controlled by children/parents) • Frequent respiratory infection (community-acquired respiratory tract infections) • Exposure during antenatal life also results in possible congenital conditions (like newborns small for gestational age) that increase the risk of exposure to air pollution later in life • Low-income settings (both in urban and in rural areas) • Exposure to passive smoking • Exposure to different pollutant in relation also to their socioeconomic status • Level of education of parents • Air pollutants type and concentration vary during the day/year resulting in different exposure depending on children's activities that differ from the ones of the adults	• Small stature: children breathe air closer to the ground that can be potentially more contaminated • Higher respiratory rate • Poor filtering through the nose • Oral breathing (predominant in infants and in small children) • Higher volume of air per minute compared to the smaller body size • Smaller airways that result in a higher impact on the lung function • Limited metabolic detoxification • Maturation and growth of organ and apparatus, including lungs, brain, immune system • Different susceptibility across the life stages • Children develop in spurts • Dosing/toxicity changes according to weight and surface area • Change in the homeostasis during later childhood, puberty and adolescence • Interference of pollutants with sexual maturation • Cumulative effects of the pollutants: toxins remain in the lungs for more time • More time to develop tumours for cells affected by carcinogenic pollutants • Blood-brain barrier not fully developed • Immaturity of the immune system

Children attending school are also exposed to community-acquired respiratory tract infection that easily spread among crowded indoor areas, like a classroom. Air pollution augments the risk of respiratory infection, and respiratory infections reduce the protection against microorganism (e.g. compromising the ciliary motility in the respiratory tract, hence reducing the mucociliary clearance [13]).

Small children that do not attend nursery and those that live at home during the entire childhood can be also affected by environmental exposure. This is especially observed in very-poor-income area where children are constantly exposed to household air pollution. This can be controversial as in very rural locations, children do not necessarily spend time inside houses, or houses are often open so that there are no clear boundaries in terms of the microenvironment between the inside and the outside. In some settings, infants and toddlers spend the whole of their times with their mums, often tied to them, also during household activities like cleaning and cooking. That could expose children from high to toxic concentration of chemicals, exhaust or fumes. The same considerations can be applied to children who spend most of their time inside a house also in higher-income countries, when the quality of the indoor air is poor.

In a developing body, in which organs and systems are not mature, any possible harm can result in long-term consequences.

It had to be noted that children can be exposed to air pollution starting from their antenatal life. This can have important implications, as pre-birth exposure lower the threshold and increases susceptibility later in childhood to the harmful effects of air pollution in an already susceptible population, namely the children (see also Chap. 26).

10.2 Epidemiology and Impact

10.2.1 Exposure

As we have seen, children can be exposed at home or at school, while commuting or when engaged in other inside and open-air activities. It implies that they can be exposed to indoor and outdoor pollution or to a mixture of both. Growing evidences link the areas in which children live or go to school with short- or long-term health problems or developing impairments. Traffic-related pollution seems accountable of most of the health effects during the time children spend at school, especially when school is located near busy roads. But effect on health can be noted also in children coming from different urban areas with different level of pollutants and attending the same school [14]. This gives a picture of the complexity in mapping the risk of air pollution exposure. It has been noted that in big cities the areas with a lower socioeconomic level are normally more polluted and children living in those neighbours are the ones that are mostly affected by air pollution [6, 15, 16]. Nonetheless children can have a more dynamic lifestyle when compared to adults: it means that they can more frequently move to different areas of a city spending several hours of the weeks in more or less polluted zones.

Although regions with a high level of industrialization are allegedly more polluted, the rate of urbanization has also a role especially in rapidly developing countries, meaning that children living in low- and middle-income areas are potentially at higher risk [4]. In those same settings, there may not be policies in place for air pollution contention; at the same time, the percentage of children in the population is higher.

The amount, the length of exposure and the type of pollutants children can be exposed to, for those reasons, can vary. Individual exposure can hardly be measured, but a map of exposure can be estimated. This can be done by monitoring the quality of the air near the places a child spend most of the time (like home, school, playground, sport field), monitoring the average of pollutants concentration in a study area, mapping the nearest busy roads, monitoring the ambient with ad hoc measurements and placing pollutant monitor outside and inside schools or at home. Personal exposure can be also measured with the use of personal equipment that children can carry (mainly for a limited time) everywhere they go. Biomarkers as well have been studied in order to establish whether they can be used as expression of early effects and exposure to given pollutants [17].

It is however difficult to establish which pollutants are more dangerous for children. Particulate matter (PM), ozone (O_3), carbon monoxide (CO), sulfur dioxide (SO_2), nitrogen oxides (NO_x) and lead [3–5] have been identified as a major cause of health problem in children. Carbon monoxide, PM2.5, nitrogen oxides, hydrocarbons, other hazardous air pollutants (HAPs) and ozone are related to traffic pollution. In rural areas fumes from combustion can be a major cause of indoor and outdoor air pollution.

The concentration of those pollutants can change among micro- and macro-areas, during season and during different times of the day or of the week. Exposure can consequently change according to daily, weekly and seasonal routines in children's lives.

10.2.2 Age and Gender

Children are an inhomogeneous group of people whose characteristics vary according to age. Because children grow over the years, the same individual changes more or less quickly, altering his or her personal features and evolving to new and different stages. Those simple assumptions cause two major consequences:

1. Air pollution can affect the way children grow.
2. Air pollution can have different effects in different stages of life.

The number of study assessing the possible implications is still limited; although it appears clear that air pollution can have effect on the antenatal life and on the early stage of life, the response to different pollutants in the different stage of life is still unclear, as well as the long-term implications that exposure to toxic substance can have in small children when compared to the most grown-up.

As we will see, evidences suggest that the whole developing process can be nonetheless put at risk, with the maturation of every organ and system potentially involved. Besides, early exposures seem to increase the risk of adverse outcomes as they (a) affect very immature systems that are rapidly changing, (b) affect systems that do not have mature protective systems, (c) compromise the normal developmental pathways, (d) create cumulative effects and (e) affect systems with peculiar physiology also related to their smaller size.

Boys and girls can have different response to the effect of air pollution, because of differences in terms of exposure and of response to pollutants. Again behavioural, environmental and physiological factors have to be considered. During puberty the differences become more clear, when the effect of growth clearly differentiates the two sex also in terms of behaviour and activities. Some studies suggest still different effects of air pollution also in younger children, especially when respiratory symptoms and allergies are considered [18, 19], with healthy boys being more affected than girls and already symptomatic girls being more at risk for relapses [20]. Those preliminary results however do not clarify the physiopathology underneath the observed effect and cannot exclude that behavioural factors alone can justify those differences. Thus, further studies are needed to proper assess the risk of air pollution in children of different ages and sex.

10.2.3 Air Pollution and Healthcare

The effects of air pollution are not always so straightforward. Most of the time, the request for help is secondary to the effects that are not clearly related to air pollution or that apparently have no link to it. So it is difficult to estimate the real impact of air pollution in children although is likely very underestimated. Although some pathologies are more frequently observed according to seasonal patterns, air pollution is not normally monitored when it comes to emergency department attendance or hospital admission.

Another limitation to assess the real impact of air pollution on the healthcare is the fact that different organs and systems can be directly or indirectly affected, and most of the time, the spatial and temporal association is difficult to understand even for those that analyse epidemical data. Primary care and hospital statistics therefore cannot easily correlate the incidence of a disease with the environmental factors.

For this reason, most of the data related to hospital admission and the consequent burden of air pollution for the health systems are based on study designs that study the effect of air pollution on hospitalization and respiratory symptoms that are more easily associated with environmental triggers.

Data strongly suggest that the number of admission and relapsed strongly related to the exposure to air pollutants and traffic-related air pollution [21].

It is in any case reasonable to think that a high number of admission to children's wards, even when apparently not directly linked to air pollution exposure, can be nonetheless been caused by environmental factors.

Those are fundamental data, for the implementation of efficient and cost-effective health policies.

10.2.4 Hospitalization

Although it is becoming clear that there is a strong association between the level of exposure to air pollution and the children's health, it is difficult to make real estimates of its burden in terms of hospitalization and expenditure for the different health systems. A main reason is the fact that some of the effects cannot be directly linked to single episodes; in other cases, acute episodes can or cannot be an expression of a chronic exposure to air pollutants or may be triggered by other factors (e.g. virus or bacteria) in children chronically exposed to the effect of air pollution. Besides, in some children the quality of the air they are breathing can increase the frequency of relapses of chronic diseases that need medical attention.

For those reasons, the exact impact on hospitalization is largely unknown. Most of the studies in addition focus on respiratory causes only of hospitalization, as those are the ones that more clearly can be related to acute exposure—or to variations in terms of exposure—to air pollutants. Even in these cases, the association is still unclear: although some studies show strong evidences that support the idea that air pollution increases the number of hospitalizations and visits to emergency departments in children [22, 23] especially for those exposed to traffic-related pollution [21] and those with a diagnosis of asthma [24, 25], others found no significant association between exposure to some particulate and hospital admissions [26]. This same study suggests that exacerbation of respiratory symptoms may reflect more on primary care or outpatients' interventions, rather than in admission, and that different mechanism and different pollutants can directly or indirectly affect the health conditions of children, meaning that only in some cases a hospital intervention may be needed. That increments the difficulty in planning epidemiological studies that can help to understand the real impact of air pollution on the health systems.

10.2.5 Morbidity and Mortality

Being the medium- and long-term effects of air pollution on children's health difficult to evaluate, it is still not clear what are the numbers in terms of morbidity and mortality directly or indirectly related to air pollution. Children respond to environment differently; they can be more vulnerable during the first months or years of life; the cumulative effect can show its signs in the adult life; air pollution can cause the exacerbation of an underlying condition and that can, in turn, result in serious or lethal damages. In the latter case, it may be difficult to ascribe disability or death to air pollution.

In particularly polluted environment, nonetheless, a substantial increment of mortality under the age of five and in neonates has been noted [27–29], although in the same settings, other studies find a more significant association with mortality and older age [30]. On the other hand, different pollutants may be more or less poisoning and have effects on children's health that are comparable to the ones on the adults; for some of them, the hazards are known [3], and it has been possible to evaluate the adverse outcomes (viz. disability and mortality) also in terms of cost for the society [31].

In selected cases, the toxic effects of air pollutants are more clear, especially when indoor pollution is considered: for instance, the household solid fuel used in low-income areas is strictly related to acute respiratory infections, chronic obstructive pulmonary disease, tuberculosis, asthma, lung cancer, ischaemic heart disease, blindness and death [32].

Further assessments are needed to better understand the real impact of air pollutants on short- and long-term disabilities and deaths in children.

10.2.6 Quality of Life

Social and health factors affect the quality of life for children that live in polluted areas. We have seen that, even if probably underestimated, the effect of air pollution on health conditions is clear and children may live with disabilities, have exacerbation of chronic conditions, experience acute events or develop morbidities that eventually impact on their growth and on their future adult life. The burden in terms of direct effect is difficult to establish, but the relationship remains clear.

The resulting effect then is not different from the one created by similar health and medical conditions in children.

On the other hand, to survive in a polluted environment, families and children have to adopt different lifestyles or take precautions that may result in an impairment of the overall quality of life. Those families that are particularly aware may decide to change neighbourhood or school, and that may result in more time spent commuting. Children may be asked to remain at home in the afternoon or during summertime not only when the level of pollution can be higher but also when children are more likely to be engaged in outdoor and recreational activities. Time spent indoor may not be necessarily safer: inactivity can be associated with risky or unhealthy behaviour in children and can influence relationships and children's social lives [33].

Behavioural changes may be fostered, but taking action against air pollution can be also stressful for children that do not necessarily understand the implications and the need for change. This is particularly true when they are asked to wear active protection equipment or to avoid particular outdoor or indoor areas.

10.2.7 Society

Estimating the impact of air pollution on society is challenging. At the same time, it is hard to understand how the exposure to air pollution and its related effects on human health depends on the differences in society and in the socioeconomic status of families and individuals.

Some studies in fact suggest that ethnicity and social background can have a role in terms of exposure and outcomes on children's health [6, 16, 34]. It remains unclear if the relationship is related to genetics, to the air quality in poorer neighbourhoods (that is supposedly worse when compared to greener or residential zones) or to lifestyles. Those are also to be considered confounding factors in designing a study to evaluate the consequences of air pollution on children's health [15].

When it comes to children, the burden of air pollution has to consider a number of factors: the direct cost for the evaluation, investigation and treatment of medical conditions directly or indirectly interrelated to the exposure to air pollutants; these impact on the expenditure of the different health systems and on the available resources. Some reports give an idea of the impact on economy of air pollution-related disease in children: selected conditions due to toxins in the environment in the sole United States are responsible of about 3% of the whole US care cost [31]. This of course largely underestimates the problem, as long as diseases not directly attributable to air pollution are not included in the surveys.

Pollution-related health problems also represent a direct cost for families, in terms of money used to reach the health centre, to pay for healthcare and for investigation and treatment. The indirect costs that in children involve the same child and his or her adult carers; for the child a possible esteem can be based on school attendance and on the number of missed days of school that is significantly increased in children exposed to air pollution especially when they suffer of respiratory conditions [31, 35, 36]. Parents are also involved in terms of days of work lost in order to provide care for ill children.

As we have seen, the quality of life as well can be compromised. It results in costs for family and the society and in the use of resources that could be used for other aims.

10.2.8 Developing Countries

Two major issues are related to the effect of air pollution on children in very poor settings: first, the percentage of children in the population is generally higher in developing countries, where on the other hand the level of assistance for families and underage people is lower. Second, although rural area may be less polluted when compared to high-income countries, in developing countries' metropolitan areas, the quality of the air is poorer secondary to a number of factors. Besides, the awareness of the problem is lower, and people are less concerned and less interested in taking personal precaution or in advocating changes. In addition, children suffer often of chronic selective malnutrition or acute malnutrition ranging from mild to severe; nutritional status can reduce the level of protection to air pollutants, especially in these areas where environmental conditions are poorer [37].

In rural areas, the level of indoor pollution, especially secondary to the use of household solid fuel mainly for cooking, is a serious concern. The smoke coming from combustion produces a number of toxins that are known to provoke damage on human health and particularly on children that are the most vulnerable ones [38]. Among them are volatile organic compounds (VOCs), carbon monoxide (CO), carbon dioxide (CO_2), nitrogen dioxide (NO_2), ozone (O_3), hydrocarbons, particulate matter (PM) and dust particles [39, 40]. Very young children in these settings spend most of their time with mums or adults engaged in household activities, often on their lap or tied on their backs. The effects are mainly respiratory symptoms, but they can range up to life-threatening conditions and death [40].

The quality of outdoor air should be supposedly better, but combustion-related pollutants come from wood or manure's fires especially where livestock are hold and fumigation is constant. In addition, the level of dust in the air can be higher, as well as the concentration of biological pollutants; even if the level of traffic is very low when compared to urban settings, the use of outdated vehicle increases the level of pollutants in those area close to the main roads.

Unfortunately, it is very difficult to assess the level of exposure in these settings [41–43], and consequently developing a strategy to reduce the health effects of air pollution on children may be extremely challenging. More studies are therefore needed to better understand the entity of the problem and to correctly address it [43], in order to produce effective policies also in those countries with a low level of awareness in the government and in the population.

10.3 Effect of Air Pollution on Organs and Systems in Children

Most of the studies on the effect of air pollution on children's health focused in the past on respiratory diseases. This was due both to the logic link among air toxins and respiration, as well as the empiric observation of exacerbation of respiratory symptoms in children with chronic conditions as asthma and exposure to pollutants. As we have seen, children can be particularly exposed to air pollutants, and the rising awareness that all the systems can be potentially affected by air pollution created new lines of investigation in the last decade. Research therefore focused as well on other health issues, more related to the growing processes which naturally are typical for children. New and consistent findings demonstrate that the whole development can be affected, starting from the antenatal life up to the puberty and the teen years. These studies gave new insights to the problem. The peculiarities of children's organism in fact made them particularly vulnerable to pollutants, so that is today difficult to separate the different health effect on system and organs.

It is reasonable to think that the whole body can be affected by air pollution especially the rapidly evolving one of children. Yet, most of the consequences are largely unknown, also because of the limited number of studies undertaken so far. It is nonetheless reasonable to think that the problem in children is largely underestimated. What we know so far is that pollutants play a decisive role in the developing of diseases in children starting from weight at birth and perinatal harms up to life-threatening events and chronic life-changing conditions.

The analysis of the possible diseases that follows is not meant to be exhaustive, because of the limited number of studies in childhood. Nonetheless it gives a thorough overview of what today we know about air pollution and diseases in children and supposedly opens the doors to further lines of research.

10.3.1 Birth Weight and Antenatal Exposure to Air Pollution

The weight at birth is a reliable marker of foetal development, as every harm received during the antenatal life can present itself with a slow rate of growth [44]. This of course is not the only feature of damages occurred during pregnancy, but it is easy to observe, to measure and to compare. In addition, children small for gestational age and foetus with intrauterine growth restriction can present with serious consequences at birth and potentially life-threatening complications during delivery and the first minutes or hours of life.

Children with a low weight at birth consequently are at higher risk; it is therefore essential to ascertain if air pollution can be among its cause: a low weight could become expression of exposure to pollutants during pregnancy, at the same time can predict possible complication or damages to the foetus due to pollution and predict possible developmental issues or chronic conditions later during childhood.

The association of inhaled toxins and antenatal development is already known, as the many studies on maternal smoke and birth weight or weight for gestational age show. In fact, observation of the effect of smoking on maternal health has been noted as early as the mid-30s of the twentieth century [45]. A number of potentially harmful molecules that can pass the placental barrier are present in tobacco's smoke, and those may be similar to chemicals that can be retrieved in indoor or outdoor polluted places. What remained unclear thus was the relationship with environmental pollutants passively breathed by mums during pregnancy and their effects on the newborn. Furthermore, changes in the antenatal grow rate may or may not correlate to long-term medical conditions or to processes that could affect the normal development of a child. A new body of evidence demonstrated that air pollution even in outdoor concentrations can affect the normal development of the foetus and consequently his or her weight at birth. Physiological mechanisms are still not clear, but causality correlation has been found for different pollutants as PM_{10}, NO_2, SO_2 and CO [46–49] and low weight at birth or preterm deliveries [49].

Other effects—mainly respiratory ones—on later childhood also have been observed with a correlation with adverse outcomes later in children's lives and maternal contact or inhalation of air toxins during pregnancy [47, 50]. Some surveys seem to confirm in fact that exposure to indoor and outdoor pollutants (pesticides, polycyclic aromatic hydrocarbon, environmental tobacco smoke, PM, NO_2, O_3, CO, airborne endotoxins, dust, SO_4, NO_3, NH_4, CI, acid vapour as HNO3, formic, acetic) in the gestational life does not affect only the same pregnancy outcomes but also has a role in later onset of neurodevelopment, cognitive, asthma, obesity and pubertal development issues [51] as well as on the birth deformities, sudden infant death syndrome, cognitive impairment and reproductive outcomes [52] and autism spectrum disorder [53, 54]. It is unclear, yet, how the timing, the amount and the acute/high versus constant/low exposure to pollutants during pregnancy lead to the development of diseases in childhood and adolescence.

The number of studies is limited, so definitive associations are difficult to describe, but it is becoming clear that air pollution can be particularly harmful in pregnancy, and adequate measure in terms of prevention and reduction of the exposure should be undertaken.

10.3.2 Growth

The development process is a key issue in children. Children change dramatically during a time span of months and years. Growth and weight gain are consequently a reliable marker of a child's wellbeing. Nevertheless, evaluating the way children are growing and establish relations between grow rate and external or environmental stimuli is extremely challenging. The main reason is that every child follows an own growth curve that can be only partially estimated with the different models of growth charts currently available. Because exposure to pollutants changes over the years and goes through phases of acute exposure to likely more modest chronic background exposure, understanding what and in which amount causes discrepancies between the normal/expected growth curve of a child and the actual rate of growing. Moreover, the number of confounders that can influence the development of a child is incredibly high and varies from family to family. Creating cluster of children that could be included in longitudinal studies is challenging, and because of the confounders, there are no guarantees that the children in a comparer group could give a reliable model in terms of exposed and not exposed group.

As long as children are quickly growing, it is reasonable to think that childhood may be a critical exposure time, and it would be logical to speculate that toxins could interfere with this rapidly ongoing process.

Very few studies for these reasons focused on the effects of air pollution on the physical development of children; most of them however agreed, even with some limitation, that children exposed to air pollution have a slower growing rate [55–57], although is not clear when, for how long, how many times and what type of exposure can be considered cause of that.

To partially overcome those issues, to assess possible growth's delay or impairment in children exposed to air pollution, the makers of bone turnover have been also used. Osteocalcin and the C-terminal telopeptide of type I collagen are related to reduction in mineral bone density, and the increase of the level of these substances is a reliable marker of increased bone turnover. Children exposed to air pollution show an increase of both, meaning a negative effect of environmental toxins on their bone turnover [58].

Bone age as well can be compared with chronological age to demonstrate delay or discrepancy in the growth rate in children. Some findings suggest that the peak bone mass can be prejudiced by air pollutants [59].

This is no surprise as growth in children is determined mostly by modification of the musculoskeletal system; the same adult height depends on the lengthening of bones during childhood, puberty and adolescence. Skeletal problems secondary to the exposure to air pollutant can therefore affect the physical development of children, as bones are still maturing and rapidly changing.

Ozone pollution seems to negatively influence the cutaneous production of vitamin D; O_3 works as a screen on sunrays, reducing the level of ultraviolet B (UVB) photons that can reach the skin [59, 60]. Besides, children that live in polluted areas are less likely to spend more time outdoor, meaning a reduced exposure to sun light. UVB have a prominent role in the production of cholecalciferol that is crucial for the development of healthy bones in children and for the prevention of rickets.

Skeletal fluorosis—a chronic metabolic condition that depends mainly on an excessive absorption of fluoride and that can lead to osteosclerosis, osteomalacia, osteoporosis and secondary hyperparathyroidism, with in some cases severe deformation—seems to have a strong relationship with indoor pollution, in those areas where burning coal is used as household solid fuel [61, 62].

Although mainly ingested with food, elemental toxicants are also present in air pollution and in environmental tobacco smoke. Among them lead, cadmium and aluminium are known to influence the skeletal metabolism, damaging directly the bone tissue or indirectly due to renal toxicity or acting on the metabolism of calcium and vitamin D. Long-term exposure and cumulative effect are related to osteodystrophy and bone disease [63–66] that eventually lead to growth impairment and disruption of the physiological physical development of children.

10.3.3 Neurological Development

Growing evidences correlate the exposure to air pollution and brain functions in children. Not surprisingly the toxic effects of environmental pollutants interact with the physiological development of the particularly plastic nervous system of children.

Air pollution in fact can affect the brain development as early as the antenatal age as long nervous cells' multiplication and structural organization are particularly active in this stage of life, while the immaturity of the barrier makes the central nervous system more exposed to toxins; exposure to different gestational ages is supposed to lead to different outcomes in terms of cognitive development later in childhood [67, 68].

The mechanism of neuro damage in children and the consequences later in the adult life are still under study, and also in this field, it is still unclear what type and what timing of exposition are mainly related to impairment of the cognitive function and the physiological maturation of the brain. A major role may have the diffuse neuro-inflammation; children chronically exposed to pollutants have been noted an increment of the oxidative stress and an increase of the inflammatory response, also expressed in terms of change in the modulation of serum cytokines and chemokines [69]. That seems to be related to modification of the central nervous system structural and volumetric responses [70].

The inflammation response in the lungs, which are the organs primarily affected by air pollutants, also produces circulating cytokines that act on the cyclooxygenase 2 in the vessels of the brain endothelium, leading eventually to central nervous system inflammation [68].

The accumulation of misfolded proteins, the harmful effect of pollutants on the neurovascular unit and the production of autoantibodies to neural and tight-junction proteins are another possible mechanism of damage. Those findings have been observed in children chronically exposed to air pollution. Particulate matter ($PM_{2.5}$ and PM_{10}), O_3 and NO exposure in childhood may be linked to development of multiple sclerosis and Parkinson's and Alzheimer's disease in adulthood [71, 72]. This is mainly due to the oxidative stress and to the brain inflammatory imbalance, also involving the gene responsible for the inflammatory response in the central nervous system, and the reduced protective response of prion protein that is altered by chronic exposure to air pollutants [69]. That processes could lead to neurodegenerative process, as well as to psychological distress and psychiatric conditions, able to give short- and long-term consequences, affecting also the performance of children at school and their behaviour.

Interestingly, some of the studies that tried to evaluate the progresses at school of children constantly exposed to pollutants (at home, during commuting, at school or in a combination of those) gave alarming results. Also when confounding factors (environmental and behavioural variables that are commonly associated with air pollution and that are known to interfere with cognitive maturation in children as social environment at home and at school; parental psychological status; parents' education, unemployment and occupation; breastfeeding; diet; parental smoking at home and/or during pregnancy; and noise [73]) were excluded, children studying in more polluted schools show a significant reduction of the cognitive development particularly affecting working memory, superior working memory and inattentiveness [74], while children exposed to black carbon levels show a deterioration in their vocabulary, composite intelligence, memory construct and learning [75].

Cognitive delay can potentially lead to life-changing consequences especially during primary school age; if air pollution compromises the process of learning, it is unclear if brain impairment can be recovered in a later stage of life. Besides, at this age, children receive key information and develop mental abilities that are crucial in their future educational life. Aside from the clinical manifestations that related air pollution to brain health, youngsters would miss opportunities that are part of their wellbeing and impact on their future quality of life.

10.3.4 Psychological Effect

Living in a polluted environment can be a cause of stress for a number or reasons. For those families that are aware of the problem, pollution can impose behavioural changes aimed to protect. Life change in itself nowadays is not considered a source of emotional concern; starting new, healthy routine can result on the contrary in an improved mood and better attitude towards life. Undesirable events, yet, are more likely to produce distress, and in this sense a change in the routine can result in psychological distress [76]. Reducing the hours spent in outdoor activities, keeping windows or doors closed even during the warmest months, avoiding busy road or travelling during rush hours and wearing personal protective equipment are some of possible behaviours that can affect children's quality of life. What may be useful to prevent environmental damages, in this case, can impact on the mental wellbeing of a child. In addition, air pollution can be a cause of physiological concerns in adults also resulting in change in personal behaviours [77], and that ultimately can lead to distress in children.

Although it is difficult to evaluate the psychological burden of air pollution (because of the many confounders that often are associated with it and that can be cause of psychological distress in themselves and because "psychosocial stress can cause symptoms similar to those of organic mental disorders" [78]), organic causes have been investigated and correlated with mental health issues in children exposed to air pollutants.

A direct link between oxidative stress in nervous tissues and brain inflammation—that can be also attributable to air pollution (see Sect. 10.3.3)—are possible findings in patients affected by anxiety. In adults short-term exposure to $PM_{2.5}$ was strongly correlated to exacerbation of anxiety [79]. Higher concentration of interleukin-6 as the one that can be found in children chronically exposed to air pollution can cause the reduction of the volume of hippocampus that has been observed in major depressive disorders [69]. Some findings suggest an increment of the use in children under the age of 18 of medication for psychiatric disorder, including sedative, sleeping pills and treatments for schizophrenia and severe acute psychosis, during increment of the level of PM_{10} and NO_2 [80]. Exposure to airborne toxins and traffic-related pollution is also correlated with onset of autism spectrum disorder [81]. Moreover, air pollution can aggravate underlying chronic condition, creating an indirect cause of stress and possible mental disorders [78] like anxiety or depression.

Because of the high number of confounders, the relationships between air pollution and behavioural problems in children or organic causes that can be attributable to those remain also unclear. Yet, exposure to traffic-related pollutants during the early childhood and the perinatal life has been allegedly associated with higher hyperactivity scores in children, as well as anxiety, depression, attention disorder and antisocial/delinquent behaviours [12, 82, 83].

10.3.5 Immunity and Infections

Although it is becoming clear that air pollution can increase the risk of respiratory infections (see Sect. 10.3.10), it remains unclear whether air toxins can act on the active and passive immunological mechanisms in children, so to lead to higher risk of infections from other focuses. Airborne toxins are known to interact with the immune system on the human body and to trigger the inflammatory response [8, 84, 85]. Air pollution is known to increase level of IgA, IgM and C3c, to reduce level of IgG and to contribute to the weakening of the immune system [86], which can eventually lead to the development of systemic infections.

Pollutants impair the activity of the cilia and reduce the normal clearance of secretions from the airways and inhibiting the activity of the macrophages in the alveoli [87, 88], increasing the likelihood of penetration of microorganisms in the respiratory tree. Although this mechanism is among the causes of air pollution-related respiratory infections, it opens also the doors to microorganisms, like the mycobacterium species, responsible of systemic infections in immunocompetent and immunosuppressed individuals.

Air pollution increases also the risk of infections related to the upper respiratory tract like tonsillitis, pharyngitis, eustachian tube dysfunction and otitis media [89–91]; those are very common conditions in children and among the first causes of medical interventions in paediatrics. The mechanisms are still unknown although it has been hypothesized that chemical irritation, the impairment of the mucociliary clearance secondary to exposure to toxins, the allergic inflammation and the immunosuppressive effect of airborne pollutants may have a major role. [92].

Another possible cause can be the alteration of the normal oral bacterial flora [93] that eventually leads to the development of pathological strains that increased the number of carriers and of individuals with active symptoms of upper respiratory infection. In fact, studies suggest that air pollution can be associated with an increased colonization of bacteria in the upper airways like the *Staphylococcus* species [94] and group A *Streptococcus* [95]. The latter one can indirectly lead to systemic diseases in children as rheumatic fever, postinfectious nephritis and sepsis.

10.3.6 Gastroenterological

Poorly soluble contaminants are ubiquitous in polluted environment. These pollutants can deposit in the oral cavity and in the upper respiratory tract. In adults and older children, cough and expectoration reduce the systemic adsorption of chemical and particulate matter. Part of these particles however is swallowed into the gastrointestinal tract as part of their clearance process. This is particularly true in small children that are not able to control the cough reflex, to cough out the sputum coming from expectoration or to properly blow the mucus out of their nose, resulting in an increased swallowing of secretion from the respiratory tract.

Pollutants can interfere with the natural homeostasis of the bowel and of the other organs of the human digestive system, resulting in an increased level of gastrointestinal disorders.

The association between ingestion of pollutants and diseases in children remains unclear. However, some studies suggest an increased number of episodes of gastroenteritis and gastrointestinal disorders in children exposed to air pollution [96, 97] and an impairment of the liver detoxification function in children coming from heavily polluted areas [98].

It is reasonable to think that children can be more affected than adults, as the effects of gastroenteritis are normally more severe in this age group that is normally at higher risk of dehydration and metabolic imbalances as ketosis. Besides the higher concentration of pollutants that can be swallowed or inhaled by small children can results in an increased level of toxins, whose clearance is reduced by the allegedly changed detoxification capacity of the liver.

Chemicals and particulate coming from air pollution deposit in the soil. Pica behaviours—common in children—can increase the risk of gastrointestinal exposure to these toxins and therefore of gastroenteritis, both by augmenting the overall amount of air pollutants ingested and increasing the likelihood of enteric infections secondary to microorganism ingestion.

10.3.7 Cardiovascular

Although the effects of air pollution on the cardiovascular system in the generic population are well known, the effect on children has not been exhaustively studied. Cardiovascular diseases are not common in this age group, and the long-term effects of minor alteration of heart and vessels are difficult to assess. A major predictor for future cardiovascular damages in children is blood pressure. Increments in the normal values of blood pressure in children are unusual and are mainly due to secondary causes as renal or endocrine diseases.

Children exposed to air pollution seem to present with higher values of systolic pressure. The number of studies available is limited, and some results are controversial [99]. This may be related also to the fact that the type of pollutants considered varies, as well as the timeframes considered in the studies. However, children exposed to ultrafine particles (UFP), PM_{10}, SO_2, NO_2, O_3 and

CO seem to be associated with increased level of arterial blood pressure and with hypertension [100, 101], both secondary to short/same-day exposure and chronic/long-term exposure.

Alterations of the cardiovascular system due to exposure to air pollution, which can potentially lead to major diseases in the adulthood, have not been proven. Nonetheless, some indirect indicators as the carotid arterial stiffness [102], increase in the level of plasma endothelin-1 [103] and endothelial dysfunction [104, 105] have been observed in children subjected to air pollutants. The presence of other risk factors as obesity can amplify the effect of air pollution on children's blood pressure [106].

Besides, as the association between long-term exposure to air pollution and cardiovascular conditions is known [107], it is reasonable to think that exposure starting from the early life can augment the risk of heart diseases in the future life of exposed children, triggering as well the same process of atherogenesis since childhood [108, 109] (see also Chap. 17).

10.3.8 Metabolism, Endocrinology and Obesity

The oxidative stress secondary to exposure to air pollutants and the poisoning interferences of toxins inhaled with pollution can alter the fine balance of the endocrine system. Because of the complexity and the heterogeneity of the hormonal signals, disruptions can happen at any level. Although the endocrinological effect of air pollution is known and well documented (see Chaps. 20, 21, and 22), there are a limited number of studies in children.

On the other hand, children are particularly affected by the effect of hormones that regulate the different phases of growth, from the first stages of life up to the transformations that happen during puberty and the transition to the adulthood. In particular, the sexual maturation of children depends on a series of chemical signals that can be easily disrupted. Interestingly one of the few studies available in children demonstrates the effect of air pollution on the adrenal cortex function [110] in schoolchildren, which may be related to a reduced spermatogenesis later in life, suggesting a possible correlation with disorder of the reproductive tract later and air pollution exposure during childhood. The same adrenal hormones are involved in the bone metabolism and mineral accumulation, suggesting possible effect on the bone resorption that may be crucial in a rapidly developing body.

The same response to insulin in children can be affected by air pollution. Insulin is mainly known to be involved in the regulation of carbohydrates and lipids metabolism, but its effects indirectly shape and design the architecture of the human body. Insulin resistance has been proven in primary schoolchildren exposed to long-term effect of NO_2, $PM_{2.5}$ and PM_{10} [111]. Although it is difficult to establish the real connection with growth restriction and insulin resistance, this finding may suggest an increased risk of development of diabetes later in life. A similar association has been observed in children with new onset of type 1 diabetes [112], suggesting a role of ozone and SO_4 in the triggering of this disease. As diabetes is a multifactorial disease, such associations are difficult to establish, and the number of studies is still

Table 10.2 Mechanism that could lead to obesity in children exposed to air pollution and road traffic

Direct	Indirect
– Direct obesogenic effect of air pollutants – Direct endocrine effect of air pollutants (can trigger obesogenic pathways) – Systemic inflammation (can trigger obesogenic pathways)	– Reduction of physical activities • Less time spent outdoor because of air pollution • Chronic disease related to air pollution • Stress related to air pollution – Reduction of walking or bicycling for transportation • Increased perceived danger coming from Traffic accidents Effect of air pollution – Change in diet • Higher intake of carbohydrates secondary to stress

limited. However, some papers point out how pollution can deteriorate the control of glucose metabolism and the insulin sensitivity secondary to oxidative stress and inflammation, although the effect on metabolic control in children has not been proven so far [113].

On the other hand, air pollution can indirectly lead to alteration of the diet and of food assumption (see Table 10.2) and to reduced physical activities that can eventually lead to increment of the body weight and of BMI in children [114]. Overweight and obesity are a known risk factor for metabolic dysfunctions and a possible cause of type 2 diabetes.

Moreover, studies suggest that some pollutants can have a direct role in the onset of obesity in children: some toxins as the phthalates can be directly obesogenic [115], while others can affect the endocrine system creating disruptions that can eventually lead to obesity.

To close the circle, obesity can worsen the respiratory symptoms of respiratory conditions (as wheeze or respiratory tract infections) triggered or caused by air pollution [116]. These pathways are highly demonstrative of the complex interaction between air pollution and the endocrine system in children, as well as the multitude of factors that can affect their growth and the development of diseases in the adult life.

10.3.9 Haematology

Impact on air pollution in schoolchildren has been described as early as the second half of the 1970s [117]. Although the relationship between exposure to pollutants and changes in the cells line in the bloods remains unclear, some studies suggest a possible reduction of the level of red cells and haemoglobin, with an increment of the number of white cells and platelets in the blood [118]. This may be in line with an inflammatory response secondary to air pollutants or to an increased incidence of respiratory, non-respiratory or systemic infections. On the other hand, it has been observed that exposure to lead in ambient air can increase the level of lead in children's blood and it has been associated in this age group with anaemia [119, 120].

Black smoke, NO_2 and SO_2 have also been considered possible causes of anaemia in children [121].

Although the data are limited, the observed association with air pollution and leukaemia in children [122–124] suggests that all the cell lines can be affected by air pollutants.

Haematological diseases can have a strong impact on young people affecting their growth and their quality of life; further studies in this field are therefore needed.

10.3.10 Respiratory

The respiratory system is a major target for air pollutions at all ages. The combination of environmental air, chemicals and particulate is actively inhaled in the pulmonary system. Gases are exchanged in the alveoli lumen; soluble and insoluble substances can be absorbed or can be deposited in the respiratory tract. Toxic components can cause damages and inflammation locally or can enter the bloodstream and diffuse systemically. Those same toxins can interact with the normal physiology, genetic expression, biochemical reactions and metabolism of the organism.

Children's lungs are physiologically more exposed to air pollutants because of the immaturity of their respiratory system and because of the higher exchange rate secondary to the higher respiratory frequency. It means that they are at higher risk for respiratory symptoms and conditions related to air pollution and for systemic diseases related to inhalation of air pollutants.

The effect of air pollution in children has been largely investigated [3, 5, 8, 9, 12, 84, 125–137]. The major effects and the possible mechanisms are summarized in Table 10.3.

The main effects observed in children are on development and exacerbation of asthma, on airway inflammation, on lung development and function and on frequency and recurrences of chest infection. Children can be exposed to outdoor and indoor pollution; different toxins seem to be associated with respiratory outcomes. In addition, respiratory symptoms in children strongly correlate with the type of exposure. The number of studies available in this field allows a grossly classification secondary to the different parameters considered:

a. *Early-life exposure*: It considers mainly the level of pollution at the birth address and at the time of birth.
b. *Long-term exposure*: It considers the overall exposure in children that lived for in a polluted area for a long period. The timeframe varies in the different studies and can be associated with the exposure since birth, differentiating those children that have been always exposed to air pollution from those that moved from a less polluted area to a more polluted area and vice versa. It is normally associated with a continuous exposure to pollutants, although in most studies, where continuous/daily monitoring of the pollution was not available, it may refer to annual, seasonal, monthly or weekly mean of exposure.

Table 10.3 Effects of air pollution on the respiratory system in children

Respiratory conditions or symptoms	Alleged responsible pollutants	Alleged type of exposure
Asthma onset	NO_2	Early life
	SO_2	
	$PM_{2.5-10}$	
	O_3	
	Traffic-related air pollution	Continuous
	$PM_{2.5-10}$	Short exposure
	O_3	
Wheeze and exacerbation of asthma[a]	NO_2	Higher peak in previous hours/day/week
	NO_x	
	SO_2	
	$PM_{2.5}$	
	O_3	
	CO	
	NO_2	Facilitate viral infections to trigger exacerbation after 1 week from exposure
Nonallergic asthma	Traffic-related air pollution	Early exposure
Airway inflammation	$PM_{2.5}$	
Progression to adult COPD	$PM_{2.5}$	Chronic
	O_3	
Decrease in lung function	SO_2, NO_2, NO_x, $PM_{2.5-10}$, $PM_{2.5}$ absorbance, O_3	Current and long-term exposure
	$PM_{2.5}$	Subchronic exposure
	O_3	
	Black smoke, PM_{10}, NO, CO	Acute exposure, affecting baseline function
	O_3	Acute exposure
Respiratory tract infection	NO_2	Short-term exposure
	$PM_{2.5}$	
	O_3	
	Solid and biomass fuel	Indoor, continuous/recurrent
	NO_2	Long-term exposure, annual average
	$PM_{2.5}$	
	$PM_{2.1}$	
Otitis media	Tobacco smoke	Passive environmental smoke; continuous (recurrent episode of otitis and earache)
Pharyngitis	Fungi	Indoor, short term
	Tobacco smoke	Passive and active smoking
Rhinitis and olfactory function	NO_2	Early life
	SO_2	
	PM_{10}	

[a]Include increased use/frequency of relief medication

c. *Short-term exposure*: It refers to children exposed to air pollution for a limited number of hours, days or months, according to different studies. It can refer also to children partially exposed to air pollution as those that commute from a non-polluted area to a highly polluted area (e.g. school, outdoors activities) and therefore are exposed to pollutants only for few hours per day/week.

d. *Same-day exposure*: It considers the immediate effects of air pollution on the respiratory system. Respiratory symptoms in this case are expression of an acute exposure to pollutants
e. *Home address exposure*: Consider the exposure at home. Children are likely to spend evenings, nights and weekend at home, when the level of pollution is normally lower. However, the number of hours spent at home is normally considered more than the one spent in other locations. Most studies consider the level of pollution in the different neighbour to evaluate the baseline exposure, especially when outdoor and school's hours' exposure is considered. It also helps to differentiate among those children that are constantly exposed to pollution (at home, because their home is in a polluted area, and at school) from those exposed to pollution only for a limited number of hours a day (at school and in other activities, because they live in a less polluted area). A common parameter to evaluate traffic-related air pollution is the distance from a busy road, while different monitoring techniques are considered to evaluate the mean exposure in different town's areas
f. *Commuting exposure*: Evaluate the time children spend commuting. Children living and going to school in relatively non-polluted areas can be even so exposed to pollution while moving on the traffic (car or public transport) or while walking or biking.
g. *School address exposure*: Children spend a considerable number of hours at school; children that attend school located in more polluted areas are more likely exposed to the effects of outdoor air pollution. In addition, the quality of indoor air in schools can be poor, because of the possible contaminants in the air (e.g. chalk's powder, fumes from the kitchen, heating), the number of people attending the building (increase the formation of dust, the spreading of respiratory infections) and the spreading of chemicals (e.g. cleaning solutions, printers' toners). Schools are more likely located near busy road than residential areas.
h. *Change of address*: Children can change address one or more times over the year. Most studies consider whether children included in their research have constantly lived in the same area for a given amount of time. It can help to differentiate from those children that have been constantly exposed to the same mean amount of pollution to those subjected to different concentrations of pollutants over the time.

Respiratory symptoms easily related to the different exposure to pollutants, as they are mainly expression of a direct damage of air pollution on the respiratory tract. Those same typologies of exposure are possibly linked also to other medical conditions associated with air pollution in children. However, the restricted number of studies in children limits this possible conclusion.

10.3.10.1 Asthma and Airway Inflammation

Asthma is a chronic inflammation of the respiratory tract characterized by reversible airway obstruction and hyper-responsiveness of the airways. The main symptoms of asthma are wheeze, increased work of breathing, dyspnoea, cough and increased secretion and mucus production. It is a relatively common condition in children even if its clinical presentation can be mismatched with other common cause of wheeze especially in infants and toddlers.

Air pollution is known to create a persistent inflammation of the respiratory tract, with an increment of the indicators of airway inflammation and oxidative stress, like pH, 8-isoprostane and cytokines [138]. Although the molecular mechanisms remain unclear, the correlation of chronic inflammation induced by indoor-, outdoor- and traffic-related air pollution with asthma is strong and consistent.

The burden of asthma in children is substantial. Symptoms can deeply affect the quality of life; asthma can affect school attendance, curricular and non-curricular activities, impacting thus on a child's career development; symptoms can present at night disrupting the sleep and therefore the daytime life; parents taking care of children with asthma may neglect their social and work life or other children not affected [139].

Asthma is a multifactorial condition. Genetic predisposition has a major role, but environmental triggers remain a key factor in its onset and relapses. Any cause of inflammation or irritation of the airways can potentially trigger exacerbations of asthma, including viral and bacterial infection and exposure to chemicals.

Considering the number of possible confounders, the real impact of air pollution on the development and exacerbation of asthma remains unclear, despite the number of studies available on this subject. Yet, although complex, the association with inhalation of air pollutants and asthma in children remains clear.

The complexity depends on the several factors that must be considered, as different pollutants and types of exposure can act on the different mechanisms involved in the onset and exacerbation of asthma in children: oxidative stress, IgE-mediated sensitisation, airway inflammation, lung function, airflow obstruction, airway responsiveness and the use of medications and preventers [133, 140, 141]. On the other hand, cleaner air seems to be associated with a reduction of the airway inflammation and to an improvement of the lung function [142].

From a clinical point of view, it may be difficult to isolate a single chemical or particulate related to asthma: exposure to air pollution is rarely selective in children, and specific prophylactic measures able to reduce the exposure to single cadre of pollutants are limited.

However, understanding the cause of onset and exacerbation of airway inflammation and asthma could help clinicians in their clinical management of children with wheeze improving: the likelihood of diagnosis, the decision to start a pharmaceutical prophylaxis, the management of acute asthma also in community-based health facilities, the health education of children affected or presumably affected by asthma and the lifelong consequences of asthma like the development of chronic obstructive pulmonary disease in adulthood, which may be aggravated, caused or accelerated by the same air pollutants [143].

Onset of Asthma

Small children presenting with recurrent wheeze may represent a clinical challenge for paediatricians who should decide if and how to treat acute symptoms and whether to start a chronic treatment and prophylaxis.

As a multifactorial disease, asthma may remain latent in children or may present with blurred clinical symptoms that may mimic—especially in the youngest—other common clinical conditions as upper and lower respiratory tract infections. Children presenting with wheeze may or may not be affected by asthma, and children with asthma can present with wheeze with or without a common viral infection. For this reason, it is difficult to evaluate the effect of air pollution in children below the age of three when studying the incidence and prevalence of asthma [144] and consequently the role of air pollution in the onset of new cases.

Many studies however show a possible association with air pollution exposure and onset of asthma in susceptible children. Early-life exposure to NO_2, SO_2, $PM_{2.5–10}$ and O_3, early-life [145] and continuous exposure to traffic-related air pollution [146, 147] and maternal smoking during pregnancy where the foetus is successively exposed to air pollutants [148] may be the major factors, with an association of 1.5–2-fold increased risk of lifetime asthma for children exposed to NO_x [127], with children living close to busy road and freeways being at higher risk [134].

New onsets of asthma in children coming from lower-income urban areas consequently to early-life exposure to NO_2 [149] and to short exposures to high concentration of O_3 and $P_{2.5}$ have been also noted [150]. However, exposure to NO_2 remains a major suspected cause for the development of asthma observed in schoolchildren [151].

These findings suggest that environmental hygiene remains a key factor in the prevention of the onset of asthma in children [152, 153]. Moreover, studying and monitoring [154] the type and amount of exposure to air pollution can help to map the population of children at higher risk to develop asthma and to create individualized risk profiles.

An environmental anamnesis is crucial, and clinician should consider information coming from a child's social history and from public health sources when considering a diagnosis of asthma and in designing a therapeutic and prophylactic plan for their little patients.

Exacerbation of Asthma

Several studies underline the relation among short-term and chronic exposure to air pollution with the relapses of episodes of acute asthma [125], highlighting how acute exposure is a stronger cause of exacerbations than long-term and early-life exposure [155]. Children with asthma already suffer from a chronic inflammation of the respiratory tract and for this reason are more vulnerable to the irritant, inflammatory and oxidant effects of air pollutants. Air pollution contributes to increment the level of airways inflammation, lowering the threshold of bronchi response and increasing the likelihood of relapses: pollution can adjuvate external or internal stimuli known to trigger asthma like viral infections or exercise. In addition, fluctuations of the concentration of chemicals and particulate in the indoor or outdoor air can trigger per se asthma attacks in children.

As genetic polymorphism may be related to lifetime asthma [156], it is not surprising that children with concomitant conditions can be at higher risk of relapses: in fact, children with atopy seem to be more prone to develop wheeze secondary to short-term exposure to air pollution [157].

Exacerbations of asthma in children known to be asthmatic are mainly correlated with short-term exposure to pollutants and with an increase of the concentration of chemicals and particulate in the air. Interval symptoms or chronic symptoms of bronchitis—namely, daily cough for three consecutive months and congestion or phlegm for at least three consecutive months—have been reported in children exposed to air pollution [158], suggesting a stretching of the acute effects related to short-term exposure and exacerbation of asthma, possibly related to the oxidative stress and continuous inflammatory response.

Thus, children constantly subjected to pollution may present a chronic inflammation of the airways, but fluctuations from baseline concentration of toxins are likely the main cause of acute episodes of asthma: studies show that daily high peak of NO_2, NO_x, SO_2, $PM_{2.5–10}$, O_3 and CO is related to an increment of relapses of the symptoms of asthma the days following the exposure [159], increments of CO, NO_2 and partially SO_2 concentration are often related to episodes of exacerbation 2–3 days after the exposure [160–162] and a cumulative effect lasting over the days after the exposure has been described [163].

Seasonal variations of the concentration of pollutants also relate to an increment of acute cases and hospitalizations [164, 165]. Seasonal and short-term rise of pollutants results in a higher number of emergency department visits and also in big cities with normally low level of pollution [166]; the overall number of relapses, asthma medication use and hospital admissions for acute asthma remains higher in children acutely or chronically exposed to air pollution [160, 162, 167–169].

Other factors influence the reactivation of the disease in asthmatic exposed children, as the use of preventers. Children using inhaled or nebulized corticosteroids may be at higher risk of exacerbation secondary to acute (daily or hourly) increment of concentration of $PM_{2.5}$, PM_{10} and O_3 [170]; NO_2-facilitated viral infections [171], seasonal climate variation, air pollen concentration and second-hand smoke are known to trigger acute asthma [172, 173], the latter being outdoor and indoor air pollutants themselves.

For these reasons, environmental prophylaxis can reduce the number of relapses, emergency departments and hospital admissions. Health education in primary care settings can help to reduce the risk of indoor pollution that may lead to acute cases of asthma. Public health policies can help in reducing the fluctuation of air pollutants, decreasing the chance of inhaling higher concentrations of toxins. Mapping and monitoring the time and areas with peak in concentration of pollutants can help in creating forecast of request of medical attention in primary, secondary and tertiary care for acute asthma—grading from mild to severe. This can be useful both to plan and optimize resources and interventions and to develop public health policies aimed to reduce the number of relapses and hospitalizations.

10.3.10.2 Respiratory Tract Infections

Respiratory tract infection is a group of very common medical conditions in children. Those are communicable diseases directly caused by microorganisms. Some like pneumonia or bronchiolitis are strongly related to increments in hospital admissions and emergency departments' visits, ranging from mild to severe and life-threatening presentations. Several factors contribute to the development of a disease following a contact with the pathogens, among which are the type of organism, its virulence, the concentration of pathogens that enter in contact with the host and the active and passive defences of the host.

Although the microorganism in itself by definition is the sole direct cause of an infectious disease, these factors can act on the chain of events that lead from exposure to a pathogen to the development of an infectious disease in the host; those are indirect causes of the disease and may be the precipitating factors that lead to the clinical manifestation of the contagion. In other words, some pathogens may infect the host, which overcome its active and passive defences and have clinical manifestations, because of an external concomitant synergic cause.

Traffic-related air pollution and indoor fumes coming from the use of solid biomasses have been related to increased risk of upper and lower respiratory tract infection [5, 131, 171, 174–181]. The exact mechanisms are not known, although the most likely factors are listed in Table 10.4.

As discussed in Sect. 10.3.10.2, viral infections have been observed in children exposed to air pollution shortly before the onset of the symptoms, in some cases leading to secondary effects as exacerbation of asthma in sensitized subjects, being the two conditions strictly related.

Table 10.4 Possible mechanisms that contribute to the development of infectious diseases in children exposed to air pollution

Most likely factor	Possible mechanism or causing condition
Increment of the virulence	Pathogens as part of air pollution in themselves (e.g. fungi)
	Pollutants acting as carrier for pathogens
	Poor environmental hygiene leading to increased concentration of pathogenic organisms and strains
	Pollutants selecting pathogenic organisms' strains
	Selection of resistant and multiresistant pathogenic strains secondary to the dissemination of chemicals in the air (e.g. disinfectants)
Increased likelihood of contagion	Increased indoor activities with poor ventilation in crowd locations (schools, recreational areas) secondary to high outdoor pollution
Impairment of passive defences in the host	Irritation of the upper airways
	Reduced or impaired ciliary motility caused by pollutants
	Thickening of airways secretions
	Impairment of the clearance of the airways
	Exacerbation of asthma or other subacute condition that facilitates penetration and expression of the pathogen
Impairment of active defences in the host	Impairment or reduction of the immune system response (cellular and humoral mediated) in children exposed to the effect of air pollution
	Interaction with the immune response pathway in the epithelial cells of the respiratory tract
	Impaired activity of alveolar macrophages as reduced adherence to surfaces, ability to phagocytize bacteria and intracellular bactericidal processes
	Increased inflammation response
	Increased oxidative stress

Atypical infections related to air pollution as well have been described: this is relevant as microorganisms like *Mycoplasma pneumoniae* causing this type of disease are resistant to first-line drugs normally used in children to treat chest infections. *M. pneumoniae* is supposed to interact with metals present in fine particulate ($PM_{2.5}$), activating the inflammatory pathway and increasing the oxidative stress [182], which eventually lead to a defective response of the immune response. Therefore, children exposed to air pollution may be at higher risk, and second and third lines of treatment have always to be considered in symptomatic patients with poor clinical response to standard antibiotics.

Chronic exposure to traffic-related pollution and household pollution can be also at higher risk of opportunistic infections: tuberculosis has been positively associated with exposure to CO and NO_2 and solid fuel fumes in adults [183, 184] and in children exposed to tobacco smoke [87]. This is not surprising as tuberculosis can take advantage of the impairment of immune response in people exposed to air pollutants; it also means that in patients with latent or subclinical tuberculosis infection, air pollution can trigger relapse and acute up to life-threatening exacerbations of the disease. On the other hand, tobacco smoke can reduce the activity of pulmonary alveolar macrophages that are a major barrier to *Mycobacterium tuberculosis* infection.

Fungi can be also a component of air pollution, and higher concentration can be observed in areas with poor environmental hygiene like schools [185], especially those with a higher concentration of dust and mould like the ones with carpeted floors [186]. This can lead to minor symptoms as stuffy sinuses, sore throats, respiratory illnesses, lethargy, itchy eyes and runny noses [187] to a higher rate of serious pulmonary infections in susceptible children.

For these reasons, children at high risk for opportunistic infection, as *HIV*-positive patients, should be particularly monitored and protected from air pollution exposure.

10.3.10.3 Lung Function

Organs and tissues belonging to the airways are the more exposed to air pollutants and to their negative effects, in terms of growth and development. As we have seen, a chronic inflammation of the respiratory tract secondary to early, long-term and short-term exposure may be responsible for chronic obstruction and hyper-responsiveness, leading to increased episode of wheeze, cough and respiratory symptoms in general. However, long-term exposure may be responsible of permanent or long-lasting effects, in terms of maturation of the respiratory tract, like lung growth and alveoli differentiation [188].

The study of lung function in children gives reliable information on the development of the respiratory system in children. Forced expiratory volume in 1 s (FEV1), forced vital capacity (FVC) and peak expiratory flow (PEF) are the parameters most commonly used to measure the lung function in children [130]; their modification from reference range or from previous individual baseline recordings in children exposed to air pollution can demonstrate acute and chronic changes.

Several studies agree that exposure to outdoor and traffic-related pollution, as well as indoor pollution, produces negative effects in terms of lung function and consequently on respiratory tract development and maturation both in asthmatic and in non-asthmatic children, with some evidences of worse outcomes in children with concomitant conditions as atopy [126, 127, 130, 135, 189–203].

Because of the heterogeneity of the studies, the possible confounders and the different type of pollutants and exposure, the real extent of the damages on developing airways remains unclear.

Further studies are needed to evaluate: what are the population at higher risk; what chemicals and particulate are more likely to interact with a developing respiratory system; what damages are caused by which toxin; which damages are permanent and which are reversible and by which extent and what type of exposure to the single component (e.g. high peak, short term, long term, continuous, sporadic, early) causes permanent or reversible damage.

As the respiratory effect of air pollution in childhood can persist in the adult life increasing also the risk of respiratory disease of the older age [5], these information are crucial. Especially in children, they can be useful to establish which actions—in terms of individual prevention and public health interventions—can be more effective to reduce the risk of lifelong effects on the respiratory tract.

10.3.10.4 Upper Respiratory Effects: Otitis, Rhinitis and Olfactory Function

The upper airways are directly exposed to the higher concentration of pollutants in the breathed air; moreover, the mechanic defences like nasal cilia and mucous, although preventing the deep inhalation of particulate and toxins, increase the chance of prolonged interactions between toxins and tissues in nose, throat and ears.

As we have seen (see Sect. 10.3.5), children exposed to ambient air pollution and particularly to indoor passive tobacco smoke show an increased risk of eustachian tube disorders that can lead to recurrent episodes of earache and otitis media, one of the commonest requests for medical attention in children. Chemical irritation as well as the effect of biological pollutants like fungi of the upper respiratory tract can also result in acute episodes of pharyngitis, while environmental tobacco smoke has been associated with chronic sinusitis, chronic disease of tonsils and adenoids and chronic laryngitis [204].

Irritation of nose and throat mucosae is a common and widely experienced effect. However, a growing body of evidence link the early-life exposure to pollutants as NO_2, SO_2 and PM_{10} to the onset of lifelong allergic rhinitis [205] and olfactory dysfunction. Moreover, the neurological damages of air pollution (see also Sect. 10.3.3) have been observed also in the olfactory bulbs, which may represent a main target for pollutants. Olfactory bulbs exposed to pollutants present endothelial hyperplasia and accumulation of ultrafine particles in the endothelial cytoplasm and basement membranes [206]; these findings are consistent with a damage of the olfactory function in children and support the idea of severe neurological damages in subject exposed to air pollution since early life.

10.3.11 Allergies

As we have seen (see Sect. 10.3.10.1), air pollution is strongly associated with asthma onset and exacerbation of asthma, particularly in children with atopy. Allergic asthma is a relatively common condition, and it is often associated with other allergic conditions in the youngest as rhinitis, hay fever, eczema and dermatitis. Airborne particles, like pollens, spore and fungi, can cause allergic reactions. Although most of them come from natural source, biological contaminants are pollutants in themselves; human activities can increase their concentration [207] in urban and rural areas or increase the chances of contact for sensitive children. Example of that are cultivation and agriculture or industrial food processing up to the alleged climate changes due to the same human activities and demographic expansion.

Pollutants can bind pollens and airborne biological allergens, working as carriers and increasing their effects on human beings; air pollution on the other hand can increase the allergenicity of pollens, secondary to morphological changes, increased allergenic proteins release and exposure and pollen release in the atmosphere as a response of the vegetation to environmental conditions [208–210] that facilitate the IgE-mediated reaction in sensitized children.

Allergens coming from natural sources are directly linked to allergic response, but there is also a strong association between long-term exposure to chemicals and air pollutants like particulate matter, NO_2, SO_2 and ozone and atopic diseases like asthmatic bronchitis, bronchial hyper-responsiveness, rhinitis, hay fever, eczema and sensitization [211–213].

Air pollutants cause airway mucosal damage and impaired mucociliary clearance increasing the concentration of inhaled allergens and therefore their presentation to the cells of the immune system, promoting consequently airway sensitisation. Particulate matter and ozone are also related to oxidative stress that leads to airway inflammation, hyperreactivity and reduction of the mechanisms that are known to prevent allergies. In addition, diesel exhausts coming from traffic-related air pollution seem to directly stimulate the IgE synthesis, which is the basis of the allergic reactions [132, 212, 214, 215].

It is still unclear if air pollutants can increase the chance of allergies' onset in children not known to have atopy, directly causing the development of the allergic sensitization [216], although strong association has been found [211, 217]. Some children may be more susceptible to the oxidative effects of air pollution secondary to their genotypes [84]; that may explain why some individuals develop allergic diseases later on in life, but it also explicates why the effect of air pollutants on the modulation of the immune response is so complex and difficult to ascertain.

10.3.12 Cancer

The mutagenic action of some of the many components of air pollution is well known and well documented, as well as the link with carcinogenesis (see Chap. 23).

Several airborne chemicals and particulate matter have been associated with respiratory and non-respiratory malignancies; for some of them, the causal effect has been demonstrated, while exposure to others shows an increased risk and odd ratio for the development of cancer.

Yet, it remains unclear whether the exposition to air pollutants during early life and infancy can increase the risk of developing cancer in later childhood and in adulthood. Most of the studies try to link traffic-related air pollution or environmental tobacco smoke exposure in childhood to some types of cancer. Although the data are extremely limited, there are mild indications of second-hand tobacco smoke and lung cancer in adult life and non-respiratory childhood cancers, particularly leukaemia, lymphoma, brain and central nervous system tumours, Wilms' tumours, acute lymphatic leukaemia and central nervous system cancers [5, 218]. The strongest correlation seems to be related to motor traffic emissions, which may be involved in the aetiology of childhood leukaemia [124, 219], Hodgkin's disease [220] and other childhood cancers [123], although results are still controversial due to the partial information available and the possible confounders [220, 221].

On the other hand, inhalation of chemicals and particulate matter is only a possible way of exposure, and some of the toxins present also in air pollution may give cancer in children or adults otherwise exposed. This assumption implies also that air pollutants deposited on surfaces can enter in contact with exposed children (e.g. skin contact, ingestion) potentially leading to DNA damages and mutations during the same childhood or later in life.

For some components of air pollution, a possible mutagenic activity has been observed, with increased level of biomarkers of genotoxic substances exposure, even if no clear association with cancer in childhood has been found [222].

However, epidemiological studies in this field are defective, as the complex mixture of toxins presents in air pollution represents a major limitation. The sole products of combustion are representative of the complexity in studying correlation between pollutants and cancer development [223], as identifying critical components in the mixture, their source and their exposure and the related effects is a challenge. This is particularly true when exposition happens during childhood, and the effects are observed decades later, when an assessment of all the possible confounders is not always possible or reliable.

Further studies are therefore needed, to better clarify the role of the numerous pollutants and their possible short- and long-term carcinogenic effects in children and young adults.

10.4 Diagnosis

From a clinical point of view, because of the numerous implications of early-life, short- and long-term exposure to air pollution in children and the possible effect on health both during childhood and later in the adult life, it may be crucial to evaluate which children are or have been exposed to pollutants and in which terms. There are no clear models at the moment that can be used in clinical scenarios and in clinical settings. The different methods used in observational studies may be appropriate in research, but can be difficult to use in the clinical practice.

Continuous monitoring of air pollution with station in proximity of residential areas and schools can offer a possible source of information, but data should always be available, and paediatrician may not be able to use that information in clinical decision-making. Community doctors should nonetheless be able to access to environmental data and to crosscheck individual demographic information to assess risk of individual exposure in single children.

As we have seen, this information may be critical in clinical practice to better define profiles of risk, to assist doctors and paediatricians in differential diagnosis, to promote a proactive approach aimed to prevention and prophylaxis of selected conditions, to promote healthy lifestyles and reduce the burden of air pollution and air pollution-related diseases in children.

Further information may come to laboratory tests, able to determine if a child has been exposed to air pollution and to specific pollutants, as well as the entity of that same exposure.

10.4.1 Possible Markers of Air Pollution Exposure in Children

The definition of markers of air pollution exposure could open a whole new chapter in the clinical practice, especially in children. As it is becoming more clear that air pollutants are involved or are direct cause of several and severe medical conditions, the assessment of exposure should become a common practice in clinical settings.

However, although the effects of pollution on human health have been known or suspected for a long time, this is a relatively new field of medicine, in terms of clinical and practical applications. The different time and form of exposures to air pollutants in children are a further limitation, also because of the many possible ways in which children can get in touch with toxins (see also Table 10.1).

For some pollutants, it may be possible to have a direct or indirect reading of the level of contamination, as the measure of concentration of heavy metals in water or in food commonly ingested by children of specific population, of the actual individual exposure to given pollutants or of the tissue or blood concentration in living individuals. While the first gives an esteem of the absorption and interaction with toxins, the former can give a direct reading of the level of contamination and therefore of the possible health effects.

Other approaches try to find those markers—as product of catabolism or effect-related molecules—that may be dosed in bloods and other body sources or samples and that may be direct or indirect expression of dose-related exposure to air pollutions and to specific toxins.

A number of chemicals and techniques have been studied; some are summarized in Table 10.5 [69, 83, 91, 98, 138, 141, 154, 224, 225].

It should be noted that some markers may result sensitive but not specific, as they merely represent the final effect of the exposure; this is particularly true for the marker of inflammation or of immune response or of oxidative stress.

Table 10.5 Some of the markers of exposure that may be used in research and in the clinical practice to ascertain if a child has been exposed to air pollution or to its effects

Marker or technique	Source	Pollutants and effect
Cotinine concentration	Serum	Environmental tobacco smoke exposure
Interleukin-6 Interleukin-8 Interleukin-10 Monocyte chemotactic protein-1 Tumour necrosis factor-alpha Interferon-gamma	Whole blood	Traffic-related air pollution exposure
Apolipoprotein E ε4 allele	Genetic test	Increased risk of brain function impairment in children with the allele exposed to air pollution
pH	Breath	Airway inflammation secondary to traffic-related air pollution
8-Isoprostane	Breath	Oxidative stress secondary to traffic-related air pollution
Interleukin-1 Interleukin-1 receptor antagonist Interleukin-2	Cerebrospinal fluids	Oxidative stress, inflammation, innate and adaptive immune responses in the central nervous system of children exposed to urban air pollution
Macrophage inhibitory factor	Serum	Inflammation in the central nervous system of children exposed to urban air pollution
Cellular prion protein	Cerebrospinal fluids	Accumulation of misfolded proteins in the central nervous system of children exposed to urban air pollution
Polycyclic aromatic hydrocarbons–DNA adducts	Umbilical cord white blood cells	Behavioural effects on children exposed to airborne polycyclic aromatic hydrocarbons
Lung radiology (X-ray and computed tomography)	Radio-imaging	Non-specific modifications in children chronically exposed to urban air pollution
[15N]Methacetin test	Urine	Reduction of liver detoxification capacity in children exposed to urban air pollution

This data, therefore, should be merged with information coming from individual demographic records and local epidemiological measurements.

10.5 Management and Prevention

The identification of those subjects at higher risk for air pollutant exposure is crucial in the management of the air pollution-related diseases in children, as well as the assessment of the type and level of exposition. For this reason, paediatricians should always consider the possible effect of air pollution when dealing with their little patients and ask about environmental anamnesis when taking a medical history. The number of tests available so far to routinely evaluate the level and the effect of exposure to air pollutants is limited, and as we have seen, they are not specific for given pollutants. Data on local level of pollution are available only in selected areas and that same information may not be useful to properly evaluate the individual exposure.

The use of mobile devices may be useful to map the daily changes in exposure, tracking children movement in the different urban areas (e.g. home, school, playground) and the air quality index in that same places. That would give a punctual picture of the hour-by-hour individual exposure, provided that children can carry smartphones and technologies with them and that a reliable monitoring of the quality of air is obtainable. Those technologies are already potentially available, but a widespread use of them is still far to reach and has practical implications that need to be overcome.

On the other hand, children living or studying in metropolitan areas, in urban areas and in places with a high level of emissions (urban, rural or semirural areas close to factories or plants) should always be considered potentially at risk.

Because of the number of acute, chronic and lifelong consequences of air pollution, the patient at risk has to be monitored during the time, and the correct information should be obtained and given in order to reduce the health risk associated with air pollution. In a patient-centred approach, doctors should improve the level of communication with children and families, aiming to a long-term relationship that could help to improve the outcome of health education.

The role of prevention, in fact, is essential, as some of the effect of air pollution may not be reversible, while others may emerge later in the adult life, when the damage has already been done.

Besides, prophylaxis and prevention at the moment are the real clinical measure that paediatrician can use to reduce the risk of adverse health effect secondary to air pollution in children.

In order to achieve that, the key actions for the management of air pollution-related diseases in children are:

a. Detecting, measuring and monitoring air pollution
b. Protecting children from pollutants
c. Cleaning up of the areas where children live or spend most of their time
d. Reducing indoor and outdoor exposure promoting behavioural changes and public health interventions

The monitoring of outdoor and indoor air pollution in those areas where children live, study and play [154] is essential. As we have seen, children are peculiar and are particularly exposed to air pollution and to its effects when compared to adults. Detection of sources of air pollution and of the characteristics of the pollutants and their diffusion in the air should be taken in account, to produce reliable information to assess individual exposures. For those reasons, specifically designed tools, tailored to children's social, behavioural and physiological specificities, should be engendered. Indoor and outdoor detectors for the main pollutants should be in use at home and at school, and those data should be available to community paediatricians, public health doctors and policy-makers. Children at higher risk as the ones with atopy or respiratory conditions should be constantly monitored so to create individualized plan of intervention aimed to reduce the chance of relapses and severe exacerbations. Alerts on individual bases should be sent during peaks of concentration of chemicals, particulates and biological pollutants. Those same alerts should take into account the place where the child lives, where he or she attends school and extra-scholar activities and the route he or she follows while commuting. The average level of exposure (weekly, monthly and annually) should be known and recorded on personal files.

In children at high risk and in those acutely or chronically exposed to high level of air pollution, personal protection devices should be considered. Face masks are a possible option, provided that their technical specificities are adequate so to effectively block fine and ultrafine particulate and hazardous chemicals. Because they may be not well tolerated by children, this equipment has to be specifically designed for children, taking in account their behavioural and physiological peculiarities.

Schoolrooms and bedrooms at home can also be screened to reduce the amount of pollutants coming from outdoor sources. At the same time, the ventilation of those ambient where children spend most of their time, especially schools [187], has to be adequate, so that clean air can clear the ambient reducing individual exposure and therefore the need of personal protection devices.

Because of the many health hazards secondary to indoor pollution, the best prevention in this sense is to reduce the source of pollutants. Second-hand tobacco smoke is a major cause of indoor pollution, and for this reason, adults should never smoke in places with children, should avoid smoking at least 10 min before entering a place where children are and should not wear or bring in clothes they had on while smoking (exposure to third-hand smoke) [226–228].

Cleaning up the air in ambient where children live, reducing the level of pollutants and consequently the amount of chronic and acute/peak exposure, is proven to improve the health condition of children and their quality of life reducing at the same time school absences, the number of hospital admission and medical intervention [136]. The reduction even of small level of air pollution indeed reduces the burden and the respiratory symptoms in children [229]. This can be achieved locally with the implementation of air cleaner devices that may be mounted at school and those indoor places where children spent their time. Personal devices can also be used at home, especially at night to guarantee a clean ambient during sleeping time. Home-based devices can be used in single rooms, like children's bedrooms, to assure a good quality of air especially for those houses close to busy road even in the night-time or near industrial areas.

Although social, extra-scholar and sport activities are extremely important in children, the amount of pollutants that they can inhale during the time they spend outdoor can have dangerous consequences on their health. Reducing the exposure, limiting the hours spent outdoor when the level of traffic and of pollutants in general is higher, should be considered. This also implies that traffic and emission should be strongly regulated in areas where children study, play or practise outdoor activities, especially during the time of the day and the year when those places are busier.

Conclusions

Air pollution is a major issue in children as they are particularly exposed to its effects when compared to adults. Although several studies highlight the damages of air pollutants to the airways, little is known about the other effect on children's health and on the lifelong consequences of early exposures to these elements.

A growing body of evidences yet show how the damages are not limited to the respiratory system and some of the effects can cause serious and permanent damages.

As children grow, the cumulative effect of toxins present in air compound can emerge developing chronic conditions that may not be treated in the adult age.

Because of the peculiarities of children's lifestyles, the chance of outdoor and indoor exposure spreads across the different areas where they live, study or play, across the different times of the night or the day and across the different seasons of the year. In other words, children can be continuously exposed to background pollution and the effect of short-term peak in concentration of toxins. Some children may be more vulnerable than others because of underlying pathologies, sensitization, responsiveness to pollutants or genetic profile. On the other hand, every child can be seriously affected by the effect of air pollution, and it is currently not possible to estimate which level of exposure to which children and to which component can bring to acute, subacute or chronic consequences in susceptible and in healthy children. That has practical consequences in paediatrics and in public health, resulting in an increased number of hospital and emergency department admissions, help-seeking behaviours and ultimately health policies development and implementation.

Changes in air quality should be monitored, and individual exposure should be estimated for every child living in highly polluted areas or in places close to source of pollution like busy road, plants or industrial areas.

For these same reasons, information on air quality should be easily accessible to doctors, as peak of concentration may result in an augmented request of intervention and hospitalization; those same children known to have recurrent relapses should be alerted, so that adequate precautions can be taken to reduce the risk of exacerbation of chronic or subacute diseases.

Because any organ or system can be potentially affected also with serious and lasting consequences, doctors should always consider air pollution exposure while taking medical history in children; at the same time, children and families should be sensitized on the effect of pollution on health, and effective measure should be suggested and taken to reduce the chance of exposure.

A number of factors contribute to an increased vulnerability in children. Some are physiological, and no individual action is possible. Others can be modified in order to reduce the risk on health: behavioural changes are important, and health education can help to reduce the children's exposure to air pollution. Nevertheless, decision-makers should engage themselves in policies aimed to radically improve the quality of the air that children breathe in residential areas, at school and outdoor.

Besides, actions to reduce the impact of air pollution on children's health have been proven to be cost-effective [136], reducing the burden for families and for the society and diminishing the costs of acute medical interventions and those secondary to the management of chronic and potentially lifelong conditions.

These considerations are not limited to high-income countries with an elevated technological development profile. The burden in developing countries is still not completely known, although the high growth rate of big cities and small centres are likely to produce a poorly regulated level of pollution that may primarily impact on children which remain a major component of those societies. In addition, even in rural areas, children may be particularly subjected especially to indoor pollution, and action to reduce the exposure had to be fostered: behavioural changes in fact can highly reduce the devastating effects of these highly poisoning mixtures on children's health [230].

Policy-makers, paediatricians, emergency department and public health doctors have to be aware of the clinical consequences of air pollution in children, so that programme of prophylaxis and prevention can be planned on individual and community level. Families as well have to be informed of the risks of air pollution on their children's health and the potential sources of exposure, which are not limited to busy or undeveloped urban areas or to external factor on which they are unable to act. Parents should be in fact aware that outdoor pollution can affect also the quality of the air in their homes and of children's bedroom during the hours of sleep. Health education programmes should also emphasize the role of indoor air pollution and of second- and third-hand smoke, so that behavioural changes can reduce the chance of exposure to toxins coming from everyday activities.

Children should be protected when appropriate with personal protection equipment, and parents and educators should always make sure that the air children are breathing is good and the ambient in which they live, study and play cleaned up from pollutants.

Simple local actions, as well as major change in policies at national and international level, are needed in order to ensure a reduction of emission and of exposure, improving the air and the ambient where children live, their health and wellbeing and eventually their quality of life and of that of their families.

References

1. Doll R. Epidemiology of cancer: current perspectives. Am J Epidemiol. 1976;104(4):396–404.
2. Green RM. Cancer of the scrotum. Boston Med Surg J. 1910;163(20):755–9.
3. Kim JJ, American Academy of Pediatrics Committee on Environmental H. Ambient air pollution: health hazards to children. Pediatrics. 2004;114(6):1699–707.
4. Kelishadi R, Poursafa P. Air pollution and non-respiratory health hazards for children. Arch Med Sci. 2010;6(4):483–95.
5. Goldizen FC, Sly PD, Knibbs LD. Respiratory effects of air pollution on children. Pediatr Pulmonol. 2016;51(1):94–108.
6. Pinault L, Crouse D, Jerrett M, Brauer M, Tjepkema M. Socioeconomic differences in nitrogen dioxide ambient air pollution exposure among children in the three largest Canadian cities. Health Rep. 2016;27(7):3–9.
7. Ritz B, Wilhelm M Air pollution impact on infants and children: UCLA Institute of the Environment and Sustainability; 2013 [cited 12 Aug 2016]. Available from: http://www.environment.ucla.edu/reportcard/article1700.html.
8. Buka I, Koranteng S, Osornio-Vargas AR. The effects of air pollution on the health of children. Paediatr Child Health. 2006;11(8):513–6.
9. Rudnik J, Herman S, Pisiewicz K, Roszkowska H. Long-term effects on health of air pollution in children. Z Erkr Atmungsorgane. 1983;161(2):177–91.

10. Karr C. Addressing environmental contaminants in pediatric practice. Pediatr Rev. 2011;32(5):190–200.
11. Moya J, Bearer CF, Etzel RA. Children's behavior and physiology and how it affects exposure to environmental contaminants. Pediatrics. 2004;113(Supplement 3):996–1006.
12. Organization WH. Effects of air pollution on children's health and development: a review of the evidence. Copenhagen: WHO; 2005.
13. Pedersen M. Ciliary activity and pollution. Lung. 1990;168(1):368–76.
14. van Roosbroeck S, Wichmann J, Janssen NA, Hoek G, van Wijnen JH, Lebret E, et al. Long-term personal exposure to traffic-related air pollution among school children, a validation study. Sci Total Environ. 2006;368(2–3):565–73.
15. Lee JT, Son JY, Kim H, Kim SY. Effect of air pollution on asthma-related hospital admissions for children by socioeconomic status associated with area of residence. Arch Environ Occup Health. 2006;61(3):123–30.
16. Chakraborty J, Zandbergen PA. Children at risk: measuring racial/ethnic disparities in potential exposure to air pollution at school and home. J Epidemiol Community Health. 2007;61(12):1074–9.
17. Feretti D, Ceretti E, De Donno A, Moretti M, Carducci A, Bonetta S, et al. Monitoring air pollution effects on children for supporting public health policy: the protocol of the prospective cohort MAPEC study. BMJ Open. 2014;4(9):e006096.
18. Demoulin-Alexikova S, Plevkova J, Mazurova L, Zatko T, Alexik M, Hanacek J, et al. Impact of air pollution on age and gender related increase in cough reflex sensitivity of healthy children in Slovakia. Front Physiol. 2016;7:54.
19. Spencer-Hwang R, Soret S, Ghamsary M, Rizzo N, Baum M, Juma D, et al. Gender differences in respiratory health of school children exposed to rail yard-generated air pollution: the ENRRICH study. J Environ Health. 2016;78(6):8–16.
20. Dong GH, Chen T, Liu MM, Wang D, Ma YN, Ren WH, et al. Gender differences and effect of air pollution on asthma in children with and without allergic predisposition: northeast Chinese children health study. PLoS One. 2011;6(7):e22470.
21. Newman NC, Ryan PH, Huang B, Beck AF, Sauers HS, Kahn RS. Traffic-related air pollution and asthma hospital readmission in children: a longitudinal cohort study. J Pediatr. 2014;164(6):1396–402.e1.
22. Atkinson RW, Ross Anderson H, Sunyer J, Ayres J, Baccini M, Vonk JM, et al. Acute effects of particulate air pollution on respiratory admissions: results from APHEA 2 project. Am J Respir Crit Care Med. 2001;164(10):1860–6.
23. Sun HL, Chou MC, Lue KH. The relationship of air pollution to ED visits for asthma differ between children and adults. Am J Emerg Med. 2006;24(6):709–13.
24. Delfino RJ, Chang J, Wu J, Ren C, Tjoa T, Nickerson B, et al. Repeated hospital encounters for asthma in children and exposure to traffic-related air pollution near the home. Ann Allergy Asthma Immunol. 2009;102(2):138–44.
25. Villeneuve PJ, Chen L, Rowe BH, Coates F. Outdoor air pollution and emergency department visits for asthma among children and adults: a case-crossover study in northern Alberta, Canada. Environ Health. 2007;6:40.
26. Liu Y, Liu J, Chen F, Bilal Haider S, Wang Q, Jiao F, et al. Coal mine air pollution and number of children hospitalizations because of respiratory tract infection: a time series analysis. J Environ Public Health. 2015;2015:649706.
27. Bobak M, Leon DA. Air pollution and infant mortality in the Czech Republic, 1986–88. Lancet. 1992;340(8826):1010–4.
28. Saldiva PH, Lichtenfels A, Paiva P, Barone I, Martins M, Massad E, et al. Association between air pollution and mortality due to respiratory diseases in children in São Paulo, Brazil: a preliminary report. Environ Res. 1994;65(2):218–25.
29. Conceicao G, Miraglia S, Kishi HS, Saldiva P, Singer JM. Air pollution and child mortality: a time-series study in São Paulo, Brazil. Environ Health Perspect. 2001;109(Suppl 3):347.
30. Gouveia N, Fletcher T. Time series analysis of air pollution and mortality: effects by cause, age and socioeconomic status. J Epidemiol Community Health. 2000;54(10):750–5.

31. Landrigan PJ, Schechter CB, Lipton JM, Fahs MC, Schwartz J. Environmental pollutants and disease in American children: estimates of morbidity, mortality, and costs for lead poisoning, asthma, cancer, and developmental disabilities. Environ Health Perspect. 2002;110(7):721.
32. Smith KR, Mehta S. The burden of disease from indoor air pollution in developing countries: comparison of estimates. Int J Hyg Environ Health. 2003;206(4):279–89.
33. Aarnio M, Winter T, Kujala U, Kaprio J. Associations of health related behaviour, social relationships, and health status with persistent physical activity and inactivity: a study of Finnish adolescent twins. Br J Sports Med. 2002;36(5):360–4.
34. O'Lenick CR, Winquist A, Mulholland JA, Friberg MD, Chang HH, Kramer MR, et al. Assessment of neighbourhood-level socioeconomic status as a modifier of air pollution-asthma associations among children in Atlanta. J Epidemiol Community Health. 2016;71(2):129–36.
35. Bener A, Kamal M, Shanks NJ. Impact of asthma and air pollution on school attendance of primary school children: are they at increased risk of school absenteeism? J Asthma. 2007;44(4):249–52.
36. Romieu I, Lugo MC, Velasco SR, Sanchez S, Meneses F, Hernandez M. Air pollution and school absenteeism among children in Mexico City. Am J Epidemiol. 1992;136(12):1524–31.
37. Romieu I, Samet JM, Smith KR, Bruce N. Outdoor air pollution and acute respiratory infections among children in developing countries. J Occup Environ Med. 2002;44(7):640–9.
38. Mishra V. Indoor air pollution from biomass combustion and acute respiratory illness in preschool age children in Zimbabwe. Int J Epidemiol. 2003;32(5):847–53.
39. Khalequzzaman M, Kamijima M, Sakai K, Chowdhury NA, Hamajima N, Nakajima T. Indoor air pollution and its impact on children under five years old in Bangladesh. Indoor Air. 2007;17(4):297–304.
40. Nandasena S, Wickremasinghe AR, Sathiakumar N. Indoor air pollution and respiratory health of children in the developing world. World J Clin Pediatr. 2013;2(2):6–15.
41. Dionisio KL, Howie S, Fornace KM, Chimah O, Adegbola RA, Ezzati M. Measuring the exposure of infants and children to indoor air pollution from biomass fuels in the Gambia. Indoor Air. 2008;18(4):317–27.
42. Baumgartner J, Schauer JJ, Ezzati M, Lu L, Cheng C, Patz J, et al. Patterns and predictors of personal exposure to indoor air pollution from biomass combustion among women and children in rural China. Indoor Air. 2011;21(6):479–88.
43. Emmelin A, Wall S. Indoor air pollution: a poverty-related cause of mortality among the children of the world. Chest. 2007;132(5):1615–23.
44. Knopik VS, Marceau K, Palmer RH, Smith TF, Heath AC. Maternal smoking during pregnancy and offspring birth weight: a genetically-informed approach comparing multiple raters. Behav Genet. 2016;46(3):353–64.
45. Campbell AM. The effect of excessive cigaret smoking on maternal health. Am J Obstet Gynecol. 1936;31(3):502–8.
46. Huang JV, Leung GM, Schooling CM. The association of air pollution with birthweight and gestational age: evidence from Hong Kong's 'children of 1997' birth cohort. J Public Health (Oxf). 2016:1–9.
47. Mortimer K, Neugebauer R, Lurmann F, Alcorn S, Balmes J, Tager I. Air pollution and pulmonary function in asthmatic children: effects of prenatal and lifetime exposures. Epidemiology. 2008;19(4):550–7. discussion 61-2
48. Yang CY, Tseng YT, Chang CC. Effects of air pollution on birth weight among children born between 1995 and 1997 in Kaohsiung, Taiwan. J Toxicol Environ Health A. 2003;66(9):807–16.
49. Ritz B, Yu F, Chapa G, Fruin S. Effect of air pollution on preterm birth among children born in Southern California between 1989 and 1993. Epidemiology. 2000;11(5):502–11.
50. Rey-Ares L, Irazola VE, Althabe F, Sobrino E, Mazzoni A, Seron P, et al. Lower respiratory tract illness in children younger than five years of age and adverse pregnancy outcomes related to household air pollution in Bariloche (Argentina) and Temuco (Chile). Indoor Air. 2015;26(6):964–75.
51. Gilliland F, Avol E, Kinney P, Jerrett M, Dvonch T, Lurmann F, et al. Air pollution exposure assessment for epidemiologic studies of pregnant women and children: lessons learned

from the centers for Children's environmental health and disease prevention research. Environ Health Perspect. 2005;113(10):1447–54.
52. Röösli M. Non-cancer effects of chemical agents on children's health. Prog Biophys Mol Biol. 2011;107(3):315–22.
53. Kalkbrenner AE, Daniels JL, Chen J-C, Poole C, Emch M, Morrissey J. Perinatal exposure to hazardous air pollutants and autism spectrum disorders at age 8. Epidemiology. 2010;21(5):631.
54. von Ehrenstein OS, Aralis H, Cockburn M, Ritz B. In utero exposure to toxic air pollutants and risk of childhood autism. Epidemiology. 2014;25(6):851.
55. Netriova Y, Brezina M, Hruskovic I, Kacenakova M. Growth in children living in areas with air pollution. Cesk Pediatr. 1990;45(8):456–8.
56. Schlipkoter HW, Rosicky B, Dolgner R, Pelech L. Growth and bone maturation in children from two regions of the F.R.G. Differing in the degree of air pollution: results of the 1974 and 1984 surveys. J Hyg Epidemiol Microbiol Immunol. 1986;30(4):353–8.
57. Thielebeule U, Pelech L, Grosser PJ, Horn K. Body height and bone age of school children in areas of different air pollution concentrations. (a repeat study conducted in Bitterfeld and Berlin, capital of the GDR). Z Gesamte Hyg. 1980;26(10):771–4.
58. Liu C, Fuertes E, Flexeder C, Hofbauer LC, Berdel D, Hoffmann B, et al. Associations between ambient air pollution and bone turnover markers in 10-year old children: results from the GINIplus and LISAplus studies. Int J Hyg Environ Health. 2015;218(1):58–65.
59. Calderon-Garciduenas L, Mora-Tiscareno A, Francolira M, Torres-Jardon R, Pena-Cruz B, Palacios-Lopez C, et al. Exposure to urban air pollution and bone health in clinically healthy six-year-old children. Arh Hig Rada Toksikol. 2013;64(1):23–34.
60. Holick MF. Environmental factors that influence the cutaneous production of vitamin D. Am J Clin Nutr. 1995;61(3):638S–45S.
61. Ando M, Tadano M, Yamamoto S, Tamura K, Asanuma S, Watanabe T, et al. Health effects of fluoride pollution caused by coal burning. Sci Total Environ. 2001;271(1):107–16.
62. Qin X, Wang S, Yu M, Zhang L, Li X, Zuo Z, et al. Child skeletal fluorosis from indoor burning of coal in southwestern China. J Environ Public Health. 2009;2009:7.
63. Goyer RA. Toxic and essential metal interactions. Annu Rev. Nutr. 1997;17(1):37–50.
64. Berglund M, Åkesson A, Bjellerup P, Vahter M. Metal–bone interactions. Toxicol Lett. 2000;112:219–25.
65. Järup L, Åkesson A. Current status of cadmium as an environmental health problem. Toxicol Appl Pharmacol. 2009;238(3):201–8.
66. Ejaz S, Camer GA, Anwar K, Ashraf M. Monitoring impacts of air pollution: PIXE analysis and histopathological modalities in evaluating relative risks of elemental contamination. Ecotoxicology. 2014;23(3):357–69.
67. Chiu YH, Hsu HH, Coull BA, Bellinger DC, Kloog I, Schwartz J, et al. Prenatal particulate air pollution and neurodevelopment in urban children: examining sensitive windows and sex-specific associations. Environ Int. 2016;87:56–65.
68. Sunyer J. The neurological effects of air pollution in children. Eur Respir J. 2008;32(3):535–7.
69. Calderon-Garciduenas L, Cross JV, Franco-Lira M, Aragon-Flores M, Kavanaugh M, Torres-Jardon R, et al. Brain immune interactions and air pollution: macrophage inhibitory factor (MIF), prion cellular protein (PrP(C)), interleukin-6 (IL-6), interleukin 1 receptor antagonist (IL-1Ra), and interleukin-2 (IL-2) in cerebrospinal fluid and MIF in serum differentiate urban children exposed to severe vs. low air pollution. Front Neurosci. 2013;7:183.
70. Calderon-Garciduenas L, Mora-Tiscareno A, Styner M, Gomez-Garza G, Zhu H, Torres-Jardon R, et al. White matter hyperintensities, systemic inflammation, brain growth, and cognitive functions in children exposed to air pollution. J Alzheimers Dis. 2012;31(1):183–91.
71. Calderon-Garciduenas L, Leray E, Heydarpour P, Torres-Jardon R, Reis J. Air pollution, a rising environmental risk factor for cognition, neuroinflammation and neurodegeneration: the clinical impact on children and beyond. Rev Neurol. 2016;172(1):69–80.

72. Calderon-Garciduenas L, Vojdani A, Blaurock-Busch E, Busch Y, Friedle A, Franco-Lira M, et al. Air pollution and children: neural and tight junction antibodies and combustion metals, the role of barrier breakdown and brain immunity in neurodegeneration. J Alzheimers Dis. 2015;43(3):1039–58.
73. Clark C, Martin R, Van Kempen E, Alfred T, Head J, Davies HW, et al. Exposure-effect relations between aircraft and road traffic noise exposure at school and reading comprehension the RANCH project. Am J Epidemiol. 2006;163(1):27–37.
74. Sunyer J, Esnaola M, Alvarez-Pedrerol M, Forns J, Rivas I, Lopez-Vicente M, et al. Association between traffic-related air pollution in schools and cognitive development in primary school children: a prospective cohort study. PLoS Med. 2015;12(3):e1001792.
75. Suglia SF, Gryparis A, Wright R, Schwartz J, Wright R. Association of black carbon with cognition among children in a prospective birth cohort study. Am J Epidemiol. 2008;167(3):280–6.
76. Mirowsky J, Ross CE. Life change: an abandoned explanation. In: Social causes of psychological distress. Piscataway, NJ: Transaction Publishers; 2003. p. 159–65.
77. Evans GW, Colome SD, Shearer DF. Psychological reactions to air pollution. Environ Res. 1988;45(1):1–15.
78. Lundberg A. Psychiatric aspects of air pollution. Otolaryngol Head Neck Surg. 1996;114(2): 227–31.
79. Power MC, Kioumourtzoglou M-A, Hart JE, Okereke OI, Laden F, Weisskopf MG. The relation between past exposure to fine particulate air pollution and prevalent anxiety: observational cohort study. BMJ. 2015;350:h1111.
80. Oudin A, Braback L, Astrom DO, Stromgren M, Forsberg B. Association between neighbourhood air pollution concentrations and dispensed medication for psychiatric disorders in a large longitudinal cohort of Swedish children and adolescents. BMJ Open. 2016;6(6):e010004.
81. Volk HE, Hertz-Picciotto I, Delwiche L, Lurmann F, McConnell R. Residential proximity to freeways and autism in the CHARGE study. Environ Health Perspect. 2011;119(6):873.
82. Newman NC, Ryan P, LeMasters G, Levin L, Bernstein D, Hershey GKK, et al. Traffic-related air pollution exposure in the first year of life and behavioral scores at 7 years of age. Environ Health Perspect. 2013;121(6):731.
83. Perera FP, Wang S, Vishnevetsky J, Zhang B, Cole KJ, Tang D, et al. Polycyclic aromatic hydrocarbons-aromatic DNA adducts in cord blood and behavior scores in New York city children. Environ Health Perspect. 2011;119(8):1176.
84. Saxon A, Diaz-Sanchez D. Air pollution and allergy: you are what you breathe. Nat Immunol. 2005;6(3):223–6.
85. Hadnagy W, Stiller-Winkler R, Idel H. Immunological alterations in sera of persons living in areas with different air pollution. Toxicol Lett. 1996;88(1):147–53.
86. Zhang J, Smith KR. Household air pollution from coal and biomass fuels in China: measurements, health impacts, and interventions. Environ Health Perspect. 2007;115(6):848–55.
87. Lin H-H, Ezzati M, Murray M. Tobacco smoke, indoor air pollution and tuberculosis: a systematic review and meta-analysis. PLoS Med. 2007;4(1):e20.
88. Houtmeyers E, Gosselink R, Gayan-Ramirez G, Decramer M. Regulation of mucociliary clearance in health and disease. Eur Respir J. 1999;13(5):1177–88.
89. Rovers MM, de Kok IM, Schilder AG. Risk factors for otitis media: an international perspective. Int J Pediatr Otorhinolaryngol. 2006;70(7):1251–6.
90. Heinrich J, Raghuyamshi VS. Air pollution and otitis media: a review of evidence from epidemiologic studies. Curr Allergy Asthma Rep. 2004;4(4):302–9.
91. Patel MA, Mener DJ, Garcia-Esquinas E, Navas-Acien A, Agrawal Y, Lin SY. Tobacco smoke exposure and eustachian tube disorders in US children and adolescents. PLoS One. 2016;11(10):e0163926.
92. Gulya AJ. Environmental tobacco smoke and otitis media. Otolaryngol Head Neck Surg. 1994;111(1):6–8.
93. Brook I. The impact of smoking on oral and nasopharyngeal bacterial flora. J Dent Res. 2011;90(6):704–10.

94. Bukharin OV, Deriabin DG, Nemtseva NV. Staphylococcal contamination of the upper respiratory tract mucosa in children exposed to air pollution. Gig Sanit. 1996;(6):6–8.
95. Nandi S, Kumar R, Ray P, Vohra H, Ganguly NK. Group a streptococcal sore throat in a peri-urban population of northern India: a one-year prospective study. Bull World Health Organ. 2001;79(6):528–33.
96. Orazzo F, Nespoli L, Ito K, Tassinari D, Giardina D, Funis M, et al. Air pollution, aeroallergens, and emergency room visits for acute respiratory diseases and gastroenteric disorders among young children in six Italian cities. Environ Health Perspect. 2009;117(11):1780–5.
97. Savilov ED, Shcherbakova EB. Acute intestinal infections in children in areas with industrial environmental air pollution. Gig Sanit. 2003;(4):6–8.
98. Herbarth O, Franck U, Krumbiegel P, Rehwagen M, Rolle-Kampczyk U, Weiss H. Noninvasive assessment of liver detoxification capacity of children, observed in children from heavily polluted industrial and clean control areas, together with assessments of air pollution and chloroorganic body burden. Environ Toxicol. 2004;19(2):103–8.
99. Liu C, Fuertes E, Tiesler CM, Birk M, Babisch W, Bauer CP, et al. The associations between traffic-related air pollution and noise with blood pressure in children: results from the GINIplus and LISAplus studies. Int J Hyg Environ Health. 2014;217(4–5):499–505.
100. Dong GH, Qian ZM, Trevathan E, Zeng XW, Vaughn MG, Wang J, et al. Air pollution associated hypertension and increased blood pressure may be reduced by breastfeeding in Chinese children: the seven northeastern cities Chinese Children's study. Int J Cardiol. 2014;176(3):956–61.
101. Pieters N, Koppen G, Van Poppel M, De Prins S, Cox B, Dons E, et al. Blood pressure and same-day exposure to air pollution at school: associations with nano-sized to coarse PM in children. Environ Health Perspect. 2015;123(7):737–42.
102. Iannuzzi A, Verga MC, Renis M, Schiavo A, Salvatore V, Santoriello C, et al. Air pollution and carotid arterial stiffness in children. Cardiol Young. 2010;20(2):186–90.
103. Calderón-Garcidueñas L, Vincent R, Mora-Tiscareño A, Franco-Lira M, Henríquez-Roldán C, Barragán-Mejía G, et al. Elevated plasma endothelin-1 and pulmonary arterial pressure in children exposed to air pollution. Environ Health Perspect. 2007;115(8):1248–53.
104. Poursafa P, Kelishadi R, Lahijanzadeh A, Modaresi M, Javanmard SH, Assari R, et al. The relationship of air pollution and surrogate markers of endothelial dysfunction in a population-based sample of children. BMC Public Health. 2011;11(1):1.
105. Kelishadi R, Hashemi M, Javanmard SH, Mansourian M, Afshani M, Poursafa P, et al. Effect of particulate air pollution and passive smoking on surrogate biomarkers of endothelial dysfunction in healthy children. Paediatr Int Child Health. 2014;34(3):165–9.
106. Dong GH, Wang J, Zeng XW, Chen L, Qin XD, Zhou Y, et al. Interactions between air pollution and obesity on blood pressure and hypertension in chinese children. Epidemiology. 2015;26(5):740–7.
107. Atkinson RW, Carey IM, Kent AJ, van Staa TP, Anderson HR, Cook DG. Long-term exposure to outdoor air pollution and incidence of cardiovascular diseases. Epidemiology. 2013;24(1):44–53.
108. Poursafa P, Kelishadi R, Haghjooy-Javanmard S, Rafiei L, Keramatian K. Synergistic effects of genetic polymorphism and air pollution on markers of endothelial dysfunction in children. J Res Med Sci. 2012;17(8):718–23.
109. Hashemi M, Afshani MR, Mansourian M, Poursafa P, Kelishadi R. Association of particulate air pollution and secondhand smoke on endothelium-dependent brachial artery dilation in healthy children. J Res Med Sci. 2012;17(4):317–21.
110. Watanabe N. Urinary 17-KS and 17-OHCS as markers of endocrine disruption by air pollution in primary school children. Environ Health Prev Med. 2000;4(4):221–3.
111. Thiering E, Cyrys J, Kratzsch J, Meisinger C, Hoffmann B, Berdel D, et al. Long-term exposure to traffic-related air pollution and insulin resistance in children: results from the GINIplus and LISAplus birth cohorts. Diabetologia. 2013;56(8):1696–704.
112. Hathout EH, Beeson WL, Ischander M, Rao R, Mace JW. Air pollution and type 1 diabetes in children. Pediatr Diabetes. 2006;7(2):81–7.

113. Tamayo T, Rathmann W, Stahl-Pehe A, Landwehr S, Sugiri D, Kramer U, et al. No adverse effect of outdoor air pollution on HbA1c in children and young adults with type 1 diabetes. Int J Hyg Environ Health. 2016;219(4–5):349–55.
114. Jerrett M, McConnell R, Wolch J, Chang R, Lam C, Dunton G, et al. Traffic-related air pollution and obesity formation in children: a longitudinal, multilevel analysis. Environ Health. 2014;13:49.
115. Grün F, Blumberg B. Endocrine disrupters as obesogens. Mol Cell Endocrinol. 2009;304(1):19–29.
116. Dong GH, Qian Z, Liu MM, Wang D, Ren WH, Fu Q, et al. Obesity enhanced respiratory health effects of ambient air pollution in Chinese children: the seven northeastern cities study. Int J Obes. 2013;37(1):94–100.
117. Roels H, Bruaux P, Buchet JP, Claeys-Thoreau F, Lauwerys R, Lafontaine A, et al. Impact of air pollution by lead on the heme biosynthetic pathway in school-age children. Arch Environ Health. 1976;31(6):310–6.
118. Poursafa P, Kelishadi R, Amini A, Amini A, Amin MM, Lahijanzadeh M, et al. Association of air pollution and hematologic parameters in children and adolescents. J Pediatr. 2011;87(4):350–6.
119. Landrigan PJ, Baker EL, Feldman RG, Cox DH, Eden KV, Orenstein WA, et al. Increased lead absorption with anemia and slowed nerve conduction in children near a lead smelter. J Pediatr. 1976;89(6):904–10.
120. Pönkä A. Lead in the ambient air and blood of children in Helsinki. Sci Total Environ. 1998;219(1):1–5.
121. Nikolić M, Nikić D, Stanković A. Effects of air pollution on red blood cells in children. Pol J Environ Stud. 2008;17(28):267.
122. Weng H-H, Tsai S-S, Chen C-C, Chiu H-F, Wu T-N, Yang C-Y. Childhood leukemia development and correlation with traffic air pollution in Taiwan using nitrogen dioxide as an air pollutant marker. J Toxicol Environ Health. 2008;71(7):434–8.
123. Pearson RL, Wachtel H, Ebi KL. Distance-weighted traffic density in proximity to a home is a risk factor for leukemia and other childhood cancers. J Air Waste Manage Assoc. 2000;50(2):175–80.
124. Crosignani P, Tittarelli A, Borgini A, Codazzi T, Rovelli A, Porro E, et al. Childhood leukemia and road traffic: a population-based case-control study. Int J Cancer. 2004;108(4): 596–9.
125. Tzivian L. Outdoor air pollution and asthma in children. J Asthma. 2011;48(5):470–81.
126. Rodriguez-Villamizar LA, Magico A, Osornio-Vargas A, Rowe BH. The effects of outdoor air pollution on the respiratory health of Canadian children: a systematic review of epidemiological studies. Can Respir J. 2015;22(5):282–92.
127. Chen Z, Salam MT, Eckel SP, Breton CV, Gilliland FD. Chronic effects of air pollution on respiratory health in Southern California children: findings from the Southern California Children's health study. J Thorac Dis. 2015;7(1):46–58.
128. Salvi S. Health effects of ambient air pollution in children. Paediatr Respir Rev. 2007;8(4):275–80.
129. Ward DJ, Ayres JG. Particulate air pollution and panel studies in children: a systematic review. Occup Environ Med. 2004;61(4):e13.
130. Gehring U, Gruzieva O, Agius RM, Beelen R, Custovic A, Cyrys J, et al. Air pollution exposure and lung function in children: the ESCAPE project. Environ Health Perspect. 2013;121(11–12):1357–64.
131. Dherani M, Pope D, Mascarenhas M, Smith KR, Weber M, Bruce N. Indoor air pollution from unprocessed solid fuel use and pneumonia risk in children aged under five years: a systematic review and meta-analysis. Bull World Health Organ. 2008;86(5):390–8C.
132. Kim BJ, Hong SJ. Ambient air pollution and allergic diseases in children. Korean J Pediatr. 2012;55(6):185–92.
133. Esposito S, Tenconi R, Lelii M, Preti V, Nazzari E, Consolo S, et al. Possible molecular mechanisms linking air pollution and asthma in children. BMC Pulm Med. 2014;14:31.

134. Gasana J, Dillikar D, Mendy A, Forno E, Ramos VE. Motor vehicle air pollution and asthma in children: a meta-analysis. Environ Res. 2012;117:36–45.
135. Chen CH, Chan CC, Chen BY, Cheng TJ, Leon GY. Effects of particulate air pollution and ozone on lung function in non-asthmatic children. Environ Res. 2015;137:40–8.
136. Wong EY, Gohlke J, Griffith WC, Farrow S, Faustman EM. Assessing the health benefits of air pollution reduction for children. Environ Health Perspect. 2004;112(2):226–32.
137. Dockery DW, Cunningham J, Damokosh AI, Neas LM, Spengler JD, Koutrakis P, et al. Health effects of acid aerosols on north American children: respiratory symptoms. Environ Health Perspect. 1996;104(5):500–5.
138. Patel MM, Chillrud SN, Deepti K, Ross JM, Kinney PL. Traffic-related air pollutants and exhaled markers of airway inflammation and oxidative stress in new York City adolescents. Environ Res. 2013;121:71–8.
139. Lenney W. The burden of pediatric asthma. Pediatr Pulmonol. 1997;24(S15):13–6.
140. Ierodiakonou D, Zanobetti A, Coull BA, Melly S, Postma DS, Boezen HM, et al. Ambient air pollution, lung function, and airway responsiveness in asthmatic children. J Allergy Clin Immunol. 2016;137(2):390–9.
141. Klumper C, Kramer U, Lehmann I, von Berg A, Berdel D, Herberth G, et al. Air pollution and cytokine responsiveness in asthmatic and non-asthmatic children. Environ Res. 2015;138:381–90.
142. Renzetti G, Silvestre G, D'Amario C, Bottini E, Gloria-Bottini F, Bottini N, et al. Less air pollution leads to rapid reduction of airway inflammation and improved airway function in asthmatic children. Pediatrics. 2009;123(3):1051–8.
143. To T, Zhu J, Larsen K, Simatovic J, Feldman L, Ryckman K, et al. Progression from asthma to chronic obstructive pulmonary disease (COPD): is air pollution a risk factor? Am J Respir Crit Care Med. 2016;194(4):429–38.
144. Hasunuma H, Sato T, Iwata T, Kohno Y, Nitta H, Odajima H, et al. Association between traffic-related air pollution and asthma in preschool children in a national Japanese nested case-control study. BMJ Open. 2016;6(2):–e010410.
145. Patel MM, Quinn JW, Jung KH, Hoepner L, Diaz D, Perzanowski M, et al. Traffic density and stationary sources of air pollution associated with wheeze, asthma, and immunoglobulin E from birth to age 5 years among New York City children. Environ Res. 2011;111(8): 1222–9.
146. Jerrett M, Shankardass K, Berhane K, Gauderman WJ, Kunzli N, Avol E, et al. Traffic-related air pollution and asthma onset in children: a prospective cohort study with individual exposure measurement. Environ Health Perspect. 2008;116(10):1433–8.
147. Shima M, Nitta Y, Adachi M. Traffic-related air pollution and respiratory symptoms in children living along trunk roads in Chiba prefecture, Japan. J Epidemiol. 2003;13(2):108–19.
148. Sonnenschein-van der Voort AM, de Kluizenaar Y, Jaddoe VW, Gabriele C, Raat H, Moll HA, et al. Air pollution, fetal and infant tobacco smoke exposure, and wheezing in preschool children: a population-based prospective birth cohort. Environ Health. 2012;11:91.
149. Nishimura KK, Galanter JM, Roth LA, Oh SS, Thakur N, Nguyen EA, et al. Early-life air pollution and asthma risk in minority children. The GALA II and SAGE II studies. Am J Respir Crit Care Med. 2013;188(3):309–18.
150. Wendt JK, Symanski E, Stock TH, Chan W, Du XL. Association of short-term increases in ambient air pollution and timing of initial asthma diagnosis among Medicaid-enrolled children in a metropolitan area. Environ Res. 2014;131:50–8.
151. Shima M, Nitta Y, Ando M, Adachi M. Effects of air pollution on the prevalence and incidence of asthma in children. Arch Environ Health. 2002;57(6):529–35.
152. Esposito S, Galeone C, Lelii M, Longhi B, Ascolese B, Senatore L, et al. Impact of air pollution on respiratory diseases in children with recurrent wheezing or asthma. BMC Pulm Med. 2014;14:130.
153. Breysse PN, Diette GB, Matsui EC, Butz AM, Hansel NN, McCormack MC. Indoor air pollution and asthma in children. Proc Am Thorac Soc. 2010;7(2):102–6.

154. Sarnat SE, Raysoni AU, Li WW, Holguin F, Johnson BA, Flores Luevano S, et al. Air pollution and acute respiratory response in a panel of asthmatic children along the U.S.-Mexico border. Environ Health Perspect. 2012;120(3):437–44.
155. Tétreault L-F, Doucet M, Gamache P, Fournier M, Brand A, Kosatsky T, et al. Severe and moderate asthma exacerbations in asthmatic children and exposure to ambient air pollutants. Int J Environ Res Public Health. 2016;13(8):771.
156. Sandstrom T, Kelly FJ. Traffic-related air pollution, genetics and asthma development in children. Thorax. 2009;64(2):98–9.
157. Mann JK, Balmes JR, Bruckner TA, Mortimer KM, Margolis HG, Pratt B, et al. Short-term effects of air pollution on wheeze in asthmatic children in Fresno, California. Environ Health Perspect. 2010;118(10):1497–502.
158. McConnell R, Berhane K, Gilliland F, Molitor J, Thomas D, Lurmann F, et al. Prospective study of air pollution and bronchitic symptoms in children with asthma. Am J Respir Crit Care Med. 2003;168(7):790–7.
159. YoussefAgha AH, Jayawardene WP, Lohrmann DK, El Afandi GS. Air pollution indicators predict outbreaks of asthma exacerbations among elementary school children: integration of daily environmental and school health surveillance systems in Pennsylvania. J Environ Monit. 2012;14(12):3202–10.
160. Schildcrout JS, Sheppard L, Lumley T, Slaughter JC, Koenig JQ, Shapiro GG. Ambient air pollution and asthma exacerbations in children: an eight-city analysis. Am J Epidemiol. 2006;164(6):505–17.
161. Liu L, Poon R, Chen L, Frescura AM, Montuschi P, Ciabattoni G, et al. Acute effects of air pollution on pulmonary function, airway inflammation, and oxidative stress in asthmatic children. Environ Health Perspect. 2009;117(4):668–74.
162. Lee SL, Wong WH, Lau YL. Association between air pollution and asthma admission among children in Hong Kong. Clin Exp Allergy. 2006;36(9):1138–46.
163. Escamilla-Nunez MC, Barraza-Villarreal A, Hernandez-Cadena L, Moreno-Macias H, Ramirez-Aguilar M, Sienra-Monge JJ, et al. Traffic-related air pollution and respiratory symptoms among asthmatic children, resident in Mexico City: the EVA cohort study. Respir Res. 2008;9:74.
164. Yeh KW, Chang CJ, Huang JL. The association of seasonal variations of asthma hospitalization with air pollution among children in Taiwan. Asian Pac J Allergy Immunol. 2011;29(1):34–41.
165. Just J, Segala C, Sahraoui F, Priol G, Grimfeld A, Neukirch F. Short-term health effects of particulate and photochemical air pollution in asthmatic children. Eur Respir J. 2002;20(4):899–906.
166. Jalaludin B, Khalaj B, Sheppeard V, Morgan G. Air pollution and ED visits for asthma in Australian children: a case-crossover analysis. Int Arch Occup Environ Health. 2008;81(8):967–74.
167. Migliaretti G, Cavallo F. Urban air pollution and asthma in children. Pediatr Pulmonol. 2004;38(3):198–203.
168. Slaughter JC, Lumley T, Sheppard L, Koenig JQ, Shapiro GG. Effects of ambient air pollution on symptom severity and medication use in children with asthma. Ann Allergy Asthma Immunol. 2003;91(4):346–53.
169. Penard-Morand C, Raherison C, Charpin D, Kopferschmitt C, Lavaud F, Caillaud D, et al. Long-term exposure to close-proximity air pollution and asthma and allergies in urban children. Eur Respir J. 2010;36(1):33–40.
170. Lewis TC, Robins TG, Mentz GB, Zhang X, Mukherjee B, Lin X, et al. Air pollution and respiratory symptoms among children with asthma: vulnerability by corticosteroid use and residence area. Sci Total Environ. 2013;448:48–55.
171. Chauhan A, Inskip HM, Linaker CH, Smith S, Schreiber J, Johnston SL, et al. Personal exposure to nitrogen dioxide (NO 2) and the severity of virus-induced asthma in children. Lancet. 2003;361(9373):1939–44.

172. Puranik S, Forno E, Bush A, Celedón JC. Predicting severe asthma exacerbations in children. Am J Respir Crit Care Med. 2016;195(7):854–9.
173. Wisniewski JA, McLaughlin AP, Stenger PJ, Patrie J, Brown MA, El-Dahr JM, et al. A comparison of seasonal trends in asthma exacerbations among children from geographic regions with different climates. Allergy Asthma Proc. 2016;37(6):475–81.
174. Rey-Ares L, Irazola V, Althabe F, Sobrino E, Mazzoni A, Serón P, et al. Lower tract respiratory infection in children younger than 5 years of age and adverse pregnancy outcomes related to household air pollution in Bariloche (Argentina) and Temuco (Chile). Indoor Air. 2016;26(6):964–75.
175. Zhu YD, Wei JR, Huang L, Wang SH, Tian HM, Guo XB. Comparison of respiratory diseases and symptoms among school-age children in areas with different levels of air pollution. Beijing Da Xue Xue Bao. 2015;47(3):395–9.
176. Upadhyay AK, Singh A, Kumar K, Singh A. Impact of indoor air pollution from the use of solid fuels on the incidence of life threatening respiratory illnesses in children in India. BMC Public Health. 2015;15:300.
177. Darrow LA, Klein M, Flanders WD, Mulholland JA, Tolbert PE, Strickland MJ. Air pollution and acute respiratory infections among children 0-4 years of age: an 18-year time-series study. Am J Epidemiol. 2014;180(10):968–77.
178. Ramesh Bhat Y, Manjunath N, Sanjay D, Dhanya Y. Association of indoor air pollution with acute lower respiratory tract infections in children under 5 years of age. Paediatr Int Child Health. 2012;32(3):132–5.
179. Brauer M, Hoek G, Van Vliet P, Meliefste K, Fischer PH, Wijga A, et al. Air pollution from traffic and the development of respiratory infections and asthmatic and allergic symptoms in children. Am J Respir Crit Care Med. 2002;166(8):1092–8.
180. Cliff J. Indoor air pollution and acute respiratory infections in children. Lancet. 1992;339(8790):396–8.
181. McCarthy CE, Duffney PF, Thatcher TH, Phipps RP, Sime PJ. Dung biomass smoke attenuates innate immune responses to a viral mimetic and influenza a virus in human small airway epithelial cells. In: B15. Health effects of indoor/outdoor pollution in childhood. New York: American Thoracic Society; 2016. p. A2874.
182. Hou W, Xu X, Lei Y, Cao J, Zhang Y, Chen L, et al. The role of the PM2. 5-associated metals in pathogenesis of child Mycoplasma Pneumoniae infections: a systematic review. Environ Sci Pollut Res. 2016;23(11):10604–14.
183. Smith GS, Van Den Eeden SK, Garcia C, Shan J, Baxter R, Herring AH, et al. Air pollution and pulmonary tuberculosis: a nested case-control study among members of a northern California health plan. Environ Health Perspect. 2016;124(6):761.
184. Mishra VK, Retherford RD, Smith KR. Biomass cooking fuels and prevalence of tuberculosis in India. Int J Infect Dis. 1999;3(3):119–29.
185. Bates JM, Mahaffy DJ. Relationships of reported allergy symptoms, relative humidity and airborne biologicals in thirteen Florida classrooms. Proc Indoor Air. 1996;96:551–6.
186. Gravesen S, Larsen L, Gyntelberg F, Skov P. Demonstration of microorganisms and dust in schools and offices. Allergy. 1986;41(7):520–5.
187. Daisey JM, Angell WJ, Apte MG. Indoor air quality, ventilation and health symptoms in schools: an analysis of existing information. Indoor Air. 2003;13(1):53–64.
188. Calogero C, Sly P. 1 developmental physiology: lung function during growth and development from birth to old age. Eur Respir Monogr. 2010;47:1.
189. Schultz ES, Gruzieva O, Bellander T, Bottai M, Hallberg J, Kull I, et al. Traffic-related air pollution and lung function in children at 8 years of age: a birth cohort study. Am J Respir Crit Care Med. 2012;186(12):1286–91.
190. Roy A, Chapman RS, Hu W, Wei F, Liu X, Zhang J. Indoor air pollution and lung function growth among children in four Chinese cities. Indoor Air. 2012;22(1):3–11.
191. Gao Y, Chan EY, Li LP, He QQ, Wong TW. Chronic effects of ambient air pollution on lung function among Chinese children. Arch Dis Child. 2013;98(2):128–35.

192. Eenhuizen E, Gehring U, Wijga AH, Smit HA, Fischer PH, Brauer M, et al. Traffic-related air pollution is related to interrupter resistance in 4-year-old children. Eur Respir J. 2013;41(6):1257–63.
193. Altug H, Gaga EO, Dogeroglu T, Ozden O, Ornektekin S, Brunekreef B, et al. Effects of air pollution on lung function and symptoms of asthma, rhinitis and eczema in primary school children. Environ Sci Pollut Res Int. 2013;20(9):6455–67.
194. Martins PC, Valente J, Papoila AL, Caires I, Araujo-Martins J, Mata P, et al. Airways changes related to air pollution exposure in wheezing children. Eur Respir J. 2012;39(2):246–53.
195. Correia-Deur JE, Claudio L, Imazawa AT, Eluf-Neto J. Variations in peak expiratory flow measurements associated to air pollution and allergic sensitization in children in Sao Paulo, Brazil. Am J Ind Med. 2012;55(12):1087–98.
196. Wiwatanadate P, Trakultivakorn M. Air pollution-related peak expiratory flow rates among asthmatic children in Chiang Mai, Thailand. Inhal Toxicol. 2010;22(4):301–8.
197. Linares B, Guizar JM, Amador N, Garcia A, Miranda V, Perez JR, et al. Impact of air pollution on pulmonary function and respiratory symptoms in children. Longitudinal repeated-measures study. BMC Pulm Med. 2010;10:62.
198. Cara AC, Degryse J, van den Akker M, Dinant GJ, Manolovici C, Buntinx F. Impact of early childhood air pollution on respiratory status of school children. Eur J Gen Pract. 2010;16(3):133–8.
199. Fuentes-Leonarte V, Ballester F, Tenias JM. Sources of indoor air pollution and respiratory health in preschool children. J Environ Public Health. 2009;2009:727516.
200. Delfino RJ, Staimer N, Tjoa T, Gillen D, Kleinman MT, Sioutas C, et al. Personal and ambient air pollution exposures and lung function decrements in children with asthma. Environ Health Perspect. 2008;116(4):550–8.
201. Kasamatsu J, Shima M, Yamazaki S, Tamura K, Sun G. Effects of winter air pollution on pulmonary function of school children in Shenyang, China. Int J Hyg Environ Health. 2006;209(5):435–44.
202. Timonen KL, Pekkanen J, Tiittanen P, Salonen RO. Effects of air pollution on changes in lung function induced by exercise in children with chronic respiratory symptoms. Occup Environ Med. 2002;59(2):129–34.
203. Gauderman WJ, Gilliland GF, Vora H, Avol E, Stram D, McConnell R, et al. Association between air pollution and lung function growth in southern California children: results from a second cohort. Am J Respir Crit Care Med. 2002;166(1):76–84.
204. Mannino DM, Siegel M, Husten C, Rose D, Etzel R. Environmental tobacco smoke exposure and health effects in children: results from the 1991 National Health Interview Survey. Tob Control. 1996;5(1):13–8.
205. Deng Q, Lu C, Yu Y, Li Y, Sundell J, Norback D. Early life exposure to traffic-related air pollution and allergic rhinitis in preschool children. Respir Med. 2016;121:67–73.
206. Calderon-Garciduenas L, Franco-Lira M, Henriquez-Roldan C, Osnaya N, Gonzalez-Maciel A, Reynoso-Robles R, et al. Urban air pollution: influences on olfactory function and pathology in exposed children and young adults. Exp Toxicol Pathol. 2010;62(1):91–102.
207. Smith WH. Air pollution and forests: interactions between air contaminants and forest ecosystems. Berlin: Springer; 2012.
208. Behrendt H, Becker W, Fritzsche C, Sliwa-Tomczok W, Tomczok J, Friedrichs K, et al. Air pollution and allergy: experimental studies on modulation of allergen release from pollen by air pollutants. Int Arch Allergy Immunol. 1997;113(1–3):69–74.
209. Majd A, Chehregani A, Moin M, Gholami M, Kohno S, Nabe T, et al. The effects of air pollution on structures, proteins and allergenicity of pollen grains. Aerobiologia. 2004;20(2):111–8.
210. Bartra J, Mullol J, Del Cuvillo A, Davila I, Ferrer M, Jauregui I, et al. Air pollution and allergens. J Investig Allergol Clin Immunol. 2007;17(Suppl 2):3–8.
211. Morgenstern V, Zutavern A, Cyrys J, Brockow I, Koletzko S, Kramer U, et al. Atopic diseases, allergic sensitization, and exposure to traffic-related air pollution in children. Am J Respir Crit Care Med. 2008;177(12):1331–7.

212. D'Amato G. Urban air pollution and plant-derived respiratory allergy. Clin Exp Allergy. 2000;30(5):628–36.
213. Janssen NA, Brunekreef B, van Vliet P, Aarts F, Meliefste K, Harssema H, et al. The relationship between air pollution from heavy traffic and allergic sensitization, bronchial hyperresponsiveness, and respiratory symptoms in Dutch schoolchildren. Environ Health Perspect. 2003;111(12):1512.
214. D'amato G, Liccardi G, D'amato M, Cazzola M. The role of outdoor air pollution and climatic changes on the rising trends in respiratory allergy. Respir Med. 2001;95(7):606–11.
215. D'Amato G, Liccardi G, D'amato M, Cazzola M. Outdoor air pollution, climatic changes and allergic bronchial asthma. Eur Respir J. 2002;20(3):763–76.
216. Gruzieva O, Bellander T, Eneroth K, Kull I, Melen E, Nordling E, et al. Traffic-related air pollution and development of allergic sensitization in children during the first 8 years of life. J Allergy Clin Immunol. 2012;129(1):240–6.
217. Kramer U, Koch T, Ranft U, Ring J, Behrendt H. Traffic-related air pollution is associated with atopy in children living in urban areas. Epidemiology. 2000;11(1):64–70.
218. Boffetta P, Tredaniel J, Greco A. Risk of childhood cancer and adult lung cancer after childhood exposure to passive smoke: a meta-analysis. Environ Health Perspect. 2000;108(1):73–82.
219. Reynolds P, Von Behren J, Gunier RB, Goldberg DE, Hertz A, Smith DF. Childhood cancer incidence rates and hazardous air pollutants in California: an exploratory analysis. Environ Health Perspect. 2003;111(4):663–8.
220. Raaschou-Nielsen O, Hertel O, Thomsen BL, Olsen JH. Air pollution from traffic at the residence of children with cancer. Am J Epidemiol. 2001;153(5):433–43.
221. Raaschou-Nielsen O, Hertel O, Thomsen BL, Olsen JH. Traffic-related air pollution at the place of residence and risk of cancer among children. Ugeskr Laeger. 2002;164(17):2283–7.
222. Ruchirawat M, Settachan D, Navasumrit P, Tuntawiroon J, Autrup H. Assessment of potential cancer risk in children exposed to urban air pollution in Bangkok, Thailand. Toxicol Lett. 2007;168(3):200–9.
223. Lewtas J. Complex mixtures of air pollutants: characterizing the cancer risk of polycyclic organic matter. Environ Health Perspect. 1993;100:211.
224. Calderon-Garciduenas L, Mora-Tiscareno A, Franco-Lira M, Zhu H, Lu Z, Solorio E, et al. Decreases in short term memory, IQ, and altered brain metabolic ratios in urban apolipoprotein epsilon4 children exposed to air pollution. J Alzheimers Dis. 2015;45(3):757–70.
225. Calderon-Garciduenas L, Mora-Tiscareno A, Fordham LA, Chung CJ, Valencia-Salazar G, Flores-Gomez S, et al. Lung radiology and pulmonary function of children chronically exposed to air pollution. Environ Health Perspect. 2006;114(9):1432–7.
226. Scotland A. Third-hand smoke. 2011.
227. Merritt TA, Mazela J, Adamczak A, Merritt T. The impact of second-hand tobacco smoke exposure on pregnancy outcomes, infant health, and the threat of third-hand smoke exposure to our environment and to our children. Przeglad lekarski. 2012;69(10):717–720.
228. Ueta I, Saito Y, Teraoka K, Miura T, Jinno K. Determination of volatile organic compounds for a systematic evaluation of third-hand smoking. Anal Sci. 2010;26(5):569–74.
229. Bayer-Oglesby L, Grize L, Gassner M, Takken-Sahli K, Sennhauser FH, Neu U, et al. Decline of ambient air pollution levels and improved respiratory health in Swiss children. Environ Health Perspect. 2005;113(11):1632–7.
230. Barnes BR, Mathee A, Krieger L, Shafritz L, Favin M, Sherburne L. Testing selected behaviors to reduce indoor air pollution exposure in young children. Health Educ Res. 2004;19(5):543–50.

Air Pollution and Elderly

11

Domenico Sabatini

11.1 News Regarding the Epidemiology of Ageing

The percentage of senior citizens aged 80 years or over made up 14% of the total population of senior citizens in 2013, and it is predicted that this could increase to 19% in 2050. If this prediction is confirmed to be true, there will be 392 million people aged 80 years or over by 2050, meaning there will be three times the current figure.

The increase in the population aged 80 years or over is occurring at a faster pace in less developed countries than it is in more developed ones (Fig. 11.1). In 1950, there were 6 million senior citizens aged 80 years or over in less developed countries and 8 million in more developed ones; in 2013, senior citizens aged 80 years or over were already more numerous in less developed countries than any others. By 2050, it is predicted that the number of people aged 80 years or over will reach 268 million in less developed countries compared to a mere 124 million in more developed countries.

Further, in addition to experiencing the demographic phenomenon of an ageing population, developing countries are also experiencing other socio-economic phenomena tied to "civilisation", such as that of urbanisation, which is relevant to our discussion. Urbanisation is "the demographic transition from rural to urban location" [2], which on one hand offers the opportunity to increase the population's health, but on the other hand increases conditions that put health at risk, including, and most notably, air pollution.

In fact, outdoor air quality in the principal cities of low- and medium-income countries is rapidly deteriorating. With the steady growth of traffic in countries such as China and India, air pollution has surpassed the standard limits on vehicle emissions. The number of deaths due to air pollution between 2005 and 2010 increased by roughly 5% in China and by roughly 12% in India.

D. Sabatini
Department of Medicine, Madonna del Soccorso" Hospital,
S. Benedetto del Tronto, Italy, London, UK
e-mail: weekend.ger@libero.it

F. Capello, A.V. Gaddi (eds.), *Clinical Handbook of Air Pollution-Related Diseases*,
https://doi.org/10.1007/978-3-319-62731-1_11

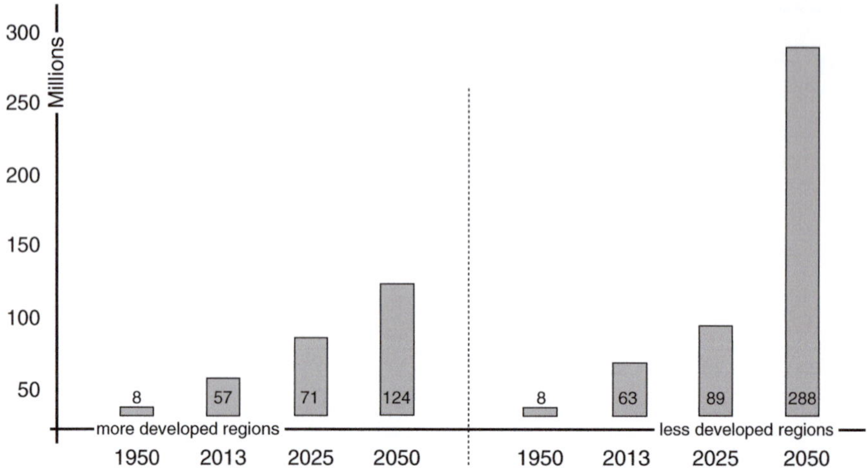

Fig. 11.1 Population aged 80 years or over in the less and more developed regions, 1950, 2013, 2025 and 2050 (taken from the Department of Economic and Social Affairs, Population Division. World Population Ageing 2013. Available online: http://www.un.org/en/development/desa/population/publications/pdf/ageing/WorldPopulationAgeing2013.pdf) [1]

Although rapid economic growth has had numerous benefits on these countries, the adverse consequences of urbanisation on health present significant policy challenges. In rapidly changing countries, such as China and India, frequently updated urbanisation data is crucial in order for health policies to mitigate these adverse effects on health, primarily within highly vulnerable populations, such as senior citizens and children.

Beyond the quantitative justifications supplied by epidemiology (or independently of them), the specific interest for the effects of air pollution on senior citizens stems also from other angles.

11.2 Frailty, Susceptibility and Vulnerability

"Ageing" is a process of progressive decline of bodily function that leads to the biological frailty of individuals and therefore to the maintenance of the functions of organs and systems to the limits of their capacity.

In the "oldest old" individuals, the functional indicators of the cardiovascular, respiratory and urinary systems approach pathological values and represent objective measures of frailty [3–5].

In the context of biological frailty, alterations of the immune system (and the progressive deterioration of the capacity to respond to infections and vaccinations) often mentioned in literature represent just one aspect—certainly an important one—of a problem that is much vaster [6]. Nevertheless, these alterations are the only reference found in literature (together with some minor references to the "senile lung") concerning the physiology of ageing.

General and specific conditions of senior citizens' susceptibility to pollutants are shown in Table 11.1.

Various pathologies and clinical conditions seem to be the best indicators that the subject is "fragile" and susceptible to the effects of air pollution. These are asthma,

Table 11.1 Factors influencing air pollution responses in the elderly (taken from a T. SANDSTROM) [7]

Factors
Smoking history
Occupational exposure history
Environmental exposure history
Infections history
Altered immune response with age
Impaired homeostasis
Altered deposition of PM in diseased airways
Antioxidant and nutritional status
Respirators, cardiovascular and other concurrent diseases
Medication

chronic obstructive pulmonary disease (COPD), coronary heart diseases, hypercoagulable conditions, heart failures and disorders of the heart's rhythm. In senior citizens, these conditions frequently overlap. Thus, in the absence of definite biological references, there is no general consensus that age can be an independent indicator of greater susceptibility to the effects of air pollution; in fact, greater susceptibility of senior citizens to pollution-related diseases is generally better explained by comorbidity. The other side of the coin shows that comorbidity (like pharmacological treatments, history of occupational exposure, etc.) can constitute an important confounding factor [8].

"Susceptibility" is defined as an elevated risk of a particular disease or event that verifies itself within a population that is subjected, together with others, to the same atmospheric exposure. Generally, susceptibility indicates a basic medical condition (e.g. respiratory diseases, cardiovascular diseases, diabetes, obesity, etc.) or a personal biological characteristic (e.g. age, female gender, race, etc.) capable of determining an increase of risk.

"Vulnerability" is different. It refers to a population at greater risk of either more frequent exposure to pollutants or exposure to higher levels of pollutants [9].

There is increasing evidence that the effect of toxic contents in the air varies from one individual to the next based on their level of sensibility, meaning the degree of vulnerability, frailty or sensitivity of the individual to exposures, stimuli and influences.

Susceptibility results from the complex interrelationships of various mechanisms. It has been suggested that those who suffer most from air pollution are people, mainly females and/or senior citizens, who are already at risk due to serious cardiovascular or pulmonary diseases [10]. Equally, lower socio-economic standing can influence susceptibility to illnesses. Mortality due to exposure to atmospheric pollutants—as mentioned in many texts—mainly concerns senior citizens, individuals with lower levels of education [11, 12], individuals who live in low-income areas and individuals who do not have access to health care. This suggests that air quality guidelines designed to protect the general population may be insufficient to actually protect older and more disadvantaged populations [9, 13, 14].

The importance of social factors concerns every country, from developing countries [15] to North America to Europe [16, 17]. A diagram summarising the conditions of susceptibility discussed so far can be found in Fig. 11.2.

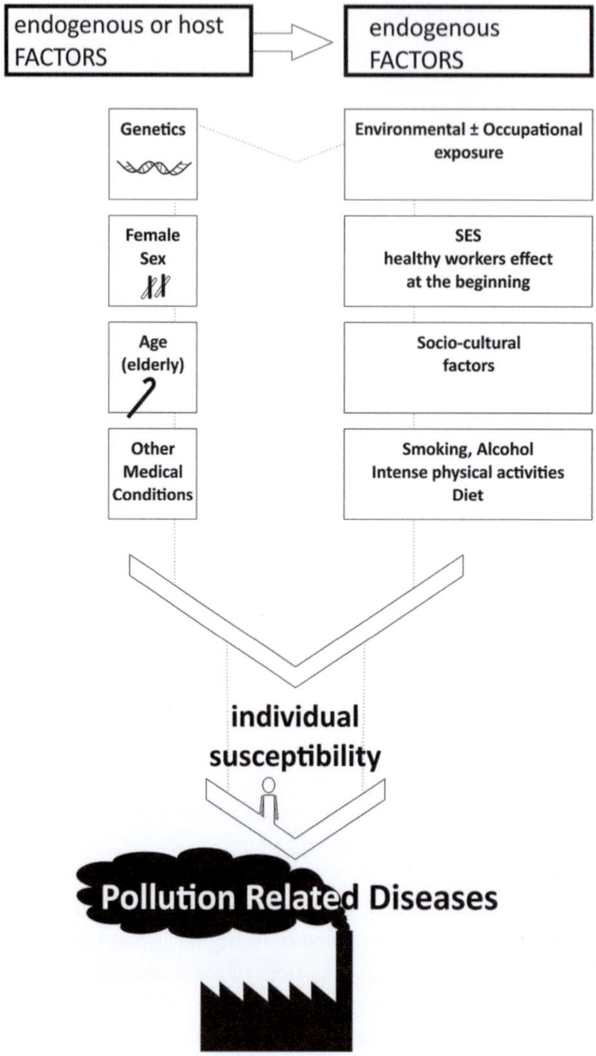

Fig. 11.2 Potential risk factors for individual susceptibility to pollution-related diseases (taken from ANNESI-MAESANO) [10]

11.3 Outdoor Environments and Senior Citizens

Epidemiological studies on the adverse effects of air pollution on senior citizens focus on the more commonly monitored air pollutants (primarily SO_2, CO, NO_2, O_3 and PM as a primary and secondary particulate), but they often estimate, based on the individual case, different levels of intensity in outcome [2, 9].

In addition, the growing importance of other outdoor air pollutants (e.g. volatile organic compounds (VOC) including benzene) is signalled.

Lastly, an emerging problem is the unregulated pollution from ultrafine particles (UFPs, diameter <100 nm). UFPs are omnipresent and it has been hypothesised that they may have greater potential for adverse effects on health with respect to their larger counterparts, although studies regarding long-term exposure are needed in order to confirm this. The primary source of outdoor urban concentrations of UFPs is motor traffic. Recent analysis has suggested that the average exposure to UFPs in Asian cities is roughly four times as much as that in European cities.

What remains unclear is the toxicological importance of various metals known as transition metals (e.g. iron, vanadium, nickel, copper and zinc) and of specific PM-related components capable of triggering oxidative stress and inflammation and thus probably capable of inflicting biological damage. Data, even merely estimative, regarding the effect of each of these substances on the health of senior citizens has not been found within the relevant literature [9].

The fact that pollutants, even in concentrations inferior to the standards that are normally tolerated, represent a serious public health problem for senior citizens has been signalled for some time and by numerous sources [18]. This citation has been willingly extracted from research performed in Switzerland, which, in addition to its historical neutrality, should also represent the cleanest environment.

On the other hand, there is no concentration threshold for pollution from particulates below which there will be no observation of a greater incidence of mortality due to cardiovascular and respiratory causes. In other words, harmful effects of air pollution on health exist even when the level of pollutants is low, and this is due to the interference of other factors, including seasonal climate change and temperature [19]. Few works have studied the variability of air pollution in diverse weather and seasonal conditions or temperatures. Some contradicting results have been found [20]. More recent works, some specifically regarding senior citizens, have identified air pollution data that is both threatening and reassuring for health in connection with even daily climatic variations. From this stems an indication to consider, even from such perspective, the effect of exposure to air pollution, particularly on the populations that are considered more susceptible, in order to better formulate recommendations and advice for prevention [21, 22].

11.4 Indoor Environments and Senior Citizens

Indoor air quality (IAQ) refers to internal air that we breathe within confined spaces, including housing, public and private offices, communal structures, environments intended for recreation and public transport. Industrial environments are excluded, since the quality of the air contained in these environments is strictly correlated with the type of manufacturing activity taking place and is subject to controls and specific laws [23].

From the WHO's communication in 2000 (in which the theme appeared to be confined to developing countries and their closed and poor dwellings as well as their utilisation of coal and biomass in the form of wood, manure and crop residues for the purpose of domestic energy) [24, 25], we have suddenly, and in very little time, come to research this same problem within our own environments to find numerous polluting substances stemming from "civilised" sources and from the return to the burning of biomass, for economic reasons or for fashion.

The agents that pollute indoor environments, as well as their sources, are numerous and variously referred to according to the criterion of nomenclature. Depending on their origin, they are divided into derivatives from biological contaminants, from combustion processes and from construction materials; or, on the basis of their characteristics, they are divided into chemical, physical and biological products.

The knowledge of each polluting agent is certainly important in order to understand its effects and to find its prevention. See, for example, research that specifically concerns formaldehyde [26] and volatile organic components (VOCs) [27, 28]. Even more important for the understanding of this phenomenon and for the deepening of our knowledge is the idea that indoor air pollution—just like outdoor air pollution—is a dynamic process. Therefore, the usual approach to indoor air pollution, which treats the indoor environment as a closed box, leads not only to mistakes but also to reading errors; in short, it produces prudent estimates for the concentration of primary indoor air pollutants, but it ignores secondary pollutants. A kinetic framework of the problem would be better, one that considers indoor production, outdoor-to-indoor transport, indoor-to-outdoor transport and indoor chemistry [29].

For our purposes, indoor pollutants, detected by three meta-analyses [29–31], are in good practice the following (summarised along with their possible sources in Table 11.2): suspended particles (TSP), i.e. PM10 (particles of aerodiameter

Table 11.2 Main indoor pollutants and potential internal sources (taken from the Italian Ministry of Health [23]

Sources	Pollutants
Gas or coal combustion processes for heating and/or cooking, fireplaces and wood stoves, vehicle exhaust fumes	Combustion products (CO, NOx, SO_2, particulates)
Building materials and insulation	Asbestos, artificial vitreous fibres, particulates, radon; biological agents (due to the presence of moisture and/or dust)
Coating materials and carpets	Formaldehyde, acrylates, VOCs and biological agents (due to the presence of moisture and/or dust)
Furnishings	Formaldehyde, VOCs and biological agents (due to the presence of moisture and/or dust)
Liquids and products for cleaning	Alcohols, phenols, VOCs
Photocopiers	Ozone (O_3), toner powder, volatile hydrocarbons (VOCs)
Cigarette smoke	Polycyclic hydrocarbons, VOC, formaldehyde, CO, fine particulates
Air conditioning systems	CO2 and VOCs (due to shortage of hourly changes or excess recycling); biological agents (due to lack of cleaning/maintenance)
Dust	Biological agents (indoor allergens, dust mites)
Individuals	Co_2 and biological agents (bacteria, viruses, etc.)
Animals	Indoor allergens (hair, etc.)
Natural sources (lava, tuffs, granite, etc.)	Radon

< 10 μm) and PM2.5 (fine particles, particles of aerodiameter < 2.5 μm), biomass, O_3, NO_2, CO, SO_2, VOCs (volatile organic compounds such as aldehydes, ketones, esters, etc.), allergens, microorganisms and ETS (environmental tobacco smoke).

Sources of indoor air pollution include fuel combustion for cooking, heating, tobacco smoking, paints, glues, polishes, pesticides and building products.

Amongst the VOCs, formaldehyde is found in numerous construction materials and commonly used products (some insulation materials, certain fitted carpets and textiles, some glues, disinfectants and other household products). Cooking is also a source of immediate VOC emissions according to the cooking method (fossil fuel, oil, wood or kerosene stove) and the food cooked (acrolein derived from oil and fat heating reactions).

Some notes on VOCs. Although VOCs exist in outdoor environments, their concentration indoors is two to five times superior than their concentration outdoors (for formaldehyde it is 10–20 times) [26], to the point that they are generally considered "indoor pollutants".

Normally, the occupiers of these buildings are exposed not to a single substance but to a mix of polluting substances, in varying concentrations within time and space, emitted by sources that differ in quantity and type (see Table 11.3).

The adverse effects on health caused by breathing in these chemical substances can range from temporary irritation of eyes and throat, nausea, headaches and severe respiratory symptoms to long-term diseases such as cancer; damage to the liver, kidneys and central nervous system; or immune system dysfunctions [30].

Some notes on organic pollutants. Few studies have addressed the effects of indoor allergens on respiratory health of senior citizens, probably because allergic sensitisation diminishes with age. Amongst the studies cited in a recent meta-analysis [33] on the effects of indoor damp and mould contamination on health, a

Table 11.3 Classes of VOCs and their possible emission sources (taken from Shaobin Wang [32])

VOC class	Environment and sources
Aliphatic and cycle hydrocarbons	1, 2, 4, 5, 7, 9–11
Aromatic hydrocarbons	1–7, 9, 11, 12
Aldehydes	1–12
Terpenes	1–4, 7–10
Alcohols	1–9, 11
Esters	1, 2, 4, 7–9
Halocarbons	1, 2, 7, 11
Glycols/glycol ethers/glycol esters	1–4, 7, 9
Ketones	1–4, 6–12
Siloxanes	11
Alkene	2, 7
Organic acids	2, 3, 7–9, 11
Ethers	9
Other VOCs	1, 2, 4, 7, 11

(1) Established buildings. (2) New and renovated buildings. (3) School. (4) New car interiors. (5) Carpets. (6) Floor coverings. (7) Wood-based panel and furniture. (8) Solid woods. (9) Paints. (10) Cleaning products. (11) Unflued gas heaters and electric ovens. (12) Office equipment

Norwegian study [34] examined the relationship between mould and respiratory symptoms in individuals aged 61 to 82. The study showed that there is a significant association with respect to phlegm, chronic cough and dyspnea.

In general—as is reported in the prologues of several works—people spend 80–90% of their time indoors. Senior citizens (and in particular very old senior citizens) spend an even greater proportion of their time indoors, due to their reduced activity and their decline in physical autonomy, or simply because of a lack of stimulation. As such, they are potentially more exposed to indoor air pollution compared to the rest of the population [35, 36]. In other words, senior citizens risk spending more time at home and, as such, risk being exposed to indoor pollutants for longer periods of time than the remainder of the population, thus making them more vulnerable to these pollutants [31]. As such, independently of the type of indoor pollutant, even the presence of substances at low concentrations can have an important biological impact on senior citizens, due to the long periods of exposure.

Senior citizens suffering from breathing problems are at an augmented risk of infection of the respiratory tract, particularly if they smoke or have developed advanced COPD. In senior citizens with COPD, higher levels of PM2.5 are associated with worse health conditions; in addition, indoor levels of PM2.5 are significantly higher in a smoker's household [37].

Moreover, it should be considered that physiological, age-dependent anatomo-functional changes can contribute in a significant manner to a predisposition for lower respiratory tract infections in healthy senior citizens. In addition, ageing comes with a gradual decline in many aspects of immune function, thus generally increasing the risk of respiratory infections [31].

Current findings do now allow us to define a causal relationship between exposure and respiratory health outcomes in senior citizens, due to a lack of information regarding lifetime exposure to air pollution and particularly regarding the intensity and duration of the exposure or the age of exposure. Nevertheless, we may be granted a conclusion: senior citizens spend more time indoors, where they are easily exposed to bad air quality; the data (although there is few of it and it is often derived from very heterogeneous research or from inconclusive results) regarding the effects of exposure to indoor air pollutants on the respiratory health of senior citizens shows that health is threatened by ETS, PM, VOCs, use of biomass and possibly by mould.

11.5 General Effects of Air Pollution on Health

Exposure to air pollution has impacts on the health of senior citizens that have been variously analysed in scientific work. These studies take into consideration both short- and long-term effects and utilise different outcomes for each of these, generally mortality and numerous indicators of morbidity.

A review conducted by Bentayeb, which reinterprets 18 studies on morbidity and 11 on mortality [38], as well as another review by this group, which adds further 3 and 5 studies, respectively [2], concluded that the effects of various outdoor air pollutants on the health of senior citizens are solidly proven with respect to short-term exposure and less supported for long-term exposure. These effects concern morbidity and cardiopulmonary mortality, but consistent evidence exists to support the fact that senior citizens exposed to outdoor air pollution experience more hospital admissions for asthma and COPD and more COPD mortality than the rest of the population.

11.5.1 Short-Term Effects

The relationship between short-term exposure to air pollution and its acute effects on health is determined primarily through time-series studies, studying the effects on a population defined by a period of exposure. For senior citizens, morbidity in the acute phase is well documented through hospital admissions and/or medical appoints both at home and in the surgery.

There is strong evidence of an association between short-term exposure to air pollutants and respiratory morbidity, primarily represented by exacerbations of COPD, acute dyspnea and respiratory infections. Symptoms in subjects aged over 65 may appear several days after an increase in levels of exposure (and anyway with a short best lag time to develop asthma exacerbations with respect to those aged below 65) [39] and can persist for a number of days.

In summary, the relationship between short-term exposure to air pollution and morbidity in senior citizens is well documented. In general, research also shows sufficient proof of mortality in the short term, which appears to be attributable in particular to pulmonary and cardiopulmonary causes. In older people (>65 years), research has found a prevalence of pulmonary pathologies related to PM10, whereas cardiopulmonary mortality seems better related to O_3 [2].

The same linear relationship between PM10 and mortality is found by Simoni in other works and in other parts of the world, even with other pollutants (PM 2.5, SO_2, NO_2, O_3, CO).

The summary in Fig. 11.3 associates pollutants with effects on diseases.

11.5.2 Long-Term Effects

Mortality tied to exposure in the short and medium term is substantially lower than that tied to long-term exposure, when observed in the same cohorts using survival analysis. These observations indicate the importance of intra-urban variations in long-term pollution for the purpose of estimating an association between exposure and mortality. They also suggest that the impact of air pollution on public health may be dominated by long-term exposure.

Generally, it is said (and written) [9] that exposure to elevated levels of pollution in the long term can reduce life expectancy by a few years. The long-term effects on the health of senior citizens have been investigated by longitudinal studies, but these are not numerous [38]. Chronic exposure to elevated levels of air pollution is especially tied to the respiratory health of senior citizens and has been related to incidence of COPD, chronic bronchitis, asthma and emphysema [40, 41].

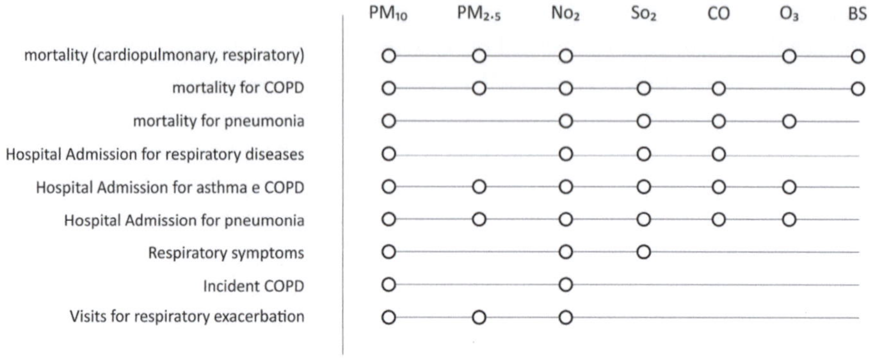

Fig. 11.3 Main respiratory health effects due to outdoor pollution exposure in the elderly: CO (carbon monoxide), NO_2 (nitrogen dioxide), PM (particulate matter), SO_2 (sulphur dioxide), O_3 (ozone), BS (black smoke) and COPD (chronic obstructive pulmonary disease). Taken from Marzia Simoni [2]

11.6 Effects of Air Pollution on Organs and Systems

11.6.1 Respiratory System

Arguments regarding respiratory physiology of ageing indicate that some measure of pulmonary function (such as forced vital capacity or FEV1) increases until one reaches adulthood and then declines for the remainder of one's life. Smoking contributes to the more rapid progression of airflow obstruction, as demonstrated by a decrease in FEV1; it is hypothesised that PM air pollution has similar, but weaker, effects. As such, in non-smokers, exposure to long-term PM results in pulmonary retention of fine particles and small airway remodelling that contribute to causing CODP [13].

There is scientific evidence that, due to pulmonary (and cardiovascular) comorbidity, senior citizens have reduced ventilatory capacity and are more affected by the damages that stem from exposure to air pollutants [31].

Forbes and colleagues, in a study performed in order to explain the relationship between chronic exposure to certain pollutants and pulmonary function in adults, found that the association between reduced FEV1 (but not FVC) and PM10, NO2 and SO2 was greater in senior citizens. The explanation, according to the authors, could be a cohort effect; in other words, senior citizens who live in more polluted areas have worse pulmonary function due to the greater exposure to air pollution that they have had in the past. However, another explanation should be considered, and that is that senior citizens are more susceptible to the effects of air pollution on the lungs [42].

There is an evident link between ageing and respiratory diseases. Senior citizens are at higher risk of suffering from various respiratory diseases, such as CODP, asthma, pneumonia and tuberculosis, compared to other age groups. Women appear to be at even higher risk than men. In old age, various forms of airway obstruction, such as asthma, emphysema and chronic bronchitis, tend to overlap. As such, it could be argued that, in the older population, respiratory diseases depend on higher intensity and longer duration of exposure to environmental risk factors. Therefore, senior citizens are at higher risk of respiratory diseases, even after controlling, for their entire lifetime, active smoke and occupation [31].

Air pollution induces subclinical pulmonary inflammation and leads to the release of humoral mediators in pathways, and later in systemic circulation, contributing to the functional deterioration of other tissues and organs. Long-term exposure to air pollution could also lead to changes in profiles of inflammation markers in lower pulmonary airways, thus worsening the anatomofunctional conditions of the lungs [43]. In other words, air pollution acts as a trigger for exacerbations of CODP by increasing oxidative stress and therefore aggravates exacerbations of CODP; many studies have correlated PM-related exacerbations of respiratory symptoms with transient declines in pulmonary function [13].

11.6.2 Cardiovascular System

Epidemiological and clinical studies have always shown that air pollution is associated not only with respiratory diseases but also with cardiovascular ones. The most common heart disease in senior citizens following exposure to air pollution is shown in Table 11.4.

Cardiovascular morbidity and mortality caused by short-term or prolonged exposure to air pollution are summarised in Table 11.5.

11.6.2.1 Blood Pressure

Although several studies have evaluated the response of blood pressure (BP) to exposure to acute air pollution, many contradictions can be found within their results.

Table 11.4 Cardiovascular disease categories and International Classification of Diseases (ICD) codes

Disease category	ICD-9	ICD-10
Arrhythmia	427	146–149
Cardiac disease	390–429	100–152, 197.0,197.1, 198.1
Cardiac failure	428	150
Ischaemic heart disease	410–413	120, 121, 122, 124, 125.2
Myocardial infarction	410	121, 122
Stroke	430–438	160–166, 167 (excluding 167.0, 167.3), 168 (excluding 168.0), 169, G45 (excluding G45.3), 646
Total cardiovascular disease	390–459	100–199 (excluding 167.3, 168.0, 188, 197.8, 197.9, 198.0), G45 (excluding G45.3), G46, M30, M31, R58

ICD-9, 9th Revision, used January–June 1998; ICD-10, 10th Revision, used July 1998–December 2001

Taken from Ref. [58]

Table 11.5 Overall summary of epidemiological evidence of the cardiovascular effects of PM2.5, traffic-related, or combustion-related air pollution exposure at ambient levels (taken from Brooks, 12)

Health outcomes	Short-term exposure (days)	Longer-term exposure (months[a], years)
Clinical cardiovascular end points from epidemiological studies at ambient pollution concentrations		
Cardiovascular mortality	↑↑↑	↑↑↑
Cardiovascular hospitalisations	↑↑↑	↑
Ischaemic heart disease[a]	↑↑↑	↑↑↑
Heart failure[a]	↑↑	↑
Ischaemic stroke[a]	↑↑	↑
Vascular diseases	↑	↑[b]
Cardiac arrhythmia/cardiac arrest	↑	↑
Subclinical cardiovascular end points and/or surrogate measure in human studies		
Surrogate makers of atherosclerosis	N/A	↑
Systemic inflammation	↑↑	↑
Systemic oxidative stress	↑	↑
Endothelial cell activation/ blood coagulation	↑↑	↑
Vascular/endothelial dysfunction	↑↑	
BP	↑↑	
Altered HRV	↑↑↑	↑
Cardiac ischaemia	↑	
Arrhythmias	↑	

The arrows are not indicators of the relative size of the association but represent a qualitative assessment based on the consensus of the writing group of the strength of the epidemiological evidence based on the number and/or quality, as well as the consistency, of the relevant epidemiological studies
↑↑↑ indicates strong overall epidemiological evidence
↑↑ indicates moderate overall epidemiological evidence
↑ indicates some but limited or weak available epidemiological evidence
Blank indicates lack of evidence
N/A indicates not applicable
[a]Categories include fatal and nonfatal events
[b]Deep venous thrombosis only

Some studies show positive relationships between recent exposure to PM and BP, while others have found negative results and insignificant results. The reasons behind these contrasting inconsistencies stem from different characteristics within the examined populations or from different methods used to measure BP or perhaps from the fact that the study of BP was not the primary outcome of interest for many of these studies [44–46].

Research specifically concerning senior citizens reveals the same inconsistencies and even adds further ones, such as the detection of increased BP (or increased differential pressure) especially in subjects on antihypertensive medication [46, 47]. The reason for this could be an increased vulnerability of hypertensives towards particulates or towards the presence of confounding risk factors within the examined populations, such as obesity. On the other hand, we note the protective value of antihypertensive drugs, following evidence that not taking antihypertensive drugs is a strong predictor of increased BP (both systolic and differential)

[45], which leads us to report a preference—though purely conceptual—for beta blockers, ACE inhibitors and ARBs due to their respective ability to block sympathetic nervous system responses and their antioxidant and anti-inflammatory responses.

To this day, available data suggests that short- and long-term exposure to PM2.5 can cause an elevation in arterial BP—at least in more sensible subjects and in certain situations—in addition, the effects of short-term exposure on health vary depending on the susceptibility/vulnerability of the patient. For example, a young or healthy individual (and therefore not a senior citizen) will probably be only slightly affected by short-term exposure to even higher levels of PM2.5 [44].

On the other hand, a patient with hypertension (especially if it is badly controlled) and/or with underlying coronary disease (even if he is asymptomatic and seemingly healthy) can be much more sensitive to haemodynamic changes stemming from short-term exposure due to their trigger effect [48].

Rapid increase in BP and/or arterial vasoconstriction could conspire to provoke instability/rupture of atherosclerotic plaque and therefore promote an acute CV event. This event could manifest itself as myocardial ischaemia, heart attack (with underlying coronary atherosclerosis) or stroke (in the presence of unstable carotid atherosclerosis). BP and haemodynamic changes, together with myocardial ischaemia of any cause, can increase the risk of heart failure in patients whose left ventricular function is compromised. So far, the most likely mechanism for a rapid haemodynamic response appears to be an imbalance of the autonomic nervous system (ANS). Nevertheless, it is possible that reductions of bioavailability of nitric oxide, which modulates basal arterial tone towards vasoconstriction or vasodilatation of ET in other haemodynamically active molecules (such as angiotensin II), also play a role in certain circumstances [44].

We are still unsure about whether long-term PM exposure is capable of promoting the development of chronically elevated BP levels and even of overt hypertension. This research has to be strengthened by additional proof. In the MESA study [47], participants who lived close to highways had a significant increase in the left ventricular mass index, analogous to that which is obtained with 5.6 mm Hg increase in long-term systolic blood pressure.

On the other hand, it is less clear whether long-term PM exposure increases the risk of developing overt hypertension. And anyhow, even a slight elevation (meaning 10/5 mm Hg) in chronic BP can increase the risk of future cardiovascular events by up to 50% within a large population [44].

11.6.2.2 Arrhythmias: Heart Rate Variability

Air pollution has been associated with anomalies in repolarisation of ECG, with ventricular arrhythmias and with atrial fibrillation [49, 50]. In accordance with these observations, elevated cardiovascular morbidity and mortality deriving from exposure to PM seems to be, in large part, caused by cardiac arrhythmias and by sudden cardiac arrest. The mechanism responsible would be the alteration of the autonomic cardiac balance and/or the impairment of cardiac ion channels induced by systemic oxidative stress.

Even still, many issues remain unclear, such as the importance of gaseous co-pollutants or the role played by pollution sources with respect to cardiovascular disease; equally unclear is the type of susceptible patient [51].

Heart rate variability (HRV), dependent on physiological factors such as postural changes, movement, respiratory rhythm, etc., is the answer to the cardiac autonomic control, which reflects the modulation of the rhythmic activity of the sinus node [52].

Reduction in HVR is a factor that demonstrates risk of cardiovascular morbidity and mortality [52]. Numerous studies, and within these there have been recent meta-analyses, regarding the effects of short-term exposure to air pollution (particularly PM2.5) have demonstrated a decrease in HVR; the existing relationships are particularly strong in senior citizens, people with persistent cardiovascular disease or diabetes or people with reduced antioxidant defences [53]. Relevant mechanisms are, from time to time, identified as systemic inflammation, oxidative stress and modifications in ion channel function or in cardiac autonomic function.

From the first studies specifically concerning senior citizens, which correlated HVR with long-term exposure to air pollution, literature on the subject has notably improved, and recent meta-analyses have concluded that long-term exposure can substantially impact autonomic cardiac function in the same way short-term exposure can [9, 54].

There are few reports on the possibility that drugs may, in some way, carry out a "protective" cardiac function, or otherwise an "aggravating" function; some studies have limited themselves to touching upon this issue by simply communicating experimental feedback, for example, that people who utilise calcium channel blockers and beta blockers have lower relationships between O3 and PM2.5 with LF heart rate [55]. Others have signalled (and sometimes even emphasised) the antioxidant and anti-inflammatory action of certain drugs, from which we can extract a "protective" action, whether demonstrated or hypothetical [56].

11.6.2.3 Ischaemic Heart Disease

Long-term exposure to elevated levels of PM2.5 increases risk of death by ischaemic heart disease, equally, so does exposure to PM10. Short-term exposure to air pollution increases hospital admissions due to ischaemic heart disease when concentration of PM2.5 and/or PM10 is elevated. Exposure can concern a period of a few hours or a couple of days [9].

In triggering an acute cardiac event, the patient's susceptibility (e.g. the presence of persistent heart disease) is very important, as suggested by Pope's angiographic study [57]. Numerous studies of various dimensions have looked at general morbidity and mortality with particular attention towards cardiovascular morbidity and mortality. Some studies have focused specifically on senior citizens; for example, Barnett's study [58] looked exclusively at cardiovascular disease and compared the results of populations aged under and over 65. Analysis of individual cardiac pathologies showed that, in senior citizens, there is a significant relationship between increase in air pollution levels and increase in hospital admissions due to ischaemic heart disease and myocardial heart attacks, whereas the same relationship was not found for the younger group.

The immediate effect of air pollution on the duration of ventricular repolarisation, on its morphology and on its variability (QTc duration, T-wave complexity and T-wave amplitude) is reported by multiple sources as representing the substrate of vulnerability of the myocardium, a key factor in the mechanisms of sudden death [59].

Numerous studies have demonstrated the association between ischaemic morbidity/mortality and short-term exposure. On the contrary, few studies have found the same relationship for long-term exposure. Van Hee's study concerned the adult population and showed that long-term exposure to air pollution altered ventricular repolarisation even in subjects without a history of cardiovascular disease and without calcification of coronary arteries [60]. Mordukhovich's study, which specifically looked at senior citizens (males of an average age of 76), reported a positive relationship between sub-chronic long-term exposure to PM2.5 and QT interval duration (which was even stronger for long-term exposure in subjects with higher oxidative stress allelic profiles). Mordukhovich defers to future high-powered studies the role of further evaluating the association between long-term exposure to polluting substances and QT interval duration, especially for the purpose of identifying susceptible subgroups [61].

Even ST-segment depression is noted as being associated with increased exposure to PM2.5 and to black carbon in cardiac patients. The risk of ST-segment depression associated with air pollution may be even greater in subjects with myocardial injury in the first month following the cardiac event [62]. Alterations in ECG repolarisation, even to the point of obvious ST-segment depression, associated with air pollution, as detected in numerous studies, including those specifically concerning very old individuals (up to 88 years old), are signs of subclinical ischaemia [63]. Together these could indicate an effect on the heart of a systemic inflammation.

11.6.2.4 Heart Failure

Heart failure is the most common cause for hospital admissions and readmissions in senior citizens; it makes up 5% of all diagnoses behind hospital discharges [64, 65].

The data collected by Healthcare Cost and Utilization Project (HCUP), which collects information regarding diseases with which patients are admitted to hospitals in the United States [66], shows that in the period between 1997 and 2008, heart failure was the most common cause for hospital admissions in the population aged between 65 and 84 and aged above 85.

The mechanisms behind myocardial infarctions, which are best known and described by their relationship with air pollution, are incredibly different from those that cause heart failure. These mechanisms can concern an increased demand for cardiac function, or an impairment of cardiac performance (such as reduced contractility), or increased myocardial injury [53]. Think, for example, about the additional work the heart has to perform when exposure to air pollution particulates becomes associated with hyperkinetic arrhythmia or with an increase in systemic arterial pressure with vasoconstriction, or with an increase in pulmonary and right ventricular pressure, etc. [64].

In addition, in the case of senior citizens, think of the physiological modifications of the heart that normally compromise ventricular diastolic filling by reducing compliance, therefore causing the heart to work in a condition that can be termed as diastolic heart failure. To this we must then add any type of concomitant myocardial, ischaemic or degenerative disease. Based on this consideration, evidence that air pollution correlates closely with diastolic function in elderly women appears particularly interesting [67].

11.6.3 The Brain

11.6.3.1 Brain Stroke

Exposure to air pollution is certainly correlated with strokes. Unfortunately, we do not currently have any more useful information that would help us better understand this problem and would therefore help us develop advice or recommendations.

In the prospective study addressed only to postmenopausal women (Women's Health Initiative Observational Study), long-term exposure to PM2.5 was associated with cardiovascular events (including strokes) and mortality [68].

On the other hand, in a study contained in the Cancer Prevention Study II carried out by the American Cancer Society, no significant association was found between stroke mortality and PM air pollution [69]. A Korean study found that PM10 and gaseous pollutants (nitrogen dioxide, sulphur dioxide, carbon monoxide) are important risk factors for stroke mortality, especially for senior citizens and women, who were found to be more sensitive towards the effect of these particulates [70]. When different types of strokes were analysed separately, it was revealed that air pollution was associated with ischaemic strokes but not with haemorrhagic strokes. The authors explained these findings through the stimulating effect on blood coagulation/viscosity stemming from free radicals produced by air pollutants, which can cause ischaemia but not haemorrhaging, and quoted other previous studies that had obtained similar results [71].

A Finnish study demonstrated that toxicological and epidemiological research suggests that smallest particles cause greater systemic damage. The study also evaluated the effects of various types of particles, including ultrafine particles (<0.1 μm), on strokes. The conclusions reached in this study were that the relationship between daily levels of air pollution and death caused by strokes concerned particularly people aged over 65 and that PM2.5, ultrafine particles and carbon monoxide are associated with an increased risk of fatal strokes, but only during warmer seasons. The effect of the season is attributable to seasonal differences in exposure or air pollution mixture.

A large recent study (the first European multicentre study) looked at long-term exposure to air pollution and the incidence of strokes and found suggestive evidence of an association between PM2.5 and strokes, though the main estimate did not reach statistical significance. Stronger associations were found in subjects ≥ 60 years old and in non-smokers. The association was also observed below the current European limit values, which indicated that, as often reported for other diseases, the harmful effects of fine particles exist even at low concentrations [72].

Lastly, a Danish study looked at the effects of short-term air pollution (particularly ultrafine particles) on stroke-related hospital admissions, clinically classifying them in a very precise fashion. The results indicated possible effects of air pollution stemming from traffic, mainly UFPs, on ischaemic stroke-related hospital admissions, especially for mild ischaemic strokes of likely thrombotic original (without AF) [73].

11.6.3.2 Cognitive Decay

The neurotoxicity of exposure to air pollution has been well known for a while and is widely demonstrated by experiments carried out on animals and by autopsy studies carried out on humans.

From the beginning of this century, Calderón-Garcidueñas has carried out studies on dogs and people and has signalled in the brain increased levels of inflammatory mediators, deposition of β-amyloid, presence of markers of oxidative damage to the DNA as well as evidence of interruption in the blood-brain barrier in subjects that lived in cities with high levels of air pollution compared to those exposed to lower levels.

Note: the reconstruction of the studies that led to researching the relationship between pollutants and cognition in humans is taken from the prologue of the works cited at numbers [74].

However, despite the many known actions of pollutants (cardiovascular and systematic), the potential relationship between exposure to common environmental toxicants and cognitive capacity in senior citizens has received relatively little consideration compared to exposure to other non-environmental risk factors. The first study that found a consistent relationship between black carbon and impairment of cognitive functions in humans dates back to 2008 and was carried out on children [75]. The first epidemiological study on adults occurred the following year and examined the effects of PM and ozone on cognition [76]. The first epidemiological study that looked at the relationship between black carbon and cognition of older men (mean age 71 ± 7) over a period of 11 years began in 2011 [77]. The conclusion was that traffic-related air pollution can have a negative effect on cognitive ability. In addition, the effect can be greater in smokers or in obese and overweight subjects, that is, in conditions defined as pro-inflammatory.

The study gave reasonable advice: "Given the ubiquitous nature of the exposure, if traffic-related air pollution is causally related to cognitive impairment in older adults, implementation of interventions to reduce exposure, including establishment of more stringent emissions standards, would be expected to have substantial benefits".

The pollutant-cognition relationship (studied in adults and older subjects) has expanded in recent years in order to consider other substances (O3, PM2.5 and NO_2) and has better explained the cognitive deficits within specific domains [78]. This has also opened the door to the possibility that subclinical atherosclerosis has a mediating role in the pathogenesis of cognitive impairment.

11.6.3.3 Depressive Syndrome

From Bullinger's first observations 1989 [79], that exposure to low levels of pollutants (in this case SO2) could influence subjective psychological well-being without causing depression, to Szyszkowicz's observations (2010) [80], that there is a possible relationship between air pollution and attempted suicide (until more recent confirmation) [81], to numerous occasional reports that relate, particularly in hypothetic manner, air pollution with mood, interest in this subject has been consolidated in recent years. This interest immediately directed itself towards the older population, within which depression is one of the most common mental health issues.

A 2012 Korean study found that growing concentrations of PM10, NO_2 and O_3 were significantly associated with depressive symptoms (measured by the Geriatric Depression Scale, short form SGDS-K) in an older population (average age 71). The individual items that explore emotional symptoms were more likely to be associated with these three polluting substances than the items that explore somatic or affective symptoms [82].

Other researches followed this same train of thought; for example, a relationship between pollution from biomass and depression in premenopausal women was found [83] and a collection of advice regarding the association of air pollution and depression in adults was reported [84]. No positive relationship was found by a group of Boston researchers between depressive symptoms and short- or long-term exposure to air pollution in a population of 732 senior citizens (aged >65, average age 78.1), when utilising the Revised Centre for Epidemiological Studies Depression Scale (CESD-R) [85].

The research was challenged on methodological grounds and has been compared with other recent works, though not numerous, which had communicated perfectly opposite results, for example, by stating that the air pollution/depression relationship does exist [86]. Wellenius [87] responded to these dissenting observations by saying that, "…it is worth noting that there are very few other studies available for direct comparison, and thus this remains very much an open research question. Additional studies in diverse populations are clearly needed to confirm or refute the presence of an association between air pollution and depressive symptoms".

On an end point expanded to include mental health (and not specifically directed at depression), it appears that, in China, senior citizens that live in cities have better mental health than senior citizens that live in rural areas [88]. On the contrary, in Finland, senior citizens in the countryside experience better mental health conditions [89]. Beyond geographical differences and diverse methodologies (in the Finnish study, mental health was included in the general health status), we believe that depression, in addition to its symptoms and the complexity of its nosology (the last DSM V classification dates to 2013), presents many socio-economic components that have varying effects on senior citizens, which should be explained in relation to their predominant meaning and thus not in relation to air pollution.

While waiting to learn more about this topic, scientific studies on depression and air pollution have turned to more complex areas of research. Madrigano, in a study of much broader interest, tried to find a mechanistic correlation, through DNA methylation, between air pollution and morbidity. In this context, he found that older males (average age 72) had a positive correlation between PM2.5 and mood, grossly divided into optimistic or pessimistic character [90].

11.7 Conclusions and Suggestions

Numerous scientific papers on the topic of air pollution have paid particular attention to the older population or have been specifically addressed to it. However, these refer to senior citizens as subjects, varying from case to case, who are over 60 years of age, or aged between 65 and 75, or over 85; this definition of an age range never refers to an objective biological fact or to a consolidated convention. It follows that current information on senior citizens is not comparable, and any attempt at generalising risks is naïve and misleading.

The affirmation that senior citizens are "susceptible" (see paragraph 2) is generic; the application of biological mechanisms of susceptibility (see, e.g. paragraph 2 on the immune response to infectious diseases and vaccinations and paragraph 4) should contain more specific references.

It is also true that, on the subject on fragility, geriatrics has written the impossible, and from that philosophy, we have never obtained a practical notion. However, the assumption that 80 years of age represents the biological and epidemiological transition to old age [3] is a long-established notion and deserves better consideration.

The safeguarding of physical autonomy is one of the cardinal principles of clinical geriatrics, and, as such, the fight against immobility and hypokinesia is its most important therapeutic practice.

Physical activity is one of the few non-pharmacological therapies that has, for years, achieved consolidated success even in official medicine. Cardiologists were the first to believe in the effectiveness of physical activity, and the indispensability of its prescription, with respect to primary and secondary prevention of myocardial infarctions. The guidelines of the two most prestigious American societies of cardiology (AHA/ACC) on secondary cardiovascular prevention (contained in the 2006 update) [91] recommend doctors to encourage everyone to undertake moderate intensity aerobic activity, such as fast walking for 30–60 min, 7 days a week (minimum 5 days a week).

In geriatrics, this recommendation has encountered too many arbitrary translations, which oscillate between two opposing principles: the first, "old age does not exist" and, the second, "elders should be guarded, not treated". Physical activity should be prescribed to senior citizens with the same conviction and the same scientific principles with which we prescribe drugs, with the addition of the art of adapting trials and guidelines specifically to that person.

The decision of whether or not to direct someone towards physical activity, in the present context, should be followed by considerations of where and when, in accordance with all the outdoor/indoor valuations within our knowledge [92]. It is therefore important to understand indoor environmental conditions of these senior citizens, such as their house and their retirement residence [93, 94] as well as the neighbouring outdoor conditions and any small quarters in which their whole life usually takes place [8, 95].

One last important consideration regarding clinical geriatrics is that older senior citizens (aged >80), normally, breathe deeply (without arriving to dyspnea) and therefore introduce greater amounts of pollutants into their systems even when at rest or carrying out minor daily activities [96]. Therefore, we should give senior citizens (and their assistants) the same advice we would give individuals who, in the same environment, would carry out intense aerobic activity.

The biopsychosocial model was a new way of looking at the person developed by Engel in the 1980s [97] and variously proposed in subsequent years on the basis of a definition of health established by the WHO (World Health Organization) in 1947: "Health is a state of complete physical, mental and social well-being and not merely the absence of disease or infirmity". Geriatrics has applied this model to its own methodology and has derived from it a need to take care of senior citizens in a multidimensional manner, because physical, mental and environmental health—together—form the senior citizen's health.

Our topic—independently of geriatrics and old age—constantly repeats this holistic vision of the person and of health. And anyway, research often shows an impression (when there isn't clearly a statement) that a strictly biomedical vision of the problem is scarcely effective, or even unrealistic (see paragraph 2).

The supervision of individual organs and systems for the purpose of primary and secondary prevention is certainly useful, if specifically addressed to pollution-dependent symptoms and signs and if it considers the overall health of a person within their environment; otherwise, it would entail the multiplication of investigations and prevention of further diseases in a world that is already overly committed to preventing old age.

Polypharmacy, defined by some as the "modern epidemic" [98], is a predominant geriatric term addressed in recent years for reasons of economic expenditure and, sometimes, for concern of safeguarding public health and as a response to the absence of scientific nature (to the point of complete irrationality) of pharmacological choices and prescriptions. Polypharmacy of senior citizens depends on multiple factors, of which comorbidity is merely the easiest to consider and to explain, but not the most important. Other factors are equally important, such as:

(a.) multi-specialist medicine for the elderly, which amplifies prescriptions and often creates overlaps and redundancies;
(b.) drugs that, according to Rochon, fall within the "prescribing cascade" [99], meaning they are prescribed to alleviate the side effects of the principal drug, with the risk of creating new side effects;
(c.) drugs that fall within the expression "medicalisation of old age" (taken from Ivan Illich), meaning they are utilised—scientifically—according to guidelines, and mocked by Tinetti centuries ago [100], or they are shifted from scientific research onto adults-elderlies and applied to centenarians;
(d.) self-prescribed drugs, which have been recommended by friends and family or by trusted barkers or are "welfare drugs", whose only scientifically proven advantage is the benefit to pharmacists.

The effect of air pollutants on drugs and the reciprocal interferences would be an interesting subject, but it is not yet legible within scientific literature. In the meantime, the antipollution pharmacological proposal for primary prevention should be carefully considered. Drugs that show contradictory result (such as arterial hypertension and strokes) are certainly not justifiable; "general" drugs, such as those with antioxidant or anti-inflammatory actions, are probably useful but should be recommended with caution [56, 101]. Drugs that follow reasonableness but not evidence (e.g. beta blockers and ARBs as better anti-hypertension drugs due to their antioxidant and hypotensive actions, or calcium channel blockers since they are protection of arrhythmias incurred by possible alterations of the myocardial calcium channels—see paragraph 7B.2) are only good as topics of discussion.

Susceptibility to air pollution brought about by individual diseases (paragraph 6) is continuously reported in scientific research; in general, it concerns respiratory and cardiovascular diseases as well as diabetes.

The allocation of individual diseases to senior citizens in the name of comorbidity is certainly a complex theme, and one that cannot be studied in depth here, but it must be at clarified that comorbidity is not generally a list of diseases, but rather the sum of pathology/gravity, where gravity is a measure of the single pathology and moves the importance of the problem from the disease to the function [102, 103].

The effects of air pollution on individual organs and systems in senior citizens were described in paragraph 6. The need for a structured medical examination is the foundation of clinical geriatrics and has been suggested sporadically in scientific works as being an effective intervention method. The doctor (and here we think of general practitioners, leaving specialist to their sub-fragmented entities) should all know the diseases, and their causes, in order to identify their effects on this or that system. But in our case (in a population that notoriously struggles to communicate general malaise) diseases, often with no declared symptoms, are attributed to misleading causes, or worse, are interpreted by assistants of any kind.

The suggestions for primary and secondary prevention aimed at senior citizens are the same that are aimed towards the rest of the population, therefore utilising well-made tools that can be found in absolutely reliable papers, websites, brochures, groups and entities. Reports will have to indicate the quality of air considering that, in good account, the older population is particularly susceptible and therefor sensible even to low concentration pollutants.

In addition, messages must be refined by taking into account the individual's overall exposure including the protection from simultaneous exposure to heat and smog, the promotion (or the exclusion) of physical activity, advice regarding known interferences with drugs and diet, etc. Therefore, any attempt to "geriatrise" general advice will only reinterpret it along the lines of reasonableness and common sense.

Eighty years of age—as we have said multiple times, even in the introduction to this same paragraph—indicated the boundary between the biology of adult age and that of old age, meaning it is the beginning of biological frailty, as defined in paragraph 2. Above and below that boundary, we will still have to consider comorbidity and the degree of physical autonomy (as well as psychological and social).

Dividing the older population into two age groups—one from 65 to 80 (which includes younger and medium-aged senior citizens and which is biologically closer to adults) and one over 80—would make sense, in that it would follow almost objective criteria and could be advantageous; for example, we could provide the population below 80 with advice regarding long- and short-term exposure to air pollution, and to that over 80 advice regarding short-term morbidity and mortality. However, in medical practice we will always find ourselves working with exceptions (which in these cases would become the rules), such as 65-year-old subjects with debilitating diseases or other comorbidities and 85-year-old patients that are healthy and almost omnipotent. From this stems a need (and an obligation) to alter and readapt general principles on the basis of each patient.

Paraphrasing Brook [9], a single valid reason pushes us to focus our attention and resources towards the primary prevention of damage caused by air pollution, and this reason is the absolute efficacy of the protection of groups at risk, with obvious short-term benefits. Other reasons, such as the reversibility of some functional damage, are less valid when applied to the over 80 population. Beyond the fact that reductions of PM levels exclusively decrease cardiovascular mortality, the possibility of reversible alterations to the respiratory system has not been proven (indeed it has been denied) [104].

Recommendations should be made to family doctors [105] and, more generally, to health professionals who interact with senior citizens as well as to the patients themselves, so as to put them in a position where they can accurately perceive the threat of exposure to air pollution on their health and they can implement protective behaviour. It is a matter of using intervention strategies capable not only of providing knowledge and understanding its benefits, so as to increase awareness regarding the seriousness of the problem, but also of modifying behaviour and aiming it towards the preservation of health [106].

Recommendations that tend to modify habits and behaviours should be viewed in terms of their effectiveness and should always be personalised to the specific age and functional status of the individual. Essentially, recommendations should be based predominantly on pragmatic logic (and not naivety), in accordance with which if the greatest proportion of pollution-related diseases stem from long-term exposure, then all interventions that keep people away from sources of long-term exposure should be associated with the greatest health benefits.

Recommendations must come from evidence-based scientific demonstrations. The treatment of traditional cardiovascular risk factors, when adequately proposed, could diminish the vulnerability of patients towards air pollution. This is equally true for patients suffering with respiratory diseases and in general for those with illnesses and biological conditions that increase the risk of pollution-based pathologies [17] and therefore for all senior citizens aged above 65 with a cardiorespiratory disease as well as those over 80 with no somatic pathology. Even the recommendations that have not been shown to reduce adverse effects should anyway constitute a series of prudent and feasible measures [9].

The conscious knowledge amongst people of air quality indexes, of warnings issued by informational organisations and of consultations with health workers can significantly influence change in outdoor activity in a way that will avoid exposure to air pollution. This is to say that it is possible, through information, to educate people—regardless of their old age—towards new and protective behaviour [107].

Note M. Simoni [2] cites two Latin American studies, which, although conducted in different times, show similar results:

Saldiva PH, Pope CA, Schwartz J, Dockery DW, Lichtenfels AJ, Salge JM, Barone I, Bohm GM: *Air pollution and mortality in elderly people: a time-series study in Sao Paulo, Brazil*, Arch Environ Health 1995; 50:159–63.

Romieu I, Gouveia N, Cifuentes LA, De Leon AP, Junger W, Vera J, STRAPPA V, Hurtado-Díaz M, Miranda-Soberanis V, Rojas-Bracho L, Carbajal-Arroyo L, Tzintzun-Cervantes G: *Multicity study of air pollution and mortality in Latin America (the ESCALA study)*, Res Rep Health Eff Inst 2012;171:5–86.

References

1. Department of Economic and Social Affairs, Population Division. Available online: http://www.un.org/en/development/desa/population/publications/pdf/ageing/WorldPopulationAgeing2013.pdf; 2013.
2. Simoni M, Baldacci S, Maio S, Cerrai S, Sarno G, Viegi G. Adverse effects of outdoor pollution in the elderly. J Thorac Dis. 2015;7(1):34–45.
3. Vergani C, Grassi M. La nuova longevità. Milano: Oscar Mondadori; 1997.
4. Fried LP, Tangen CM, Walston J, Newman AB, Hirsch C, Gottdiener J, Seeman T, Tracy R, Kop WJ, Burke G, Mcburnie MA. Frailty in older adults: evidence for a phenotype. J Gerontol A Biol Sci Med Sci. 2001;56(3):M146–56.
5. Chen X, Mao G, Leng SX. Frailty syndrome: an overview. Clin Interv Aging. 2014;9:433–41.
6. Poland GA, Ovsyannikova IG, Kennedy RB, Lambert ND, Kirkland JL. A systems biology approach to the effect of aging, immunosenescence and vaccine response. Curr Opin Immunol. 2014;29:62–8.
7. Sandstrom T, Frew AJ, Svartengren M, Viegiz G. The need for a focus on air pollution research in the elderly. Eur Respir J. 2003;21(suppl 40):92s–5s.
8. Yen IH, Michael Y, Perdue L. Neighborhood environment in studies of health of older adults a systematic review. Am J Prev Med. 2009;37:455–63.
9. Brook RD, Rajagopalan S, Arden Pope C III, Brook JR, Bhatnagar A, Diez-Roux AV, Holguin F, Hong Y, Luepker RV, Mittleman MA, Peters A, Siscovick D, Smith SC, Whitsel L, Kaufman JD. Particulate matter air pollution and cardiovascular disease an update to the scientific statement from the American heart association. Circulation. 2010;121:2331–78.
10. Annesi-Maesano I, Agabiti N, Pistelli R, Couilliot M-F, Forastiere F. Subpopulations at increased risk of adverse health outcomes from air pollution. Eur Respir J. 2003;21(Suppl. 40):57s–63s.
11. Ostro BD, Feng W-Y, Broadwin R, Malig BJ, Green RS, Lipsett MJ. The impact of components of fine particulate matter on cardiovascular mortality in susceptible subpopulations. Occup Environ Med. 2008;65:750–6.
12. Cakmak S, Dales RE, Rubio MA, Vidal CB. The risk of dying on days of higher air pollution among the socially disadvantaged elderly. Environ Res. 2011;111(3):388–93.
13. Arden Pope C III, Dockery DW. Health effects of fine particulate air pollution: lines that connect. J Air Waste Manage Assoc. 2006;56(6):709–42.
14. Martins MCH, Fatigati FL, Vespoli TC, Martins LC, Pereira LAA, Martins MA, Saldiva PHN, Braga ALF. Influence of socioeconomic conditions on air pollution adverse health effects in elderly people: an analysis of six regions in Sao Paulo, Brazil. J Epidemiol Community Health. 2004;58:41–6.
15. Kan H, London SJ, Chen G, Zhang Y, Song G, Zhao N, Jiang L, Chen B. Season, sex, age, and education as modifiers of the effects of outdoor air pollution on daily mortality in Shanghai, China: the Public Health and Air Pollution in Asia (PAPA) study. Environ Health Perspect. 2008;116:1183–8.

16. Ponce NA, Hoggatt KJ, Wilhelm M, Ritz B. Preterm birth: the interaction of traffic related air pollution with economic hardship in Los Angeles neighborhoods. Am J Epidemiol. 2005;162(2):140–8.
17. Samoli E, Peng R, Ramsay T, Pipikou M, Touloumi G, Dominici F, Burnett R, Cohen A, Krewski D, Samet J, Katsouyanni K. Acute effects of ambient particulate matter on mortality in Europe and North America: results from the APHENA study. Environ Health Perspect. 2008;116:1480–6.
18. Ackermann-Liebrich U, Leuenberger P, Schwartz J, Schindler C, Monn C, Bolognini G, Bongard JP, Brändli O, Domenighetti G, Elsasser S, Grize L, Karrer W, Keller R, Keller-Wossidlo H, Künzli N, Martin BW, Medici TC, Perruchoud AP, Schöni MH, Tschopp JM, Villiger B, Wüthrich B, Zellweger JP, Zemp E. Lung function and long term exposure to air pollutants in Switzerland. Study on air pollution and lung diseases in adults (SAPALDIA) team. Am J Respir Crit Care Med. 1997;155(1):122–9.
19. Krstić G. Apparent temperature and air pollution vs. elderly population mortality in metro Vancouver. PLoS One. 2011;6(9):e25101. https://doi.org/10.1371/journal.pone.0025101.
20. Koken PJM, Piver WT, Frank YE, Elixhauser A, Olsen LM, Portier CJ. Temperature, air pollution, and hospitalization for cardiovascular diseases among elderly people in Denver. Environ Health Perspect. 2003;111:1312–7.
21. Vanos JK, Cakmak S, Bristow C, Brion V, Tremblay N, Martin SL, Sheridan SS. Synoptic weather typing applied to air pollution mortality among the elderly in 10 Canadian cities. Environ Res. October 2013;126:66–75.
22. Yorifuj T, Suzuki E, Kashima S. Hourly differences in air pollution and risk of respiratory disease in the elderly: a time-stratified case-crossover study. Environ Health. 2014;13:67.
23. Ministero Della Salute, www.salute.gov.it/portale/temi/p2_4.jsp?lingua=italiano&tema=Ambiente%20e%20salute&area=indor.
24. Bruce N, Perez-Padilla R, Albalak R. Indoor air pollution in developing countries: a major environmental and public health challenge. Bull World Health Organ. 2000;78(9):1–15.
25. Mishra V. Effect of indoor air pollution from biomass combustion on prevalence of asthma in the elderly. Environ Health Perspect. 2003;111:71–7.
26. Salthammer T, Mentese S, Marutzky R. Formaldehyde in the indoor environment. Chem Rev. 2010;110:2536–72.
27. Yoon HI, Hong Y-C, Cho S-H, Kim H, Kim YH, Sohn JR, Kwone M, Park S-H, Cho M-H, Cheong H-K. Exposure to volatile organic compounds and loss of pulmonary function in the elderly. Eur Respir J. 2010;36:1270–6.
28. Bentayeb M, Billionnet C, Baiz N, Derbez M, Kirchner S, Annesi-Maesano I. Higher prevalence of breathlessness in elderly exposed to indoor aldehydes and VOCs in a representative sample of French dwellings. Respir Med. 2013;107:1598–607.
29. Mitchell CS, Zhang J, Sigsgaard T, Jantunen M, Lioy PJ, Samson R, Karol MH. Current state of the science: health effects and indoor environmental quality. Environ Health Perspect. 2007;115:958–64.
30. Bentayeb M, Simoni M, Norback D, Baldacci S, Maio S, Viegi G, Annesi-Maesano I. Indoor air pollution and respiratory health in elderly. J Environ Sci Health A. 2013;48:1–7.
31. Viegi M, Maio S, Simoni M, Baldacci S, Annesi-Maesano I. The epidemiological link between ageing and respiratory diseases. Eur Respir Mon. 2009;43:1–17.
32. Wang S, Ang HM, Tade MO. Volatile organic compounds in indoor environment and photocatalytic oxidation: state of the art. Environ Int. 2007;33:694–705.
33. Fisk WJ, Lei-Gomez Q, Mendell MJ. Meta-analyses of the associations of respiratory health effects with dampness and mold in homes. Indoor Air. 2007;17:284–96.
34. Duelien Skorge T, Eagan TML, Eide GE, Gulsvik A, Bakke PS. Indoor exposures and respiratory symptoms in a Norwegian community sample. Thorax. 2005;60:937–42.
35. Jaakkola MS. Environmental tobacco smoke and health in the elderly. Eur Respir J. 2002;19:172–81.
36. Simoni M, Jaakkola MS, Carrozzi L, Baldacci S, Di Pede F, Viegi G. Indoor air pollution and respiratory health in the elderly. Eur Respir J. 2003;21(Suppl. 40):15s–20s.

37. Osman LM, Douglas JG, Garden C, Reglitz K, Lyon J, Gordon S, Ayres JG. In*door air quality in homes of patients with chronic obstructive pulmonary disease*. Am J Respir Crit Care Med. 2007;176:465–72.
38. Bentayeb M, Simoni M, Baiz N, Norback D, Baldacci S, Maio S, Viegi G, Annesi-Maesano I. Adverse respiratory effects of outdoor air pollution in the elderly. Int J Tuberc Lung Dis. 2012;16:1149–61.
39. Ko FWS, Tam W, Wong TW, Lai CKW, Wong GWK, Leung T-F, Ng SSS, Hui DSC. Effects of air pollution on asthma hospitalization rates in different age groups in Hong Kong. Clin Exp Allergy. 2007;37:1312–9.
40. Tanaka T, Asai M, Yanagita Y, Nishinakagawa T, Miyamoto N, Kotaki K, Yano Y, Kozu R, Honda S, Senjyu H. Longitudinal study of respiratory function and symptoms in a non-smoking group of long-term officially-acknowledged victims of pollution-related illness. BMC Public Health. 2013;13:766.
41. Gotschi T, Heinrich J, Sunyer J, Kunzlia N. Long-term effects of ambient air pollution on lung function: a review. Epidemiology. 2008;19:690–701.
42. Forbes LJL, Kapetanakis V, Rudnicka AR, Cook DG, Bush T, Stedman JR, Whincup PH, Strachan DP, Anderson HR. Chronic exposure to outdoor air pollution and lung function in adults. Thorax. 2009;64:657–63.
43. Vossoughi M, Schikowski T, Vierkötter A, Sugiri D, Hoffmann B, Teichert T, Herder C, Schulte T, Luckhaus C, Raulf-Heimsoth M, Casjens S, Brüning T, Krämer U. Air pollution and subclinical airway inflammation in the SALIA cohort study. Immun Ageing. 2014;11:5.
44. Brook RD, Rajagopalan S. Particulate matter, air pollution, and blood pressure. J Am Soc Hypertens. 2009;3(5):332–50.
45. Dvonch JT, Kannan S, Schulz AJ, Keeler GJ, Mentz G, House J, Benjamin A, Max P, Bard RL, Brook RD. Acute effects of ambient particulate matter on blood pressure: differential effects across urban communities. Hypertension. 2009;53:853–9.
46. Jacobs L, Buczyńska AJ, Walgraeve C, Nawrot TS. Acute changes in pulse pressure in relation to constituents of particulate air pollution in elderly persons. Environ Res. 2012;117:60–7.
47. Auchincloss AH, Diez Roux AV, Timothy Dvonch J, Brown PL, Graham Barr R, Daviglus ML, Goff DC Jr, Kaufman JD, O'neill MS. Associations between recent exposure to ambient fine particulate matter and blood pressure in the multi-ethnic study of atherosclerosis (MESA). Environ Health Perspect. 2008;116:486–91.
48. Tofler GH, Muller JE. Triggering of acute cardiovascular disease and potential preventive strategies. Circulation. 2006;114:1863–72.
49. Brook RD. Cardiovascular effects of air pollution. Clin Sci. 2008;115:175–87.
50. Rich DQ, Mittleman MA, Link MS, Schwartz J, Luttmann-Gibson H, Catalano PJ, Speizer FE, Gold DR, Dockery DW. Increased risk of paroxysmal atrial fibrillation episodes associated with acute increases in ambient air pollution. Environ Health Perspect. 2006;114:120–3.
51. Lanki T, De Hartog JJ, Heinrich J, Hoek G, Janssen NAH, Peters A, Stölzel M, Timonen KL, Vallius M, Vanninen E, Pekkanen J. Can we identify sources of fine particles responsible for exercise-induced ischemia on days with elevated air pollution? The ULTRA study. Environ Health Perspect. 2006;114(5):655–60.
52. Tsuji H, Larson MG, Venditti FJ, Manders E, Evans JC, Feldman CL, Levy D. Impact of reduced heart rate variability on risk for cardiac events. The Framingham Heart study. Circulation. 1996;94(11):2850–5.
53. Pieters N, Plusquin M, Cox B, Kicinski M, Vangronsveld J, Nawrot TS. An epidemiological appraisal of the association between heart rate variability and particulate air pollution: a meta-analysis. Heart. 2012;98:1127–35.
54. Mordukhovich I, Coull B, Kloog I, Koutrakis P, Vokonas P, Schwartz J. Exposure to sub-chronic and long-term particulate air pollution and heart rate variability in an elderly cohort: the Normative Aging Study. Environ Health. 2015;14:87.
55. Park SK, O'neill MS, Vokonas PS, Sparrow D, Schwartz J. Effects of air pollution on heart rate variability: the VA Normative Aging Study. Environ Health Perspect. 2005;113(3):304–9.

56. Schwartz J, Park SK, O'Neill MS, Vokonas PS, Sparrow D, Weiss S, Kelsey K. Glutathione-S-transferase M1, obesity, statins, and autonomic effects of particles gene-by-drug-by-environment interaction. Am J Respir Crit Care Med. 2005;172:1529–33.
57. Pope CA III, Muhlestein JB, May HT, Renlund DG, Anderson JL, Horne BD. Ischemic heart disease events triggered by short-term exposure to fine particulate air pollution. Circulation. 2006;114:2443–8.
58. Barnett AG, Williams GM, Schwartz J, Best TL, Neller AH, Petroeschevsky AL, Simpson RW. The effects of air pollution on hospitalizations for cardiovascular disease in elderly people in Australian and New Zealand cities. Environ Health Perspect. 2006;114:1018–23.
59. Henneberger A, Zareba W, Ibald-Mulli A, Rückerl R, Cyrys J, Couderc J-P, Mykins B, Woelke G, Erich Wichmann H, Peters A. Repolarization changes induced by air pollution in ischemic heart disease patients. Environ Health Perspect. 2005;113:440–6.
60. Victor C, Van HEE, Szpiro AA, Prineas R, Neyer J, Watson K, Siscovick D, Park SK, Kaufman JD. Association of long-term air pollution with ventricular conduction and repolarization abnormalities. Epidemiology. 2011;22(6):773–80.
61. Mordukhovich I, Kloog I, Coull B, Koutrakis P, Vokonas P, Schwartz J. Association between particulate air pollution and qt interval duration in an elderly cohort. Epidemiology. 2016;27(2):284–90.
62. Chuang KJ, Coull BA, Zanobetti A, Suh H, Schwartz J, Stone PH, Litonjua A, Speizer FE, Gold DR. Particulate air pollution as a risk factor for st-segment depression in patients with coronary artery disease. Circulation. 2008;118:1314–20.
63. Gold Diane R, Litonjua AA, Zanobetti A, Coull BA, Schwartz J, Maccallum G, Verrier RL. Air pollution and ST-segment depression in elderly subjects. Environ Health Perspect. 2005;113(7):883–7.
64. Shah ASV, Langrish JP, Nair H, Mcallister DA, Hunter AL, Donaldson K, Newby DE, Mills NL. Global association of air pollution and heart failure: a systematic review and meta-analysis. Lancet. 2013;382:1039–48.
65. Jencks SF, Williams MV, Coleman EA. Rehospitalizations among patients in the Medicare fee-for-service program. N Engl J Med. 2009;360:1418–28.
66. HCUP. https://www.hcup-us.ahrq.gov/
67. Ohlwein S, Klümper C, Vossoughi M, Sugiri D, Stolz S, Vierkötter A, Schikowski T, Kara K, Germing A, Quass U, Krämer U, Hoffmann B. Air pollution and diastolic function in elderly women—results from the SALIA study cohort. Int J Hyg Environ Health. 2016;219(4–5):356–63.
68. Miller KA, Siscovick DS, Sheppard L, Shepherd K, Sullivan JH, Anderson GL, Kaufman JD. Long-term exposure to air pollution and incidence of cardiovascular events in women. N Engl J Med. 2007;356:447–58.
69. Arden Pope C, Burnett RT, Thurston GD, Thun MJ, Calle EE, Krewski D, Godleski JJ. Cardiovascular mortality and long-term exposure to particulate air pollution: epidemiological evidence of general pathophysiological pathways of disease. Circulation. 2004;109:71–7.
70. Hong Y-C, Lee J-T, Kim H, Eun-Hee HA, Schwartz J, Christiani DC. Effects of air pollutants on acute stroke mortality. Environ Health Perspect. 2002;110:187–91.
71. Yun-Chul H, Jong-Tae L, Kim H, Kwon H-J. Air pollution a new risk factor in ischemic stroke mortality. Stroke. 2002;33:2165–9.
72. Stafoggia M, Cesaroni G, Peters A, Andersen ZJ, Badaloni C, Beelen R, Caracciolo B, Cyrys J, De Faire U, De Hoogh K, Eriksen KT, Fratiglioni L, Galassi C, Gigante B, Havulinna AS, Hennig F, Hilding A, Hoek G, Hoffmann B, Houthuijs D, Korek M, Lanki T, Leander K, Magnusson PK, Meisinger C, Migliore E, Overvad K, Östenson C-G, Pedersen NL, Pekkanen J, Penell J, Pershagen G, Pundt N, Pyko A, Raaschou-Nielsen O, Ranzi A, Ricceri F, Sacerdote C, Swart WJR, Turunen AW, Vineis P, Weimar C, Weinmayr G, Wolf K, Brunekreef B, Forastiere F. Long-term exposure to ambient air pollution and incidence of cerebrovascular events: results from 11 European cohorts within the ESCAPE project. Environ Health Perspect. 2014;122:919–25.

73. Andersen ZJ, Olsen TS, Andersen KK, Loft S, Ketzel M, Raaschou-Nielsen O. Association between short-term exposure to ultrafine particles and hospital admissions for stroke in Copenhagen, Denmark. Eur Heart J. 2010;31(16):2034–40.
74. Chen J-C, Schwartz J. Neurobehavioral effects of ambient air pollution on cognitive performance in US adults. Neurotoxicology. 2009;30(2):231–9. https://doi.org/10.1016/j.neuro.2008.12.011.
75. Franco Suglia S, Gryparis A, Wright RO, Schwartz J, Wright RJ. Association of black carbon with cognition among children in a prospective birth cohort. Am J Epidemiol. 2008;167(3):280–6.
76. Jiu-Chiuan C, Schwartz J. Neurobehavioral effects of ambient air pollution on cognitive performance in US adults. Neurotoxicology. 2009;30(2):231–9.
77. Power MC, Weisskopf MG, Alexeeff SE, Coull BA, Spiro A III, Schwartz J. Traffic-related air pollution and cognitive function in a cohort of older men. Environ Health Perspect. 2011;119:682–7.
78. Gatto NM, Henderson VW, Hodis HN, St John JA, Lurmannd F, Chen J-C, Mack WJ. Components of air pollution and cognitive function in middleaged and older adults in Los Angeles. Neurotoxicology. 2014;40:1–7. https://doi.org/10.1016/j.neuro.2013.09.004.
79. Bullinger M. Psychological effects of air pollution on healthy residents—a time-series approach. J Environ Psychol. 1989;9(2):103–18.
80. Szyszkowicz M, Willey JB, Grafstein E, Rowe BH, Colman I. Air pollution and emergency department visits for suicide attempts in Vancouver, Canada. Environ Health Insights. 2010;4:79–86.
81. Bakian AV, Huber RS, Coon H, Gray D, Wilson P, McMahon WM, Renshaw PF. Acute air pollution exposure and risk of suicide completion. Am J Epidemiol. 2015;181(5):295–303.
82. Youn-Hee L, Kim H, Kim JH, Bae S, Park HY, Hong Y-C. Air pollution and symptoms of depression in elderly adults. Environ Health Perspect. 2012;120:1023–8.
83. Banerjee M, Siddique S, Dutta A, Mukherjee B, Ray MR. Cooking with biomass increases the risk of depression in pre-menopausal women in India. Soc Sci Med. 2012;75(3):565–72.
84. Calderón-Garcidueñas L, Calderón-Garcidueñas A, Torres-Jardón R, Avila-Ramírez J, Kulesza RJ, Angiulli AD. Air pollution and your brain: what do you need to know right now. Prim Health Care Res Develop. 2015;16(4):329–45.
85. Wang Y, Eliot MN, Koutrakis P, Gryparis A, Schwartz JD, Coull BA, Mittleman MA, Milberg WP, Lipsitz LA, Wellenius GA. Ambient air pollution and depressive symptoms in older adults: results from the MOBILIZE Boston study. Environ Health Perspect. 2014;122(6):553–8.
86. Gao Y, Xu T, Sun W. Ambient air pollution and depressive symptoms in older adults. Environ Health Perspect. 2015;123(5):A114. http://ehp.niehs.nih.gov/1409657/
87. Wellenius GA, Koutrakis P, Wang Y. Ambient air pollution and depressive symptoms in older adults: Wellenius et al. respond. Environ Health Perspect. 2015;123(5):A114–5. http://ehp.niehs.nih.gov/1409657R/
88. Tian T, Chen Y, Zhu J, Liu P. Effect of air pollution and rural-urban difference on mental health of the elderly in China. Iran J Public Health. 2015;44(8):1084–94.
89. Fogelholm M, Valve R, Absetz P, Heinonen H, Uutela A, Patja K, Karisto A, Konttinen R, Makela T, Nissinen A, Jallinoja P, Nummela O, Talja M. Rural-urban differences in health and health behaviour: a baseline description of a community health-promotion programme for the elderly. Scand J Public Healt. 2006;34:632–40.
90. Madrigano J, Baccarelli A, Mittleman MA, Sparrow D. Air pollution and DNA methylation: interaction by psychological factors in the VA normative aging study. Am J Epidemiol. 2012;176(3):224–32.
91. Smith SC Jr, Allen J, Blair SN, Bonow RO. AHA/ACC guidelines for secondary prevention for patients with coronary and other atherosclerotic vascular disease: 2006 update. J Am Coll Cardiol. 2006;47(10):2130–9.

92. Andersen ZJ, De Nazelle A, Mendez MA, Garcia-Aymerich J, Hertel O, Tjønneland A, Overvad K, Raaschou-Nielsen O, Nieuwenhuijsen MJ. A study of the combined effects of physical activity and air pollution on mortality in elderly urban residents: the danish diet, cancer, and health cohort. Environ Health Perspect. 2015;123(6):557–63.
93. Annesi-Maesano ND, Zielinski J, Bernard A, Gratziou C, Sigsgaard T, Sestini P, Viegi G, Gerie Study. Geriatric study in Europe on health effects of air quality in nursing homes (GERIE study) profile: objectives, study protocol and descriptive data. Multidiscip Respir Med. 2013;8(1):71.
94. Bentayeb M, Norback D, Bednarek M, Bernard A, Cai G, Cerrai S, Eleftheriou KK, Gratziou C, Holst GJ, Lavaud F, Nasilowski J, Sestini P, Sarno G, Sigsgaard T, Wieslander G, Zielinski J, Viegi G, Gerie Study. Indoor air quality, ventilation and respiratory health in elderly residents living in nursing homes in Europe. Eur Respir J. 2015;45(5):1228–38.
95. Polidori A, Arhami M, Sioutas C, Delfino RJ, Allen R, Fraser S. Indoor/outdoor relationships, trends, and carbonaceous content of fine particulate matter in retirement homes of the Los Angeles Basin. Air Waste Manage Assoc. 2007;57:366–79.
96. Bentayeb M, Helmer C, Raherison C, Dartigues JF, Tessier J-F, Annesi-Maesano I. Bronchitis-like symptoms and proximity air pollution in French elderly. Respir Med. 2010;104:880–8.
97. Engel GL. The need for a new medical model. A challenge for biomedicine. Science. 1977;196:129–36.
98. Nyborg G, Straand J, Brekke M. Inappropriate prescribing for the elderly—a modern epidemic? Eur J Clin Pharmacol. 2012;68(7):1085–94. https://doi.org/10.1007/s00228-012-1223-8.
99. Rochon PA, Gurwitz JH. Optimising drug treatment for elderly people: the prescribing cascade. BMJ. Oct. 1997;315:25.
100. Tinetti ME, Bogardus ST Jr, Agostini JV. Potential pitfalls of disease-specific guidelines for patients with multiple conditions. N Engl J Med. 2004;351:27.
101. Romieu I, Te Llez-Rojo MM, Lazo M, Manzano-Patino A, Cortez-Lugo M, Julien P, BeLanger MC, Hernandez-Avila M, Holguin F. Omega-3 fatty acid prevents heart rate variability. Reductions associated with particulate matter. Am J Respir Crit Care Med. 2005;172:1534–40.
102. Valderas JM, Starfield B, Sibbald B, Salisbury C, Roland M. Defining comorbidity: implications for understanding health and health services. Ann Fam Med. 2012;10:142–51. https://doi.org/10.1370/afm.1337.
103. Marengoni A, Anglemana S, Melis R, Mangialaschea F, Karpa A, Garmena A, Meinowa B, Fratiglioni L. Aging with multimorbidity: A systematic review of the literature. Ageing Res Rev. 2011;10:430–9.
104. Yanagita Y, Senjyu H, Asai MT, Tanaka T, Yano Y, Miyamoto N, Nishinakagawa T, Kotaki K, Kozu R, Tabusadani M, Honda S. Air pollution irreversibly impairs lung function: a twenty-year follow-up of officially acknowledged victims in Japan. Tohoku J Exp Med. 2013;230:177–84.
105. Abelsohn A, Stieb DM. Health effects of outdoor air pollution. Approach to counseling patients using the Air Quality Health Index. Can Fam Physician. 2011;57:881–7.
106. Radisic S, Bruce Newbold K, Eyles J, Williams A. Factors influencing health behaviours in response to the air quality health index: a cross-sectional study in Hamilton, Canada. EHR. 2016;59(1):17–29.
107. Wen XJ, Balluz L, Mokdad A. Association between media alerts of air quality index and change of outdoor activity among adult asthma in six states, BRFSS, 2005. J Community Health. 2009;34:40–6.

Pollutants and Drugs: Interactions and Human Health

12

Vincenzo Tumiatti, Carmela Fimognari, Andrea Milelli, and Davide Manucra

12.1 Introduction

Drug consumption in our modern society has reached unequaled proportions and it is still increasing. Recent data on the revenue of the worldwide pharmaceutical market pointed out a potent increase in the global drug market starting from the 390.1 billion US dollars for 2001 up to 1057.2 billion US dollars of revenue for 2014 (data taken from statistical website). We consider this data only for their general significance, and we cannot examine the market of the different classes of drugs, but we report these data only to realize the massive diffusion of drugs in the world and the dimensions of this phenomenon. On the other hand, pollution represents another reliable problem due to its global dimensions. These two issues are not distinct, but they overlap one to another and they can influence heavily the human health. In fact, drugs and pollutants are metabolized by the same enzymatic systems, which will no longer be able to process these massive doses of substrates. In this chapter, we will consider the interactions between drugs and environmental pollutants based on scientific evidence which appeared recently on the literature.

12.2 Drug Metabolism

To better understand the possible interactions between drugs and pollutants, it is important to briefly revise the different phases and machinery of the human drug metabolism. It is important to point out that the main aim of the human metabolism

V. Tumiatti (✉) • C. Fimognari • A. Milelli
Dipartimento di Scienze per la Qualità della Vita, Alma Mater Studiorum-Università di Bologna, corso d'Augusto 237, Rimini, Italy
e-mail: vincenzo.tumiatti@unibo.it; carmela.fimognari@unibo.it

D. Manucra
ARPA Emilia Romagna, Via F. Rocchi, 19 - Via Triachini, 17, 40138 Bologna, Italy

F. Capello, A.V. Gaddi (eds.), *Clinical Handbook of Air Pollution-Related Diseases*,
https://doi.org/10.1007/978-3-319-62731-1_12

is to eliminate all xenobiotics introduced by different ways into the human body and able to interfere with its homeostasis; this process is called detoxification. This important goal should be reached very fast in order to avoid any effects on the body machinery. The main strategy to maximize the xenobiotic elimination is generally based on the increase of their hydrophilicity in order to improve their elimination and/or excretion through the key system represented by kidneys into the urine. Different chemical reactions can contribute to this important process, and they are usually enzymatic reactions that take place mainly in the liver, although several enzymatic systems are ubiquitous in the body. The kidney is the major organ for drug excretion. It excretes hydrophilic drugs and drug metabolites through glomerular filtration. Detoxification process is related to different chemical modifications carried out on xenobiotic compounds to facilitate their elimination; however, the same chemical reactions responsible for these important processes might transform nontoxic xenobiotics into a toxic one by unmasking and/or activating some reactive chemical functions. More details about these processes will be reported below.

12.3 Metabolic Processes

Drug metabolism via the cytochrome P450 system has emerged as an important determinant in the occurrence of several drug-drug interactions. A greater degree of interaction predictability has been achieved through the identification of P450 isozymes and some of the drugs that share them. Six different P450 isozymes—CYP1A2, CYP2C19, CYP2C9, CYP2D6, CYP2E1, and CYP3A4—that play important roles in drug metabolism have been identified. Of these six isozymes, shared metabolism by the CYP3A4 isozyme has resulted in several clinically significant drug-drug interactions. More information about the effects of certain drugs on enzyme-mediated biotransformation has led to identification of enzyme inducers and inhibitors, providing even greater insight into the nature of the interactions. Cytochrome P450 represents a family of isozymes responsible for biotransformation of many drugs via oxidation. The enzymes are heme-containing membrane proteins, which are located in the smooth endoplasmic reticulum of several tissues. Although a majority of the isozymes are located in the liver, extrahepatic metabolism also occurs in the kidneys, skin, gastrointestinal tract, and lungs. Significant inactivation of some orally administered drugs is due to the extensive first-pass metabolism in the gastrointestinal tract by the CYP3A4 isozyme.

12.4 Metabolic Reactions

Toxicants or their metabolic precursors, called protoxicants, are characterized by a common behavior with other compounds assumed by the human body (i.e., drugs), and in particular, they are subject to different processes as absorption, metabolism, temporary storage, distribution, and finally excretion. These compounds may generally undergo chemical modifications by enzymatic systems which normally process endogenous substrates. For example, the monoaminooxidase enzymatic system may process endogenous catecholamines as well as basic xenobiotic compounds leading to deaminated and/or to oxidized xenobiotic compounds. This kind of reactions is normally referred to as biotransformations. To note, that exist also some biotransformation reactions not linked to an enzymatic system such as hydrolysis and some oxidation/reduction processes. However, the following metabolic phases classified as phase I and phase II reactions and reported herein are enzymatic.

The chemical nature of the xenobiotic is heavily related to the probability to be processed by an enzymatic system. In particular, compounds characterized by a high polarity (i.e., carboxylic acids) have a low degree of probability to undergo enzymatic processes, and also they hardly enter in the body (we simplify the situation not considering the presence of dedicated transport systems in the intestinal lumen). Nevertheless, if this occurs, they are quickly excreted. Also volatile compounds are hardly subjected to enzymatic modifications due to their short stay in the body. The xenobiotics subjected more frequently to enzymatic reactions are those characterized by a lipophilic nonpolar backbone. These species present low hydrosolubility and, if not metabolized, the ability to bioaccumulate in the lipid tissues.

12.5 Phase I Reactions

The metabolic reactions are classified in two main classes and normally reported as reactions of phase I and phase II. They can occur at the same time, although they act in different ways on the xenobiotic structures. As reported, they are important to mitigate the effects of toxic substances.

Phase I reactions are those able to introduce reactive and polar functional groups into lipophilic xenobiotic backbone. The products are promptly conjugated with other hydrophilic compounds naturally present in the body to form a substance which can be readily excreted. The most important reactions of this phase are normally oxidation of different chemical groups characterized by carbon, nitrogen, sulfur, and phosphorous atoms. Reduction reactions also occur on reducible chemical groups by addition of hydrogen or removal of the oxygen atom. This class of metabolic reaction can include also some hydrolysis processes if a hydrolysable group exists on the xenobiotic chemical structure (i.e., ester, amides). This kind of reactions normally occurs in *endoplasmic reticulum*. Herein, we report a list of the different reactions of phase I with a brief description of their characteristics.

Epoxidation Normally it involves an addition of an oxygen atom to an unsaturated system. It occurs also on aromatic rings activating the reactivity of such compounds toward nucleophilic species (i.e., DNA) in a process called intoxication. The most famous example is related to the epoxidation of benzene ring introducing an epoxide group on the aromatic structure and increasing its reactivity. The resultant epoxide can also undergo to hydration by addition of water to the epoxide ring, resulting in a phenol function. If this function is located on a benzene ring, the new functionality is called phenolic group and this process is very important to eliminate the aromatic compounds. This route is defined as aromatic hydroxylation.

Hydroxylation This process consists on the introduction of a hydroxyl group on the carbon chain and occurs in two typical positions, the first one called Ω and the second one $\Omega-1$, which correspond to the last or the penultimate carbon atom of the chain, respectively. If an aromatic ring is present, it can occur also on the first carbon atom linked to the aromatic ring (benzylic position).

Oxidative reactions also occur on non-carbon atoms such as oxygen, nitrogen, sulfur, and phosphorous leading sometimes to more toxic compounds. In addition, some oxidative reactions can be mediated by flavin-containing monooxygenase (FMO) which is specialized in the oxidation of amino derivatives such as neurotransmitters (adrenaline, noradrenaline, serotonin, dopamine, and others) to carboxylic acid. This kind of reactions normally occur in the synaptic cleft just to stop the activity of neurotransmitters, but it can also take place on different substrates such basic xenobiotics, leading to more polar derivatives which can be easily excreted.

Oxygenated compounds, characterized by an alcoholic group, can be oxidized to aldehyde or carboxylic acids. This reaction is very important to metabolize ethanol but also to eliminate some toxic aldehydes (detoxification process) deriving from the body metabolism which can be oxidized to carboxylic acids. These compounds can undergo to phase II reactions and promptly eliminate (Table 12.1).

In addition to the oxidative reactions, also reductive metabolic reactions can occur on the different substrates. This kind of reactions is less common than the oxidative ones. Reductase enzymes are mainly located in the liver and to a lesser extent also in the lungs and kidneys. In addition, some reductions of xenobiotics can also occur in the intestines due to the presence of gut flora. In particular, aldehydes and ketones can be reduced to alcohol, sulfoxides to sulfides, disulfides to thiols, alkenes to alkanes, and finally azo and nitro groups to amines (see Table 12.2).

Finally, another class of metabolic reactions able to modify the structure of xenobiotics is represented by hydrolysis reactions. They may occur when on the xenobiotic structure hydrolysable functional groups such as esters and amides are present. This process consists on the addition of a water molecule to the hydrolysable group leading to a cleavage of the molecule in two species. In the case of ester, a molecule of carboxylic acid and another of an alcohol will be provided and, in the case of an amide, a carboxylic acid and an amine. The two enzymes responsible of these two processes are called esterases and amidases, respectively, and normally they are ubiquitous in the body. The resulting compounds might be more or less toxic than the parent compounds.

Table 12.1 Main oxidative metabolic reactions

Functional groups	Process	Products
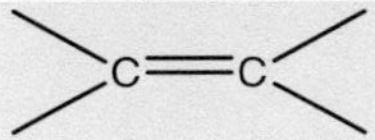	Epoxidation	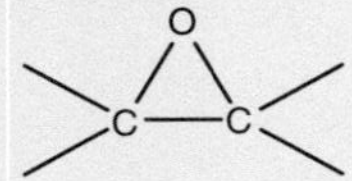
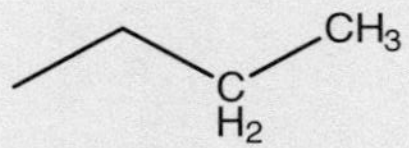	Hydroxylation	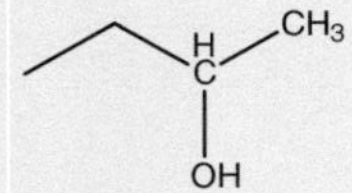
	Aromatic hydroxylation	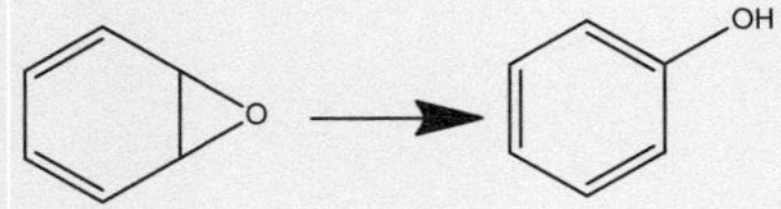
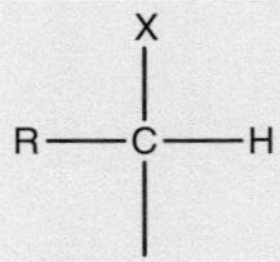	Dehalogenation	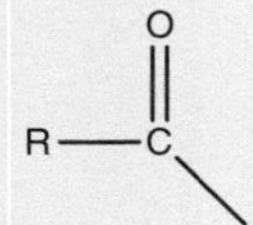
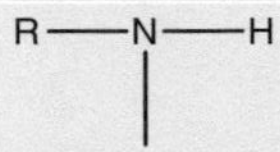	*N*-Hydroxylation	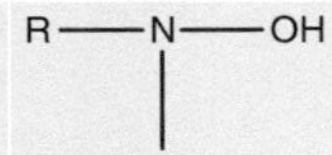
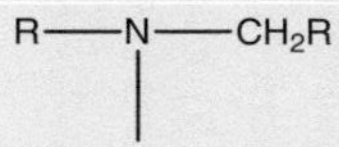	*N*-Dealkylation	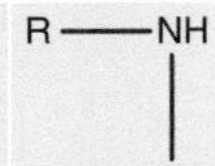
R—O—CH_3	*O*-Dealkylation	R—OH
R—S—CH_3	*S*-Dealkylation	R—SH

Table 12.2 Main reductive metabolic reactions

Functional groups	Process	Products
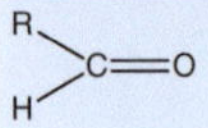	Aldehyde reduction	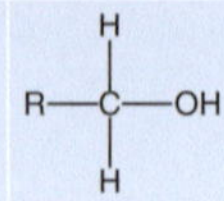
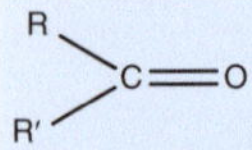	Ketone reduction	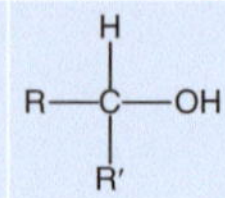
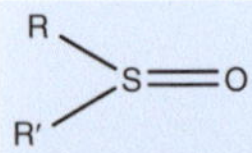	Sulfoxide reduction	R—S—R′
R—S—S—R′	Disulfide reduction	R—SH + HS—R′
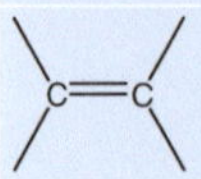	Alkene reduction	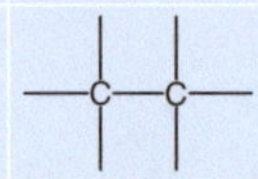
R—N═N—R′	Azo-reduction	R—NH_2 + H_2N—R′
R—NO_2	Nitro-reduction	R—NH_2

12.6 Phase II Reactions

Phase II reactions are also defined as conjugation reactions because they link together two molecules, the xenobiotic compounds or their metabolites, with a second one which normally occurs in the organism. Xenobiotics, metabolites, or other chemical species interact with an activated conjugation agent leading to new compounds characterized by an increased hydrophilicity and thus more easily removable. In this process, reactions of phase I are useful to introduce some "chemical handles" on the xenobiotic backbone in order to facilitate the conjugation process.

The classification of these important reactions is based on the different conjugating agent. One of the most common is the glucuronic acid which derived from a molecule of glucose with its sixth carbon atom oxidized to a carboxylic acid. It is activated in position 1 by a UDP group, where it can be attached by a nucleophilic group. Different atoms present on xenobiotics and endowed with nucleophilic ability are able to react with UDP-activated glucuronic acid (UDPGA), in particular nitrogen, sulfur, and oxygen atoms. The carboxylic group is normally ionized at physiological pH increasing the water solubility of the conjugated metabolites.

A second type of conjugation reaction involves glutathione (GSH) as conjugating agent. This molecule is characterized by a thiol group able to react with electrophilic species. These compounds are very dangerous for the cell organization because they can easily interact with electron-rich structures such as DNA and proteins causing serious damage to the cells. Different xenobiotic species such as epoxides, aromatic hydrocarbons, aromatic and alkyl halides, aromatic nitro compounds, and others are very reactive and can react with GSH. GSH is formed by the condensation of three amino acids like glycine, cysteine, and glutamic acid. It acts as a strong nucleophilic agent thanks to the loss of a proton on thiol group. It interacts at the beginning with the electrophilic species giving a GSH conjugate, and later there is a loss of the glutamyl and glycinyl residues with a final acetylation leading to a mercapturic acid conjugate. This final compound is readily excreted thanks to its high water solubility (see Table 12.3).

Another strategy to improve the hydrophilicity of xenobiotics is represented by the transfer of a sulfate on their structure, thanks to the enzymatic activity of sulfotransferases using 3′-phpsphoadenosine-5′-phosphosulfate (PAPS) as donating group. The presence of this residue facilitates the excretion of the conjugated xenobiotic due to the improvement of its water solubility. Other minor reactions of conjugation are also reported in literature, and they comprise condensation with some amino acids such as glycine, glutamine, taurine, and serine. Finally, the migration of an acetyl or a methyl group on nucleophilic residues is reported as strategy to firstly inactivate xenobiotic toxicity and secondly to eliminate them from the organism. The donors of the two chemical groups are acetyl coenzyme A and S-adenosylmethionine (SAM), respectively.

Table 12.3 Main activated conjugating agents

Conjugating agents	Name	Products
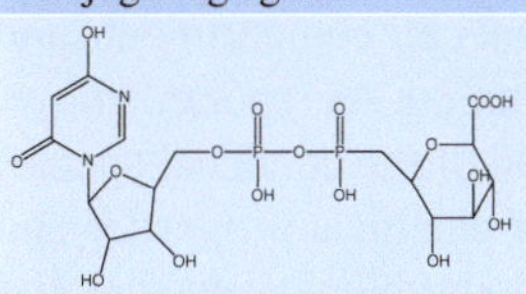	Uridine-5′-diphospho-α-D-glucuronic acid (UDPGA)	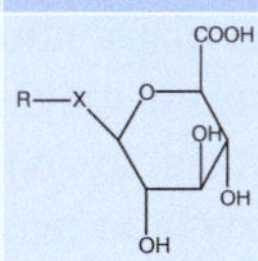
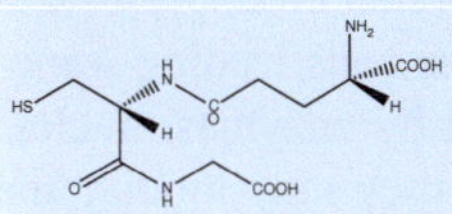	Glutathione (GSH)	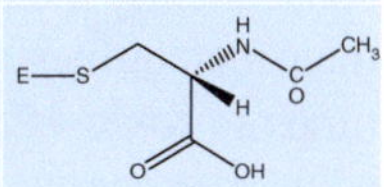
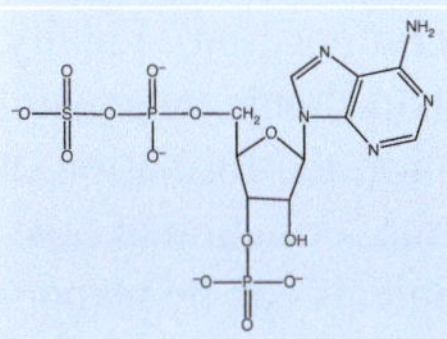	3′-Phosphoadenosine-5′-phosphosulfate (PAPS)	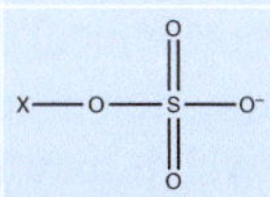
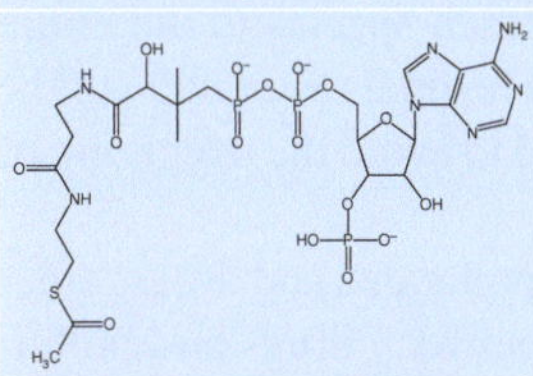	Acetyl CoA	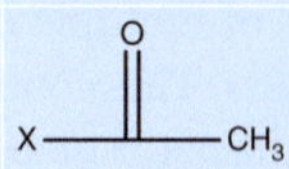
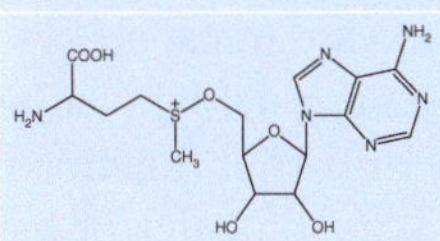	*S*-Adenosylmethionine (SAM)	X—CH_3

12.7 Bases of Interactions Between Drugs and Pollutants

As reported in literature by Ariens [1] some years ago, interactions between chemicals may occur during the different phases of the toxication and detoxication processes in the human body, in particular, in the pre-absorption (exposure), chemokinetic (distribution), and chemodynamic (action) phases. Only in the last years it has been considered the interactions between drugs and pollutants, although the concentration values of the latter cannot reach remarkable values. This result has been possible thanks to the development of new techniques of detection and cellular investigations. The interactions can be of different types: noninteractive (independent) and interactive (synergistic or antagonistic). Normally, the small molecules may enter into the cells and/or excreted by one or more than the following basic mechanisms: passive diffusion, active and facilitated transport, and endo- and exocytosis. The first mechanism is the most common of uptake and excretion of lipophilic chemicals. Most of the interactive reactions may occur during the facilitated and active transport phase considering that chemicals able to use these processes should resemble to the endogenous compounds. Both modes of transport have an important role in the excretion of exogenous chemicals, including common metabolites deriving from their conjugation with glucuronic acid. A lot of examples for competition between the different metabolites of exogenous chemicals for the active excretion system at renal level have been so far reported. Little is known about the cellular distribution of exogenous chemicals in the different cell compartments and this process may play an important role in their cell distribution. As reported above, it is crucial to consider the possibility of biotransformation of the chemical species by enzymatic reactions which may deeply influence their chemical properties and consequently their behavior inside the cells. These exogenous chemical species and their metabolites may interact with different targets inside the cells leading sometimes to toxic effects. These targets are usually represented by enzymes, nucleic acids, membrane receptors, and cofactors, but in addition, they may modify the physicochemical environment (i.e., pH, redox potential, ionic composition). Finally, the formation of reactive species from the xenobiotic metabolism should be considered, which may lead to massive cell damages. We can mention the formation of highly reactive electrophilic or free radical metabolites and/or the formation of reactive oxygen species (ROS). The presence of two or more chemicals can be responsible of multiple cell damage due to the contribution of each species to the entire process. This contribution can be additive or synergistic. The final result will still be a damage to some cell structures which should be repaired to avoid the triggering of the apoptotic pathway, which leads to the cell to death.

In addition to common pathways of metabolization of pollutants and drugs, other different mechanisms, involved in possible interactions between drugs and pollutants, were very recently taken into account [2]. A particular outward transport system is represented by P-glycoprotein system (P-gp), which is a natural detoxification system expressed in normal tissues characterized by secretory or barrier functions. It is found in the small and large bowel, biliary canaliculi, proximal tubules of the kidney, vascular endothelial cells of the central nervous system, placenta, adrenal glands, and testicles. This system consists of an efflux pump through which a drug or other substances can be (r)ejected outside the cell. It is presently accepted as an important factor of distribution and excretion of drugs. Possible interactions of pollutants with this system can reduce its efficacy in the outward drugs transport thereby sensitizing the human body to toxic chemicals that would otherwise be effluxed. Such system is also responsible for many cases of drug resistance in bacteria, cancer cells, and plasmodia due to its ability to reject the drug, once diffused in the cell compartment through the cell membrane, outside the cell again. More generally, it is reported on the ability of different pollutants to act as inhibitors, substrates, or inducers of different drug transporters systems. In a review published on Expert Opinion [3], the authors reported a brief overview of the different drug transport systems involved in the interaction with drugs or other substances. These systems may also competitively interact not only with pollutants but also with different xenobiotic conjugates, preventing the drug efflux and leading to an increase in their concentrations inside the cells, which is responsible for their toxic effects. Now a list of some examples related to interactions between pollutants and drugs will be reported below. It will be given a brief description of the interaction between pollutants and drugs just to better understand the context and the dimension of the problem, referring to the cited articles for more details.

12.8 Examples of Interactions Between Drug(s) and Pollutant(s)

We have reported the general mechanisms of human body detoxification. It is clear that cross interactions between different chemicals (pollutants and/or drugs) may lead to a general inhibition of the chemical metabolism resulting, in the case of drugs, in an increase of their concentration in human body and consequent amplification of their side effects. It is the case reported by Li et al. [4] about the inhibition of drug metabolism due to the presence of endocrine-disrupting chemicals (EDCs) in the body. In particular, they investigated the effect of EDCs on the metabolism of zidovudine (also known as azidothymidine (AZT)), a special drug used in the protocol for AIDS treatment. EDCs are well-known chemicals widely distributed and able to interact with the human hormone system by mimicking or blocking hormone activity. One of the most famous EDCs is bisphenol A (BPA) used as monomer for the production of polycarbonate and also as precursor of epoxy resins. There are scientific evidences about the contamination of babies by BPA, because they can be frequently exposed to this contaminant. In fact, they normally assume liquid milk by cans which can be contaminated by BPA. The exposure to this contaminant has been correlated with different diseases such as cardiovascular disease and diabetes as well as to an increase of inflammation and oxidative stress. BPA is mainly eliminated through its conjugation with activated glucuronic acid (UDP-glucuronosyltransferases). The interaction of this pathway with other chemicals can inhibit the elimination of different drugs causing an increase of their staying and levels in the human body with relative intensification of their side effects.

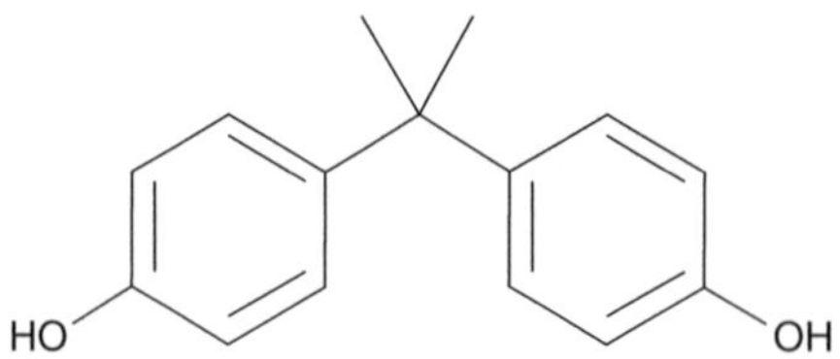

Bisphenol A

Investigations about the influence of BPA on the elimination of azidothymidine (AZT) were carried out, through in vitro experiments by Li et al. [4]. These experiments verified a concentration-dependent inhibition of BPA toward AZT glucuronidation. On these bases, the authors conclude that metabolic elimination of some clinical drugs can be affected by the presence of BPA through the influence on the UDP-glucuronosyltransferases.

Another paper on the same topic published by Verner et al. [5] confirmed these results by determining the influence of different drugs on glucuronidation of BPA and nonylphenol in the hepatocytes of rats. BPA and nonylphenol are endowed with endocrine-disrupting properties. BPA is a component of polycarbonate plastics; nonylphenol derives from the biodegradation of nonylphenol polyethoxylates, used in many products such as latex paints and cosmetics. Their toxic effects are linked to the ability of these compounds to interact with estrogen receptors, and their traces were found in human blood, breast milk, and fetal serum samples. They are extensively metabolized in the rat liver and intestine mainly as glucuronic acid conjugates. The simultaneous exposure to these chemicals together with drugs can potentially cause pharmacokinetic and pharmacodynamic interactions.

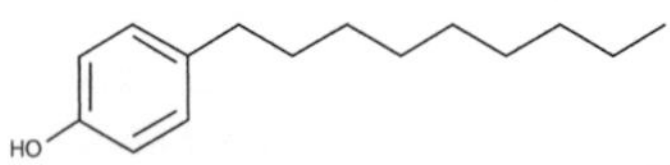

nonylphenol

The elimination of these chemicals is normally very fast as inactive glucuronidated metabolites. A metabolic inhibition of this important pathway by drugs such as naproxen, salicylic acid, ibuprofen, and diclofenac can lead to a dramatic increase of the internal levels of the chemicals due to a potential competitive interaction by drugs with the conjugation system. Furthermore, the conjugation of some drugs with glucuronic acid may lead to a depletion of cofactor UDPGA, thus influencing the biotransformation of BPA and nonylphenol and impacting on the toxicity of these chemicals. The results of this investigation demonstrated that the biotransformation of these chemicals through glucuronic acid conjugation was significantly inhibited by many drugs in freshly isolated rat hepatocytes [5].

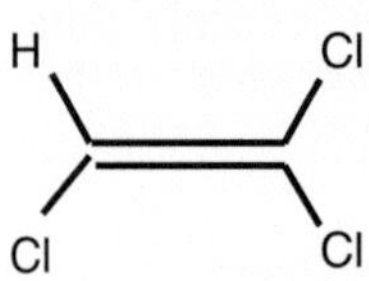

trichloroethylene

The same group more recently has investigated the role of naproxen, salicylic acid, and valproic acid on the pharmacokinetics of trichloroethylene (TCE) and its metabolites in rats. TCE is a volatile chlorinated hydrocarbon solvent widely used in a large spectrum of industrial applications. It is estimated that about 3.5 million of workers are exposed to TCE each year and about 60% of the worldwide production is released in the atmosphere and in oceans. It represents a serious environmental hazard since its recycling and disposal is problematic. TCE can be easily inhaled by the workers, and also general population can be exposed to this pollutant which was detected in the ambient air in a concentration range of 0.01–0.3 ppb or higher concentrations in drinking water. Several health risks were reported about its assumption by the humans. In the website of Agency for Toxic Substances and Disease Registry, it is reported that TCE is carcinogenic to humans (evidence for cancer is based on kidney cancer, limited evidence for non-Hodgkin lymphoma and liver cancer, as well as various tumors in animals). The co-exposure to TCE and drugs can present more potential dangerous effects due to a possible cross interaction. This can be due to the pharmaceuticals metabolism which can affect those of TCE in the liver. In particular, its pharmacokinetics and pharmacodynamics and in general its toxicity in target tissue can be improved by the simultaneous drug consumption. This investigation was carried out on rats treated with the different drugs (bolus dose equivalent to tenfold greater than the recommended daily dose) and exposed to TCE inhalation (50 ppm for 6 h). It is worth to mention the metabolism of TCA because it can define its possible interactions with drug metabolism. TCE is oxidized in the first step to chloral hydrate and then in trichloroacetic acid or by reduction in trichloroethanol. In parallel, it can be conjugated with GSH leading to a cysteine derivative. Haddad and coworkers considered three common drugs in this study which can undergo GSH conjugation as well as other metabolic CYP-dependent pathways, demonstrating an influence of naproxen on TCE metabolism. This interaction should be taken into account for adequate risk evaluation and management related to TCE exposure [6].

The same group investigated the effect on TCE metabolism by the presence of 14 widely used drugs in human-suspended hepatocytes. The drugs comprised in this investigation are mainly excreted as glucuronic acid conjugates in order to maximize their elimination from the human body. Also TCE is metabolized by the same pathway, and the competition for the same process can lead to an increase of the drugs and/or TCE levels in the human body resulting in an increase of side or toxic effects. As reported above, it is important to note that TCE is classified as carcinogenic substance, but also its main phase-I metabolites (chemical structure modifications) are associated with harmful effects. The results of this investigation revealed that of the entire group, only two drugs are endowed with the ability to interact with TCE metabolism. In particular, naproxen and carbamazepine are able to inhibit the formation of the conjugate between trichloroethanol (a metabolite of TCE) and glucuronic acid. The authors conclude that it is necessary to carry out an in vivo investigation for an adequate assessment of the risk to human health due to the metabolic interactions between TCE and the 14 investigated drugs [7].

The drugs are administered usually when necessary or for short period, unlike drugs of abuse that are taken for long period, and in this case, it is interesting to investigate the interactions between these compounds with pollutants. Carvajal et al. [8] reported some experimental evidence on this subject. Starting from the continuing exposure of humans to a variety of environmental neurotoxicants including herbicides, insecticides, fungicides, and pesticides, which alone can already cause health problems, they reported the effects of these chemicals when administered simultaneously with drugs of abuse such as heroin, cocaine, methamphetamine, and alcohol. The first report was on formaldehyde, a volatile compound widely used in the wood industry, which demonstrated the capacity to exhibit cross-sensitization with cocaine depending both on the dose and on the pattern of exposure to this volatile chemical. A second report is based on the possible interactions between heavy metals and drugs of abuse. The human exposure to these agents is very common through a variety of sources, and it can alone produce deleterious effects on human health. The different data reported in the literature contribute to forming a framework not very clear on the possible interactions between heavy metals and drugs of abuse. In fact, some studies demonstrated a possible potentiation of the effects of the different drugs of abuse in the presence of different heavy metals; on the other hand, other investigations showed an attenuation of the effects. The situation is still unclear, and the authors concluded that exposure to lead during the development can cause long-term changes in the response that these individuals give to drugs of abuse in adulthood. A lot of investigations were carried out on the effects of ethanol with neurotoxicants. Ethanol is the most famous and common drug of abuse used worldwide. In Europe it has been estimated that the pro capita consumption of alcohol is about 10 l of pure ethanol per year, with the consequent problems related to health and socioeconomic effects. On this basis, we can assume a more likely exposure to pollutants and neurotoxicants by alcohol consumers. One of the effects observed in animals exposed to ethanol for 8 weeks and lead is an increase of the latter in the blood and brain. Other effects were observed after lead or aluminum exposure together with alcohol, demonstrating the possible interactions between them. These effects range from the decrease of dopamine levels to reduced behavioral sensitivity to cocaine, indicating an intriguing framework of interactions. In the case of pesticides, it has been firstly considered the direct effects of these chemicals on the different target organs and/or systemic pathways and subsequently the possible interactions with the drugs of abuse. In particular, several investigations were devoted to elucidate the interactions between organophosphate and nicotine or ethanol. Epidemiological studies described a persistent intolerance to these two drugs of abuse in German workers involved in the manufacturing of chemical weapons during the World War II. The same was observed more recently in veterans of Gulf War, who presented intolerance to low doses of ethanol. Unfortunately, no more studies were carried out on the direct exposure of humans to these agents. Investigations on animals showed the enhanced effects of nicotine, partially explained through the inhibition of the enzyme acetylcholinesterase by the organophosphates, leading to acute cholinergic overstimulation at nicotinic and muscarinic synapses of peripheral autonomic and central nervous system. In addition, other receptor systems were considered in order to explain the acute and lethal action of exposure to organophosphate agents such as glutamatergic and GABAergic receptor systems. Several more data were collected relatively to a possible interaction between alcohol and organophosphates deriving from the observation that the exposure to organophosphate reduced voluntary ethanol drinking and increases its sedative effect. In conclusion, the authors suggest that organophosphates can affect the brain systems at the central level involved in neurobehavioral response to ethanol. However much studies need to be developed in order to further identify and characterize the mechanisms involved in such interactions.

As reported above, Nicklisch et al. [2] reported on the ability of pollutants to interact with P-gp system; in particular, they identified specific congeners of organochlorine pesticides, polychlorinated biphenyls, and polybrominated diphenyl ethers able to inhibit mouse and human P-gp (Fig. 12.1). They have approached the problem in the classical way starting from the structure of P-gp and verifying the ability of different pollutants to interact with it. To this aim, they carried out a series of biochemical and cellular assays on human and mouse P-gp in order to identify specific POPs able to inhibit this transport system. Interestingly, they validated the results on one of these congeners (polybrominated diphenyl ether) by using X-ray crystallography and verifying the binding of this chemical in the ligand pocket of the transporter, providing the first snapshot of a pollutant binding to P-gp. Finally, to provide more details about the environmental relevance of the different POPs on P-gp, they measured their level in yellowfin tuna (*Thunnus albacares*) and used these data to examine the effects on the human health of a representative POP mixture influencing the transport activity of human P-gp. They found relevant values of POPs in some samples of tuna tissues (total POPs 12.6 μM) which could affect vulnerable populations such as neonates, in which the xenobiotic metabolism is still not fully active. We have reported in detail the results of this paper just to demonstrate that interactions between drugs and pollutants are not only hypotheses. Scientific evidence coming from ongoing toxicological studies reveal the potential toxic effects of pollutants-drugs interaction.

Furthermore, it was observed that other chemicals can interfere with other drug transport systems. Fardel et al. [3] reviewed environmental chemicals as substrates, inhibitors, or inducers of drug transporters. They reported that a long series of

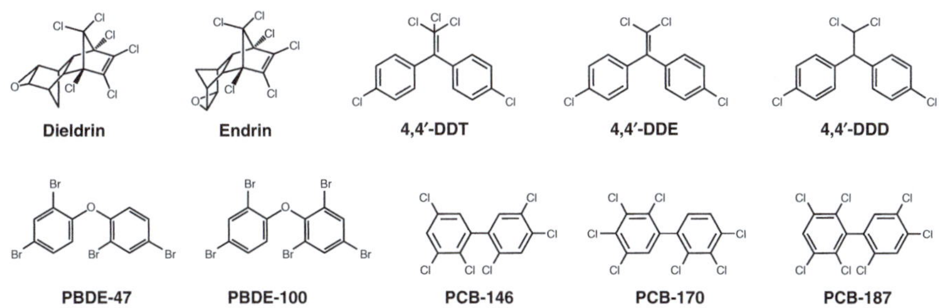

Fig. 12.1 Chemical structures of organochlorine pesticides and other polychlorinated biphenyls

environmental contaminants interact with different transporter causing an increase in the toxic effects of the drugs due to the lower rate of elimination. In addition, the authors also reported the ability of some environmental chemicals to modulate the drug transporter expression. This means that pollutants can interact with intracellular receptors, which translocate into the nucleus where they dimerize with receptor nuclear translocators and interact with specific responsive elements. Based on this mechanism, many contaminants can interact with xenobiotic-sensing receptors such as aryl hydrocarbon receptor (AhR), pregnane X receptor (PXR), constitutive androstane receptor (CAR), and peroxisome proliferator-activated receptor alpha (PPAR alpha) and, after their activation (agonistic behavior), alter the expression of these transporters. Besides this mechanism, some environmental chemicals are able to modulate transport systems without drug-sensitizing receptor interactions. For example, low doses of ochratoxin A (mycotoxin) to rats induce expression of organic anion transporter ($Oat_{1,3,4,5}$), which causes an accumulation of this substance in the kidney leading to a probable nephrotoxicity. The opposite effect was observed with high doses of ochratoxin A. Finally, an organophosphate called diazinon showed to induce the P-gp expression in intestinal Caco-2 cells, without affecting the levels of RNA messenger of multidrug resistance protein (MDR1). Other indications arise from transporter knockout mice which reveal the role of the transport systems in the deposition of the chemicals in the different body districts. The presence of such transport at intestinal level may have a role in carcinogenesis processes because they can determine the intake or not of particular substances which can contribute to cancer formation or can contribute to defeat the body from carcinogens. The role of transporters was also demonstrated at the level of blood-brain barrier, which is responsible for the penetration of pesticides in the brain leading to neurologic disease. However, it is worthy to note that the deposition of different chemicals inside the body sometimes depends on passive transport system based on the concentration gradient between the outer and inner part of the cell membrane.

At the end of this overview on the effects of the pollutants on the pharmacokinetics and pharmacodynamics of the different drugs, we would like to show the results from a paper of Ahmad and coworkers. They reported one of the first examples of the direct effect of pollutants on the conformation and biological activity of a protein such as human serum albumin (HSA) [9]. In particular, they analyzed the structural changes in HSA induced by different pollutants such as 1- and 2-naphtol and its isostere 8-quinolinol by circular dichroism, fluorescence spectroscopy, and dynamic light scattering. From these investigations, the authors discovered that the interactions between pollutants and HSA determined changes in HSA tertiary structure, enhancing its thermal stability and enzymatic activity as beta-lactamase. The authors postulated and concluded that in humans exposed to these pollutants, (a) the determination of HSA concentration by enzymatic assay can be erroneous, (b) the beta-lactams' resistance can be enhanced, and (c) the dose regime of beta-lactams antibiotics, and those of other drugs able to interact with HSA, should be carefully considered.

Other papers were published concerning the possible interactions of different pollutants with drugs. Scientific evidence has arisen from these investigations about potential interactions between drugs and pollutants, which however need to be confirmed through in vivo experiments to better mimic the conditions which may occur in the human body when exposed to different kind of pollutants at different concentrations. Finally, chronic effects should be also considered to have a whole vision of all effects.

Conclusions

Environmental pollutants may theoretically affect the pharmacokinetics of several drugs prescribed to and used by humans. The result is a contaminant-drug interaction that can involve drug transporter activity or expression. Since the exposure to environmental pollutants can induce toxic effects on the liver or kidney, i.e., organs involved in drug disposition, they can alter the functional expression of hepatic or renal transporters and decrease biliary or renal clearance. Recent studies has evidenced that several pollutants can act as endocrine disruptors and, by this way, may impact on hormone-regulated drug transporters and modify carrier-mediated transport of drugs. However, the consequences of those interactions are scarcely defined. Further studies should be performed to characterize the in vitro interactions for a wide range of pollutants through the use of high-throughput screening technologies. In a second step, in vivo studies should be realized to define the repercussion of transporter modulation on the pharmacokinetics of drugs that are substrates of pollutant-targeted transporters. In a third step, the results from the in vitro and in vivo toxicological studies should be integrated with data from exposure science. This step is critical. Experimental studies are often conducted with dose levels much higher than those achievable in human population. Furthermore, human population is exposed to a mixture of environmental pollutants. The components of the mixture can interact with each other, and the effects of the pollutants in the mixture toward transporters can be synergistic, additive, or antagonistic. An aspect deserving attention is that the biological responses in humans throughout the life span can be influenced by the exposures from the environment, diet, behavior, and endogenous processes. In 2005, Dr. Christopher Wild coined the term "exposome" and suggested that the exposome "encompasses life-course environmental exposures (including lifestyle factors), from the prenatal period onwards" [10]. The concept of exposome can play a pivotal role in defining the human exposure to environmental pollutants because it places that exposure within the broader context of diet, behavior, and other exogenous and endogenous agents and incorporates how human body reacts to environment pressures, including epigenetic changes and mutations, as well as its biochemical reactions.

References

1. Ariens EJ. Adverse drug interactions. Proc Eur Soc Study Drug Toxicol. 1972;13:37–63.
2. Nicklisch SC, Rees SD, McGrath AP, et al. Global marine pollutants inhibit P-glycoprotein: environmental levels, inhibitory effects, and cocrystal structure. Sci Adv. 2016;2:e1600001.
3. Fardel O, Kolasa E, Le Vee M. Environmental chemicals as substrates, inhibitors or inducers of drug transporters: implication for toxicokinetics, toxicity and pharmacokinetics. Expert Opin Drug Metab Toxicol. 2012;8:29–46.
4. Li WP, Wang YF, Gao J, et al. In vitro evidence for endocrine-disrupting chemical (EDC)'s inhibition of drug metabolism. Afr Health Sci. 2014;14:185–8.
5. Verner MA, Magher T, Haddad S. High concentrations of commonly used drugs can inhibit the in vitro glucuronidation of bisphenol a and nonylphenol in rats. Xenobiotica. 2010;40:83–92.
6. Rouhou MC, Charest-Tardif G, Haddad S. In vivo effects of naproxen, salicylic acid, and valproic acid on the pharmacokinetics of trichloroethylene and metabolites in rats. J Toxicol Environ Health A. 2015;78:671–84.
7. Cheikh Rouhou M, Haddad S. Modulation of trichloroethylene in vitro metabolism by different drugs in human. Toxicol In Vitro. 2014;28:732–41.
8. Carvajal F, Sanchez-Amate M, Lerma-Cabrera JM, Cubero I. Interaction between exposure to neurotoxicants and drug abuse. In: Oosthuizen J, editor. Environmental health-emerging issues and practice. Rijeka: InTech; 2012. p. 273–88.
9. Ahmad E, Rabbani G, Zaidi N, et al. Pollutant-induced modulation in conformation and β-lactamase activity of human serum albumin. PLoS One. 2012;7:e38372.
10. Wild CP. Complementing the genome with an "Exposome": the outstanding challenge of environmental exposure measurement in molecular epidemiology. Cancer Epidemiol Biomark Prev. 2005;14:1847–50.

Epigenetic Effects of Air Pollution

13

Ernesto Burgio and Agostino Di Ciaula

13.1 Introduction

According to the WHO, 12.6 million deaths globally (23% of all deaths) were attributable in the 2012 to the environment, and, considering both death and disability, the global burden of environmental-related disease is 22% (95%CI 12–32) [1]. A considerable part of this burden is due to atmospheric pollution and principally to fine particulate matter (PM) that could be responsible for 3.2 million deaths per year and 76 million years of healthy life lost [2]. These assessments are certainly and strongly underestimated, since these did not consider a number of diseases (i.e., diabetes and other insulin-resistance-related disease, chronic neurological diseases, endocrine and immune disorders, perinatal and pregnancy disorders, miscarriage, etc.) not "classically" related to environmental pollution but having well-demonstrated pathogenic links with it.

In particular, the incidence of a number of chronic diseases including cancer (and, principally, childhood cancer [3]), neurodevelopmental disorders [4], psychiatric and neurodegenerative diseases (i.e., Alzheimer disease [5]), metabolic disorders such as obesity [6–8], type 2 [9] and type 1 diabetes [10, 11] is rapidly rising worldwide, being currently recognized as the *twenty-first-century epidemiological transition* [12].

Genetic factors seem to have a limited impact on the continuous increase of all these diseases, taking into account the relatively short period in which these

E. Burgio (✉)
European Cancer and Environment Research Institute, Bruxelles, Belgium

ISDE (International Society of Doctors for Environment), Basel, Switzerland
e-mail: erburg@libero.it

A. Di Ciaula
Division of Internal Medicine, Hospital of Bisceglie (ASL BAT), Bisceglie, Italy

ISDE Italia (International Society of Doctors for Environment), Arezzo, Italy

F. Capello, A.V. Gaddi (eds.), *Clinical Handbook of Air Pollution-Related Diseases*,
https://doi.org/10.1007/978-3-319-62731-1_13

variations are occurring, as compared with the large time window requested to establish genetic changes at a population level.

On the other hand, in relation to the origins of noncommunicable diseases, a critical role is emerging for environmentally induced *epigenetic changes* (heritable variations in gene activity and expression that occur without alteration in DNA sequence [13–16]), principally for the ones that could negatively affect the *fetal programming* (which relates to the concept of *developmental plasticity,* where genes can express different ranges of physiological or morphological states in response to the environmental conditions during fetal development) [17], with later health consequences of in utero and/or perinatal exposure and even transgenerational transmission of risk [18, 19].

On the other hand, it is important to emphasize that most of these epigenetic changes are potentially modifiable by environmental and therapeutic interventions, which could make it possible to put forward definite *primary prevention* measures.

13.2 What Air Pollution Really Means

According to the WHO, "air pollution" is a heterogeneous mixture of gases and solid particles, each component having potential effect on the human body [20]. But is this definition really exhaustive?

- Air pollution is not only a more or less significant increase of indefinite pollutants able to promote respiratory injuries, heart attacks, and strokes in vulnerable people for constitutional reasons or age. It is a mixture of toxic gases, heavy metals, ultrafine particulate, and persistent and not biodegradable toxic molecules able to enter the human body through airways and/or by foods, to reach the blood, to easily cross all biological barriers (including placenta, blood-brain barrier, cell and nuclear membranes), to access the central nervous system and embryo-fetal tissues during their early development, to interfere with all the inter- and intracellular signaling pathways and with hormonal systems, and, therefore, with all the major biochemical and molecular mechanisms, including the main steps of gene expression: chromatin variations, transcription and translation of gene sequences, protein folding, and activation of mobile genetic elements, in particular in the brain tissue.
- Air pollution does not only produce thousands of deaths per day (more than 7 million preventable deaths/year according to the WHO) but is also one of the major determinants of chronic systemic inflammation (or subacute low-grade inflammation) which accompanies us for most of our lives opening the way to atherosclerosis, cardiovascular and cerebrovascular diseases, immunological disorders such as allergies and autoimmune diseases, endocrine and metabolic diseases including obesity and diabetes, neurodevelopmental and neurodegenerative disorders, and cancer. Not coincidentally, all these diseases are largely increasing, as already mentioned (*the epidemiological twenty-first-century transition*) [12].

– Above all, air pollution interferes with cell differentiation and proliferation of:
  (1) *Stem cells* in various adult tissues (particularly during chronic stress or tissue damage, interfering with cell differentiation [21] and/or opening the way to proliferative processes [22])
  (2) *Embryo-fetal cells* during the whole period of ontogenetic development, strongly conditioning (throughout life) cell differentiation and, therefore, the epigenetic programming of different tissues and organs and of the entire system [23]
  (3) The *gametes*, with potential transgenerational effects, as demonstrated by numerous toxicological and epidemiological studies [24, 25]

13.3 A Polluting Agent Widely Pervasive and Toxic to the Development: The Ultrafine Particulate

Particulate matter (PM) is composed of particles of different composition and size ranging from nanometric scale up to tens of micrometers. PM10 (particles with a diameter less than or equal to 10 μm) are usually retained and transformed at the level of upper airways/lung. PM2.5 (particles with a diameter less than or equal to 2.5 μm) penetrate deeply (lower airways) and are able to cross the blood-lung barrier.

There is a direct relationship between the air concentration of PM and human health [20], with damages starting at a subcellular and cellular level [26–31]. The major pathogenic mechanisms induced by acute and chronic exposure to PM include oxidative stress [32], vasoconstriction [33], pro-thrombotic [34] and antifibrinolytic [33] activity, ischemic damage induction [35], genotoxicity [36], insufficient DNA repair [37], and altered gene expression.

No apparent threshold exists below which the association between the air concentration of PM and adverse health effects no longer applies [38], and there is a direct relationship between PM10, PM2.5, and various diseases for concentration levels ranging from 0 to up to 200 $\mu g/m^3$ [39].

Of note, the health hazards of PM are inversely proportional to the diameter of particles, being maximal in the case of ultrafine particles (UFP, < 0.1 μm), which are able to cross all anatomic and functional barriers (including the blood-brain barrier [40] and the placenta [41]) and to penetrate cells (starting from lung macrophages, directly interfering with local [42, 43]), transporting toxics everywhere throughout the body. Moreover, and above all, ultrafine and fine particles are very high in number. They have have greater total surface area than particles with larger size and are able to absorb and retain a number of toxic substances due to their spongy surface [44].

UFP health impacts are quite different than PM10 effects [45] and unfortunately they are not currently regulated by national air quality standards, although they represent a major threat to human health [46–50], principally due to their high capacity to quickly evade the lungs and to deposit in extrapulmonary tissues [51, 52].

As compared with greater particles, UFP have a largely wide surface in relation with volume, are able to transport a large amount of adsorbed toxics and have much more pronounced pro-oxidative and pro-inflammatory properties [53, 54].

UFP are not routinely measured in any part of the world and are able to remain suspended in the atmosphere for days and to be transported by wind traveling hundreds of kilometers [55]. All toxic particles (including UFP) are generated by combustion processes, including motor vehicles [56] (with an emission rate of about 10^{15} UFPs/km [57]) and industrial procedures processing fossil fuels [58–61], biomass [60, 62], or waste [63, 64]. Of note, besides direct emission, particles (including UFP) can also take origin in the ambient air (i.e., *secondary particulate*), following chemical reactions between precursors (i.e., sulfur dioxide, nitrogen oxides, ammonia) [64–71] and having the same harmful effects of primary particulate [71].

Particularly critical targets for UFP are pregnant women and the fetal development. Actually the fetal development is a vulnerable period for many toxic substances, in particular considering the immune system development, which might be strongly secondarily affected by maternal exposure [72]. UFP can induce apoptosis in cultured human umbilical endothelial cells [73], and it has been shown that nanosized particles are able to cross the human placenta [41, 74], moving easily in fetal tissues and in particular translocating in the fetal brain [75], contributing significantly to the negative effects of road traffic on brain development [76].

In animal models it has been shown that nanoparticles may be cytotoxic to trophoblasts [77] and are able, in offspring, to alter postnatal growth [78], to damage kidneys [79], to induce allergic or inflammatory effects [80] and neurobehavioral disorders [81–83], and to affect the reproductive function [84].

13.4 The Epigenome as the Main Link Between Air Pollution and Disease Occurrence

Epigenetic mechanisms have been indicated as the key factor in the pathogenesis of several diseases, due to exposure to toxic agents [85] during fetal development [86, 87]. The epigenome should be considered as a dynamic molecular network surrounding the DNA, opened to the stimuli deriving from the environment that will induce (for life) the *programming* of the phenotype, without changing the DNA sequence [87].

Many studies indicate the main epigenetic mechanisms (DNA methylation [88], histone acetylation/deacetylation, and noncoding microRNAs mediating gene-silencing [89]) having structural or regulatory functions [90] as the essential intersection between environmental factors and the inherited nucleotide polymorphisms in the determination of noncommunicable diseases, including endocrine and metabolic [16, 91], neurobiological [92], and immunological [93, 94] diseases and cancer [14].

Black carbon and PM2.5 are able to decrease LINE-1 and Alu element methylation, which increases their activity as retrotransposons in human blood cells according to time-related variables [95]. A decreased global DNA methylation in whole blood from non-smoking adults has also been associated with environmental exposure to NO_2, PM10, and ozone [96]. Altered DNA methylation secondary to environmental pollution has been linked (also in the short term) with blood pressure [97], heart rate variability [98], alterations in cardiovascular biomarkers [99], higher fasting blood glucose levels in nondiabetic subjects [98], asthma [100], impaired systemic immunity [101], and metabolic diseases [91].

Of note, epigenetic mechanisms may also be involved in transgenerational transmission of risk through exposed pregnant women and/or heritable germline epimutations.

It has been recently demonstrated the presence of differential offspring DNA methylation in mitochondria-related genes following NO_2 exposure during pregnancy, with a concurrent altered expression of genes involved in antioxidant protective pathways [102].

On the other hand, animal studies have suggested that exposure to chemical toxics in adult age or during early life (prenatal exposure) is able to alter the DNA methylation patterns in germ cells, with transgenerational transmission of these alterations and subsequent phenotypic effects [103]. The epigenetic transgenerational inheritance has been shown in a variety of species, including humans [104]. It has been also suggested that induction of epigenetic transgenerational inheritance of sperm epimutations by environmental factors is able to promote genome instability and acquisition of genetic mutations in later generations, pointing to a combination of environmental-induced epigenetic and genetic variations in the transgenerational determination of phenotypes [105].

13.5 Environment, Epigenome, and Cancer

Unlike the majority of cancers in adult age, childhood cancer has a short latency (months or years) before occurrence, which emphasizes the importance of environmental factors causing the disease, acting in a limited and well-recognizable time window through specific mechanisms (gametal/transplacental exposure, epigenome-mediated alterations in gene expression, epigenetic transgenerational inheritance).

In the last decades (1970–2012), a slow but progressive rise in the incidence of some leukemias and brain cancers (the most frequent types of pediatric cancer) has been reported in US children and adolescents (0–19 years), with brain cancer surpassing leukemia as the leading cause of cancer death in this age group [106].

An increasing trend has also been noticed, in US children, for differentiated thyroid carcinomas [107]. Besides US children, a significant increase in cancer incidence rate has been reported (years 2000–2010) among Chinese children [108], and an increasing trend for pediatric hepatoblastoma (an embryonal tumor) has been reported, in the period 1991–2012, in Germany [109]. Data from 33 population-based cancer registries in 15 European countries (ACCIS, period 1978–1997) showed an increasing trend (average yearly percentage change of 1.1%) including the most common cancer types (soft tissue sarcomas, brain tumors, tumors of the sympathetic nervous system, germ-cell tumors, carcinomas, lymphomas, renal tumors, and leukemias [3]). Moreover, ACCIS data showed the maximum increase in the first year of age, suggesting a critical role for transplacental and/or transgenerational (epigenetic/germline) exposure to environmental toxics.

These data, taken together, would suggest a critical role for environmental factors and for their epigenetic effects, which seem to be of major importance as compared with the hypothesis of stochastic genetic mutations, whose fast increase in a few decades would be difficult to explain. Several observations demonstrated an increased occurrence of cancer in children exposed to pollution and/or living in contaminated sites [110–115].

An increased risk of neuroblastoma has been detected in children whose mother was exposed to solvents (in particular aromatic hydrocarbons) in the preconceptional period [116], and paternal preconceptional occupational exposure to polycyclic aromatic hydrocarbons is linked with increased risks of all childhood brain tumors [117].

In children, specific variations in DNA methylation have been shown as a potential mediator of environmental risks in the onset of acute lymphoblastic leukemia [118], have been described in malignant gliomas [119], and are associated with tumor progression and poor prognosis in brain tumors [120].

Specific epigenetic mechanisms seem to have a critical role also in the determination of cancer in adults [121–124]. In particular, disturbed epigenetic regulation of gene transcription has been described in hematopoietic malignancies [121], microRNAs might have a specific role in the promotion of lung inflammation and carcinogenesis [125], and DNA methylation seems to have direct relationships with carcinogenesis [123, 126]. Also in adults, epigenetic changes are linked with environmental exposure to toxics. In fact, it has been shown that traffic-related air pollution is able to induce alteration of the DNA methylation pattern in human lung cells [127], Downregulation of miR-144 has been shown critical for air pollution-related lung cancer [128], and exposure to polycyclic aromatic hydrocarbons is linked with hypo- or hyper-methylation at multiple promoter regions in women with breast cancer [129].

Some years ago the English epidemiologist Ernest Knox analyzed the addresses of children who died of cancer in Britain between 1955 and 1980, showing that the risk increased significantly not so much for children born and/or raised near busy roads, gas stations, bus stations, and industrial facilities as for the children of women living in these areas during the period of pregnancy [130, 131]. Both the ACCIS data and the Knox studies that focused on the relationship between air pollution and cancer (according to Knox a quarter of childhood cancer has this origin) brought a significant contribution to the thesis of a possible transplacental or transgenerational transmission of children cancer [14], which is also confirmed by the presence of the common leukemic fusion genes, TEL-AML1 t (12; 21) (p12; q22) or AML1-ETO t (8; 21) (q22q22) in the cord blood of 1:100 children [132]. More recently, a national registry-based case-control study [ESCALE (Etude Sur les Cancers et les Leucémies de l'Enfant)], carried out in France, led to assess the effect of exposure to road traffic exhaust fumes on the risk of childhood leukemia and supported the hypothesis that living close to heavy-traffic roads may increase the risk [133].

13.6 The Link Between Air Pollution and Chronic Neurological Diseases

The brain is a specific target of air pollution, which might induce neurotoxicity mainly through oxidative stress and neuro-inflammation, leading to chronic neurodevelopmental disorders [134] and neurodegenerative diseases [135].

Environmental factors are able to affect the epigenetic mechanisms involved in brain development during fetal life, also through the activation of the maternal immune system during pregnancy [136]. Prenatal exposure to PAH air pollutants contributes to ADHD and other neurodevelopmental disorders by disrupting the development of left hemisphere white matter, whereas postnatal PAH exposure disturbs white matter expansion in dorsal prefrontal regions [137].

Air pollution is responsible for cognitive disorders during childhood mainly due to oxidative stress [138], and several studies linked autism spectrum disorders (ASD) with pollution in particular for exposures occurring during periods of marked neuro- and glio-genesis [139].

The risk of autism significantly increases in the case of fetal exposure to air pollutants produced by urban traffic and industrial emissions [140, 141], in particular during the third trimester [140, 142].

Gene-environment interaction through epigenetic mechanisms seems to have a major role in increasing the risk of ASD in susceptible children [143], and definite inherited alterations of DNA methylation could affect gene expression, contributing to ASD susceptibility [144].

Growing interest also exists in the association between environmental pollution and neurodegenerative disorders such as Alzheimer and Parkinson diseases [145], considering in particular the role of early (i.e., prenatal and postnatal) exposures in the occurrence of these diseases in later life [146] and the epigenetic mechanisms (i.e., DNA methylation, chromatin remodeling, miRNAs) operating in this way from prenatal life throughout lifetime [147–150].

Of great impact were the studies conducted in the last decade by pediatricians and epidemiologists in one of the most polluted megacities of the world, Mexico City, which showed that air pollution (especially UFPs and heavy metals) may favor the accumulation of misfolded proteins (hyperphosphorylated tau (HPπ), alpha-synuclein, and beta-amyloid) and their early deposition in the brain of dogs [151], adolescents [152], and even children [153] in cerebral areas exactly coinciding with the anatomical distribution observed in both Alzheimer and Parkinson diseases.

Of particular relevance in this context appears to be the *Latent EARly-life associated regulation model (LEARn)*, in which many early exposures to environmental agents remain latent for decades to manifest themselves again at maturity or senescence, increasing the production of Aβ and opening the way to Alzheimer disease (AD) [154, 155]. These models attest the great role that the environmental pollution seems to have in the genesis of neurodevelopmental and neurodegenerative diseases that are in great increase all over the world, by interfering epigenetically on the programming of the nervous system [156, 157]. Many scientists and researchers have recently pointed out that this awareness could, at the same time, open the way to effective strategies of primary prevention based on the reduction of atmospheric pollution and more generally on the detoxification of our environment [158, 159].

13.7 The Epigenetic Relationships Between the Progressive Rise in Metabolic Diseases and Environmental Pollution

Obesity has more than doubled since 1980 worldwide [160, 161], and the obesity epidemic continues to spread [162] with no evident reversal trend, although a growing number of countries have adopted specific policies to prevent the obesity epidemic [162].

Similar to obesity, in the past few decades, the incidence of type 2 diabetes has doubled, with a rising rate also observed in children [163, 164]. Increasing epidemiological trends have also been shown, in pediatric age, for type 1 diabetes [10, 11, 165].

Gene-environment interactions [91, 166–171] and epigenetic mechanisms [16] have a crucial role in the determination of obesity [171]. Maternal exposure to toxics during pregnancy is able to influence the metabolic phenotype of the offspring through in utero programming of later obesity [172], mainly by epigenetic mechanisms [173], and elevated maternal serum concentrations of persistent organic pollutants in the first trimester of pregnancy are related to the onset of obesity in offspring [174].

Specific variations in the DNA methylation patterns are associated with body mass index (BMI) and waist circumference [175] and, remarkably, in monozygotic twin pairs discordant for BMI [176].

The US National Institute of Environmental Health Sciences/National Toxicology Program (NIEHS/NTP) listed, in the year 2012, more than 200 human studies which described specific relationships between environmental pollutants, obesity, and type 2 diabetes [177]. Nowadays, a growing role is emerging for the effects of endocrine-disrupting chemicals (EDCs) in the onset of obesity both in children and in adults [178, 179]. EDCs, in particular, dysregulate epigenetic mechanisms [180, 181] (mainly by increasing histone deacetylation and altering DNA methylation [181–183], inducing an increased adipocyte formation and an increased fat storage [184–186]. Besides environmental toxics introduced with contaminated food (i.e., pesticides, BPA, phthalates, POPs), a major role in the determination of metabolic diseases seems to be played by air pollutants.

Environmental exposure to PM10 and NO_2 has been linked with insulin resistance in German school-aged children [187], with a critical role for oxidative stress induced by air pollution [188]. DNA methylation is strongly influenced by air concentrations of PM2.5, which are able to influence, in the short medium term, fasting blood glucose levels [98].

Furthermore, an association between chronic exposure to pollutants, type 2 diabetes [189], insulin resistance [187, 188], and type 1 diabetes (T1D) [190] has been also shown in children.

The MIREC study showed that maternal exposure to nitrogen dioxide and PM2.5 during pregnancy increases adiponectin levels in umbilical cord blood, underscoring the effects of air pollution on fetal metabolic function and on the potential development of childhood obesity [191]. Maternal exposure during pregnancy seems to be relevant also in the case of T1D, since an association has been described between elevated levels of ozone and NOx during pregnancy and the disease occurrence in offspring [192]. The epigenome has a central role in the onset of T1D [94, 193], as suggested by a number of studies linking DNA methylation [194], histone post-translational modifications [195], and microRNA dysregulation [196, 197] with the onset of this chronic disease in children [190]. As the incidence of type 1 diabetes has doubled every 20 years over the last few decades, a trend that could not be explained by genetic factors and suggests strong environmental effects, it is important to note that the different DNA methylation profiles in twins point to strong epigenetic effects in the etiology of the disease [198].

Conclusions

The effects of air pollution on the onset and clinical history of both acute and chronic (in particular noncommunicable) diseases are a pressing public health issue, considering the progressive epidemiological rise of widely diffused conditions such as metabolic, allergic and autoimmune, and neurodevelopmental and neurodegenerative diseases and cancer both in adults and in children.

All these health effects are potentially avoidable through primary prevention measures, and the growing knowledge of epigenetic mechanisms actually represents a powerful tool to reduce their incidence and their costs, in particular pointing to the safeguard of pregnancy and the perinatal period.

However, there is also an urgent need for research protocols focusing on environmental exposures and early nutrition, in order to provide better insights into pathogenesis.

In this context it will be necessary, above all, to focus on the epigenetic regulation of inflammatory pathways that are a likely candidate for persistent changes in metabolic and brain function, as a consequence of the perinatal exposure to environmental toxicants. The results of such studies could provide the opportunity for more and more effective strategies of primary prevention, necessary to reduce the increasing burden of chronic diseases linked to environmental pollution and especially to the atmospheric pollution of our cities.

References

1. Prüss-Ustün A, Wolf J, Corvalán C, Bos R, Neira M. Preventing disease through healthy environments. A global assessment of the burden of disease from environmental risks. Geneva: WHO; 2016.

2. Lim SS, Vos T, Flaxman AD, Danaei G, Shibuya K, Adair-Rohani H, Amann M, Anderson HR, Andrews KG, Aryee M, Atkinson C, Bacchus LJ, Bahalim AN, Balakrishnan K, Balmes J, Barker-Collo S, Baxter A, Bell ML, Blore JD, Blyth F, Bonner C, Borges G, Bourne R, Boussinesq M, Brauer M, Brooks P, Bruce NG, Brunekreef B, Bryan-Hancock C, Bucello C, Buchbinder R, Bull F, Burnett RT, Byers TE, Calabria B, Carapetis J, Carnahan E, Chafe Z, Charlson F, Chen H, Chen JS, Cheng AT, Child JC, Cohen A, Colson KE, Cowie BC, Darby S, Darling S, Davis A, Degenhardt L, Dentener F, Des Jarlais DC, Devries K, Dherani M, Ding EL, Dorsey ER, Driscoll T, Edmond K, Ali SE, Engell RE, Erwin PJ, Fahimi S, Falder G, Farzadfar F, Ferrari A, Finucane MM, Flaxman S, Fowkes FG, Freedman G, Freeman MK, Gakidou E, Ghosh S, Giovannucci E, Gmel G, Graham K, Grainger R, Grant B, Gunnell D, Gutierrez HR, Hall W, Hoek HW, Hogan A, Hosgood HD III, Hoy D, Hu H, Hubbell BJ, Hutchings SJ, Ibeanusi SE, Jacklyn GL, Jasrasaria R, Jonas JB, Kan H, Kanis JA, Kassebaum N, Kawakami N, Khang YH, Khatibzadeh S, Khoo JP, Kok C, Laden F, Lalloo R, Lan Q, Lathlean T, Leasher JL, Leigh J, Li Y, Lin JK, Lipshultz SE, London S, Lozano R, Lu Y, Mak J, Malekzadeh R, Mallinger L, Marcenes W, March L, Marks R, Martin R, McGale P, McGrath J, Mehta S, Mensah GA, Merriman TR, Micha R, Michaud C, Mishra V, Mohd Hanafiah K, Mokdad AA, Morawska L, Mozaffarian D, Murphy T, Naghavi M, Neal B, Nelson PK, Nolla JM, Norman R, Olives C, Omer SB, Orchard J, Osborne R, Ostro B, Page A, Pandey KD, Parry CD, Passmore E, Patra J, Pearce N, Pelizzari PM, Petzold M, Phillips MR, Pope D, Pope CA III, Powles J, Rao M, Razavi H, Rehfuess EA, Rehm JT, Ritz B, Rivara FP, Roberts T, Robinson C, Rodriguez-Portales JA, Romieu I, Room R, Rosenfeld LC, Roy A, Rushton L, Salomon JA, Sampson U, Sanchez-Riera L, Sanman E, Sapkota A, Seedat S, Shi P, Shield K, Shivakoti R, Singh GM, Sleet DA, Smith E, Smith KR, Stapelberg NJ, Steenland K, Stockl H, Stovner LJ, Straif K, Straney L, Thurston GD, Tran JH, Van Dingenen R, van Donkelaar A, Veerman JL, Vijayakumar L, Weintraub R, Weissman MM, White RA, Whiteford H, Wiersma ST, Wilkinson JD, Williams HC, Williams W, Wilson N, Woolf AD, Yip P, Zielinski JM, Lopez AD, Murray CJ, Ezzati M, MA AM, Memish ZA. A comparative risk assessment of burden of disease and injury attributable to 67 risk factors and risk factor clusters in 21 regions, 1990-2010: a systematic analysis for the global burden of disease study 2010. Lancet. 2012;380(9859):2224–60. https://doi.org/10.1016/S0140-6736(12)61766-8.
3. Kaatsch P, Steliarova-Foucher E, Crocetti E, Magnani C, Spix C, Zambon P. Time trends of cancer incidence in European children (1978-1997): report from the automated childhood cancer information system project. Eur J Cancer. 2006;42(13):1961–71. https://doi.org/10.1016/j.ejca.2006.05.014.
4. Fombonne E. Epidemiology of pervasive developmental disorders. Pediatr Res. 2009;65(6):591–8. https://doi.org/10.1203/PDR.0b013e31819e7203.
5. Brookmeyer R, Johnson E, Ziegler-Graham K, Arrighi HM. Forecasting the global burden of Alzheimer's disease. Alzheimers Dement. 2007;3(3):186–91. https://doi.org/10.1016/j.jalz.2007.04.381.
6. Roth J, Qiang X, Marban SL, Redelt H, Lowell BC. The obesity pandemic: where have we been and where are we going? Obes Res. 2004;12(Suppl 2):88S–101S. https://doi.org/10.1038/oby.2004.273.
7. Stein CJ, Colditz GA. The epidemic of obesity. J Clin Endocrinol Metab. 2004;89(6):2522–5. https://doi.org/10.1210/jc.2004-0288.
8. Stevens GA, Singh GM, Lu Y, Danaei G, Lin JK, Finucane MM, Bahalim AN, McIntire RK, Gutierrez HR, Cowan M, Paciorek CJ, Farzadfar F, Riley L, Ezzati M, Global Burden of Metabolic Risk Factors of Chronic Diseases Collaborating G. National, regional, and global trends in adult overweight and obesity prevalences. Popul Health Metrics. 2012;10(1):22. https://doi.org/10.1186/1478-7954-10-22.
9. Shaw JE, Sicree RA, Zimmet PZ. Global estimates of the prevalence of diabetes for 2010 and 2030. Diabetes Res Clin Pract. 2010;87(1):4–14. https://doi.org/10.1016/j.diabres.2009.10.007.
10. Di Ciaula A. Association between air pollutant emissions and type 1 diabetes incidence in European countries. Adv Res. 2014;2(7):409–25.

11. Patterson CC, Gyurus E, Rosenbauer J, Cinek O, Neu A, Schober E, Parslow RC, Joner G, Svensson J, Castell C, Bingley PJ, Schoenle E, Jarosz-Chobot P, Urbonaite B, Rothe U, Krzisnik C, Ionescu-Tirgoviste C, Weets I, Kocova M, Stipancic G, Samardzic M, de Beaufort CE, Green A, Dahlquist GG, Soltesz G. Trends in childhood type 1 diabetes incidence in Europe during 1989-2008: evidence of non-uniformity over time in rates of increase. Diabetologia. 2012;55(8):2142–7. https://doi.org/10.1007/s00125-012-2571-8.
12. Burgio E. Environmental and fetal programming: the origins of some current "pandemics". J Pediatr Neonat Individual Med. 2015;4(2):e040237. https://doi.org/10.7363/040237.
13. Bird A. Perceptions of epigenetics. Nature. 2007;447(7143):396–8. https://doi.org/10.1038/nature05913.
14. Burgio E. Notes on the epigenetic origins of childhood cancer. Epidemiol Prev. 2013;37(1 Suppl 1):261–5.
15. Burgio E, Migliore L. Towards a systemic paradigm in carcinogenesis: linking epigenetics and genetics. Mol Biol Rep. 2015;42(4):777–90. https://doi.org/10.1007/s11033-014-3804-3.
16. Lopomo A, Burgio E, Migliore L. Epigenetics of obesity. Prog Mol Biol Transl Sci. 2016;140:151–84. https://doi.org/10.1016/bs.pmbts.2016.02.002.
17. Byrne CD, Phillips DI. Fetal origins of adult disease: epidemiology and mechanisms. J Clin Pathol. 2000;53(11):822–8.
18. Gluckman PD, Hanson MA. Developmental origins of disease paradigm: a mechanistic and evolutionary perspective. Pediatr Res. 2004;56(3):311–7. https://doi.org/10.1203/01.PDR.0000135998.08025.FB.
19. Landrigan PJ, Sonawane B, Butler RN, Trasande L, Callan R, Droller D. Early environmental origins of neurodegenerative disease in later life. Environ Health Perspect. 2005;113(9):1230–3.
20. World Health Organization. Review of evidence on health aspects of air pollution - REVIHAAP project- first results. Denmark: World Health Organization–Regional Office for Europe; 2013.
21. Cui Y, Jia F, He J, Xie X, Li Z, Fu M, Hao H, Liu Y, Liu DZ, Cowan PJ, Zhu H, Sun Q, Liu Z. Ambient fine particulate matter suppresses in vivo proliferation of bone marrow stem cells through reactive oxygen species formation. PLoS One. 2015;10(6):e0127309. https://doi.org/10.1371/journal.pone.0127309.
22. Cui Y, Sun Q, Liu Z. Ambient particulate matter exposure and cardiovascular diseases: a focus on progenitor and stem cells. J Cell Mol Med. 2016;20(5):782–93. https://doi.org/10.1111/jcmm.12822.
23. Wadhwa PD, Buss C, Entringer S, Swanson JM. Developmental origins of health and disease: brief history of the approach and current focus on epigenetic mechanisms. Semin Reprod Med. 2009;27(5):358–68. https://doi.org/10.1055/s-0029-1237424.
24. Somers CM. Ambient air pollution exposure and damage to male gametes: human studies and in situ 'sentinel' animal experiments. Syst Biol Reprod Med. 2011;57(1–2):63–71. https://doi.org/10.3109/19396368.2010.500440.
25. Yauk CL, Lucas Argueso J, Auerbach SS, Awadalla P, Davis SR, Demarini DM, Douglas GR, Dubrova YE, Elespuru RK, Glover TW, Hales BF, Hurles ME, Klein CB, Lupski JR, Manchester DK, Marchetti F, Montpetit A, Mulvihill JJ, Robaire B, Robbins WA, Rouleau GA, Shaughnessy DT, Somers CM, Taylor JG, Trasler J, Waters MD, Wilson TE, Witt KL, Bishop JB. Harnessing genomics to identify environmental determinants of heritable disease. Mutat Res. 2013;752(1):6–9. https://doi.org/10.1016/j.mrrev.2012.08.002.
26. Dick CA, Brown DM, Donaldson K, Stone V. The role of free radicals in the toxic and inflammatory effects of four different ultrafine particle types. Inhal Toxicol. 2003;15(1):39–52.
27. Donaldson K, Brown DM, Mitchell C, Dineva M, Beswick PH, Gilmour P, MacNee W. Free radical activity of PM10: iron-mediated generation of hydroxyl radicals. Environ Health Perspect. 1997;105(Suppl 5):1285–9.
28. Eiguren-Fernandez A, Shinyashiki M, Schmitz DA, DiStefano E, Hinds W, Kumagai Y, Cho AK, Froines JR. Redox and electrophilic properties of vapor- and particle-phase components of ambient aerosols. Environ Res. 2010;110(3):207–12.

29. Laing S, Wang G, Briazova T, Zhang C, Wang A, Zheng Z, Gow A, Chen AF, Rajagopalan S, Chen LC, Sun Q, Zhang K. Airborne particulate matter selectively activates endoplasmic reticulum stress response in the lung and liver tissues. Am J Phys Cell Phys. 2010;299(4):C736–49.
30. Osornio-Vargas AR, Serrano J, Rojas-Bracho L, Miranda J, Garcia-Cuellar C, Reyna MA, Flores G, Zuk M, Quintero M, Vazquez I, Sanchez-Perez Y, Lopez T, Rosas I. In vitro biological effects of airborne PM(2).(5) and PM(1)(0) from a semi-desert city on the Mexico-US border. Chemosphere. 2011;83(4):618–26. https://doi.org/10.1016/j.chemosphere.2010.11.073.
31. Velasco G. Endoplasmic reticulum stressed by pollution. Focus on "airborne particulate matter selectively activates endoplasmic reticulum stress response in the lung and liver tissues". Am J Phys Cell Phys. 2010;299(4):C727–8.
32. Miller MR, Shaw CA, Langrish JP. From particles to patients: oxidative stress and the cardiovascular effects of air pollution. Futur Cardiol. 2012;8(4):577–602. https://doi.org/10.2217/fca.12.43.
33. Mills NL, Tornqvist H, Robinson SD, Gonzalez M, Darnley K, MacNee W, Boon NA, Donaldson K, Blomberg A, Sandstrom T, Newby DE. Diesel exhaust inhalation causes vascular dysfunction and impaired endogenous fibrinolysis. Circulation. 2005;112(25):3930–6. https://doi.org/10.1161/CIRCULATIONAHA.105.588962.
34. Lucking AJ, Lundback M, Mills NL, Faratian D, Barath SL, Pourazar J, Cassee FR, Donaldson K, Boon NA, Badimon JJ, Sandstrom T, Blomberg A, Newby DE. Diesel exhaust inhalation increases thrombus formation in man. Eur Heart J. 2008;29(24):3043–51. https://doi.org/10.1093/eurheartj/ehn464.
35. Mills NL, Tornqvist H, Gonzalez MC, Vink E, Robinson SD, Soderberg S, Boon NA, Donaldson K, Sandstrom T, Blomberg A, Newby DE. Ischemic and thrombotic effects of dilute diesel-exhaust inhalation in men with coronary heart disease. N Engl J Med. 2007;357(11):1075–82.
36. Rittinghausen S, Bellmann B, Creutzenberg O, Ernst H, Kolling A, Mangelsdorf I, Kellner R, Beneke S, Ziemann C. Evaluation of immunohistochemical markers to detect the genotoxic mode of action of fine and ultrafine dusts in rat lungs. Toxicology. 2013;303:177–86. https://doi.org/10.1016/j.tox.2012.11.007.
37. Hartwig A. Role of DNA repair in particle- and fiber-induced lung injury. Inhal Toxicol. 2002;14(1):91–100.
38. Ware JH. Particulate air pollution and mortality--clearing the air. N Engl J Med. 2000;343(24):1798–9.
39. World Health Organization. Guidelines for air quality. WHO ed., Geneva, Switzerland; 1999.
40. Elder A, Gelein R, Silva V, Feikert T, Opanashuk L, Carter J, Potter R, Maynard A, Ito Y, Finkelstein J, Oberdorster G. Translocation of inhaled ultrafine manganese oxide particles to the central nervous system. Environ Health Perspect. 2006;114(8):1172–8.
41. Muoth C, Aengenheister L, Kucki M, Wick P, Buerki-Thurnherr T. Nanoparticle transport across the placental barrier: pushing the field forward! Nanomedicine. 2016;11(8):941–57. https://doi.org/10.2217/nnm-2015-0012.
42. Chang C. The immune effects of naturally occurring and synthetic nanoparticles. J Autoimmun. 2010;34(3):J234–46. https://doi.org/10.1016/j.jaut.2009.11.009.
43. Renwick LC, Donaldson K, Clouter A. Impairment of alveolar macrophage phagocytosis by ultrafine particles. Toxicol Appl Pharmacol. 2001;172(2):119–27. https://doi.org/10.1006/taap.2001.9128.
44. World Health Organization Air quality guidelines, global update 2005. Report on a Working Group meeting, Bonn, Germany; 2005.
45. Ibald-Mulli A, Wichmann HE, Kreyling W, Peters A. Epidemiological evidence on health effects of ultrafine particles. J Aerosol Med. 2002;15(2):189–201.
46. Donaldson K, Brown D, Clouter A, Duffin R, MacNee W, Renwick L, Tran L, Stone V. The pulmonary toxicology of ultrafine particles. J Aerosol Med. 2002;15(2):213–20.
47. Donaldson K, Seaton A. The Janus faces of nanoparticles. J Nanosci Nanotechnol. 2007;7(12):4607–11.

48. Donaldson K, Tran L, Jimenez LA, Duffin R, Newby DE, Mills N, MacNee W, Stone V. Combustion-derived nanoparticles: a review of their toxicology following inhalation exposure. Part Fibre Toxicol. 2005;2:10.
49. Duffin R, Mills NL, Donaldson K. Nanoparticles-a thoracic toxicology perspective. Yonsei Med J. 2007;48(4):561–72.
50. Seaton A, MacNee W, Donaldson K, Godden D. Particulate air pollution and acute health effects. Lancet. 1995;345(8943):176–8.
51. Kreyling WG, Semmler M, Erbe F, Mayer P, Takenaka S, Schulz H, Oberdorster G, Ziesenis A. Translocation of ultrafine insoluble iridium particles from lung epithelium to extrapulmonary organs is size dependent but very low. J Toxicol Environ Health A. 2002;65(20):1513–30.
52. Kreyling WG, Semmler-Behnke M, Moller W. Ultrafine particle-lung interactions: does size matter? J Aerosol Med. 2006;19(1):74–83.
53. Schmid O, Moller W, Semmler-Behnke M, Ferron GA, Karg E, Lipka J, Schulz H, Kreyling WG, Stoeger T. Dosimetry and toxicology of inhaled ultrafine particles. Biomarkers. 2009;14(Suppl 1):67–73. https://doi.org/10.1080/13547500902965617.
54. Yang H, Liu C, Yang D, Zhang H, Xi Z. Comparative study of cytotoxicity, oxidative stress and genotoxicity induced by four typical nanomaterials: the role of particle size, shape and composition. J Appl Toxicol. 2009;29(1):69–78. https://doi.org/10.1002/jat.1385.
55. Hinds WC. Aerosol technology: behavior and measurements of airborne particles. 2nd ed. New York: Wiley Interscience; 1999.
56. Young LH, Keeler GJ. Characterization of ultrafine particle number concentration and size distribution during a summer campaign in southwest Detroit. J Air Waste Manage Assoc. 2004;54(9):1079–90.
57. Morawska L, Ristovski Z, Jayaratne ER, Keogh DU, Ling X. Ambient nano and ultrafine particles from motor vehicle emissions: characteristics, ambient processing and implications on human exposure. Atmos Environ. 2008;42:8113–38.
58. Chen R, Meng X, Zhao A, Wang C, Yang C, Li H, Cai J, Zhao Z, Kan H. DNA hypomethylation and its mediation in the effects of fine particulate air pollution on cardiovascular biomarkers: a randomized crossover trial. Environ Int. 2016a;94:614–9. https://doi.org/10.1016/j.envint.2016.06.026.
59. Jayasekher T. Aerosols near by a coal fired thermal power plant: chemical composition and toxic evaluation. Chemosphere. 2009;75(11):1525–30. https://doi.org/10.1016/j.chemosphere.2009.02.001.
60. Rogula-Kozlowska W. Size-segregated urban particulate matter: mass closure, chemical composition, and primary and secondary matter content. Air Qual Atmos Health. 2016;9:533–50. https://doi.org/10.1007/s11869-015-0359-y.
61. Sambandam B, Devasena T, Islam VI, Prakhya BM. Characterization of coal fly ash nanoparticles and their induced in vitro cellular toxicity and oxidative DNA damage in different cell lines. Indian J Exp Biol. 2015;53(9):585–93.
62. Hata M, Chomanee J, Thongyen T, Bao L, Tekasakul S, Tekasakul P, Otani Y, Furuuchi M. Characteristics of nanoparticles emitted from burning of biomass fuels. J Environ Sci. 2014;26(9):1913–20. https://doi.org/10.1016/j.jes.2014.07.005.
63. Cernuschi S, Giugliano M, Ozgen S, Consonni S. Number concentration and chemical composition of ultrafine and nanoparticles from WTE (waste to energy) plants. Sci Total Environ. 2012;420:319–26. https://doi.org/10.1016/j.scitotenv.2012.01.024.
64. Maguhn J, Karg E, Kettrup A, Zimmermann R. On-line analysis of the size distribution of fine and ultrafine aerosol particles in flue and stack gas of a municipal waste incineration plant: effects of dynamic process control measures and emission reduction devices. Environ Sci Technol. 2003;37(20):4761–70.
65. Cassee FR, Boere AJ, Bos J, Fokkens PH, Dormans JA, Van LH. Effects of diesel exhaust enriched concentrated PM2.5 in ozone preexposed or monocrotaline-treated rats. Inhal Toxicol. 2002;14(7):721–43.
66. Seinfeld JH, Pandis SN. Atmospheric chemistry and physics: from air pollution to climate change. 2nd ed. New York: Wiley; 2006.

67. European Commission D-GfSr, Development. ExternE–externalities of energy, Oil & gas, vol. 4: Office for Official Publications of the European Communities; 1995. http://www.externe.info/oldvolumes/vol4.pdf
68. Grosjean D, Seinfeld JH. Parameterization of the formation potential of secondary organic aerosols. Atmos Environ. 1989;23:1733–47.
69. Janssen NA, Schwartz J, Zanobetti A, Suh HH. Air conditioning and source-specific particles as modifiers of the effect of PM(10) on hospital admissions for heart and lung disease. Environ Health Perspect. 2002;110(1):43–9.
70. Laden F, Neas LM, Dockery DW, Schwartz J. Association of fine particulate matter from different sources with daily mortality in six U.S. cities. Environ Health Perspect. 2000;108(10):941–7.
71. Mangia C, Cervino M, Gianicolo EA. Secondary particulate matter originating from an industrial source and its impact on population health. Int J Environ Res Public Health. 2015;12(7):7667–81. https://doi.org/10.3390/ijerph120707667.
72. Hertz-Picciotto I, Park HY, Dostal M, Kocan A, Trnovec T, Sram R. Prenatal exposures to persistent and non-persistent organic compounds and effects on immune system development. Basic Clin Pharmacol Toxicol. 2008;102(2):146–54. https://doi.org/10.1111/j.1742-7843.2007.00190.x.
73. Zhu MT, Wang Y, Feng WY, Wang B, Wang M, Ouyang H, Chai ZF. Oxidative stress and apoptosis induced by iron oxide nanoparticles in cultured human umbilical endothelial cells. J Nanosci Nanotechnol. 2010;10(12):8584–90.
74. Wick P, Malek A, Manser P, Meili D, Maeder-Althaus X, Diener L, Diener PA, Zisch A, Krug HF, von Mandach U. Barrier capacity of human placenta for nanosized materials. Environ Health Perspect. 2010;118(3):432–6. https://doi.org/10.1289/ehp.0901200.
75. Oberdorster G, Sharp Z, Atudorei V, Elder A, Gelein R, Kreyling W, Cox C. Translocation of inhaled ultrafine particles to the brain. Inhal Toxicol. 2004;16(6–7):437–45. https://doi.org/10.1080/08958370490439597.
76. Woodward N, Finch CE, Morgan TE. Traffic-related air pollution and brain development. AIMS Environ Sci. 2015;2(2):353–73. https://doi.org/10.3934/environsci.2015.2.353.
77. Huang JP, Hsieh PC, Chen CY, Wang TY, Chen PC, Liu CC, Chen CC, Chen CP. Nanoparticles can cross mouse placenta and induce trophoblast apoptosis. Placenta. 2015;36(12):1433–41. https://doi.org/10.1016/j.placenta.2015.10.007.
78. Blum JL, Xiong JQ, Hoffman C, Zelikoff JT. Cadmium associated with inhaled cadmium oxide nanoparticles impacts fetal and neonatal development and growth. Toxicol Sci. 2012;126(2):478–86. https://doi.org/10.1093/toxsci/kfs008.
79. Blum JL, Edwards JR, Prozialeck WC, Xiong JQ, Zelikoff JT. Effects of maternal exposure to cadmium oxide nanoparticles during pregnancy on maternal and offspring kidney injury markers using a murine model. J Toxicol Environ Health Part A. 2015;78(12):711–24. https://doi.org/10.1080/15287394.2015.1026622.
80. El-Sayed YS, Shimizu R, Onoda A, Takeda K, Umezawa M. Carbon black nanoparticle exposure during middle and late fetal development induces immune activation in male offspring mice. Toxicology. 2015;327:53–61. https://doi.org/10.1016/j.tox.2014.11.005.
81. Davis DA, Bortolato M, Godar SC, Sander TK, Iwata N, Pakbin P, Shih JC, Berhane K, McConnell R, Sioutas C, Finch CE, Morgan TE. Prenatal exposure to urban air nanoparticles in mice causes altered neuronal differentiation and depression-like responses. PLoS One. 2013;8(5):e64128. https://doi.org/10.1371/journal.pone.0064128.
82. Ghaderi S, Tabatabaei SR, Varzi HN, Rashno M. Induced adverse effects of prenatal exposure to silver nanoparticles on neurobehavioral development of offspring of mice. J Toxicol Sci. 2015;40(2):263–75. https://doi.org/10.2131/jts.40.263.
83. Mohammadipour A, Fazel A, Haghir H, Motejaded F, Rafatpanah H, Zabihi H, Hosseini M, Bideskan AE. Maternal exposure to titanium dioxide nanoparticles during pregnancy; impaired memory and decreased hippocampal cell proliferation in rat offspring. Environ Toxicol Pharmacol. 2014;37(2):617–25. https://doi.org/10.1016/j.etap.2014.01.014.

84. Yoshida S, Hiyoshi K, Oshio S, Takano H, Takeda K, Ichinose T. Effects of fetal exposure to carbon nanoparticles on reproductive function in male offspring. Fertil Steril. 2010;93(5):1695–9. https://doi.org/10.1016/j.fertnstert.2009.03.094.
85. Christensen BC, Marsit CJ. Epigenomics in environmental health. Front Genet. 2011;2:84. https://doi.org/10.3389/fgene.2011.00084.
86. Loke YJ, Novakovic B, Ollikainen M, Wallace EM, Umstad MP, Permezel M, Morley R, Ponsonby AL, Gordon L, Galati JC, Saffery R, Craig JM. The peri/postnatal epigenetic twins study (PETS). Twin Res Hum Genet. 2012;16(1):1–8. https://doi.org/10.1017/thg.2012.114.
87. Mazzio EA, Soliman KF. Basic concepts of epigenetics: impact of environmental signals on gene expression. Epigenetics. 2012;7(2):119–30. https://doi.org/10.4161/epi.7.2.18764.
88. Heijmans BT, Tobi EW, Stein AD, Putter H, Blauw GJ, Susser ES, Slagboom PE, Lumey LH. Persistent epigenetic differences associated with prenatal exposure to famine in humans. Proc Natl Acad Sci U S A. 2008;105(44):17046–9. https://doi.org/10.1073/pnas.0806560105.
89. Yajnik CS. Transmission of obesity-adiposity and related disorders from the mother to the baby. Ann Nutr Metab. 2014;64(Suppl 1):8–17. https://doi.org/10.1159/000362608.
90. Mattick JS, Taft RJ, Faulkner GJ. A global view of genomic information--moving beyond the gene and the master regulator. Trends Genet. 2010;26(1):21–8. https://doi.org/10.1016/j.tig.2009.11.002.
91. Di Ciaula A, Portincasa P. Fat, epigenome and pancreatic diseases. Interplay and common pathways from a toxic and obesogenic environment. Eur J Intern Med. 2014;25(10):865–73. https://doi.org/10.1016/j.ejim.2014.10.012.
92. Lahiri DK, Maloney B, Zawia NH. The LEARn model: an epigenetic explanation for idiopathic neurobiological diseases. Mol Psychiatry. 2009;14(11):992–1003. https://doi.org/10.1038/mp.2009.82.
93. Dang MN, Buzzetti R, Pozzilli P. Epigenetics in autoimmune diseases with focus on type 1 diabetes. Diabetes Metab Res Rev. 2013;29(1):8–18. https://doi.org/10.1002/dmrr.2375.
94. MacFarlane AJ, Strom A, Scott FW. Epigenetics: deciphering how environmental factors may modify autoimmune type 1 diabetes. Mamm Genome. 2009;20(9–10):624–32. https://doi.org/10.1007/s00335-009-9213-6.
95. Baccarelli A, Wright RO, Bollati V, Tarantini L, Litonjua AA, Suh HH, Zanobetti A, Sparrow D, Vokonas PS, Schwartz J. Rapid DNA methylation changes after exposure to traffic particles. Am J Respir Crit Care Med. 2009;179(7):572–8. https://doi.org/10.1164/rccm.200807-1097OC.
96. De Prins S, Koppen G, Jacobs G, Dons E, Van de Mieroop E, Nelen V, Fierens F, Int Panis L, De Boever P, Cox B, Nawrot TS, Schoeters G. Influence of ambient air pollution on global DNA methylation in healthy adults: a seasonal follow-up. Environ Int. 2013;59:418–24. https://doi.org/10.1016/j.envint.2013.07.007.
97. Wang C, Chen R, Cai J, Shi J, Yang C, Tse LA, Li H, Lin Z, Meng X, Liu C, Niu Y, Xia Y, Zhao Z, Kan H. Personal exposure to fine particulate matter and blood pressure: a role of angiotensin converting enzyme and its DNA methylation. Environ Int. 2016. https://doi.org/10.1016/j.envint.2016.07.001.
98. Peng C, Bind MC, Colicino E, Kloog I, Byun HM, Cantone L, Trevisi L, Zhong J, Brennan K, Dereix AE, Vokonas PS, Coull BA, Schwartz JD, Baccarelli AA. Particulate air pollution and fasting blood glucose in non-diabetic individuals: associations and epigenetic mediation in the normative aging study, 2000–2011. Environ Health Perspect. 2016;124(11):1715–21. https://doi.org/10.1289/EHP183.
99. Chen R, Hu B, Liu Y, Xu J, Yang G, Xu D, Chen C. Beyond PM2.5: the role of ultrafine particles on adverse health effects of air pollution. Biochim Biophys Acta. 2016b;1860(12):2844–55. https://doi.org/10.1016/j.bbagen.2016.03.019.
100. Somineni HK, Zhang X, Biagini Myers JM, Kovacic MB, Ulm A, Jurcak N, Ryan PH, Khurana Hershey GK, Ji H. Ten-eleven translocation 1 (TET1) methylation is associated with childhood asthma and traffic-related air pollution. J Allergy Clin Immunol. 2016;137(3):797–805.e795. https://doi.org/10.1016/j.jaci.2015.10.021.

101. Hew KM, Walker AI, Kohli A, Garcia M, Syed A, McDonald-Hyman C, Noth EM, Mann JK, Pratt B, Balmes J, Hammond SK, Eisen EA, Nadeau KC. Childhood exposure to ambient polycyclic aromatic hydrocarbons is linked to epigenetic modifications and impaired systemic immunity in T cells. Clin Exp Allergy. 2015;45(1):238–48. https://doi.org/10.1111/cea.12377.
102. Gruzieva O, Xu CJ, Breton CV, Annesi-Maesano I, Anto JM, Auffray C, Ballereau S, Bellander T, Bousquet J, Bustamante M, Charles MA, de Kluizenaar Y, den Dekker HT, Duijts L, Felix JF, Gehring U, Guxens M, Jaddoe VV, Jankipersadsing SA, Merid SK, Kere J, Kumar A, Lemonnier N, Lepeule J, Nystad W, Page CM, Panasevich S, Postma D, Slama R, Sunyer J, Soderhall C, Yao J, London SJ, Pershagen G, Koppelman GH, Melen E. Epigenome-wide meta-analysis of methylation in children related to prenatal NO2 air pollution exposure. Environ Health Perspect. 2016;125(1):104–10. https://doi.org/10.1289/EHP36.
103. Pacchierotti F, Spano M. Environmental impact on DNA methylation in the germline: state of the art and gaps of knowledge. Biomed Res Int. 2015;2015:123484. https://doi.org/10.1155/2015/123484.
104. Hanson MA, Skinner MK. Developmental origins of epigenetic transgenerational inheritance. Environ Epigenet. 2016;2(1):dvw002. https://doi.org/10.1093/eep/dvw002.
105. Skinner MK, Guerrero-Bosagna C, Haque MM. Environmentally induced epigenetic transgenerational inheritance of sperm epimutations promote genetic mutations. Epigenetics. 2015;10(8):762–71. https://doi.org/10.1080/15592294.2015.1062207.
106. Siegel RL, Miller KD, Jemal A. Cancer statistics, 2016. CA. 2016;66(1):7–30. https://doi.org/10.3322/caac.21332.
107. Vergamini LB, Frazier AL, Abrantes FL, Ribeiro KB, Rodriguez-Galindo C. Increase in the incidence of differentiated thyroid carcinoma in children, adolescents, and young adults: a population-based study. J Pediatr. 2014;164(6):1481–5. https://doi.org/10.1016/j.jpeds.2014.01.059.
108. Zheng R, Peng X, Zeng H, Zhang S, Chen T, Wang H, Chen W. Incidence, mortality and survival of childhood cancer in China during 2000-2010 period: a population-based study. Cancer Lett. 2015;363(2):176–80. https://doi.org/10.1016/j.canlet.2015.04.021.
109. Tulla M, Berthold F, Graf N, Rutkowski S, von Schweinitz D, Spix C, Kaatsch P. Incidence, trends, and survival of children with embryonal tumors. Pediatrics. 2015;136(3):e623–32. https://doi.org/10.1542/peds.2015-0224.
110. Cannone GA, Lo Monaco MG, Carone S, Melcarne A, Tanzarella M, Bisceglia L, Ardizzone A, Minerba S, Quarta F, Coviello V, Gruppo di lavoro Registro tumori P. Childhood cancer incidence in Apulia region (southern Italy), 2003–2008. Epidemiol Prev. 2015;39(3):176–82.
111. Danysh HE, Zhang K, Mitchell LE, Scheurer ME, Lupo PJ. Maternal residential proximity to major roadways at delivery and childhood central nervous system tumors. Environ Res. 2016;146:315–22. https://doi.org/10.1016/j.envres.2016.01.012.
112. Garcia-Perez J, Lopez-Abente G, Gomez-Barroso D, Morales-Piga A, Romaguera EP, Tamayo I, Fernandez-Navarro P, Ramis R. Childhood leukemia and residential proximity to industrial and urban sites. Environ Res. 2015;140:542–53. https://doi.org/10.1016/j.envres.2015.05.014.
113. Garcia-Perez J, Morales-Piga A, Gomez J, Gomez-Barroso D, Tamayo-Uria I, Pardo Romaguera E, Fernandez-Navarro P, Lopez-Abente G, Ramis R. Association between residential proximity to environmental pollution sources and childhood renal tumors. Environ Res. 2016b;147:405–14. https://doi.org/10.1016/j.envres.2016.02.036.
114. Garcia-Perez J, Morales-Piga A, Gomez-Barroso D, Tamayo-Uria I, Pardo Romaguera E, Fernandez-Navarro P, Lopez-Abente G, Ramis R. Risk of neuroblastoma and residential proximity to industrial and urban sites: a case-control study. Environ Int. 2016a;92-93:269–75. https://doi.org/10.1016/j.envint.2016.04.023.
115. Spycher BD, Feller M, Roosli M, Ammann RA, Diezi M, Egger M, Kuehni CE. Childhood cancer and residential exposure to highways: a nationwide cohort study. Eur J Epidemiol. 2015;30(12):1263–75. https://doi.org/10.1007/s10654-015-0091-9.

116. Parodi S, Merlo DF, Ranucci A, Miligi L, Benvenuti A, Rondelli R, Magnani C, Haupt R, Group SW. Risk of neuroblastoma, maternal characteristics and perinatal exposures: the SETIL study. Cancer Epidemiol. 2014;38(6):686–94. https://doi.org/10.1016/j.canep.2014.09.007.
117. Cordier S, Monfort C, Filippini G, Preston-Martin S, Lubin F, Mueller BA, Holly EA, Peris-Bonet R, McCredie M, Choi W, Little J, Arslan A. Parental exposure to polycyclic aromatic hydrocarbons and the risk of childhood brain tumors: the SEARCH international childhood brain tumor study. Am J Epidemiol. 2004;159(12):1109–16. https://doi.org/10.1093/aje/kwh154.
118. Timms JA, Relton CL, Rankin J, Strathdee G, McKay JA. DNA methylation as a potential mediator of environmental risks in the development of childhood acute lymphoblastic leukemia. Epigenomics. 2016;8(4):519–36. https://doi.org/10.2217/epi-2015-0011.
119. Vladimirova V, Mikeska T, Waha A, Soerensen N, Xu J, Reynolds PC, Pietsch T. Aberrant methylation and reduced expression of LHX9 in malignant gliomas of childhood. Neoplasia. 2009;11(7):700–11.
120. Castelo-Branco P, Choufani S, Mack S, Gallagher D, Zhang C, Lipman T, Zhukova N, Walker EJ, Martin D, Merino D, Wasserman JD, Elizabeth C, Alon N, Zhang L, Hovestadt V, Kool M, Jones DT, Zadeh G, Croul S, Hawkins C, Hitzler J, Wang JC, Baruchel S, Dirks PB, Malkin D, Pfister S, Taylor MD, Weksberg R, Tabori U. Methylation of the TERT promoter and risk stratification of childhood brain tumours: an integrative genomic and molecular study. Lancet Oncol. 2013;14(6):534–42. https://doi.org/10.1016/S1470-2045(13)70110-4.
121. Chung YR, Schatoff E, Abdel-Wahab O. Epigenetic alterations in hematopoietic malignancies. Int J Hematol. 2012;96(4):413–27. https://doi.org/10.1007/s12185-012-1181-z.
122. Dawson MA, Kouzarides T. Cancer epigenetics: from mechanism to therapy. Cell. 2012;150(1):12–27. https://doi.org/10.1016/j.cell.2012.06.013.
123. Marsit C, Christensen B. Blood-derived DNA methylation markers of cancer risk. Adv Exp Med Biol. 2013;754:233–52. https://doi.org/10.1007/978-1-4419-9967-2_12.
124. Zheng SC, Widschwendter M, Teschendorff AE. Epigenetic drift, epigenetic clocks and cancer risk. Epigenomics. 2016;8(5):705–19. https://doi.org/10.2217/epi-2015-0017.
125. Wei J, Li F, Yang J, Liu X, Cho WC. MicroRNAs as regulators of airborne pollution-induced lung inflammation and carcinogenesis. Arch Toxicol. 2015;89(5):677–85. https://doi.org/10.1007/s00204-015-1462-4.
126. Cao Y. Environmental pollution and DNA methylation: carcinogenesis, clinical significance, and practical applications. Front Med. 2015;9(3):261–74. https://doi.org/10.1007/s11684-015-0406-y.
127. Ding R, Jin Y, Liu X, Zhu Z, Zhang Y, Wang T, Xu Y. Characteristics of DNA methylation changes induced by traffic-related air pollution. Mutat Res. 2016;796:46–53. https://doi.org/10.1016/j.mrgentox.2015.12.002.
128. Pan HL, Wen ZS, Huang YC, Cheng X, Wang GZ, Zhou YC, Wang ZY, Guo YQ, Cao Y, Zhou GB. Down-regulation of microRNA-144 in air pollution-related lung cancer. Sci Rep. 2015;5:14331. https://doi.org/10.1038/srep14331.
129. White AJ, Chen J, Teitelbaum SL, McCullough LE, Xu X, Hee Cho Y, Conway K, Beyea J, Stellman SD, Steck SE, Mordukhovich I, Eng SM, Beth Terry M, Engel LS, Hatch M, Neugut AI, Hibshoosh H, Santella RM, Gammon MD. Sources of polycyclic aromatic hydrocarbons are associated with gene-specific promoter methylation in women with breast cancer. Environ Res. 2016;145:93–100. https://doi.org/10.1016/j.envres.2015.11.033.
130. Knox EG. Childhood cancers and atmospheric carcinogens. J Epidemiol Community Health. 2005;59(2):101–5. https://doi.org/10.1136/jech.2004.021675.
131. Knox EG. Roads, railways, and childhood cancers. J Epidemiol Community Health. 2006;60(2):136–41. https://doi.org/10.1136/jech.2005.042036.
132. Greaves M. Pre-natal origins of childhood leukemia. Rev Clin Exp Hematol. 2003;7(3):233–45.
133. Amigou A, Sermage-Faure C, Orsi L, Leverger G, Baruchel A, Bertrand Y, Nelken B, Robert A, Michel G, Margueritte G, Perel Y, Mechinaud F, Bordigoni P, Hemon D, Clavel J. Road traffic and childhood leukemia: the ESCALE study (SFCE). Environ Health Perspect. 2011;119(4):566–72. https://doi.org/10.1289/ehp.1002429.

134. Weisskopf MG, Kioumourtzoglou MA, Roberts AL. Air pollution and autism spectrum disorders: causal or confounded? Curr Environ Health Rep. 2015;2(4):430–9. https://doi.org/10.1007/s40572-015-0073-9.
135. Costa LG, Cole TB, Coburn J, Chang YC, Dao K, Roque PJ. Neurotoxicity of traffic-related air pollution. Neurotoxicology. 2015;59:133–9. https://doi.org/10.1016/j.neuro.2015.11.008.
136. Nardone S, Elliott E. The interaction between the immune system and epigenetics in the etiology of autism spectrum disorders. Front Neurosci. 2016;10:329. https://doi.org/10.3389/fnins.2016.00329.
137. Peterson BS, Rauh VA, Bansal R, Hao X, Toth Z, Nati G, Walsh K, Miller RL, Arias F, Semanek D, Perera F. Effects of prenatal exposure to air pollutants (polycyclic aromatic hydrocarbons) on the development of brain white matter, cognition, and behavior in later childhood. JAMA Psychiat. 2015;72(6):531–40. https://doi.org/10.1001/jamapsychiatry.2015.57.
138. Annavarapu RN, Kathi S. Cognitive disorders in children associated with urban vehicular emissions. Environ Pollut. 2016;208(Pt A):74–8. https://doi.org/10.1016/j.envpol.2015.09.036.
139. Allen JL, Oberdorster G, Morris-Schaffer K, Wong C, Klocke C, Sobolewski M, Conrad K, Mayer-Proschel M, Cory-Slechta DA. Developmental neurotoxicity of inhaled ambient ultrafine particle air pollution: parallels with neuropathological and behavioral features of autism and other neurodevelopmental disorders. Neurotoxicology. 2015;59:140–54. https://doi.org/10.1016/j.neuro.2015.12.014.
140. Raz R, Roberts AL, Lyall K, Hart JE, Just AC, Laden F, Weisskopf MG. Autism spectrum disorder and particulate matter air pollution before, during, and after pregnancy: a nested case-control analysis within the nurses' health study II cohort. Environ Health Perspect. 2015;123(3):264–70. https://doi.org/10.1289/ehp.1408133.
141. von Ehrenstein OS, Aralis H, Cockburn M, Ritz B. In utero exposure to toxic air pollutants and risk of childhood autism. Epidemiology. 2014;25(6):851–8. https://doi.org/10.1097/EDE.0000000000000150.
142. Kalkbrenner AE, Windham GC, Serre ML, Akita Y, Wang X, Hoffman K, Thayer BP, Daniels JL. Particulate matter exposure, prenatal and postnatal windows of susceptibility, and autism spectrum disorders. Epidemiology. 2015;26(1):30–42. https://doi.org/10.1097/EDE.0000000000000173.
143. Keil KP, Lein PJ. DNA methylation: a mechanism linking environmental chemical exposures to risk of autism spectrum disorders? Environ Epigenet. 2016;2(1). https://doi.org/10.1093/eep/dvv012.
144. Homs A, Codina-Sola M, Rodriguez-Santiago B, Villanueva CM, Monk D, Cusco I, Perez-Jurado LA. Genetic and epigenetic methylation defects and implication of the ERMN gene in autism spectrum disorders. Transl Psychiatry. 2016;6(7):e855. https://doi.org/10.1038/tp.2016.120.
145. Moulton PV, Yang W. Air pollution, oxidative stress, and Alzheimer's disease. J Environ Public Health. 2012;2012:472751. https://doi.org/10.1155/2012/472751.
146. Chin-Chan M, Navarro-Yepes J, Quintanilla-Vega B. Environmental pollutants as risk factors for neurodegenerative disorders: Alzheimer and Parkinson diseases. Front Cell Neurosci. 2015;9:124. https://doi.org/10.3389/fncel.2015.00124.
147. Bihaqi SW, Schumacher A, Maloney B, Lahiri DK, Zawia NH. Do epigenetic pathways initiate late onset Alzheimer disease (LOAD): towards a new paradigm. Curr Alzheimer Res. 2012;9(5):574–88.
148. Kwok JB. Role of epigenetics in Alzheimer's and Parkinson's disease. Epigenomics. 2010;2(5):671–82. https://doi.org/10.2217/epi.10.43.
149. Marques S, Outeiro TF. Epigenetics in Parkinson's and Alzheimer's diseases. Subcell Biochem. 2013;61:507–25. https://doi.org/10.1007/978-94-007-4525-4_22.
150. Marques SC, Oliveira CR, Pereira CM, Outeiro TF. Epigenetics in neurodegeneration: a new layer of complexity. Prog Neuro-Psychopharmacol Biol Psychiatry. 2011;35(2):348–55. https://doi.org/10.1016/j.pnpbp.2010.08.008.
151. Calderon-Garciduenas L, Maronpot RR, Torres-Jardon R, Henriquez-Roldan C, Schoonhoven R, Acuna-Ayala H, Villarreal-Calderon A, Nakamura J, Fernando R, Reed W, Azzarelli B,

Swenberg JA. DNA damage in nasal and brain tissues of canines exposed to air pollutants is associated with evidence of chronic brain inflammation and neurodegeneration. Toxicol Pathol. 2003;31(5):524–38.

152. Calderon-Garciduenas L, Franco-Lira M, Torres-Jardon R, Henriquez-Roldan C, Barragan-Mejia G, Valencia-Salazar G, Gonzalez-Maciel A, Reynoso-Robles R, Villarreal-Calderon R, Reed W. Pediatric respiratory and systemic effects of chronic air pollution exposure: nose, lung, heart, and brain pathology. Toxicol Pathol. 2007;35(1):154–62. https://doi.org/10.1080/01926230601059985.
153. Calderon-Garciduenas L, Kavanaugh M, Block M, D'Angiulli A, Delgado-Chavez R, Torres-Jardon R, Gonzalez-Maciel A, Reynoso-Robles R, Osnaya N, Villarreal-Calderon R, Guo R, Hua Z, Zhu H, Perry G, Diaz P. Neuroinflammation, hyperphosphorylated tau, diffuse amyloid plaques, and down-regulation of the cellular prion protein in air pollution exposed children and young adults. J Alzheimers Dis. 2012;28(1):93–107. https://doi.org/10.3233/JAD-2011-110722.
154. Lahiri DK, Maloney B. The "LEARn" (latent early-life associated regulation) model integrates environmental risk factors and the developmental basis of Alzheimer's disease, and proposes remedial steps. Exp Gerontol. 2010;45(4):291–6. https://doi.org/10.1016/j.exger.2010.01.001.
155. Lahiri DK, Zawia NH, Greig NH, Sambamurti K, Maloney B. Early-life events may trigger biochemical pathways for Alzheimer's disease: the "LEARn" model. Biogerontology. 2008;9(6):375–9. https://doi.org/10.1007/s10522-008-9162-6.
156. Grandjean P, Landrigan PJ. Developmental neurotoxicity of industrial chemicals. Lancet. 2006;368(9553):2167–78. https://doi.org/10.1016/S0140-6736(06)69665-7.
157. Grandjean P, Landrigan PJ. Neurobehavioural effects of developmental toxicity. Lancet Neurol. 2014;13(3):330–8. https://doi.org/10.1016/S1474-4422(13)70278-3.
158. Calderon-Garciduenas L, Torres-Jardon R. The impact of air pollutants on the brain. JAMA Psychiat. 2015;72(6):529–30. https://doi.org/10.1001/jamapsychiatry.2015.0192.
159. Maloney B, Sambamurti K, Zawia N, Lahiri DK. Applying epigenetics to Alzheimer's disease via the latent early-life associated regulation (LEARn) model. Curr Alzheimer Res. 2012;9(5):589–99.
160. World Health Organization. Obesity and overweight. Fact sheet n.311. Geneva: WHO; 2015.
161. Ogden CL, Carroll MD, Fryar CD, Flegal KM. Prevalence of obesity among adults and youth: United States, 2011-2014. NCHS Data Brief. 2015;219:1–8.
162. OECD. Obesity update 2014. Paris: OECD Directorate dor Employement, Labour and Social Affairs; 2014.
163. Bhupathiraju SN, Hu FB. Epidemiology of obesity and diabetes and their cardiovascular complications. Circ Res. 2016;118(11):1723–35. https://doi.org/10.1161/CIRCRESAHA.115.306825.
164. Zimmet PZ, Alberti KG. Epidemiology of diabetes-status of a pandemic and issues around metabolic surgery. Diabetes Care. 2016;39(6):878–83. https://doi.org/10.2337/dc16-0273.
165. Fazeli Farsani S, Souverein PC, van der Vorst MM, Knibbe CA, Herings RM, de Boer A, Mantel-Teeuwisse AK. Increasing trends in the incidence and prevalence rates of type 1 diabetes among children and adolescents in the Netherlands. Pediatr Diabetes. 2014;17(1):44–52. https://doi.org/10.1111/pedi.12232.
166. Demerath EW, Choh AC, Johnson W, Curran JE, Lee M, Bellis C, Dyer TD, Czerwinski SA, Blangero J, Towne B. The positive association of obesity variants with adulthood adiposity strengthens over an 80-year period: a gene-by-birth year interaction. Hum Hered. 2013;75(2–4):175–85. https://doi.org/10.1159/000351742.
167. Kilpelainen TO, Qi L, Brage S, Sharp SJ, Sonestedt E, Demerath E, Ahmad T, Mora S, Kaakinen M, Sandholt CH, Holzapfel C, Autenrieth CS, Hypponen E, Cauchi S, He M, Kutalik Z, Kumari M, Stancakova A, Meidtner K, Balkau B, Tan JT, Mangino M, Timpson NJ, Song Y, Zillikens MC, Jablonski KA, Garcia ME, Johansson S, Bragg-Gresham JL, Wu Y, van Vliet-Ostaptchouk JV, Onland-Moret NC, Zimmermann E, Rivera NV, Tanaka T, Stringham HM, Silbernagel G, Kanoni S, Feitosa MF, Snitker S, Ruiz JR, Metter J, Larrad

MT, Atalay M, Hakanen M, Amin N, Cavalcanti-Proenca C, Grontved A, Hallmans G, Jansson JO, Kuusisto J, Kahonen M, Lutsey PL, Nolan JJ, Palla L, Pedersen O, Perusse L, Renstrom F, Scott RA, Shungin D, Sovio U, Tammelin TH, Ronnemaa T, Lakka TA, Uusitupa M, Rios MS, Ferrucci L, Bouchard C, Meirhaeghe A, Fu M, Walker M, Borecki IB, Dedoussis GV, Fritsche A, Ohlsson C, Boehnke M, Bandinelli S, van Duijn CM, Ebrahim S, Lawlor DA, Gudnason V, Harris TB, Sorensen TI, Mohlke KL, Hofman A, Uitterlinden AG, Tuomilehto J, Lehtimaki T, Raitakari O, Isomaa B, Njolstad PR, Florez JC, Liu S, Ness A, Spector TD, Tai ES, Froguel P, Boeing H, Laakso M, Marmot M, Bergmann S, Power C, Khaw KT, Chasman D, Ridker P, Hansen T, Monda KL, Illig T, Jarvelin MR, Wareham NJ, Hu FB, Groop LC, Orho-Melander M, Ekelund U, Franks PW, Loos RJ. Physical activity attenuates the influence of FTO variants on obesity risk: a meta-analysis of 218,166 adults and 19,268 children. PLoS Med. 2011;8(11):e1001116. https://doi.org/10.1371/journal.pmed.1001116.

168. Painter RC, Osmond C, Gluckman P, Hanson M, Phillips DI, Roseboom TJ. Transgenerational effects of prenatal exposure to the Dutch famine on neonatal adiposity and health in later life. BJOG. 2008;115(10):1243–9. https://doi.org/10.1111/j.1471-0528.2008.01822.x.
169. Shi H, Su B. Molecular adaptation of modern human populations. Int J Evol Biol. 2011;2011:484769. https://doi.org/10.4061/2011/484769.
170. Stanner SA, Yudkin JS. Fetal programming and the Leningrad siege study. Twin Res. 2001;4(5):287–92. https://doi.org/10.1375/1369052012498.
171. Veerman JL. On the futility of screening for genes that make you fat. PLoS Med. 2011;8(11):e1001114. https://doi.org/10.1371/journal.pmed.1001114.
172. Sarr O, Yang K, Regnault TR. In utero programming of later adiposity: the role of fetal growth restriction. J Pregnancy. 2012;2012:134758. https://doi.org/10.1155/2012/134758.
173. Ravelli AC, van Der Meulen JH, Osmond C, Barker DJ, Bleker OP. Obesity at the age of 50 y in men and women exposed to famine prenatally. Am J Clin Nutr. 1999;70(5):811–6.
174. Vafeiadi M, Georgiou V, Chalkiadaki G, Rantakokko P, Kiviranta H, Karachaliou M, Fthenou E, Venihaki M, Sarri K, Vassilaki M, Kyrtopoulos SA, Oken E, Kogevinas M, Chatzi L. Association of prenatal exposure to persistent organic pollutants with obesity and cardiometabolic traits in early childhood: the rhea mother-child cohort (Crete, Greece). Environ Health Perspect. 2015;123(10):1015–21. https://doi.org/10.1289/ehp.1409062.
175. Aslibekyan S, Demerath EW, Mendelson M, Zhi D, Guan W, Liang L, Sha J, Pankow JS, Liu C, Irvin MR, Fornage M, Hidalgo B, Lin LA, Thibeault KS, Bressler J, Tsai MY, Grove ML, Hopkins PN, Boerwinkle E, Borecki IB, Ordovas JM, Levy D, Tiwari HK, Absher DM, Arnett DK. Epigenome-wide study identifies novel methylation loci associated with body mass index and waist circumference. Obesity. 2015;23(7):1493–501. https://doi.org/10.1002/oby.21111.
176. Pietilainen KH, Ismail K, Jarvinen E, Heinonen S, Tummers M, Bollepalli S, Lyle R, Muniandy M, Moilanen E, Hakkarainen A, Lundbom J, Lundbom N, Rissanen A, Kaprio J, Ollikainen M. DNA methylation and gene expression patterns in adipose tissue differ significantly within young adult monozygotic BMI-discordant twin pairs. Int J Obes. 2016;40(4):654–61. https://doi.org/10.1038/ijo.2015.221.
177. Thayer KA, Heindel JJ, Bucher JR, Gallo MA. Role of environmental chemicals in diabetes and obesity: a National Toxicology Program workshop review. Environ Health Perspect. 2012;120(6):779–89. https://doi.org/10.1289/ehp.1104597.
178. Gore AC, Chappell VA, Fenton SE, Flaws JA, Nadal A, Prins GS, Toppari J, Zoeller RT. Executive summary to EDC-2: the endocrine society's second scientific statement on endocrine-disrupting chemicals. Endocr Rev. 2015;36(6):593–602. https://doi.org/10.1210/er.2015-1093.
179. Janesick AS, Blumberg B. Obesogens: an emerging threat to public health. Am J Obstet Gynecol. 2016;214(5):559–65. https://doi.org/10.1016/j.ajog.2016.01.182.

180. Janesick A, Blumberg B. Obesogens, stem cells and the developmental programming of obesity. Int J Androl. 2012;35(3):437–48. https://doi.org/10.1111/j.1365-2605.2012.01247.x.
181. Stel J, Legler J. The role of epigenetics in the latent effects of early life exposure to obesogenic endocrine disrupting chemicals. Endocrinology. 2015;156(10):3466–72. https://doi.org/10.1210/en.2015-1434.
182. Rajesh P, Balasubramanian K. Phthalate exposure in utero causes epigenetic changes and impairs insulin signalling. J Endocrinol. 2014;223(1):47–66. https://doi.org/10.1530/JOE-14-0111.
183. Tabb MM, Blumberg B. New modes of action for endocrine-disrupting chemicals. Mol Endocrinol. 2006;20(3):475–82. https://doi.org/10.1210/me.2004-0513.
184. Biemann R, Fischer B, Navarrete Santos A. Adipogenic effects of a combination of the endocrine-disrupting compounds bisphenol a, diethylhexylphthalate, and tributyltin. Obes Facts. 2014;7(1):48–56. https://doi.org/10.1159/000358913.
185. Blumberg B. Obesogens, stem cells and the maternal programming of obesity. J Dev Orig Health Dis. 2011;2(1):3–8. https://doi.org/10.1017/S2040174410000589.
186. Janesick A, Blumberg B. Endocrine disrupting chemicals and the developmental programming of adipogenesis and obesity. Birth Defects Res Part C. 2011;93(1):34–50. https://doi.org/10.1002/bdrc.20197.
187. Thiering E, Cyrys J, Kratzsch J, Meisinger C, Hoffmann B, Berdel D, von Berg A, Koletzko S, Bauer CP, Heinrich J. Long-term exposure to traffic-related air pollution and insulin resistance in children: results from the GINIplus and LISAplus birth cohorts. Diabetologia. 2013;56(8):1696–704. https://doi.org/10.1007/s00125-013-2925-x.
188. Kelishadi R, Mirghaffari N, Poursafa P, Gidding SS. Lifestyle and environmental factors associated with inflammation, oxidative stress and insulin resistance in children. Atherosclerosis. 2009;203(1):311–9. https://doi.org/10.1016/j.atherosclerosis.2008.06.022.
189. Kramer U, Herder C, Sugiri D, Strassburger K, Schikowski T, Ranft U, Rathmann W. Traffic-related air pollution and incident type 2 diabetes: results from the SALIA cohort study. Environ Health Perspect. 2010;118(9):1273–9.
190. Di Ciaula A. Type I diabetes in paediatric age in Apulia (Italy): incidence and associations with outdoor air pollutants. Diabetes Res Clin Pract. 2016;111:36–43. https://doi.org/10.1016/j.diabres.2015.10.016.
191. Lavigne E, Ashley-Martin J, Dodds L, Arbuckle TE, Hystad P, Johnson M, Crouse DL, Ettinger AS, Shapiro GD, Fisher M, Morisset AS, Taback S, Bouchard MF, Sun L, Monnier P, Dallaire R, Fraser WD. Air pollution exposure during pregnancy and fetal markers of metabolic function: the MIREC study. Am J Epidemiol. 2016;183(9):842–51. https://doi.org/10.1093/aje/kwv256.
192. Malmqvist E, Larsson HE, Jonsson I, Rignell-Hydbom A, Ivarsson SA, Tinnerberg H, Stroh E, Rittner R, Jakobsson K, Swietlicki E, Rylander L. Maternal exposure to air pollution and type 1 diabetes–accounting for genetic factors. Environ Res. 2015;140:268–74. https://doi.org/10.1016/j.envres.2015.03.024.
193. Ke X, Cortina-Borja M, Silva BC, Lowe R, Rakyan V, Balding D. Integrated analysis of genome-wide genetic and epigenetic association data for identification of disease mechanisms. Epigenetics. 2013;8(11):1236–44.
194. Stefan M, Zhang W, Concepcion E, Yi Z, Tomer Y. DNA methylation profiles in type 1 diabetes twins point to strong epigenetic effects on etiology. J Autoimmun. 2014;50:33–7. https://doi.org/10.1016/j.jaut.2013.10.001.
195. Miao F, Chen Z, Zhang L, Liu Z, Wu X, Yuan YC, Natarajan R. Profiles of epigenetic histone post-translational modifications at type 1 diabetes susceptible genes. J Biol Chem. 2012;287(20):16335–45. https://doi.org/10.1074/jbc.M111.330373.
196. Hezova R, Slaby O, Faltejskova P, Mikulkova Z, Buresova I, Raja KR, Hodek J, Ovesna J, Michalek J. microRNA-342, microRNA-191 and microRNA-510 are differentially expressed in T regulatory cells of type 1 diabetic patients. Cell Immunol. 2010;260(2):70–4. https://doi.org/10.1016/j.cellimm.2009.10.012.

197. Nielsen LB, Wang C, Sorensen K, Bang-Berthelsen CH, Hansen L, Andersen ML, Hougaard P, Juul A, Zhang CY, Pociot F, Mortensen HB. Circulating levels of microRNA from children with newly diagnosed type 1 diabetes and healthy controls: evidence that miR-25 associates to residual beta-cell function and glycaemic control during disease progression. Exp Diabetes Res. 2012;2012:896362. https://doi.org/10.1155/2012/896362.
198. Stefan M, Zhang W, Concepcion E, Yi Z, Tomer Y. DNA methylation profiles in type 1 diabetes twins point to strong epigenetic effects on etiology. J Autoimmun. 2013;50:33–7. https://doi.org/10.1016/j.jaut.2013.10.001.

Complexity in Pollution/Disease Management 14

Antonio Vittorino Gaddi, Ibar Sabatini, and Fabio Capello

14.1 Foreword: How Are Diseases Caused by Pollution Classified?

In order to understand complexity when applied to air pollution-related diseases, it may be useful to understand the implication of classification in medicine. In this case, the problem is to set and classify diseases caused by pollution; therefore, in this chapter we focus on the principles of classification passing through historic profiles.

The most ancient classifications were approximate and methodical, based on body topography or aetiology. The traditional distribution *a capite ad calcem* was more an order of didactic exposition than a real classification, as explained by Mirko D. Grmek [1].

Thanks to the development of a new medical vision regarding the modern clinical concept of disease in particular, the hope of a natural disease classification rose in the eighteenth century. The branch of medicine which is in charge of the aforementioned systematic classification is nosology that is based on the idea that it is possible to identify a disease in an unambiguous way. In fact, it depends on aetiology, pathogenetic mechanism or symptoms. The classification based on organ and anatomical systems, even though still being used, is too complicated because of the fact that some pathologies hit more than one organ.

Since the nineteenth century, however, pathology took the place of nosology in order to study living system changes by biochemical and biophysical

A.V. Gaddi (✉)
Scientific Director of Health Lab of GTechnology Foundation, Modena, Italy
e-mail: profgaddi@gmail.com

I. Sabatini
Proeon, San Benedetto del Tronto, Italy

F. Capello
Alder Hey Children's Hospital, Liverpool, UK

F. Capello, A.V. Gaddi (eds.), *Clinical Handbook of Air Pollution-Related Diseases*, https://doi.org/10.1007/978-3-319-62731-1_14

mechanisms, passing from trigger causes to clinical symptoms. This was due to the impossibility of elaborating a coherent clinical system and a natural classification of diseases.

In spite of this, a compromise has been reached because of the necessity of an epidemiologic control: this is an artificial classification composed by different areas of clinical coherence. The beginning of international statistic classification of diseases goes back in 1893, when the international nomenclature of death cause ideated by Jacques Bertillon was adopted at the Chicago Conference. After some years in which lots of revisions were made, in 1993 the tenth revision of the classification of diseases was publicized, the first one that seemed to be completed [2]. It is a unified system of codex ICD and is called the International Statistical Classification of Diseases and Related Health Problems, Tenth Revision (CIM-10). Since it was the result of some agreements of different experts, it presented various contradictions. For this reason, the traditional system of numeric codification was substituted by an alphanumeric system. The number of categories passed from 17 to 21, where the majority of categories are defined by the localization of disturbs. To make the CIM-10 completely operative, it was completed with an international nomenclature of diseases and a group of related healthcare classification.

Despite this need of classifying diseases, however, we can understand that it is difficult to mark clear boundaries. This is owed to the fact that diseases are the consequence of an extremely complex interaction of several factors which operate contemporarily or in sequence on an organism. This new concept becomes detached from the old reductionist vision which had represented a failure in the clinical thinking, trying to simplify everything, and is the reason of systems medicine: the science of complexity.

Systems medicine is an interdisciplinary field of study that looks at the systems of the human body as part of an integrated whole, incorporating biochemical, physiological and environment interactions. It is based on systems science and systems biology and considers complex interactions within the human body in light of a patient's genomics, behaviour and environment.

In 1992 the earliest uses of the term systems medicine appeared in an article on systems medicine and pharmacology by B. J. Zeng [3] and in a paper on systems biomedicine by T. Kamada [4].

> A key feature of systems medicine is that existing networks, through dynamic (time-dependent) interactions, manifests 'emergent properties' that define the whole and that these properties are not simply the sum of the features of its component parts. […] Nearly all major classes of human diseases are genetically complex, with more than a single gene contributing risk for disease development.

An important theme in systems medicine and systems biomedicine, as just mentioned, is the development of computational models.

Those models are one of the keys and at the same time the problem and the solution that can help to understand what complexity is when applied to medicine.

Thus, classification opens the door to this major field of research, still largely unexplored.

14.2 Complexity: An Overview

Speaking about pollution is a hard task. As we will see, air pollution fits perfectly the model of complexity when applied to medicine and human health. Air pollution, in fact, is composed by a mix of different airborne chemicals, particulate, organic molecules and microorganisms, which have different interactions with human tissues and cells. Different components cause direct or indirect effect alone, in combination or in synergy with other chemicals and microorganisms. Different sources, concentrations, time and ways of exposure produce different effects visible immediately or anytime afterwards (hours, days, weeks, months, years, decades).

We need some definitions for a better comprehension of the idea of complexity. First of all, we need to separate concepts such as *complicated* and *complex*. These terms have different meanings based on the context. They may concern decision making, forecasting methods and/or system's composition.

The adjective *complex* comes from the Latin *complexus* (from the verb *complecti* meaning "tighten, understand, embrace"), and it is defined as "consisting of many different and connected parts", while the noun *complexity* is defined as "characteristic of a system (hence called complex), conceived as an organic and structured aggregate of parts interacting with each other, according to which the global behaviour of the system is not immediately attributable to that of the individual constituents, depending on the way in which they interact". The adjective *complicated* (comes from the Latin verb *complicare* meaning "fold together, wrap") can be defined as "neither simple nor easy; confusing, intricate" or "involving many different and confusing aspects" [5, 6].

These definitions help us to understand that a system may be defined by one of these terms, none or both.

We can use the model described in the Stacey matrix to give a clearer definition [7] (Fig. 14.1).

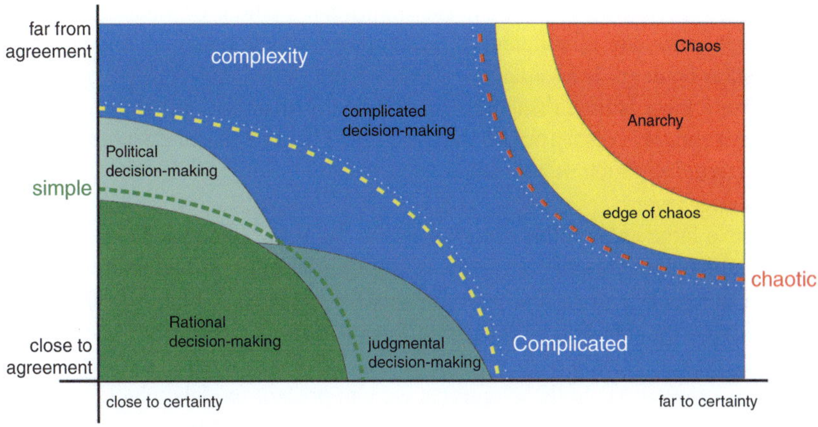

Fig. 14.1 The Stacey matrix

In this model, we have all the types of decision making based on two variables: *certainty* and *agreement*. *Certainty* concerns about the knowledge of the problem and its causes, links and possible outcomes. *Agreement* means "harmony or accordance in opinion or feeling", but also "The absence of incompatibility between two things; consistency" [6]. The decision makers then may be close to agreement or not. What kinds of decisions exist?

1. *Rational decision making*
   It lays in the region on the matrix which is close to certainty and close to agreement. In this region, decision makers use techniques which gather data from the past and use that to predict the future. They plan specific paths of action to achieve outcomes and monitor the actual behaviour by comparing it against these plans. The goal is to repeat what works to improve efficiency and effectiveness.
2. *Political decision making*
   Some issues have a great deal of certainty about how outcomes are created but high levels of disagreement about which outcomes are desirable. Neither plans nor shared mission is likely to work in this context. Instead, politics become more important. Coalition building, negotiation and compromise are used to create the organization's agenda and direction.
3. *Judgemental decision making*
   Some issues have a high level of agreement but not much certainty as to the cause and effect linkages to create the desired outcomes. In these cases, monitoring against a preset plan will not work. A strong sense of shared mission or vision may substitute for a plan in these cases. Comparisons are made not against plans but against the mission and vision for the organization. In this region, the goal is to head towards an agreed upon future state even though the specific paths cannot be predetermined.
4. *Chaos*
   Situations where there are very high levels of uncertainty and disagreement often result in a breakdown or anarchy. The traditional methods of planning, visioning and negotiation are insufficient in these contexts. One personal strategy to deal with such contexts is avoidance—avoiding the issues that are highly uncertain and where there is little disagreement. While this may be a protective strategy in the short run, it is disastrous in the long run. This is a region that organizations should avoid as much as possible.
5. *Complexity zone*
   There is a large area on this diagram which lies between the anarchy region and regions of the traditional management approaches. Stacey calls this large central region the zone of complexity—others call it the edge of chaos. In the zone of complexity, the traditional management approaches are not very effective, but it is the zone of high creativity, innovation and breaking with the past to create new modes of operation. In these regions, we can present models which extrapolate from past experience and thereby can be used to forecast the future. This is the

Table 14.1 Complexity in decision making according to the definition of scales of certainty and agreement

	Decision makers	Data	Forecast
Simple	Agree	Clear and collectible	Systematic
Complicated	Not agree	Clear and collectible	Systematic
	Agree	Unclear and/or uncollectible	Systematic
Complex	Not agree	Clear and collectible	Unsystematic
	Agree	Unclear and/or uncollectible	Unsystematic

hallmark of good science in the traditional mode. It is useful that decision makers have a diversity of approaches to deal with the diversity of contexts.

We have five zones in the diagram. One is the zone of simple decisions. Two and three are zones of complicated decisions. Four and five is the zone of complex decisions. We can extract parameters and values to define these concepts (see Table 14.1).

These distinctions used to describe decision making can be used to talk about systems too. We must consider how much we know about a system and if it possible to forecast outcomes based on our model. Systems do not have the agreement aspect strictly, but science is a matter of study and discussion, and this aspect may generate other problems or solutions.

14.3 Complexity in Medicine and Healthcare

Complexity is part of the nature including what is visible, measurable, invisible and not measurable (nevertheless theoretically possible), intended as a whole system that surrounds us and our existence. This axiom is crucial as it contains in itself the meaning of complexity when it is applied to biological systems based on the elusive concept of life.

A first step towards complexity is the idea that what is complex for fundamental systems (studied by physics and chemistry) cannot be applied to life (prokaryote and eukaryote cells and cells' clusters). The second step is that complexity applied to basic biological system cannot be applied to upper life forms according to a nonlinear and noncontinuous complexity scale (invertebrate, vertebrate, mammals). The third step is that complexity applied to the previous system cannot be applied to the human being.

A simple observation suggests that the augmented complexity in humans depends on larger and unquantifiable sets of variables where the whole life experience is considered. It means that memories, opinions, emotions, relationship, instincts, dreams and hopes are gathered with the whole system of biological events and functions with a concatenation of causes and effects. This chain of events and interactions is also spread on a timescale which varies according to subjective and objective scales (for instance, the chronological age may not correspond to the actual age of an individual: a person can look and appear older than another one of the same age, presenting also the same biological features of people older than him).

This complexity is summarized by a number of studies which try to disclose the fine biological mechanisms and laws that rule the human physiology or biochemistry, as well as the ones that lead to the development of diseases and pathological conditions. Each subject treated and researched may consist of a large number of publications, showing some sides of the truth and many gaps in the knowledge of the human nature. Every discovery is not definitive, and it shall be valuated with others which come from similar (or non-similar) studies.

This process generates big data, and complex computational systems may try to handle them, or reductionist models can try to synthetize them. The results are simulations which miss information in any case and oversimplify (or overcomplicate) a reality not known precisely.

To understand better the idea of complexity in upper biological life forms, it is important to think that biological systems are made of the whole of the physical, chemical and organic systems that are on the lower side of the scale: quantum systems with their own complexity organized in hadrons and molecules from subatomic structures, chemical interactions which rule simple reactions and subcellular structures with their own level of complexity, and cellular clusters organized in colonies or tissues and eventually in organs, apparatuses and organisms. Every level also has its own complexity, so with an added complexity that is the combination of these complex systems in a whole.

This idea can be summarized also in the same origin of life, where limited information coming from two gametes generate, during ontogenesis, an exponential mass of information that theoretically did not exist when the cell was born [8]. In addition, the spiritual essence of the human race should be mentioned, as a plausible and controversial factor able to augment the complexity.

Planet Earth is part of a complex system that selected different life forms through the millenniums together with their ecosystem with its fine balance. Human beings are a selection of a number of events that led to the rise and fall of species and environments. The fine-tuning of what the human organism is in every single individual today is part of this selection that can possibly push complexity ahead.

This is not only a problem that can be resolved with the integration of the knowledge coming from different fields, as in the idea of the systems medicine. It can be considered a starting point but it shows only a partial truth because it is not knowable totally.

In addition, human beings interact and create relationships and interactions, creating complex social systems where roles, obligations and responsibilities are created, accepted and discharged [9]. These systems create behaviours, push policies, implement decisions and build societies, and they are surrounded by an environment which humans have mutual exchanges with.

Considering air pollution, the whole idea of complexity is crucial, because it implies the interaction of a very complex system (the multitude of components of air pollution, their interaction between themselves and with the environment according to different spatial and temporal scales) with another very complex system (the human body and in general the human life).

A medical phenomenon is said complex [10] when—although measurable and observable—its evolution (which conventionally and schematically we can call disease) follows not entirely predictable rules, those that depend on a number of ($n => 1$) interaction/iterations of relatively simpler systems. The result or the effect cannot be predicted by summing the single effects coming from each subsystem. In addition, the logic of individual interactions (or their sum) is not fully known and this information may remain unknown.

Airborne pollutants (whether still floating in the atmosphere, deposited or became part of new ecological cycles) or their combination creates changes to the human body at different levels (subatomic particles composing the atoms of the body, up to the full organism with organs and apparatuses). This interaction depends on exposures which happen in different fashions and periods with single, cumulative or synergic effects. These effects could manifest their consequences in different stages and in different times (immediate outcomes, short-, medium- and long-term outcomes) or with a combination of all the previous.

From a medical point of view that creates unpredictable experimental models, the population, the effect or intervention, the comparison and the same outcomes may remain completely unknown and unmeasurable with the current logic of research.

14.4 Complexity in Air Pollution and Human Health

Why do we need such definitions for pollution? Because it's a complex problem which we may study in order to make it complicated. Daniel Vallero talked about the risk of disease due to pollution. He began the chapter introducing an equation:

$$R = H \times E$$

R (risk): it's the risk of a specific disease due to a substance.
H (hazard): they are chemical and physical properties of a substance which may cause negative effects.
E (exposure): it's the time of exposure to that substance.

The risk of disease increases based on the type of pollutant and time of exposure. We may say we have a function with two variables. We can imagine an *x*-axis which is "type of pollutants", a *y*-axis "time of exposure" and a *z*-axis "risk of disease" (Fig. 14.2).

There is an important note. This graph doesn't represent an exact, correct and mathematic function. It has the goal to clarify some valid concepts. Imagine we have two types of pollutants (*a, b*). We can see that pollutant *a* has low level of risk [7] at the beginning and it increases to a higher level [11]. We can see pollutant *b* has already a high level of risk at the beginning and it increases not as well as pollutant *a*. We can extract some information from this model. The two pollutants are different because the outcomes are not the same in the same period. Pollutant *a* is less dangerous than pollutant *b* because its risk of disease is similar to the other in the

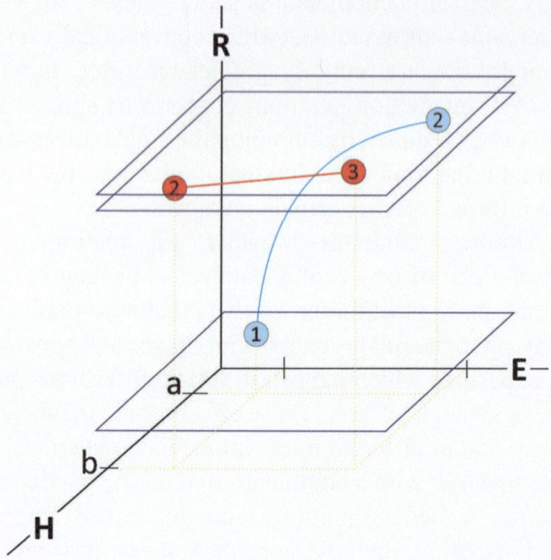

Fig. 14.2 Risk of disease based on the type of pollutant and time of exposure (see text for details)

worst case only. If a pollutant starts with a high level of risk, it may increase but not so much. How much can a high risk become higher? On the contrary, a pollutant with a low level of risk may have different outcomes based on its type. We may assume we can create models for every kind of pollutant in order to create solutions but it's not true. We began talking about concepts such as complicated and complex, and you may find that graph quite simple. There are other variables we didn't mention. Every person has different parameters such as age, sex, previous diseases, etc. Pollutants don't act singularly often. A pollutant mixed with others may create new outcomes. It's like the pollutants *a* and *b* create a new one with different consequences. The list can increase so the number of variables concurs in the risk factor. The human eye cannot see further than the third dimension, so we cannot put a graph of this function with all these variables. This part shows the complexity of the problem because there are so many variables that results may not be predictable accurately. We know the obstacles if we face the whole problem, but we don't say a solution cannot be found.

We may assume we can choose specific variables and see how R changes. We have to maintain not chosen variables constant in order to put a noticeable graph. Let us make an example (Fig. 14.3).

These graphs belong to Goran Krstic's work [12] about the relation between apparent temperature and mortality due to $PM_{2.5}$ in Metro Vancouver. In this scenario, he shows mortality levels based on apparent temperature values. In this case, the constants are exposure time, people's age, type of disease, place and type of PM. The two graphs are different because one of these constants has changed and it is the exposure time. Other variations might occur if we imagine to change another constant. The idea is that we are able to plot scenarios based on our purposes. We know the variables and the constants we need if we're interested on studying

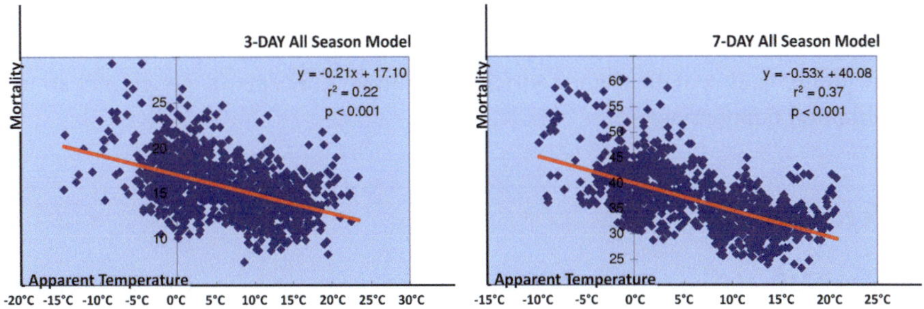

Fig. 14.3 Mortality per 100 K elderly population from circulatory and respiratory causes associated with apparent temperature (3-day and 7-day all-season models)

diseases on elderly people in specific places. This method is objective oriented and it enables to focus on the relevant information cutting off the useless ones. This example suggests a way of studying different cases. What can we expect for the future? Will we be able to forecast outcomes accurately? We said that this problem is composed by many variables, so we can expect even more related data exist. Big data studies are important tools for a solid solution. Artificial intelligences are another important tool in our opinion. IBM's Watson [13] is inspiring. Watson was created as a question-answering computer system capable of answering questions posed in natural language. The formidable feature of this system is its ability to adapt step by step. We won't describe the whole functions of this machine but the idea it suggests is quite interesting. We know that complex problems don't allow a systematic way of forecasting because they can change or evolve. Artificial intelligences such as Watson could do it too, so they might be the complex solution for complex problems.

Conclusions

The acknowledgement of the idea of complexity may be the only actual fact that we can know about life. Air pollution in this sense adds a new level of uncertainty, because it is a complex phenomenon in itself and because the real effect on humans is unknown.

Understanding how air pollutant interacts with the human cells and tissues implies the research of new knowledge which needs an open-minded approach, able to overcome the restraints of research models where the form is more important than the discoveries. The risk may be science trespassing into philosophy and pure speculation. On the other side, the ideas are the driving force of research which may lead to innovation and not confirmations of known theories.

Thus, air pollution can instil new insights in science, inspiring a new generation of researchers. This is because air pollution will remain a growing concern in the decades to come, and most of its effects will probably continue to show up, even in the desirable case of reduced emissions with a proper environmental administration.

At the same time, no good will come from approaches which ignore complexity and moreover the human beings intended as individuals who live and interact in groups and societies, not restricted in simple concepts of geographical and temporal boundaries.

References

1. Grmek MD. Malattia e salute. Enciclopedia della Scienza e della Tecnica; 2007. Available from: http://www.treccani.it/enciclopedia/malattia-e-salute_(Enciclopedia-della-Scienza-e-della-Tecnica)/
2. Capocci M. Specie nosologica. Enciclopedia della Scienza e della Tecnica; 2008. Available from: http://www.treccani.it/enciclopedia/specie-nosologica_(Enciclopedia-della-Scienza-e-della-Tecnica)/
3. Zeng BJ. On the holographic model of human body. In: 1st national conference of comparative studies traditional Chinese medicine and west medicine, medicine and philosophy, Guangzhou, April 1992
4. Kamada T. System biomedicine: a new paradigm in bio-medical engineering. Front Med Biol Eng. 1992;4(1):1–2.
5. Simone R. Treccani: dizionario della lingua italiana: Istituto italiano della enciclopedia italiana; 2013.
6. Stevenson A. Oxford dictionary of English. Oxford: OUP; 2010.
7. Stacey RD. Strategic management and organisational dynamics: the challenge of complexity to ways of thinking about organisations. London: Pearson Education; 2007.
8. Barbieri M. Code biology–a new science of life. Biosemiotics. 2012;5(3):411–37.
9. Martin M, Dobson J, Bonatti M. Conversation theory: reasoning about significance and mutuality. Available from: http://www.ncl.ac.uk/kite/assets/downloads/conversation1.htm; Accessed on 10 Jan 2017.
10. Manca M. Handbook of complexity in medicine. Cham: Springer; 2017.
11. Vallero D. Fundamentals of air pollution. Cambridge, MA: Academic Press; 2014.
12. Krstić G. Apparent temperature and air pollution vs. elderly population mortality in metro Vancouver. PLoS One. 2011;6(9):e25101.
13. IBM. What is Watson? 2017 [14/01/2017]. Available from: http://www.ibm.com/watson/what-is-watson.html

Air Pollution and the Nervous System: Diseases of the Brain Caused by Air Pollutants and Air Pollution

15

G.L. Lenzi and F. Bonsanto

15.1 Definition of Air Pollution in 2017: Historical Background

The WHO estimates that air pollution causes no less than 2.5% of all deaths in developed countries, corresponding in 2012 to seven million people [1] and representing one eighth of total global mortality. It would seem logical that both mortality and its correlate morbidity, as expressed in hospitalization, should be a worldwide, growing, public health concern [2]. The damages due to air pollution regard not only humans but also fish, animals, trees, etc. Air pollutants destroy human artefacts, buildings and masterpieces as well. Corrosion of monuments (now known to be due to acid rains) is a long-lasting acknowledged problem, as witnessed by the poor remains of the "Fonte Gaia" in Siena's, Piazza del Campo, completed in 1419 by Jacopo della Quercia but placed under the roof of the town hall for protection in 1858. An identical fountain replaced the original in the square (Fig. 15.1).

We quote here the definition from Sermin et al. "Air pollution collectively describes the presence of a diverse and complex mixture of chemical, particulate matter (PM), or of biological material in the ambient air which can cause harm or discomfort to humans or other living animals. The sources of air pollution can either be natural (e.g. volcanic eruptions) or manmade (e.g. industrial activities), and air pollution emerge as a serious health problem especially in rapidly growing countries"[3].

G.L. Lenzi (✉)
University of Rome Sapienza, Rome, Italy
e-mail: gianluigi.lenzi@gmail.com

F. Bonsanto
Alma Mater Studiorum, University of Bologna, Bologna, Italy

F. Capello, A.V. Gaddi (eds.), *Clinical Handbook of Air Pollution-Related Diseases*,
https://doi.org/10.1007/978-3-319-62731-1_15

Fig. 15.1 (**a**) Effects of air pollution and rain over four centuries on the Fonte Gaia by Jacopo della Quercia in Piazza del Campo in Siena (from Eugenio Müntz; http//www.ilpalio.org/siena471.htm). (**b**) Copy by the sculptor Tito Sarrocchi (1858), from Siena

The harm and discomfort potentially may affect all living aspects of biology, from grass to crops to trees to invertebrate animals, and, as said above, human artefacts as well. Finally, PM may be solid or liquid, be transported with the winds and/or diffused by the rain. It may be deposited on or absorbed by skin.

This definition appears to consider *air* as a passive, neutral carrier: the pollutant's damages are due to pollutants carried in the air, PMs or biological material. This approach seems to avoid evaluation of:

(a) The effects of PMs (and biological material) on some basic aspects of air that are relevant for life and well-being, such as transparency and clarity, which allow for full permeability of solar photons.

(b) The hazards of air pollutants such as ozone, CO_2, CO and other humanly generated gases.

Furthermore, air transports other important hazards for human beings as well as animals and vegetables. One important example is environmental noise, so often linked to airports and/or nearby highways: its damages may just superimpose those of engine's exhaust. Electromagnetic waves may be one other example of something that is "contained" in the air and that could be harmful. We will discuss environmental noise in the more general panorama of air pollution effects on superior cerebral activities, due to its potential major contribution to the striking increase in the last years of psychiatric presentations such as depressions and general anxiety syndromes with panic attacks.

Electromagnetic waves will not be discussed, as the continuing use of cellular phones seems to be impossible to be prohibited on the bases that it is "unhealthy".

It is important to underline that in the Sermin et al. definition (of air pollution) reported above, nothing is said about the mechanisms involved in air pollutions causing the final harm or discomfort: these may be direct or indirect on the target individual or organ, and consequently the pollutant may act directly and/or indirectly. That is, the air pollution per se may be a risk factor just because the normal aspects of air in the atmosphere are impaired even before entering in contact with the target individual and/or organ. In other words, the air pollutants may act by directly impairing the nervous system or through impairment(s) of human body functions (respiratory, circulatory, endocrine, metabolic, etc.) which then indirectly (in a secondary way) affect the nervous system. Possibly they may act just because they are present in the air that surrounds the individual. All these mechanisms may be more or less harmful depending on the genetics and the age of the target. Socioeconomic factors are also claimed to influence the degree of final damage.

Continuing, while it seems *politically correct* to speak about air pollution in the western countries, it is considered *politically incorrect* to apply these same words to the developing, third-world countries. In particular in developing countries and rapidly growing economics, air pollution will continue to increase as a major health problem, but these countries so far do not agree to international treaties for placing pollution under control, thus causing a dangerous domino effect.

In my personal experience in the 1960s, the "… dense smog that engulfed the city < of London > …" which was the core for one of the first reports highlighting, in 1953, the effects on health of air pollution [4] was very similar to the dense fog that accompanied me in 2010 during a trip from New Delhi to Agra, comparable now to the yellow air that engulfs Shanghai (Fig. 15.2). It is noteworthy that on March 5, 2017, the government of China officially announced their intention to give blue skies back to the Chinese population.

We are strongly convinced that the global hazard to health, as defined by OMS, is due not only to the PM inhaled or entering in contact with the nasal mucosa and the skin but also to the very fact of living in a turbid non-transparent non-clear environment, with fading colours and a disorderly yellow or grey overcast descending on the town.

While air pollutant's effects on cardiovascular and respiratory diseases are well recognized since quite a few decades, and currently under important scrutiny and

Fig. 15.2 Air pollution in China

investigation (cfr [5]), the central nervous system as a target for air pollutants is a more recent issue. The available data are not very numerous, as shown in Table 15.1 [from Ref. [12] with references up to 1 year ago. In this same important chapter from Ref. [4], out of the 78 references, *only one* regarding air pollution versus CNS *and published before the year 2000* is listed.

In the new millennium, the immunology of the CNS and all the different aspects of immune functions appear to be "hot" topics. Accordingly, the effects of air pollution on the immune system and, through these effects, on the CNS, with pathological changes called chronic inflammation or other definitions (such as oxidative stress, microglia activation and so on), are also a "hot" topic: scientists working on multiple sclerosis and related neuro-immunological diseases claim an important harmful influence from air pollutants (see below).

However, all considered, an important hiatus appears to remain between basic scientific studies on air pollution and clinical neurosciences and in particular between air pollution and brain diseases, being these diseases labelled as neurological and/or cognitive and/or psychiatric.

15.2 Definition of the Nervous System at 2017

Before initiating an overview on the topic of this chapter "Air pollution and Nervous System Diseases", it is important to define what "nervous system diseases" means in the year 2017.

Table 15.1 Particulate exposure and CNS effects

Animal/human experimental:	
Calderón-Garcidueñas et al. (1999, 2003, 2004, 2008b)	Mexico City air pollution associated with CNS inflammatory and neurodegenerative changes increased brain COX-2 and β-amyloid (humans), DNA damage and neuronal tangles (dogs)
Campbell et al. (2005)	Two-week ultrafine/fine PM near-highway exposure in ovalbumin-sensitized mice led to increased brain NF-κB and IL-1α
Crüts et al. (2005)	Diesel exhaust exposure caused an increase in fast-wave EEG activity in humans
Veronesi et al. (2005)	Five-month exposure to ultrafine/fine PM induced loss of dopaminergic neurons in substantia nigra of $ApoE^{-/-}$ mice
Elder et al. (2006)	Twelve-day exposure to Mn oxide ultrafine particles led to inflammatory cell activation and oxidative stress in olfactory bulb and other brain regions
Kleinman et al. (2008)	Six-week exposure to concentrated near-highway ultrafine PM in ApoE-null mice led to activation of inflammatory mediator transcription factors
Gerlofs-Nijland et al. (2010)	Four-week diesel exhaust exposure induced elevations in rat striatal TNF-α and IL-1α
Suzuki et al. (2010)	Mice exposed in utero to diesel PM had reduced locomotor activity
Fonken et al. (2011)	Ten-month $PM_{2.5}$ exposure led to oxidative stress and inflammatory changes in mouse hippocampus and to decreased learning and memory
Allen et al. (2013)	Mice exposed as neonates or adults to ultrafine PM have preference for immediate reward upon behavioural testing
Guerra et al. (2013)	Two-month coarse/fine/ultrafine Mexico City ambient air exposures in rats led to region- and PM size-specific increases in oxidative stress, inflammation and unfolded protein responses
Epidemiological	
Rauh et al. [6]	Second-hand tobacco smoke exposure in pre- and post-natal periods associated with decreased cognitive function in children
Calderón-Garcidueñas et al. (2008a, 2011)	Mexico City air pollution associated with cognitive changes in children
Suglia et al. [7]	Decrease in verbal and non-verbal intelligence and memory in children in association with traffic-related particles
Zeng et al. [8]	Poor air quality in China associated with poor cognitive function in older adults
Power et al. [9]	Cognitive function declines in older men in association with black carbon exposure
Volk et al. (2011, 2013)	Proximity to roadways and exposure to traffic-related pollutants or $PM_{2.5}$ in utero or in early life predict higher likelihood of autism
Weuve et al. [10]	$PM_{2.5}$ and PM_{10} levels associated with cognitive function decline in adult females
Wellenius et al. [11]	Association between proximity to roadways and cognitive function changes in elderly
Becerra et al. (2013)	$PM_{2.5}$ and ozone exposure during pregnancy increase odds of autistic disorder in children

From: Ref. [12], p. 270

The indication "nervous system diseases" is in general reserved for the organic diseases of the nervous system, i.e. neurology.

One of the most respected textbooks in the international panorama of neurology is *Brain's Diseases of the Nervous System*, first published by Prof. W. Russell Brain in 1933, which reached its twelfth edition in 2009, and was edited by Prof. Michael Donaghy [13]. In accordance with our premise, the title of this textbook indicates that the contents are exactly the diseases of the nervous system. And in his preface to the first edition, W. Russell Brain wrote "… Neurology is more dependent than many other branches of medicine upon anatomy and physiology". That was in 1933. Eighty-three years later, we would like to add *neuroimaging*. However, in the 2009 edition index, there is no trace of "air pollution" or "pollution" or similar references. Chapter 5 "Toxic and Environmental Disorders" (Ref. [13], p. 141–163) explores many CNS disease, from alcohol toxicity to metal toxicity to environmental and physical insults. However, besides the Parkinsonian syndromes attributed to carbon disulphide and carbon tetrachloride mixtures in grain workers (Ref. [13], p. 152), air pollution is not taken into consideration.

The same is true for the textbook edited by Anthony Schapira in 2006 [14]. Nothing is mentioned about air pollution. In addition, the relationships between pollutants and neurodegenerative disorders such as ALS (see below) and Parkinson's disease (PD) (see below) receive little attention.

However, in 2017, it is acknowledged that there are *diseases* different from those listed in the neurological textbooks indicated above, as well as and including many or all other neurological textbooks referring to the nervous system (cfr Table 2.1 in Ref. [13], p. 19), and where clinical presentations from *the brain*, such as dementia, are not present. The absence of anatomical and/or pathological changes at post-mortem and cultural influences as well (as an example, the development of psycho-analysis) has brought about (in the second half of the last century) a regrouping of these other brain diseases under the labels: psychiatry, psychopathology, psychology or others.

A potential link between neurology and psychiatry (or similar psychological disciplines) may be the identification of "psychologically determined disorders" (see Chap. 4 in Ref. [13]), with sections on pseudo-neurotoxic disorders and on mass sociogenic illnesses presenting as neurological diseases. These may be mass anxiety hysteria or mass motor hysteria, following mass exposures to substances, usually odorous, that act through an inhalation mechanism [15]. Sometimes, these sociogenic illnesses concern potential harmful mechanisms such as pollutants contained in the chemtrails that are observed in the clear sky.

In the "classical" neurology, when "… physical manifestations predominate, the underlying psychological cause may be ignored" ([13], p. 129). However, the data shown in Sects. 15.9 and 15.10 suggest that these underlying psychological causes may have deep roots in the physiology of our central nervous system.

To be completely fair, one must acknowledge that neurologists, and neurology, are well aware that in many neurological illnesses there are "associated psychiatric states"([13], p. 133) such as anxiety, depression and symptoms such as fatigue, clumsiness, gait disorders, headache, memory loss, cognitive impairment and others.

Clinical presentations with a dramatic cognitive impairment and a complete memory loss may be due to depression (the depressive pseudo dementia—see below) and therefore have to be recognized and treated accordingly. The physician assigned to the job has to have a complete neurological and psychiatric background and formation. In the same way, neurologists are often confronted with psychiatric adverse events, sometimes serious, due to neurological-neurosurgical therapies, such as in deep brain stimulation in Parkinson's disease [16] or L-dopa therapy side effects, as hallucinations and delusions.

The Webster's dictionary definition of psychiatry is "a branch of medicine that deals with mental, emotional or behavioural disorders ". As it will be shown later on, many nervous system's diseases due to air pollution concern or comprehend mental, emotional and behavioural impairments, in addition to classical-neurological signs and symptoms. However, in the psychiatric field, very few traces are to be found of problems linked or due to air pollution and air pollutants. In a possibly related area, the clinical features of "Substance Use Disorders" have been studied and described, but always or mainly from the point of view of a voluntary use of a "… substance despite significant substance-related problems" [14].

It appears that the brain's disorders caused and/or linked to air pollution are scientifically and culturally in a limbo, where perhaps the conflicts of interest between the different medical branches play a major role.

Let us make a quick example: dementia is a clinical concept, but it is widely acknowledged that dementia could be the result of:

(a) A neurodegenerative disease (such as Alzheimer's disease, Pick's disease, Huntington's chorea, etc.)
(b) A cerebrovascular disease (vascular dementia)
(c) A cerebrospinal fluid disease (idiopathic normal pressure hydrocephalus)
(d) Many others

Neurologists, psychiatrists, GPs and physicians are often confronted with the difficulty of differential diagnosis and with the several multifaceted problems of management of demented patients in their everyday routines, considering that in 2017 these diseases affect *millions* of elderly people. In a nearly complete *undervaluation* of these hard data on pathogenesis, "dementia" has been inside the psychiatric panel of WHO up to a few years ago. And now in many western countries, geriatricians are claiming their central role in the treatment of these disorders (perhaps again with some conflict of interest, *to be politically incorrect*).

Therefore, the *diseases of the brain* have been selected as the headline of this chapter. Accordingly, a full description will be made of the different clinical presentations related to air pollution and air pollutants actions on the nervous system, be it the CNS (the brain and the spinal cord) or the PNS (roots and peripheral nerves), with neurological signs and symptoms or without, with psychiatric signs and symptoms or without and with cognitive signs and symptoms or without. A consideration will also be made of all these signs and symptoms due to malfunctioning of nerve cells, of their fibres and connections, of their networks, of their glial relationships and so on: the malfunctioning of the brain's normal work.

15.3 The Diseases of the Brain

In mammals, and therefore in man, the nervous system is formed by a central part (CNS) and by a peripheral part (PNS). CNS comprehends the brain which is composed of the cerebrum, cerebellum and brainstem inside the skull, the spinal cord and the vertebral channel: the largest amount of neurons and of glial cells lies within the CNS. The PNS is composed of the spinal roots, with their neurons, and of the peripheral nerves, with the glial cells surrounding the nerve fibres.

The human brain reaches a volume of 1300–1500 cm^3. It has been calculated that in 1 mm^3 (= 0.001 cm^3) of the brain, there are 100,000 neurons (and three times more glial cells) and 4 km of nerve fibres (the axons) that connect neurons with other neurons and with glial cells as well. The total number of neurons in the human brain is approximately 85 billions. The neuron-to-neuron main contact is the synapse. Attributing to a single neuron 10,000 synapses, the estimated total number of synapses reaches upwards to 8.5 hundred trillions that furthermore may work in different allosteric positions (such as the receptors). All this bring us up to many billions of biological transistors or chips that are not something stable and/or fixed, but that renew every day, in a mere 1 mm^3. If we add the multiple connections with glial cells, and the many potential (still largely unknown), the system appears amazingly complex. Every microsecond, there are micro-parts that became obsolete, or damaged, in need of fast repair, or fast renewing, in particular at the level of the synapse. With 2788 unique proteins already identified as integral to the composition of each synapse, the systems controlling the microenvironment of the synapses and of these proteins should be both complex and delicate. One of these systems is the ubiquitin-proteasome system (UPS) that works to ensure that old proteins are discarded to make way for new ones and that the proper components of building materials are available [17]. We will again address this ubiquitin system in Sect. 15.7.

This amazingly complex nervous system works 24 h/day, full-time, throughout one's lifespan. To work properly (in quantitative terms, without any relationship to the quality of the CNS performances!!) CNS needs a continuous, very delicate, hydro-electrolytic and energetic homeostasis, a continuous influx of glucose and oxygen, through the blood supply (CBF, cerebral blood flow), to maintain constant for both the CMRO2 (cerebral metabolic rate for oxygen) and the CMRGlu (cerebral metabolic rate for glucose), at general as well as regional levels. In addition, the brain needs a *regular alternation* of sleep and wakefulness.

Global or regional impairments in these basic aspects of the brain's work (cfr [18]) result in modifications of physiological mechanism that, up to a certain extent, are coped within the normal possibilities of the neuron(s) and/or fibre(s) and/or network(s). Cognitive aspects such as consciousness or mental activities are not impaired in a noticeable way, but it is possible that poor motor performances (to be out of practice) as well as decreased creativity (lack of creativity) may reflect these modifications, still within normal limits.

When compensatory mechanisms can no longer maintain a function within the considered normal limits (for a particular individual, but possibly for a group exposed to an air pollutants), neuropsychiatric signs and symptoms appear. These clinical presentations will be named "diseases of the brain" collecting together the different clinical presentations that the nervous system offers in response to air pollution and air pollutants actions, be it on the CNS (the brain and the spinal cord) or on the PNS (roots and peripheral nerves) and in general all the signs and symptoms due to malfunctioning of nerve cells, of their fibres and connections, of their networks and of their glial relationships.

With this in mind, a central paresis will be considered to be caused by a decreased function of the central motor system, mainly of the pyramidal cells (Area 4 in the Fig. 15.3), while an impairment in decision making and in problem solving as caused by an impairment of the lateral-frontal association areas and of their interrelationships (Areas 6, 8, 9, 10 in the Fig. 15.3).

It is well known that the studies on correlations between neurological presentations and anatomical localization of a lesion began with the observations of Paul Broca on Napoleonic war veterans in 1861, nearly two centuries ago. Therefore, they are already accepted in classical neurology. However, the correlations between problem solving and activation of particular cortical areas are definitely more recent and still largely debated (cfr [19]). In addition, the anatomical and physiological bases of autism, neurosis, panic attacks and psychoses are completely uncertain. Diseases due to a neurotransmitter deficit, such as Parkinson's disease, could be modified by therapies such as statins presumably acting on completely different organs and systems outside CNS (cfr [20]), showing that it's "a long way to Tipperary". Perhaps technologies correlating neuroimaging and function may be helpful to understanding the human brain functional connectome (cfr [21]) and therefore the diseases of the brain.

I am not sure what this has to do with diseases of the brain?

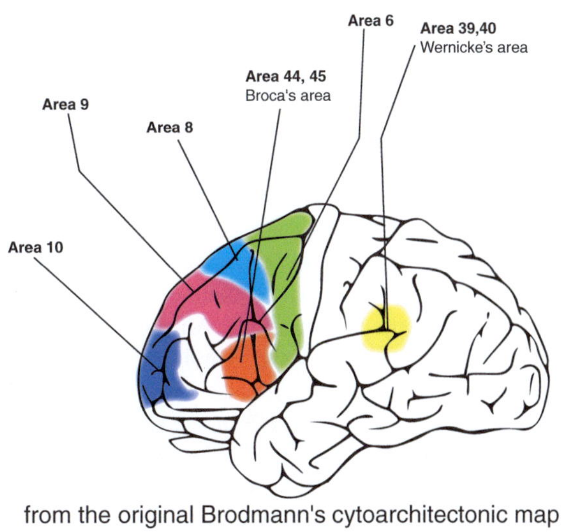

Fig. 15.3 Brodmann areas in the cerebral cortex of man

15.4 Penetration of Air Pollutants into the Nervous System

A range of very different particles such as dust, pollen, soot, smoke, liquid droplets and others can contaminate air. They may have their origin in nature, or they may be due to human activities.

How they enter the human body through the respiratory tract has been described in a different chapter in this handbook.

The principal air pollutants may be divided in groups respecting the areas in which they are produced: (a) urban and (b) non-urban (rural). The urban airborne particulate matter (PM) can be divided into three main groups: PM_{10} (coarse particles), $PM_{2.5}$ (fine particles) and $PM_{0.1}$ (ultrafine particles, UFP), depending on their properties (less than 10 μm to less than 2.5 μm in aerodynamic diameter and less than 0.1 μm in thermodynamic diameter.

Urban particles are generated by mechanical break-up of larger solid particles (PM_{10}), by combustion sources and heating processes including various precursor gases ($PM_{2.5}$), by natural processes, such as volcanoes, or by industrial activities, such as internal combustion engines, power plants and fumes (UFP).

Non-urban particles are often coarse, due to agricultural processes as well as to pollen grains, mould spores and plant and insect parts, but they may be due to evaporation of sea spray. Since fine particles are largely formed by gases, their presence in non-urban environment is minor, and it is different from zone to zone (highway borders are not different from urban areas). However, we have to consider in the non-urban air pollution effects as those due to acid deposition, impaired visibility and ozone, which are caused by fine particles.

As to the direct effects on the human body (and, the concern here is on the nervous system, directly or indirectly), environmental effects must be included that act both directly and indirectly on the brain. Further on and in more detail, the psychological effects of living in a contaminated and dilapidated environment, which appears not only to increase the individual's tendency to depression but also to decrease immunological reactions, will be considered. In addition to the effects of the main components (directly emitted and secondary compounds) themselves, some compounds react with other particles to form reaction products with important effects. One of these important reaction products is tropospheric ozone.

Inhaled larger PM, from their site of deposition in the respiratory tract, is cleared by macrophage phagocytosis in the alveoli and transported by mucociliary clearance to the naso-oro-pharynx and then, via swallowing, to the gastrointestinal tract. The smaller PM, the UFP, can have a similar clearance, or they may relocate into the blood or lymph, enter in contact with peripheral vegetative nerve fibres and thus penetrate the brain directly. Air pollutants may enter the human body through the skin, both via blood and lymphatic vessels as well as through the peripheral sensory nerve fibres. A particular aspect of this penetration mechanism lies in the olfactory bulb: up to the 11% of the UFP deposited on the olfactory mucosa may be detected in the olfactory bulb, that is, in the brain [22]: it could be relevant to remember that this is the same pathway followed by polio virus (around 30 nm) to reach and infect the brain. Magnetic nanoparticles, formed by combustion and/or friction derived heating, <200 nm in diameter, can enter the brain directly through the olfactory nerve [23]. A similar direct penetration may occur in the trigeminal ganglion, where it is possible for the PM and in particular the UFP to enter in contact with peripheral sensory axons through the skin. Interestingly, a similar mechanism, through an enteric neurotropic pathogen pathway, has recently been demonstrated in Parkinson's disease, since the risk of developing this classical neurodegenerative disorder appears to be decreased in those patients who underwent vagotomy decades before the neurological presentation [24]. Furthermore, it has been reported that nanoparticles (gold, 40–80 nm in diameter) may cross synapses (cfr [4], p. 277). This same mechanism may be active throughout our skin, both directly (penetration of pollutants) and indirectly (impaired signals to nerve endings).

The uptake into the brain from blood and lymph circulations is deterred by the presence of the blood-brain barrier (BBB), which, with its tight junctions, restrict this penetration pathways, as it does for many substances and therapeutic drugs, such as antibiotics. Inflammations, as well as osmotic agents, disrupt the BBB and open it to a free passage of components.

Another translocation pathway to CNS (the brain and spinal cord) is represented by the cerebrospinal fluid (CSF) [25], through the choroid plexus, which forms the blood-CSF barrier (not the BBB, which is formed by the microvascular endothelium of the brain). PM entering the CSF (as well as the drugs eventually injected) is quickly washed away through the CSF circulation and Pacchioni's granulations.

Figure 15.4 illustrates the potential translocation pathways for UFP to the brain.

Particle deposition depends on the particle's size, density, shape, hygroscopicity and ambient humidity. This is evident for the nasal absorption of the different PM,

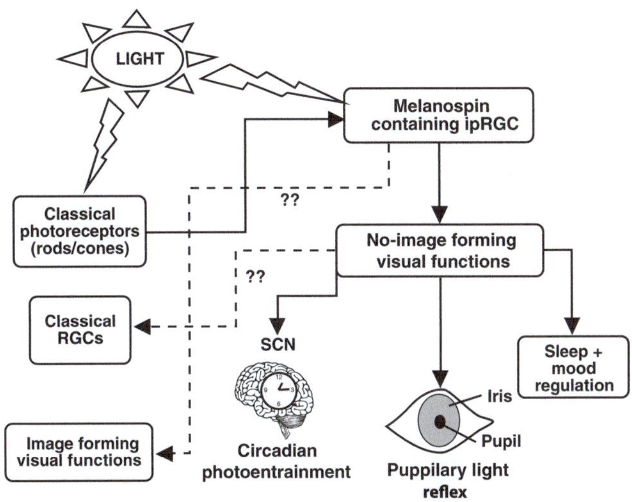

Fig. 15.4 The potential translocation pathways for UFP to the brain. Modified from Ref. [12], Fig. 10.4, p. 275

and the experimental data support this concept (cfr [4]). It could be important to add to these aspects the lipophilic properties of a pollutant. Hexanes are a striking example of severe work intoxication that may be due to skin absorption as well as to inhalation. Hexanes are utilized as cheap solvents to obtain glues for shoes and leather products, as well as to extract cooking oils (such as soy oil) and in food-based soybean oil extraction. Since hexane has considerable vapor pressure at room temperature, inhalation may cause both acute and chronic exposure. Inhalation of n-hexane at 2500–1000 ppm for 12 h produces drowsiness, fatigue, loss of appetite and paraesthesias in the distal extremities. Occupational hexane poisoning has been described in many countries, including Italy, as far back as 1974 [26], as well as trigeminal and/or peripheral neuropathies which have been reported in glue sniffers (cfr [27]).

It has been affirmed that the entrance of the nervous system into the field of air-borne pollutant's damages has to be pinpointed in 2002 [28]. However, the epidemiological studies on idiopathic PD have detected over the years quite a few risk factors that may strongly be connected with air pollutants, such as rural residence, exposure to pesticides and herbicides, farming, wood pulp mills and iron and steel production. Many of these pollutants may enter the nervous system through an enteric neurotropic mechanism as well (cfr [24]). Regional differences in the PD prevalence are well documented over more than 30 years (cfr [28]), however, with differences and variations in the collection and analysis of data that have often made the meta-analysis contradictory. The final results of a meta-analysis, performed on all the epidemiological studies on PD published from 1985 to 2010 [23] indicated that "In our analysis of the prevalence of PD by age and geographic location, individuals 70–79 years of age in Asia had a significantly lower prevalence of PD (646 per 100,000) compared with individuals of the same age in Europe, North America, and Australia (1.602 per 100.000; $p < 0.05$)....*The data were insufficient to make comparisons between geographic locations* ..." ([28], p. 1587). In contrast, the meta-analysis performed by Pezzoli and Cereda in 2013 [29] concluded that "The literature supports the hypothesis that exposure to pesticides or solvents is a risk factor for PD". Therefore, if studies on clinical consequences of the inhalation of heavy metals in workers of industries employing aluminium, copper, iron, mercury, zinc and manganese have mostly been negative (or—to be more accurate—not so positive as to induce courts or institutions to offer a compensation to the affected workers), the inhalation of their solvents is an evident risk factor for PD.

Similar data have been published for Alzheimer's disease (AD) well before 2002: a study in the Seattle area on 23.000 persons aged 60 years or more, who entered the study between 1987 and 1992, had shown that organic solvents, such as benzene, toluene, phenols and alcohol, increased AD odd's ratio in males to 6.0 [30].

Stroke too has been associated with air pollution on epidemiological, clinical and experimental backgrounds. Both short-term and long-term air pollution exposures have been linked to hospitalizations for ischaemic stroke and—with some discrepancies—to ischaemic stroke mortality [31]. The link with haemorrhagic stroke appears less strong. All these relationships may be viewed in the general panorama of air pollution and cardiovascular (or, better, cerebro-cardiovascular) diseases

It will be interesting to see if the same group (Ref. [5]) will present similar data on carotid and vertebral arteries, twin sisters of the coronaries.

Nervous system damages resulting from pollutants entering the nervous system through other pathways have been demonstrated, as in the cases of heavy metals or chemical compounds, such as in the "frozen cases" of Parkinsonism caused by methyl-phenyl-tetrahydropyridine (MPTP), injected as a synthetic substitute for heroin [32]. In the same way, lead intoxication has been linked to the aetiology of Alzheimer's disease (AD), as well as to accelerated declines in cognition in adults [33]. It is debated as to whether these toxic substances may act through air inhalation, but it probably is true for heavy metal powders derived, for instance, from engine combustion. Environmental NPs have been shown to increase the rate of protein fibrillation, thus linking neurodegenerative disorders to air pollution [34]. The development of nanotechnologies and nanomaterials (NMs), with their capacity to bypass or cross the BBB, may therefore be the cause of a new generation of air pollutants with CNS damage, through mechanisms such as oxidative stress, autophagy and lysosome dysfunction [35]. However, the access pathways to the CNS of nanomolecules such as titanium dioxide, silicon dioxide, zinc oxide, copper oxide, iron NPs, and carbon nanotubes as well as their presence in the environment might pose this neurotoxicity problem only in particular conditions, such as their production or utilization sites (04-15). Recent reports on both combustion-derived NPs and manufactured or engineered NPs (04-16) stress the hazard of these compounds, which may penetrate the CNS through inhalation or the nasal mucosa (cfr Fig. 15.4), as magnetite nanoparticles do. Clinical and behavioural aspects of neurotoxicity induced by nanoparticles will be discussed in Sect. 15.6.

15.5 The Effects of the "Pollutants" on the Biology of the Nervous System

Nervous system damage due to air pollution may be secondary to respiratory, cardiovascular, inflammatory and autoimmune diseases, described elsewhere in this book. Thus, the adverse effects on the CNS may be due to triggering the release of soluble inflammatory mediators from primary entry organs or secondary deposition sites. In addition to these secondary mechanisms, we have to consider a primary, direct effect, taking into account that "Although the exact mechanisms underlying brain pathology induced by air pollution are not fully understood, several lines of current evidence point out that neuroinflammation, oxidative stress, glial activation, and cerebrovascular damage might be the primary pathway" ([3], p. 1). All these damages cause a final cellular death, a common hallmark in neurodegenerative diseases.

These processes (neuroinflammation, oxidative stress, glial activation, cerebrovascular damage, BBB breakdown) may be linked and/or interact in multiple ways. For multiple sclerosis (MS), the important role of oxidative stress has been suggested by several AA, on findings from cerebrospinal fluid samples, blood samples, tissue homogenates and animal models (cfr [36]). In physiological conditions, neurons, astrocytes and oligodendroglia express molecules that bind to receptors on microglia and inhibit microglia activation. The inflammation causes the formation of an initial phagocytic lesion that contains few CD8-positive T-cells, with an initial tissue injury that induces activation of microglia, infiltration of the nervous tissue, secondary recruitment of T-cells and B-cells, macrophages for the removal of debris, further tissue injury and so on. It has been shown that the preactivated microglia is more easily converted into a cytotoxic form when exposed to a pro-inflammatory cytokine milieu, which happens in systemic inflammation. For these processes, the effects of the release of iron from myelin debris, which occurs in active MS lesions, amplify the first wave of oxidative stress and thus create new waves and progression of the disease. A similar harmful effect may be due to mitochondrial injury and the consequent ATP deficiency, leading to the failure to remove sodium from the inside of the neuronal membrane to the extracellular space. Accumulated sodium is replaced by calcium ions, with activation of calpains, and the beginning of the proteolytic degeneration of cytoskeletal proteins [37]. Finally, several factors related to local supply-to-demand oxygen balance may add their negative effects on the final picture. An explanation of the old data is seen in oedematous tissue swelling and increased oxygen consumption due to inflammatory infiltrates, found on particular locations of MS plaques on the boundary zones between major cerebral arteries, the so-called watersheds [38], more recently confirmed with MR [39]. These concepts have, very recently, been further expanded by the demonstration of the presence of meningeal lymphocytic follicula and of their possible pathogenic role both in the primary progressive MS (P1MS) and in the late transformation of relapsing-remitting MS (RRMS) into a secondary progressive MS (P2MS) (cfr [40]). Oxidative stress is indicative of an imbalance between the production of reactive oxygen species (ROS) and the cells ability to detoxify reactive intermediates or to repair cellular damage due to ROS. ROS are highly hazardous because of their unpaired electrons, thus impairing the normal redox state of the tissues with final production of peroxides and free radicals that may lead to neuronal death. The therapeutic war against free radicals is one of the main targets in the therapy for ageing and age-related diseases such as stroke and neurodegenerative pathologies. Under normal conditions, ROS are easily neutralized by cellular antioxidants defences such as glutathione and antioxidant enzymes; however, if the ROS production increases and/or if the defence mechanisms are diminished, a pro-inflammatory response is generated. This is particularly true for the brain, due to its high oxidative metabolic activity, its low antioxidant enzyme concentrations (such as superoxide dismutase and catalase), its low endogenous content in antioxidants and its high content in redox metals such as iron and copper; the nanomaterials (cfr [41]), that are increasingly engineered and utilized, are a potential source of pollutants that penetrate the nervous system through inhalation and then induce an increased production of ROS. Recently, the presence of biologically formed nanoparticles of the strongly magnetic mineral magnetite (well known to have redox activity, surface charge and a powerful magnetic behaviour) has been found in the human brains (they may enter through the olfactory bulb [23] or through other nerves as well (cfr [24])) of people living in Mexico City (cfr [42–44]).

Experimental data on the activation of several inflammation-related transcription factors to astrocyte activation in CNS have been reported by several AA, with related cortical damage as well as with loss of dopaminergic neurons in the substantia nigra, with dose-related relationships (cfr [5]). Neuronal and microglial cell lines or primary cultures of these cells have been exposed to concentrated ambient air particles, diesel exhaust particles and toxic gases such as ozone, to bacterial endotoxins or toxic elements (heavy metals such as manganese), with results still to be digested by the scientists specifically working on these aspects. This is a huge problem due also to the "…long list of confounding parameters experimentally …" obscuring the results [5] nonetheless by clinicians. The same "long list" appears to apply to clinical data, such as on stroke; this will be discussed later on. In our opinion, it is important to distinguish scientifically proven hypotheses from chemtrails. This is particularly true for brain's diseases, affecting large populations of human being, such as stroke, dementia, anxiousness and depression.

Direct damage to the nervous tissue, with accumulation of beta-amyloid (a hallmark for AD) and of alpha-synuclein (a hallmark for PD), as well as deficits in cognition, memory and executive function and increased prefrontal lesion (as measured with MR imaging), has been demonstrated in several studies by Calderòn-Garciduenas et al., from the University of Montana (cfr [42–44]), without, unfortunately, pollutant's characterization. If it was noteworthy that animals living in Mexico City, and exposed to the same air pollutants as the inhabitants, showed damage to the nasal mucosa and the olfactory bulb, as well as in subcortical and cortical structures, in comparison to control groups of mongrel canines living in less polluted rural areas, the correlation between a particular air pollutant and the final CNS damage was outside the experimental protocol. Similar findings have been shown in the brains of patients living in Mexico City, who were presenting a high density of nanoparticles formed by the highly magnetic iron salt magnetite [23].

The developing brain appears to be a relevant target for air pollutants, because of its high neuronal proliferation and differentiation rates, its hypermetabolism (a child shows a CMRO2 many times higher than an 18-year-old) and its imperfect BBB [45, 46]. Placenta may serve as a powerful barrier against environmental hazardous substances, but it is not enough protection against all or many components of air pollution. Maternal smoking, alcohol and drugs consumption and abuse, as well as exposure to industrial chemicals, are well documented hazards. Out of the air pollutants, ozone has been one of the best investigated both in animals and in humans, showing disruption of the monoaminergic systems, in particular in the cerebellum. The hazards on the developing brain are well demonstrated by the reports on cognitive functions in exposed children, which will be reviewed below.

15.6 The Effects of the Principal "Pollutants" on the Functions of the Nervous System

So far the basic mechanisms through which the air pollutants may impair the biological functioning of both nerve cells and glia, of the brain vessels and of the BBB have been summarized. It must be underlined that many aspects of the normal physiological interactions between neurons and glia and BBB and vessels are still far from being fully understood and clarified. It is possible to imagine a black box in which we know some of the inputs, some of the physiological relationships and some of the pathophysiological changes induced by pollutants. But the greater part of what is happening inside the black box is still unknown. However, the outputs may be measured in terms of particular aspects of the "brain work", comparing normal functions with impaired functions. This section will include these impairments (of the physiological functioning of the "brain") in terms of the interconnected nets and pathways leading to a final result of the "brain work" (cfr [17]). In a broad pathophysiological sense, impairments of the (basic) neurological functions (motor and sensory, consciousness) and of the cognitive activities (superior cortical/cerebral activities, content of consciousness, memory, intelligence, attention, problem solving, etc.), leading to psychological and finally psychiatric aspects and impairments of these cerebral activities (mood, behaviour, autism, schizophrenia and so on), will be examined.

Air pollutants that enter the human body through the skin affect peripheral nerve endings, with a dying back mechanism that produces a decreased function of motor or sensory fibres or of both, as well as central effects since the debris travels onto the neuronal bodies, then to the spinal cord and finally to the other parts of the CNS.

It is seen not only in a polyneuropathy indicating the PNS impairment (cfr Ref. [25]) or the impairments of cranial nerves [27] but also in the CNS with damage which may be seen as lesions of the spinal cord, with a clinical presentation of spastic paraparesis, with sensory deficits in the same territories and with EEG alterations, as described by our group in glue workers [26].

Healthy young male volunteers were exposed for 1 h to full diesel exhaust, and the electrical activity of their brains was monitored by means of quantitative electroencephalography in a crossover study [47]. A significant change in the median power frequency in the frontal lobes, with an increased beta-2 EEG wave activity, was observed after 30 min of exposure, indicating that cortical stress could lead to functional changes.

Experimental studies on zinc oxide nanoparticles, that may reach the brain through inhalation or translocation along the nasal mucosa and the olfactory nerve pathways, have shown in old animals an increased oxidative stress level, impaired learning and memory abilities. The experimental changes induced by zinc nanoparticles appear to be very similar to those involved in the onset and progression of neurodegeneration [48].

A recent meta-analysis of papers dealing with the effects of outdoor air pollution and noise on cognitive and psychological functions in adults has shown that "both exposures were separately shown to be associated with one or several measures of global cognitive function, verbal and nonverbal learning and memory, activities of daily living, depressive symptoms, elevated anxiety, and nuisance" [49]. However, no consistent data are available on the relationship between length of exposure and progression of the deficits, caused by a *nonpollutant related air pollution*, as in the case of noise. As said above, it appears that embryos and children are more sensitive to these mechanisms of pollution; however, little is known on the possible recovery of brain functions after removal of the pollutants or moving the subjects into cleaner, unpolluted environments ("more suitable airs" Manzoni A; Cinque Maggio). School-aged children living in Boston (USA) and chronically exposed to traffic particles showed a decrease in cognitive functions of the order of two IQ points [7]. Similar data were later obtained in older men living in the Boston area, with an inverse correlation between levels of exposure and cognitive functions [9]. Accordingly, data on larger cohorts, both in 15.973 older adults in China [8] and in 19.409 nurses in the USA [49] have confirmed these damages in cognitive functions. Furthermore, the higher baseline exposure was correlated with a faster rate in decline over time [10]. Again in Boston, the residential proximity to major roadways has been correlated with impaired cognitive functions ([11]; cfr [24]). The same is for long-term particle exposure [50] and for ozone and PM_{10} [51].

In 2011, WHO published an important report on the "Burden of disease from environmental noise" [52], reviewing the evidence available to obtain a quantification of the damage, in terms of lost DALYs, on the relationship between environmental noise (END) and specific health effects, including cardiovascular disease (61,000 years lost), cognitive impairment (45,000 years lost), sleep disturbances (903,000 years lost), annoyance (654,000 years lost) and tinnitus (22,000 years lost). These results indicate that at least 1 million healthy life years are lost every year due to traffic-related noise within Western Europe. This environmental noise pollution—in our opinion—has to be considered within the different mechanisms of the more general air pollution. The complex clinical presentations linked to annoyance have to be considered as resulting from impairments of the cortical circuitries regulating behaviour and mood, and it is important to underline that these impairments are more evident in children [53, 54] and in the elderly: however, these aspects are still debated and await better nosology. Annoyance may lead to anxiousness, to depression and to cognitive impairment, depending on the initial definition and on the protocols utilized. The WHO publication includes stroke within the chapter on cardiovascular diseases, beside the fact that worldwide nearly half of the deaths labelled cardiovascular are due to stroke. Cognitive impairment is particularly evident in children and seen in tasks involving central processing and language, such as learning, attention and memory. It is important to report the studies concerning the relocation of Munich's airport in 1992. Prior to relocation, high noise exposure was associated with deficits in long-term memory and reading comprehension. The deficit disappeared when the study was repeated 2 years after the closure of the old airport and its relocation. This good news, meaning the possibility of complete recovery if the exposure ceases, was counterbalanced by the (quite obvious) report that the same cognitive damages were found in children who again became exposed to noise near the new airport (cfr [52]). How these aspects may be linked to the annoyance reported in the exposed children as well as in the exposed adults, and how they may be interpreted with the growing number of individuals (let's not call them patients so far !) showing severe serotoninergic problems, will be discussed further, in Sect. 15.10.

Tobacco smoke particles have been shown to be very similar to ambient particles. Primary tobacco smoke has more gases and less combustion particles than second-hand smoke that may result in a decrease in cognitive and academic abilities when the exposure happens during childhood [6] or when there is prenatal exposure and conditions of material hardship [55]. Higher current exposure to second-hand smoke is correlated with deficits in superior cognitive abilities [6]. The number of scientific reports is very high, and we refer to the report from the Centers for Disease Control and Prevention, Atlanta, 2010 [56]. In fact, there is an important inconsistency in general assessments of children's cognition and intelligence, which in our opinion reflects the debate on the real nature of cognition and of intelligence, in spite of how we may measure them. The topic is hot, and commercial interests may well be a major factor in confounding a sound and safe approach.

Neurodevelopmental disorders have been linked to many factors: medical, genetic and environmental as well. A very low birth weight and foetal alcohol exposure (even in absence of stigmata of foetal alcoholic syndrome) are certainly associated with these disorders that may have different clinical presentations, such as autism (or, better, autism spectrum disorders (ASD), including Asperger's disease), attention-deficit/hyperactivity disorder (ADHD), intellectual disability, social communication disorders, stuttering and many others. In a population-based case-control study in Southwestern Pennsylvania, it has been reported that the "Living in areas with higher levels of styrene and chromium during pregnancy was associated with increased risk of ASD, with borderline effects for PAHs and methylene chloride. ….It is unclear … whether these chemicals are by themselves risk factors or if they reflect the effect of a mixture of pollutants. Future work should include spatio-temporal estimates of exposure to air toxics, taking into account the dynamic movements of individuals during daily life" [57]. However, we have to remember that, for many decades, the first seminal papers in the 1940s by the child psychiatrist Leo Kanner, showing the first identified autism, made the hypothesis that infantile autism resulted from rejection of the infant by emotionally cold parents ("refrigerator mothers"). This put a lot of pain and guilt on the already unhappy parents, until further research disclaimed this hypothesis. Modern society is affected by *politically incorrect* (in our opinion) attitudes such as NIMBY and compensation. The multiple varietals of hypotheses on autism may just reflect the inconsistency of the data, and we have to accept that unproven hypotheses remain simply hypotheses. The debate on autism may be considered a reflection of similar debates on many psychiatric diseases. Exposures to toxins, chemicals, poisons and other substances have been hypothesized to cause autism, only because anecdotal case reports suggested that such exposures may play a role in isolated cases of autistic disorder, giving support for a request of compensation, not taking into account the fact that a causative role for toxins in the development of autism in general has not yet been demonstrated. The same is true for organochlorine pesticides dicofol and endosulfan during the first trimester of pregnancy. In some parts of the world, exposure to specific toxins may influence local autism rates. For example, the high incidence of autism in areas of Japan has been hypothesized to be due to a toxic effect of certain fish. Although toxins may play a role in the development of isolated cases of autism in Japan, they have not been proved to be the principal cause of autism there.

Another possible explanation for the high autism rates in Japan is the reported excellent training of Japanese clinicians; low rates elsewhere may reflect the limited abilities of those clinicians to correctly diagnose autism (cfr [58, 59]). This consideration reflects the difficulties involved in comparing data obtained in different countries using different methods and protocols by different groups, a problem affecting experimental research since Galileo.

The same concerns should apply to the interpretation of experimental data and their application to clinical problems, regarding recent experimental data [60]: these AA have described how exposure to air pollution early in life produces harmful changes in the brains of mice, producing a widespread neuroinflammatory response, changes in CNS neurotransmitters and diffuse glial activation, leading finally to an enlargement of the lateral ventricles, "… a neuropathology that has been associated with poor neurodevelopmental outcome, autism, and schizophrenia". However, these changes, as well as an increased hippocampal glutamate, occurred predominantly or only in male mice: a gender difference that is not observed in humans. Furthermore, enlargement of lateral ventricles is frequently observed in normal ageing, without clinical counterparts. It seems without doubt that exposure to air pollutants and to noise is something to always be avoided, in particular for children. People should make better use of their good common sense in accepting the interpretations of the available experimental data, to avoid—for example—what happened to the unfortunate "refrigerators mothers", incriminated, even before the theory was approved scientifically, for they are responsible for their babies' autism. It is true that emotional development of mothers has a major effect on the emotional development of the new baby (cfr [61]), but the risks are in terms of further difficulties in emotional relationships and not in causing the presentation of ASDs. Beware of chemtrails! And of the "no-vax" hysteria !

15.7 Air Pollutants and the Pathogenesis of Particular Neurological and Psychiatric Diseases: Neurodegenerative-Neurocognitive Disorders (SDAT, PD, ALS), Stroke and Depression

The last 70 years after the end of the World War II have shown a continuous general increase in life expectancy worldwide, in parallel with an increase in the incidence and prevalence of many diseases, due to their correlation with age. For the brain, neurodegenerative-neurocognitive disorders (including dementia) and stroke are, from an epidemiological point of view, the two most important health problems which refer nosologically to neurology, while depression (including anxiety) is the largest health problem referred to psychiatry. The burden of the "mental disorders and other disorders of the brain" has been expressed, for Europe in 2010, in terms of the size of expenses for their treatment (Table 15.2) [62], which are proportional to the lost DALYs. How much of this burden is due to "growing older", and how strong may be the contribution of other aspects of modern life that have shown a similar increase after World War II?

Table 15.2 Revised rankings of DALY estimates (age > 15) by selected mental and neurological disorders for the EU-27 population

Men				Women				Both			
Diagnosis	Total DALYs	Rate per 10,000 DALYs	Rank	Diagnosis	Total DALYs	Rate per 10,000 DALYs	Rank	Diagnosis	Total DALYs	Rate per 10,000 DALYs	Rank
Alcohol use disorders	1,668,597	82.8	1	Unipol. depr. dis.	2,891,945	134.4	1	Unipol. depr. dis.	4,320,400	1033	1
Unipol. depr. dis.	1,428,455	70.9	2	Dementias	1,476,756	68.6	2	Dementias	2,216,514	53.7	2
Stroke	783,449	38.9	3	Stroke	793,389	36.9	3	Alcohol use disorders	2,039,741	48.9	3
Dementias	759,758	37.7	4	Migraine	490,744	22.8	4	Stroke	1,576,838	37.8	4
Drug use disorders	578,236	28.7	5	Alcohol use disorders	371,143	17.2	5	Drug use disorders	756,548	18.2	5
Bipolar affective disorder	367,638	182	6	Bipolar affective disorder	360,204	16.7	6	Bipolar affective disorder	727,841	17.5	6
Schizophrenia	329,945	16.4	7	Schizophrenia	307,748	14.3	7	Migraine	642,677	15.4	7
Parkinson's disease	174,037	8.6	8	Panic disorder	256,932	11.9	8	Schizophrenia	637,693	15.3	8
Insomnia (primary)	168,845	8.4	9	Insomnia (primary)	220,908	10.3	9	Insomnia (primary)	389,753	9.4	9
Migraine	151,933	75	10	OCD	181,777	8.4	10	Panic disorder	383,783	9.2	10
Epilepsy	150,145	75	11	PTSD	181,422	8.4	11	Parkinson's disease	334,446	8.0	11
OCD	147,907	7 3	12	Drug use disorders	178,312	8.3	12	OCD	329,684	7.9	12
Panic disorder	126,851	63	13	Parkinson's disease	160,409	7.5	13	Epilepsy	260,424	6%	13
Multiple sclerosis	72,722	3,6	14	Epilepsy	110,279	5.1	14	PTSD	245,475	5.9	14
PTSD	64,054	3.2	15	Multiple sclerosis	1011,104	4.7	15	Multiple sclerosis	172,826	4.1	15
Mental retardation, lead caused	2657	0.1	16*	Mental retardation, lead caused	2736	0.1	16*	Mental retardation, lead caused	5393	0.1	16*
Total mental dis.	5,642,944	280.0		Total mental dis.	6429,1182	298.8		Total mental dis.	12,072,826	289.7	
Total neurological dis.	1,332,286	66.1		Total neurological dis.	1,654,926	76.9		Total neurological dis.	2,987,212	71.7	
Total other neuropsychiatric	423,544	21.0		Total other neuropsychiatric	382,981	17.8		Total other neuropsychiatric	806,525	19.4	
TOTAL	7,398,774	367.1		TOTAL	8,467,718	3935		Total	15,865,563	380.7	
Total all cause DALY	31,616,455			Total all cause DALYs	28,125,319			Total all cause DALYs	59,741,774		
Proportion of neuropsychiatric		23.4%		Proportion of neuropsychiatric		30.1%		Proportion of neuropsychiatric		26.6%	

*DALYs due to lead lamed mental retardation represent only a small fraction of the burden attributable to mental retardation

Considering that air pollution and pollution of the air (including noise) have certainly increased in this same temporal span, with some geographical dislocation of particular air pollutions, as seen above, in regard to the Munich airport, we could be legitimized to consider these pollutants as an additional component to the pathological mechanisms of these disorders. Again, it is not easy to discriminate sound data from hypotheses.

One of the more important human pathologies linked to ageing is *dementia*, due in a majority of cases to a neurodegenerative disorder. Dementia however is only a clinical syndrome, affecting 5% of the world population older than 65 years of age, with a prevalence that doubles every 5 years: in subjects older than 80 years, it reaches 40%. It has been calculated that in 2015 there are 46.8 million people affected by dementia.

The cost of dementia in 2015 has been calculated in 818 billion US dollars. Nearly 60% of patients affected by dementia live in medium- or low-income countries [63]. It has been not easy to distinguish between the effects of "normal" ageing and dementia as a disease, since the time of the "De Senectute" written by Cato Major in 44 BC. Modern dementologists have developed the concept of mild cognitive impairment (MCI), to indicate a patient at risk of being affected by a dementia, whatever the form may be.

For neurological textbooks, neurodegenerative disorder is an old definition, based on pathological classification proposed by Van Bogaert in 1949 [65] and it included all these various neurological presentations with a primary involvement of a neuronal system or of large cortical areas, with unknown aetiology and with clear-cut post-mortem findings of degeneration of those neuronal systems and/or cortical areas. The time course was a progressive worsening, leading to death. Therapy was only palliative. This concept of neurodegenerative disorder is no longer accepted by modern neurology (cfr [13]), but it is maintained in other international textbooks (cfr [14]). The majority of the neurodegenerative disorders are accompanied by cognitive impairment, as in primary dementias. In psychiatric textbooks (cfr 02-05), all these clinical presentations are included in a large nosological group, the neurocognitive disorders (NCDs), with some caveats, such as "Although cognitive deficits are present in many if not all mental disorders (e.g. schizophrenia, bipolar disorders), only disorders whose core features are cognitive are included in the neurocognitive disorders" or such as "The NCDs are unique among DSM-V categories in that these are syndromes for which the underlying pathology, and frequently the aetiology as well, can potentially be determined (02-05, p. 591). In general, in psychiatry the term *dementia* is reserved to older adults, and it indicates an acquired cognitive deficit. The term *neurocognitive disorder* is utilized in younger individuals, and it may reflect the progression of a variety of damages to the brain, be it congenital, toxic, traumatic or others.

Primary neurodegenerative dementias include Alzheimer's disease (AD; 2.12% in the general population; 50% of all cases with dementia; 190/100.000 aged 65 or older/year; 46.8 million worldwide), Lewy body dementia (LBD) together with Parkinson-dementia complex (PDD; 4–30% of all cases with dementia; 4–7/100.000/year; 15–20 million worldwide) and frontotemporal dementia (FTD;

5–10% of all cases with dementia; 4–15/100.000/year; 2–5 million worldwide). These four clinical presentations represent 60–70% of all the presentations labelled as "dementia", that is a "…clinical syndrome of impairment in multiple domains of cognitive function, which must include impairment of episodic memory, in a patient who remains alert with normal arousal" ([13], p. 978). The rarer presentations, such as subcortical dementias, prion dementias, cerebral infections and inflammatory disorders, inherited metabolic and storage disorders will not be discussed. Vascular dementia, which accounts for 30% of all the dementias, and which may be present in many cases of primary neurodegenerative dementia as well, will be discussed later on in this section. On the whole, clinical presentation commonly labelled as dementias is now a real epidemic, which may be due not only to the increased lifespan which has increased in the last 70 years but also to changes in the environment that have really exploded during this same time period. These "epidemics" will be discussed in Sect. 15.10.

Primary dementias pathologies differ from AD to LBD to FTD to PDD, leading to the actual differentiation between taupathies and synucleinopathies. However, all these clinical presentations show an accumulation of protein catabolites in the tissue (extracellular space the taupathies, intracellular space the synucleinopathies), their precipitates assuming particular shapes and forms.

In AD, neuropathology shows the presence of senile plaques and neurofibrillary tangles. The senile plaques are observed in the association cortices and are composed of beta-amyloid peptide (a protein derived from a much larger transmembrane molecule, the amyloid precursor protein). The neurofibrillary tangles are intraneuronal depositions consisting of paired helical filaments with the microtubule-associated protein tau, being the major component. AD is accompanied by a severe neuronal cellular loss, maximal in the hippocampus (a core center for memory) and in associated cortical areas. Neuronal loss is evident in subcortical nuclei, in particular in the amygdala (a relay center for integrate visual inputs with emotional experiences [64]) and in the nucleus basalis of Meynert, which is the origin of the cholinergic subcortical projection system (therefore, replacement of the cholinergic deficit is actually the core of the AD therapy).

LBD pathology affects both cortex and midbrain, and it is characterized by the intraneuronal presence of Lewy bodies and by Lewy neurites, reported also in Parkinson's disease and in PDD. Actually, many AA do not see a major difference between LBD and PDD. Lewy bodies are intraneuronal inclusions mainly composed of alpha-synuclein [65], and they are diffused throughout the CNS, with different severity from region to region in different patients, which is the reason for the many different clinical presentations. Synuclein and tau proteins are linked in some way, since the toxicity of beta-amyloid seems to depend on the presence of microtubule-associated protein tau, the hyperphosphorilated forms of which aggregate and deposit in AD as neurofibrillary tangles; however, this "hot" topic is outside the targets of this chapter (cfr [66]).

FTD is composed of different disorders which are considered together since they share the common involvement of the frontal and/or the temporal lobes, with the clinical presentations reflecting the predominant anatomical distribution of the neurodegeneration. Neuropathological backgrounds are variable, ranging from Pick's disease, aspecific frontal lobe atrophy, asymmetrical cortical degenerations, hereditary taupathies and some particular motoneuron disease. Neurons may show tau inclusions and ubiquitin inclusions, and sometimes links with prion dementias have been described (cfr [67]).

Aetiology of neurodegenerative dementias is multifactorial, a synonym often used to mask poor knowledge. Risk factors include genetic predisposition as well as environmental factors. A history of cerebral trauma or of general anaesthesia has also been cited. The Apolipoprotein E (ApoE) has been demonstrated to be deeply involved in AD pathogenesis. ApoE is a major cholesterol carrier, which supports lipid transport and injury repair in the brain. ApoE has polymorphic alleles, and the individuals carrying the e4 allele are at increased risk of AD compared with those carrying the more common e3 allele. Subjects with an e4/e4 genotype have an OR 14.9 in comparison to e2/e4 (OR, 2.6) or e3/e4 (OR, 3.2), while e2/e2 seems to be protective against AD, with OR, 0.6.

Genetic studies have indicated that many chromosomes may be implicated in AD pathogenesis (as well as for the other neurodegenerative dementias), but it is presumed that genetics account for 70% of the risk, and the remaining 30% include environmental factors and human lifestyle patterns [68, 69]. Air pollutants (PMs), organic solvents [29] and in particular long-term exposure with bioaccumulation over an individual's lifetime are speculated to affect the nervous tissue. Accordingly, although different neurodegenerative diseases have somehow different pathologies and clinical presentations, as said above, they appear to share common mechanisms such as protein aggregation, oxidative stress injury, neuroinflammation, microglial activation, apoptosis and mitochondrial dysfunction, which have been described above as due (also) to air pollutants. A model (LEARn, latent early-life associated regulation) with an underlying "two-hit" theory, which combines genetic and environmental risk factors, has been proposed, suggesting that AD risk is established early in life and exacerbated by many humanly generated factors, such as obesity, lack of instruction, lack of exercise, hypertension, diabetes, air pollutants exposure and many others. The interaction between the genetic risk factors and the environmental-personal risk factors is counteracted by the capacity of the brain to cope with injury and with ageing processes, the so-called "reserve", that may be divided into a "brain reserve" and a "cognitive reserve". Brain reserve refers to differences in the brain structure that may increase tolerance to pathology; cognitive reserve refers to differences between individuals in how tasks are performed that might enable some people to be more resilient to brain changes than others. This concept (of brain or cognitive reserve) provides an explanation for differences between individuals in susceptibility to age-related brain changes or pathology related to AD and to the other neurodegenerative dementias. Epidemiological studies suggest that lifelong experiences, including educational and occupational attainment, and leisure activities in later life, can increase this reserve. For example, the risk of developing AD is reduced in individuals with higher educational or occupational attainment. However, we have seen the cognitive decline of presidents and premiers, who certainly had a higher educational and/or occupational attainments. Greater understanding of the concept of cognitive reserve could lead to interventions aimed at slowing down cognitive ageing and/or reducing the risk of dementia, but it is crystal clear that education and lifestyles are two of the main components of this "brain-bank account" where individuals deposit their individual patrimony to later be spent in ageing. It could be said that socioeconomic factors are therefore quite important. Rentals in proximity of airports or highways may be more affordable than rentals located near parks, and thus children from low-income families may be condemned from birth to an increased risk factor for later developing a neurodegenerative disorder. The findings reported in many paper referred (cfr Refs. [7, 29, 42–45, 49, 68] and other) to in this chapter may be reinterpreted under the light of equal chances of an individual's to develop or not brain or cognitive reserves.

For PD, many of the concepts and findings reported above are fully valid. PD is the second largest neurodegenerative disease, with an annual incidence of 15–20/100.000 people and a prevalence of 1–2/1000/year; it is slightly more frequent in men (2.0%) than in women (1.3%). Approximately, 4–6 million people worldwide are affected by PD.

It is due to a neuronal degeneration localized in the *pars compacta* of the mesencephalic substantia nigra, resulting in a dopaminergic failure in the striatum, the main target of the nigral neurons; these circuits are crucial for the execution of movements and postural functions, explaining the nature of PD symptoms. Histological examination of these affected regions show the presence of both Lewy bodies and of Lewy neuritis, with definite alpha-synuclein pathology. As for AD, also in PD, age appears to be the leading risk factor, but epidemiological studies have detected a variety of other risk factors, including rural residence, exposure to well water, pesticides and herbicides, farming, wood pulp mills and iron and steel production, indicating that the exposure to unidentified environmental toxin(s) may have a noticeable contribution in the development of PD. Medical doctors appear to be at higher risk for PD in respect to other professions. Protective factors for PD may be detrimental for other human diseases, such as smoking or excess coffee drinking. Vagotomy has been shown to decrease the risk of developing, decades later, PD (cfr [24]). The important role of environmental factors is supported by the link with encephalitis, by post-encephalitic Parkinsonism and by a low concordance in monozygotic twins. The 1983 epidemic of Parkinsonism caused by MPDP (cfr [32]) has been briefly described in Sect. 15.4, and it represents a clear-cut demonstration of how a toxin, acting on mitochondrial and oxidative factors, may cause CNS damage that mimics closely human PD. One similar finding may be considered in the reduced activity of the ubiquitin proteasome system due to epoxomicin, a product of actinomycetes found in soil and in some root vegetables [70]. We have discussed these aspects in Sect. 15.4, and there is an ongoing debate on the relationship between PD and rural activities as well as with activities employing organic solvents. Heterogeneity in risk estimates makes final statements unsafe. To illustrate just one example, the route of exposure to air pollutants (e.g. inhaled or transcutaneous) has never been investigated.

Risk appears to increase as the duration of exposure increases, but this is also true for age. Synergy between different pollutants and toxins could also be relevant, making the situation even more complicated. Well worth taking in high consideration is the concern of Pezzoli and Cereda [27] who state "Although some compounds have been withdrawn from the market in industrialized countries, they are still in use in developing parts of the world". However, PD incidence seems to decrease with therapy acting on different targets, such as statins (cfr [20]), as said above. Therefore, as multifactorial as the risk factors may be, multifactorial may be the strategies to counteract the pathology/ies. New hypotheses on PD causes are emerging in these last years (cfr [24]).

Amyotrophic lateral sclerosis (ALS) (motoneuron disease; Lou Gehrig disease; ALS/MND) is the most common fatal neurodegenerative disease affecting motor neurons, both at cortical level as well as in the cranial nerve motor neurons and in the spinal cord. Incidence (2/100.000) and prevalence (6–8/100.000/year) are fairly uniform throughout the world, with the exception of a few high-incidence foci, such as the @Kii Peninsula of Japan and the island of Guam. The disease predominantly affects middle-aged and elderly individuals, though younger individuals can also be affected, and frequently individuals with a previous intense and/or professional sports activity, such as Lou Gehrig, who gave the disease one of its eponyms. The clinical presentation is dramatic, with progressive paralysis and death within a few years of onset. Cognitive defects can also accompany motor deficits. At histology, ubiquitin-rich cytoplasmic inclusions are demonstrated in the remaining motor neurons, signaling enhanced repair activity.

All the mechanisms described for neurodegenerative dementias and PD have been claimed in ALS, with the (usual) interplay of genetic factors, oxidative stress, excitotoxicity, mitochondrial damages as well as injury from neighbouring non-neuronal glial cells. Recent data have reported an increased risk of developing motor neuron disease in military personnel, in airline pilots and in Italian professional football players [71, 72]. Current data suggest that about 40% of ALS cases are environmentally determined. The data on war veterans also suggest that, if the final trigger remains unknown, risk factors may well include multiple vaccinations, physical activity and traumatic injuries, besides environmental exposure. A very recent study on occupational history has confirmed the consistent association between occupational pesticide exposure and ALS. In contrast, lead exposure showed a protective effect against ALS [73].

Multiple sclerosis is generally considered an autoimmune disorder, but many other factors have been claimed to be present since the early description by Charcot in 1868. Geographical distribution of MS shows that the disease has a higher prevalence in medium-temperate countries, but it is difficult to attribute these differences to the presence of some air component or pollutant. Since MS is linked to neuroinflammation, the neuroinflammatory properties of some air pollutant have been claimed to be part of the mechanisms leading to MS (cfr Sect. 15.5; [74]). In recent years, acrolein, a beta-unsaturated aldehyde, that is both a product and a catalyst of lipid peroxidation, has been shown to induce demyelination in an animal model of MS, and that sequestering acrolein promoted functional recovery and a decrease amount of demyelination [75]. Acroleine, as an air pollutant, has many exogenous sources, since it is emitted during various manufacturing processes, during burning cigarettes, from the exhaust of combustion engines, as well as vapours from overheated cooking oil. Acrolein is also generated by endogenous sources, through the oxidation of various compounds, in particular in conditions such as trauma, stress and ageing, at concentrations 40 times greater than normal, and it is 100 times more reactive in respect to similar compounds generated in the oxidative stress. Furthermore, acrolein remains active in the body much longer than commonly studied oxidative species; most oxygen radicals decay within fractions of a second, while acrolein persists for 7–10 days. It has been suggested [75] that acrolein could sustain or even further aggravate the inflammatory environment presumably liable for functional deterioration in MS, as well as in other pathological processes of the nervous tissue where oxidative stress may play a role.

Stroke and cerebrovascular diseases (CVD) represent the second cause of death worldwide, and they are the first cause of invalidity. Stroke is strongly correlated with ageing. In the western world, below aged 45, the incidence is 13/100.000, and prevalence is 65/100.000/year, to climb to 2224 and 8796, respectively, for those over 75 years old.

In the third-world and generally in the developing and underdeveloped countries, these data are completely different, due to the fact that the mean lifespan often does not surpass 50 years of age. Mortality is correlated with ageing, as the severity at onset of symptoms. In parallel, the therapeutic response with early thrombolysis is better in lightly affected patients treated earlier, definitely worse in severe patients treated towards the end of the therapeutic window (around 5–6 h from onset). Males are more affected (1.25×) in respect to women. 70–80% of strokes are ischaemic, and 20% are reoccurences (often are preceded by a TIA or a very minor stroke). 10–12% are haemorrhages, and 2–5% are subarachnoid haemorrhages (these are due in general to the rupture of an aneurysm). Besides age and gender, that are unmodifiable risk factors so far, the other main but modifiable risk factors are obesity, lifestyle (such as social isolation and loneliness) [76]), hypertension, nonvalvular atrial fibrillation and diabetes. Also stress and socioeconomic factors have, in our opinion, a major influence, often influencing lifestyle. Stroke and CVD in general produce depressive symptoms unrelated to the severity of the neurological deficit [77], and both stroke and CVD are correlated with air pollution. All these aspects, in relation to air pollution, are dealt in different Chapters of this Handbook (cfr Ref. [1]).

Stroke is strictly correlated with vascular dementia, representing the clinical result of multiple strokes or of strokes affecting critical regions. The clinical progression from the first stroke to the stage of dementia is classically stepwise, each step signalling a new stroke. Vascular dementia (VD) represents 30% of all presentations of dementia, but it should be remembered that many cases of AD (or other neurodegenerative dementias) show, at a careful neuroimaging, the presence of old vascular (mainly ischaemic) lesions. Accordingly, the first case described by Alois Alzheimer [77], and later described in greater detail by Gaetano Perusini [78], shows, at the histological examination, the presence of several vascular lesions as well. That is, cases with a typical VD progression will show at the pathological examination the presence of AD beta-amyloid depositions and vice-versa [79]. It follows that all the above considerations regarding the relationships between air pollution and air pollutants with both dementia and stroke are equally valid for VD and for the many forms of mixed dementia; notwithstanding how they are finally labeled.

Depression (mood disorder) affects, in Europe, 33.3 million people; in comparison to dementia, 6.3 million and in stroke 8.2 million are affected. It may be considered to be linked to anxiety disorders that affect 69.1 million people in Europe in 2010 [80]. Depression is—in our opinion—a symptom and not a disease. It has to be evaluated within the history of the patient, both familiar, past medical and recent medical. It is important to consider if it is the patient himself who is willing to tell the physician about his depression or if the situation has to be evaluated through a relative of the "silent depressed patient". DSM-V (cfr 02-05) correctly separates the depressive phase of a bipolar disorder (where the patient is very silent and a bipolar history quite evident from the relative's comments) from the depressive disorders due to disruptive mood disorder, dysthymia, premenstrual, dysphoric disorder, substance-/medication-induced depressive disorder and many others (where the patient is very keen to pour onto the ear of the poor physician the effects of a continuous *negative rumination*, with an avalanche of symptoms, complaints, criticisms, bereavements, sadness, incomprehensiveness (obviously quite all due to bad behaviour of other persons!!), which are the core and the motivation for their depressive mood). The common features of both these clinical presentations (which are very different in their pathogenesis, prognosis, treatment) are the presence of sad, empty or irritable moods, by somatic symptoms and by cognitive impairment. However, in the patient who talks, talks, talks, the cognitive impairment is limited to a maladjustment with the situations of his/her life, while in the silent, bipolar patient, it may be severe enough to appear to be dementia (the so-called depressive pseudodementia). Children may be affected by these diseases, showing persistent irritability and frequent episodes of their behaviour being extremely out of control. However, while they will seldom develop into typical bipolar disorders, they may very frequently develop into unease, then anxiety and maladjustment with possible unipolar depression later on.

Later in Sect. 15.10, the possible environmental causes of these clinical presentations, associated with discomfort, bereavements, annoyance, anxiousness, negative ruminations, panic attacks and so on, and their relationships to air pollution will be discussed. Bipolar disorders, with their important genetic background, appear to be less prone to be influenced by air pollutants.

However, the neurotransmitter system that appears to be the underlying mechanism for depression is the aminergic system. A serotoninergic involvement is also found, particularly in ruminative patients, while dopamine appears to be more specifically involved in schizophrenia. There is evident that a monoaminergic depletion in the CNS could be correlated with depression, and it has been supported by many reports, including studies in cell cultures ([79]; cfr Sect. 15.5). It follows that air pollution due to diesel engine exhausts could bring on depression since these pollutants induce aminergic neurotoxicity, possibly through oxidative stress mechanisms. In Canada, researchers have reported short-term effects of air pollution from combustion of fossil fuels in motor vehicles, coinciding with sharp increases in emergency department consultations because of depression and suicide attempts [81, 82]. Ozone too has been shown to be disruptive for the monoaminergic systems.

In the elderly, depression is one of the common mental health problems, and it is correlated with increased mortality and suicide. Known risk factors are alcohol and substance abuse, sleep disturbances, bereavement, concomitant severe medical conditions and gender (female). AA from Korea reported that increasing concentrations of PM_{10}, NO_2 and O_3 were significantly associated with depressive symptoms in the elderly population [83]. Mechanistic studies linking air pollution to depression are lacking, and in general, studies on depression in relation to air pollutants and air pollution are not numerous, and it could be *politically incorrect* to endeavour to pursue the causes for this low interest.

15.8 When the Story Began? La "Ville Lumière": A Hypothesis from the Impressionists

Epidemiological data presented above ([1, 52]; cfr Table 15.2) have underlined the burden (human, economical, etc.) of the "diseases of the brain", with depression, anxiousness and dementia in the pole positions. These figures are prone to increase in the next decades and may be considered as the epidemics, the black death of the third millennium. Humanity—especially in western countries—has witnessed a decline in deaths caused by infectious diseases (such as pneumonia, typhoid fever, tbc, polio, etc.), and the winners of these battles have been awarded with the most prestigious prizes. Society is actually supporting both scientists fighting against cancer and cardio-cerebrovascular diseases, as well as the research for genetically determined diseases. However, it seems that very little is done to understand (We will not use the verb "to fight") one of the simplest questions about these third millennium epidemics: why now? Why are we seeing these epidemics just now after the end of the World War II? One of the reasons for this lack of insight could be the answer, more or less satisfactory, that the human lifespan has increased and is viewed as a major achievement of the modern society. Humans have succeeded in living longer. That is all very nice, yet one must accept that half of the oldest elderly population will be demented. That is not nice at all. Why is humanity suffering from depression and why are we seeing an increasing number of young people, from the teen years onwards, showing anxiousness with panic attacks, negative ruminations and phobias? Why are these questions left to philosophers and to sociologists and why are they not examined with a biomedical neuroscientific approach?

A simple examination of the major changes in lifestyle worldwide would have shown that, parallel to a prolonged lifespan, many other (perhaps negative) changes have occurred since 1950.

Placing the beginning of this story with an anecdote from the time of the impressionists, Paris in 1900 (as well as London) was suffering from devastating air pollution. This phenomena is well described in Puccini's La Bohème song "Nei cieli bigi guardo fumar dai mille comignoli Parigi". (In grey skies I watch the smoke of a thousands pariginian chimneys) [84]. The (then unknown) impressionist painter Van Gogh was advised by his physician to move to Provence to get rid of a very annoying eye inflammation, due to this terrible air pollution. When he arrived in Arles, he discovered a new world full of clear light and of bright colors, and he wrote to his close friend Gauguin to leave everything in Paris behind and to join him where the light was light and the color was colorful. Paris in 1900 was called *la ville lumière*, but this *ville lumière* had killed the natural light and the colors: the *lumière* had poisoned the air and the light.

The killing of the natural light quickly changed the habits of the (civilized) men, no longer subject to the light of the sun for their daily activities and productivity. For centuries, all human activities stopped during winter, even wars, due to the difficulties and/or impossibilities for moving. In a way, it was as an obliged winter rest. Nowadays, after the discovery of artificial light and electricity, the hours for working (or enjoying life) are theoretically infinite. All these activities could be spent completely away from solar light and from solar photons, and humanity does not worry about the loss of the principal beneficial properties of air and solar light. Before analysing what this air pollution was causing if only just through the filtering of the solar photons, let us analyse the causes of the air pollution: that is the air pollutants themselves and how some of them have increased significantly during the last 70 years.

Since 1950, the following sources of air pollutants have shown an exponential increase: car engines, diesel engines, factories engines, highways, airports and aircrafts, found everywhere. In towns, warming engines, cooling engines, engines, engines, engines.

In rural areas, thousands of products have been created to increase agricultural production, regardless—or sometime just because—of their killing capabilities on other biological species, including men. Could all these put together be called a pollution/pollutant epidemic?

As said above, the epidemiological data reported above are showing that the diseases affecting millions of people and with a dramatic increase since 1950 are depression, anxiousness with mood disorder (and panic attacks) and dementia. Out of these three epidemics, only dementia appears linked to brain ageing. Depression and more so anxiousness with mood disorder and panic attacks are increasingly detected in young people. So the interpretation that these increases in incidence and prevalence are just due to the ageing of the brain appears unsatisfactory. Certainly, elderly people are more prone to depression, but we have reported data showing how these depressive aspects are increasingly reported in young people, and how they appear to be correlated with environmental factors, including air pollutants, as well as with socioeconomics factors and gender. Experimental observations on the effects of these pollutants on neurons, glia, synaptic activities, metabolic activities and other aspects of the brain's work have been published by many AA, but they are still awaiting a systematic collocation (statistical analysis?). What is still absolutely unknown so far is the future consequences of the damages that appear so much more readily produced in children by the air pollutants and by the air pollution in general, including noise. With the two curves, pollution and disease, both growing exponentially, could it not be a possible abstract working hypothesis to study the possible links between the growth of pollution and the growth of these important diseases of the brain (depression and anxiousness syndromes with mood disorders and panic attacks)?

15.9 About the Qualities of the "Air": Air Pollution Versus Pollution of the Air

Scientific literature on the effects of air pollutants and air pollution on cognitive and functional aspects of the brain, including mood changes, anxiousness, panic attacks, bereavement, discomfort, annoyance, negative ruminations, intrusive thoughts and so on, have been described above. The main pathway of penetration into the nervous tissue has been identified as inhalation and therefore the lungs. However, other biological pathways are also the nasal mucosa and the skin. Previous sections have reviewed some of the available data.

Now, one different aspect of air pollution will be discussed one that it is not strictly connected with the quality and specifications of the different inhaled pollutants that constitute the bulk of air pollution. The aspect of air pollution presented in this section is not the cause of the loss in luminosity, in transparency and in vivacity of the colours seen in Paris in 1900 but the loss itself of these normal characteristics of solar light (which Van Gogh and Gauguin rediscovered in Provence) which occurs when *air is polluted*. The presence of pollutants in the air as shown in Fig. 15.2 causes, besides many other medical problems, a physical change, that is, the loss of absorption of billions of solar photons, which are just stopped short in the polluted atmosphere. Humanity is deprived of billions of photons, and, as far as the human body is concerned, the eyes and the skin are the two human organs most affected. It could be correct to include in this concept "the pollution of air" also "non-particulate" pollutants such as noise or electromagnetic waves: this is outside however the borders of this chapter. It will be enough to underline that both noise and electromagnetic waves are particulate [85].

It was commonly assumed that human (and mammalian) retina contained only two classes of photoreceptors, rods and cones, with two different pigments, respectively, the rhodopsins and the photopsins (which are really three different kinds). In the last 20 years, many studies have pointed out that a third class of mammalian photoreceptors exists, working with a new light-sensitive molecule, called melanopsin, assigned not to the classical vision of objects but to signal the ambient light levels (*irradiance*) to the brain [86]. Surprisingly, these photoreceptors are the retinal ganglion cells (RGCs): a part of them shows the unique ability to communicate directly, and not, as for all the other receptors of mammalian brain, with the higher visual cortices. A smaller part (<5%; the intrinsically photosensitive ipRGSs) of the ganglion cells, characterized by a very large dendritic tree diameter, is signaling to SCN, and in particular to the suprachiasmatic nucleus (the site of the mammalian central circadian clock) and to the olivary pretectal nucleus, other aspects of the irradiance, the level of pupillary contraction, the

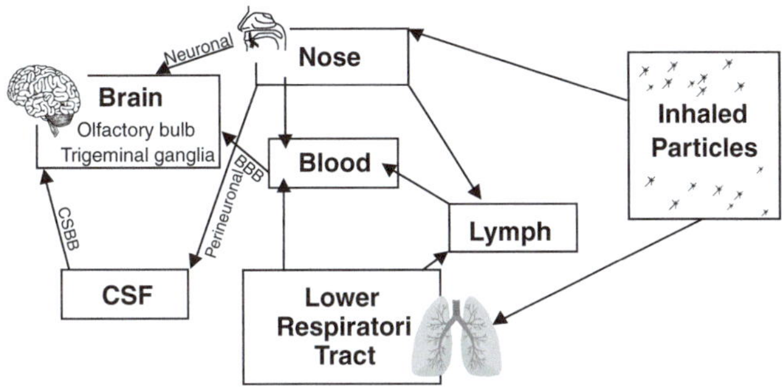

Fig. 15.5 *Regulation of image and nonimage forming visual functions of the eye*

neuroendocrine regulation and the circadian physiological rhythms to the light/dark cycle, all behaviours that are remarkably tonic and require prolonged integration of ambient light levels [87]. Production of melanopsin is under control of a specific gene (opn4), and in transgenic mice lacking this gene, it has been shown that the normal circadian behaviours are impaired, with loss of the photoentrainment (Fig. 15.5, from Ref. [88]).

The visual functions of the human eye are regulated by the classical photosensory pigments (rhodopsin and photopsins) present in rods and cones receptors that activate the classical (non-melanopsin) RGCs. The melanopsin positive ipRGCs (<5%) respond directly to ambient light and mediate a variety of nonimage forming visual functions, namely, circadian photoentrainment of the SCN, pupillary light response and regulation of sleep and mood. Classical RGCs may have a role in these nonimage forming visual functions, as well as the ipRGCs may contribute to image forming. *RGC* retinal ganglion cells, *ipRGCs* intrinsically photosensitive retinal ganglion cells, *SCN* suprachiasmatic nucleus

It should be remembered that the circadian rhythms of our physiology and behaviour—biological aspects fundamental for our well-being—are regulated by the hypothalamic SCN, the master circadian oscillator, which requires regular synchronization with the daily and seasonal fluctuations of the environmental photoperiod, and its different levels of irradiance. Current studies are addressing the role of melanopsin in signaling to the brainstem's neurosecretory nuclei the level of irradiance. Its impairment may possibly be connected with the pathogenesis of circadian and sleep abnormalities observed in neurodegenerative disorders, also in their early phase [89, 90]. Furthermore, melanopsin levels, determined by irradiance, may be the biological basis for the recurrent depressions in fall and winter that are so common in patients presenting with seasonal affective disorder (SAD), now classified (cfr Ref. 02-05) as "*seasonal pattern for recurrent major depressive disorder that occurs at a specific time of the year and fully remits otherwise*". Recurrent depressions are common also in patients who do not present with a major depressive disorder but who present with the phobic-obsessive syndromes with panic attack or with generalized anxiousness; observed more and more often in our outpatient clinics and also seen with disruptive mood dysregulation disorders, as the hormones regulating mood and sleep become unbalanced due to seasonal changes, giving origin to these "winter blues". Individuals with low levels of melanopsin have been found to frequently suffer from mood and sleep disorder [91, 92], and melanopsin gene variants (a single missense variant, P10L of Opn4 gene) are also associated with increased risk of SAD in humans [93]. It is not possible today to exclude that other factors (or other genes) may contribute to this imbalance; together with the melanopsin mechanisms, other signals of a decreased arrival of solar photons may reach the mood regulating centers in the SCN. The seasonality of depressive episodes in a large cohort of these patients and the return to normal non-depressive behaviours in response to therapies showing a resetting of the body's internal clock (thus restoring the sleep-wake cycle and mood issues in SAD patients) are clearly pointing out the positive effects of light on the human brain's clock and photoentrainments. Many psychiatric groups claim a positive effect with light therapy, in accordance with the data on melanopsin. Finally, it could be supposed that, as a Darwinian inheritance, also human skin may not be so indifferent to changes in solar light, and not only as summer's sun-burns. It would not be surprising if future research might demonstrate melanopsin-like properties in some skin receptor, as well as other photo-pigments acting on the SCN and circadian activities.

Serotonin (5-HT) and the 5-HT agonists can modify the response of the SCN in mammals [94], altering the photic effects on the SCN circadian oscillator, which is due to a transcription-/translation-based molecular clock that autonomously regulates activity patterns in near 24-h rhythms. It appears that there is an endogenous inhibitory serotoninergic input to the biological clock that generates circadian rhythms in mammals, as well as behavioural changes related to the seasonal irradiance or photoperiod, arriving to the SCN from the serotoninergic nuclei of the brainstem, namely, the dorsal raphe nucleus of the mesencephalon [95]. The circuit therefore is formed by the ipRGCs (activated by light) that project directly to the SCN and to the DRN [96], with DRN neurons firing back on the SCN, through a serotoninergic pathway. In addition, SCN has to be reset periodically so that circadian rhythms are synchronized to the light/dark cycle. Light is by far the most potent synchronizing agent, and this action is mediated through the ipRGCs.

The relevance of this tuning of the human body in general and of the human brain connectome in particular to daily and seasonal irradiance is signaled by reports indicating the effects of ipRGCs on modulation of cognitive performance [97] and enhancement of alertness [98] and depression, with a decrease of depressive behaviour when the activity of ipRGCs is enhanced [99].

15.10 Effects of the "Pollution of the Air" on the CNS: The New Pandemics: Number of People Affected and Costs

Section 15.6 (cfr Ref. [52]) and Sect. 15.7 (cfr: Ref. [79]) dealt with numbers regarding the diseases of the brain, quantified in term of lost DALYs and affected individuals. The 2011 CDBE study group report has shown that sleep disturbances and annoyance are ten times more costly (as lost DALYs), then cardiovascular diseases or cognitive impairment. The 2010 EEC CDBE report has shown [80] that depression affects, in Europe, 33.3 million people, to be compared with dementia (6.3 millions) and stroke (1.3 millions). However, anxiety affects 61.3 million people in Europe today. Physicians may be more interested in carotid occlusions due to air pollution [5], but economists and politicians, key persons in the allocation of funds both for research and medical care, may be more interested in the larger numbers.

The same CDBE report (Table 15.3, from Ref. [81]) compares data obtained—with the same methodology—in 2010 and in 2004, showing striking increases for some disease, less or no significant changes for other brain diseases. Anxiety disorders rise 60% in costs (+24 billions €); mood disorders rise 7% (+7 billions €); dementia rises 90% (+ 50 billions €); stroke rises 22% (+ 5 billions); Parkinson's disease rises 30% (+3 billions €). On the contrary, brain tumours, epilepsy and migraine show a decrease in the number of subjects affected and of related costs.

Table 15.3 Comparison of 2010 and 2004 estimates, excluding diagnoses and indirect costs that were not included in the 133C2005 study

	Estimates in 2010			Estimates in 2004		
	Number of subjects[a] (million)	Costs per subject[b] (€PPP, 2010)	Total costs (million €PPP, 2010)	Number of subjects[a] (million)	Costs per subject (€PPP, 2010)	Total costs (million €PPP, 2004)
Addiction	15.5	4227	65,684	9.2	6229	57,275
Anxiety disorders[c]	61.3	1076	65,995	41.4	999	41,372
Brain tumour	0.24	21,590	5174	0.14	33,907	4586
Dementia	6.3	16,584	105,163	4.9	11,292	55,176
E3110754	2.6	5221	13,800	2.7	5778	15,546
Migraine	49.9	370	18,463	40.8	662	27,002
Mood disorders[d]	33.3	3406	113,405	20.9	5066	105,666
Multiple sclerosis	0.54	26,974	14,559	0.38	23,101	8769
Parkinson's disease	1.2	11,153	13,933	1.2	9251	10,722
Psychotic disorders[e]	5.0	5805	29,007	3.7	9554	35,229
Stroke[f]	1.3	21,000	26,641	1.1	19,394	21,895
Traumatic brain injury[f,e]	1.2	4209	5085	0.71	4143	2937
Total	178.5	2672	476,911	127.0	3040	386,175

[a]Including also persons with zero costs
[b]Weighted mean from all countries and diagnoses
[c]Excluding PTSD
[d]Referred to as "affective disorders" in 2005
[e]Excluding indirect costs
[f]Includes only incident cases in 2010
From Ref. [81], Table 4

If a threshold is allowed to be drawn, calling "normal" an increase below 50% (for these 6 years), very possibly due to better scrutiny, better diagnostics, better knowledge of the problem and so on (thus explaining the +66% increase in costs for MS, with only 0.5 million people affected), then the increases in anxiety disorders and dementias appear to achieve a different dimension in respect to all the other diseases of the brain: dimensions due to the number of affected people (61.3 and 6.3 million people, respectively) and to the total costs (65,995 million € and 105,163 million €, respectively) for care.

Only mood disorders achieve a comparable cost (33.3 million people affected with 113,405 million € being the costs), but the border with anxiety disorders is quite foggy (remaining within the field of air pollution!!).

These brain diseases seem today to achieve the dimension of a new III millennium pandemic or new epidemics; furthermore, in our clinical experience, there are no indications that they are slowing down. Outpatient clinics are increasingly populated by these patients, complaining, ruminating, hyperventilating and panicking.

Some of the increases from 2004 to 2010 reported in Table 15.3 may be—at least partially—explained by the increase of lifespan, as, for example, the 22% increase in stroke costs with a +18% in people affected. However, when the increase in number of people affected by dementia (+90%) is compared with the increase in people affected by cerebrovascular disease and stroke (+22%), taking into consideration that for one patient affected by CVD and stroke there are five to six patients affected by dementias, one must realize that the simple increase in lifespan cannot be the considered main factor. Accordingly, it should be remembered that many reports have indicated that an increased lifespan also represents a prolonged exposition to air pollutants and air pollution, with all the harms and hazards that this prolonged exposure may produce on the nervous system. Experimental and clinical data, in adults, in elderly and in children, have demonstrated that these pollutants and this pollution damage neurons, glia and neuronal functions.

It was underlined in the previous section how in the last 70 years many normal habits in human lives have changed, with respect to what they were up until World War II. In particular, humanity has moved indoors from outdoors, freeing itself from slavery to solar light, of weather conditions and of seasonal conditions. These changes combined with billions of engines working and billions of tons of fuel burning have provided men these "freedoms". Human beings are growing accustomed to a decrease of light and a decrease in the purity of solar light, accustomed to grey skies and accustomed to increased background noise. Lifespan is increased, but often elderly individuals "survive" in complete loneliness and social isolation, risk factors for stroke, for coronary heart disease and for depression (cfr [75]). Data on worldwide fuel consumption and data on worldwide air pollution are showing that their yearly increases continue, aside from all the concerns for the environment. The two curves, increase in fuel consumption and increase in people affected by anxiousness or by dementia, have a noteworthy similarity. The trend of these curves to be parallel could have meant nothing, if a consistent bulk of data, only a small amount of which is quoted here, had not shown how truly correlated they are. Pollutants, pollution and noise all cause anxiousness; sleep disturbances; decreased capacities in learning, memory and intelligence; a decrease in the brain reserve and in the cognition reserve that are correlated with increased risk of cognitive impairment (MCI) and dementia. All of them are increasing exactly in the same time period. It must be underlined, once again, how pollutants and pollution change the natural aspects of solar light that are fundamental for our internal clocks related to circadian rhythm, to seasonal rhythm and to human endeavours and efforts.

The working hypothesis proposed here at the end of our chapter is that the brain's malfunctions and diseases of its connectome are the combined product of many risk factors. Some are unmodifiable (such as genetics and age); some are modifiable if better recognized (such as air pollutants and noise); others are still largely unrecognized (such as pathogenetic for brain diseases, such as depression, mood disorders, anxiousness with panic attacks, MCI and dementia). It has been shown that one of these brain diseases, that affects nearly 100 million people in Europe in the present decade, is represented by the disorders of mood, including anxiety, negative ruminations, panic attacks, discomfort, annoyance, complaints, bereavements and SAD, and is also clearly different from bipolar disorders and by unipolar depression. Data has been presented showing how the solar light influences, through hyper-specialized receptors in the human retina, human beings' adjustments of circadian rhythms, to weather conditions and to seasonal conditions. Serotonin appears to be the neurotransmitter that from the raphe nuclei and other serotoninergic brainstem nuclei modulates the sensitivity of the CNS clock.

Experimentally induced serotonin depletion causes a profound alteration of the daily activities of mammals that mimics human depression. Therapies with inhibitors of the serotonin uptake, that prolong the activity of the normally produced serotonin in the human brain, appear to be successful in helping patients to move out of their brooding, their negative ruminations, their anxiousness, and also they appear to decrease both the frequency and the intensity of the panic attacks. Accordingly, many groups claim that a therapy with blue light achieves similar results (cfr [98, 99]).

In our working hypothesis, the continuous prolonged decrease of solar photons activation of the ipRGCs, due to the pollution of air, may result in a decreased secondary stimulation of the SCN, with hazards for the endogenous clock and circadian rhythms, therefore requesting an increased serotoninergic activation from the mesencephalic raphe nuclei. For a few years, this compensatory mechanism may be able to cope with the situation, but at a certain moment, and possibly this "when" is genetically and/or socially determined, it cannot maintain this increased production of serotonin. Sleep disturbances and mood depression begin to reappear. Physical manifestations that may possibly have an underlying psychological cause proceed in parallel with psychological bereavement, with an inexplicable unhappiness.

Epidemiological data show that the moment when this imbalance occurs is consistently earlier in one's lifespan: accordingly, one may not forget all the reports (cfr [6, 53–55]) indicating how children are particularly sensitive to damages on their cognition, memory and intelligence, due to air pollutants.

If not treated, these clinical presentations—with anxiousness and with negative ruminations and panic attacks, often with SAD (the winter blues) - will increase their burden with depressive feelings and, in elderly people, with decreased cognitive abilities.

Chemtrails may be a useful representation of the *pollution of the air* that the IIIth millennium is now increasingly producing and experiencing: when chemtrails will become thousands upon thousands, totally covering the sky, there will no longer be room for the sun's light to reach humanity. Prometheus is dead.

References

1. Mills NL, et al. Adverse cardiovascular effects of air pollution. Nat Clin Pract Cardiovasc Med. 2009;6(1):36–44.
2. Chen H, Goldberg MS, et al. A systematic review of the relation between long-term exposure to ambient air pollution and chronic diseases. Rev Environ Health. 2008;23(4):243–97.
3. Sermin G, et al. The adverse effects of air pollution on the nervous system. J Toxicol. 2012;2012:782462. pp. 23
4. Nanadur SS, Hollingsworth JW. Preface. In: Nanadur SS, Hollingsworth JW, editors. Air pollution and health effects. Heidelberg: Humana Press; 2015. p. V.
5. Kaufman JD, et al. Association between air pollution and coronary artery calcification within six metropolitan areas in the USA (Multi-Ethnic Study of Atherosclerosis and Air Pollution): a longitudinal cohort study. Lancet. 2016;388:696–704.
6. Rauh VA, et al. Developmental effects of exposure to environmental tobacco smoke and material hardship among inner-city children. Neurotoxicol Teratol. 2004;26:373–85.
7. Suglia SF, et al. Association of black carbon with cognition among children in a prospective birth cohort study. Am J Epidemiol. 2008;167:280–6.
8. Zeng Y, et al. Association of environmental factors with elderly health and mortality in China. Am J Public Health. 2010;100:298–305.
9. Power MS, et al. Traffic-related air pollution and cognitve function in a cohort of older men. Environ Health Perspect. 2011;119:682–7.
10. Weuve J, et al. Exposure to particulate air pollution and cognitive decline in older women. Arch Intern Med. 2012;172:219–27.
11. Wellenius GA, et al. Residential proximity to major roadways and cognitive function in community dwelling seniors: results from the MOBILIZE Boston study. Arch Intern Med. 2012;60:2075–80.
12. Elder A, et al. Particulate air pollution and CNS health. In: Nanadur SS, Hollingsworth JW, editors. Air pollution and health effects. Heidelberg: Humana Press; 2015. p. 269–88.
13. Donaghy M, editor. Brain's diseases of the nervous system. 12th ed. Oxford: Oxford University Press; 2009.
14. Schapira AHV. Neurology and clinical neuroscience. Philadelphia: Mosby Elsevier; 2007.
15. Jones TF, et al. Mass psychogenic illness attributed to toxic exposure at a high school. N Engl J Med. 2000;342:96–100.
16. Cozac VV, et al. Older candidates for subthalamic deep brain stimulation in Parkinson's disease have a higher incidence of psychiatric serious adverse events. Front Aging Neurosci. 2016;8:132. https://doi.org/10.3389/fnagi.2016.00132.
17. Atkin G, Paulson H. Ubiquitin pathways in neurodegenerative disease. Front Mol Neurosci. 2014;7:63. https://doi.org/10.3389/fnmol.2014.00063.
18. Lenzi GL. Brain work: function, metabolism and blood flow. Arch Ital Biol. 1982;120:189–200.
19. Kleibeuker SW, et al. Prefrontal cortex involvement in creative probelm solving in middle adolescence and adulthood. Dev Cogn Neurosci. 2013;3:197–206.
20. Lin K, et al. Statin therapy prevents the onset of Parkinson disease in patients with diabetes. Ann Neurol. 2016;80:532. https://doi.org/10.1002/ana.24751.

21. Abbott DF. Probing the human brain functional connectome with simultaneous EEG and fMRI. Front Neurosci. 2016;10:302. https://doi.org/10.3389/fnins.2016.00302.
22. Elder A, et al. Translocation of inhaled ultrafine manganese oxide particles to the central nervous system. Environ Health Perspect. 2006;114:1172–8.
23. Maher B, et al. Magnetite pollution nanoparticles in the human brain. Proc Natl Acad Sci. 2016;113:10797. https://doi.org/10.1073/pnas 1605941113.
24. Svensson E, et al. Vagotomy and subsequent risk of Parkinson's disease. Ann Neurol. 2015;78:522–9.
25. Ramcharan K, et al. Encephalopathy and neuropathy due to glue, paint thinner, and gasoline sniffing in Trinidad and Tobago-MTI findings. Case Reports in Neurological Medicine. 2014;2014:850109. 4pp
26. Pringsheim T, et al. The prevalence of Parkinsos's disease: a systematic review and meta-analysis. Mov Disord. 2014;29:1583–90.
27. Pezzoli G, et al. Exposure to pesticides or solvents and risk of Parkinson disease. Neurology. 2013;80:2035–41.
28. Kukull WA, et al. Solvent exposure as a risk factor for Alzheimer's disease : a case-control study. Am J Epidemiol. 1995;141:1059–71.
29. Ljungman PL, et al. Ambient air pollution and stroke. Stroke. 2014;45:3734–41.
30. Langston JW, et al. MPTP-induced parkinsonism in human and non-human primates – clinical and experimental aspects. Acta Neurol Scand Suppl. 1984;100:49–54.
31. Bakulski KM, et al. Alzheimer's disease and environmental exposure to lead: the epidemiologic evidence and potential role of epigenetics. Curr Alzheimer Res. 2012;9:563–73.
32. Colvin VL, Kulinowski KM. Nanoparticles as catalysts for protein fibrillation. Proc Natl Acad Sci. 2007;104:8679–80.
33. Feng X, et al. Central nervous system toxicity of metallic nanoparticles. Int J Nanomedicine. 2015;10:4321–40.
34. Migliore L, et al. Nanomaterials and neurodegeneration. Environ Mol Mutagen. 2015;56:149–70.
35. Win-Shwe TT, Fujimaki H. Nanoparticles and neurotoxicity. Int J Mol Sci. 2011;12:6267–80.
36. Haider L. Inflammation, iron, energy failure, and oxidative stress in the pathogenesis of multiple sclerosis. Oxid Med Cell Longev. 2015;2015:1. https://doi.org/10.1155/2015/725370.
37. Trapp BD, Stys PK. Virtual hypoxia and chronic necrosis of demyelinated axons in multiple sclerosis. Lancet Neurol. 2009;8:280–91.
38. Brownell B, Hughes JY. The distribution of plaques in the cerebrum in multiple sclerosis. J Neurol Neurosurg Psychiatry. 1962;25:315–20.
39. Holland CM, et al. The relationship between normal cerebral perfusion patterns and white matter lesion distribution in 1,249 patients with multiplke sclerosis. J Neuroimaging. 2012;22:129–36.
40. Ciccarelli O, et al. Pathogenesis of multiple sclerosis. Insights from molecular and metabolic imaging. Lancet Neurol. 2015;13:807–22.
41. Nel A, et al. Toxic potential of materials at the nanolevel. Science. 2006;311:622–7.
42. Calderon-Garciduenas L, et al. Brain inflammation and Alzheimer's-like pathology in individuals exposed to severe air pollution. Toxicol Pathol. 2004;32:650–8.
43. Calderòn-Garciaduenas L, et al. Exposure to severe urban air pollution influences cognitive outcomes, brain volume and systemic inflammation in clinically healthy children. Brain Cogn. 2011;77:345–55.
44. Calderon-Garciduenas L, et al. Air pollution and children: neural and thigh junction antibodies and combustion metals, the role of barrier breakdown and brain immunity in neurodegeneration. J Alzheimers Dis. 2014;43:1039. https://doi.org/10.3233/JAD-141365.
45. Sunyer J. The neurological effects of air pollution in children. Eur Respir J. 2008;32:535–7.
46. Costa LG, et al. Neurotoxicants are in the air: convergence of human, animal, and in vitro studies on the effects of air pollution on the brain. j. 2014;2014:736385, 8pp.
47. Crüts B, et al. Exposure to diesel exhaust induces changes in EEG in human volunteers. Part Fibre Toxicol. 2008;5:1–6. Int J Mol Sci. 2011, 12

48. Tian L, et al. Neurotoxicity induced by zinc oxide nanoparticles: age-related differences and interaction. Sci Rep. 2015;5:16117.
49. Tzivian L, et al. Effect of long-term outdoor air pollution and noise on cognitive and psychological functions in adults. Int J Hyg Environ Health. 2015;1:1–11.
50. Ranft U, et al. Long-term exposure to traffic-related particulate matter impairs cognitive function in the elferly. Environ Res. 2009;109:1004–11.
51. Chen J, Schwartz J. Neurobehavioral effects of ambient air pollution on cognitive performance in US adults. Neurotoxicology. 2009;30:231–9.
52. World Health Organization, Regional Office for Europe. Burden of disease from Environmental noise: quantification of healthy life years lost in Europe. Copenaghen: WHO Regional Office for Europe; 2011.
53. Fields JM, et al. Standardized general-purpose noise reaction questions for community noise surveys: research and recommendation. J Sound Vib. 2001;242:641–79.
54. van Kempen EE, et al. Children's annoyance reactions to aircraft and road traffic noise. J Acoust Soc Am. 2009;125:895–904.
55. Yolton K, et al. Exposure to environmental tobacco smoke and cognitive abilities among U.S. children and adolescents. Environ Health Perspect. 2005;113:98–103.
56. Office on Smoking and Health; Centers for Disease Control and Prevention. How tobacco smoke causes disease: the biology and behavioral basis for smoking-attributable disease: a report of the surgeon general. Atlanta. 2010.
57. Talbott EO, et al. Air toxics and the risk of autism spectrum disorder: the results of a population based case-control study in southwestern Pennsylvania. Environ Health. 2015;14:80.
58. Larsson HJ, et al. Risk factors for autism: perinatal factors, parental psychiatric history, and socioeconomic status. Am J Epidemiol. 2005;161:916–25. discussion 926-8
59. Lawler CP, et al. Identifying environmental contributions to autism: provocative clues and false leads. Ment Retard Dev Disabil Res Rev. 2004;10:292–302.
60. Allen JL, et al. Early postnatal exposure to ultrafine particulate matter air pollution: persistent ventriculomegaly, neurochemical disruption, and glial activation preferentially in male mice. Environ Health Perspect. 2014;122(9):939–45.
61. Lenzi D, et al. Neural basis of attachment-caregiving systems interaction: insights from neuroimaging studies. Front Psychol. 2015;6:1241.
62. Wittchen HU, et al. The size and burden of mental disorders and other disorders of the brain in Europe 2010. Eur Neuropsychopharmacol. 2011;21:655–79.
63. World Alzheimer Report 2015, published by Alzheimer's Disease International (ADI). London. 2015.
64. Whitenhouse P, et al., editors. Concepts of Alzheimer disease: biological, clinical, and cultural perspectives. Baltimore: Johns Hopkins University Press; 2000.
65. Spillantini MG, Goedert M. Tau pathology and neurodegeneration. Lancet Neurol. 2013;12:609–22.
66. Moussaud S, et al. Alpha-synuclein and tau : teanmates in neurodegeneration ? Mol Neurodegener. 2014;9:43–51.
67. Forman MS, et al. Frontotemporal dementia: clinicopathological correlations. Ann Neurol. 2006;59:952–62.
68. Chen H, et al. Living near major roads and the incidence of dementia, Parkinson's disease, and multiple sclerosis : a population-based cohort study. Lancet. 2017;389:718–26.
69. Manivannan Y, et al. Role of environmental contaminants in the etiology of Alzheimer's disease: a review. Curr Alzheimer Res. 2015;12:116–46.
70. McNaught KS, et al. Systemic exposure to proteasome inhibitors causes a progressive model of Parkinson's disease. Ann Neurol. 2004;56:149–62.
71. Weisskopf MGP, et al. Prospective study of military service and mortality from ALS. Neurology. 2005;64:32–7.
72. Chio A, et al. Severely increased risk of amyotrophic lateral sclerosis among Italian professional foot-ball players. Brain. 2005;128:172–6.

73. Su FC, et al. Association of environmental toxins with amyotrophic lateral sclerosis. JAMA Neurol. 2016;73:803–11.
74. Tully M, Shi R. New insights in the pathogenesis of multiple sclerosis – role of acrolein in neuronal and myelin damage. Int J Mol Sci. 2013;14:20037–47.
75. Leung G, et al. Anti-acrolein treatment improves behavioural outcome and alleviates myelin damage in experimental autoimmune encephalomyelitis mouse. Neuroscience. 2011;173:150–5.
76. Valtorta NK, et al. Loneliness and social isolation as risk factors for coronary heart disease and stroke: systematic review and meta-analysis of longitudinal observational studies. Heart. 2016;102:1009. https://doi.org/10.1136/2015-308790.
77. Altieri M, et al. Depression after minor stroke: prevalence and predictors. Eur J Neurol. 2012;19:517–21.
78. Alzheimer A. Über eigenartige Krankheitsfälle des späteren Alters. Z Gesamte Neurol Psychiatr. 1911;4:356–85.
79. Block ML, et al. Nanometer size diesel exhaust particles are selectively toxic to dopaminergic neurons: the role of microglia, phagocytosis, and NADPH oxidase. FASEB J. 2004;18:1618–20.
80. Gustavsson A, et al. Cost of disorders of the brain in Europe 2010. Eur Neuropsychopharmacol. 2011;21:718–79.
81. Szyszkowicz M, et al. Air pollution and daily emergency department visits for depression. Int J Occup Med Environ Health. 2009;22:355–62.
82. Szyszkowicz M, et al. Air pollution and emergency department visits for suicide attempts in Vancouver, Canada. Environ Health Insights. 2010;4:79–86.
83. Lim Y-H, et al. Air pollution and symptoms of depression in elderly adults. Environ Health Perspect. 2012;120:1023–8.
84. Illica L & Giacosa G. La Bohème. Libretto from "La vie de Bohème". Murger H. 1896.
85. Gribbin J. In search of Schrodinger's cat—quantum physics and reality. New York: Bantam Books; 1984.
86. Provencio I, et al. Melanopsin: an opsin in melanophores, brain and eye. Proc Nat Acad Sci U S A. 1998;95:340–5.
87. Hannibal J, et al. Central projections of intrinsically photosensitive retinal ganglion cells in the macaque monkey. J Comp Neurol. 2014;522:2231–48.
88. Sengupta A, et al. Localization of melatonin receptor 1 in mouse retina and ita role in the circadian regulation of the electroretinogram and dopamine levels. PLoS One. 2011;6(9):e24483.
89. La Morgia C, et al. Melanopsin-expressing retinal ganglion cells: implications for human diseases. Vis Res. 2011;51:296–302.
90. LeGates TA, et al. Aberrant light directly impairs mood and learning through melanopsin-expressing neurons. Nature. 2012;491:594–8.
91. Panda S, et al. Melanopsin (Opn4) requirement for normal light-induced circadian phase shifting. Science. 2002;298:2213–6.
92. Roecklein K, et al. Melanopsin, photosensitive ganglion cells, and seasonal affective disorder. Neurosci Biobehav Rev. 2013;37:229–39.
93. Roecklein KA, et al. A missense variant (P10L) of the melanopsin (OPN4) gene in seasonal affective disorder. J Affect Disord. 2009;114:279–85.
94. Smith BN, et al. Serotoninergic modulation of retinal input to the mouse suprachiasmatic nucleus mediated by 5-HT1B and 5-HT7 receptors. J Biol Rhythm. 2001;16:25–38.
95. Gannon RL. Serotoninergic serotonin (1A) mixed agonists/antagonista elicit large-magnitude phase shifts in hamster circadian wheel-running rhythms. Neuroscience. 2003;119:567–76.
96. Luan L, et al. Y-like retinal ganglion cells innervate the dorsal raphe nucleus in the mongolian gerbil (Meriones unguiculatus). PLoS One. 2011;6:e18938. https://doi.org/10.1371/journal.pone.0018938.
97. Vanderwalle C, et al. Blue light stimulates cognitive brain activity in visually blind individuals. J Cogn Neurosci. 2013;25:2072–85.

98. Viola AU, et al. Blue-enriched white light in the workplace improves self-reported alertness, performance and sleep quality. Scand J Work Environ Health. 2008;34:297–306.
99. Meesters Y, et al. Low-intensity blue-enriched white light (750 lux) and standard bright light (10,000 lux) are equally effective in treating SAD. A randomized controlled study. BMC Psychiatry. 2011;11:17.
100. Pardridge WM. CSF, blood-brain barrier, and brain drug delivery. Expert Opin Drug Deliv. 2016;13:963–75.
101. Battistini N, et al. Osservazione di un gruppo di polineuropatie da collanti. 1. Contributo clinico ed elettromiografico e contributo sperimentale. Riv Pat Nerv Ment. 1974;95:1–14.
102. Van Bogaert L. Les maladies systématisées. In: Lamierre A, et al., editors. Traité de medicine, vol. XVI. Paris: Masson; 1949.
103. Perusini G. Über klinisch und histologisch eigenartige psychische Erkrankungen des späteren Lebensalter. Histologische und histopathologische Arbeiten über die Großhirnrinde mit besonderer Berücksichtigung der pathologischen Anatomie des Geisteskranken. Plates XVII–XXIII. Jena: G. Fischer. 1909–1910; 297–358.
104. Snowdon DA. Healthy aging and dementia: findings from the nun study. Ann Intern Med. 2003;139:450–4.

Air Pollution and Cardiovascular Diseases (Risk Factors and the Myocardial Cell Defence)

16

Antonio Vittorino Gaddi

16.1 Cardiovascular and Cardiopulmonary Diseases Caused by Air Pollution: Integrated Prevention

Several researches provide compelling evidences on the cardiovascular effects of fine particulate matter (PM) for the development of atherosclerosis and cardiovascular disease and in triggering acute cardiac events [1]. PM_{10}, $PM_{2.5}$ and $PM_{0.3}$ are strongly associated with increased incidence of myocardial infarction (MI), stroke, arrhythmias and vascular dysfunction; fine and ultrafine particles showed direct toxicity on both endothelial and myocardial cells, whilst all PMs (including gross dust) showed indirect toxic effects that are able to promote systemic inflammation [2].

The more recent systematic review on the effects of indoor pollution (IP) on the cardiovascular system [3] outlines the effects of IP on coronary heart disease (CHD) and describes possible effects on heart rate (HR) and HR variability, on hypertension and lipid profile and on vascular pathology. This paper, published on *British Medical Bulletin*—like many others that we omit for brevity—is strongly influenced by the conventional although arguable selection criteria adopted in meta-analysis and systematic reviews.[1] Moreover, this scientific approach is appropriate (feasible) for those clinical conditions characterized by a fatal or nonfatal "new event" and therefore with an inception point easily knowable, which corresponds to a "late" symptom in the context of severe events (e.g., a coronary acute syndrome).

[1] That is, of 580 published publications, only 26 were analysed. It is very difficult to prove that these highly selective processes and these criteria will lead to a level of information and knowledge greater than that of other modern methodological approaches.

A.V. Gaddi
Scientific Director of Health Lab of GTechnology Foundation,
Modena, Italy
e-mail: profgaddi@gmail.com

F. Capello, A.V. Gaddi (eds.), *Clinical Handbook of Air Pollution-Related Diseases*,
https://doi.org/10.1007/978-3-319-62731-1_16

It is therefore impossible to select the appropriate inception cohorts or follow the evolutionary history of a disease. Not surprisingly, a recent study on 47,000 emergency department critical care accesses at Shanghai Hospital clarifies the effects of the outdoor pollution on all subtypes of coronary heart disease (symptomatic or only detectable by instruments), demonstrating strong effects of PM_{10}, SO_2 and NO_2 on sudden cardiac death; moderate effects on acute myocardial infarction and angina; weak effects on ischaemic cardiomyopathy; and "no effect on occult CHD" (defined as EKG ST depression or other signs of ischaemia).

A recent estimate of total cardiovascular death's burden in China demonstrated that PM_{10} is responsible for a total of 1252 cardiovascular death out of 28,365 with a total loss of about 27,000 years of life due to cardiovascular diseases, mainly in women [4]. These data do not consider the AP role as cofactor of other genetic and environmental risk factors; thus, probably, the total burden is higher, as suggested by several researches demonstrating that the impact of fine and ultrafine particulate matter on cardiovascular diseases generally is equal or exceeds those on pulmonary diseases.

The authors' conclusion is that "air quality improvement has become necessary to protect public health". This is certainly true; but to protect the health, we need to know "how to do it" and then to have other studies on the interactions between traditional risk factors, pollution (of air, water and food) and the history of the evolution of the coronary disease in each individual.

In fact, the slow effects and the linkages between type and concentration of pollutants and the slow evolution of chronic degenerative diseases (also characterized by steps of remission and phases of progression) are less defined; this is due to the intrinsic difficulties of this type of studies and to the high number of variables that can influence the evolution of the disease.

Uzoigwe et al., in a review on the effect of indoor and outdoor pollutants on the cardiovascular and the pulmonary system [5], stress the need to elucidate the biological mechanisms underlying the link between air pollution and cardiovascular diseases in order to develop new therapeutic strategies and diet prevention of AP-induced diseases.

A similar position was proposed recently by Cascio et al. [6] that underline the need of knowing the pathophysiologic framework underneath the association between exposure to AP and premature cardiovascular mortality, pragmatically suggesting the "insights for public health translation".

Probably, this is one of the possible ways to plan studies aimed to understand the real effect of different pollutants on the cardiovascular system, even if it does not face the problem of complexity (see Chaps. 1 and 14) that is particularly relevant in case of AP and the cardiovascular system (both complex systems interacting in a complex way).

In general, the possible mechanisms of damage of the pollutants on the cardiovascular system may depend from the strengthening or synergy between long-term exposures to air pollutants and the traditional cardiovascular risk factors (genetic, environmental and behavioural) and on the effect of pollutants on vascular thrombosis and electrical dysfunction. Because chronic and acute exposure is usually concomitant with systemic inflammation, with changes in the autonomic nervous

Table 16.1 Summary of the risk profiles as per the available research on air pollution and cardiovascular diseases

No.	Disease	Acute	Chronic	Indoor	Outdoor	Chapter
1	Coronary atherosclerosis			+	+	Ghiselli G
2	Carotid atherosclerosis			+	+	Ghiselli G
3	Atherosclerosis (other)			D	+	Ghiselli G
4	Vessel diseases (other than atherosclerosis)			D	D	
5	Myocardial infarction			++	++	Present
6	Other ischaemic heart diseases			+D	+	Present
7	Heart failure			+		Present
8	Arrhythmias		+	++		Present
9	Pulmonary hypertension	+D	+	++	D	Present/Cipolla M
10	CODP/lung diseases					Cipolla M
11	Heart and lung infectious diseases	++	+	++	++	
12	Pulmonary embolism and deep vein thrombosis					Present
13	Stroke/other cerebrovascular					Lenzi GL
14	Hypertension	++	+D	+	++	Present

system balance and in the pro/antithrombotic and pro/anti-fibrinolytic balance [6], it seems logical that a preventive approach directed to minimize the exposure to all risk factors and the pollutant remains a key factor.

A summary of the profile of the available studies, which is also an index of the topics treated, is reported in Table 16.1.

16.2 Hypertension

Hypertension is one of the main risk factor for both cardio- and cerebrovascular, fatal and nonfatal new events. Air pollution is now regarded as an independent risk factor[2] for cardiovascular morbidity and mortality and not only as an "etiologic component" of diseases. Some authors estimate that air pollution explains at least the 10–25% of incidence of coronary disease that is unexplained by traditional risk factors [1].

From the public health's point of view, a recent systematic revision underlines that even a small mean effect of air pollution on blood pressure (increments of few mmHg) would have "enormous global public health implications" [7], taking into account the ubiquity of air pollution and the high prevalence, worldwide, of hypertension.

A recent research performed in Canada demonstrated that recent exposure to several pollutants, SO_2, NO_2 and $PM_{2.5}$ (during cold season) and SO_2 and O_3 (during

[2]The individual components of the air pollution have toxic effects directly on the cells, tissues and organs and must therefore be regarded as etiologic factors of disease. For other, in the recent literature, many "non-specific" measures of air pollution (as the total PM or the Air Quality Index) are directly associated with other risk factors and to the incidence of cardiovascular new events, irrespective of causal link.

the warm season), has a statistically significant positive association with the number of emergency department visits for hypertension or hypertensive crisis. The authors conclude that several air pollutants can be capable of elevating blood pressure leading to an increase of emergency visits [8].

Several researches suggested a direct effect of pollutant on blood pressure in the short-term period in men and women (both with and without previous known hypertension) [9–11] rising the request for medical attention [12]. The hypertensive effect is demonstrated also in children [13] and seems related to all the tested pollutant (O_3, NO_2, SO_2, CO and PM_{10}) (see also Chap. 10).

Moreover, the acute effect of exposition at extremely high concentrations of both fine particles and black carbon particles rises systolic and diastolic blood pressure and insulin resistance: The authors underline that these results provide important global public health warnings that "air pollution may pose a risk to cardio metabolic health" of billions of people, particularly in China [14].

The long-term exposure also rises blood pressure values; however, in the long term, the correlation seemed less evident [15], perhaps secondary to methodological reasons, because of several interfering cofactors. However, NO_2 traffic-related emissions[3] increase systolic blood pressure [16], although in a Danish cohort, the NOx exposition was not associated with incident hypertension [17]. On the contrary, in a study on 24,000 Chinese men, a positive association between blood pressure values and hypertension was found for SO_2, PM_{10} and O_3 [18].

Probably, the presence of diabetes, blood lipids and obesity strongly influences the relationships between pollutant and hypertension [19, 20].

Several physiopathological mechanisms were proposed to explain the relationship between the exposure to some air pollutants and the rise in systolic blood pressure: probably, the factors involved are many. The literature has highlighted, in experimental animals, an effect on air pollutant reactive oxygen species (ROS) mediated by upregulation of the Rho-associated protein kinase (ROCK) pathway [21]. Recently, Cascio et al. also suggested an interesting hypothesis on the role for the biological component of air pollution, like endotoxin or β-1,3-d-glucan contained in particulate matter [22], thus opening a new scenario in air pollution research.

16.3 Pulmonary Hypertension and Heart-Lung Diseases

Pulmonary hypertension, chronic obstructive lung diseases and right heart failure mutually reinforce each other: they are not only "coexisting" comorbidity but on the contrary often constitute a single alteration of the heart-lung system; they also have common multifactorial causes.

We think that in the future this is bound to become one of the main areas of air pollution-related pathologies; however, today integrated researches on these topics (or, better, on this type of patients) are not available. Particularly in low- and

[3] NO_2 is used as a marker of traffic exposition.

medium-income countries, billions of people are exposed to indoor pollutant, and the incidence of the deaths and physical disabilities induced by these diseases rose in the last decades.

Bloomfield et al. in a publication with the evocative title "Waiting to Inhale…." reviewed the recent articles on this matter and suggested a joined approach to these diseases. In fact, the common causes of pulmonary hypertension[4] potentiate the effect of indoor air pollution [23]; moreover, chronic exposure to air pollutants probably accelerates and worsens the clinical course of pulmonary hypertension, of COPD (see Chap. 14) and of right heart failure.

Moreover, Bloomfield suggests that there may be a "direct link between exposure to household air pollution and right heart failure": in other words, that indoor air pollution is a direct cause of right heart failure, independent of pre-existing diseases or comorbidities [23].

The role of tobacco smoking on pulmonary hypertension is well known [24], and probably the smoke plays a double role: it has a direct action on smokers, but it also increases the level of CO, CO_2 and PM in the house air (second- and third-hand smoking; see also Chap. 9), even if the effect of second-hand smoking was not confirmed by all studies [25].

Indoor air pollution probably worsens the high-altitude-induced pulmonary hypertension [26].

The associated burden is significant: the social cost of pulmonary hypertension is very high, whilst the quality of life of patients is very low [27]; however, the patient awareness of air pollution-related risk for the heart and lung remains scarce [28].

Possibly, the physiopathology of pulmonary hypertension secondary to air pollution depends on the thickening of the pulmonary arteries and on an increased systolic pressure in right heart via T-cell-produced cytokines, interleukin (IL)-13 and IL-17A [29, 30]; PM decreases endothelium-dependent relaxation and eNOS production in pulmonary arteries [31].

16.4 Heart Failure

Experimental researches on healthy volunteers demonstrate the inotropic effect of some pollutants on the heart; the endothelial dysfunction and BNP increase after exposition to diesel exhaust [32–34] and confirm the efficacy of diesel exhaust filter in lowering the toxic effects [34].

The air pollution effects (PM and NO_2) on several cardiovascular diseases and on congestive heart failure are also linked to heterogeneous social factors and to race, with some preliminary evidences that confirm the so-called Hispanic health paradox for sensitivity to NO_2 ethnicity [35]. In patients with cardiomyopathy and also in

[4] Human immunodeficiency virus, schistosomiasis, haemoglobinopathies, interstitial lung disease and COPD, healed tuberculosis infection, high altitudes and chronic pulmonary hypertension (thromboembolic)

Table 16.2 Association between single pollutant, concentration and cardiovascular risk

Carbon monoxide	1 ppm	3.52% (95% CI = 2.52–4.54)
Sulphur dioxide	10 ppb	2.36% (95% CI = 1.35–3.38)
Nitrogen dioxide	10 ppb	1.70% (95% CI = 1.25–2.16)
$PM_{2.5}$	10 mg/mc	2.12% (95% CI = 1.42–2.82)
PM_{10}	10 mg/mc	1.63% (95% CI = 1.20–2.07)

nonhospitalized patients affected by heart failure, the short-term exposition to $PM_{2.5}$ and CO affects both perceived (symptoms) and factual parameters of heart function [36]. Cardiomyopathy confers susceptibility to the inflammatory and the rhythmogenesis effects of acute $PM_{2.5}$ inhalation [37]

Few years ago, a large study performed in Guangzhou (China) [38] provided a good epidemiologic evidence that outdoor pollution exacerbates heart failure symptoms and probably rises the access to emergency departments in general free-living population. Similar conclusions were derived from a study performed in the UK with a different experimental approach [39]. An interesting [40] research performed in Kenia demonstrates an effect on isolated right heart failures in women exposed to dust (occupational) or poor kitchen ventilation and HIV [40].

However, we have few information on long-term exposure to different pollutants and on progression, therapy response and outcomes of different types of heart failures.

A meta-analysis published by Shah et al. [41] only confirmed the effects of acute of pollution on mortality and on access to hospital for heart failure. This relevant paper demonstrated an association between single pollutants and cardiovascular risk. The details of these studies are summarized in Table 16.2 (values refer to heart failure hospitalization plus death).

Ozone resulted to be not associated with both hospitalization and death by heart failure. Quantitative reviews fail to demonstrate clear and definite correlations between ozone exposition in long period and death from lung pathologies and other diseases [42].

However, several authors suggested the inclusion of the above (or similar) values into recommendatory acts of the European Union and of other countries [43].

Probably, the toxic effects of air pollution on the heart are relevant in the long term and must be added to all other causes of myocardial damage; the outcome depends on the sum (and the synergic action) of all the injuries among which air pollution affects a patient during his or her entire life; an example of this type of pathological interaction comes from studies that demonstrate specific geographical associations between air pollution exposition in youth and in early stage of life and lung infection with consequent rheumatic heart disease complications [44].

16.5 Ischaemic Heart Diseases (IHD)

The effect of air pollution on coronary heart disease (CHD) and coronary artery disease, mainly due to early onset of coronary atherosclerosis, is described also in another chapter of this book (see Chap. 17). In this paragraph, we summarize the relationships between pollutants and ischaemic heart disease (like myocardial infarction and its complications as angina pectoris, effort-induced ischaemia, etc.).

A recent time-series study performed in Beijing on 369,469 IHD cases and 53,247 deaths demonstrated a non-linear, significant, dose-response relationships between $PM_{2.5}$ and both morbidity and mortality. During the 3 years of the study, the researchers observed that "there were 7703 cases and 1475 deaths advanced by $PM_{2.5}$ pollution over expected rates if daily levels had not exceeded the WHO target" [45]. A successive study carried out in Shanghai on over 600,000 CHD outpatients confirmed that exposure to PM_{10} and $PM_{2.5}$ was associated with an increased risk of and leads to a marked increase in emergency department visits. Several factors, including season, gender and age, influence the association [46], and their associations were effect modifiers.

The association between cardiovascular diseases, including myocardial infarction, and $PM_{2.5}$ and PM_{10} was also demonstrated in a very exhaustive meta-analysis on 59 articles (out of 1464 published papers): short-term exposures to particulate matter (between 0.3 and 10 μm) are associated with increases in mortality. However, the author concludes that evidence of constituent-associated health effects, long-term effects and morbidity in China is still inadequate [47]. These conclusions are very cautious and rigorous from a scientific point of view. We can add that that applies to all over the world and not only for China.

The meta-analysis supports an inverse relationship between HRV, a marker for a worse cardiovascular prognosis, and particulate air pollution [48].

The sum of more pollutants (in particular, the sum of $PM_{2.5}$, NO_2 and O_3) is correlated with the incidence and severity of the myocardial infarction [49]; the same author demonstrated an association between myocardial infarction and pollens [50] that are a new emerging environmental risk factor and a biological constituent of pollution.

Conclusions

A number of studies demonstrate that air pollution has a central role in the development of cardiovascular diseases.

In relation to hypertension, from the scientific point of view, there is a great confusion: The number of studies is insufficient, whilst the number of variables is very high. We are in the field of complexity, and we cannot consider satisfactory nor conclusive the available data.

However, from the clinical-pragmatic point of view, we can argue that air pollution is a worsening factor when the values of the arterial pressure are considered, so that we need to alert people (children, adults and the elderly) about the risk and the possible outcomes. Fortunately, the arterial hypertension can be effectively cured.

The association with pulmonary hypertension and heart-lung diseases is also being researched. The studies carried out so far are few and sectorial. The literature lacks the holistic view of the individual; in addition, the same idea of air pollution affecting the cardiopulmonary system is missing or scarcely expressed.

From the clinical point of view, however, an interesting scenario is presented because on the one hand there is enough evidence to support that exposure (chronic and indoor) to pollution increases the pulmonary hypertension and on the other hand such effects of pollution can be prevented, both on the general population (primary prevention) and on subjects at high risk or already suffering from pulmonary hypertension.

The clinical findings and preliminary scientific conclusions on cardiac failure and IHD are common, not only because of synthesis, but also because the clinical picture of the two "diseases" is often confused in the literature; especially in the elderly, the two conditions overlap; this is interesting as most of fatal and nonfatal cases studied in clinical trials come from this cadre (see also Chap. 11). From a scientific point of view, the studies supporting the idea of a relationship between air pollution and heart failure are numerous and are more than the one presented. We have quoted only part of those, as detailed reviews on these topics are easily available. There are still not enough data for the long-term effects, but we have solid evidences to conclude that the PM and some gases substantially increase the risk of ischaemic heart disease and cardiac failure and, in a wider sense of all the other cardiac diseases, including cardiac arrhythmias [51, 52], cardio-metabolic risk factors, like obesity [53, 54], hyperlipidaemias (few studies available; see also Chap. 16) [55] and diabetes [56–60]. The latter has been studied in the long term too.

References

1. Franklin BA, Brook R, Arden Pope C III. Air pollution and cardiovascular disease. Curr Probl Cardiol. 2015;40(5):207–38.
2. Du Y, Xu X, Chu M, Guo Y, Wang J. Air particulate matter and cardiovascular disease: the epidemiological, biomedical and clinical evidence. J Thorac Disease. 2016;8(1):E8–E19.
3. Fatmi Z, Coggon D. Coronary heart disease and household air pollution from use of solid fuel: a systematic review. Br Med Bull. 2016;118(1):91–109.
4. Li G, Zeng Q, Pan X. Disease burden of ischaemic heart disease from short-term outdoor air pollution exposure in Tianjin, 2002-2006. Eur J Prev Cardiol. 2016;23(16):1774–82.
5. Uzoigwe JC, Prum T, Bresnahan E, Garelnabi M. The emerging role of outdoor and indoor air pollution in cardiovascular disease. N Am J Med Sci. 2013;5(8):445–53.
6. Cascio WE. Proposed pathophysiologic framework to explain some excess cardiovascular death associated with ambient air particle pollution: insights for public health translation. Biochim Biophys Acta. 2016;1860(12):2869–79.
7. Giorgini P, Di Giosia P, Grassi D, Rubenfire M, Brook RD, Ferri C. Air pollution exposure and blood pressure: an updated review of the literature. Curr Pharm Des. 2016;22(1):28–51.
8. Brook RD, Kousha T. Air pollution and emergency department visits for hypertension in Edmonton and Calgary, Canada: a case-crossover study. Am J Hypertens. 2015;28(9):1121–6.
9. Mannisto T, Mendola P, Liu D, Leishear K, Sherman S, Laughon SK. Acute air pollution exposure and blood pressure at delivery among women with and without hypertension. Am J Hypertens. 2015;28(1):58–72.
10. Shan J, Li HY, Liu GF, Yang X, Dong W, Jian WY, et al. Effect of air pollution on health service demand of the elderly and middle-age patients with hypertension, cardiovascular and cerebrovascular diseases: based on analysis of data from CHARLS. J Peking Univ Health Sci. 2016;48(3):460–4.

11. Guo Y, Tong S, Zhang Y, Barnett AG, Jia Y, Pan X. The relationship between particulate air pollution and emergency hospital visits for hypertension in Beijing, China. Sci Total Environ. 2010;408(20):4446–50.
12. Arbex MA, Saldiva PH, Pereira LA, Braga AL. Impact of outdoor biomass air pollution on hypertension hospital admissions. J Epidemiol Community Health. 2010;64(7):573–9.
13. Dong GH, Qian ZM, Trevathan E, Zeng XW, Vaughn MG, Wang J, et al. Air pollution associated hypertension and increased blood pressure may be reduced by breastfeeding in Chinese children: the seven northeastern cities Chinese Children's study. Int J Cardiol. 2014;176(3):956–61.
14. Brook RD, Sun Z, Brook JR, Zhao X, Ruan Y, Yan J, et al. Extreme air pollution conditions adversely affect blood pressure and insulin resistance: the air pollution and cardiometabolic disease study. Hypertension. 2016;67(1):77–85.
15. Coogan PF, White LF, Jerrett M, Brook RD, Su JG, Seto E, et al. Air pollution and incidence of hypertension and diabetes mellitus in black women living in Los Angeles. Circulation. 2012;125(6):767–72.
16. Foraster M, Basagana X, Aguilera I, Rivera M, Agis D, Bouso L, et al. Association of long-term exposure to traffic-related air pollution with blood pressure and hypertension in an adult population-based cohort in Spain (the REGICOR study). Environ Health Perspect. 2014;122(4):404–11.
17. Sorensen M, Hoffmann B, Hvidberg M, Ketzel M, Jensen SS, Andersen ZJ, et al. Long-term exposure to traffic-related air pollution associated with blood pressure and self-reported hypertension in a Danish cohort. Environ Health Perspect. 2012;120(3):418–24.
18. Dong GH, Qian ZM, Xaverius PK, Trevathan E, Maalouf S, Parker J, et al. Association between long-term air pollution and increased blood pressure and hypertension in China. Hypertension. 2013;61(3):578–84.
19. Zhao Y, Qian ZM, Wang J, Vaughn MG, Liu YQ, Ren WH, et al. Does obesity amplify the association between ambient air pollution and increased blood pressure and hypertension in adults? Findings from the 33 communities Chinese health study. Int J Cardiol. 2013;168(5):e148–50.
20. Basile JN, Bloch MJ. Exposure to air pollution increases the incidence of hypertension and diabetes in black women living in Los Angeles. J Clin Hypertens (Greenwich). 2012;14(11):819–20.
21. Sun Q, Yue P, Ying Z, Cardounel AJ, Brook RD, Devlin R, et al. Air pollution exposure potentiates hypertension through reactive oxygen species-mediated activation of rho/ROCK. Arterioscler Thromb Vasc Biol. 2008;28(10):1760–6.
22. Cascio WE, Gilmour MI, Peden DB. Ambient air pollution and increases in blood pressure: role for biological constituents of particulate matter. Hypertension. 2015;66(3):469–71.
23. Bloomfield GS, Lagat DK, Akwanalo OC, Carter EJ, Lugogo N, Vedanthan R, et al. Waiting to inhale: an exploratory review of conditions that may predispose to pulmonary hypertension and right heart failure in persons exposed to household air pollution in low- and middle-income countries. Glob Heart. 2012;7(3):249–59.
24. Keusch S, Hildenbrand FF, Bollmann T, Halank M, Held M, Kaiser R, et al. Tobacco smoke exposure in pulmonary arterial and thromboembolic pulmonary hypertension. Respiration. 2014;88(1):38–45.
25. Ren X, Hsu PY, Dulbecco FL, Fleischmann KE, Gold WM, Redberg RF, et al. Remote second-hand tobacco exposure in flight attendants is associated with systemic but not pulmonary hypertension. Cardiol J. 2008;15(4):338–43.
26. Mirrakhimov AE, Strohl KP. High-altitude pulmonary hypertension: an update on disease pathogenesis and management. Open Cardiovasc Med J. 2016;10:19–27.
27. Rubenfire M, Lippo G, Bodini BD, Blasi F, Allegra L, Bossone E. Evaluating health-related quality of life, work ability, and disability in pulmonary arterial hypertension: an unmet need. Chest. 2009;136(2):597–603.
28. Nowka MR, Bard RL, Rubenfire M, Jackson EA, Brook RD. Patient awareness of the risks for heart disease posed by air pollution. Prog Cardiovasc Dis. 2011;53(5):379–84.

29. Park SH, Chen WC, Durmus N, Bleck B, Reibman J, Riemekasten G, et al. The effects of antigen-specific IgG1 antibody for the pulmonary-hypertension-phenotype and B cells for inflammation in mice exposed to antigen and fine particles from air pollution. PLoS One. 2015;10(6):e0129910.
30. Park SH, Chen WC, Esmaeil N, Lucas B, Marsh LM, Reibman J, et al. Interleukin 13- and interleukin 17A-induced pulmonary hypertension phenotype due to inhalation of antigen and fine particles from air pollution. Pulmon Circul. 2014;4(4):654–68.
31. Davel AP, Lemos M, Pastro LM, Pedro SC, de Andre PA, Hebeda C, et al. Endothelial dysfunction in the pulmonary artery induced by concentrated fine particulate matter exposure is associated with local but not systemic inflammation. Toxicology. 2012;295(1–3):39–46.
32. Vieira JL, Guimaraes GV, de Andre PA, Saldiva PH, Bocchi EA. Effects of reducing exposure to air pollution on submaximal cardiopulmonary test in patients with heart failure: analysis of the randomized, double-blind and controlled FILTER-HF trial. Int J Cardiol. 2016;215:92–7.
33. Vieira JL, Guimaraes GV, de Andre PA, Cruz FD, Saldiva PH, Bocchi EA. Respiratory filter reduces the cardiovascular effects associated with diesel exhaust exposure: a randomized, prospective, double-blind, controlled study of heart failure: the FILTER-HF trial. JACC Heart Fail. 2016;4(1):55–64.
34. Mentz RJ, O'Brien EC. Air pollution in patients with heart failure: lessons from a mechanistic pilot study of a filter intervention. JACC Heart Fail. 2016;4(1):65–7.
35. Grineski SE, Herrera JM, Bulathsinhala P, Staniswalis JG. Is there a Hispanic health paradox in sensitivity to air pollution? Hospital admissions for asthma, chronic obstructive pulmonary disease and congestive heart failure associated with NO2 and PM2.5 in El Paso, TX, 2005–2010. Atmos Environ. 2015;119:314–21.
36. Goldberg MS, Wheeler AJ, Burnett RT, Mayo NE, Valois MF, Brophy JM, et al. Physiological and perceived health effects from daily changes in air pollution and weather among persons with heart failure: a panel study. J Expos Sci Environ Epidemiol. 2015;25(2):187–99.
37. Carll AP, Haykal-Coates N, Winsett DW, Hazari MS, Ledbetter AD, Richards JH, et al. Cardiomyopathy confers susceptibility to particulate matter-induced oxidative stress, vagal dominance, arrhythmia and pulmonary inflammation in heart failure-prone rats. Inhal Toxicol. 2015;27(2):100–12.
38. Yang C, Chen A, Chen R, Qi Y, Ye J, Li S, et al. Acute effect of ambient air pollution on heart failure in Guangzhou, China. Int J Cardiol. 2014;177(2):436–41.
39. Bennett O, Kandala NB, Ji C, Linnane J, Clarke A. Spatial variation of heart failure and air pollution in Warwickshire, UK: an investigation of small scale variation at the ward-level. BMJ Open. 2014;4(12):e006028.
40. Lagat DK, DeLong AK, Wellenius GA, Carter EJ, Bloomfield GS, Velazquez EJ, et al. Factors associated with isolated right heart failure in women: a pilot study from western Kenya. Glob Heart. 2014;9(2):249–54.
41. Shah AS, Langrish JP, Nair H, McAllister DA, Hunter AL, Donaldson K, et al. Global association of air pollution and heart failure: a systematic review and meta-analysis. Lancet. 2013;382(9897):1039–48.
42. Atkinson RW, Butland BK, Dimitroulopoulou C, Heal MR, Stedman JR, Carslaw N, et al. Long-term exposure to ambient ozone and mortality: a quantitative systematic review and meta-analysis of evidence from cohort studies. BMJ Open. 2016;6(2):e009493.
43. Forastiere F, Agabiti N. Assessing the link between air pollution and heart failure. Lancet. 2013;382(9897):1008–10.
44. Phillips DI, Osmond C, Williams ML, Jones A. Air pollution in early life and adult mortality from chronic rheumatic heart disease. Int J Epidemiol. 2016. https://doi.org/10.1093/ije/dyw249.
45. Xie W, Li G, Zhao D, Xie X, Wei Z, Wang W, et al. Relationship between fine particulate air pollution and ischaemic heart disease morbidity and mortality. Heart. 2015;101(4):257–63.
46. Ye X, Peng L, Kan H, Wang W, Geng F, Mu Z, et al. Acute effects of particulate air pollution on the incidence of coronary heart disease in shanghai, China. PLoS One. 2016;11(3):e0151119.

47. Lu F, Xu D, Cheng Y, Dong S, Guo C, Jiang X, et al. Systematic review and meta-analysis of the adverse health effects of ambient PM2.5 and PM10 pollution in the Chinese population. Environ Res. 2015;136:196–204.
48. Pieters N, Plusquin M, Cox B, Kicinski M, Vangronsveld J, Nawrot TS. An epidemiological appraisal of the association between heart rate variability and particulate air pollution: a meta-analysis. Heart. 2012;98(15):1127–35.
49. Weichenthal S, Lavigne E, Evans G, Pollitt K, Burnett RT. Ambient PM2.5 and risk of emergency room visits for myocardial infarction: impact of regional PM2.5 oxidative potential: a case-crossover study. Environ Health. 2016;15(1):46.
50. Weichenthal S, Lavigne E, Villeneuve PJ, Reeves F. Airborne pollen concentrations and emergency room visits for myocardial infarction: a multicity case-crossover study in Ontario, Canada. Am J Epidemiol. 2016;183(7):613–21.
51. Zhao A, Chen R, Kuang X, Kan H. Ambient air pollution and daily outpatient visits for cardiac arrhythmia in shanghai, China. J Epidemiol. 2014;24(4):321–6.
52. Chiu HF, Tsai SS, Weng HH, Yang CY. Short-term effects of fine particulate air pollution on emergency room visits for cardiac arrhythmias: a case-crossover study in Taipei. J Toxicol Environ Health Part A. 2013;76(10):614–23.
53. Wei Y, Zhang JJ, Li Z, Gow A, Chung KF, Hu M, et al. Chronic exposure to air pollution particles increases the risk of obesity and metabolic syndrome: findings from a natural experiment in Beijing. FASEB J. 2016;30(6):2115–22.
54. Qin XD, Qian Z, Vaughn MG, Trevathan E, Emo B, Paul G, et al. Gender-specific differences of interaction between obesity and air pollution on stroke and cardiovascular diseases in Chinese adults from a high pollution range area: a large population based cross sectional study. Sci Total Environ. 2015;529:243–8.
55. Xiao S, Liu R, Wei Y, Feng L, Lv X, Tang F. Air pollution and blood lipid markers levels: estimating short and long-term effects on elderly hypertension inpatients complicated with or without type 2 diabetes. Environ Pollut. 2016;215:135–40.
56. Zhao A, Chen R, Wang C, Zhao Z, Yang C, Lu J, et al. Associations between size-fractionated particulate air pollution and blood pressure in a panel of type II diabetes mellitus patients. Environ Int. 2015;80:19–25.
57. Wang B, Xu D, Jing Z, Liu D, Yan S, Wang Y. Effect of long-term exposure to air pollution on type 2 diabetes mellitus risk: a systemic review and meta-analysis of cohort studies. Eur J Endocrinol. 2014;171(5):R173–82.
58. Rosenbauer J, Tamayo T, Bachle C, Stahl-Pehe A, Landwehr S, Sugiri D, et al. Ambient air pollution and early manifestation of type 1 diabetes. Epidemiology. 2016;26(3):e31–2.
59. Weinmayr G, Hennig F, Fuks K, Nonnemacher M, Jakobs H, Mohlenkamp S, et al. Long-term exposure to fine particulate matter and incidence of type 2 diabetes mellitus in a cohort study: effects of total and traffic-specific air pollution. Environ Health. 2015;14(1):53.
60. Rao X, Montresor-Lopez J, Puett R, Rajagopalan S, Brook RD. Ambient air pollution: an emerging risk factor for diabetes mellitus. Curr Diabet Rep. 2015;15(6):603.

Air Pollution and Atherosclerosis

17

Marilisa Bove and Giancarlo Ghiselli

17.1 Introduction

There is increasing concern around the impact that air pollution has on human health. The Global Health Observatory of the WHO has determined that worldwide air pollution contributes to 5.4% of all deaths from any causes [1, 2]. The higher impact has been recorded in highly industrialized countries as well as in developing countries with large population residing in urban areas. Along with cigarette smoking and secondhand smoke, the exposure to polluted air is now recognized as an important environmental contributor to cardiovascular morbidity and mortality. Indeed, ambient air pollution ranks among the top ten modifiable disease risk factors along with low physical activity, high-sodium diet, high cholesterol, and drug use. Studies reporting an association between high levels of air pollution and health deterioration had been published with increasing frequency since the end of the last century [3–8]. However, convincing epidemiological evidence has become available only recently [9–18] and has called the attention of the medical community, regulatory agencies, and public administrators on air pollution as a main risk factor for cardiovascular disease morbidity and mortality in the population at large.

M. Bove
Medicine and Surgery Sciences Department, Sant'Orsola Hospital, University of Bologna, Bologna, Italy

G. Ghiselli (✉)
Meetsci, Spruce Street, Suite 706, Philadelphia, PA 19107, USA
e-mail: gghiselli@meetsci.com

F. Capello, A.V. Gaddi (eds.), *Clinical Handbook of Air Pollution-Related Diseases*,
https://doi.org/10.1007/978-3-319-62731-1_17

17.2 Air Pollution and Atherosclerosis: The Epidemiological Evidence

In the past two decades, a number of human cross-sectional, time-series, and cohort studies have suggested a strong link between the level of particulate matter (PM) from industrial, farming, household, or mobile exhaustions and the progression of atherosclerosis [19, 20]. The first focused epidemiological study published in 2005 investigated 798 residents of Los Angeles [21]. Air pollution exposures were estimated individually based on PM values recorded by local pollution-monitoring stations. The study concluded that for every 10 μg/m^3 rise in PM, aortic intima-media thickness (ITM) increases by 4% after adjustment for potential confounding variables. Similar conclusions were reached by assessing the effect of living close to a main urban road [22]. The coronary artery calcium, used as a biomarker of coronary atherosclerosis, was found to increase by 60% in the sample exposed to the higher PM burden. Halving the distance between the residence and a main road resulted in a 10.2% increase in coronary artery calcification. In a large prospective multicenter study (Multi-Ethnic Study of Atherosclerosis and Air Pollution, aka MESA Air), the relationship between long-term air pollution exposure and the progression of subclinical atherosclerosis has been investigated over several years [23–25]. In the study, long-term exposure to PM 2.5 was found to be significantly associated with decreased endothelial function and increased IMT progression even during a relatively short follow-up period. Importantly, the trial demonstrated that reducing exposure to PM 2.5 can slow IMT progression. Recently, the German Heinz Nixdorf Recall Study group has reported that long-term exposure to fine PM is an independent risk factor associated with subclinical atherosclerosis [26, 27]. The study included a population-based cohort of 4814 randomly selected participants, and thoracic aortic calcification was used to monitor the progression of atherosclerosis. The studies cited and others have been the subject of recent excellent review articles [24, 28–33]. The reader is directed to this literature for an in-depth examination of the study design and the conclusions reached in different population cohorts. A consensus panel has recently concluded that increased levels and exposure to PM are strongly and positively associated to accelerated progression of atherosclerosis [34]. The findings have been a significant driving factor for researchers to identify mechanism(s) by which PM affects the cardiovascular system and particularly the progression of atherosclerosis. In this review article, the experimental evidences linking exposure to polluted air to the development of atherosclerosis are presented.

17.3 Effect of Air Pollutants on Atherogenesis in Experimental Animals

Air pollutants are heterogeneous both physically and chemically. The complex mixture includes particles of different size (mainly PM 10, PM 2.5, and PM of size lower than 1 μm), fumes, and gases such as carbon monoxide (CO), nitrogen dioxide (NO_2), sulfur dioxide (SO), and ozone (O_3). The particles are a mix of dust and liquid droplets composed of chemicals, acids, metals, and soil. Air pollutants from different sources are endowed with distinct health risk profiles and each may be preferentially linked to pulmonary diseases, systemic inflammation and oxidative stress, endothelial dysfunction, pro-thrombotic and coagulant changes, or to the progression of atherosclerosis.

Investigations on the pathogenesis mechanism underlying the pro-atherogenic effect of various air pollutants have made use of atherosclerosis-prone animals. To this end, Suwa et al. have utilized Watanabe heritable hyperlipidemic (WHHL) rabbits that naturally develop systemic atherosclerosis [35]. The animals were exposed to PM 10 concentrate suspended in saline by intrapharyngeal instillation twice a week for 4 weeks. The results indicated that repeated exposure to urban air PM 10 elicits a systemic inflammatory response and accelerate atherosclerosis in the coronary arteries and aorta of the animals. The severity of the atherosclerotic lesions correlated with the extent of PM 10 phagocytosed by alveolar macrophages in the lung. The qualitative histological observations demonstrated extensive atherosclerosis in the aorta, increase of the lipid-laden areas, and increased cellular lipid turnover. In another study [36], female WHHL were exposed to PM 10 by intratracheal instillation twice a week for 4 weeks, and the recruitment of BrdU-labeled monocytes into the vessel walls and the atherosclerotic plaques were measured through quantitative histology. The exposure to PM 10 increased the number of BrdU-labeled monocytes adherent to the endothelium covering the plaques and promoted the subendothelial migration of the monocytes at the site of plaque formation. In order to investigate the effect of air pollution on the ultrastructural properties of atherosclerotic plaques induced by PM 10 in WHHL rabbits, concentrated ambient particulate matter was instilled into the lungs of the rabbits twice per week for 4 weeks [37]. When examined by transmission electron microscopy, atherosclerotic plaques of PM-treated rabbits displayed increased accumulation of macrophage-derived foam cells compared to saline-treated animals. In addition, type IV collagen was present in the thickened extracellular matrix material beneath the endothelium at the site of plaque formation. These ultrastructural changes are likely to increase the probability of plaque rupture that may trigger a thrombotic event.

Atherosclerosis-prone mice (apoE−/−) fed with a high-fat diet [38], exposed to PM 2.5 at tenfold ambient concentrations for 6 h/day, 5 days per week for a total of 6 months, had a significant increase of lipid content in the aortic arch compared to animal fed with the same diet and breathing filtered air. The exposed animals displayed marked increases in macrophage infiltration, expression of the inducible isoform of nitric oxide synthase, increased generation of reactive oxygen species, and greater immunostaining for the protein nitration product 3-nitrotyrosine. In a similar study, apoE−/− mice were exposed to concentrated PM 2.5 ultrafine particles, and the progression of their atherosclerosis is compared to that in animals breathing filtered air [39]. The exposure to PM 2.5 resulted in a much earlier appearance of large atherosclerotic lesions. A deterioration of the HDL anti-inflammatory capacity and a significant increase of the level of genetic markers for systemic oxidative stress were also recorded. To investigate the effect of diesel exhaustion particles on the progression of atherosclerosis induced by a pro-atherogenic diet [40], apoE−/− mice fed with a Western diet received a twice-a-week oropharyngeal instillation of diesel exhaust particulate or alternatively saline for 4 weeks. A larger number of atherosclerotic lesions per vessel and more buried fibrous caps were observed in animals instilled with the solution containing diesel particulate matter. The pro-atherosclerotic effect was concomitant with pulmonary inflammation and systemic oxidative stress. In a recent study [41], atherosclerosis development accelerated in apoE−/− mice exposed to ambient ultrafine particles (UFP) with diameter smaller than 180 nm. UFP contain pro-oxidant or otherwise toxic organic chemicals. When these agents were removed by heating the UFP, the level of biomarkers of oxidative stress and the sizes of arterial plaques were significantly reduced. This has led the authors to conclude that removal of organic constituents from ambient particles affords a significant reduction of toxic cardiovascular effects of air pollution exposure. In order to investigate the gene expression alterations leading to accelerated atherosclerosis, global gene expression analysis was performed on atherosclerotic plaques from apoE−/− mice exposed for 6 h/day, 5 days/week for 5 months, to filtered air or concentrated (tenfold) ambient air from an area located 40 km north of New York City [42]. The gene expression profiling showed upregulation of genes linked to inflammation, proliferation, matrix remodeling, and oxidative stress, involved in classical mechanisms of atherosclerosis and plaque progression.

The impact of chronic exposure to urban air on the susceptibility of LDL to oxidative modifications and the development of anti-oxLDL antibodies in blood has been investigated in hyperlipemic mice (LDLR−/−) exposed to ambient or filtered air for 4 months [43]. Exposure to polluted air led to a significant increase of circulating oxidized LDL shown as well by the immune response to the modified lipoproteins. When ApoE(−/−) or LDLR(−/−) mice were exposed to filtered air or concentrated ambient PM 2.5 for 6 months [44], PM 2.5 increased the quantity of oxidized cholesterol (7-ketocholesterol) carried by low-density lipoproteins leading to its accumulation in atherosclerotic plaques. At the cellular level, macrophages from mice exposed to the particulate matter displayed increased uptake of oxidized lipoproteins. In apoE−/− mice, both acute and subchronic inhalation of traffic-generated air pollution lead to increased plasma oxLDL as well as the expression of the oxLDL receptor LOX-1 in vascular endothelial cells [45]. Expression of LOX-1 is increased in systemic arteries following exposure to diesel and gasoline emissions and ozone. Blocking LOX-1 through injection in animals of specific anti-LOX1 antibodies prevents exhaust-induced aortic lipid peroxidation and inflammation. Diesel exhaust emissions contain a large number of ultrafine particles, enriched in organic content such as polycyclic aromatic hydrocarbons. The exposure of apoE−/− mice to diesel exhaust for 2 weeks reduced HLD anti-inflammatory and antioxidant activity. These changes were negatively correlated with paraoxonase enzymatic activity in plasma and to activation of the 5-lipoxygenase pathway in the liver, leading to the formation of peroxidated lipids [46].

Animal studies have allowed researchers to address the key question of the relationship between the dose of exposure to pollutants and the progression and magnitude of atherosclerosis effect based on specific end points. To this regard, particularly instructive experiments have been carried out to compare the effect produced by PM to that produced by passive tobacco smoke on the atherosclerosis progression in apoE−/− mice [47]. By performing noninvasive (high-resolution ultrasounds) sequential measurements of the plaques in the aortic arch, PM at doses one-third those of tobacco smoke induced comparable degrees of plaque at similar time points. The results demonstrate that PM induces atherosclerosis at a pace comparable to that of passive smoke inhalation, but at much lower $\mu g/mm^3$ concentration. The results also demonstrated a nonlinear dose-response effect of PM and the progression of atherosclerosis. Whereas exposure to low levels of PM is associated to a rapid increase in cardiovascular risk, the strength of the correlation lessen as the concentration of inhaled PM increases. To investigate this effect further, the dose-response effect of the inhalation of increasing concentration of diesel engine emissions on atherosclerosis has been evaluated in apoE−/− mice [48]. In this experiment, the animals were exposed for 6 h/day for 50 days to increasing concentrations of diesel-exhausted material or to particulate-filtered diesel exhaust. The exposure to the emissions induced dose-related alterations in gene markers of vascular remodeling and aortic lipid peroxidation. An increase number of macrophages and accumulation of collagen was observed, consistent with the presence of advanced and more fragile atherosclerotic plaques prone to rupture.

17.4 Atherogenic Activity of the Particulate Matter and Potential Mechanisms of Action

An important question relates to the mechanisms underlying the transference of toxicity from pulmonary exposures to PM to the development of atherosclerosis. The inflammation hypothesis [49–51] proposes that inhaled particles bind to or are taken up by the lung alveolar macrophages and epithelial cells and activate the cells to an extent that induces a marked pulmonary inflammation. Inflammatory and pro-oxidative mediators (e.g., cytokines and activated immune cells) then enter the circulation and alter cardiovascular function. There is ample experimental evidence that exposure to air PM initiates an inflammatory response and oxidative stress in the lung [52–54]. The inhaled air pollutants can trigger local oxidative stress by reacting with protective secretions of the airways, thus generating reactive oxygen species (ROS) through the Fenton reaction [55, 56]. Ultrafine particles may also penetrate the cells of the alveolar surface and upon reaching the mitochondria affect the cell respiratory cycle [57]. When the PM-generated ROS overwhelm the pulmonary stress response system, an inflammatory reaction is triggered which releases mediators and further activates pro-inflammatory transcription factors. The released mediators (e.g., cytokines and ROS) readily enter the circulation, cause systemic injury to endothelial cells, and enhance the endothelial barrier permeability [58]. This facilitates the entry of fine and ultrafine PM into the blood stream where they can reach peripheral tissues and elicit further inflammatory and toxic effects [59–61]. A plethora of pro-inflammatory cytokines are released from alveolar macrophages following exposure to PM [62, 63]. PM 2.5 particles are recognized by intimal macrophages via Toll-like receptors (TLRs), TLR2 and TLR4. The binding of the particulate to macrophages activates the NF-κB pathway which in turn leads to the releases of cytokines and chemotactic agents [64, 65]. The involvement of TLR4 in the PM 2.5-mediated reaction of macrophages has been demonstrated in TLR4-deficient mice. In spite of being chronically challenged with PM 2.5, the animals displayed a normal cytokine profile [66]. There is compelling evidence that PM 2.5-derived ROS at level found in the body following exposure to polluted air have the potential to trigger an atherogenic cascade by generating oxidized LDL and altering the function of vascular cells [59, 67, 68]. The oxidized lipoproteins are taken up by vessel-resident macrophages, which in the process are converted into lipid-laden foam cells. Macrophages can also phagocytize PM 2.5 and, as a consequence, undergo apoptosis [69]. Necrotic remnants of macrophages contribute to the formation of the atheroma necrotic core which is surrounded by foam cells, smooth muscle cells, and extracellular matrix. In endothelial cells, PM 2.5 increases ROS level through p38 mitogen-activated protein kinase and heat shock protein 27-dependent pathways [58]. ROS may be also generated through a cell intrinsic pathway via the nicotinamide adenine dinucleotide phosphate (NADPH) oxidase pathway. PM exposure induces the production of NADPH oxidase-derived superoxide in monocytes and aortic tissue thus contributing through an alternative pathway to atherosclerosis progression [66]. Finally, ROS by scavenging the vasodilator NO produced by the endothelial cells can trigger vascular constriction leading to hypertension [70]. The pro-atherogenic effect of air PM mediated by the induction of a pro-inflammatory and pro-oxidant state is enhanced when the pollutant itself has high oxidizing potential as in the case of ozone or PM 2.5, which contains organic chemicals, transition metals, and high surface areas, all of which can contribute to local generation of ROS.

References

1. Brook RD, Bard RL, Burnett RT, Shin HH, Vette A, Croghan C, Phillips M, Rodes C, Thornburg J, Williams R. Differences in blood pressure and vascular responses associated with ambient fine particulate matter exposures measured at the personal versus community level. Occup Environ Med. 2011;68(3):224–30.
2. Lim SS, Vos T, Flaxman AD, Danaei G, Shibuya K, Adair-Rohani H, Amann M, Anderson HR, Andrews KG, Aryee M, Atkinson C, Bacchus LJ, Bahalim AN, Balakrishnan K, Balmes J, Barker-Collo S, Baxter A, Bell ML, Blore JD, Blyth F, Bonner C, Borges G, Bourne R, Boussinesq M, Brauer M, Brooks P, Bruce NG, Brunekreef B, Bryan-Hancock C, Bucello C, Buchbinder R, Bull F, Burnett RT, Byers TE, Calabria B, Carapetis J, Carnahan E, Chafe Z, Charlson F, Chen H, Chen JS, Cheng AT, Child JC, Cohen A, Colson KE, Cowie BC, Darby S, Darling S, Davis A, Degenhardt L, Dentener F, Des J, Devries K, Dherani M, Ding EL, Dorsey ER, Driscoll T, Edmond K, Ali SE, Engell RE, Erwin PJ, Fahimi S, Falder G, Farzadfar F, Ferrari A, Finucane MM, Flaxman S, Fowkes FG, Freedman G, Freeman MK, Gakidou E, Ghosh S, Giovannucci E, Gmel G, Graham K, Grainger R, Grant B, Gunnell D, Gutierrez HR, Hall W, Hoek HW, Hogan A, Hosgood HD III, Hoy D, Hu H, Hubbell BJ, Hutchings SJ, Ibeanusi SE, Jacklyn GL, Jasrasaria R, Jonas JB, Kan H, Kanis JA, Kassebaum N, Kawakami N, Khang YH, Khatibzadeh S, Khoo JP, Kok C, Laden F, Lalloo R, Lan Q, Lathlean T, Leasher JL, Leigh J, Li Y, Lin JK, Lipshultz SE, London S, Lozano R, Lu, Y,

Mak J, Malekzadeh R, Mallinger L, Marcenes W, March L, Marks R, Martin R, McGale P, McGrath J, Mehta S, Mensah GA, Merriman TR, Micha R, Michaud C, Mishra V, Mohd HK, Mokdad AA, Morawska L, Mozaffarian D, Murphy T, Naghavi M, Neal B, Nelson PK, Nolla JM, Norman R, Olives C, Omer SB, Orchard J, Osborne R, Ostro B, Page A, Pandey KD, Parry CD, Passmore E, Patra J, Pearce N, Pelizzari PM, Petzold M, Phillips MR, Pope D, Pope CA III, Powles J, Rao M, Razavi H, Rehfuess EA, Rehm JT, Ritz B, Rivara FP, Roberts T, Robinson C, Rodriguez-Portales JA, Romieu I, Room R, Rosenfeld LC, Roy A, Rushton L, Salomon JA, Sampson U, Sanchez-Riera L, Sanman E, Sapkota A, Seedat S, Shi P, Shield K, Shivakoti R, Singh GM, Sleet DA, Smith E, Smith KR, Stapelberg NJ, Steenland K, Stockl H, Stovner LJ, Straif K, Straney L, Thurston GD, Tran JH, Van DR, van DA, Veerman JL, Vijayakumar L, Weintraub R, Weissman MM, White RA, Whiteford H, Wiersma ST, Wilkinson JD, Williams HC, Williams W, Wilson N, Woolf AD, Yip P, Zielinski JM, Lopez AD, Murray CJ, Ezzati M, AlMazroa MA, and Memish ZA. A comparative risk assessment of burden of disease and injury attributable to 67 risk factors and risk factor clusters in 21 regions, 1990–2010: a systematic analysis for the Global Burden of Disease Study 2010. Lancet. 2012;380(9859):2224–2260.

3. Dockery DW, Pope CA III, Xu X, Spengler JD, Ware JH, Fay ME, Ferris BG Jr, Speizer FE. An association between air pollution and mortality in six U.S. cities. N Engl J Med. 1993;329(24):1753–9.
4. Pope CA III, Thun MJ, Namboodiri MM, Dockery DW, Evans JS, Speizer FE, Heath CW Jr. Particulate air pollution as a predictor of mortality in a prospective study of U.S. adults. Am J Respir Crit Care Med. 1995;151(3 Pt 1):669–74.
5. Pope CA III, Verrier RL, Lovett EG, Larson AC, Raizenne ME, Kanner RE, Schwartz J, Villegas GM, Gold DR, Dockery DW. Heart rate variability associated with particulate air pollution. Am Heart J. 1999;138(5 Pt 1):890–9.
6. Rossi G, Vigotti MA, Zanobetti A, Repetto F, Gianelle V, Schwartz J. Air pollution and cause-specific mortality in Milan, Italy, 1980-1989. Arch Environ Health. 1999;54(3):158–64.
7. Schwartz J. Air pollution and daily mortality: a review and meta analysis. Environ Res. 1994;64(1):36–52.
8. Schwartz J. Air pollution and hospital admissions for heart disease in eight U.S. counties. Epidemiology. 1999;10(1):17–22.
9. Analitis A, Katsouyanni K, Dimakopoulou K, Samoli E, Nikoloulopoulos AK, Petasakis Y, Touloumi G, Schwartz J, Anderson HR, Cambra K, Forastiere F, Zmirou D, Vonk JM, Clancy L, Kriz B, Bobvos J, Pekkanen J. Short-term effects of ambient particles on cardiovascular and respiratory mortality. Epidemiology. 2006;17(2):230–3.
10. Beelen R, Raaschou-Nielsen O, Stafoggia M, Andersen ZJ, Weinmayr G, Hoffmann B, Wolf K, Samoli E, Fischer P, Nieuwenhuijsen M, Vineis P, Xun WW, Katsouyanni K, Dimakopoulou K, Oudin A, Forsberg B, Modig L, Havulinna AS, Lanki T, Turunen A, Oftedal B, Nystad W, Nafstad P, de Faire U, Pedersen NL, Ostenson CG, Fratiglioni L, Penell J, Korek M, Pershagen G, Eriksen KT, Overvad K, Ellermann T, Eeftens M, Peeters PH, Meliefste K, Wang M, Bueno-de-Mesquita B, Sugiri D, Kramer U, Heinrich J, de Hoogh K, Key T, Peters A, Hampel R, Concin H, Nagel G, Ineichen A, Schaffner E, Probst-Hensch N, Kunzli N, Schindler C, Schikowski T, Adam M, Phuleria H, Vilier A, Clavel-Chapelon F, Declercq C, Grioni S, Krogh V, Tsai MY, Ricceri F, Sacerdote C, Galassi C, Migliore E, Ranzi A, Cesaroni G, Badaloni C, Forastiere F, Tamayo I, Amiano P, Dorronsoro M, Katsoulis M, Trichopoulou A, Brunekreef B, Hoek G. Effects of long-term exposure to air pollution on natural-cause mortality: an analysis of 22 European cohorts within the multicentre ESCAPE project. Lancet. 2014;383(9919):785–95.
11. Brook RD, Franklin B, Cascio W, Hong Y, Howard G, Lipsett M, Luepker R, Mittleman M, Samet J, Smith SC Jr, Tager I. Air pollution and cardiovascular disease: a statement for healthcare professionals from the expert panel on population and prevention science of the American Heart Association. Circulation. 2004;109(21):2655–71.
12. Brook RD, Rajagopalan S, Pope CA III, Brook JR, Bhatnagar A, Diez-Roux AV, Holguin F, Hong Y, Luepker RV, Mittleman MA, Peters A, Siscovick D, Smith SC Jr, Whitsel L, Kaufman

JD. Particulate matter air pollution and cardiovascular disease: an update to the scientific statement from the American Heart Association. Circulation. 2010;121(21):2331–78.

13. Cesaroni G, Forastiere F, Stafoggia M, Andersen ZJ, Badaloni C, Beelen R, Caracciolo B, de Faire U, Erbel R, Eriksen KT, Fratiglioni L, Galassi C, Hampel R, Heier M, Hennig F, Hilding A, Hoffmann B, Houthuijs D, Jockel KH, Korek M, Lanki T, Leander K, Magnusson PK, Migliore E, Ostenson CG, Overvad K, Pedersen NL, Penell J, Pershagen G, Pyko A, Raaschou-Nielsen O, Ranzi A, Ricceri F, Sacerdote C, Salomaa V, Swart W, Turunen AW, Vineis P, Weinmayr G, Wolf K, de Hoogh K, Hoek G, Brunekreef B, Peters A. Long term exposure to ambient air pollution and incidence of acute coronary events: prospective cohort study and meta-analysis in 11 European cohorts from the ESCAPE project. BMJ. 2014;348:f7412.
14. D'Ippoliti D, Forastiere F, Ancona C, Agabiti N, Fusco D, Michelozzi P, Perucci CA. Air pollution and myocardial infarction in Rome: a case-crossover analysis. Epidemiology. 2003;14(5):528–35.
15. Laden F, Schwartz J, Speizer FE, Dockery DW. Reduction in fine particulate air pollution and mortality: extended follow-up of the Harvard six cities study. Am J Respir Crit Care Med. 2006;173(6):667–72.
16. Miller KA, Siscovick DS, Sheppard L, Shepherd K, Sullivan JH, Anderson GL, Kaufman JD. Long-term exposure to air pollution and incidence of cardiovascular events in women. N Engl J Med. 2007;356(5):447–58.
17. Peters A, Dockery DW, Muller JE, Mittleman MA. Increased particulate air pollution and the triggering of myocardial infarction. Circulation. 2001;103(23):2810–5.
18. Pope CA III, Muhlestein JB, May HT, Renlund DG, Anderson JL, Horne BD. Ischemic heart disease events triggered by short-term exposure to fine particulate air pollution. Circulation. 2006;114(23):2443–8.
19. Adar SD, Sheppard L, Vedal S, Polak JF, Sampson PD, Diez Roux AV, Budoff M, Jacobs DR Jr, Barr RG, Watson K, Kaufman JD. Fine particulate air pollution and the progression of carotid intima-medial thickness: a prospective cohort study from the multi-ethnic study of atherosclerosis and air pollution. PLoS Med. 2013;10(4):e1001430.
20. Kaufman JD, Adar SD, Barr RG, Budoff M, Burke GL, Curl CL, Daviglus ML, Diez Roux AV, Gassett AJ, Jacobs DR Jr, Kronmal R, Larson TV, Navas-Acien A, Olives C, Sampson PD, Sheppard L, Siscovick DS, Stein JH, Szpiro AA, Watson KE. Association between air pollution and coronary artery calcification within six metropolitan areas in the USA (the multi-ethnic study of atherosclerosis and air pollution): a longitudinal cohort study. Lancet. 2016a;388(10045):696–704.
21. Kunzli N, Jerrett M, Mack WJ, Beckerman B, LaBree L, Gilliland F, Thomas D, Peters J, Hodis HN. Ambient air pollution and atherosclerosis in Los Angeles. Environ Health Perspect. 2005;113(2):201–6.
22. Hoffmann B, Moebus S, Mohlenkamp S, Stang A, Lehmann N, Dragano N, Schmermund A, Memmesheimer M, Mann K, Erbel R, Jockel KH. Residential exposure to traffic is associated with coronary atherosclerosis. Circulation. 2007;116(5):489–96.
23. Allen RW, Criqui MH, Diez Roux AV, Allison M, Shea S, Detrano R, Sheppard L, Wong ND, Stukovsky KH, Kaufman JD. Fine particulate matter air pollution, proximity to traffic, and aortic atherosclerosis. Epidemiology. 2009;20(2):254–64.
24. Gill EA, Curl CL, Adar SD, Allen RW, Auchincloss AH, O'Neill MS, Park SK, Van H, Diez Roux AV, Kaufman JD. Air pollution and cardiovascular disease in the multi-ethnic study of atherosclerosis. Prog Cardiovasc Dis. 2011;53(5):353–60.
25. Kaufman JD, Adar SD, Allen RW, Barr RG, Budoff MJ, Burke GL, Casillas AM, Cohen MA, Curl CL, Daviglus ML, Dieez Roux AV, Jacobs DR Jr, Kronmal RA, Larson TV, Liu SL, Lumley T, Navas-Acien A, O'Leary DH, Rotter JI, Sampson PD, Sheppard L, Siscovick DS, Stein JH, Szpiro AA, Tracy RP. Prospective study of particulate air pollution exposures, subclinical atherosclerosis, and clinical cardiovascular disease: the multi-ethnic study of atherosclerosis and air pollution (MESA air). Am J Epidemiol. 2012;176(9):825–37.
26. Bauer M, Moebus S, Mohlenkamp S, Dragano N, Nonnemacher M, Fuchsluger M, Kessler C, Jakobs H, Memmesheimer M, Erbel R, Jockel KH, Hoffmann B. Urban particulate matter air

pollution is associated with subclinical atherosclerosis: results from the HNR (Heinz Nixdorf recall) study. J Am Coll Cardiol. 2010;56(22):1803–8.

27. Kalsch H, Hennig F, Moebus S, Mohlenkamp S, Dragano N, Jakobs H, Memmesheimer M, Erbel R, Jockel KH, Hoffmann B. Are air pollution and traffic noise independently associated with atherosclerosis: the Heinz Nixdorf recall study. Eur Heart J. 2014;35(13):853–60.
28. Fernandez-Ruiz I. Risk factors: air pollution accelerates progression of atherosclerosis. Nat Rev Cardiol. 2016;13(7):379.
29. Franklin BA, Brook R, Arden PC III. Air pollution and cardiovascular disease. Curr Probl Cardiol. 2015;40(5):207–38.
30. Kaufman JD, Spalt EW, Curl CL, Hajat A, Jones MR, Kim SY, Vedal S, Szpiro AA, Gassett A, Sheppard L, Daviglus ML, Adar SD. Advances in understanding air pollution and CVD. Glob Heart. 2016b;11(3):343–52.
31. Meo SA, Suraya F. Effect of environmental air pollution on cardiovascular diseases. Eur Rev Med Pharmacol Sci. 2015;19(24):4890–7.
32. Shrey K, Suchit A, Deepika D, Shruti K, Vibha R. Air pollutants: the key stages in the pathway towards the development of cardiovascular disorders. Environ Toxicol Pharmacol. 2011;31(1):1–9.
33. Sun Q, Hong X, Wold LE. Cardiovascular effects of ambient particulate air pollution exposure. Circulation. 2010;121(25):2755–65.
34. Newby DE, Mannucci PM, Tell GS, Baccarelli AA, Brook RD, Donaldson K, Forastiere F, Franchini M, Franco OH, Graham I, Hoek G, Hoffmann B, Hoylaerts MF, Kunzli N, Mills N, Pekkanen J, Peters A, Piepoli MF, Rajagopalan S, Storey RF. Expert position paper on air pollution and cardiovascular disease. Eur Heart J. 2015;36(2):83–93b.
35. Suwa T, Hogg JC, Quinlan KB, Ohgami A, Vincent R, van Eeden SF. Particulate air pollution induces progression of atherosclerosis. J Am Coll Cardiol. 2002;39(6):935–42.
36. Yatera K, Hsieh J, Hogg JC, Tranfield E, Suzuki H, Shih CH, Behzad AR, Vincent R, van Eeden SF. Particulate matter air pollution exposure promotes recruitment of monocytes into atherosclerotic plaques. Am J Physiol Heart Circ Physiol. 2008;294(2):H944–53.
37. Tranfield EM, van Eeden SF, Yatera K, Hogg JC, Walker DC. Ultrastructural changes in atherosclerotic plaques following the instillation of airborne particulate matter into the lungs of rabbits. Can J Cardiol. 2010;26(7):e258–69.
38. Sun Q, Wang A, Jin X, Natanzon A, Duquaine D, Brook RD, Aguinaldo JG, Fayad ZA, Fuster V, Lippmann M, Chen LC, Rajagopalan S. Long-term air pollution exposure and acceleration of atherosclerosis and vascular inflammation in an animal model. JAMA. 2005;294(23):3003–10.
39. Araujo JA, Barajas B, Kleinman M, Wang X, Bennett BJ, Gong KW, Navab M, Harkema J, Sioutas C, Lusis AJ, Nel AE. Ambient particulate pollutants in the ultrafine range promote early atherosclerosis and systemic oxidative stress. Circ Res. 2008;102(5):589–96.
40. Miller MR, McLean SG, Duffin R, Lawal AO, Araujo JA, Shaw CA, Mills NL, Donaldson K, Newby DE, Hadoke PW. Diesel exhaust particulate increases the size and complexity of lesions in atherosclerotic mice. Part Fibre Toxicol. 2013;10:61.
41. Keebaugh AJ, Sioutas C, Pakbin P, Schauer JJ, Mendez LB, Kleinman MT. Is atherosclerotic disease associated with organic components of ambient fine particles? Sci Total Environ. 2015;533:69–75.
42. Floyd HS, Chen LC, Vallanat B, Dreher K. Fine ambient air particulate matter exposure induces molecular alterations associated with vascular disease progression within plaques of atherosclerotic susceptible mice. Inhal Toxicol. 2009;21(5):394–403.
43. Soares SR, Carvalho-Oliveira R, Ramos-Sanchez E, Catanozi S, da Silva LF, Mauad T, Gidlund M, Goto H, Garcia ML. Air pollution and antibodies against modified lipoproteins are associated with atherosclerosis and vascular remodeling in hyperlipemic mice. Atherosclerosis. 2009;207(2):368–73.
44. Rao X, Zhong J, Maiseyeu A, Gopalakrishnan B, Villamena FA, Chen LC, Harkema JR, Sun Q, Rajagopalan S. CD36-dependent 7-ketocholesterol accumulation in macrophages mediates progression of atherosclerosis in response to chronic air pollution exposure. Circ Res. 2014;115(9):770–80.

45. Lund AK, Lucero J, Harman M, Madden MC, McDonald JD, Seagrave JC, Campen MJ. The oxidized low-density lipoprotein receptor mediates vascular effects of inhaled vehicle emissions. Am J Respir Crit Care Med. 2011;184(1):82–91.
46. Yin F, Lawal A, Ricks J, Fox JR, Larson T, Navab M, Fogelman AM, Rosenfeld ME, Araujo JA. Diesel exhaust induces systemic lipid peroxidation and development of dysfunctional pro-oxidant and pro-inflammatory high-density lipoprotein. Arterioscler Thromb Vasc Biol. 2013;33(6):1153–61.
47. Chen LC, Quan C, Hwang JS, Jin X, Li Q, Zhong M, Rajagopalan S, Sun Q. Atherosclerosis lesion progression during inhalation exposure to environmental tobacco smoke: a comparison to concentrated ambient air fine particles exposure. Inhal Toxicol. 2010;22(6):449–59.
48. Campen MJ, Lund AK, Knuckles TL, Conklin DJ, Bishop B, Young D, Seilkop S, Seagrave J, Reed MD, McDonald JD. Inhaled diesel emissions alter atherosclerotic plaque composition in ApoE(−/−) mice. Toxicol Appl Pharmacol. 2010;242(3):310–7.
49. Bai Y, Sun Q. Fine particulate matter air pollution and atherosclerosis: mechanistic insights. Biochim Biophys Acta. 2016;1860(12):2863–8.
50. Campen MJ, Lund A, Rosenfeld M. Mechanisms linking traffic-related air pollution and atherosclerosis. Curr Opin Pulm Med. 2012;18(2):155–60.
51. Du Y, Xu X, Chu M, Guo Y, Wang J. Air particulate matter and cardiovascular disease: the epidemiological, biomedical and clinical evidence. J Thorac Dis. 2016;8(1):E8–E19.
52. Fujii T, Hayashi S, Hogg JC, Mukae H, Suwa T, Goto Y, Vincent R, van Eeden SF. Interaction of alveolar macrophages and airway epithelial cells following exposure to particulate matter produces mediators that stimulate the bone marrow. Am J Respir Cell Mol Biol. 2002;27(1):34–41.
53. Gurgueira SA, Lawrence J, Coull B, Murthy GG, Gonzalez-Flecha B. Rapid increases in the steady-state concentration of reactive oxygen species in the lungs and heart after particulate air pollution inhalation. Environ Health Perspect. 2002;110(8):749–55.
54. Wang G, Zhao J, Jiang R, Song W. Rat lung response to ozone and fine particulate matter (PM2.5) exposures. Environ Toxicol. 2015;30(3):343–56.
55. Ghio AJ, Cohen MD. Disruption of iron homeostasis as a mechanism of biologic effect by ambient air pollution particles. Inhal Toxicol. 2005;17(13):709–16.
56. Lodovici M, Bigagli E. Oxidative stress and air pollution exposure. J Toxicol. 2011;2011:487074.
57. Muhlfeld C, Rothen-Rutishauser B, Blank F, Vanhecke D, Ochs M, Gehr P. Interactions of nanoparticles with pulmonary structures and cellular responses. Am J Physiol Lung Cell Mol Physiol. 2008;294(5):L817–29.
58. Wang T, Chiang ET, Moreno-Vinasco L, Lang GD, Pendyala S, Samet JM, Geyh AS, Breysse PN, Chillrud SN, Natarajan V, Garcia JG. Particulate matter disrupts human lung endothelial barrier integrity via ROS- and p38 MAPK-dependent pathways. Am J Respir Cell Mol Biol. 2010;42(4):442–9.
59. Cui Y, Xie X, Jia F, He J, Li Z, Fu M, Hao H, Liu Y, Liu JZ, Cowan PJ, Zhu H, Sun Q, Liu Z. Ambient fine particulate matter induces apoptosis of endothelial progenitor cells through reactive oxygen species formation. Cell Physiol Biochem. 2015;35(1):353–63.
60. Moller P, Jacobsen NR, Folkmann JK, Danielsen PH, Mikkelsen L, Hemmingsen JG, Vesterdal LK, Forchhammer L, Wallin H, Loft S. Role of oxidative damage in toxicity of particulates. Free Radic Res. 2010;44(1):1–46.
61. Nemmar A, Hoet PH, Vanquickenborne B, Dinsdale D, Thomeer M, Hoylaerts MF, Vanbilloen H, Mortelmans L, Nemery B. Passage of inhaled particles into the blood circulation in humans. Circulation. 2002;105(4):411–4.
62. Hetland RB, Cassee FR, Lag M, Refsnes M, Dybing E, Schwarze PE. Cytokine release from alveolar macrophages exposed to ambient particulate matter: heterogeneity in relation to size, city and season. Part Fibre Toxicol. 2005;2:4.
63. Hiraiwa K, van Eeden SF. Contribution of lung macrophages to the inflammatory responses induced by exposure to air pollutants. Mediat Inflamm. 2013;2013:619523.
64. Dagher Z, Garcon G, Billet S, Verdin A, Ledoux F, Courcot D, Aboukais A, Shirali P. Role of nuclear factor-kappa B activation in the adverse effects induced by air pollution

particulate matter (PM2.5) in human epithelial lung cells (L132) in culture. J Appl Toxicol. 2007;27(3):284–90.

65. Shoenfelt J, Mitkus RJ, Zeisler R, Spatz RO, Powell J, Fenton MJ, Squibb KA, Medvedev AE. Involvement of TLR2 and TLR4 in inflammatory immune responses induced by fine and coarse ambient air particulate matter. J Leukoc Biol. 2009;86(2):303–12.
66. Kampfrath T, Maiseyeu A, Ying Z, Shah Z, Deiuliis JA, Xu X, Kherada N, Brook RD, Reddy KM, Padture NP, Parthasarathy S, Chen LC, Moffatt-Bruce S, Sun Q, Morawietz H, Rajagopalan S. Chronic fine particulate matter exposure induces systemic vascular dysfunction via NADPH oxidase and TLR4 pathways. Circ Res. 2011;108(6):716–26.
67. Araujo JA, Nel AE. Particulate matter and atherosclerosis: role of particle size, composition and oxidative stress. Part Fibre Toxicol. 2009;6:24.
68. Miller MR. The role of oxidative stress in the cardiovascular actions of particulate air pollution. Biochem Soc Trans. 2014;42(4):1006–11.
69. Hardwick SJ, Hegyi L, Clare K, Law NS, Carpenter KL, Mitchinson MJ, Skepper JN. Apoptosis in human monocyte-macrophages exposed to oxidized low density lipoprotein. J Pathol. 1996;179(3):294–302.
70. Moller P, Mikkelsen L, Vesterdal LK, Folkmann JK, Forchhammer L, Roursgaard M, Danielsen PH, Loft S. Hazard identification of particulate matter on vasomotor dysfunction and progression of atherosclerosis. Crit Rev Toxicol. 2011;41(4):339–68.

Air Pollution and Lung Diseases

18

Maurizio Cipolla, Mario Sorgenti, Carlo Gentile, and Marwa Maria Bishara

18.1 Introduction

During the twentieth century, major innovations made by man, in the industrial, agricultural, technological, and transport fields, have led to the appearance of numerous atmospheric, climatic, and biophysical changes. Air pollutants can be defined as a heterogeneous group of substances able to alter the natural chemical composition of air, leading to an impact on human health and the environment. According to recent estimates of the World Health Organization (WHO), more than 90% of European citizens are exposed to annual levels of contaminants (higher than those indicated by WHO in the Europe report of 2 May 2015 [1]), resulting in approximately 600,000 deaths per year. Air pollution may contribute to the development, concern, or exacerbation of some of the most common respiratory diseases, particularly bronchial asthma and chronic obstructive pulmonary disease (COPD). The most vulnerable population groups are represented by children and elderly people suffering from chronic diseases. Respiratory diseases, caused by pollutants, are known to prevail in industrialized areas, and their severity is correlated to time of exposure to pollution and atmospheric pollutants. Numerous studies have been conducted in pulmonology and allergology fields, with the aim of improving levels of treatment and prevention of respiratory diseases caused by pollutants. In this chapter, we will analyze the direct and indirect effects of pollution on the lungs, the major pollutants responsible, the diseases they cause, and possible treatments and prevention strategies (Table 18.1).

M. Cipolla (✉)
Division of general pathology, Public Health Authority at Catanzaro, Calabria, Italy
e-mail: cipolla.maurizio@virgilio.it

M. Sorgenti
Division of Pneumology, Public Health Authority at Catanzaro, Calabria, Italy

C. Gentile • M.M. Bishara
"Magna Graecia" University of Catanzaro, Calabria, Italy

F. Capello, A.V. Gaddi (eds.), *Clinical Handbook of Air Pollution-Related Diseases*,
https://doi.org/10.1007/978-3-319-62731-1_18

Table 18.1 Exposition limit value, averaging time, and % population over exposed in urban area to many common pollutants

	WHO exposition limit values (μg/m^3)	Averaging period	% population exposed beyond limits
PM_{25}	50	1 day	96
PM_{10}	25	1 day	88
O_3	100	8 h	98
NO_2	200	1 h	94
SO_2	20	1 day	46

18.2 Direct Effect of Air Pollutants on the Respiratory System

In the early 1970s, it was noted that air pollution can cause adverse effects on human health. Thanks to developments in medical research and application of new technologies, which have enabled more sensitive and accurate measurements; it was possible to identify air pollutants and determinate their impact on human health. Numerous epidemiological studies have demonstrated the significance, as well as the direct and indirect impact of air pollution on the lungs. The presence of solids, liquids, and gases, in the air we breathe, is a constant irritant, which could result in short-term and long-term effects on the respiratory system, some of which are not yet fully known.

18.2.1 Indoor and Outdoor Pollution

Contaminants that damage the respiratory system are present both "outdoors," particularly in cities with a high rate of motor vehicle traffic, and "indoors," meaning inside buildings or housing. Main outdoor pollution sources are represented by the burning of coal and petroleum products from traffic, heating and cooling, and industrial activities. Several studies [2–4] have demonstrated the correlation between exposure to outdoor air pollution and the functional decline of respiratory system among adults [5]. At the same time, it was also shown that a reduced exposure to pollutants can partly restore lung function. The main sources of indoor pollution are represented by people or animals, cigarette smoke, dust (receptacle of microorganisms), structures, building materials, furniture, equipment (air conditioners, humidifiers, plumbing), consumer products, household activities, and outdoor air.

18.2.2 Chemical Composition and Sources of Gaseous Pollutants

To better define pollution's impact on the respiratory system, the different contamination should be divided into solids, liquids, and gaseous fuels based on their chemical composition.

Gaseous pollutants, which are known to have a greater impact on the health of the respiratory tract, are nitrogen and sulfur oxides, ozone, carbon monoxide, benzene, and polycyclic hydrocarbons (3–4 benzopyrene, benzoanthracene).

These substances are able to damage the bronchial and pulmonary mucosa, causing numerous respiratory diseases. The harmful effects of nitrogen oxides on the respiratory system can occur either acutely (e.g., inflammation and irritation of the mucous membranes of the upper and lower respiratory tract: decreased respiratory function and pulmonary edema) or chronically (pulmonary changes at cellular level and increased susceptibility to bacteria). Nitrogen monoxide (NO) not only causes inflammation of the upper and lower airways and pulmonary edema but also possesses an eye irritant that can lead to hematomas, resulting in formation of methemoglobinemia. In severe cases, exposure to this pollutant can also lead to death. Following its emission, nitrogen oxide gradually oxidizes to nitrogen dioxide (NO2), which is about four times more toxic than NO. To date, the biochemical mechanisms through which the NO2 induces its toxic effects have not been fully elucidated.

However, it is known that the oxidation of NO causes severe damage to cell membranes, as a result of protein and lipid. Acute effects include inflammation of the mucous membranes, decreased lung function, and pulmonary edema. The long-term effects include increased incidences of respiratory diseases, lung disorders at cellular and tissue level, and increased susceptibility to bacterial and viral lung infections. The group most at risk is people with asthma and children. The main sources of NO are related to industrial civilization and involve combustion processes, such as the use of vehicles with diesel engines, gasoline, or LPG and the production of heat and electricity. In the presence of other pollutants, for example, hydrocarbons, ozone, and other free radicals produced by photo dissociation reactions, they can trigger a complex of chemical reactions that lead to the formation of photochemical smog. Also sulfur oxides, better known as sulfur dioxide (SO_2) and sulfur trioxide (SO_3), are hazardous air pollutants. In particular, sulfur dioxide is extremely soluble in water and is easily absorbed by the mucous membranes of the nose and respiratory tract, causing a strong irritating effect. Prolonged exposure to even very low concentrations (a few parts per billion (ppb)) can cause pharyngitis, fatigue, and sensory tract disorders (eyes, nose, etc.) or, more rarely, tracheitis, bronchial spasms, difficulty in breathing, asthma, chronic bronchitis, and emphysema.

This phenomenon allows SO_2, carried by the particles, to reach deeper into the lungs, with serious medium- and long-term effects. The main source of these pollutants is fossil fuels (coal and petroleum derivatives), in which sulfur of a high impurity is present. Atmospheric concentrations are greater in winter, owing to the addition of home heating. A well-known element, ozone (O_3), also damages the respiratory system, acting at intracellular level and oxidizing the sulfhydryl groups in enzymes, coenzymes, proteins, and unsaturated fatty acids. In this way, the O_3 causes damage to the membranes of cellular organisms. The main target is the respiratory system, in which the main damage is borne by the macrophages on the walls of the small pulmonary arteries. Acute effects include dryness of the throat and nose, increased mucus production, cough, pharyngitis, bronchitis, decreased lung function, chest pain, decreased lung bactericidal capacity, eye irritation, and headaches. The consequences of chronic exposure are fibrosis, teratogenic effects, and effects on the parathyroid and on the reproductive organs. The role of O3 in the etiology of lung tumors has not yet been completely clarified: it is a typical secondary pollutant, which is formed in the atmosphere, as a result of photochemical reactions in load of precursor pollutants produced by combustion processes, such as NO_2, NO_3, hydrocarbons, and aldehydes. Air pollution contributes to the increase of ozone's concentration in the air. Predicted environmental concentrations tend to increase during the summer, as heat favors the production of such gases. For the same reason, during the day, levels are lower in the morning, and when the photochemical process begins, it reaches a peak in the early afternoon and then falls gradually toward the evening, owing to the decrease of solar radiation. Carbon monoxide is another pollutant whose negative effects are well known. Its ability to compete with oxygen in the bond with hemoglobin in blood, for which it has higher affinity, forms carboxyhemoglobin (COHb). In this way, what occurs is a reduction in the ability of the blood to carry oxygen; consequently, the amount that is released in the tissues is less. The COHb is 75 times more stable than oxyhemoglobin, and it was shown that the concentration present in the blood is related to the amount of CO present in inhaled air [6]. The damage caused by COHb to human health occurs mainly at the expense of the cardiovascular and nervous systems, and the most sensitive individuals are anemic patients and pregnant women. However, it cannot rule out a direct or indirect involvement of the respiratory tract. The main sources of carbon monoxide production are exhaust fumes, especially gasoline, discharge from industries, involved in the treatment and disposal of waste, and oil refineries and smelters. Among aromatic pollutants, the most important structure is the benzene ring (C_6H_6), which acts primarily on the central nervous system and cardiovascular system. In fact, in cases of acute poisoning, following an initial period of agitation and dizzy spells, the main symptoms are shortness of breath, wheezing, and chest tightness, accompanied by physical weakness and headaches. Higher concentrations of aromatic pollutants can lead to symptoms such as excitement, euphoria, and exhilaration, followed by fatigue, drowsiness, and respiratory arrest. Among the long-term effects, it is important to highlight the relationship between exposure to this agent and the development of hemopoietic tumors and lung cancer. Motor vehicles represent the main sources of benzene emissions, i.e., about 85% is released into the atmosphere through exhaust gases, while the remaining 15% gets released via fuel evaporation during refueling operations. Airborne concentrations increase during the winter period, where stagnation conditions of air pollutants are more frequent. Other harmful aromatic compounds, affecting the respiratory system, are the polycyclic aromatic hydrocarbons (PAHs), which, besides irritating the nose, throat, and eyes, are recognized for their mutagenic and neoplastic properties. PAHs are considered to be the most carcinogenic of all airborne pollutants, including benzopyrene (BaP). PAHs have been included in group 1 of the IARC classification: they are present everywhere in the atmosphere, are produced by the incomplete combustion of organic material, and are derived from the use of oil fuels, gas, coal, and wood in the production of energy. Emissions are more common in winter, as it is favored by the stagnation of air pollutants. The most important source are motor vehicles, followed by thermal plants, thermal power plants, and incinerators. Polycyclic aromatic hydrocarbons are most often associated with airborne dust. In this case, the size of the airborne particles is the main parameter which influences entry and deposition in the respiratory system and, therefore, relative toxicity. Present in the urban cycle, they are generally associated with particles having an aerodynamic diameter of 2 microns and, therefore, easily reach the alveolar regions of the lungs, hence the blood and then body tissue. Finally, it is important to mention the presence of Radon (Rn), a dangerous inert gas, which could be an important predisposing factor in the development of lung cancer, second only to cigarette smoking. The main source of this gas is the terrain (other sources, of lesser extent, may be construction materials, especially if volcanic such as tuff or granites and water), from which it is dispersed into the environment, accumulating in enclosed areas, where it becomes dangerous.

18.2.3 Chemical Composition and Production Sources of Solid and Liquid Pollutants

As for solid and liquid pollutants, total suspended particles are of great importance. This term encompasses a heterogeneous group of micro-solid and micro-liquid particles released into the atmosphere. Suspended particles, of less than 10 μm in diameter (PM10), are of particular danger to the respiratory tract and the cardiovascular system. To identify and quantify the effects that these particles play on lung health, it is very important to determine the size and chemical composition of the particles. In fact, the degree of penetration into the respiratory tract is inversely proportional to the diameter of such particles and is also influenced by their ability to react with other pollutants, such as PAHs, heavy metals, and SO2. Obviously symptoms are closely related to the area of deposition: particles deposited in the upper airways (nasal cavity, pharynx, and larynx) mainly cause local irritation, such as dryness and inflammation.

Particles deposited in the bronchial trachea tract may cause constriction, decreased respiratory capacity, exacerbation of chronic respiratory diseases (asthma, bronchitis, and emphysema), and possibly cancer. Particles, with a diameter less than 5.6 μm, can be deposited in the respiratory bronchioli, while those with a diameter of less than 1.1 μm could even reach the pulmonary alveoli. This can cause inflammation, fibrosis, interstitial spaces, or even neoplasms. The fine particles can also indirectly induce systemic effects on specific target organs, following the release of pollutants in biological fluids conveyed by it. The most sensitive groups are patients with asthma and chronic bronchitis. Synergistic effects were also discovered when combining the exposure of suspended particles with SO_2. The main sources of such particles are emission from industrial plants and motor vehicles. Sources of heat are found to contribute to a lesser extent. We must specify that a fraction of the particles present in the atmosphere are of secondary origin or a result

Table 18.2 Correlations behind size of the MPS, areas of deposition, and symptoms

Dimensions (μm)	Areas of deposition	Symptoms
9–5.8	Nose	Dryness and rhinitis
5.8–4.7	Pharynx	Dryness and pharyngitis
4.7–3.3	Trachea and primary bronchioli	Bronchoconstriction, bronchitis, asthma, pulmonary emphysema, pulmonary fibrosis, interstitial, lung cancer
3.3–2.1	Secondary bronchioli	
2.1–1.1	Final bronchioli	
1.1–0.65	Alveoli	
0.65–0.43	Alveoli	

of chemical reactions that start from primary gaseous pollutants (emitted by the atmosphere itself) and secondary gaseous pollutants (as a result of human activities). These dust particles are more common in highly urbanized areas, but even rural ones are not negligible. The concentration is higher in winter. In Table 18.2, the diameter of microparticles is associated with the capacity of penetration and diseases caused.

18.2.4 Chemical Composition and Production Sources of Metal Pollutants

The most toxic metals for lung function are aluminum (Al), cadmium (Cd), nickel (Ni), mercury (Me), lead (Pb), chromium (Cr), arsenic (As), and asbestos. These pollutants, in addition to developing specific respiratory diseases, can lead to lung cancer. Trace elements, such as arsenic, cadmium, and nickel, often exist as air pollutants.

As a result of industrial emission, airborne concentrations of some metals, in urban and industrial areas, can reach values of 10–100 times higher than those of rural areas. The exposure to trace elements, associated with multiple effects on the health of the lungs, is still not entirely known. Responsible sources, for the natural increase in the amounts of metals, are mainly mining, smelters and refineries, energy production, waste incineration, and farming. Heavy metals are present in the atmosphere in the form of airborne particles; the size of the particles, to which they are associated and their chemical composition, strongly depends on the type of emission source. In winter, stagnation conditions of air pollutants are more frequent. In particular, people subjected to the constant inhalation of air containing cadmium, beryllium, or those metals happy alloys are at higher risk of developing lung cancer. People exposed to nickel and chromium are not only at risk of developing lung cancer but are also at risk of developing neoplasms of nasal and paranasal sinuses. Finally, those subjected to arsenic, in addition to developing cancers of the lung, can be subject to skin cancer. Lead (Pb) is a highly toxic element, found in traces in the air. Not only it may affect the respiratory system, but all the other organs are potential targets. Effects are extremely varied and include anemia and damage to the central and peripheral nervous system, kidneys, reproductive system, and cardiovascular, hepatic, endocrine, gastrointestinal, and immune system. Lead (Pb) absorbed through the lung epithelium enters the bloodstream and is distributed in decreasing amounts in the bones, liver, kidneys, muscles, and brain. Acute intoxication is rare and only occurs after ingestion or inhalation of large quantities of Pb. Lead's toxicity could be partly explained by binding to sulfhydryl groups of proteins or replacing essential metal ions that interfere with various enzyme systems. Groups most at risk are children and pregnant women. The main source of air pollution is vehicle exhaust fumes during winter.

18.3 Indirect Effects of Pollution to the Respiratory System

18.3.1 Atopical Increase in Lung Disease

Environmental changes that have occurred over the last century have had a significant impact on the presence of pollen and onset appearances of respiratory allergic diseases. Climate change has had an impact on the presence and variety of vegetations and, subsequently, the spread of pollen, both qualitatively and quantitatively. In particular, over the last 40 years, there has been a lengthening of the flowering seasons in the Northern Hemisphere at higher latitudes, which was not common before.

There has also been an increase in flora and fauna migrating toward the poles and higher altitudes. In addition to the damaging oxidative pollution, "stressed" woody plants have been stimulated to produce pollen, containing a higher concentration of allergenic proteins "pan-allergens," widely found in plant and identified as "pathogenesis-related proteins (PRP)." This phenomenon is a direct result of premature, prolonged exposure to pollen and fungal spores, resulting in an increased sensitization.

18.3.2 The Spread of Neophyte Plants and the Appearance of New Allergens

Over the last 500 years, approximately 12,000 plant species, especially decorative garden plants, have been imported to Europe, from every corner of the globe. Most of these plants have remained in commercial and private gardens, while 300 species have, over time, spread to other habitats (the so-called neophytes). About a dozen (the so-called invasive plants) of them prevents the growth of other plants and have thereby become a serious problem. The infestation of weeds is partly man-made but in part also depends on weather changes. An early example of an invasive plant is the *Robinia pseudoacacia*, also known as acacia. It settles among all other native tree species, causing them to disappear, along with many herbaceous and shrub species. Since this is a species of legumes, it may play an active role in allergen sensitization. Another major plant pest is the *Ailanthus*, the "Tree of Heaven," kind of

Simaroubaceae, whose leaves give off an unpleasant smell. It is able to colonize bare or semi-fertile land, often found along railroad tracks. The pollen of the *Ailanthus* can also produce allergic reactions. In addition there is also the *Buddleja davidii*, a type of Scrophulariaceae, which is a shrub native to the Far East, known for its large, purple, florescent cone. Another plant species also able to give allergic reactions is the American pokeweed, of the Phytolaccaceae family, curiously called "Turkish grape," native to the American continent, which invades all uncultivated areas and heaps of rubble. The *Solidago canadensis* and *S. gigantea*, of the Compositae family, are also potentially allergenic. These plants, with their thick, florescent yellow coloring, are used for decoration purposes. The Jerusalem artichoke (*Helianthus tuberosus*), imported for food as edible tubers, belongs to the composite family that can also create an allergic reaction. The *Carpobrotus acinaciformis* or "Hottentot figs," a type of Aizoaceae, is a succulent plant and weed, which grows in the sand or on rocks and blooms from April to May. Other species include *Ambrosia*, an inconspicuous herb that flowers in the fields and along the paths and creates allergic reactions or even severe forms of asthma in patients over 60. Pollen generation is highest between the second half of August and the first half of September.

Alterations of local flora, caused by numerous invasive plants, are not easy to remedy. The best solution would be to find a natural predator or parasite for each plant type; the effects of introducing another organism, however, are often unpredictable and potentially harmful.

18.3.3 Climate Changes and Increased Infectious Diseases

Climatic changes caused by air pollution have required amendments in the management of infectious diseases. It has in fact changed the incidence and severity of specific climate-related infectious diseases. The impact on air temperature, caused by pollution, makes for easier transmission of infectious diseases such as Legionnaires' disease, tuberculosis, malaria, and yellow fever. Climatic variations and changes in humidity are increasing incidence of bacterial infections not transmitted via human to human but through the inhalation of the aerosol flows and/or contaminated water, such as Legionnaires' disease. These diseases are common in air-conditioned areas, humidifiers, cooling towers, etc. *Legionella* bacteria, in particular, are able to penetrate and adhere to the walls of the respiratory tract. The mucociliary clearance may be reduced as a result of exposure to air pollutants. Once they have penetrated, these bacteria tend to induce the activation of ineffective, cell-mediated immunity. This results in bacterial proliferation and subsequently cellular breakdown. The infection that follows may manifest in the form of pneumonia. Macroscopic abscesses are frequently observed on the central necrosis. Most common clinical manifestations include symptoms such as high fever, headaches, myalgia, cough, hemoptysis, and chills or less frequently gastrointestinal symptoms such as abdominal pains, nausea, vomiting, and diarrhea.

18.4 The Effects of Pollution on the Respiratory System

18.4.1 Main Diseases Caused by Pollution

Epidemiological evidence, both clinical and experimental, demonstrated a correlation between the current levels of air pollution, both indoors and outdoors, and the development of diseases and respiratory symptoms. The respiratory system is the first line of defense against the onset and progression of diseases caused by air pollutants. The intensity and level of damage depend on the amount of pollutants inhaled and their level of cellular deposition. Short-term effects of outdoor air pollution include changes in the level of lung function, respiratory symptoms, and mortality from respiratory causes. Long-term effects, on the other hand, include damage to lung growth, chronic obstructive pulmonary disease (COPD), lung cancer, and possible development of asthma and allergies [7]. In addition to lung disease, directly caused by environmental pollutants, an increase of some infectious diseases, related to the impact of air pollution on climate, flora, and fauna, can also be detected.

18.4.2 Effects of Air Pollution on Health

Table 18.3 summarizes the main effects of air pollution on society. It is evident that these effects can vary, depending on nature and amount of contaminant, period of exposure, and individual's susceptibility. Symptoms shift from mere annoyance that cease with time, without the use of therapies, to hospitalization or, in severe cases, death of the person exposed to pollutants.

18.4.3 Clinical Symptoms

The effects of air pollution on the respiratory system vary according to the nature of inhaled substances, concentration in the air, and amount entered into the lungs. Acute respiratory symptoms due to high concentration of air pollution may appear shortly after exposure. The symptoms may include irritation of the airways, wheezing, coughing, chest tightness, and asthma attacks. During these attacks, which can

Table 18.3 Effects of air pollution on society

Effects of air pollution on society
Nuisance manifestation
Development of symptoms
Medicinal use
Limitation of activities
Prescribed specialist consultations
Admissions to emergency room
Hospital admission
Death

be sudden or gradual, one may witness a deterioration of lung function in patients with pre-existing lung disease. Chronic effects may occur following exposure to low or moderate levels of pollution for long periods or repeated exposure to high levels of pollution. The symptoms include coughing and chronic bronchitis, increased incidences of chronic obstructive pulmonary disease, increased incidences of lung tumors, and ultimately increased mortality from cardiopulmonary causes.

18.4.4 Lung Damage

The respiratory system possesses important defensive mechanisms to counter mechanical and chemical effects of inhaled particles. Some pollutants have an irritant effect on the respiratory system. In particular, cough receptors found in the airways trigger the reflex to eject it when stimulated by a foreign body. The polluting particles are a stimulus to these receptors, causing the onset of cough. A second mechanism is the one related to inflammations generated by oxidative stress, involving epithelial cells and alveolar macrophages. This toxic stress may cause degeneration of the epithelial cells with various effects. On the one hand, the excretion of excess mucus and other secretions becomes more difficult in the presence of pollution. On the other hand, the release of vasoconstrictor and vasodilator substances causes changes to vascular permeability, resulting in edema. Oxidative stress plays a central role in the mechanisms by which air pollutants are damaging to health. For example, microparticles contain transition metal atoms that may mediate chemical reactions, resulting in the synthesis of free radicals and oxidizing substances. The respiratory system is equipped with defenses to counteract oxidative stress, such as vitamins C, A, and E and the tripeptide glutathione. When subjected to excessive amounts of air pollutants penetrating into the lungs, there is a reduction of antioxidant defenses and a high prevalence of free radicals that damage lung cells. Even nitric oxide undergoes changes, as a result of air pollution exposure. This substance is an endogenous regulator that gets produced through the action of enzyme systems called nitric oxide synthase (NOS) and endothelial, epithelial, and some inflammatory cells. Pulmonary diseases, enhanced with nitric oxide (FENO), are an indicator of airways inflammation. The measurement of FENO distinguishes asthmatic patients from non-asthmatics.

Human macrophages play a role in the functional changes brought about to lungs damaged by air pollution. These immune system cells, when exposed to inhaled pollution particles inside the lungs, release a series of cytokines, including tumor necrosis factor α, interleukins 6 and 1β of the inflammatory protein 1α macrophages, granulocyte-colony-stimulating factor, and other mediators, leading to the activation of nuclear factor kB or activator protein 1.The result of this signaling is a massive migration of neutrophils in the lungs and the recruitment and activation of T lymphocytes [8] which damages the epithelium of the pulmonary alveoli, resulting in the manifestation of other diseases.

18.4.5 Investigations

The diagnosis of lung diseases following exposure to air pollution is based mostly on direct or significant medical history and physical examination, followed by functional diagnosis through basic spirometry and after administration of a single dose of bronchodilator or after a short asthma therapy cycle. Such measurements are noninvasive tests that evaluate the presence or absence of obstructive alteration compatible with bronchial asthma framework, or other diseases, such as chronic obstructive pulmonary disease or other lung diseases, and assess the extent and their reversibility. To run the test using a spirometer, the patient is advised to avoid taking bronchodilators, 8–12 hours prior to testing. One of the most important values tasted is the FEV1 (forced expiratory volume in 1 s). This parameter is reduced following exposure to large amounts of air pollutants. The spirometric values will reflect the modifications of the respiratory volumes, recorded by the spirometer, related to lung damage induced by the contaminants. Of course in the event of fibrosis, a restrictive framework will occur, with a reduction in the forced vital capacity (FVC) and FEV a Tiffeneau-Pinelli index (FEV/FVC) unaltered. In the case of COPD instead, there will be an obstructive framework, with decreased FVC, FEV, and FVC. A patient with acute asthma may show reduced volumes as in COPD; however, via pharmacodynamic tests, it will be possible to demonstrate the reversibility of bronchial obstruction. This test is carried out by running a spirometry baseline on a patient, before and after administration of the bronchodilator. There is also a second option, the bronchial provocation test with meta-choline (TBPM), using methacholine (a muscarinic antagonist), to evaluate the bronchial hyper reactivity and then the possible presence of asthma in remission. Patients, chronically exposed to air pollution, such as PM10 particles, have shown a reduction in vital lung capacity during performing spirometry tests. Chest radiography is a useful examination and very common in lung diseases. It involves the use of direct X-ray chest, made in two projections (anteroposterior and lateral side). In the event of pneumonia, caused by legionella, it is possible to identify, but not specify, the presence of pulmonary infiltrates.

The high-resolution CT scan of the chest (HRCT) allows you to evaluate the anatomical and functional alterations of various diseases. The test is highly sensitive and specific and can be especially useful in the diagnosis and staging of cancers of the respiratory tract. It is also frequently used in the analysis of bronchus obstructive pulmonary diseases. The percutaneous oximetry apparatus is a noninvasive examination, useful for assessing cardiorespiratory functions, able to detect the percentage (%) oxyhemoglobin saturation. The results provide a measurement of the amount of dissolved oxygen present in the arterial blood. In this way it is possible not only to determine the level of available oxygen in a patient, at a certain time, but also to update the data in real time. The arterial blood gas analysis is more invasive and enables the assessment of the individual's lung function, acid-base balance, and electrolyte content in the blood. The sputum test is a very important test in evaluating respiratory infections. It allows one to make an etiological agent responsible for diagnosing infections. Identifying the elements responsible for the main allergies requires a blood or dermal test checking for signs of allergies. The PEF (peak expiratory flow) is a more specific test used in the field of pneumology in assessing adverse effects caused by air pollution. This test evaluates the maximum speed at which one is able to expel air from the lungs. Tests for carbon monoxide diffusion (DLCO), which study the pulmonary gas exchange measuring the alveolar-capillary diffusion of carbon monoxide, reveal the presence of various lung diseases and in particular those characterized by interstitial. In the end, eNO test (exhaled nitric oxide) can help in diagnosis and treatment of asthma. It measures the level of nitric oxide gas in an exhaled sample of your breath and is helpful in distinction to asthmatic patients from non-asthmatics.

18.5 Prevention of Diseases Caused by Air Pollution

18.5.1 Behavioral Norms Useful in Minimizing Exposure to Pollutants

Considering the short- and long-term effects of exposure to several airborne pollutants, it is advisable to adopt behavioral patterns that could diminish the inhalation of such pollutants. Some ways to curb respiratory tract infections could be avoiding long-term exposure to air conditioning in polluted areas, practicing sports in green areas, and of course abstaining from smoking. For those who live on the ground floor in polluted areas, it would be appropriate to install airtight windows and arrange, where possible, windows in covered courtyards, rather than on the road. Needless to say it would be useful to open the windows that face the street only at times when traffic is less. Other measures include avoid placing the jets of air conditioners and ventilation systems on streets carrying heavy traffic. Recirculate the air inside ones car when driving in traffic or through tunnels. Equip one's home and appliances with thermostats, in order to maintain a constant, comfortable ambient temperature, thus minimizing unnecessary energy losses. Finally, use appropriately graduated face masks for adequate protection against various types of airborne pollutants.

Something that requires underscoring is that the more sensitive groups such as children, the elderly, people with chronic respiratory diseases, and pregnant women need particular care.

18.5.2 Allergen Monitoring Systems for the Cure and Prevention of New Allergies

Undoubtedly the relationship between climate change, allergens, and allergic diseases requires further study. However, in order to improve the prevention and treatment of these diseases, it has become necessary to monitor the activities of allergens in the atmosphere. Improved forecasting of pollen and fungal spores is paramount, in order to promptly implement preventative remedies necessary in combatting diversified allergens. The scenario that lies ahead includes likely future climate changes, with significant variations in allergic and other disease types, which will need to be addressed with appropriate mitigation measures and in an adequate manner.

References

1. Air quality in Europe—EEA report No. 5/2015.
2. Andersen ZJ, Hvidberg M, Jensen SS. Chronic obstructive pulmonary disease and long-term exposure to traffic-related air pollution: a cohort study. Am J Respir Crit Care Med. 2011;183(4):455–61.
3. Forbes LJ, Kapetanakis V, Rudnicka AR, et al. Chronic exposure to outdoor air pollution and lung function in adults. Thorax. 2009;64:657–63.
4. Kan H, Heiss G, Rose KM, Whitsel E, et al. Traffic exposure and lung function in adults: the atherosclerosis risk in communities study. Thorax. 2007;62:873–9.
5. Boogaard H, Fischer PH, Janssen NA, et al. Respiratory effects of a reduction in outdoor air pollution concentrations. Epidemiology. 2013;24:753–61.
6. Cobrun RF, Foster RE, Kane PB. Comparison of the physiopatological variables that determinate carboxyhemoglobin concentration in man. J Clin Invest. 1965;44(11):1899–910.
7. Annesi-Maesano I, Dab W. Air pollution and the lung: epidemiological approach. Med Sci (Paris). 2006;22(6–7):589–94.
8. Sierra-Vargas MP, Teran LM. Air pollution: impact and prevention. Respirology. 2012;17(7):1031–8.

Eye and Pollution

19

Costantino Schiavi and Giuseppe Giannaccare

19.1 The Ocular Surface

The ocular surface system consists of the cornea, conjunctiva, lacrimal and Meibomian glands, nasolacrimal duct, and their associated tear and connective tissue matrices, as well as the eyelids and eyelashes, all integrated by continuous epithelia and interconnected nervous, endocrine, immune, and vascular systems [1]. This functional unit protects the eye from the external environment and provides for an optimal refractive surface of the cornea through the production of an efficient tear film [2].

The *cornea* has the highest dioptric power of the optical complex. It needs to be avascular in order to be transparent and receive its nutrients through diffusion from the tear film and aqueous humor. The cornea is one of the body structures most densely innervated; the innervations come from axons of the sympathetic ganglion and trigeminal ganglion. The epithelium is the external layer; underneath is Bowman's layer, stroma, the recently recognized Dua's layer, Descemet's membrane, and at the most inner level the endothelial cells [3, 4]. The corneal epithelium is a stratified, non-keratinized squamous layer. It has three types of cells: the most external type of cells are the superficial epithelial cells, in the middle are the wing cells, and located on top of the inner layer are the basal epithelial cells. Due to its histological nature, the epithelium has the primary function of providing a barrier to the cornea and to the entire eyeball. Bowman's layer is composed of thin, type I, III, V, and VI collagen microfibrils. It is not an independent membrane but a modification of the most superficial portion of the stroma of the cornea. The stroma represents the main support of the corneal structure and comprises up to 90% of its volume. This compartment is about 450 μm thick and contains nerves, stromal

C. Schiavi • G. Giannaccare (✉)
Ophthalmology Unit Head Prof. EC Campos, S. Orsola-Malpighi Teaching Hospital, University of Bologna, Bologna, Italy
e-mail: giuseppe.giannaccare@gmail.com

F. Capello, A.V. Gaddi (eds.), *Clinical Handbook of Air Pollution-Related Diseases*,
https://doi.org/10.1007/978-3-319-62731-1_19

keratocytes with different morphologies, and type I and V collagen fibers. Dua's layer is a strong acellular layer in the pre-Descemet's cornea, made of 5–8 thin lamellae of tightly packed type I, VI, and VI collagen bundles running in longitudinal, transverse, and oblique directions [5]. The Descemet's membrane represents the basal membrane of the posterior epithelium. It is formed by very thin filaments of type IV collagen, which are arranged in a very regular pattern. The endothelium is a monolayer of cells that aids in keeping the corneal transparency not only by its barrier function but also by its ionic pump function.

The *tear film* covers the ocular surface and provides major refractive power of the visual system, nutrition, lubrication, and protection [6]. It forms a thin film layer of 8 μm thick. Although typically considered as formed of three layers (namely, the external lipid layer, the central aqueous layer, and the inner mucin layer), it is now recognized that the tear film is more a lipid boundary layer with aqueous phases incorporating differing concentrations of mucins throughout. Meibomian and Moll glands produce the lipid component, mainly wax esters, triglycerides, free fatty acids, as well as neutral diesters. Lacrimal glands produce the aqueous component and goblet cells which are located in the conjunctiva, secrete the mucin, and contain membrane-associated glycoproteins. Other components of the tear film are metabolites and electrolytes. Interestingly, the proteins contained in the tear film take part in other processes, for instance, they work as antimicrobials and anti-inflammatories and also help in healing processes after trauma, as well as mechanical protection to the surface of the cornea.

19.2 Ocular Surface Disease

Factors disturbing the delicate homeostatic balance of the ocular surface system can adversely affect tear film stability and osmolarity, resulting in cellular osmotic, mechanical, and inflammatory damage with dry eye disease (DED) onset [11]. DED, in general, is the most frequent disorder in ophthalmological practice, and its prevalence varies from 10.8% to 57.1% depending on the population studied [7–10].

The updated Dry Eye WorkShop (DEWS) definition states that "dry eye is a multifactorial disease of the tears and ocular surface that results in symptoms of discomfort, visual disturbance, and tear film instability with potential damage to the ocular surface. It is accompanied by increased osmolarity of the tear film and inflammation of the ocular surface." The major classes of DED are aqueous tear deficient and evaporative. The first category refers chiefly to a failure of lacrimal gland secretion. The latter has been subdivided to distinguish those causes that are dependent on intrinsic conditions of the lids and ocular surface and those that arise from extrinsic influences, for example, from environmental, occupational, and pollution.

There are a lot of tests performed to diagnose DED and to monitor the efficacy of the therapy. The main tests are summarized briefly below:

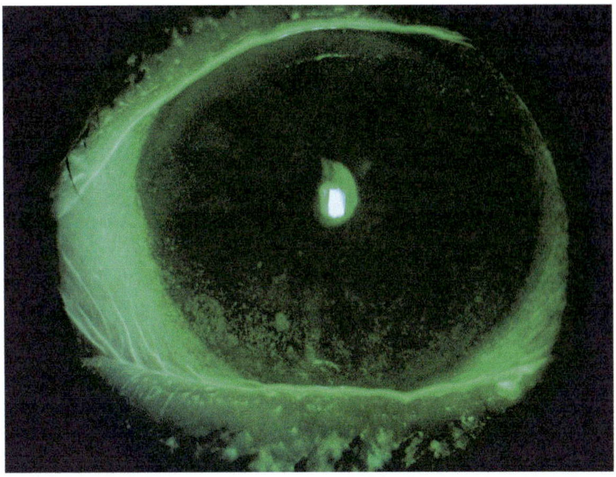

Fig. 19.1 Representative image of corneal epithelial damage. The fluorescein is deposited as green spots in the central and inferior region of the cornea. The *black spots* in the central cornea represent areas of thinning and break of the tear film

- *Schirmer test type I* is obtained by putting paper strips over the lower lid margin, midway between the middle and outer third, and by evaluating the measurement of the wet paper after 5 min. It is an estimation of tear flow production stimulated reflexly by insertion of a filter paper into the conjunctival sac.
- *Tear film breakup time (BUT)* is defined as the interval in seconds between the last complete blink and the first appearance of a dry spot or disruption in the tear film. It is an index of tear stability.
- *Ocular surface staining* is used to show the damaged areas of cornea and conjunctiva where colorants can deposit. Usually in the clinical practice, fluorescein is used to grade the staining of the cornea (Fig. 19.1) while lissamine green to grade the staining of the conjunctiva. Three systems for quantifying staining of the ocular surface are currently used, the van Bijsterveld system [12], the Oxford system [13], and a standardized version of the NEI/Industry Workshop system.
- *Tear osmolarity* is obtained by the collection of small nanoliter tear sample by a standard micropipette and then automatically transferred to a chip surface. A precise readout is obtained in seconds after the transfer.
- *Ocular Surface Disease Index (OSDI)* is the most used questionnaire about subjective symptoms employed in the clinical practice to assess the efficacy of such treatment or to grade disease severity; it consists of 12 questions about 3 major items: visual function (6 questions), ocular symptoms (3), and environmental triggers (3). It is validated in dry eye population and used as outcome measure in randomized clinical trials.

19.3 Environmental Dry Eye Disease

Human eyes are very susceptible to ill effects of air pollution. The dense innervation present at the ocular surface is extremely sensitive to the environmental agents; furthermore, the ocular mucosa is constantly exposed to the external environment protected by only a thin layer of tear film. Environmental factors are associated with ocular surface inflammatory conditions, and they cause a clinical subtype of dry eye disease that is called *environmental dry eye disease (EDED)* [14]. EDED is a single clinical entity with a unique set of symptoms and clinical findings different from the other forms of DED. The most common symptoms are sorrow eyes and visual fatigue. Clinical findings are alterations in blinking rate and tear film breakup time and corneal epithelial damage. Tear film instability, which derives from the exposure to environmental factors, compromises the corneal epithelial barrier function resulting in corneal and conjunctival epithelial chronic damage and inflammation. The individual risk factors for EDED are similar to those in other population affected by the other types of DED: aging, female sex, allergic or autoimmune conditions, and contact lens wearing. Environmental causative factors can be divided into two categories: *outdoor* risk factors, like exposure to open areas with extreme temperatures, gases, suspended particles, intense UV exposure, petrochemical industries, and urban traffic, and *indoor* environmental conditions, like closed ambient with variations in airflow, low humidity, and excessive time in front of computer and other video displays. The two categories can be often associated, for example, a desiccating environment can lead to increase in tear film evaporation, with a consequent exposure of the ocular surface to hazardous environmental elements that trigger or exacerbate EDED symptoms. Since the anterior ocular surface is the most densely innervated area of the body, it is very sensitive to irritants and adverse environmental conditions, but the cutoff limits of the most harmful pollutants and environmental risk factors to the ocular surface capable to induce EDED are still unknown. For example, airborne particle deposition velocity onto the ocular surface depends on external physical factors like turbulence, presence of electrostatic field, and temperature difference between the ocular surface and the surrounding air. The amount of particles on the ocular surface is the result of that once deposited from the air and removed by the tear film [15]. Furthermore the risk of DED is higher with the increasing values of atmospheric pressure [16].

Environmental factors cause ocular discomfort mediated by the activation of ocular surface receptors, which induces pro-inflammatory cytokines, chemoattractant expression, and elaboration of extracellular stromal matrix. These mechanisms induce alterations in tear film composition and ocular surface components, which may persist even after the individuals are no longer exposed to the related environmental factors: slower blinking rate and larger lid opening, eyelid margin alterations, corneal epithelial damage, loss of goblet cells and hyperplasia, lower mucous production, and low lysozyme levels in tears. Tear film osmolarity is an important indicator of ocular surface injury, and an increase in tear osmolarity is considered the single best marker of DED. It has been shown that a significant association between air pollution levels and tear film osmolarity exists [17]. Mucins play a critical role in ocular surface homeostasis because they provide the clearance of allergens, pathogens and debris, lubrification, and antimicrobial activity. Responses of the ocular surface to irritation include an increase in mucin levels and goblet cells hyperplasia. These changes are common in taxi drivers and traffic controllers exposed to ambient levels of particular matter (PM) and nitrogen dioxide (NO_2).

They represent an adaptive ocular surface response to ambient air pollution which causes a relative decrease in tear film osmolarity. However a chronic exposure to high levels of environmental pollution induces a long-term reduction in goblet cell count and mucin levels and consequent tear film hyperosmolarity [18].

There is a significant association between traffic-derived air pollution exposure and ocular discomfort symptoms measured by OSDI, and in these subjects, there is a significant increase in frequency of meibomitis and low BUT values [19].

19.4 Conjunctivitis

The ill effects of air pollution on human eyes are mostly irritation and inflammation, with conjunctivitis being a major clinical picture. The major signs are conjunctival hyperemia and papillary reaction of the superior and inferior tarsal conjunctiva (Fig. 19.2), while the pathognomonic symptom is itching.

A study by Chang and co-workers [20] investigated the associations between outpatient visits for nonspecific conjunctivitis and air pollution levels. This analysis found that ozone (O_3) has the greatest impact on causing conjunctivitis, followed by nitrogen dioxide (NO_2) and sulfur dioxide (SO_2). Their effects were more prominent in winter. The exact pathophysiological mechanisms linking air pollutants and conjunctivitis are not well known. However possible causes have been identified in the changes in the lacrimal pH, caused by the acidifications of tears in an atmosphere with a high oxidant power (NO_2, O_2).

Furthermore, air pollutants seem to play a key role in determining allergic reactions in ocular mucosa as already demonstrated in the lung. A study by Riediker and coauthors [21] showed that rhinoconjunctival symptoms in pollen-allergic patients seem to be strongly influenced by air pollutants represented by nitrogen oxide (NOx) and ozone (O_3) throughout the pollen season. Indeed, it showed the existence of a threshold level below which pollen or their allergens do not provoke any symptoms and that this

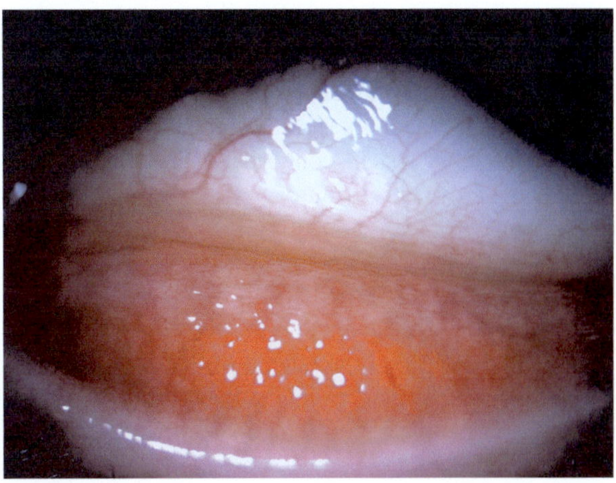

Fig. 19.2 Papillary reaction of the inferior tarsal conjunctiva

threshold seems to be lowered by increasing air pollution. Therefore, susceptibility toward allergens might be increased in areas with increased levels of air pollutants. Symptoms were related to even low levels of pollutants, suggesting that rhinoconjunctival tissue is very sensitive to irritant stimuli during an ongoing allergic inflammation.

Another study showed that outpatient visits for allergic conjunctivitis were significantly associated also with ambient particulate matter with an aerodynamic diameter of 2.5 μm or less (PM 2.5) during the non-pollen season [22]. There are several possible reasons why PM 2.5 has a stronger association with allergic conjunctivitis than other air pollutants. The first reason is the physical form in the atmosphere, since PM 2.5 are particles and the other air pollutants are gases. Conjunctivitis induced by PM 2.5 is probably due to inflammation related to friction caused by these particles on the conjunctiva. PM 2.5 may also affect tear film stability, causing breakup and thinning of the film by disrupting the tear lipid layer. Furthermore, ultrafine PM 2.5 may also increase the levels of inflammatory cytokines in tear fluid. In addition, PM 2.5 has been noted to have an adjuvant effect containing various airborne particles, such as pollutants, aerosols, pollens, bacteria, and fungi. It is likely that components of PM such as pollen and fungi are related to its adjuvant effect on allergic reactions in patients with allergic conjunctivitis.

19.5 Blepharitis

Air pollution has also proven to have deleterious impact on eyelid margin disease, and these effects were observed even though air pollutant concentration levels met the established normal ranges of air quality. In a recent study, all cases of changes in the eyelids were recorded, and clinical findings were rated on a scale from zero (normal) to two (severe alterations) [23]. These changes were then correlated to air pollution. Eyelid debris were significantly correlated with air pollutants measured, while an increase in Meibomian gland secretion was significantly correlated with NO_2 levels only. Increases of 28.8 μg/m^3 in the concentrations of particulate matter and 1.1 ppm in the concentrations of carbon monoxide (CO) were associated with increases in cases of blepharitis on the day of exposure. An increase of 51.1 μg/m^3 of NO_2 levels caused a raise in the number of cases of augmented Meibomian gland secretion at the same day of the clinical examination. There are several postulated mechanisms involved in the effects of air pollutants on the eyelid margin disease: these mechanisms involve oxidative stress on the site of particulate matter deposition, through overloading of the antioxidative defense system, conformational changes in the structure of antioxidants such as essential fatty acids by nitrogen dioxide, and chronic inflammation [24].

19.6 Cataract

Age-related cataract is another eye disease, probably influenced by outdoor and indoor air pollution. According to the World Health Organization (WHO), more than one million people become blind worldwide each year, and cataract accounts for more than 50% of these cases. A broad range of factors may influence cataract onset. These include aging, cigarette smoking, exposure to UV radiation (sunlight), ionizing radiation, microwave radiation, high temperatures, and metabolic conditions such as malnutrition, diabetes, and chronic severe diarrhea. Exposure to indoor air pollution generated by combustion of traditional biomass fuels (wood, charcoal, animal dung, and crop wastes) and coal is a significant environmental hazard, predominantly affecting populations of developing countries [25–27]. Recently, several epidemiological studies have provided evidence of an association between cooking with solid fuels and cataract development [28–30]. Wood combustion was found to be an important factor in the etiology of age-related cataract. Typical occupational cataracts were thoroughly described in ore-melting oven (steel) workers and workers exposed to organic solvents (TNT) and to ionizing radiation (rarely seen nowadays due to efficient protection). The therapy for environmental-related cataract is surgical regardless its cause because the clinical picture is not reversible.

19.7 Retinal Diseases

Environmental tobacco smoke (ETS) has been proposed to be a risk factor for *age-related macular degeneration* [31]. In vitro studies showed that benzo(e)pyrene (B(e)P), a toxic element contained in cigarette smoke, is toxic for human retinal pigment epithelial cells [32]. It causes cell death and induces apoptosis by the involvement of multiple caspase pathways.

Remky et al. [33] measured *retinal vessel* diameters using non-mydriatic fundus photography in a medical and ergonomic field study of occupational exposure to carbon disulfide (CS2). They found that chronic exposure to carbon disulfide may lead to retinal vein changes with larger caliber and retinopathy. Resch and coauthors [34] investigated the effect of inhaled CO on retinal and choroidal blood flow and found that CO inhalation caused the widening of retinal arteries and veins independent of the significant increase of carboxyhemoglobin. Retinal blood flow, subfoveal choroidal blood flow, and fundus pulsation amplitude increased significantly in response to CO inhalation. Recent studies have suggested that exposure to exhaust particles and ambient air pollution increases carotid intima-media thickness and reduces ocular blood flow velocity. Memişoğullari et al. [35] assessed the relationship between serum homocysteine, a potential parameter for atherosclerosis, and ocular blood flow velocity and resistivity index in 22 highway toll collectors and 24 control subjects using color Doppler ultrasonography. The authors showed significant correlations between serum homocysteine level and ophthalmic artery resistivity index in both highway toll collectors and controls and suggested that exposure to exhaust particles might increase serum homocysteine level, which in turn could lead to reduced ocular blood flow and increased resistivity index.

19.8 Retinoblastoma

Retinoblastoma is a malignant tumor of the retina that occurs most commonly in young children. Approximately two third of cases are diagnosed before the age of 2 years and almost the totality before the age of 5 years, suggesting that exposures occurring during the perinatal period are likely to be important to retinoblastoma development. Retinoblastoma occurs due to a loss or mutation of both alleles of the RB1 gene. In hereditary retinoblastoma, a defective allele is inherited from a parent because of a hereditary or de novo mutation in parental germline cells; most of these cases present as bilateral disease. In sporadic retinoblastoma, both alleles are inactive somatically at some point after conception, and these cases most frequently present as unilateral disease. Little is known about retinoblastoma etiology. Few studies analyzed the possible association between air pollution and retinoblastoma, but no clear relation was observed. In a recent study, the ambient exposure to specific air toxics in the perinatal period was analyzed in relation to retinoblastoma development [36]. One hundred three cases ascertained from California Cancer Registry and 30,601 controls were included in the study. Using logistic regression analyses, authors modeled the risk of retinoblastoma due to air toxic exposure, separately for exposures in pregnancy and the first year of life. Retinoblastoma risk was found to be increased with pregnancy exposure to benzene (Odds ratio [OR] = 1.67) and other toxics which primarily arise from gasoline and diesel combustion: toluene, 1,3-butadiene, ethyl benzene, ortho-xylene, and meta/para-xylene; these six toxics were high correlated. Retinoblastoma risk was also increased with pregnancy exposure to chloroform (OR = 1.35), chromium (OR = 1.29), paradichlorobenzene (OR = 1.24), nickel (OR = 1.48), and in the first year of life, acetaldehyde (OR = 1.62). Overall survival, eye salvage and preservation of vision are largely dependent on the stage of presentation, but the mortality remains still high worldwide [37].

19.9 Ultraviolet Light Damage

The human eye is constantly exposed to sunlight and artificial lighting. Light transmission through the eye is fundamental to its unique biological functions of directing vision and circadian rhythm [38]. The sun supports life on our planet, but its life-giving rays also pose dangers. The sun's primary danger is in the form of ultraviolet (UV) radiation. Artificial sources, like welding machines, tanning beds, and lasers, can also produce UV radiation. There are three types of UV radiation: UV-C is absorbed by the ozone layer and does not present any threat, while UV-A and UV-B radiation can have long- and short-term negative effects on the eyes and vision.

If the eyes are exposed to excessive amounts of UV radiation over a short period of time, *photokeratitis* can likely occur. Symptoms include pain, red eyes, foreign body sensation, gritty feeling in the eyes, extreme sensitivity to light, and excessive tearing. Fortunately, these symptoms are usually temporary and rarely cause permanent damage to the eyes. On the other hand, the longer the eyes are exposed to solar radiation, the greater is the risk of developing *cataract* or *macular degeneration* later in life. The removal of these wavelengths from ocular exposure would greatly reduce the risk of these diseases. One way this may be easily done is by wearing sunglasses that block wavelengths below 400 nm (marked 400 on the glasses).

However, because of the geometry of the eye, these glasses must be wraparound sunglasses to prevent reflective UV radiation from reaching the eye. Additional protection may be offered by contact lenses that absorb significant amounts of UV radiation.

In addition to UV radiation, short blue visible light (400–440 nm) that is part of the visible light spectrum is a risk factor for the adult human retina. This wavelength of light is not essential for sight and not necessary for a circadian rhythm response. Many digital devices emit this shorter-wavelength visible light. The sun emits blue light, as do artificial light sources, such as LEDs, computers, and smartphones. Lenses that absorb harmful blue light but allow beneficial blue light transmission are entering the marketplace. It is also possible to apply a special clear coating to traditional lenses to enhance their ability to block these harmful rays while using computers and smartphones.

To provide adequate protection for your eyes, sunglasses should:

- Block out 99 to 100 percent of both UV-A and UV-B radiation
- Screen out 75 to 90 percent of visible light
- Have lenses that are perfectly matched in color and free of distortion and imperfection
- Have lenses that are gray for proper color recognition

The bad protection from UV light provided by cheap sunglasses is a current issue. When it's really bright out, squint, and pupils' constriction to tiny dots occurs to limit the amount of light getting in. When people put on sunglasses, the effect is the same as being in darkened room, the pupil dilates to let in more light. Cheap glasses that don't block UV rays may offer some relief from visible light and reduce the need to squint, but the additional exposure to UVA and UVB due to pupil dilation can allow much more harmful light into the eyes than if the cheap sunglasses are not worn at all.

19.10 Management of Environmental Dry Eye Disease

Environmental dry eye disease chore treatment includes the same medications and interventions available for the other types of DED. In addition the golden rule in case of eye irritation is *to avoid vigorous eye rubbing* so as to prevent increased allergy and chance of contacting infection.

Few additional helpful tips include:

- Cool compress to closed eyes for 5 min twice per day
- Frequent use of tear substitutes prescribed by the ophthalmologist to dilute allergens and air pollutants
- Use of appropriate sunglasses outdoors
- Avoid direct splashing of water to open eyes
- Avoid contact lens wearing and eye makeup if eyes are feeling sore

References

1. Gipson IK. The ocular surface: the challenge to enable and protect vision: the Friedenwald lecture. Invest Ophthalmol Vis Sci. 2007;48:4391–8.
2. Aragona P, Rolando M. Towards a dynamic customised therapy for ocular surface dysfunctions. Br J Ophthalmol. 2013;97:955–60.
3. Böhnke M, Masters BR. Confocal microscopy of the cornea. Prog Retin Eye Res. 1999;18:553–628.
4. DelMonte DW, Kim T. Anatomy and physiology of the cornea. J Cataract Refract Surg. 2011;37:588–98.
5. Dua HS, Faraj LA, Said DG, et al. Human corneal anatomy redefined: a novel pre-Descemet's layer (Dua's layer). Ophthalmology. 2013;120:1778–85.
6. Holly FJ, Lemp MA. Tear physiology and dry eyes. Surv Ophthalmol. 1977;22:69–87.
7. Albietz JM. Prevalence of dry eye subtypes in clinical optometry practice. Optom Vis Sci. 2000;77:357–63.
8. Hikichi T, Yoshida A, Fukui Y, Hamano T, Ri M, Araki K, Horimoto K, Takamura E, Kitagawa K, Oyama M. Prevalence of dry eye in Japanese eye centers. Graefes Arch Clin Exp Ophthalmol. 1995;233:555–8.
9. Versura P, Cellini M, Torreggiani A, Profazio V, Bernabini B, Caramazza R. Dryness symptoms, diagnostic protocol and therapeutic management: a report on 1,200 patients. Ophthalmic Res. 2001;33:221–7.
10. Giannaccare G, Bonifazi F, Sessa M, Fresina M, Arpinati M, Bandini G, Versura P. Dry eye is already present in hematological patients before hematopoietic stem cell transplantation. Cornea. 2016;35:638–43.
11. Dry Eye WorkShop. Research in dry eye: report of the Research Subcommittee of the International Dry Eye WorkShop. Ocul Surf. 2007;5:179–93.
12. van Bijsterveld OP. Diagnostic tests in the sicca syndrome. Arch Ophthalmol. 1969;82:10–4.
13. Bron AJ, Evans VE, Smith JA. Grading of corneal and conjunctival staining in the context of other dry eye tests. Cornea. 2003;22:640–50.
14. Alves M, Novaes P, Morraye Mde A, Reinach PS, Rocha EM. Is dry eye an environmental disease? Arq Bras Oftalmol. 2014;77:193–200.
15. Schneider T, Bohgard M. Airborne particle deposition onto the ocular surface. Indoor Air. 2005;15:215–9.
16. Galor A, Kumar N, Feuer W, Lee DJ. Environmental factors affect the risk of dry eye syndrome in a United States veteran population. Ophthalmology. 2014;121:972–3.
17. Torricelli AA, Novaes P, Matsuda M, Braga A, Saldiva PH, Alves MR, Monteiro ML. Correlation between signs and symptoms of ocular surface dysfunction and tear osmolarity with ambient levels of air pollution in a large metropolitan area. Cornea. 2013;32:11–5.
18. Torricelli AA, Matsuda M, Novaes P, Braga AL, Saldiva PH, Alves MR, Monteiro ML. Effects of ambient levels of traffic-derives air pollution on the ocular surface. Environ Res. 2014;131:59–63.
19. Novaes P, Saldiva PH, Matsuda M, Macchione M, Rangel MP, Kara-José N, Berra A. The effects of chronic exposure to traffic derived air pollution on the ocular surface. Environ Res. 2010;110:372–4.
20. Chang CJ, Yang HH, Chang CA, Tsai HY. Relationship between air pollution and outpatient visits for nonspecific conjunctivitis. Invest Ophthalmol Vis Sci. 2012;53:429–33.
21. Riediker M, Monn C, Koller T, Stahel WA, Wuthrich B. Air pollutants enhance rhinoconjunctivitis symptoms in pollen-allergic individuals. Ann Allergy Asthma Immunol. 2001;87:311–8.
22. Mimura T, Ichinose T, Yamagami S, Fujishima H, Kamei Y, Goto M, Takada S, Matsubara S. Airborne particulate matter (PM2.5) and the prevalence of allergic conjunctivitis in Japan. Sci Total Environ. 2014;487:493–9.
23. Malerbi FK, Martins LC, Saldiva PH, Braga AL. Ambient levels of air pollution induce clinical worsening of blepharitis. Environ Res. 2012;112:199–203.

24. Wolkoff P. Ocular discomfort by environmental and personal risk factors altering the precorneal tear film. Toxicol Lett. 2010;199:203–12.
25. Sreenivas V, Prabhakar AK, Badrinath SS, Fernandez T, Roy IS, Sharma T, Shah B. A rural population based case-control study of senile cataract in India. J Epidemiol. 1999;9:327–36.
26. Ughade SN, Zodpey SP, Khanolkar VA. Risk factors for cataract: a case control study. Indian J Ophthalmol. 1998;46:221–7.
27. Ellegard A. Tears while cooking: an indicator of indoor air pollution and related health effects in developing countries. Environ Res. 1997;75:12–22.
28. Saha A, Kulkarni PK, Shah A, Patel M, Saiyed HN. Ocular morbility and fuel use: an experience from India. Occup Environ Med. 2005;62:66–9.
29. Shalini VK, Luthra M, Srinivas L, Rao SH, Basti S, Reddy M, Balasubramanian D. Oxidative damage to the eye lens caused by cigarette smoke and fuel smoke condensates. Indian J Biochem Biophys. 1994;31:261–6.
30. Pokhrel AK, Smith KR, Khalakdina A, Deuja A, Bates MN. Case-control study of indoor cooking smoke exposure and cataract in Nepal and India. Int J Epidemol. 2005;34:709–10.
31. Lois N, Abdelkader E, Reglitz K, Garden C, Ayres JG. Environmental tobacco smoke exposure and eye disease. Br J Ophthalmol. 2008;92:1304–10.
32. Sharma A, Neekhra A, Gramajo AL, Patil J, Chwa M, Kuppermann BD, Kenney MC. Effects of Benzo(e)Pyrene, a toxic component of cigarette smoke, on human retinal pigment epithelial cells in vitro. Invest Ophthalmol Vis Sci. 2008;49:5111–7.
33. Remky A, Dressen D, Plange N, Korinth G, Drexler H, Arend O. Retinal vessel diameters determined in an ergo- medical field study: diameter changes associated with the occupational exposure to carbon disulfide. Invest Ophthalmol Vis Sci. 2004;45:U619.
34. Resch H, Zawinka C, Weigert G, Schmetterer L, Garhöfer G. Inhaled carbon monoxide increases retinal and choroidal blood flow in healthy humans. Invest Ophthalmol Vis Sci. 2005;46:4275–80.
35. Memişoğullari R, Yüksel H, Coskun A, Yüksel HK, Yazgan O, Bilgin C. High serum homocysteine levels correlate with a decrease in the blood flow velocity of the ophthalmic artery in highway toll collectors. Tohoku J Exp Med. 2007;212:247–52.
36. Heck JE, Park AS, Qiu J, Cockburn M, Ritz B. Retinoblastoma and ambient exposure to air toxics in the perinatal period. J Expo Sci Environ Epidemiol. 2015;25:182–6.
37. Delhiwala KS, Vadakkal IP, Mulay K, Khetan V, Wick MR. Retinoblastoma: an update. Semin Diagn Pathol. 2016;33:133–40.
38. Roberts JE. Ultraviolet radiation as a risk factor for cataract and macular degeneration. Eye Contact Lens. 2011;37:246–9.

Skin and Air Pollution: State of the Art and New Frontiers 20

Alessandro Guido Cavalieri Manasse
and Antonio De Bitonto

20.1 State of the Art

Actually, effects of air pollutants on skin are generally well known and discussed [1–5].

Irritant contact dermatitis (ICD) is provoked by a direct damage of the skin caused by the irritant substances. It develops if the exposure of the substance or substances is sufficient or if the skin is particularly susceptible as is found in atopic dermatitis (AD) patients. The most common irritants found in the environment are soaps, detergents, water, solvents, and a dry atmosphere. The occurrence of irritant contact dermatitis depends upon the degree of exposure [6–8]. A strong relationship between high concentration of air pollutants (ozone, nitrogen dioxide, and fine particulate matter with a median aerodynamic diameter of less than 2.5 μm) and urticaria is confirmed by a large Canadian observational study [9, 10]. It also generally ascertained a link between climate and pollutant-induced eczema [11].

Even allergic contact dermatitis (ACD) is caused by delayed allergenic mechanisms Th1 (T helper 1) response. Common causes are nickel, fragrances, rubber additives, preservatives, plants, and medicaments.

Environmental acne, called chloracne, results from various chemical exposures, and the eruption may be mild, involving localized exposure, or covered areas of the body or severe, explosive, and disseminated with involvement of almost every follicular orifice [12]. Chloracne almost always represents a cutaneous sign of systemic exposure to highly toxic chemicals. Chloracne results from environmental

A.G. Cavalieri Manasse (✉)
R&D Department Freia Farmaceutici, Milan, Italy
e-mail: alessandro.cavalieri@freiafarmaceutici.it

A. De Bitonto
Policlinico S. Marco Zingonia, Osio Sotto, BG, Italy

F. Capello, A.V. Gaddi (eds.), *Clinical Handbook of Air Pollution-Related Diseases*,
https://doi.org/10.1007/978-3-319-62731-1_20

exposure to certain halogenated aromatic hydrocarbons, such as polyhalogenated naphthalenes, polyhalogenated biphenyls, and dibenzofurans, contaminants of polychlorophenol compounds – especially herbicides (2,4,5-T and pentachlorophenol) and herbicide intermediates (2,4,5-trichlorophenol), contaminants of 3,4-dichloroaniline and related herbicides (propanil, methazole, etc.) [13].

Chemical leukoderma is defined as pigmentation or hypopigmentation of the skin due to industrial exposure to one chemical or more chemicals known to have a destructive effect on epidermal melanocytes [14]. A connection between traffic-related soot (both volatile organic compounds and NO2) and exposure and increased lentigo formation is clear demonstrated [15–17]. The most commonly chemicals able to provoke leukoderma are hydroquinone, monobenzylether of hydroquinone, monoethyl ether of hydroquinone (*p*-ethoxyphenol), monomethyl ether of hydroquinone (*p*-methoxyphenol), *p*-cresol, *p*-isopropylcatechol, *p*-methylcatechol, *p*-nonylphenol, *p*-octylphenol, *p*-phenylphenol, *p*-tert-amylphenol, *p*-tert-butylcatechol, *p*-tert-butylphenol, *N*,*N*′,*N*″-triethylenethiophosphoramide (thio-TEPA), mercaptoamines, and physostigmine.

Scleroderma-like diseases and quartz-induced scleroderma, in addition to changes of the skin similar to those caused by scleroderma, also involve other organ systems but are not consistent with classical scleroderma [18]. Factors inducing scleroderma-like diseases are vinyl chloride, bis(4-amino-3-methylcyclohexyl) methane, quartz, bleomycin, pentazocine, L-tryptophan, and silicons.

Environmental skin cancer [19], besides effects of ionizing and UV radiations [20], is strongly associated to exposure to agents as polycyclic hydrocarbons (tar distilling, coal gas manufacturing, briquettes manufacturing, shale oil compounds, refinery exhaust compounds) [21, 22].

20.2 Edge of Knowledge

In the last two decades, scientists and clinical researchers have been committed on evaluating and protecting skin from UV damages.

Moreover, in the last years, as a result of an increasing interest and pressure from cosmetic industry, researchers' focus is shifting to evaluating effects on the skin of prolonged or repetitive exposure to high levels of air pollutants; even if, it is interesting to note that most papers are published in technical works (chemistry, environment, etc.) and not in medical journals.

Chemico-physical studies have demonstrated that volatile organic compound (VOC) dimensions – responsible of soot carbonaceous particles' formation – are on the order of nanometers, precisely in the field of hyperfine particles (<10 nm) [23] and how they can pass through skin barrier [24].

Transdermal uptake directly from air that is demonstrated could be comparable to or larger than intake via inhalation for many semivolatile VOCs [25–27].

However, today knowledge regarding the pollutants' transdermal absorption by skin is quite limited.

According to present knowledge, there are some mechanisms by which air pollutants affect skin health:

1. Skin chemical impairment
2. Production of reactive oxygen species (ROS)
3. Subsequent induction of inflammatory cascade

Epidemiological studies on the cutaneous impacts of particulate matters (PM) showed that PM affects the development and exacerbation of skin diseases [28].

Exposure to high level of PM10 and PM2.5 has been related with a squalene peroxidation [29], a lower level of vitamin E and an increase of lactic acid [30].

PM and polycyclic aromatic hydrocarbons (PAHs) enter the skin via nanoparticles and generate quinones [31], which are redox cycling chemicals that produce ROS and pro-inflammatory cytokines secretion (TNF-alpha, IL-1alpha, and IL-8) [32, 33]. Such prolonged exposure to reactive oxygen species (ROS) can be linked with a significant increase of matrix metalloproteinases (MMPs), enzymes contributing to the breakdown of collagen while inhibiting new collagen formation and increasing inflammatory skin diseases and skin aging, which includes skin hyperpigmentation [34], atopic and urticarial skins, a higher frequency of red dermographism, an important seborrheic status at the forehead level, and a lower level of dandruffs [30].

Also, ultrafine particles (UFPs) including black carbon and polycyclic aromatic hydrocarbons (PAHs) enhance the incidence of skin cancer. Mainly the water-soluble faction, as transition metals with redox potential, plays an important role in the initiation of oxidative DNA damage and membrane lipid peroxidation [35]. The link between chemical compositions and particle toxicity tends to be stronger for the fine and ultrafine PM size fractions [36, 37].

20.3 New Frontiers

Recent toxicological investigations demonstrated that VOCs are able to penetrate skin tissues [38, 39].

In order to understand the effects of new vehicles penetrating the skin, it is useful to remind how dermal absorption is a complex process [40] whereby substances are transported across the skin's biomembrane and into the body's living tissues [41–43].

Such process is strongly influenced by the skin's particular absorption characteristics. As a result of topical contact, substances which do not evaporate or are not mechanically removed may penetrate into the epidermis and subsequently reach the hypodermal vascular network via the dermis. During this process, an absorbed substance may also undergo a biotransformation [44, 45].

In addition to the skin's properties, characteristics of the substance being absorbed also influence dermal absorption, including the same substance, vehicle (for dissolved substances), occlusion, concentration, and exposure pattern. The most important substance properties affecting transdermal absorption are [46]:

1. Liposolubility
2. Molecular weight (smaller molecules penetrate more easily)
3. Dilution factor of compound or vehicle substance
4. Electron configurations and dissociation constants (the greater is the molecule's polarity and/or ionization, the less is the penetration)
5. The presence/absence of specific elements favoring/inhibiting penetration

Given the above, in our opinion, on one side, it is useful to make some assumptions, and, on the other, further work remains necessary to deepen them.

The first consideration is related to the specific pathway represented by the skin and the health consequences related to pollutants entry into the body through it.

Pollutant entrance via ingestion, actually, is partially detoxified by enzymes and acids present in the stomach, intestine, and liver before being metabolized and transferred into the blood. This detoxification process does not take place through the transdermal passage. Through this pathway further defense filters do not exist and pollutants enter intact into the blood stream: it is therefore reasonable to expect, for equal exposure, major damages to health and a larger involvement of other organs [47].

A further consideration must be made about the body area affected by the pollution skin contact. As well reported in literature, the skin on different areas of the same organism has different absorption patterns; e.g., scrotal skin is 12 times more permeable than forearm skin and forehead skin more than cheek skin [48]. In this regard, in 1775, the first cancer of any type to be linked with environmental exposure was scrotal squamous carcinomas in British chimney sweeps reported by Percivall Pott [49].

In-depth studies need to be undertaken in this topic in order to ascertain the true damage extent caused by the transcutaneous pollution passage, and empirical data should be collected in order to verify what type of clothing may constitute a defense and an adequate barrier to prevent pollution passage directly to the blood through the skin.

A further area of research should be focused on common hygiene and skincare processes when exposed to pollutants. Are commonly used detergents and cosmetics an aid to erase pollutants settled on the skin? Or do they favor a more rapid and increased absorption? If new studies should be undertaken along this route, it is possible to make some assumptions.

A first consideration can be made regarding cleansing products acting by contrast. In such products, surfactants, using electrical charges, make hydrophilic pollutants. Can these polarity changes affect the pollutants' permeability deposited onto the skin? Some studies seem to demonstrate that cleansing by affinity might be the best answer in order to eliminate pollutants without increasing absorption, just for the skin defensive capacity versus high lipophilic compounds [50].

A further consideration is related to the conveying action carried by cosmetic ingredients when linked with pollutants. Polar or nonpolar interface with skin components can affect properties of carrier substances.

Such substances also remain onto the skin for a long time, and they are developed to be more easily absorbed and penetrate into the deeper layers of the dermis. If they have the ability to link with pollutants, can they accelerate their transcutaneous absorption and ease a direct passage to the blood and lymphatic system?

New researches should investigate the full role of the pollutants transcutaneous penetration, the role of common skin cleansing and makeup may have in facilitating or preventing pollutants absorption, and the role of different tissues and clothes can play in protecting or in exposing the skin to these pollutants.

References

1. English JSC, Dawe RS, Ferguson J. Environmental effects and skin disease. Br Med Bull. 2003;68:129–42.
2. Goldsmith LA. Skin effects of air pollution. Otolaryngol Head Neck Surg. 1996;14:217–9.
3. Kampa M, Castanas E. Human health effects of air pollution. Environ Pollut. 2008;151:362–7.
4. Katsouyanni K. Ambient air pollution and health. Br Med Bull. 2003;68:143–56.
5. Dales R, Liu L, Wheeler AJ, Gilbert NL. Quality of indoor residential air and health. CMAJ. 2008;179:147–52.
6. Meding B, Swanbeck G. Occupational hand eczema in an industrial city. Contact Dermatitis. 1990;22:13–23.
7. Dotterud LK, Kvammen B, BoUe R, Balk ES. A survey of atopic diseases among schoolchildren in Sor-Varanger community. Possible effects of subarctic climate and industrial pollution from Russia. Acta Derm Venereol. 1994;74:124–8.
8. Eberlein-König B, Przybilla B, Kûhnl P, Pechak J, Gebefügi I, Kleinschmidt J, et al. Influence of airborne nitrogen dioxide or formaldehyde on parameters of skin function and cellular activation in patients with atopic eczema and control subjects. J Allergy Clin Immunol. 1998;101:141–3.
9. Kousha T, Valacchi G. The air quality health index and emergency department visits for urticaria in Windsor. Canada J Toxicol Environ Health A. 2015;78(8):524–33.
10. Bekö G, Callesen M, Weschler CJ, Toftum J, Langer S, Sigsgaard T, Høst A, Kold Jensen T, Clausen G. Phthalate exposure through different pathways and allergic sensitization in preschool children with asthma, allergic rhinoconjunctivitis and atopic dermatitis. Environ Res. 2015 Feb;137:432–9.
11. Lee YL, Su HJ, Sheu HM, Yu HS, Guo YL. Traffic-related air pollution, climate, and prevalence of eczema in Taiwanese school children. J Invest Dermatol. 2008;128:2412–20.
12. Taylor JS. Environmental chloracne: update and review. Ann N Y Acad Sci. 1979;320:295–307.
13. Tindall JP. Chloracne and chloracnegens. J Am Acad Dermatol. 1985;13:539–58.
14. Wattanakrai P, Miyamoto L, Taylor JS. Occupational pigmentary disorders. In: Kanerva L, Elsner P, Wahlberg JE, Maibach HI, editors. Handbook of occupational dermatology. Berlin: Springer; 2000. p. 280–94.
15. Vierkotter A, Schicovski T, Ranft U, Sugiri D, Matsui M, Kramer U, et al. Airobone particle exposure and extrinsic skin aging. J Invest Dermatol. 2010;130:2719–26.
16. Vierkotter TS, Ranft U, Sugiri D, Matsui M, Kramer U, J. Krutmann, MPP-1 and -3 promoter variants are indicative of a common susceptibility for skin and lung aging: results from a cohort of elderly women (SALIA). J Invest Dermatol. 2015;135:1268–74.
17. Vassoughi M, Schicovski T, Vierkotter A, Sugiri D, Hoffmann B, Teicher T, et al. Air pollution and subclinical airway inflammation in the SALIA cohort study. Immun Ageing. 2014;11:5.

18. Black CM, Welsh KI. Occupationally and environmentally induced scleroderma-like illness: etiology, pathogenesis, diagnosis, and treatment. Int Med Spec. 1988;9:135–54.
19. Baudouin C, Charveron M, Tarroux R, Gall Y. Environmental pollutants and skin cancer. Cell Biol Toxicol. 2002;18:341–8.
20. Burke KE, Wei H. Synergistic damage by UVA radiation and pollutants. Toxicol Ind Health. 2009;25:219–24.
21. Cruikshank CD, Squire JR. Skin cancer in the engineering industry from use of mineral oil. Br J Ind Med. 1950;7:1–11.
22. Epstein JH, Ormsby A, Adams RM. Occupational skin cancer. In: Adams RM, editor. Occupational skin disease. 3rd ed. Philadelphia, PA: WB Saunders Company; 1999.
23. Apicella B, et al. Size exclusion chromatography of particulate produced in fuel-rich combustion of different fuels. Energy & Fuel. 2003;17:565.
24. Gajjar RM, Miller MA, Kasting GB. Evaporation of volatile organic compounds from human skin in vitro. Ann Occup Hyg. 2013;57(7):853–65.
25. Weschler CJ, Nazaroff WW. SVOC exposure indoors: fresh look at dermal pathways. Indoor Air. 2012 Oct;22(5):356–77.
26. Weschler CJ, Nazaroff WW. Dermal uptake of organic vapors commonly found in indoor air. Environ Sci Technol. 2014;48(2):1230–7.
27. Brown HS, Bishop DR, Rowan CA. The role of skin absorption as a route of exposure for volatile organic compounds (VOCs) in drinking water. Am J Public Health. 1984;74(5):479–84.
28. Lademann J, Schaefer H, Otberg N, Teichmann A, Blume-Peytavi U, Sterry W. Penetration of microparticles into human skin. Hautarzt. 2004;55:1117–9.
29. Pham DM, et al. Oxidization of squalene, a human skin lipid: a new and reliable marker of environmental pollution studies. Int J Cosmet Sci. 2015;37:357.
30. Lefebvre MA, et al. Evaluation of the impact of urban pollution on the quality of skin: a multicentre study in Mexico. Int J Cosmet Sci. 2015;37(3):329–38.
31. Penning TM, Burcynski ME, Hung CF, McCoull KD, Palackal NT, Tsuruda LS. Dihydrodiol dehydrogenase and polycyclic aromatic hydrocarbon activation: generation of reactive and redox active o-quinones. Chem Res Toxicol. 1999;12:1–18.
32. Kim KE, et al. Air pollution and skin diseases: adverse effects of airborne particulate matter on various skin diseases. Life Sci. 2016;152:126.
33. Lefebvre MA, et al. Evaluation of the impact of urban pollution on the quality of skin: a multicentre study in Mexico. Int J Cosmet Sci. 2015;37(3):329–38.
34. Roberts WE. Pollution as a risk factor for the development of melasma and other skin disorders of facial hyperpigmentation is there a case to be made? J Drugs Dermatol. 2015;14(4):337–41.
35. Valavanidis A, Fiotakis K, Vlachogianni T. Airborne particulate matter and human health: toxicological assessment and importance of size and composition of particles for oxidative damage and carcinogenic mechanisms. J Environ Sci Health C Environ Carcinog Ecotoxicol Rev. 2008;26(4):339–62.
36. Pan TL, et al. The impact of urban particulate pollution on skin barrier function and the subsequent drug absorption. J Dermatol Sci. 2015;78(1):51–60.
37. Li M, Vierkotter A, Schicowki T, Huls A, Ding A, Matsui MS, et al. Epidemiological evidence that indoor air pollution from cooking with solid fuels accelerates skin aging in Chinese women. J Dermatol Sci. 2015;79:148–54.
38. Magnani ND, Muresan XM, Belmonte G, Cervellati F, Sticozzi C, Pecorelli A, Miracco C, Marchini T, Evelson P, Valacchi G. Skin damage mechanisms related to airborne particulate matter exposure. Toxicol Sci. 2016;149(1):227–36.
39. Wu CC, Bao LJ, Tao S, Zeng EY. Dermal uptake from airborne organics as an important route of human exposure to E-waste combustion fumes. Environ Sci Technol. 2016;50(13):6599–605.
40. Brown, R, Cofone, L, Diwan I, Mittelman A, Hoang K, Schaum J. Ability of theoretical skin permeability models to predict permeability coefficient or flux values for environmental pollutants. Presented at Principles of Route-to-Route Extrapolation Meeting; March 1990. Hilton Head, SC; 1990.

41. Pannier K, Brand RM, Jones DD. Fuzzy modeling of skin permeability coefficients. Pharm Res. 2003;20(2):143–8.
42. Katritzky R, Dobchev DA, Fara DC, Hür E, Tämm K, Kurunczi L, Karelson M, Varnek A, Solov'ev VP. Skin permeation rate as a function of chemical structure. J Med Chem. 2006;49(11):3305–14.
43. Rivas N. The impermeable facts of skin penetration and absorption. In: Personal care: information based on scientific facts. Saluda, SC: Personal Care Truth; 2011.
44. Noonan PK, Wester RC. Cutaneous metabolism of Xenobiotics. In: Bronough RL, Maibach HI, editors. Percutaneous absorption. New York, NY: Marcel Dekker; 1989. p. 53–75.
45. Kurtmann J, Liu W, Li L, Pan X, Crawford M, Sore G, et al. Pollution and skin: from epidemiological and mechanistic studies to clinical implications. J Dermatol Sci. 2014;76:163–8.
46. Schaefer H, Redelmeier TE. Skin barrier: principles of percutaneous absorption. New York, NY: S. Karger Publishers, Inc.; 1996.
47. Ushio H, Nohara K, Fujimaki H. Effect of environmental pollutants on the production of pro-inflammatory cytokines by normal human dermal keratinocytes. Toxicol Lett. 1999;105:17–24.
48. Wépierre J. L'Absorption Percutanée. Produits Pharmaceutiques. 1970;26(6):312–29.
49. Waldron HA. A brief history of scrotal cancer. Br J Ind Med. 1983;40:390–401.
50. Frederiksen M, Vorkamp K, Jensen NM, Sørensen JA, Knudsen LE, Sørensen LS, Webster TF, Nielsen JB. Dermal uptake and percutaneous penetration of ten flame retardants in a human skin ex vivo model. Chemosphere. 2016;162:308–14.

Air Pollution and Endocrine-Disrupting Chemicals

21

Annamalai Jayshree and Namasivayam Vasudevan

21.1 Air Pollution and Air Pollutants

Over the past decades, rapid industrialization and economic development have been concomitant with large migration from rural areas to cities [1]. This outlook has resulted in urban growth, modernization, and increased concentration of air pollutants, in turn ensuing increased number of exposed people [2]. Atmosphere being a geochemical reservoir of organic compounds partition's into gaseous and airborne particulate phase that interact with oceans, land, and living organisms [3]. The source and concentration of pollutants in outdoor and indoor atmosphere vary based on point of discharge. The outdoor pollutants are resultant of coal and oil combustion in power plants; disposal of industrial, medical, and municipal solid waste; motor vehicles emissions; industrial and sewage treatment plant discharges; and biomass burning. The causatives of indoor pollutants are unvented combustion, building materials, furnishings, paint, floor and wall coverings, cleaning products, cosmetics, detergents, pesticides, and electronic appliances [4]. The shift-ups in inhaling the above pollutants indeed justify the presence of toxic endocrine-disrupting chemicals (EDCs) body burden in every human being's blood, urine, and body tissues [5].

21.1.1 Air Pollutants in Terms of EDCs

EDCs are synthetic chemicals used in industrial, agricultural, and household applications, noted for specific property in disrupting endocrine system. On inhaling these chemicals as air particulates, they enter different systems of the body and disrupt normal homeostasis by either mimicking or blocking hormones. EDCs

A. Jayshree (✉) • N. Vasudevan
Centre for Environmental Studies, Anna University,
CEG Campus, Chennai 600025, Tamil Nadu, India
e-mail: jayshreeannamalai@yahoo.com

F. Capello, A.V. Gaddi (eds.), *Clinical Handbook of Air Pollution-Related Diseases*,
https://doi.org/10.1007/978-3-319-62731-1_21

Table 21.1 Solubility, partition coefficients, and half-life of major EDCs in the atmosphere

Endocrine-disrupting chemicals	Water solubility at (mg/L) 25 °C	Log KOW	Log KOC	Log KOA	Mean half-life in air (h)
Phthalates					
Di (2-ethylhexyl) phthalate	0.27	7.6	5.83	12.55	6
Dibutyl phthalate	13	4.9	3.80	8.63	7.4
Diethyl phthalate	1080	2.6	2.18	7.02	22.2
Bisphenol A	120	3.32	3.20	12.8	24
Polychlorinated biphenyls					
2,2′,3,4,5′-Pentachlorobiphenyl	0.0463	6.24	n.a.	9.16	3330
2,3′,4′-Trichlorobiphenyl	0.353	5.7	n.a.	8.92	2720
Polycyclic aromatic hydrocarbons					
Anthracene	0.045	4.540	4.47	7.34	60
Benzo[a]anthracene	0.011	5.91	5.30	10.80	300
Benzo[a]pyrene	0.0038	6.04	6.01	10.71	40
Phenanthrene	1.1	4.57	4.36	7.45	1040
Naphthalene	31.7	3.37	3.11	5.13	1210
Brominated flame retardants					
2,2′,4,4′-tetrabromodiphenyl ether	0.011	6.0	n.a.	9.8	22,416
2,2′,4,4′,5-pentabromodiphenyl ether	0.0024	6.8	n.a.	11.2	40.8
Decabromodiphenyl ether	0.02–0.03	9.9	n.a.	13.1	318
Pesticides					
Hexachlorobenzene	0.0062	5.73	3.59	4.47	4310
γ-Hexachlorocyclohexane	17	3.72	3.04	n.a.	3330
Dioxins and furans					
2,3,7,8-Tetrachloro-dibenzofuran	0.085	6.31	n.a.	n.a.	2190
2,3,7,8-Tetrachloro-dibenzo-p-dioxin	19.3 ng/L	6.8	7.15	9.70	1860
Alkylphenols					
4-Nonylphenol	7.0	5.8	4.63	n.a.	5.0
4-tert-octylphenol	5.0	5.3	3.42	n.a.	6.1
4-tert-butylphenol	700	3.3	n.a.	n.a.	6.3
Perfluorinated chemicals					
Perfluorooctane sulfonate	570	na	2.57	6.63	2736
Perfluorooctanoic acid	9.5×10^3	na	2.06	5.73	2160

Ref: HSDB [16], ATSDR [17], HSDB [18], EPA [19], Medellin-Castillo et al. [20], WHO [21], Safdari and Golmohammadi [22], Yuyin et al. [23], Mackay et al. [24], Kidd and James [25], Earl et al. [26], HSDB [27, 28], ATSDR [29], ATSDR [30], Chen and Bunce [31], WHO IPCS [32], Kelly et al. [33], Arnot et al. [34], Farmer et al. [35], Lewis [36], Hansch et al. [37], Yuyin et al. [23], Hollifield [38], Ripping [39], Mackay et al. [40], HSDB [28, 41], SRC [42], Groshart et al. [43], EPA Factsheet [44], Arnot and Gobas [45], and Zhu and Zou [46]

downstream effects start either by interacting with nuclear, hormone, or orphan receptors or by modifying enzymatic pathways involved in steroid biosynthesis or metabolism [6]. The long-range transport of EDCs via atmosphere and ocean routes makes their ubiquitous presence and source of exposure [7]. Distribution of EDCs in the atmosphere is governed by three equilibrium partitioning coefficients: air-water, water-octanol, and octanol-air (Table 21.1).

21.1.2 Major EDCs Causing Pathetic Disorders

Chemical production in the preceding 20 years correlates to the growing incidence of endocrine-associated pediatric disorders that include male reproductive disorders, early female puberty, leukemia, brain cancer, and neurobehavioral disorders [5]. Early life exposure to EDCs has also been implicated to the alteration of developmental programming, resulting in higher susceptibility to obesity, respiratory disorders, asthma, diabetes, cancer, endometriosis, neurological disorders, thyroid disorders, and cardiovascular disease [8–15]. The major EDCs involved in all of the above anomalies are phthalate esters (PEs), bisphenol A (BPA), polychlorinated biphenyls (PCBs), polyaromatic hydrocarbons (PAHs), polybrominated flame retardants (BFRs), pesticides, dioxins, alkylphenol (APs), perfluorinated compounds (PFCs), and heavy metals.

21.2 Emerging Sources of EDCs

Numerous chemicals were developed to meet wide range of medical, scientific, agricultural and industrial needs. Despite their socioeconomic benefits, the release of chemicals into the environment and exposure had led to serious health consequences irrespective of ages and species (Fig. 21.1).The source and point of discharge of major EDCs into the atmosphere are as follows:

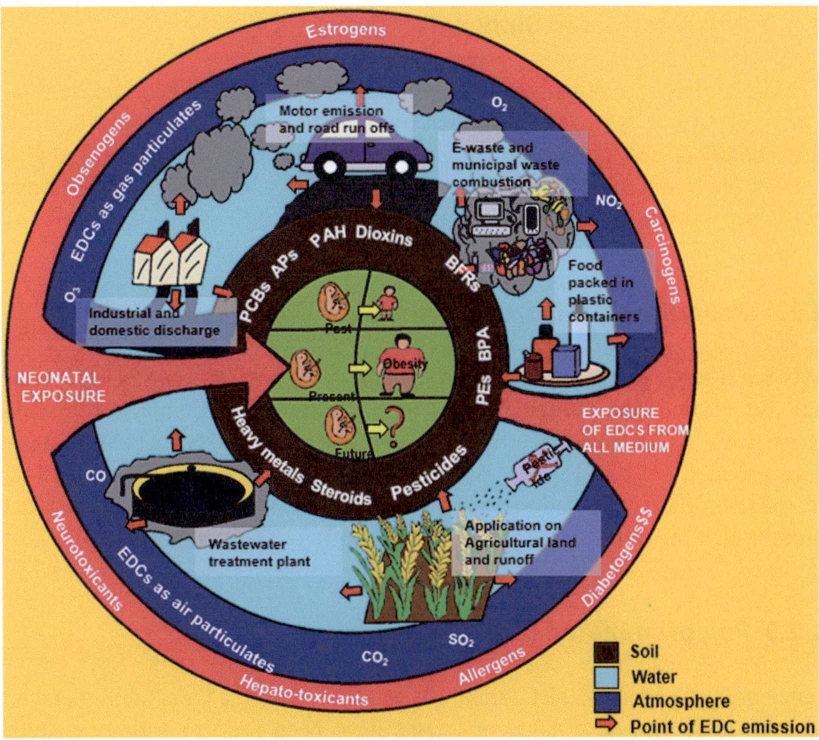

Fig. 21.1 Schematic representation of air pollution sources, emerging EDCs, and health issues

21.2.1 Phthalates

PEs are dialkyl or alkyl aryl esters of phthalic acid commonly used as plasticizers and encountered as indoor pollutant [47]. They are used in the manufacture of polyvinylchloride (PVC) products, building materials, toys, clothing, cosmetics, perfumes, food packaging, and medical appliances [48]. Phthalates are not physically bound to the polymers making their diffusion easier out of the plastics into the environment. Release of household and industrial wastewater from production and processing units and disposal of materials are sources of phthalates occurring in the atmosphere [49].

21.2.2 Bisphenol A

BPA is an organic compound composed of two phenol rings connected by a methyl bridge, with two methyl functional groups. It is a high-volume production chemical used worldwide in the manufacture of polycarbonate plastics including numerous consumer products like food and water containers and bottles. BPA is also found in the resin linings of food and beverage cans and dental sealants [50], leaching readily from many of these products lead to exposure in large segments of the population [51].

21.2.3 Polychlorinated Biphenyls

PCBs are aromatic, synthetic chemicals formed by two linked benzene rings with some or all of the hydrogen substituted by chlorine atoms. PCBs have been used commercially since 1929 as insulating fluid in transformers, capacitors, and plasticizers in open systems comprising numerous building materials including adhesives, caulk, ceiling tiles, paints, and sealants [52].

21.2.4 Polyaromatic Hydrocarbons

PAHs are a large group of organic compounds with two or more fused aromatic rings. Based on origin, pyrogenic PAHs are formed by the incomplete combustion of fossil fuels, forest fires, and tobacco smoke; petrogenic PAHs are present in crude oil, its product, and coal [53]. PAHs enter the environment primarily through sewage, road runoffs, smelter industries, and oil spills [54, 55]. The offshore PAHs enter water through oil seeps, spills, and discharges from offshore oil installations [56].

21.2.5 Polybrominated Flame Retardants

In ancient Egypt about 450 BC, alum was used to reduce the flammability of wood, and ever since that time flame retardants have been used in various materials.

The halogen-containing compounds are used today as flame retardants in electronic equipment, textiles, plastics, paints, and printed circuit boards preventing fire eruptions by capturing free radicals [57].

21.2.6 Pesticides

Pesticides are substances or chemical mixture intended for preventing, destroying, repelling, or lessening the damage of pest (Fig. 21.2). This includes herbicides, insecticides, fungicides, and rodenticides used in agriculture and public health. Occupational exposure to pesticides in agricultural workplace occurs during preparation (mixing and loading) and application (spraying) [58].

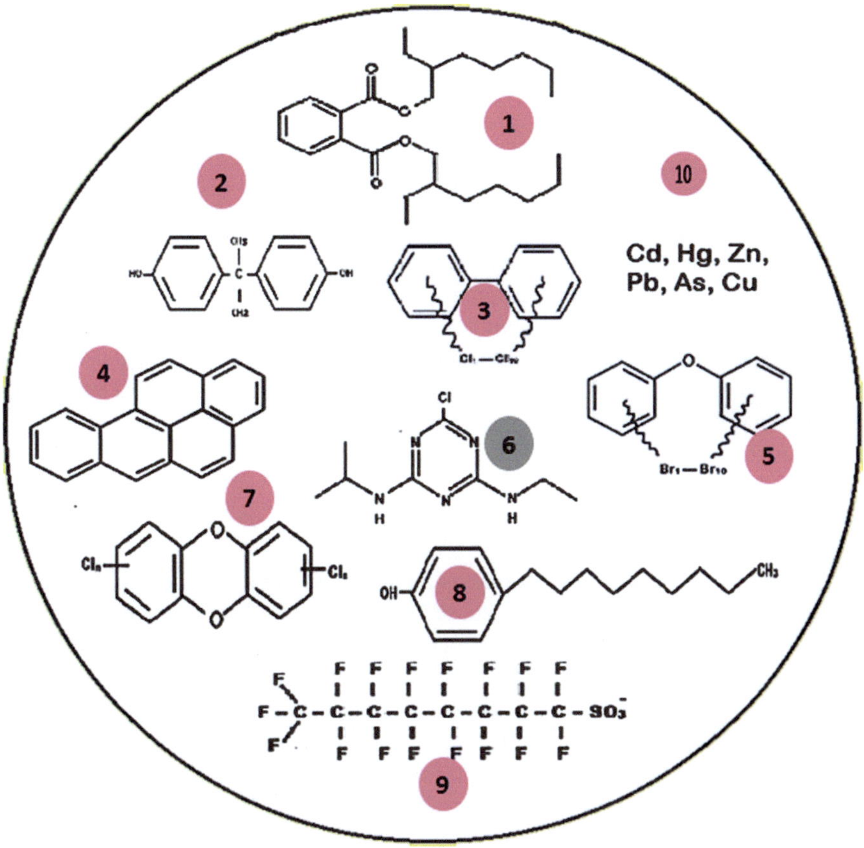

Fig. 21.2 Structure of major EDCs causing potential threat to health: (1) Di (2-ethylhexyl) phthalate, (2) bisphenol A, (3) polychlorinated biphenyl, (4) polyaromatic hydrocarbon, (5) brominated biphenyl ethers, (6) atrazine, (7) polychlorinated dibenzo-dioxins, (8) nonylphenol, and (9) perfluorooctane sulfonate

21.2.7 Dioxins

Dioxins are a group of 210 organic chemicals, among which 75 congeners are polychlorinated dibenzo-p-dioxins (PCDDs), and 135 are polychlorinated dibenzofurans (PCDFs). Prior to industrialization, low concentrations of dioxins existed in nature due to natural combustion and geological processes [59]. Presently, in industries though dioxins are not produced commercially, they are formed as by-products when reaction temperature is not well controlled in chemical manufacturing processes [60]. Dioxins are also produced in small concentrations when organic materials are burned in the presence of chlorine ions or atoms in the fuel formulae [61].

21.2.8 Alkylphenol

APs are widely used as nonionic surfactants in detergents, pesticides, herbicides, emulsifiers, paints, cosmetics, plastic ware, and even in jet fuel [62]. They are commonly found in wastewater discharges and effluents from sewage treatment plants [63].

21.2.9 Perfluorinated Compounds

PFCs are synthetic compounds characterized by long, fully fluorinated carbon chains with different functional head groups resisting them to degradation [64]. PFCs are used in variety of products to resist grease, oil, stains, and water and are also used in fire-fighting foam [65]. PFC contamination in the environment originates from direct or indirect anthropogenic sources. Direct source includes manufacture and use of perfluoroalkylated acids (PFAA), whereas indirect sources include product impurities and production of chemicals that may degrade to PFAA [66].

21.2.10 Heavy Metals

Heavy metals are commonly defined as metals those having a specific density of more than 5 g cm^3. The main threats to human health from heavy metals are associated with exposure to lead, cadmium, mercury, and arsenic. Heavy metals are being used in many different areas for thousands of years. Lead has been used for at least 5000 years, early applications including building materials, pigments for glazing ceramics, and pipes for transporting water [67]. Cigarette smoke contains about 30 metals, of which cadmium, arsenic, and lead are in the highest concentrations [68].

21.3 Mechanism Behind Endocrine Disruption

EDCs exert numerous disrupting mechanisms interfering endogenous hormonal functions depending on certain factors such as exposure level, duration, age, and susceptibility (Table 21.2). Owing to this action, EDCs are able to disrupt two or

Table 21.2 Nuclear receptors, biological half-life, and concentration of EDCs in the atmosphere and humans

Endocrine-disrupting chemical	Nuclear receptors	Biological half-life	Concentration in the atmosphere	Concentration in humans
Phthalates	PPARs, CAR/PXR, GR	From hours to days	Indoor: DEP, 2.29 mg m^{-3}; BBP, 3.97 mg m^{-3}; DEHP, 2.43 mg m^{-3} Outdoor: as particulates North Sea to Arctic, 0.03–5.03 ng m^{-3}; in gaseous phase 1.11–3.09 ng m^{-3}	Prenatal phthalate metabolite in urine of mothers, 2.54–816 μg L^{-1} Monoesters of DEHP in children's urine, 91.3 μg L^{-1}
Bisphenol A	ERs, AR, TR,GR	6 h		In blood, urine, fat, and fetal tissue, 0.1–10 ng mL^{-1}
Polychlorinated biphenyls	CAR, PXR	Longer than 90 days	Indoor: New York city schools, <50–807 ng m^{-3}; in classrooms, 236–2920 ng m^{-3} Outdoor: in Turkey–urban: gaseous phase, 23.5 ng m^{-3}; suburban: gaseous phase, 109.7 ng m^{-3}	In blood of mothers, 190 μg kg^{-1} lipid; in children-boys, 222 μg kg^{-1} lipid; girls, 153 μg kg^{-1} lipid; umbilical cord blood, 168.2–566.8 μg kg^{-1} lipid
Polyaromatic hydrocarbons	AhR, ERα, ERβ, PKCs, CYP1A	Benzo(a)pyrene, 2.2–12.4 days	Occupational region: aluminum production plants, 6 ng m^{-3}; coke ovens, 135–200,000 ng m^{-3}; iron foundries, 6400 ng m^{-3} Outdoor: France: rural, 92 ng m^{-3}; urban, 127 ng m^{-3}; suburban, 149 ng m^{-3}	In lung tissues of African-American males, 0.784 ng g^{-1}; Caucasian males, 0.407 ng g^{-1}; African-American females, 0.308 ng g^{-1}; Caucasian females, 0.279 ng g^{-1}

(continued)

Table 21.2 (continued)

Endocrine-disrupting chemical	Nuclear receptors	Biological half-life	Concentration in the atmosphere	Concentration in humans
Brominated flame retardants	PXR, ERs, TR	In serum, from weeks to months; infant, several years	Indoor: In Belgium and UK-school, 78–1741 ng g^{-1}; home, 33–1019 ng g^{-1}; office, 80–384 ng g^{-1}; e-waste dismantling hall, 20 ng m^{-3} Outdoor: e-waste site, 4.49–398 pg m^{-3}	Mean levels in adipose tissue: Europe and Asia, <5 ng (g lipid)$^{-1}$ North America, >200 ng (g lipid)$^{-1}$ In fetal liver, 4–98.5 ng (g lipid)$^{-1}$ In breast milk, 1.57–73.9 ng (g lipid)$^{-1}$
Pesticides	ERα, AR	5 years	Indoor: Outdoor: rural, 384 pg m^{-3}; urban, 379 pg m^{-3}	DDE, <5 to >15,000 mg (kgBW)$^{-1}$
Dioxins	Via AhR: PPARγ, ERs	7–11 years	Outdoor: Germany, rural, 5–50 fg m^{-3}; urban, 20–220 fg m^{-3}; suburban, 10–100 fg m^{-3}	In adipose tissue, 3.6pg (g lipid)$^{-1}$ In blood, 2.2 ppt
Alkylphenols	ERs, AR, CAR	NP in blood: 2–3 h	Indoor: in Japan, 4-t-BP, 387 ng m^{-3}; 4-t-OP, 45.7 ng m^{-3}; 4-NP 680 ng m^{-3} Outdoor: NP in USA: 0.01–81 ng m^{-3}	In urine, 0.4–13.9 ng ml^{-1} In adipose tissue, 57 ng g^{-1}(Spain)
Perfluorinated compounds	ERs, AR, PPARs	PFOS, 5.4 years	Indoor: FASAs, 20–300 pg m^{-3}; FASEs, 20–200 pg m^{-3}	Serum-level medians: PFOS, 19.9 μg L^{-1}; PFOA, 3.9 μg L^{-1}
		PFOA, 3.8 years	Outdoor: Northern hemisphere, 972 pg m^{-3}; Central Europe, 243 pg m^{-3}; Asia, 2466 pg m^{-3}; Northern America, 403 pg m^{-3}	

Table 21.2 (continued)

Endocrine-disrupting chemical	Nuclear receptors	Biological half-life	Concentration in the atmosphere	Concentration in humans
Heavy metals	ERα, ERβ, GPR30 (non nuclear)	As, 3–5 days	Cd level in outdoor and precipitation: Northern Europe, 0.05–0.2 ng m^{-3}and 0.02–0.1 μg L^{-1};	Cd, Pb, and Hg levels in blood: Urban, 0.62, 20.0 and 0.43 μg L^{-1}
		Cd, > 20 years	Central Europe, 0.2–0.5 ng m^{-3}and 0.04–0.2 μg L^{-1}; Southern Europe, 0.06–0.12 ng m^{-3}	Suburban, 0.82, 21.5, 0.47 μg L^{-1} Rural, 0.87, 22.5 and 0.33 μg L^{-1}

Ref: Pie et al. [72], Xie et al. [73], Thomas et al. [52], Demircioglu et al. [74], Morville et al. [75], Sjodin et al. [76], Tian et al. [77], Ali et al. [78], Devi et al. [79], WYGE [80], Ying [81], Saito et al. [82], Dreyer et al. [83], Barber et al. [84], Oono et al. [85], Stock et al. [86], Jahnke et al. [87], Gilbert [88], Graef [89], Ling et al. [90], Goldman et al. [91], Sponder et al. [92], Aas and Breivik [93]

more endocrinal functions with widespread consequences on the biological processes controlled by vulnerable endocrine glands. *In vivo* studies predict activation of hormonal receptors at nanomolar (nM) levels and enzymatic disruption at micromolar (mM) levels resulting in genomic instability and alteration of hormonal feedback regulation [69]. EDCs exert their actions through nuclear hormone receptors such as estrogen receptors (ERs), androgen receptors (ARs), progesterone receptors (PRs), thyroid receptors (TRs), and retinoid receptors (RXR) [6]. Indeed, EDCs are also capable of acting through nonsteroid receptors, transcriptional coactivators, enzymatic pathways, and genomic mechanism [70, 71].

21.3.1 Metabolic Disruption Through Hormone Receptors

Hormone receptors belong to a class of classic hormone receptors that recognize only one or a few molecules with high affinity. Thyroid hormone (TH), mineralocorticoid, glucocorticoid, retinoic acid, estrogen, vitamin D, progesterone, and androgen receptors belong to this class. Interaction of EDCs with these receptors results in developmental and reproductive effects, as well as metabolic alterations.

21.3.1.1 Estrogen Receptor

Estrogen receptors (ERα and β) have well-established roles in reproduction; in addition to that, they are also involved in brain development and function of many other organs such as the skin, bone, and liver (Fig. 21.3). At the molecular level, ERs and estrogens regulate glucose transport, glycolysis, mitochondrial activity, and fatty acid oxidation [94]. Based on experimental evidences, early exposure to BPA enhances adipocyte differentiation or permanently disrupt adipocyte-specific gene expression and leptin synthesis [95, 96]. Estrogenic surfactant octylphenols potentially elevate adipocyte production of resistin through activation of the ER and regulate

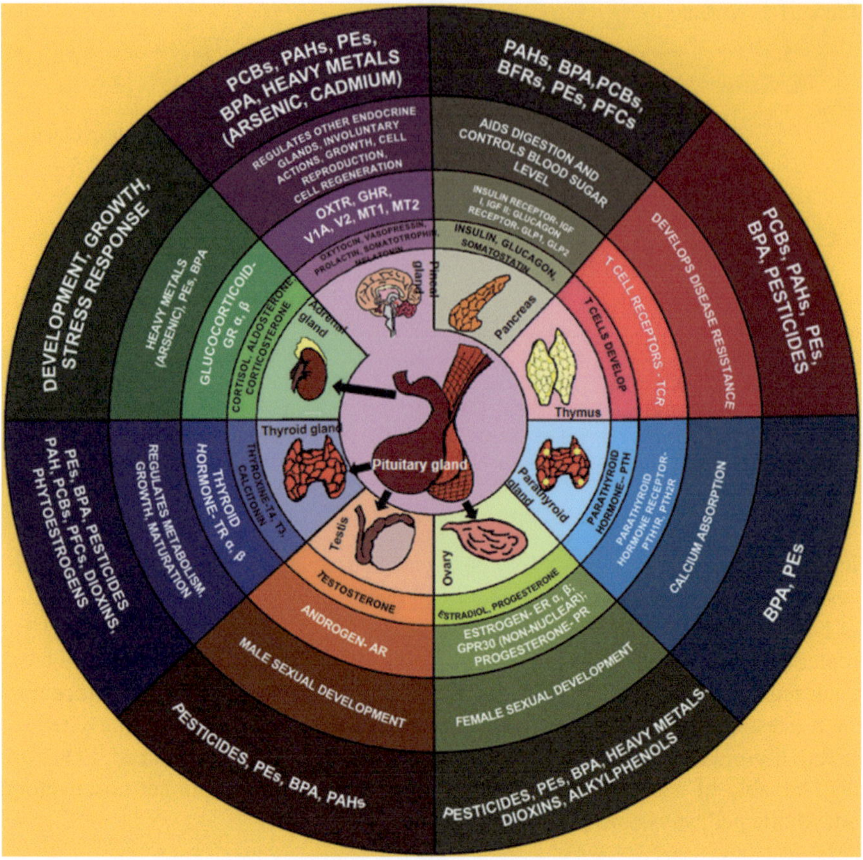

Fig. 21.3 Schematic representation of endocrine glands, hormones, receptors, significant functions, and antagonistic EDCs

extracellular signal kinase pathways [97]. Resistin secreted by adipocytes causes insulin resistance and predisposition to type 2 diabetes [98]. BPA also affects ERα activity in the pancreas and increases insulin secretion [99].

21.3.1.2 Thyroid Hormone Receptor and Glucocorticoid Receptor

TH activity is mediated by the TRs, TRα, and β which form heterodimers with RXR to bind the promoter sequences of target genes. Elevated TH levels accelerate metabolism, increase lipolysis and hepatic cholesterol biosynthesis, and provoke weight loss; the exact vice versa exists with low TH levels. In contrast to TR, GR forms homodimers and resides in the cytosol, forming complexes with molecular chaperones, and influences gene expression. Glucocorticoids act through GRs promoting gluconeogenesis, increasing blood glucose levels, and mobilizing the oxidation of fatty acids. Based on experimental evidences, BPA and PEs stimulate GR-mediated lipid accumulation and synergize with a weak GR agonist to increase expression of adipocyte-specific markers [100]. The effects of BPA on TR during development may be significant in long-term body weight increase, while BFRs decrease glucose oxidation, characteristically associated with obesity, insulin resistance, and type 2 diabetes [101, 102].

21.3.2 Metabolic Disruption Through Xenosensors

The body is protected from the accumulation of toxic chemicals by a complex strategy that takes place in the liver, regulating the expression of drug-metabolizing enzymes and transporters. This adaptive response integrates at least three xenosensors: pregnane X receptor (PXR), constitutive androstane receptor (CAR), and aryl hydrocarbon receptor (AhR).

21.3.2.1 Pregnane X Receptor and Constitutive Androstane Receptor

PXR and CAR regulate gene expression by forming heterodimers with RXR that bind to xenobiotic response sequences present in the promoters of their target genes. PXR is located primarily in the nucleus and is strongly activated upon ligand binding. In contrast, in the absence of ligand, CAR is retained in the cytoplasm through association with the cytoplasmic CAR retention protein (CCRP) and heat shock protein 90 (HSP90) [103]. PXR and CAR are highly expressed in the liver and act as master regulators of detoxification pathways [104]. Based on experimental evidence, EDCs such as nonylphenols (NPs), DEHP, MEHP, PCBs, BPA, PFCs PFOA, PFOS, and organochlorine methoxychlor are reported to activate PXR and CAR [105–107]. DEHP induces CAR-dependent activation of the nuclear receptor pathway, controlling the cellular clock and functions in energy metabolism [106]. EDC activated PXR and CAR also tends to regulate several CYP (cytochrome P450) family members involved in detoxifying pathways by metabolizing steroids and other endogenous hormones [103].

21.3.2.2 Aryl Hydrocarbon Receptor

AhR is a ligand-activated transcription factor which mainly senses and mediates the toxic effects of dioxins TCDD. The inactivated AhR protein resides in the cytosol and, upon ligand-mediated activation, translocates into the nucleus and heterodimerizes with the ubiquitously expressed aryl hydrocarbon receptor nuclear translocator (ARNT). Then AhR/ARNT complex binds to specific regulatory DNA sequences to regulate gene expression [108]. Recently, AhR has also been implicated as a regulator of energy metabolism, organogenesis, embryonic development, the cell cycle, immunosuppression, and carcinogenicity. Experimental studies report that certain EDCs such as PCBs and TCDD trigger inappropriate activation of AhR unrelated to detoxification. This affects the genes in an AhR-dependent manner linked to hepatic circadian rhythm, cholesterol biosynthesis, fatty acid synthesis, glucose metabolism, and adipocyte differentiation [109, 110]. AhR also disrupts ER signaling pathways by increasing ER proteasomal degradation and modulating estrogen levels via CYP expression and altering ER transcriptional activity [111, 112].

21.3.3 Peroxisome Proliferator-Activated Receptors

PPARs are sensor receptors with large ligand-binding domain that accommodates variety of ligands, primarily lipid derivatives. In presence of ligand, PPARs heterodimerize with RXR and bind to the PPAR response elements localized in the promoter regions of target genes [113]. PPAR is composed of three isotypes: PPARα, β/δ, and γ. PPARα is expressed predominantly in tissues characterized by a high rate of fatty acid catabolism such as the liver, kidney, heart, and muscle. PPARβ shares partially overlapping functions with PPARα and has a role in cell differentiation and survival [114, 115]. PPARγ functions in adipogenesis, lipid storage, inflammatory responses, and the control of insulin sensitivity [116]. Plasticizers, surfactants, pesticides, and dioxins modulate PPAR activity, although fairly little is known about the molecular mechanisms and the physiological outputs involved. *In utero* exposure induces alterations in fat structure and metabolism, with a disorganization of hepatic and gonadal architecture, steatosis in the liver, and an increase in lipid accumulation and mature adipocytes [117].

21.4 Endocrine System and Determination of EDC's Dose Response

Available literature and reports suggest that endogenous hormones and EDCs act at extremely low serum concentrations, typically in the picomolar to nanomolar range. But question arises whether the risk assessment studies and animal model studies are certainly able to round off such low dose. Complementally, low-dose effect is defined as any biological change that occurs in the range of typical human exposures or the dose lower than those typically used in standard protocols and toxicology assessments [118]. As an account, endocrine system displays specificity in response to endogenous hormones via hormone receptors that may be found specifically in a single cell or few cell types or throughout the body; e.g., thyroid-stimulating hormone (TSH) receptors are found specifically in thyroid gland, while thyroid hormone (TH) receptors are found throughout the body [119]. The receptors found in multiple cell types may also vary in response or effects as different co-regulators influence the behavior of target genes [120, 121].

Concentrations of active endogenous hormones vary based on the age and physiological status of the individual; for instance, plasma testosterone levels are less than 1 ng mL^{-1} in male children but increase to approximately 5–7 ng mL^{-1} in adult; during menses, estradiol levels are typically less than 100 pg mL^{-1} but, prior ovulation, spikes to 800 pg mL^{-1} [122, 123]. In addition, it is noted that active concentrations of endogenous hormones vary from species to species and even vary between strains of the same species [124]. Apart from endogenous hormone mechanism, EDCs to which exposure is increasing in day-to-day life also mimic and exert several mechanisms in binding to hormone receptors (Fig. 21.4). Single EDC is able to regulate several pathways; in contrast the atmospheric exposure remains to be

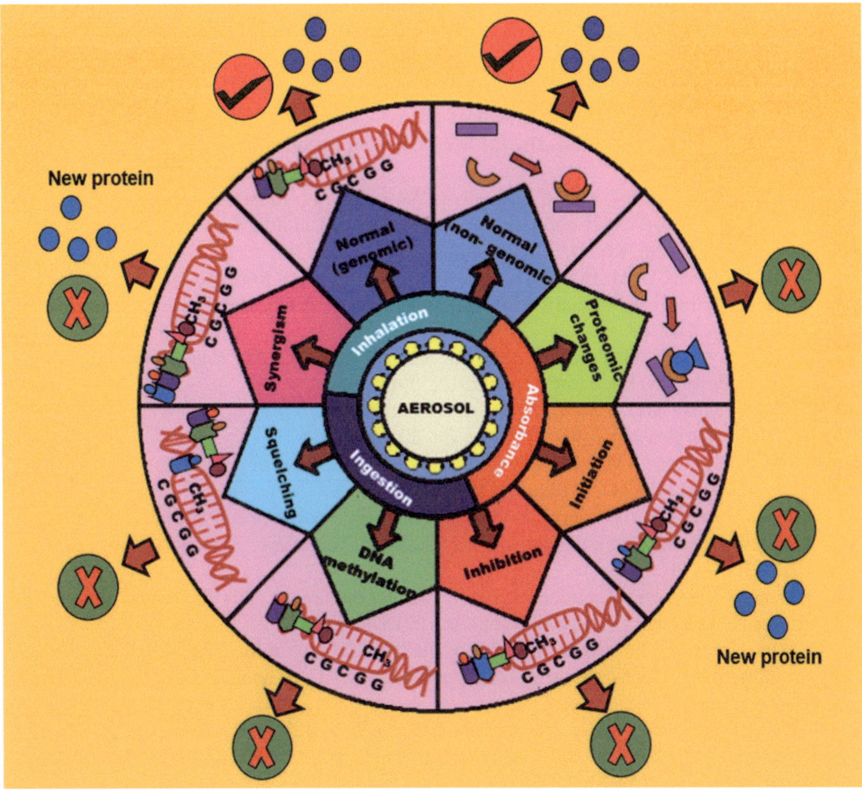

Fig. 21.4 Mechanisms exerted by EDCs in disrupting normal metabolism of a cell

mixture of EDCs, and occupational exposure may be of higher dose, the effect of which may vary from low-dose response. Considering all of the above, though oblique statement prevails in determining the mechanism, concentration, and effects of EDCs, the postulated mechanisms, determined toxic levels, and health effects may not be unnoticed.

References

1. Zhang Y, Mo J, Weschler CJ. Reducing health risks from indoor exposures in rapidly developing urban China. Environ Health Perspect. 2013;121(7):752–5.
2. United Nations, Department of Economic and Social Affairs, Population Division. World Urbanization Prospects: The 2014 Revision, Highlights (ST/ESA/SER.A/352). 2014.
3. Fu P, Kawamura K. Uniquity of bisphenol A in the atmosphere. Environ Pollut. 2010;158:3138–43.
4. Annamalai J, Namasivayam V. Endocrine disrupting chemicals in the atmosphere: their effects on humans and wildlife. Environ Int. 2015;76:78–97.
5. Gore AC, Crews D, Doan LL, Merrill ML, Patisau Hl, Zota A. Introduction to endocrine disrupting chemicals (EDCs): A guide for public interest organizations and policy-makers. Endocrine Society- Hormone to Health. 2014. p. 1–69.

6. Diamanti-Kandarakis E, Bourguignon JP, Giudice LC, Hauser R, Prins GS, Soto AM, et al. Endocrine-disrupting chemicals: an Endocrine Society scientific statement. Endocr Rev. 2009;30(4):293–342.
7. Lohmann R, Breivik K, Dachs J, Muir DG. Global fate of POPs: current and future research directions. Environ Pollut. 2007;150:150–65.
8. Bruner-Tran KL, Osteen KG. Dioxin-like PCBs and endometriosis. Syst Biol Reprod Med. 2010;56(2):132–46.
9. Clark NA, Demers PA, Karr CJ, Koehoorn M, Lencar C, Tamburic L, Brauer M. Effect of early life exposure to air pollution on development of childhood asthma. Environ Health Perspect. 2010;118:284–90.
10. Cock MD, Maas YGH, Bor MVD. Does perinatal exposure to endocrine disruptors induce autism spectrum and attention deficit hyperactivity disorders? Review. Acta Paediatr. 2012;101:811–8.
11. Garcia-Mayor RV, Vidal AL, Dicet Caamano MF, Gimenez AL. Endocrine disruptors and obesity: obesogens. Endocrinol Nutr. 2012;59(4):261–7.
12. Kim KH, Jahan SA, Kabir E. A review on human health perspective of air pollution with respect to allergies and asthma. Environ Int. 2013;59:41–52.
13. Kirkley AG, Sargis RM. Environmental endocrine disruption of energy metabolism and cardiovascular risk. Curr Diab Rep. 2014;14:494.
14. Kuo CC, Moon K, Thayer KA, Navas-Acien A. Environmental chemicals and type 2 diabetes: an updated systematic review of the epidemiologic evidence. Curr Diab Rep. 2013;13:831–49.
15. Macon MB, Fenton SE. Endocrine disruptors and the breast: early life effects and later life disease. J Mammary Gland Biol Neoplasia. 2013;18:43–61.
16. HSDB. Hazardous Substances Data Bank. National Library of Medicine. 2010. http://toxnet.nlm.nih.gov/cgi-bin/sis/htmlgen? HSDB and search on CAS number. Last Accessed 18 July 2010.
17. ATSDR (Agency for Toxic Substances and Disease Registry). Toxicological profile for di (2-ethyl hexyl) phthalate. Atlanta, GA: U.S. Department of Health and Human Services, Public Health Service; 2002.
18. HSDB (Hazardous Substances Data Bank.National). Library of medicine. 2001. http://toxnet.nlm.nih.gov/cgi-bin/sis/search/a?dbs+hsdb:@term+@DOCNO+922.
19. EPA. Toxicological review of dibutyl phthalate. External peer review draft. 2006a. http://www.epa.gov/ncea/iris/recent 2006.htm.
20. Medellin-Castillo NA, Ocampo-Perez R, Leyva-Ramos R, Sanchez-Polo M, Rivera-Utrilla J, Mendez-Diaz JD. Removal of diethyl phthalate from water solution by adsorption, photo-oxidation, ozonation and advanced oxidation process (UV/H_2O_2, O_3/H_2O_2 and O_3/activated carbon). Sci Total Environ. 2013;442:26–35.
21. WHO (World Health Organization). Diethyl phthalate. Concise International Chemical Assessment Document 52. ISBN 92-4-153052-9 (LC/NLM Classification: QV 612), ISSN 1020–6167. Geneva. 2003.
22. Safdari M, Golmohammadi H. Prediction of *n*-octanol-water partition coefficient for polychlorinated biphenyls from theoretical molecular descriptors. Eur J Chem. 2010;1(4):266–75.
23. Yuyin L, Chunsheng Y, Hongyan L, Zhongsheng Y, Yang W. 3D-QSAr study on half-lives of POPs using CoMFA and CoMSIA. J Environ Sci. 2008;20(12):1433–8.
24. Mackay D, Shiu WY, Ma KC. Illustrated handbook of physical-chemical properties and environmental fate for organic chemicals: pesticide chemicals, vol. 5. Boca Raton, FL: Lewis; 1997.
25. Kidd H, James DR. The agrochemicals hand book. 3rd ed. Cambridge: Royal Society of Chemistry Information Services; 1991. p. 6–10.
26. Earl N, Cartwright CD, Horrocks SJ, Warboys M, Swift S, Kirton JA, et al. Review of the fate and transport of selected contaminants in the soil environment. United Kingdom: Environment Agency. Technical Report P5–079/TR1. 2004.

27. HSDB (Hazardous Substances Data Bank). Arsenic, elemental. Canadian Centre for Occupational Health and Safety. U.S. National Library of Medicine. 2004. http://toxnet.nlm.nih.gov/cgi-bin/sis/htmlgen?HSDB.
28. HSDB (Hazardous Substances Data Bank). National Library of Medicine, Bethesda, MD (TOMES.CD-ROM Version). 1995. Denver, CO: Micromedex, Indlc.
29. ATSDR (Agency for Toxic Substances and Disease Registry). Toxicological profile for polycyclic aromatic hydrocarbons. Atlanta, GA: U.S. Department of Health and Human Services, Public Health Service; 1995.
30. ATSDR (Agency for Toxic Substances and Disease Registry). Toxicological profile for polybrominated biphenyls and polybrominated diphenyl ethers. Atlanta, GA: U.S. Department of Health and Human Services, Public Health Service; 2004.
31. Chen G, Bunce NJ. Polybrominated diphenyl ethers as Ah receptor agonists and antagonists. Toxicol Sci. 2003;76:310–20.
32. WHO IPCS. World Health Organization. International Programme on Chemical Safety. Brominated diphenyl ethers. Environmental Health Criteria 162. ISBN 92-4-157162. Geneva. 1994.
33. Kelly BC, Ikonomou MG, Blair JD, Morin AE, Gobas FAPC. Food web-specific biomagnification of persistent organic pollutants. Science. 2007;317:236–9.
34. Arnot J, Gouin T, Mackay D. Development and Application of Models of Chemical Fate in Canada: Practical methods for estimating environmental biodegradation rates. Canadian Environmental Modelling Network Report No. 200503. 2005.
35. Farmer WJ, Yang M, Letey J. Problems associated with the land disposal of an organic industrial hazardous waste containing HCB: Residual management by land disposal; U.S. Environmental Protection Agency, Report EPA-600/9–76-015, Cincinnatia, Ohio. 1976. p. 177–185.
36. Lewis NM. Attenuation of Polybrominated Biphenyls and Hexachlorobenzene by Earth Materials Project Summary. U. S. Environmental Protection Agency, Report EPA600/S2-81-191, Cincinnati, Ohio. 1981.
37. Hansch C, Leo A, Hoekman D. Exploring QSAR: Hydrophobic, Electronic, and Steric Constants. American. 1995.
38. Hollifield HC. Rapid nephelometric estimate of water solubility of highly insoluble organic chemicals of environmental interest. Bull Environ Contam Toxicol. 1979;23:579–86.
39. Ripping G. Screening of absorption behavior of new chemicals: natural spoils and model absorptions. Ecotoxicol Environ Saf. 1972;6:236–45.
40. Mackay D, Shiu WY, Ma KC, Lee SC. Handbook of physical-chemical properties and environmental fate for organic chemicals. In: Halogenated hydrocarbons, vol. 2. Boca Raton, FL: CRC Press/Taylor and Francis Group; 2006.
41. HSDB. Hazardous Substances Data Bank. 1999. Available online at http://sis.nlm.nih.gov.
42. SRC (Syracuse Research Corporation). Environmental Fate Data Base (EFDB): CHEMFATE. Database. SRC. Syracuse: NY. 2003. http://www.syrres.com/esc/efdb.htm.
43. Groshart CP, Okkerman PC, Wassenberg WBA, Pijnenburg, AMCM. Chemical study on Alkylphenols. RIKZ Report 2001.029 Nat. Inst. Coast and Marine Management (RIKZ). 2001.
44. EPA Factsheet Emerging Contaminants - Perfluorooctane Sulphonate (PFOS) and Perflurooctanoic Acid (PFOA). 2012. www.EPA.gov.
45. Arnot JA, Gobas FA. A review of bioconcentration factor (BCF) and bioaccumulation factor (BAF) assessments for organic chemicals in aquatic organisms. Environ Rev. 2006;14(4):257–97.
46. Zhu Z, Zuo Y. Bisphenol A and other alkylphenols in the environment - occurrence, fate, health effects and analytical techniques. Adv Environ Res. 2013;2(3):179–202.
47. Rudel RA, Camann DE, Spengler JD, Korn LR, Brody JG. Phthalates, alkylphenols, pesticides, polybrominated diphenyl ethers and other endocrine disrupting compounds in indoor air and dust. Environ Sci Technol. 2003;37(20):4543–53.

48. Wormuth M, Scheringer M, Vollenweider M, Hungerbuhler K. What are the sources of exposure to eight frequently used phthalic acid esters in Europeans? Risk Anal. 2006;26(3):803–24.
49. Cifci DI, Arikan OA. Occurrence of phthalates in sewage sludge from three wastewater treatment plants in Istanbul, Turkey. Clean (Weinh). 2013;41(9):851–5.
50. Welshons WV, Nagel SC, vom Saal FS. Large effects from small exposures. III. Endocrine mechanisms mediating effects of bisphenol A at levels of human exposure. Endocrinology. 2006;147:S56–69.
51. Vandenberg LN, Chahoud I, Heindel JJ, Padmanabhan V, Paumgartten FJ, et al. Urinary, circulating, and tissue biomonitoring studies indicate widespread exposure to bisphenol A. Environ Health Perspect. 2010;118:1055–70.
52. Thomas K, Xue J, Williams R, Jones P, Whitake D. Polychlorinated Biphenyls (PCBs) in School Buildings: Sources, Environmental Levels, and Exposures. EPA/600/R-12/051. 2012.
53. Vasudevan N, Aruazhagan P. Surfactants mediated recovery of anthracene and pyrene from contaminated soil. Asian J Microbiol Biotechnol Environ Exp Sci. 2007;9(2):237–42.
54. Mascarelli A. After the oil. Nature. 2010;467:22–4.
55. Vasudevan N, Bharathi S, Arulazhagan P. Role of plasmid in the degradation of petroleum hydrocarbon by *Pseudomonas fluorescens* NS1. J Environ Sci Health. 2007;A42:1141–6.
56. Kemsley J. Water eased oil removal in Gulf. Chem Eng News Arch. 2012;6:32–3.
57. Birnbaum LS, Staskal DF. Brominated flame retardants: cause for concern? Environ Health Perspect. 2004;112:9–17.
58. Khan AA, Ma S, Rahman SU. Occupational exposure to pesticides and its effects on health status of workers in Swat, Khyber Pakhtunkwa, Pakistan. J Biol Life Sci. 2013;4:43–55.
59. CPCB (Central pollution control board). Study of environmental contamination with polychlorinated. 2007.
60. CPCB (Central pollution control board). Dioxin (PCDDs) and Furan (PCDFs) - Critical persistent organic pollutants (POPs). Final Report. 2004.
61. EPA. An Inventory of Sources and Environmental Releases of Dioxin-Like Compounds in the United States for the Years 1987, 1995, and 2000. EPA/600/P-03/002F. Final Report. Nov 2006b.
62. Cevdet U, Mesude I, Inci T. Alkylphenols in the environment and their adverse effects on living organisms. Kocatepe Vet J. 2009;2(1):49–58.
63. Chen B, Duan JC, Mai BX, Luo XJ, Yang QS, Sheng GY, et al. Distribution of alkylphenols in the Pearl River Delta and adjacent northern South China Sea, China. Chemosphere. 2006;63(4):652–61.
64. Fromme H, Tittlemier SA, Volkel W, Wilhelm M, Twardella D. Perfluorinated compounds-exposure assessment for the general population in western countries. Int J Hyg Environ Health. 2009;212:239–70.
65. Bjorklund JA, Thuresson K, De Wit CA. Perfluoroalkyl compounds (PFCs) in indoor dust: concentrations, human exposure estimates, and sources. Environ Sci Technol. 2009;43:2276–81.
66. Prevedouros K, Cousins IT, Buck RC. Sources, fate, and transport of perfluorocarboxylates. Environ Sci Technol. 2006;40(1):32–44.
67. Jarup L. Hazards of heavy metal contamination. Br Med Bull. 2003;69:167–82.
68. Pizent A, Ariba B, Zivkovic T. Reproductive toxicity of metals in men. Arh Hig Rada Toksikol. 2012;63(1):35–46.
69. Iguchi T, Katsu Y. Commonality in Signaling of endocrine disruption from snail to human. Bioscience. 2008;58(11):1061–7.
70. Diamanti-Kandarakis E, Palioura E, Kandarakis SA, Koutsilieris M. The impact of endocrine disruptors on endocrine targets. Horm Metab Res. 2010;42(8):543–52.
71. Moral R, Wang R, Russo IH, Lamartiniere CA, Pereira J, Russo J. Effect of prenatal exposure to the endocrine disruptor bisphenol A on mammary gland morphology and gene expression signature. J Endocrinol. 2008;196(1):101–12.
72. Pie XQ, Song M, Guo M, Mo FF, Shen XY. Concentration and risk assessment of phthalates present in indoor air from newly decorated apartments. Atmos Environ. 2013;68:17–23.

73. Xie Z, Ebinghaus R, Temme C, Lohmann R, Caba A, Ruck W. Occurrence and Air-Sea exchange of phthalates in the Arctic. Environ Sci Technol. 2007;41:4555–60.
74. Demircioglu E, Sofuoglu A, Odabasi M. Atmospheric concentrations and phase partitioning of polycyclic aromatic hydrocarbons in Izmir, Turkey. Clean (Weinh). 2011;39:319–27.
75. Morville S, Delhomme O, Millet M. Seasonal and diurnal variations of atmospheric PAH concentrations between rural, suburban and urban areas. Atmos Pollut Res. 2011;2:366–73.
76. Sjodin A, Jones RS, Focant JF, Lapeza C, Wang RY, McGhaee EE, et al. Retrospective time-trend study of polybrominated diphenyl ether and polybrominated and polychlorinated biphenyl levels in human serum from the United States. Environ Health Perspect. 2004;112(6):654–8.
77. Tian M, Chen SJ, Wang J, Luo Y, Luo XJ, Mai BX. Plant uptake of atmospheric brominated flame retardants at an e-waste site in southern China. Environ Sci Technol. 2012;46:2708–14.
78. Ali N, Harrad S, Goosey E, Neels H, Covaci A. "Novel" brominated flame retardants in Belgian and UK indoor dust: implications for human exposure. Chemosphere. 2011;83(10):1360–5.
79. Devi NL, Qi S, Chakraborty P, Zhang G, Yadav IC. Passive air sampling of organochlorine pesticides in a northeastern state of India, Manipur. J Environ Sci. 2011;23(5):808–15.
80. WYGE (WYG Environmental) (Ireland) Ltd. Monitoring of Dioxin and Furan concentrations in the ambient air Haulbowline Island Island, Co. Cork 8: Resource and Environmental Consultants Ltd. 2008.
81. Ying GG. Fate, behavior and effects of surfactants and their degradation products in the environment. Environ Int. 2006;32(3):417–31.
82. Saito I, Onuki A, Seto H. Indoor air pollution by alkylphenols in Tokyo. Indoor Air. 2004;14(5):325–32.
83. Dreyer A, Weinberg I, Temme C, Ebinghaus R. Polyfluorinated compounds in the atmosphere of the Atlantic and southern oceans: evidence for a global distribution. Environ Sci Technol. 2009;43:6507–14.
84. Barber JL, Berger U, Chaemfa C, Huber S, Jahnke A, Temme C, et al. Analysis of per- and polyfluorinated alkyl substances in air samples from northwest. J Environ Monit. 2007;9:530–41.
85. Oono S, Matsubara E, Harada KH, Takagi S, Hamada S, Asakava A, et al. Survey of airborne polyfluorinated telomeres in Keihan area, Japan. Bull Environ Contam Toxicol. 2008;80:102–6.
86. Stock NL, Lau FK, Ellis DA, Martin JW, Muir DCG, Mabury SA. Polyfluorinated telomere alcohols and sulfonamides in the North American troposphere. Environ Sci Technol. 2004;38:991–6.
87. Jahnke A, Ahrens LR, Ebingham U, Berger Jl, Barber C, Tenense. An improved method for the analysis of volatile polyfluorinated alkyl substances in environmental air samples. Anal Bioanal Chem. 2007; 387: 965–975.
88. Gilbert SG. A small dose of toxicology: the health effects of common chemicals. 1st ed. Boca Raton, FL: CRC Press; 2004.
89. Graef JW. 'Cadmium' in Harrison's principles of internal medicine. In: Isselbacher KJ, et al., editors. . 13th ed. New York, NY: McGraw Hill; 1994. p. 2462–3.
90. Ling B, Han G, Xu Y. PCB levels in humans in an area of PCB transformer recycling. Ann N Y Acad Sci. 2008;1140:135–42.
91. Goldman R, Enewold L, Pellizzari E, Beach JB, Bowman ED, Krishnan SS, Shields PG. Smoking increases carcinogenic polycyclic aromatic hydrocarbons in human lung tissue. Cancer Res. 2001;61:6367–71.
92. Sponder M, Fritzer-Szekeres M, Marculescu R, Mittlbock M, Uhl M, Kohler-Vallant B, Strametz-Juranek J. Blood and urine levels of heavy metal pollutants in female and male patients with coronary artery disease. Vasc Health Risk Manag. 2014;10:311–7.
93. Aas W, Breivik K. Heavy metals and POP measurements 2003. Kjeller, Norwegian Institute for Air Research (EMEP-CCC Report #9/2005). 2005.

94. Chen JQ, Brown TR, Russo J. Regulation of energy metabolism pathways by estrogens and estrogenic chemicals and potential implications in obesity associated with increased exposure to endocrine disruptors. Biochim Biophys Acta. 2009;1793:1128–43.
95. Phrakonkham P, Viengchareun S, Belloir C, Lombes M, Artur Y, Canivenc-Lavier MC. Dietary xenoestrogens differentially impair 3T3-L1 preadipocyte differentiation and persistently affect leptin synthesis. J Steroid Biochem Mol Biol. 2008;110:95–103.
96. Wada K, Sakamoto H, Nishikawa K, Sakuma S, Nakajima A, et al. Lifestyle-related diseases of the digestive system: endocrine disruptors stimulate lipid accumulation in target cells related to metabolic syndrome. J Pharmacol Sci. 2007;105:133–7.
97. Lee MJ, Lin H, Liu CW, Wu MH, Liao WJ, et al. Octylphenol stimulates resisting gene expression in 3T3-L1 adipocytes via the estrogen receptor and extracellular signal-regulated kinase pathways. Am J Physiol Cell Physiol. 2008;294:1542–51.
98. Steppan CM, Bailey ST, Bhat S, Brown EJ, Banerjee RR, et al. The hormone resistin links obesity to diabetes. Nature. 2001;409:307–12.
99. Alonso-Magdalena P, Ropero AB, Carrera MP, Cederroth CR, Baquie M, et al. Pancreatic insulin content regulation by the estrogen receptor ERα. PLoS One. 2008;3:e2069.
100. Sargis RM, Johnson DN, Choudhury RA, Brady MJ. Environmental endocrine disruptors promote adipogenesis in the 3T3-L1 cell line through glucocorticoid receptor activation. Obesity. 2010;18:1283–8.
101. Legler J. New insights into the endocrine disrupting effects of brominated flame retardants. Chemosphere. 2008;75:216–22.
102. Zoeller RT, Bansal R, Parris C. Bisphenol-A, an environmental contaminant that acts as a thyroid hormone receptor antagonist in vitro, increases serum thyroxine, and altersRC3/neurogran in expression in the developing rat brain. Endocrinology. 2005;146:607–12.
103. Kretschmer XC, Baldwin WS. CAR and PXR: xenosensors of endocrine disrupters? Chem Biol Interact. 2005;155:111–28.
104. Wada T, Gao J, Xie W. PXR and CAR in energy metabolism. Trends Endocrinol Metab. 2009;20:273–9.
105. Eveillard A, Lasserre F, de Tayrac M, Polizzi A, Claus S, et al. Identification of potential mechanisms of toxicity after di-(2-ethylhexyl)-phthalate (DEHP) adult exposure in the liver using a systems biology approach. Toxicol Appl Pharmacol. 2009b;236:282–92.
106. Eveillard A, Mselli-Lakhal L, Mogha A, Lasserre F, Polizzi A, et al. Di-(2-ethylhexyl)-phthalate (DEHP) activates the constitutive androstane receptor (CAR): a novel signaling pathway sensitive to phthalates. Biochem Pharmacol. 2009a;77:1735–46.
107. Ren H, Aleksunes LM, Wood C, Vallanat B, George MH, et al. Characterization of peroxisome proliferator-activated receptor α–independent effects of PPARα activators in the rodent liver: di-(2- ethylhexyl) phthalate also activates the constitutive-activated receptor. Toxicol Sci. 2010;113:4–59.
108. Bock KW, Kohle C. Ah receptor: dioxin-mediated toxic responses as hints to deregulated physiologic functions. Biochem Pharmacol. 2006;72:393–404.
109. Arsenescu V, Arsenescu RI, King V, Swanson H, Cassis LA. Polychlorinated biphenyl-77 induces adipocyte differentiation and proinflammatory adipokines and promotes obesity and atherosclerosis. Environ Health Perspect. 2008;116:761–8.
110. Sato S, Shirakawa H, Tomita S, Ohsaki Y, Haketa K, et al. Low-dose dioxins alter gene expression related to cholesterol biosynthesis, lipogenesis, and glucose metabolism through the aryl hydrocarbon receptor-mediated pathway in mouse liver. Toxicol Appl Pharmacol. 2008;229:10–9.
111. Matthews J, Gustafsson JA. Estrogen receptor and aryl hydrocarbon receptor signaling pathways. Nucl Recept Signal. 2006;4:e016.
112. Swedenborg E, Ruegg J, Makela S, Pongratz I. Endocrine disruptive chemicals: mechanisms of action and involvement in metabolic disorders. J Mol Endocrinol. 2009;43:1–10.
113. Feige JN, Gelman L, Michalik L, Desvergne B, Wahli W. From molecular action to physiological outputs: peroxisome proliferator-activated receptors are nuclear receptors at the cross roads of key cellular functions. Prog Lipid Res. 2006;45:120–59.

114. Barish GD, Narkar VA, Evans RM. PPARδ: a dagger in the heart of the metabolic syndrome. J Clin Investig. 2006;14:590–7.
115. Michalik L, Wahli W. Peroxisome proliferator-activated receptors (PPARs) in skin health, repair and disease. Biochim Biophys Acta. 2007;1771:991–8.
116. Auwerx J. PPARγ, the ultimate thrifty gene. Diabetologia. 1999;42:1033–49.
117. Takeuchi S, Matsuda T, Kobayashi S, Takahashi T, Kojima H. In vitro screening of 200 pesticides for agonistic activity via mouse peroxisome proliferator-activated receptor (PPAR)α and PPARγ and quantitative analysis of in vivo induction pathway. Toxicol Appl Pharmacol. 2006;217:235–44.
118. Vandenberg LN, Colborn T, Hayes TB, Heindel JJ, Jacobs DR Jr, Lee DH, et al. Hormones and endocrine-disrupting chemicals: low-dose effects and non-monotonic dose responses. Endocr Rev. 2012;33(3):378–455.
119. Cheng SY, Leonard JL, Davis PJ. Molecular aspects of thyroid hormone actions. Endocr Rev. 2010;31:139–70.
120. Fu M, Wang C, Zhang X, Pestell RG. Acetylation of nuclear receptors in cellular growth and apoptosis. Biochem Pharmacol. 2004;68:1199–208.
121. Kress E, Samarut J, Plateroti M. Thyroid hormones and the control of cell proliferation or cell differentiation: paradox or duality? Mol Cell Endocrinol. 2009;313:36–49.
122. Jones KA. Summation of basic endocrine data. In: Gass GH, Kaplan HM, editors. Handbook of endocrinology. 2nd ed. New York, NY: CRC Press; 1996. p. 1–42.
123. Neill JD. Knobil and Neill's physiology of reproduction. 3rd ed. New York: Academic Press; 2005.
124. Stokes WS. Selecting appropriate animal models and experimental designs for endocrine disruptor research and testing studies. ILAR J. 2004;45:387–93.

Health Consequences Due to Prenatal Endocrine-Disrupting Chemical Exposure

22

Annamalai Jayshree and Namasivayam Vasudevan

22.1 Consequences on Health Secondary to EDC Exposure

There are several characteristics of the endocrine system that must be taken into account to understand the full mechanisms and consequences of exposure to EDCs. Similar to hormones, EDCs function at very low doses in a tissue-specific manner. EDCs also exert nontraditional dose responses due to the complicated dynamics of hormone receptor occupancy and saturation [1]. Thus low doses have more impact on a target tissue than higher doses, and the effects are entirely different. The age at which an individual is exposed to an EDC also has important implications on resulting health consequences. Indeed, it is clear that exposure to EDCs during development results in different effects than exposures during adulthood [2].

22.1.1 Prenatal Exposure

Studies suggest that when embryos are exposed to EDCs in terms of critical windows of *in utero* development, leading to altered programming in tissue structure and function. In return, this predisposes to a higher disease risk later in adult life [3]. Fetuses are exposed to EDCs by maternal exposure via placenta. Effects of low-dose exposure during development last long even after EDCs disappear from the body. To accentuate this concept, the term "fetal basis of adult disease" (FeBAD) has been coined to describe the interactions between the developing organism and the environment that determines the propensity of an individual to develop disease later in life. This prediction extends beyond the fetal period to include the early

A. Jayshree (✉) • N. Vasudevan
Centre for Environmental Studies, Anna University,
CEG Campus, Chennai 600025, Tamil Nadu, India
e-mail: jayshreeannamalai@yahoo.com

F. Capello, A.V. Gaddi (eds.), *Clinical Handbook of Air Pollution-Related Diseases*,
https://doi.org/10.1007/978-3-319-62731-1_22

postnatal developmental period while the organs continue to undergo substantial development [4].

Based on the experimental data and EDCs involved, consequences of prenatal exposures are discussed below.

22.1.1.1 Obesity

Obesity delineating an excessive body fat has become a fast-growing human health crisis heaving worldwide attention. Obesity and overweight are known to seriously affect human health, and it tends to be prognostic to number of diseases such as type 2 diabetes, hypertension, stroke, gout, liver disease, asthma and pulmonary problems, gall bladder disease, kidney disease, reproductive problems, osteoarthritis, and certain forms of cancer [5]. In addition to this, it is further linked to psychological problems including social discrimination, poor self-esteem, and depression [6]. Obesity is of significant health risk for children and adolescents than adults since obesity has dramatically increased the incidence of type 2 diabetes [7, 8].

EDCs Involved

Polycyclic Aromatic Hydrocarbons and Air Pollution

Epidemiological data strongly support causal association between maternal smoking and increased risk of obesity or overweight in offspring [9]. Benzo(a)pyrene has also been shown to inhibit lipolysis and cause increased fat accumulation [10]. Air pollution near roads is suggested to contribute to obesity, insulin resistance, and inflammation in later part of life on exposure during childhood or fetal development [11–13].

Bisphenol A

The chemical structure of BPA enables it to fit into the binding site of the estrogen receptor in turn activating both nuclear and cell membrane localized estrogen receptors [14]. BPA and its derivatives are suspected to act as obesogens by inducing adipocyte differentiation and the expression of genes involved in adipogenesis [15, 16]. In laboratory studies, BPA-exposed female mice fed a high-fat diet consumed more food and gained more weight than control animals on the same diet. This suggests that early-life exposure to BPA might lead to sexually dimorphic alterations in hypothalamic energy balance circuitry, resulting in increased susceptibility to developing diet and induced obesity and metabolic impairment [17].

Brominated Flame Retardants

Flame retardants have been associated with various adverse health outcomes, including obesity and altered or inappropriate thyroid function [18–20]. Other effects include early female puberty, weight gain before puberty and continued to adulthood, male cardiac hypertrophy, glucose intolerance, and increased serum levels of thyroxine, as well as reduced hepatic carboxylesterase activity in the dams [21]. Subsequent studies report that the obesogenic effects of BFR might be mediated by binding to and activation of PPARγ [22, 23].

Polychlorinated Biphenyls

Exposure to PCB remains ubiquitous because of its improper disposal and bioaccumulation in the environment, although banned in most developed countries. Studies suggest that these lipophilic pollutants which have accumulated at high levels in adipose tissue serve as factors in epidemic obesity [24]. In a survey by NHANES 1999–2002, an association between waist circumference and BMI in individuals with detectable levels of PCBs and other persistent organic pollutants also suggests contribution to the ongoing obesity epidemic [25].

Phthalates

Prenatal and neonatal exposure to DEHP in pregnant mice led to increased body weight, numbers and size of adipocytes, and activation of PPARγ in male offspring, suggesting sexually dimorphic effect. Activation of PPARγ leads to the expression of its target genes and differentiation of cells into adipocytes [26]. Thus PEs have the ability to increase adipose tissue mass both during development and in adulthood, although the mechanism and sensitivity in these periods are probably different.

Perfluorinated Compounds

Human and animal studies on the metabolic effects of PFOA compounds are inconclusive. Young adult mice exposed *in utero* to perfluorooctanoic acid were overweight and had increased serum levels of insulin and leptin [27].

Symptoms

The primary warning sign of obesity is being above-average body weight, and the symptoms include breathing disorders (e.g., sleep apnea, chronic obstructive pulmonary disease), certain types of cancers (prostate and bowel cancer in men, breast and uterine cancer in women), coronary artery (heart) disease, depression, diabetes, gallbladder or liver disease, gastroesophageal reflux disease (GERD), high blood pressure, high cholesterol levels, joint disease (osteoarthritis), and stroke. High blood pressure, high cholesterol levels, breathing problems, and joint pain (in the knees or lower back) are also common.

Diagnosis

A measurement called the body mass index (BMI) does not directly measure body fat but is a useful tool to assess the health risk associated with being overweight or obese. A BMI of 18.5–24.9 is considered within the healthy range. The BMI is calculated using kilograms (kg) and meters (m) instead of pounds (lb) and inches/feet. 1 lb equals 0.45 kg and 1 in. equals 0.0254 m. BMI = body weight (kg)/height^2 (m).

Body fat is calculated using skin calipers that measure the thickness of skin. Body shape also plays a significant role, since weight around the waist poses a greater risk of heart disease and diabetes than big hips and thighs. Waist circumference is a good measure of abdominal obesity; women with a waist more than 35 in. or men with a waist more than 40 in. are at increased risk.

Treatment

Weight reduction is achieved by consuming fewer calories and increasing activity and exercise. Certain approaches include modified diet and lifestyle, regular exercise, prescription weight loss medications and surgery. Some common types of surgeries are:

1. *Gastroplasty*—also known as stomach stapling where a small pouch is created in the stomach that allows only limited amounts of food to be eaten at one time.
2. *Laparoscopic adjustable gastric banding*—an adjustable band is placed around the stomach with minimal invasive surgery.
3. *Gastric bypass*—this is the most effective weight loss surgery. However, it also carries a greater risk of complications, both short term and long term. A small pouch in the upper part of the stomach is created, and a hole is made in the small intestine beyond the normal stomach attachment. Then the pouch is attached to the hole, bypassing the rest of the stomach and top part of the small intestine.

22.1.1.2 Asthma

Asthma is the most common chronic disease in childhood with increased prevalence over the past decades [28, 29]. In relation to this, air pollution is one of the sources consistently linked to exacerbate the prevalence of asthma via both indoor and outdoor exposures.

EDCs Involved

Based on the outdoor studies, increases in asthma incidence or asthma symptoms were observed among the children exposed to higher levels of traffic-related air pollution [30, 31]. Pre-birth and post-birth exposures to tobacco smoking are also independently associated with increased asthma incidence [32]. Although air pollution exposures before 2–3 years of age appear to have more significant role in asthma development, the effect of pre-birth or *in utero* exposure is not much studied except PAH [33, 34]. In accordance to indoor studies, household air and dust contain higher concentrations of consumer product chemicals such as pesticides, PCBs, BFRs, phthalates, and certain volatile organic compounds which are suggested to be associated with asthma [35–38]. However, inadequate information is available on the contribution of many of these EDCs in asthma. Among personal care and cosmetic products, fragrances are shown to exacerbate asthma, and PEs (di-ethylhexyl phthalate) are also reported to be associated with asthma and wheezing in children [39].

Symptoms

Shortness of breath, chest tightness or pain, trouble sleeping caused by shortness of breath, coughing or wheezing, a whistling or wheezing sound when exhaling (wheezing is a common sign of asthma in children), and coughing or wheezing attacks that are worsened by a respiratory virus, such as a cold or the flu.

Diagnosis

1. *Physical examination*—to rule out other possible conditions such as a respiratory infection or chronic obstructive pulmonary disease (COPD).
2. *Tests to measure lung function*—quantity of air moves in and out while breathing.
   (a) *Spirometry*—to check quantity air exhaled after a deep breath and rate of exhalation
   (b) *Peak flow*—to check the rate of breathe
3. *Additional tests include* methacholine test, nitric oxide test, imaging tests, allergy tests, sputum eosinophil test, and provocative testing for exercise- and cold-induced asthma.

Treatment

1. *Medications*
   (a) *Long-term asthma control medications*—generally taken daily, are the cornerstone of asthma treatment. Types of long-term control medications include inhaled corticosteroids, leukotriene modifiers, long-acting beta agonists, combination inhalers, and theophylline.
   (b) *Quick-relief (rescue) medications*—are used for rapid, short-term symptom relief during an asthma attack or before exercise. Types of quick-relief medications include short-acting beta agonists, ipratropium (Atrovent), and oral and intravenous corticosteroids.
   (c) *Allergy medications*—if asthma is triggered or worsened by allergies. This includes allergy shots (immunotherapy) and omalizumab (Xolair).
2. *Bronchial thermoplasty*—heats the insides of the airways in the lungs with an electrode, reducing the smooth muscle inside the airways. This limits the ability of the airways to tighten, making breathing easier and possibly reducing asthma attacks. Recommended in case of severe asthma.

22.1.1.3 Reproductive Disorders

Development of the mammalian reproductive systems begins in early pregnancy with specification and migration of germ cells, followed by morphogenesis of the gonads, reproductive tract structures, and external genitalia. As the reproductive tissues form, they differentiate under the influence of numerous molecules including growth factors, transcription factors, and steroid hormones. Gross morphogenesis of reproductive tissues is largely complete before birth, but slow growth and regional and cellular differentiation continue through the onset of puberty. During puberty, a rapid phase of growth and additional structural and cellular reorganization occurs, regulated in large part by steroid hormones. In this case, EDC interruption via gestational exposure is strongly correlated with increased female and male reproductive tract structural anomalies including rare forms of cancer, infertility, genital abnormalities, poor pregnancy outcomes in female offspring, and urological cancer in male offsprings [40, 41].

Male Reproductive Disorders

Testicular Dysgenesis Syndrome

Prenatal exposure to EDCs causes Leydig and Sertoli cell dysfunction due to impairment of germ cell and androgen insufficiency. This condition leads to the abnormal development of the fetal testis, causing testicular dysgenesis syndrome (TDS) that constitutes three anomalies: urogenital tract disorder, distortion of semen quality, and testicular cancer (TC) [42].

Urogenital Tract Malformation

Hypospadias

Hypospadias is a congenital abnormality of the male genitalia characterized by incomplete development of the urethra such that the external urethral opening is on the ventral surface of the penis or on the scrotum, rather than at the tip of the penis [43]. The development of the penile urethra is complete by 12 weeks after ovulation (at 14 weeks of gestation) and depends on the secretion of testosterone by the fetal testes [44]. The urethra opening in the glans or corona (sulcus) is called distal, and this mild form of hypospadias often does not necessitate any treatment. If the urethral meatus is located in the penile shaft or penoscrotal area, the hypospadias is called proximal, and these require surgical management [45].

EDCs Involved

The etiology of hypospadias is still unknown, but the most common risk factors for hypospadias include low birth weight, relatively high maternal age, genetics, maternal subfertility, smoking, and paternal exposure to solvents and pesticides [46, 47]. The *in utero* exposure to environmental estrogens, antiandrogens, phthalates, or chemicals such as 2,3,7,8-tetrachlorodibenzo-p-dioxin may contribute to increased hypospadias.

Symptoms

Symptoms include a downward urinary spray (in older children with more severe hypospadias), downward curve of the penis (chordee), extra foreskin along the top side of the penis (hooded appearance), and an abnormal appearance of the tip of the penis (the glans).

Diagnosis

It is based on a physical examination. In cases of severe hypospadias, the absence of testicles in the scrotum, sex determination evaluation is performed.

Treatment

Treatment involves surgery to reposition the urethral opening and, if necessary, straighten the shaft of the penis. During surgery, a pediatric urology surgeon uses tissue grafts from the foreskin or from the inside of the mouth to reconstruct the urinary channel in the proper position, correcting the hypospadias.

Cryptorchidism
Cryptorchidism is a condition where one or both testicles fail to descend into the scrotal sac at the time of birth. The prevalence of cryptorchidism at birth is 2.5–9% [48]. It affects 1–1.9% of 3-month-old boys and 0.8–1.5% of 18-month-old boys [49]. Mutations in the gene for insulin-like factor 3 (INSL3) and its receptors have been recognized as the causes of cryptorchidism; in some cases, chromosomal alterations and Klinefelter syndrome are also significantly involved. The role of INSL3 in testicular descent is mainly related to its effects on gubernaculum differentiation during the transabdominal and inguinoscrotal phase.

EDCs Involved
Risk factors for cryptorchidism include genetic predisposition, preterm birth, low birth weight, and prenatal exposure to EDCs or tobacco in either the mother or the father [50, 51]. The inguinoscrotal phase is androgen dependent, and exposure to EDCs may affect normal androgen production and consequently interrupt normal descent of testes. Parental exposure to pesticide and phthalates is relatively studied cause for cryptorchidism [52].

Symptoms
Undescended testicles may be palpable or unpalpable. About 80% of cases with palpable undescended testicle are felt during physical examination. The testicle is usually located at the end of the inguinal canal, a channel that carries the spermatic cord toward the penis and scrotum in males. Unpalpable undescended testicles are felt during a physical examination. There are three main types based on location and morphology: *abdominal*, undescended testicle in the abdomen; *inguinal*, the testicle moved into the inguinal canal; and *atrophic* or *absent*, the testicle is either very small or has never formed.

Diagnosis
The detection methods include:

(a) *Ultrasound scan*—to locate the testicle, if it is in the groin
(b) *MRI scan*—to show whether the testicle is in the groin or abdomen
(c) *Laparoscopy*—corrective surgery involving small incision in abdomen
(d) *Open surgery*—necessary in rare complicated cases, to explore directly inside the abdomen

Treatment
In around half of all patients with cryptorchidism, the testicle descends within 3 months. However, in 1–2 of every 100 cases, testicles do not descend by the time a child is 6 months old, and a surgery called orchidopexy or orchiopexy is needed. This surgical procedure is commonly used to free undescended testicle and implant it into the scrotum. It is usually done between the ages of 6 and 18 months, as delaying the operation may increase the long-term risk of developing testicular cancer or infertility. The testicles in the scrotum are provided with lower temperature to produce sperm than rest of the body.

Distortion of Semen Quality

Despite the potential importance and relevance of early-life exposure to EDCs, the epidemiological evidence on the relationship between semen quality and adult is limited [53].

EDCs Involved

Phthalates

PE metabolites are associated with increased DNA damage, a lower sperm concentration, and density as well as decreased motility and morphology [54, 55]. Phthalates are known to alter gene expression encoding enzymes involved in testosterone biosynthesis and PPAR-α activation in a dose-dependent manner [56].

Polychlorinated Biphenyls

In a study, the role of environmental estrogens like PCBs was evaluated in relation to the deterioration of semen parameters in men who were infertile without obvious etiology. Semen was analyzed, and PCB concentrations were measured in 21 infertile men and 32 controls. PCBs were found in the seminal plasma of infertile men, but not in that of controls. Increasing seminal PCB concentrations showed a significant negative effect on the ejaculate volume, total progressive motility, sperm vitality, and sperm osmoregulatory capacity [57].

Dioxins

In the case of pregnant mothers exposed to the Seveso disaster, they had median serum dioxin concentration of 26.0 ppt at conception, whereas the median for the comparison group was 10.0 ppt. At the age of 4–5 months—a critical time point for proliferation of Sertoli cells that determines spermatogenic potential in adulthood—21 of 39 boys who were breastfed had increased median dioxin of 40 ppt. In later life, these men had significantly decreased sperm concentration, total sperm count, and total number of motile sperm.

Pesticides

Several epidemiological studies suggest an association between pesticide exposure and altered semen quality [58].In a small study on male partners of pregnant women, urinary concentrations of pesticide biomarkers in 34 men with sperm concentration, motility, and morphology below the median were compared to 52 men with above-median semen parameters (controls). An elevated odds ratio for poorer semen quality in relation to urinary concentrations of alachlor mercapturate, 2-isopropoxy-4-methyl-pyrim-idinol (diazinon metabolite), atrazine mercapturate, 1-naphthol (carbaryl and naphthalene metabolite), and 3,5,6-trichloro-2-pyridinol (chlorpyrifos metabolite) was found [59, 60].

Symptoms

Although most men with male infertility do not notice symptoms other than inability to conceive a child, signs and symptoms associated with male infertility include problems with sexual function (difficulty with ejaculation or small volumes of fluid ejaculated and erectile dysfunction); pain, swelling, or a lump in the testicle area; recurrent respiratory infections; inability to smell; abnormal breast growth (gynecomastia); decreased facial or body hair or other signs of chromosomal or hormonal abnormality; and lower than normal sperm count (fewer than 15 million sperm per milliliter of semen or a total sperm count of less than 39 million per ejaculate).

Diagnosis

(a) *Physical examination and medical history.*
(b) *Semen analysis*—to measure the number of sperm, abnormalities in the shape (morphology), and movement (motility) of the sperm.
(c) *Scrotal ultrasound*—to find whether there is a varicocele or other problems in the testicles.
(d) *Hormone testing*—hormones produced by the pituitary gland, hypothalamus, and testicles play a key role in sexual development and sperm production. Abnormalities in other hormonal or organ systems contribute to infertility. A blood test measures the level of testosterone and other hormones.
(e) *Post-ejaculation urinalysis*—sperm in urine indicates sperm traveling backward into the bladder instead to the penis during ejaculation (retrograde ejaculation).
(f) *Genetic test*—in case of low sperm concentration, subtle changes in the Y chromosome, a sign of genetic abnormality is checked.
(g) *Testicular biopsy*—involves removal of samples from the testicle with a needle and determining sperm production.
(h) Specialized sperm function tests—to check sperm survival and penetration into the egg after ejaculation.
(i) *Transrectal ultrasound*—to check for blockages of the tubes that carry semen (ejaculatory ducts and seminal vesicles).

Treatment

(a) *Surgery*—in the case of varicocele, it is surgically corrected, or an obstructed vas deferens is repaired. In the case where no sperm are present in the ejaculate, sperms are often retrieved directly from the testicles or epididymis using sperm retrieval techniques.
(b) *Hormone treatments and medications*—hormone replacement or medications in cases where infertility is caused by high or low levels of certain hormones or problems with the way the body uses hormones.
(c) *Assisted reproductive technology (ART)*—involves obtaining sperm through normal ejaculation, surgical extraction, or from donor individuals and insertion into female genital tract or by performing *in vitro* fertilization or intracytoplasmic sperm injection.

Testicular Cancer

Testicular cancer (TC) is the most common type of malignancy in men aged 15–40 years in industrialized countries and is the most frequent cause of death from solid tumors. Approximately 95–98% of all TCs are testicular germ cell tumors (TGCT) and 1–5% of TC result from hyperplasia of testicular somatic cells (Sertoli and Leydig cells). TGCTs consist of a diverse group of neoplasms based on anatomical locations within the testis. Histopathologically, TGCT has been classified into two main categories: seminomas, which features similar to those of prostate gland cancer, and non-seminomas, which include mixed germ cell tumors, embryonal carcinomas, teratomas, choriocarcinomas, and yolk sac tumors [61, 62].

EDCs Involved

Studies report epidemiologic evidence between PCB congeners linked and TC risk [63, 64]. A recent study consisting of 125 patients and 103 controls concluded that serum concentration of PCBs and hexachlorobenzene were statistically significant for increased risk of TC and lower semen quality [65].

Symptoms

TC affects usually only one testicle; symptoms include lump or enlargement in either testicle, heaviness in the scrotum, dull ache in the abdomen or groin, sudden collection of fluid in the scrotum, pain or discomfort in a testicle or the scrotum, enlargement or tenderness of the breasts, and back pain.

Diagnosis

(a) *Physical examination.*
(b) *Ultrasound*—to determine the nature of the testicular lumps (solid or fluid filled) and position (inside or outside testicle).
(c) *Blood tests*—to determine the levels of tumor markers (in the case of TC, certain substances in blood elevates).
(d) *Determination of cancer stage*—in stage I, cancer is limited to the testicle; in stage II, cancer had spread to the lymph nodes in the abdomen; in stage III, cancer had spread to other parts of the body. Testicular cancer most commonly spreads to the lungs and liver.

Treatment

(a) *Surgery*—to remove testicle (radical inguinal orchiectomy) or nearby lymph nodes (retroperitoneal lymph node dissection)
(b) *Radiation therapy*—used in the case of seminoma type TC after surgery
(c) *Chemotherapy*—drugs to kill cancer cells

Prostate Cancer

The prostate gland develops from endodermal origins and is highly dependent on androgen receptor activity for its development [66]. Androgen-dependent expansion of both stromal and epithelial compartments of the prostate occurs throughout sexual maturation, growing from 1 g at birth to 20 g following puberty. Expression of 5-α reductase and aromatase by the prostatic mesenchyme is necessary for conversion of androgens to testosterone and estrogen by aromatase, in turn significant in cell proliferation and morphogenesis [67]. The prostatic stroma expresses ER-α while prostate epithelial cells express ER-β. Prostate cancer (PC) is the most common solid cancer in males associated with chronically elevated estrogens [68]. The prostatic zones differ in their cancer susceptibility; the inner zone is the common site of benign prostatic hyperplasia (BHP), while the outer zone is most likely the site of PCs [69].

EDCs Involved

Pesticides

In a large epidemiology study, involving more than 55,000 pesticide applicators, it revealed a direct link between increased prostate cancer rates on exposure to methyl bromide, a fungicide with unknown mechanism of action [70]. Chlorpyrifos, fonofos, coumaphos, phorate, permethrin, and butylate of 45 common agricultural pesticides showed significant correlation with exposure and increased prostate cancer rates in men with a familial history of PC [70, 71].

Bisphenol A

BPA has a significantly lower potency than endogenous estrogens *in vitro* and is agonist for both ER-α and ER-β. BPA induces ER through non-genomic pathways with an EC50 equivalent to 17 β-estradiol, suggesting that *in vivo* estrogenic activity of BPA may be due to non-genomic activation of ER [72, 73].

Polychlorinated Biphenyls

PCBs have estrogenic or antiandrogenic activity that perturbs the prostate gland. An extensive epidemiological study of capacitor-manufacturing plant workers and electric utility workers exposed to high levels of PCBs revealed a strong exposure-response relationship for prostate cancer mortality [74–76]. Evidences also suggest that PCBs inhibit estrogen sulfotransferase activity in the liver and effectively increase bioavailable estrogen in the body [77].

Cadmium

Cadmium has been shown to act as a ligand for the ER and tends to have proliferative action on human prostate cells through an ER-dependent mechanism rather androgen dependence [78]. As cadmium bioaccumulates in the body, further epidemiological analysis of cadmium and prostate cancer risk is warranted, particularly among men with occupational exposures.

Arsenic

Epidemiological studies have shown association between arsenic exposure and prostate cancer mortality in Taiwan [79]. It has also been documented that arsenic mediates endocrine disruption, specifically through interaction with ERs and activation of estrogen-regulated genes [80]

Symptoms

Prostate cancer may cause no signs or symptoms in its early stages. Prostate cancer in more advanced stage causes trouble in urinating, decreased force in the stream of urine, blood in the semen, discomfort in the pelvic area, bone pain, and erectile dysfunction.

Diagnosis

(a) *Digital rectal exam (DRE)*—to check for abnormalities in the texture, shape, or size of prostate gland
(b) *Prostate-specific antigen (PSA) test*—to check for elevated levels of PSA in blood indicating prostate infection, inflammation, enlargement, or cancer

If an abnormality is detected on a DRE or PSA test, ultrasound and prostate biopsy are done to confirm the presence of cancer. If the spread of cancer is suspected, imaging tests are recommended that include either bone scan, computerized tomography (CT) scan, magnetic resonance imaging (MRI), or positron emission tomography (PET) scan. Then cancer stage is assessed: stage I signifies very early cancer is confined to a small area of the prostate; in stage II, cancer cells may be aggressively grown on both sides of the prostate gland; in stage III, cells spread beyond the prostate to the seminal vesicles or other nearby tissues; and in stage IV, cancer has grown to invade nearby organs, such as the bladder, or spread to lymph nodes, bones, lungs, or other organs.

Treatment

For men diagnosed with very-early-stage prostate cancer, treatment may not be necessary; instead active surveillance, regular follow-up blood tests, rectal exams, and possibly biopsies are performed to monitor progression of cancer. If cancer is progressing, following treatment may be required.

(a) *Radiation therapy*—treatment of prostate gland to external radiations requiring 5 days a week for several weeks or treatment involving placement of many rice-sized radioactive seeds in prostate tissues which would deliver a low dose of radiation over a long period of time.
(b) *Hormone therapy*—medication with luteinizing hormone-releasing hormone (LH-RH) that stops body testicles from producing testosterone, medication with antiandrogens to prevent testosterone from reaching cancer cells, or surgery performed to remove testicles (orchiectomy) that in turn would reduce testosterone levels in body.
(c) *Surgery*—to remove prostate gland, some surrounding tissues, and few lymph nodes (radical prostatectomy).

(d) *Freezing prostate tissue*—cryosurgery or cryoablation involves freezing tissue to kill cancer cells.
(e) *Chemotherapy*—this treatment is opted in cases where cancer has spread to distant areas of the body and where hormonal therapy been left ineffective.
(f) *Biological therapy*—this is an immunotherapy involving the body's immune system to destroy cancer cells.

Female Reproductive Disorders

Experimental studies evidence that EDCs on exposure have the potential to impair ovarian physiology leading to morphological and functional alterations of the female reproductive system [81]. Early exposure to EDCs has been linked to reproductive tract malformations, precocious puberty [82], alterations of the hypothalamic-pituitary-ovarian axis, disorders in ovulation [83], fertility and fecundity, endometriosis, premature ovarian failure, and development of PCOS traits [4].

Polycystic Ovarian Syndrome

Polycystic ovarian syndrome, a heterogeneous syndrome characterized by persistent anovulation, amenorrhea, and hyperandrogenism in the absence of thyroid, pituitary, and adrenal disease [84, 85]. In the ovary, conscription and growth of follicles occur without selection of dominant preovulatory follicles leading to accumulation of multiple, small, antral follicles [86]. Hyperfunctioning of the theca and relative hypofunctioning of the granulosa cells accompany the acyclicity of the syndrome. PCOS has its origins both within and outside the hypothalamic-pituitary-ovarian axis contributing to feedback mechanisms and neuroendocrine functioning, ovarian steroidogenesis, insulin resistance, and obesity. Obesity and insulin resistance predominantly occur in about 50% of women with PCOS while obese women have a 12% risk of having PCOS [87].

EDCs Involved

Bisphenol A

In an epidemiological study constituting 71 PCOS and 100 healthy women, BPA levels were higher in PCOS women compared to normal ovulating non-hyperandrogenemic women [88]. In another study, female workers from BPA-exposed and unexposed factories in China add evidence that BPA has a deleterious impact on women's hormone homeostasis showing positive association between increased urine BPA concentration and higher prolactin and progesterone levels [89]. Developmental BPA exposure is reported to disrupt oogenesis, early meiosis, and normal follicle growth leading to accelerated follicle transition and increased incidence of atretic follicles [90–93]. BPAs are also reported to induce modifications in ovarian steroidogenic enzymes including 17β hydroxylase (P450c17), cholesterol side-chain cleavage enzyme (P450scc), and steroidogenic acute regulatory protein (StAR) implicating PCOS hyperandrogenism. There are two *in vivo* and *in vitro* mechanisms behind BPA-increasing androgens. The first mechanism involves interaction between BPA and sex hormone-binding globulin (SHBG) displacing androgens and disturbing androgen/estrogen balance, in turn leading to increased levels of circulating androgens. The second mechanism involves BPA-induced downregulation of androgen metabolism or hydroxylation by decreasing testosterone 2α-hydroxylase and testosterone 6β-hydroxylase activities [94].

Symptoms

(a) *Irregular periods*—the most common characteristic.
(b) *Excess androgen*—elevated levels of male hormones (androgens) may result in excess facial and body hair (hirsutism), adult acne or severe adolescent acne, and male pattern baldness (androgenic alopecia).
(c) *Polycystic ovaries*—enlarged ovaries containing numerous small fluid-filled sacs.

Diagnosis

(a) *Physical examination* of height, weight, and blood pressure
(b) *Pelvic examination* for signs of masses, growths, or other abnormalities
(c) *Blood tests*—to measure hormone levels and exclude possible causes of menstrual abnormalities or excess androgen that mimic PCOS
(d) *Ultrasound examination* of ovaries and the thickness of the lining of uterus

Treatment
PCOS treatment is generally focused on management of individual's main concerns, such as infertility, hirsutism, acne, or obesity. To treat these, lifestyle change to reduce weight through a low-calorie diet combined with moderate exercise activities is recommended. Followed by medications to regulate menstrual cycle, ovulate, and reduce excessive hair growth.

Reproductive Tract Anomalies
The differentiation of germ cells in the ovaries begins in the first trimester of intrauterine life; later, between the second and third quarter, primordial follicles are formed, and then follicles enter a latency period lasting from 15 to 50 years old in women [81]. The oocytes are non-degenerative cells with a longer life in the human body, and consequently they are exposed to the effects of EDCs during the entire period of their existence. Formation of normal ovarian follicles in the fetus depends on a balance between systemic concentrations of estrogen, inhibin, and activin. The uterus and fallopian tubes develop from two ducts known as the Müllerian ducts. During normal development, these ducts unite together, and single uterus with an open cavity and two fallopian tubes is formed [95]. Malformations of such structure to double, bicornuate, septate, or arcuate uterus are known as Müllerian anomaly, uterine anomaly, or reproductive tract anomaly. Estrogen exposure during the

critical period of formation of follicles changes follicular dynamics and leads to the above consequences.

EDC Involved

BPA crosses the placental barrier and has been linked to several adverse human reproductive effects, including recurrent miscarriage [96–98]. BPA tends to affect meiosis in ovaries, accelerate follicle transition, reduce oocyte quality in women undergoing *in vitro* fertilization, impair uterine endometrial proliferation, decrease uterine receptivity, and increase implantation failure [99, 100].

Symptoms

Women with double uterus have normal sex lives, pregnancies, and deliveries. Rarely double uterus and other abnormalities of uterine development are associated with infertility, miscarriage, premature birth, and kidney abnormalities.

Diagnosis

(a) *Pelvic examination*—to examine uterine anomalies
(b) *Ultrasound and sonohysterogram*—to check for abnormalities in the shape of uterus
(c) *Magnetic resonance imaging (MRI)*—to create cross-sectional images

Treatment

Double uterus has no signs or symptoms; treatment is rarely needed. Surgery to unite a double uterus is rarely done—to sustain a pregnancy, in case of partial division within uterus and no other previous medical explanation of pregnancy loss. In case of double vagina adding to double uterus, an operation to remove the wall of tissue separating the two vaginas is done making child birth slightly easier.

Uterine Leiomyomas

Uterine leiomyomas (UL, fibroids) are benign smooth muscle tumors of the myometrium causing morbidities including menorrhagia, abdominal pain, pelvic prolapse, infertility, and miscarriage in women. Fibroids range in size from seedlings, undetectable by the human eye, to bulky masses that can distort and enlarge the uterus. As UL occurrence is common among young women, size of leiomyomas increases with increased secretion of estrogens by ovaries and later decreases in size after menopause. It is the most common indication for hysterectomy in premenopausal women and can be effectively treated by gonadotropin-releasing hormone (GnRH) [101]. The level of ERs in leiomyomas is higher than in normal myometrium, indicating that leiomyoma is more sensitive to stimulation by estrogen [102].

EDCs Involved

BPA is suspected EDC for the increasing prevalence of uterine leiomyoma since BPA binds ERs and acts as an estrogen [103]. In an animal model study, exposure to BPA for 18 months increased the occurrence of myoma in mice [7, 8]. However, to verify the effect of BPA in growth of leiomyomas, a close sequential monitoring and epidemiological study is needed.

Symptoms

UL sometimes remains symptomless, but in some patients the symptoms include heavy menstrual bleeding, menstrual periods lasting more than a week, pelvic pressure or pain, frequent urination, difficulty in emptying the bladder, constipation, and backache or leg pain.

Diagnosis

(a) *Pelvic examination*—to check for irregularities in the shape of uterus and the presence of fibroids
(b) *Ultrasound*—to check for abnormalities transabdominally and transvaginally
(c) *Blood tests*—to check for anemia due to chronic blood loss during abnormal menstrual bleeding and other bleeding or thyroid disorders
(d) *MRI*—to identify type, size, and location of fibroids
(e) *Hysterosonography*—to observe submucosal fibroids and endometrium images
(f) *Hysterosalpingography*—to view highlighted X-ray images of uterine cavity and openings of fallopian tubes
(g) *Hysteroscopy*—to examine uterine walls and openings of fallopian tubes

Treatment

1. *Medications* are hormone based to regulate menstrual cycle, heavy menstrual bleeding, and pelvic pressure. This treatment does not eliminate fibroids but may shrink them.
2. *Noninvasive procedure*—the ultrasound transducer focuses sound waves (sonications) into the fibroid to heat and destroy small areas of fibroid tissue.
3. *Minimally invasive procedures* involve:
   (a) *Uterine artery embolization*—small embolic agents are injected into the uterus arteries, cutting off blood flow to fibroids and causing them to shrink and destroy.
   (b) *Myolysis*—a laparoscopic procedure involving radiofrequency energy, electric current, or laser destroying the fibroids by shrinking their blood vessels. A similar procedure called cryomyolysis that freezes the fibroids may also be done.
   (c) *Laparoscopic or robotic myomectomy*—fibroids are removed by breaking them into smaller pieces called morcellation.
   (d) *Hysteroscopic myomectomy*—opted in case of submucosal fibroids.
   (e) *Endometrial ablation*—uses heat, microwave energy, hot water, or electric current to destroy the lining of uterus, either ending menstruation or reducing menstrual flow.

4. *Traditional surgical procedures:*
   (a) *Abdominal myomectomy*—in case of multiple fibroids, very large fibroids or very deep fibroids, open abdominal surgical procedure is done to remove the fibroids.
   (b) *Hysterectomy*—a major surgery involving removal of the uterus, only proven permanent solution for UL.

Premature Ovarian Failure

Premature ovarian failure (POF), cessation of proper ovarian function before the age of 40, occurs in about 1% of reproductive age women [104]. In such case, ovaries fail to produce normal amount of estrogen and fail to release ova regularly, and early symptoms of menopause occur associated with other comorbidities. As the total ovarian follicle is established before birth in humans [105], EDC interference at early stage of development decreases ovarian follicle resting pool resulting in POF. The three possible mechanisms involved in causing POF include acceleration of apoptosis, blockage of follicle maturation, and premature activation of the follicle. Disruption of germ cell migration from the genital ridge into the developing gonad also results in ovarian dysgenesis [106].

EDCs Involved

Bisphenol A

Exposure to BPA and other EDCs leads to multioocyte follicles (MOF), a potential precursor of premature ovarian failure, with the process being mediated by ER-β agonist actions [107]. In animal studies, cystic adenomas were observed in the group of mice receiving higher doses (1 μg/kg) of BPA, but not in the controls. This data supports the hypothesis that BPA exposure causes adverse effects on the reproductive system during critical periods of differentiation [108]. The increased expression of genes involved in ovarian meiosis also explains the effects of BPA on female germ cells [109].

Pesticide

4-Vinylcyclohexene diepoxide (VCD) is considered as an occupational ovotoxic chemical [110]. Repeated doses of VCD may accelerate the apoptotic process of atresia and selectively destroy primordial and primary follicles in rats and mice [111]. Therefore, women exposed to VCD are considered at risk of POF. Methoxychlor (MXC), an estrogenic compound, has been demonstrated to show inhibition of growth and implantation of embryo and ovarian atrophy due to inhibition of folliculogenesis, leading to reduced atretic follicles and ovulation [112].

Symptoms

Symptoms of POF are similar to those experienced by menopaused and estrogen-deficient women. Infertility is a common complication along with irregular or skipped periods (amenorrhea), hot flashes, night sweats, vaginal dryness, irritability or difficulty in concentrating, and decreased sexual desire.

Diagnosis

(a) *Physical examination*—pelvic examination.
(b) *Pregnancy test*—to check for the possibility of an unexpected pregnancy in a woman of childbearing age.
(c) *Follicle-stimulating hormone (FSH) test*—FSH is a hormone released by the pituitary gland that stimulates the growth of follicles in ovaries. Women with premature ovarian failure have abnormally high levels of FSH in the blood.
(d) *Estradiol test*—the blood level of estradiol, a type of estrogen, is usually low in women with premature ovarian failure.
(e) *Prolactin test*—high levels of prolactin, the hormone that stimulates breast milk production, in blood can lead to problems with ovulation.
(f) *Karyotype*—to examine all 46 of chromosomes for abnormalities. Some women with premature ovarian failure may have only one X chromosome instead of two or may have other chromosomal defects.
(g) *FMR1 gene testing*—the FMR1 gene is the gene associated with fragile X syndrome, an inherited disorder that causes intellectual problems. The FMR1 test checks both X chromosomes to make sure they appear to be normal.

Treatment

Treatment for POF usually focuses on the problems that arise from estrogen deficiency. Certain recommended therapies are:

(a) *Estrogen therapy*—to prevent osteoporosis and relieve hot flashes and other symptoms of estrogen deficiency by compensating with estrogen that ovaries no longer produce
(b) *Calcium and vitamin D supplements*—important for preventing osteoporosis and are recommended if needed after bone density test.

Aneuploidy

Aneuploidy is an error in cell division that results in the daughter cells having either missing or extra number of chromosomes. Meiotic aneuploidy causes 10–20% of birth defects or results in spontaneous miscarriage of the fetus, whereas mitotic aneuploidy causes solid tumor cancers. When an extra copy of chromosome 21 is present, the fetus has Down syndrome, and if only one copy of the X chromosome is present, then the fetus suffers from Turner syndrome [113].

Turner Syndrome

It is relatively rare (about 1 out of 2500 live births), and approximately 10% of spontaneously aborted fetuses have this disorder. This indicates that most victims of Turner syndrome do not survive fetal life. In general, lacking one or more chromosomes is usually fatal [113].

Down Syndrome

Extra copies of the autosomal chromosomes also usually result in death. One extra copy is called trisomy. Trisomy of three of the smallest chromosomes (numbers 13, 18, and 21) usually does not result in fetal death but instead results in severe birth defects and, in most cases, early childhood death. An individual with extra copies of chromosomes 21 develops Down syndrome, though they are able to survive adulthood, they have significant health problems such as developmental delays and mental retardation. Other than this, an individual with two X and one Y chromosome (XXY) develops Klinefelter syndrome: a sterile male with many female body characteristics; an individual with three X chromosomes (XXX) develops into a sterile female, and an individual with extra copies of the Y chromosome (XYY) results in outwardly normal males with neurocognitive development [113].

EDCs Involved

Bisphenol A

Chromosomal arrangements in human eggs prior to fertilization have revealed a high frequency of chromosomal misalignment in relation to the spindle called congression failure. Studies suggest that variations in hormonal conditions either because of intrinsic changes related to aging in women or contamination by hormonally active compounds like BPA may lead to congression failure resulting in aneuploidy. In an animal study, BPA in pregnant dams during mid-gestation affected the developing ovaries causing anomalies in meiotic prophase including synaptic defects while in mature animals exposed *in utero* increased aneuploid oocytes and embryos. Such alterations led to cell cycle arrest and oocyte death, thus depleting the complement of normal oocytes [93].

Symptoms

Down syndrome individuals have intellectual and developmental problems ranging from mild to moderate. Though not all children with Down syndrome have the same features, some of the more common features are flattened facial features, small head, short neck, protruding tongue, upward slanting eyes, unusual for the child's ethnic group, unusually shaped or small ears, poor muscle tone, broad, short hands with a single crease in the palm, relatively short fingers and small hands and feet, excessive flexibility, tiny white spots on the colored part (iris) of the eye called Brushfield spots, and short height.

Infants with Down syndrome may be average size, but typically they grow slowly and remain shorter than other children the same age. In general, developmental milestones, such as sitting and crawling, occur at about twice the age of children without impairment. Thus, it is recommended to screen for Down syndrome in pregnant women, regardless of age.

Diagnosis

Screening and diagnostic tests are recommended to all pregnant women, regardless of age.

Screening Tests During Pregnancy

1. *Blood test*—to measure the levels of pregnancy-associated plasma protein-A (PAPP-A) and the pregnancy hormone known as human chorionic gonadotropin (HCG). Abnormal levels of PAPP-A and HCG indicate problem with the baby.
2. *Ultrasound*—to measure a specific area on the back of neonate's neck known as a nuchal translucency screening test. When abnormalities are present, more fluid than usual tends to collect in this neck tissue.
3. *Cell-free fetal DNA analysis*—done using circulating mother's blood after 10 weeks of gestation. In case of unclear results, any of the following samples is used but these samplings carry higher risk of miscarriage.
   (a) *Amniocentesis*—amniotic fluid surrounding the fetus used to analyze the chromosomes of the fetus.
   (b) *Chorionic villus sampling*—cells are taken from the placenta to analyze the fetal chromosomes.
   (c) *Cordocentesis (percutaneous umbilical blood sampling)*—fetal blood is taken from a vein in the umbilical cord and examined for chromosomal defects.

Diagnostic Tests for Newborns

After birth, the initial diagnosis of Down syndrome is often based on the baby's appearance. Chromosomal karyotype test is done to analyze the presence of extra chromosome 21 present in all or some cells.

Treatment

Children with Down syndrome are stimulated at an early age with appropriate sensory, motor, and cognitive activities involving development of motor, language, social, and self-help skills. This may involve a team of specialist including primary care pediatrician; pediatric cardiologist; pediatric gastroenterologist; pediatric endocrinologist; developmental pediatrician; pediatric neurologist; pediatric ear, nose, and throat (ENT) specialist; pediatric eye doctor (ophthalmologist); audiologist; physical therapist; speech pathologist; and occupational therapist.

Breast Cancer

More recently, epidemiological studies have revealed that the intrauterine environment may influence the risk to develop breast cancer later in life. The typical factors influencing the development of breast cancer include age at menarche, first pregnancy, menopause, lactation, and exposure to EDCs. All of these factors are related to lifetime exposures to ovarian hormones. During postnatal life, the mammary gland undergoes massive architectural changes, comparable to those usually associated with organogenesis. These changes occur in response to alterations in endogenous hormone levels such as those associated with puberty and pregnancy. Many studies also suggest developmental exposure to EDCs (exogenous hormone) mimics and alters normal patterns of tissue organization resulting in disruption of stromal-epithelial interactions [14, 114]. Thus, both epidemiological and experimental data are consistent with the hypothesis that excessive estrogen exposure during development may increase the risk of developing breast cancer.

EDCs Involved

Bisphenol A

In animal studies, BPA is shown to readily cross the placenta [115, 116] and bind α-fetoprotein which is an estrogen-binding protein that prevents maternal estrogen from entering the circulation of the fetus with negligible affinity relative to estradiol, resulting in an enhanced bioavailability during neonatal development. In this case, BPA present in fetus and amniotic fluid during maternal exposure was higher in concentrations than that of maternal blood. In another study, BPA-exposed pregnant dams exhibited altered growth parameters of the mammary gland anlagen. Changes in the appearance of the mammary epithelium such as decreased cell size, delayed lumen formation, and increased ductal area were observed. In stroma, BPA exposure promoted advanced maturation of the fat pad and altered localization of fibrous collagen [14]. On puberty, the area and number of terminal end buds relative to the gland ductal area increased, whereas cell death decreased in BPA-exposed offspring compared with controls. These effects may be attributed to an increased sensitivity to estradiol that has been observed in the BPA-exposed animals [117].

When BPA was administered to pregnant dams, fetal exposure to BPA, from gestational d9 to postnatal d1, resulted in the development of carcinomas in situ in the mammary glands of 33% of the rats at the time of puberty, whereas no neoplasias were observed in control [118]. Fetal exposure to BPA significantly increased the number of precancerous lesions (intraductal proliferation) by three to four times, and the lesions were highly proliferative and contained abundant ER-positive cells, suggesting that the proliferative activity in these lesions may be estrogen mediated [119].

Dioxins

Prenatally exposed rats to TCDD increased number of terminal buds at the time of puberty, in turn increasing the propensity of cancer by altering mammary gland morphogenesis [120]. Contrastingly, another study revealed that the prenatal exposure to TCDD resulted in impaired development of terminal end buds that remained in the gland for prolonged periods, whereas in control terminal end buds were transient structures that regressed when ductal development was complete [121].

Symptoms

Breast lump or thickening that feels different from the surrounding tissue; change in the size, shape, or appearance of a breast; changes to the skin over the breast, such as dimpling; a newly inverted nipple; peeling, scaling, or flaking of the pigmented area of skin surrounding the nipple (areola) or breast skin; and redness or pitting of the skin over the breast.

Diagnosis

(a) Breast examination—for the presence of any lymph nodes in the armpit or other abnormalities in breast
(b) Mammogram—screening for abnormalities
(c) Breast ultrasound—to produce images of breast lump (solid mass or a fluid-filled cyst) structures deep within the body
(d) Removing a sample of breast cells for biopsy
(e) Breast magnetic resonance imaging (MRI)
(f) Computerized tomography (CT) scan
(g) Positron emission tomography (PET) scan

Treatment
Breast cancer treatment options are based on type, stage, size, and sensitivity of cancer cells to hormones.

1. *Breast cancer surgery.*
   (a) *Removal of the breast cancer (lumpectomy)*—the tumor and a small margin of surrounding healthy tissue are removed.
   (b) *Removal of the entire breast (mastectomy)*—in case of skin-sparing mastectomy, the skin over the breast is left intact to improve reconstruction and appearance.
   (c) *Removal of a limited number of lymph nodes (sentinel node biopsy)*—to determine whether cancer has spread to lymph nodes in the armpit.
   (d) *Removal of several lymph nodes (axillary lymph node dissection)*—if cancer is found in the sentinel lymph nodes.
   (e) *Removal of both breasts*—in case of increased risk of cancer in the other breast because of genetic predisposition or strong family history.
2. *Radiation therapy*—external beam radiation is commonly used after lumpectomy for early-stage breast cancer.
3. *Chemotherapy*—uses drugs to destroy cancer cells and decrease the chance that the cancer will recur. This is also known as adjuvant chemotherapy.
4. *Hormone therapy*—termed hormone-blocking therapy, used to treat breast cancers that are sensitive to hormones such as estrogen receptor-positive (ER-positive) and progesterone receptor-positive (PR-positive) cancers.
5. *Targeted drug*—to treat specific abnormalities within cancer cells.
6. *Supportive (palliative) care*—focuses on providing relief from pain and other symptoms of serious illness.

Endometriosis
Endometriosis is an estrogen-dependent gynecological disorder occurring in 6–10% of women associated with pelvic pain, infertility, and dysfunctional immune system. This disorder involves an inappropriate uncontrolled growth of endometrial cells confined to the lining of the uterus.

In the case of nonmalignant tumors, cells lose their normal control, often dedifferentiate, and grow in an inappropriate place. These endometrial cells attach to the organs and tissues and are able to invade the underlying tissue. This leads to the formation of deep lesions and abnormal bands of fibrous tissue causing pelvic tissues and organs to stick to each other. The endometrial lesions are often fluid-filled cysts and are highly invasive [122].

EDCs Involved

Dioxins

As with all dioxin-like toxicants and coplanar PCBs, cell function is mediated by AhR and is well expressed in both endometrium and immune cells [123, 124]. This AhR agonistically affects the inflammatory processes by increasing estrogen levels, suppressing progesterone level, and inducing immune dysfunction, all of which in turn initiate endometrial menstruation [125]. Seveso disaster evidences the dioxin toxicity among vulnerable (*in utero* and neonates) exposure groups at the time of the accident are now detected with endometriosis, postulating early year exposure [126].

Polychlorinated Biphenyl

Studies demonstrate that adult body burden assessment studies alone are unlikely to reveal the true association of PCBs to the subsequent development of endometriosis, as the toxic effects of PCBs are greater when exposure occurs earlier in life [127, 128].

Symptoms

(a) *Painful periods (dysmenorrhea)*—pelvic pain and cramping accompanied by lower back and abdominal pain.
(b) *Pain with intercourse* is common with endometriosis.
(c) *Pain with bowel movements or urination* during period.
(d) *Excessive bleeding*—occasional heavy periods (menorrhagia) or bleeding between periods (menometrorrhagia).
(e) *Infertility*—endometriosis is first diagnosed in some women seeking treatment for infertility.
(f) *Other symptoms include* fatigue, diarrhea, constipation, bloating, or nausea, especially during menstrual periods.

Diagnosis

(a) *Pelvic examination*—to check for abnormalities, such as cysts on reproductive organs or scars behind the uterus.
(b) *Ultrasound*—to identify cysts associated with endometriosis (endometriomas).
(c) *Laparoscopy*—to check for endometrial tissue outside the uterus and find the location, extent, and size of the endometrial implants.

Treatment

Treatment for endometriosis usually involves medications or surgery. The approach depends on the severity of symptoms.

(a) Medication—pain reliever such as the nonsteroidal anti-inflammatory drugs are used to ease painful menstrual cramps.
(b) *Hormone therapy*—supplemental hormones effectively reduce and eliminate the pain of endometriosis. The rise and fall of hormones during the menstrual cycle causes endometrial implants to thicken, break down, and bleed. In case of hormone medication, though it is not a permanent fix for endometriosis, it slows the endometrial tissue growth and prevents new implants of endometrial tissues.
(c) *Conservative surgery*—surgery removes as much endometriosis as possible besides preserving the uterus and ovaries and (conservative surgery) increases chances of pregnancy success.
(d) *Hysterectomy*—in severe cases of endometriosis, surgery to remove the uterus and cervix (total hysterectomy) as well as removal of both ovaries may be the best treatment.

Precocious Puberty

Precocious puberty (PP) is defined as development of pubertal changes before the age of 8 years in girls along with an increase in gonadotropin, sex steroid, accelerated somatic development, and bone age. Advanced progression of secondary sexual characteristics occurs in children with PP leading to compromised adult height, poor social adaptability, and emotional disorders [129]. PP is of two classes.

Central precocious puberty (CPP)—due to early maturation of the hypothalamic-pituitary-gonadal (HPG) axis. The frequency of CPP is 1 in 5000–1 in 10,000. It is more common in girls, female/male ratio ranging from 3:1 to 23:1 [130]. Childhood obesity, intrauterine growth retardation, parental obesity, diabetes, and cancer are associated with early thelarche as hormones stays at an abnormal level for a long time [131]. CPP among girls is also observed with central nervous system pathologies: hydrocephalus, meningomyelocele, neurofibromatosis, and hypothalamic hamartomas [132–134].

Peripheral precocious puberty (PPP) or gonadotropin-independent precocious puberty—endogenous or exogenous estrogen in girls in excess leads to PPP manifested with recurrent follicular cyst and irregular vaginal bleeding. PPP is due to mutations in the gene encoding the alpha subunit of the Gs protein. Primary hypothyroid is obvious among thelarche in girls without accelerated growth. Some may have elevated prolactin, galactorrhea, or ovarian cysts [130].

EDCs Involved

Polybrominated Biphenyls

PBB exposure effects after an industrial accident in the state of Michigan in 2000 were studied among pregnant women and their daughters who were breastfed. The girls with intrauterine exposure to high PBB concentrations (>7 ppm) were compared with the girls who had no exposure (<1 ppm); menarche was observed to occur 1 year earlier in those who were exposed to high concentrations of PBB. Breastfed girls were also observed to have pubarche 1 year earlier than the girls who were not breastfed. No significant difference in terms of breast development was studied [135, 136].

Pesticides

Studies in relation to intrauterine exposure to high doses of pesticides reveal the incidence of early thelarche and early pubarche in girls. Effects of DDT and its metabolite DDE on pubertal development have been investigated in several studies. One year earlier menarche was observed among girls exposed to high amounts of DDT/DDE in the intrauterine period [137]. In animal studies, endosulfan showed estrogenic effect by inhibiting follicle-stimulating hormone (FSH), luteinizing hormone (LH), and testosterone production [138].

Polychlorinated Biphenyls

While there is no clear relationship detected between intrauterine or postnatal exposure to PCB and age of puberty or menarche, few studies report occurrence of menarche significantly earlier in girls exposed to PCB 52, 70, 101, +90, and 187 subgroups [139, 140].

Dioxins

On exposure to dioxins, reduction in testicular volume in boys and delayed breast development in girls but no change in age of menarche or pubarche were observed [141].

Phthalates

Serum samples from 41 girls from Puerto Rico with premature breast development and 35 control cases were analyzed for determining the possible presence of phthalate esters. Girls with premature breast development, 28 of 41 (68%), had measurable levels of phthalates [dimethyl, diethyl, dibutyl, and di-(2-ethylhexyl)] compared with 6 of the 35 (17%) control samples [142].

Bisphenol A

BPA is suggested to have estrogenic effects causing precocious puberty. In animal studies, rats exposed to BPA showed earlier signs of puberty and abnormal pattern of ovulation cycle [143].

Lead

Girls with high serum lead levels were observed to have delayed menarche and pubarche [144, 145].

Symptoms

PP symptoms include early development of breast (early thelarche) and first period (menarche) before age 8 in girls and enlarged testicles and penis, facial hair (usually grows first on the upper lip), and deepening voice before age 9 in boys. Symptoms such as pubic or underarm hair, rapid growth, acne, and adult body odor are noticed in both boys and girls.

Diagnosis

1. *Physical examination and reviewing family medical history.*
2. *Blood test*—to determine gonadotropin-releasing hormone (Gn-RH) stimulation. In children with CPP, Gn-RH causes other hormone levels to rise. In children with PPP, other hormone levels remain the same.
3. X-rays of hand and wrist to determine whether bones are growing too quickly (bone age).
4. *Additional tests for CPP*
   (a) *MRI*—to check if any brain abnormalities are causing the early start of puberty
   (b) *Thyroid test*—to check for hypothyroidism which would sign for fatigue, sluggishness, increased sensitivity to cold, constipation, a drop in school performance, or pale, dry skin
5. *Additional testing for PPP*
   (a) *Blood test*—to check hormone levels
   (b) *Ultrasound*—in girls to check for an ovarian cyst or tumor

Treatment

Treatment for precocious puberty depends on the cause. The primary goal of treatment is to enable the child to grow to a normal adult height.

(a) *Treating central precocious puberty*—most children with CPP have no underlying medical conditions and are treated effectively by Gn-RH analogue therapy. The medication is continued until normal age of puberty reaches.
(b) *Treating an underlying medical condition*—in this case, if tumor is involved in producing hormones, it is surgically removed which in turn will stop progression of puberty.

22.1.1.4 Thyroid Disorders

Thyroid hormones are involved in regulating metabolism, bone remodeling, cardiac function, and neurodevelopment. During pregnancy, fetus relies entirely on transplacental transfer of maternal thyroid hormones for the development of brain as its absence or even a marginal low thyroxine level causes reduction in cognitive functions by reducing neuronal growth and differentiation in the cerebral cortex, hippocampus, and cerebellum [146, 147].

Mechanism Involved

EDCs disturb overall activity of the thyroid gland by interfering with TSH receptor, as thyroid-stimulating hormone (TSH) stimulates all steps of the hormone production.

The function of the sodium iodide symporter (NIS) or thyroid peroxidase (TPO) is affected either by stimulation or inhibition. In human, EDCs interfere with thyroid-binding globulin (TBG) compromising the bioavailability of thyroid hormones to the nuclear thyroid hormone receptors (TR) [148]. EDCs are also known to decrease T4 half-life in serum by inducing activity of the hepatic uridine diphosphate glucuronyl transferase that glucuronidate T4 [149]. In order to compensate the increased metabolism of T4, thyroid hormone production is increased [150]. Albeit, the compensating capacity of the thyroid gland combined the long-term effects of numerous EDCs that accumulate with time resulting in euthyroid goiter or hypothyroidism.

EDCs Involved

Polychlorinated Biphenyls and Dioxins

Many of these compounds, especially the hydroxylated metabolites, have high degree of structural resemblance to thyroxine (T4). In experimental animal studies, exposure to PCBs or dioxin resulted in the reduction of serum thyroid hormone levels, especially affecting T4 in dose-dependent manner [151, 152]. In a study among US-Vietnam war veterans who had sprayed Agent Orange, high serum TCDD levels were associated with increased serum TSH levels [153]. Infants of women exposed to the Seveso industrial accident also had elevated neonatal TSH levels compared to infants whose mothers were not exposed to TCDD [154].

Polybrominated Flame Retardants

Derivatives of BFRs exhibited higher thyroid hormone activities than parental compounds, indicating that hydrogen bonds are an important factor governing thyroid hormone activities [155, 156]. BFRs interfere with thyroid function at several levels, by interacting with TR, binding proteins and hepatic clearance resulting in the impairment of TH-dependent metamorphosis [157, 158]. PBDE acts through induction of hepatic enzymes involved in glucuronidation [159] or by downregulating the transport protein transthyretin (TTR) and transmembranal thyroid hormone transport [160].

Pesticides

Among numerous chemicals constituting pesticides, both persistent organochlorine and nonpersistent such as organophosphorus, carbamates, and pyrethroids interfere with thyroid function. Methoxychlor is found to decrease hepatic deiodinase activity [161].

Perfluorinated Chemicals

High level and continuous exposure to PFOA show negative association between PFOA and T4 as the study of 506 employees from PFC manufacturer company had shown reduced thyroid function due to high level of exposure [162]. Several animal studies have found decreased levels of T4 after both short-term and long-term exposure to PFOS [163, 164].

Phthalates

A study among 76 pregnant women reported significant negative association between the metabolite of dibutyl phthalate and T4 (free and total) [165]. In another study, negative associations between DEHP exposure and free T4 and total T3 were reported in adult men. In the study of children exposed to similar concentration of PEs as adult, it showed negative association with serum levels of T3 and height attainment [148].

Bisphenol A

BPA binds to thyroid hormone receptor (TR) as a weak ligand and acts as an antagonist to T3, inhibiting TR-mediated transcriptional activity [166, 167]. Derivatives of BPA are reported to show higher affinity for the receptors [157, 168]. Developmental studies in rats suggest that BPA specifically antagonizes the β-TR and less effective antagonist to α-TR. This thereby inhibits negative feedback of thyroid hormone at pituitary level, resulting in elevated T4 levels with unchanged serum TSH similar to that of thyroid hormone resistance syndrome [169].

Lead

Lead adversely affects the pituitary-thyroid axis through an unknown mechanism. Several occupational studies have demonstrated a pattern of low peripheral T4 levels with inappropriate low serum TSH concentrations. Long-term low-level exposure among male adolescent automobile mechanics also exhibited negative correlation between blood lead levels and free T4 with no difference in serum TSH or T3 levels compared to unexposed controls [170].

Symptoms

Hypothyroidism

Symptoms based on severity of the hormone deficiency and age of occurrence are as follows:

(a) *Adults*—in adults the symptoms include fatigue, increased sensitivity to cold, constipation, dry skin, weight gain, puffy face, hoarseness, muscle weakness, elevated blood cholesterol level, muscle aches, tenderness and stiffness, swelling in joints, heavier or irregular menstrual periods, thinning hair, slowed heart rate, depression, and impaired memory.
(b) *Infants*—in offspring's born without thyroid gland or with impaired function, hypothyroidism occurs associated with other complications such as yellowing of the skin and whites of the eyes (jaundice); frequent choking; a large, protruding tongue; puffy appearance to the face; constipation; poor muscle tone; and excessive sleepiness and in case not treated leads to severe physical and mental retardation.
(c) *Children and teens*—the symptoms are almost similar to that of adults; in addition other symptoms are stunted growth, delayed development of permanent teeth, delayed puberty, and poor mental development.

Diagnosis

Blood test—to measure the level of TSH and thyroid hormone, thyroxine

Treatment

Standard treatment for hypothyroidism involves prescription of the synthetic thyroid hormone.

22.1.1.5 Developmental Neurotoxicity

Developmental Susceptibility of the Brain to EDCs

The developing human brain is inherently much more susceptible to injury caused by toxic agents than the brain of an adult. The susceptibility stems from the fact that during the 9 months of prenatal life, the human brain develops from a strip of cells along the dorsal ectoderm of the fetus into a complex organ consisting of billions of precisely located, highly interconnected, and specialized cells. Optimum brain development requires neurons that move along precise pathways from their points of origin to their assigned locations; establish connections with other cells, both nearby and distant; and learn to communicate with other cells [171].

All these processes take place within a tightly controlled time frame, in which each developmental stage has to be reached on schedule and in the correct sequence. Because of the extraordinary complexity of human brain development, windows of unique susceptibility to toxic interference that arises have no counterpart in the mature brain or in any other organ. If a developmental process in the brain is halted or inhibited, there may be little potential for later repair; otherwise the consequences are permanent [172]. During fetal development, the placenta offers some protection against unwanted chemical exposures but is not an effective barrier against environmental pollutants [173].

Neuro-Disorders

One in every six children has a developmental disability, and in most cases these disabilities affect the nervous system. The susceptibility of infants and children to EDCs is enhanced by their increased exposures, augmented absorption rates, and diminished ability to detoxify many exogenous compounds, relative to that of adults [174, 175]. The most common neurodevelopmental disorders include learning disabilities, sensory deficits, developmental delays, cerebral palsy, autism, and attention deficit hyperactivity disorder [176]. These disorders are difficult to treat, and the disabilities are permanent, exorbitant to family and society [177–180].

EDCs Involved

Polychlorinated Biphenyls

Several studies have reported negative associations between prenatal exposure of PCBs and measures of cognitive function in infancy or childhood, learning, memory, and IQ [181]. The Netherland and Michigan cohort studies have also reported associations between PCB exposures and problems such as inattention, impulsiveness, and ADHD-related behaviors [182, 183]. In addition to the cognitive and

ADHD-like effects observed, children from Taiwan exhibited sexually dimorphic behavior with boys displaying a deficit in spatial abilities [184].

Polycyclic Aromatic Hydrocarbons

A cohort study of Polish Caucasians reported adverse effects of exposure on fetal growth and cognitive development [185, 186]. Animal studies have reported neurodevelopmental and behavioral defects including depression-like symptoms and memory impairment in the absence of other overt toxicological effects on PAH exposure during prenatal and neonatal periods [187, 188]. Benzo(a)pyrene is also been reported to alter levels of noradrenaline, dopamine, and serotonin underlying the anomalies in learning and cognitive development [189].

Phthalates

Studies suggest more ADHD-like behaviors among 4–7-year-old children and autistic-like behavior among 7–9-year-old children among mothers who had higher urinary levels of low molecular weight phthalate metabolites during pregnancy [190, 191]. Prenatal exposure to PEs also tends to be associated with decreased masculine behavior in preschool boys [192].

Bisphenol A

The gestational BPA concentrations are positively associated with aggression and hyperactivity in children; this association is stronger in girls than in boys. BPA levels in maternal urine are also associated with higher hyperactivity in 2-year-old girls, but not boys [193]. These results suggest that developmental exposures to BPA may have sexually dimorphic effects on behavior later in life. In contrast, in a mixed-race mother-child cohort study, child urinary BPA concentrations at 3 years of age were associated with higher emotional reactivity scores in children [194]. Animal studies that show changes in dopaminergic system in the forebrain on exposure to BPA suggest action via estrogen receptors to alter dopamine signaling, leading to possible hyperactivity and attention deficits in humans [195, 196].

Pesticides

Prenatal exposure to chlorpyrifos (CPF) is linked to developmental cognitive effects such as reduced head circumference, lower birth weight, abnormal reflexes, attention deficits, neurodevelopmental abnormalities, and significant reduction in childhood IQ by 5–7 points [197]. Another investigation also reveals link between CPF exposure and brain structure in children ages 5.9–11.2 years from a community-based cohort [198]. Magnetic resonance imaging of the agriculture-based subjects revealed enlargement of multiple brain regions in individuals with high CPF exposure, specifically in regions that are critically involved in various cognitive and behavioral processes including attention, social cognition, reward, emotion, and inhibitory control [199–202].

Heavy Metals

Cadmium has been shown to decrease calmodulin (CaM) protein which modulates calcium/CaM-dependent kinases (CaMKs), enzymes involved in the regulation of neurotransmitter synthesis, neurotransmitter release, and synaptic plasticity [203].

Attention Deficit Hyperactivity Disorder

ADHD is a brain disorder marked by an ongoing pattern of inattention (lacks persistence, difficulty in sustaining focus, and disorganized), hyperactivity (excessive fidgets, taps, talks, and extreme restlessness), and impulsivity (hasty actions, making important decisions without considering long-term consequences) that interferes with functioning or development.

Symptoms

ADHD symptoms start before age 12 and, in some children, are noticeable as early as 3 years of age. ADHD symptoms can be mild, moderate, or severe, and they may continue into adulthood. ADHD occurs more often in males than in females. The symptoms include:

(a) *Inattention*—fails to pay close attention to details or makes careless mistakes in schoolwork, has trouble in staying focused in tasks or play, is appearing not to listen, has difficulty in following on instructions, has trouble in organizing tasks and activities, avoids tasks that require focused mental effort, loses items (toys, school assignments, pencil) needed for tasks, be easily distracted, and is forgetting to do some daily activities.
(b) *Hyperactivity and impulsivity*—fidget with or tap hands or feet or squirm in the seat, have difficulty in staying seated, be in constant motion, run around or climb in situations when it's not appropriate, have trouble playing or doing activities quietly, are talking too much, are blurting out answers, are interrupting the questioner, have inability to delay gratification, are socially intrusive, and are interrupting or intruding on others' conversations or activities.

Diagnosis

There is no specific test for ADHD, but making a diagnosis includes:

(a) *Medical exam*ination—to rule out possible causes of symptoms
(b) *Information gathering*—such as any current medical issues, personal and family medical history, and school records
(c) *Interviews* or *questionnaires* for family members
(d) *Determination of ADHD criteria* from the Diagnostic and Statistical Manual of Mental Disorders
(e) *ADHD rating scales*—to collect and evaluate information about the child

Treatment

Standard treatments for ADHD disorder in children include medications, education, training, and counseling.

Autism Spectrum Disorder

Autism spectrum disorder (ASD) is a serious neurodevelopmental disorder that impairs a child's ability to communicate and interact with others. It also includes restricted repetitive behaviors, interests, and activities. These issues cause significant impairment in social, occupational, and other areas of functioning.

Symptoms

Some children show signs of ASD in early infancy; others with ASD develop normally for the first few months or years of life but later suddenly become withdrawn or aggressive or lose language skills that they have already acquired. Every child with ASD has a unique pattern of behavior and level of severity—from low functioning to high functioning. Severity is based on their social communication impairments, restrictiveness, and repetitive nature of behaviors, along with how these impact the ability to function.

(a) *Social communication and interaction*—the child fails to respond to his or her name or appears not to hear; resists cuddling and holding; lacks facial expression and eye contact; has delayed speech; speaks in abnormal tone; repeats words or phrases verbatim; is unable to understand simple questions; is unaware of feelings; and has inappropriate social interaction by being passive, aggressive, or disruptive.
(b) *Pattern of behavior*—performs repetitive movements, such as rocking, spinning, or hand flapping; develops specific routines and gets disturbed at the slightest change; moves constantly, resistant to change; has odd movement patterns, such as clumsiness or walking on toes; has odd, stiff, or exaggerated body language; is unusually sensitive to light, sound, and touch and yet oblivious to pain; is becoming fixated on an object or activity with abnormal intensity or focus; and is having odd food preferences, such as eating only a few food or eating only foods with a certain texture.

Diagnosis

(a) ASD varies widely in severity, making diagnosis difficult. There is no specific medical test to determine the disorder. Instead, a specialist in ASD observes a child's social interactions, communication skills, and behavior that have developed and changed over time.
(b) Check whether child meets the symptom criteria in the Diagnostic and Statistical Manual of Mental Disorders (DSM-5).
(c) Other tests include genetic disorder test—to identify the presence of fragile X syndrome.

Treatment

No cure exists for autism spectrum disorder. The goal of treatment is to maximize child's ability to function by reducing ASD symptoms and supporting development and learning.

Congenital Adrenal Hyperplasia

CAH is an autosomal recessive disorder that induces elevated adrenal androgens, particularly testosterone, in the fetus. The cause is deficiency in the enzyme 21 hydroxylase. It occurs in 1 in 14,500 live births. Girls with CAH exhibit male features in the external genitalia and usually are treated with corticosteroids. Even with treatment, CAH girls, presumably due to elevated prenatal testosterone, display some masculine proclivities in toy choice, rough-and-tumble play patterns, and choice of playmates, choosing boys as often as girls, departure from normal girl choices [204]. CAH can serve as a striking model of the potency of developmental hormonal conditions to influence sexually dimorphic behaviors.

Symptoms

There are two major types of CAH.

(a) *Classic CAH*—This more severe form of the disease is usually detected in infancy or early childhood. The most obvious sign in girls is often abnormal-appearing genitals that look more male than female, which may include an enlarged clitoris—a condition called ambiguous external genitalia. The condition is not typically as easily seen in baby boys, although some affected male infants have an enlarged penis. Other symptoms in infants include poor weight gain, weight loss, dehydration, and vomiting. In children and adults, the symptoms include very early puberty, rapid growth during childhood, irregular menstrual cycles in women, and infertility in both women and men.
(b) *Nonclassic CAH*—This milder form usually becomes evident in late childhood or early adulthood. The symptoms in women include irregular or absent menstrual periods; masculine characteristics such as facial hair, excessive body hair, and a deepening voice; and infertility. The symptoms in both women and men include early puberty, rapid growth during childhood, severe acne, low bone density, and obesity.

Diagnosis

(a) *Physical examination*
(b) *Blood and urine tests*—to measure levels of cortisol, aldosterone, and androgens
(c) *Prenatal testing*—amniocentesis and chorionic villus sampling based on family history
(d) *Karyotyping*—to determine the sex of a child

Treatment

(a) *Medications*—replacement hormone medication given to boost the levels of deficient hormones and restore them to normal levels.
(b) *Surgery*-—to correct the appearance and function of the genitals. The surgery is typically performed between 2–6 months of age.
(c) *Prenatal management*—when CAH is diagnosed in the fetus, pregnant mother is given powerful corticosteroid drug that may cross placenta and

suppress the activity of the fetus's adrenal glands. By reducing the secretion of androgens, female genitals are allowed to develop normally, and masculine features that may develop later in female fetuses are also reduced. In the case of male fetus, shorter course of lower-dose corticosteroid drug treatment is recommended.

(d) *Prognosis*—many children with CAH successfully manage the condition by staying on replacement hormone medications.

22.1.1.6 Epigenetics and Disease Susceptibility

The term epigenomics generally refers to the changes in gene expression that takes place without change in the DNA sequence [205]. Epigenetic changes occur by molecular modification in both DNA and chromatin. The more-evidenced DNA methylation occurs at carbon-5 position of cytosine in CpG dinucleotide and chromatin packing in DNA by posttranslational histone modifications [206–208]. The other epigenetic mechanisms affecting gene expression are noncoding of RNAs (microRNAs) and regulation of higher-level organization chromatins within nucleus. These are the general mechanism by which prenatal and postnatal exposures are linked to phenotypic changes later in life [205]. Alteration of epigenetic marks has a central role in determining the functional output of the information that is stored in the genome. The three genomic targets that are susceptible to gene expression changes owing to EDC's perturbation are the promoter regions of some housekeeping genes, transposable elements lying adjacent to genes, and regulatory elements of imprinted genes. These target regions are rich in CpG dinucleotide sequences, which unmethylate or methylate; in some cases, histone modifications occur in these region [209, 210]. Epigenetic changes may inherit mitotically in somatic cells by providing long-term effects on gene expression or may inherit transgenerationally in germ cells, affecting the health of future generations [211–213].

Major EDCs that have been implicated in promoting toxicity for multiple generations are BPA, polycyclic hydrocarbons, and pesticides [214–218]. In a study, pregnant rat was exposed to methoxychlor (an estrogenic pesticide), and it caused defects in spermatogenic capacity that were transmitted through four subsequent generations. In addition, transgenerational phenotypes included adult-onset diseases such as kidney diseases, immune abnormalities, behavioral changes, learning disabilities, prostate lesions, and cancer [53, 211, 219–222]. The loss and subsequent reestablishment of the epigenetic profile in the developing embryo comprise a critically sensitive period during which the system is particularly vulnerable to environmental influences. Thus, though exposure to EDCs may not result in obvious phenotypes, the exposure can alter the epigenetic programming of both somatic and germ cells inducing subtle functional changes leading to disease later in life and in future generations (Fig. 22.1).

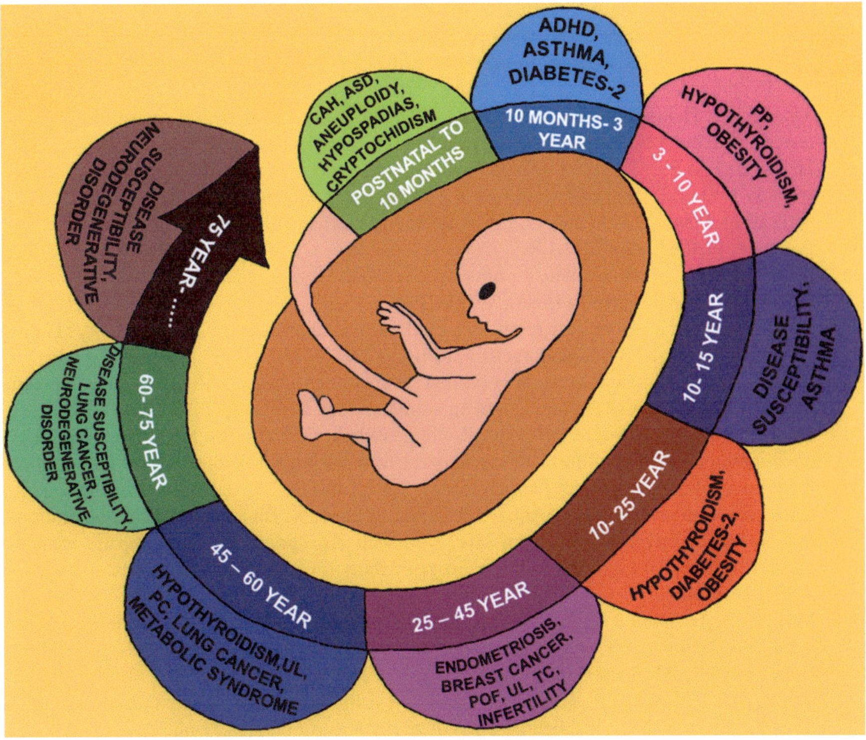

Fig. 22.1 Disease susceptibility on prenatal and postnatal exposure

References

1. Schug TT, Janesick A, Blumberg B, Heindel JJ. Endocrine disrupting chemicals and disease susceptibility. J Steroid Biochem Mol Biol. 2011;127(3-5):204–15.
2. Polyzos SA, Kountouras J, Deretzi G, Zavos C, Mantzoros CS. The emerging role of endocrine disruptors in pathogenesis of insulin resistance: a concept implicating nonalcoholic fatty liver disease. Curr Mol Med. 2012;12:68–82.
3. Sharpe RM, Drake AJ. Bisphenol A and metabolic syndrome. Endocrinology. 2010;151:2404–7.
4. Diamanti-Kandarakis E, Bourguignon JP, Giudice LC, Hauser R, Prins GS, Soto AM, et al. Endocrine-disrupting chemicals: an Endocrine Society scientific statement. Endocr Rev. 2009;30(4):293–342.
5. Baillie-Hamilton PF. Chemical toxins: a hypothesis to explain the global obesity epidemic. J Altern Complement Med. 2002;8:185–92.
6. Wardle J, Cooke L. The impact of obesity on psychological well-being. Best Pract Res Clin Endocrinol Metab. 2005;19:421–40.

7. Newbold RR, Jefferson WN, Padilla-Banks E. Long-term adverse effects of neonatal exposure to bisphenol A on the murine female reproductive tract. Reprod Toxicol. 2007a;24:253–8.
8. Newbold RR, Padilla-Banks E, Snyder RJ, Phillips TM, Jefferson WN. Developmental exposure to endocrine disruptors and the obesity epidemic. Reprod Toxicol. 2007b;23:290–6.
9. Thayer KA, Heindel JJ, Bucher JR, Gallo MA. Role of environmental chemicals in diabetes and obesity: a National Toxicology Program workshop review. Environ Health Perspect. 2012;120:779–89.
10. Irigaray P, et al. Benzo[a]pyrene impairs β-adrenergic stimulation of adipose tissue lipolysis and causes weight gain in mice. A novel molecular mechanism of toxicity for a common food pollutant. FEBS J. 2006;273:1362–72.
11. Jerrett M, et al. Traffic-related air pollution and obesity formation in children: a longitudinal, multilevel analysis. Environ Health. 2014;13:49.
12. McConnell R, et al. Does near-roadway air pollution contribute to childhood obesity? Pediatr Obes. 2015;11:1–3. https://doi.org/10.1111/ijpo.12016.
13. Zheng Z, et al. Exposure to ambient particulate matter induces a NASH-like phenotype and impairs hepatic glucose metabolism in an animal model. J Hepatol. 2013;58:148–54.
14. Vandenberg LN, Maffini MV, Wadia PR, Sonnenschein C, Rubin BS, Soto AM. Exposure to environmentally relevant doses of the xenoestrogen bisphenol-A alters development of the fetal mouse mammary gland. Endocrinology. 2007;148:116–27.
15. Chamorro-Garcia R, Sahu M, Abbey RJ, Laude J, Pham N, Blumberg B. Transgenerational inheritance of increased fat depot size, stem cell reprogramming, and hepatic steatosis elicited by prenatal exposure to the obesogen tributyltin in mice. Environ Health Perspect. 2013;121:359–66.
16. Masuno H, Iwanami J, Kidani T, Sakayama K, Honda K. Bisphenol A accelerates terminal differentiation of 3T3-L1 cells into adipocytes through the phosphatidylinositol 3-kinase pathway. Toxicol Sci. 2005;84:319–27.
17. Mackay H, Patterson ZR, Khazall R, Patel S, Tsirlin D, Abizaid A. Organizational effects of perinatal exposure to bisphenol-A and diethylstilbestrol on arcuate nucleus circuitry controlling food intake and energy expenditure in male and female CD-1 mice. Endocrinology. 2013;154:1465–75.
18. Chao HR, Wang SL, Lee WJ, Wang YF, Papke O. Levels of polybrominated diphenyl ethers (PBDEs) in breast milk from central Taiwan and their relation to infant birth outcome and maternal menstruation effects. Environ Int. 2007;33:239–45.
19. Herbstman JB, Sjodin A, Apelberg BJ, Witter FR, Halden RU, Patterson DG, et al. Birth delivery mode modifies the associations between prenatal polychlorinated biphenyl (PCB) and polybrominated diphenyl ether (PBDE) and neonatal thyroid hormone levels. Environ Health Perspect. 2008;116:1376–82.
20. Legler J, Hamers T, van Eck van der Sluijs-van de Bor M, Schoeters G, van der Ven L, Eggesbo M et al. The OBELIX project: early life exposure to endocrine disruptors and obesity. Am J Clin Nutr. 2011;94:1933S–8S.
21. Patisaul HB, Roberts SC, Mabrey KA, McCaffrey KA, Gear RB, Braun J, et al. Accumulation and endocrine disrupting effects of the flame retardant mixture Firemaster® 550 in rats: an exploratory assessment. J Biochem Mol Toxicol. 2013;27:124–36.
22. Belcher SM, Cookman CJ, Patisaul HB, Stapleton HM. In vitro assessment of human nuclear hormone receptor activity and cytotoxicity of the flame retardant mixture FM 550 and its triarylphosphate and brominated components. Toxicol Lett. 2014;228:93–102.
23. Pillai HK, Fang F, Beglov D, Kozakov D, Vajda S, Stapleton HM, et al. Ligand binding and activation of PPARγ by Firemaster® 550: effects on adipogenesis and osteogenesis in vitro. Environ Health Perspect. 2014;122:1225–32.
24. Bourez S, Van den D, Aelen C, Le Lay S, Poupaert J, Larondelle Y, Thome JP, et al. The dynamics of accumulation of PCBs in cultured adipocytes vary with the cell lipid content and the lipophilicity of the congener. Toxicol Lett. 2013;216:40–6.
25. Elobeid MA, Brock DW, AllisonDB PMA, Ruden DM. Endocrine disruptors and obesity: an examination of selected persistent organic pollutants in the NHANES 1999–2002 data. Int J Environ Res Public Health. 2010;7:2988–3005.

26. Hao C, Cheng X, Xia H, Ma X. The endocrine disruptor mono-(2-ethylhexyl) phthalate promotes adipocyte differentiation and induces obesity in mice. Biosci Rep. 2012;32:619–29.
27. Hines EP, White SS, Stanko JP, Gibbs-Flournov EA, Lau C, Fenton SE. Phenotypic dichotomy following developmental exposure to perfluorooctanoic acid (PFOA) in female CD-1 mice: low doses induce elevated serum leptin and insulin, and overweight in mid-life. Mol Cell Endocrinol. 2009;304:97–105.
28. Asher MI, Montefort S, Bjorksten B, Lai CKW, Strachan DP, Weiland SK, et al. Worldwide time trends in the prevalence of symptoms of asthma, allergic rhinoconjunctivitis, and eczema in childhood: ISAAC phases one and three repeat multicountry cross-sectional surveys. Lancet. 2006;368(9537):733–43.
29. WHO (World Health Organization). Asthma. Fact Sheet No. 307. 2006. Available http://www.who.int/mediacentre/factsheets/fs307/en/. Accessed 22 Oct 2006.
30. Brauer M, Hoek G, Smit HA, de Jongste JC, Gerritsen J, Postma DS, Kerkhof M, Brunekreef B. Air pollution and development of asthma, allergy and infections in a birth cohort. Eur Respir J. 2007;29(5):879–88.
31. Gauderman WJ, Avol E, Lurmann F, Kuenzli N, Gilliland F, Peters J, McConnell R. Childhood asthma and exposure to traffic and nitrogen dioxide. Epidemiology. 2005;16(6):737–43.
32. Haberg SE, Stigum H, Nystad W, Nafstad P. Effects of pre- and postnatal exposure to parental smoking on early childhood respiratory health. Am J Epidemiol. 2007;166(6):679–86.
33. McConnell R, Berhane K, Yao L, Jerrett M, Lurmann F, Gilliland F, et al. Traffic, susceptibility, and childhood asthma. Environ Health Perspect. 2006;114:766–72.
34. Zmirou D, Gauvin S, Pin I, Momas I, Sahraoui F, Just J, et al. Traffic related air pollution and incidence of childhood asthma: results of the Vesta case-control study. J Epidemiol Community Health. 2004;58(1):18–23.
35. Dodson RE, Levy JI, Spengler JD, Shine JP, Bennett DH. Influence of basements, garages, and common hallways on indoor residential volatile organic compound concentrations. Atmos Environ. 2008;42(7):1569–81.
36. Lorber M. Exposure of Americans to polybrominated diphenyl ethers. J Expo Sci Environ Epidemiol. 2008;18(1):2–19.
37. Rudel RA, Seryak LM, Brody JG. PCB-containing wood floor finish is a likely source of elevated PCBs in residents' blood, household air and dust: a case study of exposure. Environ Health. 2008;7:2.
38. Zota AR, Rudel RA, Morello-Frosch RA, Brody JG. Elevated house dust and serum concentrations of PBDEs in California: unintended consequences of furniture flammability standards? Environ Sci Technol. 2008;42(21):8158–64.
39. Bornehag CG, Sundell J, Weschler CJ, Sigsgaard T, Lundgren B, Hasselgren M, et al. The association between asthma and allergic symptoms in children and phthalates in house dust: a nested case–control study. Environ Health Perspect. 2004;112:1393–7.
40. Blunt E. Diethylstilbesrol exposure: it's still an issue. Holist Nurs Pract. 2004;18(4):187–91.
41. Schrager S, Potter BE. Diethylstilbestrol exposure. Am Fam Physician. 2004;69(10):2395–400.
42. Skakkebaek NE, Rajpert-De Meyts E, Main KM. Testicular dysgenesis syndrome: an increasingly common developmental disorder with environmental aspects. Hum Reprod. 2001;16:972–8.
43. Dolk H, Vrijheid M, Scott JES, Addor MC, Botting B, Vigan C, et al. Toward the effective surveillance of hypospadais. Environ Health Perspect. 2004;112(3):398–402.
44. Ledger WL, Bahathiq AOS. Historical background and functional anatomy. In: The fallopian tube in infertility and IVF practice. 1st ed. New York: Cambridge University Press; 2010. p. 1–7.
45. Brouwers MM, van der Zanden LFM, de Gier RPE, Barten EJ, Zielhuis GA, Feitz WFJ, et al. Hypospadias: risk factor patterns and different phenotypes. BJU Int. 2009;105:254–62.
46. Kurahashi N, Murakumo M, Kakizaki H, Nonomura K, Koyanagi T, Kasai S, et al. The estimated prevalence of hypospadias in Hokkaido, Japan. J Epidemiol. 2004;14(3):73–7.
47. Yoo JH. Differential diagnosis of hypospadias. J Korean Soc Pediatr Endocrinol. 2006;11(1):15–21.
48. Boisen KA, Kaleva M, Main KM, Virtanen HE, Haevisto A-M, Schmidt IM, et al. Difference in prevalence of congenital cryptorchidism in infant between two Nordic countries. Lancet. 2004;363:1264–9.

49. Preiksa RT, Zilaitiene B, Matulevicius V, Skakkebaek NE, Petersen JH, Jørgensen N, et al. Higher than expected prevalence of congenital cryptorchidism in Lithuania: a study of 1204 boys at birth and 1 year follow-up. Hum Reprod. 2005;20:1928–32.
50. Damgaard IN, Skakkebaek NE, Toppari J, Virtanen HE, Shen H, Schramm KW, et al. Persistent pesticides in human breast milk and cryptorchidism. Environ Health Perspect. 2006;114:1133–8.
51. Thorup J, Cortes D, Petersen BL. The incidence of bilateral cryptorchidism is increased and the fertility potential is reduced in sons born to mothers who have smoked during pregnancy. J Urol. 2006;176:734–7.
52. Fratric I, Zivkovic D, Vukmirovic S. Human exposure to endocrine disrupting chemicals as a prenatal risk factor for cryptorchidism. Paediatr Croat. 2015;59:19–24.
53. Anway MD, Skinner MK. Epigenetic transgenerational actions of endocrine disruptors. Endocrinology. 2006;147:S43–9.
54. Mendiola J, Jorgensen N, Andersson A, Calafat A, Silva M, Redmon J, et al. Association between urinary metabolites of di(2-ethylhexyl) phtahalate and reproductive hormones in fertile men. Int J Androl. 2010;14:1–10.
55. Pan G, Hanoka T, Yoshimura M, Zhang S, Wang P, Tsukino H, et al. Decreased serum free testosterone in workers exposed to high levels of di-n-butyl phthalate (DBP) and di-2-ethylhexyl phthalate (DEHP): a cross sectional study in China. Environ Health Perspect. 2006;114(1):1643–8.
56. Corton JC, Lapinskas PJ. Peroxisome proliferator-activated receptors: mediators of phthalate ester-induced effects in the male reproductive tract? Toxicol Sci. 2005;83(1):4–17.
57. Rozati R, Reddy PP, Reddanna P, Mujtaba R. Role of environmental estrogens in the deterioration of male factor fertility. Fertil Steril. 2002;78(6):1187–94.
58. Tan LF, Sun XZ, Li YN, Ji JM, Wang QL, Chen LS, et al. Effects of carbaryl production exposure on the sperm and semen quality of occupational male workers. Zhonghua Lao Dong Wei Sheng Zhi Ye Bing Za Zhi. 2005;23:87–90.
59. Meeker JD, Ryan L, Barr DB, Herrick RF, Bennett DH, Bravo R, et al. The relationship of urinary metabolites of carbaryl/naphthalene and chlorpyrifos with human semen quality. Environ Health Perspect. 2004;112:1665–70.
60. Swan SH, Kruse RL, Liu F, Barr DB, Drobnis EZ, Redmon JB, et al. Semen quality in relation to biomarkers of pesticide exposure. Environ Health Perspect. 2003;111:1478–84.
61. Del-Mazo J, Brieno-Enriquez MA, Garcia-Lopez J, Lopez-Fernandez LA, De-Felici M. Endocrine disruptors, gene deregulation and male germ cell tumors. Int J Dev Biol. 2013;57(2-4):225–39.
62. Giannandrea F, Paoli D, Figa-Talamanca I, Lombardo F, Lenzi A, Gandini L. Effect of endogenous and exogenous hormones on testicular cancer: the epidemiological evidence. Int J Dev Biol. 2013;57(2-4):255–63.
63. Erickson RL. Persistent organochlorine pesticides and risk of testicular germ cell tumors. J Natl Cancer Inst. 2008;100(9):663–71.
64. LeCornet C, Fervers B, Oksbjerg Dalton S, Feychting M, Pukkala E, Tynes T, et al. Testicular germ cell tumours and parental occupational exposure to pesticides: a register-based case–control study in the Nordic countries (NORD-TEST study). Occup Environ Med. 2015;72(11):805–11.
65. Paoli D, Giannandrea F, Gallo M, Turci R, Cattaruzza MS, Lombardo F, et al. Exposure to polychlorinated biphenyls and hexachlorobenzene, semen quality and testicular cancer risk. J Endocrinol Investig. 2015;38(7):745–52.
66. Sweeney MF, Hasan N, Soto AM, Sonnenschein C. Environmental endocrine disruptors: effects on the human male reproductive system. Rev Endocr Metab Disord. 2016;16:341–57.
67. Cowin PA, Foster PMD, Risbridger GP. Endocrine disruption in the male. In: Gore AC, editor. Endrocrine-disrupting chemicals: from basic research to clinical practice. Totowa: Humana; 2007. p. 33–62.
68. Modugno F, Weissfeld JL, Trump DL, Zmuda JM, Shea P, Cauley JA, et al. Allelic variants of aromatase and the androgen and estrogen receptors: toward a multigenic model of prostate cancer risk. Clin Cancer Res. 2001;7:3092–6.

69. Katzenwadel A, Wolf P. Androgen deprivation of prostate cancer: leading to a therapeutic dead end. Cancer Lett. 2015;367(1):12–7.
70. Alavanja MC, Samanic C, Dosemeci M, Lubin J, Tarone R, Lynch CF, et al. Use of agricultural pesticides and prostate cancer risk in the Agricultural Health Study cohort. Am J Epidemiol. 2003;157:800–14.
71. Mahajan R, Bonner MR, Hoppin JA, Alavanja MC. Phorate exposure and incidence of cancer in the agricultural health study. Environ Health Perspect. 2006;114:1205–9.
72. Song KH, Lee K, Choi HS. Endocrine disruptor bisphenol A induces orphan nuclear receptor Nur77 gene expression and steroidogenesis in mouse testicular Leydig cells. Endocrinology. 2002;143:2208–15.
73. Walsh DE, Dockery P, Doolan CM. Estrogenreceptor independent rapid non-genomic effects of environmental estrogens on [Ca2]i in human breast cancer cells. Mol Cell Endocrinol. 2005;230:23–30.
74. Charles LE, Loomis D, Shy CM, Newman B, Millikan R, Nylander-French LA, et al. Electromagnetic fields, polychlorinated biphenyls, and prostate cancer mortality in electric utility workers. Am J Epidemiol. 2003;157:683–91.
75. Prince MM, Ruder AM, Hein MJ, Waters MA, Whelan EA, Nilsen N, et al. Mortality and exposure response among 14,458 electrical capacitor manufacturing workers exposed to polychlorinated biphenyls (PCBs). Environ Health Perspect. 2006;114:1508–14.
76. Ritchie JM, Vial SL, Fuortes LJ, Guo H, Reedy VE, Smith EM. Organochlorines and risk of prostate cancer. J Occup Environ Med. 2003;45:692–702.
77. Kester MH, Bulduk S, van Toor H, Tibboel D, Meinl W, Glatt H, et al. Potent inhibition of estrogen sulfotransferase by hydroxylated metabolites of polyhalogenated aromatic hydrocarbons reveals alternative mechanism for estrogenic activity of endocrine disrupters. J Clin Endocrinol Metab. 2002;87:1142–50.
78. Benbrahim-Tallaa L, Liu J, Webber MM, Waalkes MP. Estrogen signaling and disruption of androgen metabolism in acquired androgen-independence during cadmium carcinogenesis in human prostate epithelial cells. Prostate. 2007a;67:135–45.
79. Davey JC, Bodwell JE, Gosse JA, Hamilton JW. Arsenic as an endocrine disruptor: effects of arsenic on estrogen receptor-mediated gene expression in vivo and in cell culture. Toxicol Sci. 2007;98:75–86.
80. Benbrahim-Tallaa L, Webber MM, Waalkes MP. Mechanisms of acquired androgen independence during arsenic-induced malignant transformation of human prostate epithelial cells. Environ Health Perspect. 2007b;115:243–7.
81. Costa EM, Spritzer PM, Hohl A, Bachega TA. Effects of endocrine disruptors in the development of the female reproductive tract. Arq Bras Endocrinol Metabol. 2014;58:153–61.
82. Ozen S, Darcan S. Effects of environmental endocrine disruptors on pubertal development. J Clin Res Pediatr Endocrinol. 2011;3:1–6.
83. Mlynarcikova A, Nagyova E, Fickova M, Scsukova S. Effects of selected endocrine disruptors on meiotic maturation, cumulus expansion, synthesis of hyaluronan and progesterone by porcine oocyte-cumulus complexes. Toxicol In Vitro. 2009;23:371–7.
84. Azziz R, Carmina E, Dewailly D, Diamanti-Kandarakis E, Escobar-Morreale HF, Futterweit W, et al. Position statement: criteria for defining polycystic ovary syndrome as a predominantly hyperandrogenic syndrome. An Androgen Excess Society guideline. J Clin Endocrinol Metab. 2006;91:4237–45.
85. Legro RS, Azziz R, Giudice L. A twenty-first century research agenda for polycystic ovary syndrome. Best Pract Res Clin Endocrinol Metab. 2006;20:331–6.
86. Franks S, Mason H, Willis D. Follicular dynamics in the polycystic ovary syndrome. Mol Cell Endocrinol. 2000;163:49–52.
87. Yildiz BO, Knochenhauer ES, Azziz R. Impact of obesity on the risk for polycystic ovary syndrome. J Clin Endocrinol Metab. 2008;93:162–8.
88. Kandaraki E, Chatzigeorgiou A, Livadas S, Palioura E, Economou F, Koutsilieris M, et al. Endocrine disruptors and polycystic ovary syndrome (PCOS): elevated serum levels of bisphenol A in women with PCOS. J Clin Endocrinol Metab. 2011;96:480–4.

89. Miao M, Yuan W, Yang F, Liang H, Zhou Z, Li R, et al. Associations between Bisphenol A exposure and reproductive hormones among female workers. Int J Environ Res Public Health. 2015;12:13240–50.
90. Eichenlaub-Ritter U, Vogt E, Cukurcam S, Sun F, Pacchierotti F, Parry J. Exposure of mouse oocytes to bisphenol A causes meiotic arrest but not aneuploidy. Mutat Res. 2008;651:82–92.
91. Li Y, Zhang W, Liu J, Wang W, Li H, Zhu J, et al. Prepubertal bisphenol A exposure interferes with ovarian follicle development and its relevant gene expression. Reprod Toxicol. 2014;44:33–40.
92. Rivera OE, Varayoud J, Rodríguez HA, Muñoz-de-Toro M, Luque EH. Neonatal exposure to bisphenol A or diethylstilbestrol alters the ovarian follicular dynamics in the lamb. Reprod Toxicol. 2011;32:304–12.
93. Susiarjo M, Hassold TJ, Freeman E, Hunt PA. Bisphenol A exposure in utero disrupts early oogenesis in the mouse. PLoS Genet. 2007;3:e5.
94. Zhou W, Liu J, Liao L, Han S, Liu J. Effect of bisphenol A on steroid hormone production in rat ovarian theca-interstitial and granulosa cells. Mol Cell Endocrinol. 2008;283(1–2):12–8.
95. Fact Sheet. The patient education website of the American Society for reproductive medicine. 2012. www.ReproductiveFacts.org.
96. Berkowitz G. Limitations of a case–control study on bisphenol A (BPA) serum levels and recurrent miscarriage. Hum Reprod. 2006;21:565–6.
97. Fenichel P, Dechaux H, Harthe C, Gal J, Ferrari P, Pacini P, et al. Unconjugated bisphenol A cord blood levels in boys with descended or undescended testes. Hum Reprod. 2012;27:983–90.
98. Sugiura-Ogasawara M, Ozaki Y, Sonta S, Makino T, Suzumori K. Exposure to bisphenol A is associated with recurrent miscarriage. Hum Reprod. 2005;20:2325–9.
99. Caserta D, Di Segni N, Mallozzi M, Giovanale V, Mantovani A, Marci R, et al. Bisphenol A and the female reproductive tract: an overview of recent laboratory evidence and epidemiological studies. Reprod Biol Endocrinol. 2014;12:37.
100. Peretz J, Vrooman L, Ricke WA, Hunt PA, Ehrlich S, Hauser R, et al. Bisphenol A and reproductive health: update of experimental and human evidence, 2007–2013. Environ Health Perspect. 2014;122:775–86.
101. Wallach EE, Vlahos NF. Uterine myomas: an overview of development, clinical features, and management. Obstet Gynecol. 2004;104:393–406.
102. Hunter DS, Hodges LC, Eagon PK, Vonier PM, Fuchs-Young R, Bergerson JS, et al. Influence of exogenous estrogen receptor ligands on uterine leiomyoma: evidence from an in vitro/in vivo animal model for uterine fibroids. Environ Health Perspect. 2000;108(Suppl 5):829–34.
103. Hiroi H, Tsutsumi O, Takeuchi T, Momoeda M, Ikezuki Y, Okamura A, et al. Differences in serum bisphenol A concentrations in premenopausal normal women and women with endometrial hyperplasia. Endocr J. 2004;51:595–600.
104. Sinha P, Kuruba N. Premature ovarian failure. J Obstet Gynaecol. 2007;27:16–9.
105. McNatty KP, Reader K, Smith P, Heath DA, Juengel JL. Control of ovarian follicular development to the go-nadotrophin-dependent phase: a 2006 perspective. Soc Reprod Fertil Suppl. 2007;64:55–68.
106. Kaipia A, Hsueh AJ. Regulation of ovarian follicle atresia. Annu Rev Physiol. 1997;59:349–63.
107. Kirigaya A, Kim H, Hayashi S, Chambon P, Watanabe H, Lguchi T, et al. Involvement of estrogen receptor beta in the induction of polyovular follicles in mouse ovaries exposed neonatally to diethylstilbestrol. Zool Sci. 2009;26(10):704–12.
108. Newbold RR, Jefferson WN, Padilla-Banks E. Prenatal exposure to bisphenol A at environmentally relevant doses adversely affects the murine female reproductive tract later in life. Environ Health Perspect. 2009;117(6):879–85.
109. Brieno-Enríquez MA, Reig-Viader R, Cabero L, Toran N, Martínez F, Roig I, et al. Gene expression is altered after bisphenol A exposure in human fetal oocytes in vitro. Mol Hum Reprod. 2012;18(4):171–83.
110. Hoyer PB, Devine PJ, Hu X, Thompson KE, Sipes IG. Ovarian toxicity of 4-vinylcyclohexene diepoxide: a mechanistic model. Toxicol Pathol. 2001;29(1):91–9.

111. Kappeler CJ, Hoyer PB. 4-vinylcyclohexene diepoxide: a model chemical for ovotoxicity. Syst Biol Reprod Med. 2012;58(1):57–62.
112. Gaido KW, Maness SC, McDonnell DP, Dehal SS, Kupfer D, Safe S. Interaction of methoxychlor and related compounds with estrogen receptor alpha and beta, and androgen receptor: structure-activity studies. Mol Pharmacol. 2000;58(4):852–8.
113. Colborn T, Demanoski D, Myers JP. Our stolen future. New York: Penguin; 1997.
114. Markey CM, Rubin BS, Soto AM, Sonnenschein C. Endocrine disruptors: from wingspread to environmental developmental biology. J Steroid Biochem Mol Biol. 2002;83:235–44.
115. Richter CA, Birnbaum LS, Farabollini F, Newbold RR, Rubin BS, Talsness CE, et al. In vivo effects of bisphenol A in laboratory rodent studies. Reprod Toxicol. 2007;24:199–224.
116. Zalko D, Soto AM, Dolo L, Dorio C, Rathahao E, Debrauwer L, et al. Biotransformations of bisphenol A in a mammalian model: answers and new questions raised by low-dose metabolic fate studies in pregnant CD1 mice. Environ Health Perspect. 2003;111:309–19.
117. Munoz-de-Toro M, Markey CM, Wadia PR, Luque EH, Rubin BS, Sonnenschein C, et al. Perinatal exposure to bisphenol-A alters peri-pubertal mammary gland development in mice. Endocrinology. 2005;146:4138–47.
118. Murray TJ, Maffini MV, Ucci AA, Sonnenschein C, Soto AM. Induction of mammary gland ductal hyperplasia and carcinoma in situ following fetal bisphenol A exposure. Reprod Toxicol. 2007;23:383–90.
119. Durando M, Kass L, Piva J, Sonnenschein C, Soto AM, Luque EH, et al. Prenatal bisphenol A exposure induces preneoplastic lesions in the mammary gland in Wistar rats. Environ Health Perspect. 2007;115:80–6.
120. Brown NM, Manzolillo PA, Zhang JX, Wang J, Lamartiniere CA. Prenatal TCDD and predisposition to mammary cancer in the rat. Carcinogenesis. 1998;19:1623–9.
121. Fenton SE. Endocrine-disrupting compounds and mammary gland development: early exposure and later life consequences. Endocrinology. 2006;147:S18–24.
122. Olive DL, Schwartz LB. Endometriosis. N Engl J Med. 1993;328:1759–69.
123. Igarashi T, Osuga U, Tsutsumi O, Momoeda M, Ando K, Matsumi H, et al. Expression of Ah receptor and dioxin-related genes in human uterine endometrium in women with or without endometriosis. Endocr J. 1999;46:765–72.
124. Kuchenhoff A, Seliger G, Klonisch T, Tscheudschilsuren G, Kaltwasser P, Seliger E, et al. Arylhydrocarbon receptor expression in the human endometrium. Fertil Steril. 1999;71:354–60.
125. Rier S, Foster WG. Environmental dioxins and endometriosis. Toxicol Sci. 2002;70:161–70.
126. Porpora MG, Ingelido AM, di Domenico A, Ferro A, Crobu M, Pallante D, et al. Increased levels of polychlorobiphenyls in Italian women with endometriosis. Chemosphere. 2006;63:1361–7.
127. Bruner-Tran KL, Osteen KG. Dioxin-like PCBs and endometriosis. Syst Biol Reprod Med. 2010;56(2):132–46.
128. Crain DA, Janssen SJ, Edwards TM, Heindel J, Ho SM, Hunt P, et al. Female reproductive disorders: the roles of endocrine-disrupting compounds and developmental timing. Fertil Steril. 2008;90:911–40.
129. Ibanez L, Ferrer A, Marcos MV. Early puberty: rapid progression and reduced final height in girls with low birth weight. Paediatrics. 2000;106:1–3.
130. Chittwar SS, Ammini AC. Precocious puberty in girls. Indian J Endocrinol Metab. 2012;16:S188–91.
131. Neely EK, Crossen SS. Precocious puberty. Curr Opin Obstet Gynecol. 2014;26:332–8.
132. Kotwal N, Yanamandra U, Menon AS, Nair V. Central precocious puberty due to hypothalamic hamartoma in a six-month-old infant girl. Indian J Endocrinol Metab. 2012;16:627–30.
133. Mogensen SS, Aksglaede L, Mouritsen A, Sorensen K, Main KM, Gideon P, et al. Pathological and incidental findings on brain MRI in a single center study of 229 consecutive girls with early or precocious puberty. PLoS One. 2012;7:e29829.
134. Trivin C, Couto-Silva AC, Sainte-Rose C, Chemaitilly W, Kalifa C, Doz F, et al. Presentation and evolution of organic central precocious puberty according to the type of CNS lesion. Clin Endocrinol. 2006;65:239–45.

135. Blanck HM, Marcus M, Hertzberg V, Tolbert PE, Rubin C, Henderson AK, et al. Determinants of polybrominated biphenyl serum decay among women in the Michigan PBB cohort. Environ Health Perspect. 2000a;108:147–52.
136. Blanck HM, Marcus M, Tolbert PE, Rubin C, Henderson AK, Hertzberg VS, et al. Age at menarche and tanner stage in girls exposed in utero and postnatally to polybrominated biphenyl. Epidemiology. 2000b;11:641–7.
137. Gladen BC, Ragan NB, Rogan WJ. Pubertal growth and development and prenatal and lactational exposure to polychlorinated biphenyls and dichlorodiphenyl dichloroethene. J Pediatr. 2000;136:490–6.
138. Wade MG, Desaulniers D, Leingartner K, Foster WG. Interaction between endosulfan and dieldrin on estrogen-mediated processes in vitro and in vivo. Reprod Toxicol. 1997;11:791–8.
139. Denham M, Schell LM, Deane G, Gallo MV, Ravenscroft J, DeCaprio AP, Task Force on the Environment, et al. Relationship of lead, mercury, mirex, dichlorodiphenyldichloroethylene, hexachlorobenzene, and polychlorinated biphenyls to timing of menarche among Akwesasne Mohawk girls. Pediatrics. 2005;115:127–34.
140. Yang CY, Yu ML, Guo HR, Lai TJ, Hsu CC, Lambert G, et al. The endocrine and reproductive function of the female Yucheng adolescents prenatally exposed to PCBs/PCDFs. Chemosphere. 2005;61:355–60.
141. Den Hond E, Roels HA, Hoppenbrouwers K, Nawrot T, Thijs L, Vandermeulen C, et al. Sexual maturation in relation to polychlorinated aromatic hydrocarbons: Sharpe and Skakkebaek's hypothesis revisited. Environ Health Perspect. 2002;110:771–6.
142. Colon I, Caro D, Bourdony CJ, Rosario O. Identification of phthalate esters in the serum of young Puerto Rican girls with premature breast development. Environ Health Perspect. 2000;108:895–900.
143. Fernández M, Bianchi M, Lux-Lantos V, Libertun C. Neonatal exposure to bisphenol A alters reproductive parameters and gonadotropin releasing hormone signaling in female rats. Environ Health Perspect. 2009;117(5):757–62.
144. Selevan SG, Rice DC, Hogan KA, Euling SY, Pfahles-Hutchens A, Bethel J. Blood lead concentration and delayed puberty in girls. N Engl J Med. 2003;348:1527–36.
145. Wu T, Buck GM, Mendola P. Blood lead levels and sexual maturation in US girls: the third National Health and Nutrition Examination Survey, 1988–1994. Environ Health Perpect. 2003;111:737–41.
146. Auso E, Lavado-Autric R, Cuevas E, Del Rey FE, Morreale DE, Berbel P. A moderate and transient deficiency of maternal thyroid function at the beginning of fetal neocorticogenesis alters neuronal migration. Endocrinology. 2004;145:4037–47.
147. Berbel P, Mestre JL, Santamaria A, Palazon I, Franco A, Graells M, et al. Delayed neurobehavioral development in children born to pregnant women with mild hypothyroxinemia during the first month of gestation: the importance of early iodine supplementation. Thyroid. 2009;19:511–9.
148. Boas M, Feldt-Rasmussen U, Main KM. Thyroid effects of endocrine disrupting chemicals. Mol Cell Endocrinol. 2012;355:240–8.
149. Howdeshell KL. A model of the development of the brain as a construct of the thyroid system. Environ Health Perspect. 2002;110(3):337–48.
150. Brucker-Davis F. Effects of environmental synthetic chemicals on thyroid function. Thyroid. 1998;8:827–56.
151. Gauger KJ, Kato Y, Haraguchi K, Lehmler HJ, Robertson LW, Bansal R, et al. Polychlorinated biphenyls (PCBs) exert thyroid hormone-like effects in the fetal rat brain but do not bind to thyroid hormone receptors. Environ Health Perspect. 2004;112:516–23.
152. Martin L, Klaassen CD. Differential effects of polychlorinated biphenyl congeners on serum thyroid hormone levels in rats. Toxicol Sci. 2010;117:36–44.
153. Pavuk M, Schecter AJ, Akhtar FZ, et al. Serum 2,3,7,8-tetrachlorodibenzo-p-dioxin (TCDD) levels and thyroid function in Air Force veterans of the Vietnam War. Ann Epidemiol. 2003;13:335–43.

154. Baccarelli A, Giacomini SM, Corbetta C, et al. Neonatal thyroid function in Seveso 25 years after maternal exposure to dioxin. PLoS Med. 2008;5:e161.
155. Li F, Xie Q, Li X, Li N, Chi P, Chen J, et al. Hormone activity of hydroxylated polybrominated diphenyl ethers on human thyroid receptor-beta: in vitro and in silico investigations. Environ Health Perspect. 2010;118:602–6.
156. Meerts IA, van Zanden JJ, Luijks EA, Leeuwen-Bol I, Marsh G, Jakobsson E. Potent competitive interactions of some brominated flame retardants and related compounds with human transthyretin in vitro. Toxicol Sci. 2000;56:95–104.
157. Fini JB, Le MS, Turque N, Palmier K, Zalko D, Cravedi JP, et al. An in vivo multiwell-based fluorescent screen for monitoring vertebrate thyroid hormone disruption. Environ Sci Technol. 2007;41:5908–14.
158. Jagnytsch O, Opitz R, Lutz I, Kloas W. Effects of tetrabromobisphenol a on larval development and thyroid hormone-regulated biomarkers of the amphibian Xenopus laevis. Environ Res. 2006;101:340–8.
159. Hallgren S, Sinjari T, Hakansson H, Darnerud PO. Effects of polybrominated diphenyl ethers (PBDEs) and polychlorinated biphenyls (PCBs) on thyroid hormone and vitamin A levels in rats and mice. Arch Toxicol. 2001;75:200–8.
160. Richardson VM, Staskal DF, Ross DG, Diliberto JJ, DeVito MJ, Birnbaum LS. Possible mechanisms of thyroid hormone disruption in mice by BDE 47, a major polybrominated diphenyl ether congener. Toxicol Appl Pharmacol. 2008;226:244–50.
161. Zhou LX, Dehal SS, Kupfer D, Morrell S, McKenzie BA, Eccleston ED Jr, et al. Cytochrome P450 catalyzed covalent binding of methoxychlor to rat hepatic, microsomal iodothyronine 50-monodeiodinase, type I: does exposure to methoxychlor disrupt thyroid hormone metabolism? Arch Biochem Biophys. 1995;322:390–4.
162. Olsen GW, Zobel LR. Assessment of lipid, hepatic, and thyroid parameters with serum perfluorooctanoate (PFOA) concentrations in fluorochemical production workers. Int Arch Occup Environ Health. 2007;81:231–46.
163. Chang SC, Thibodeaux JR, Eastvold ML, Ehresman DJ, Bjork JA, Froehlich JW, et al. Thyroid hormone status and pituitary function in adult rats given oral doses of perfluorooctanesulfonate (PFOS). Toxicology. 2008;36:330–9.
164. Yu WG, Liu W, Jin YH. Effects of perfluorooctane sulfonate on rat thyroid hormone biosynthesis and metabolism. Environ Toxicol Chem. 2009;28:990–6.
165. Huang PC, Kuo PL, Guo YL, Liao PC, Lee CC. Associations between urinary phthalate monoesters and thyroid hormones in pregnant women. Hum Reprod. 2007;22:2715–22.
166. Freitas J, Cano P, Craig-Veit C, Goodson ML, David FJ, Murk AJ. Detection of thyroid hormone receptor disruptors by a novel stable in vitro reporter gene assay. Toxicol In Vitro. 2011;25:257–66.
167. Sun W, Ban J-B, Zhang N, Zu Y-K, Sun W-X. Perinatal exposure to di-(2-ethylhexyl)-phthalate leads to cognitive dysfunction and phospho-tau level increase in aged rats. Environ Toxicol. 2014;29(5):596–603.
168. Hofmann PJ, Schomburg L, Kohrle J. Interference of endocrine disrupters with thyroid hormone receptor-dependent transactivation. Toxicol Sci. 2009;110:125–37.
169. Zoeller RT, Bansal R, Parris C. Bisphenol-A, an environmental contaminant that acts as a thyroid hormone receptor antagonist in vitro, increases serum thyroxine, and alters RC3/neurogran in expression in the developing rat brain. Endocrinology. 2005;146:607–12.
170. Dundar B, Oktem F, Arslan MK, et al. The effect of long-term low-dose lead exposure on thyroid function in adolescents. Environ Res. 2006;101:140–5.
171. Rodier PM. Developing brain as a target of toxicity. Environ Health Perspect. 1995;103(6):73–6.
172. Rice D, Barone S Jr. Critical periods of vulnerability for the developing nervous system: evidence from humans and animal models. Environ Health Perspect. 2000;108(Suppl 3):511–33.
173. Andersen HR, Nielsen JB, Grandjean P. Toxicologic evidence of developmental neurotoxicity of environmental chemicals. Toxicology. 2000;144:121–7.

174. Ginsberg G, Hattis D, Sonawane B. Incorporating pharmacokinetic differences between children and adults in assessing children's risks to environmental toxicants. Toxicol Appl Pharmacol. 2004;198:164–83.
175. National Research Council. Pesticides in the diets of infants and children. Washington: National Academy; 1993.
176. Rutter M. Incidence of autism spectrum disorders: changes over time and their meaning. Acta Paediatr. 2005;94:2–15.
177. Landrigan PJ, Schechter CB, Lipton JM, Fahs MC, Schwartz J. Environmental pollutants and disease in American children: estimates of morbidity, mortality, and costs for lead poisoning, asthma, cancer, and developmental disabilities. Environ Health Perspect. 2002;110:721–08.
178. Mendola P, Selevan SG, Gutter S, Rice D. Environmental factors associated with a spectrum of neurodevelopmental deficits. Ment Retard Dev Disabil Res Rev. 2002;8:188–97.
179. National Research Council. Scientific frontiers in developmental toxicology and risk assessment. Washington: National Academy; 2000.
180. Trasande L, Landrigan PJ, Schechter C. Public health and economic consequences of methyl mercury toxicity to the developing brain. Environ Health Perspect. 2005;113:590–6.
181. Schantz SL, Widholm JJ, Rice DC. Effects of PCB exposure on neuropsychological function in children. Environ Health Perspect. 2003;111(3):357–576.
182. Jacobson LJ, Jacobson SW. Prenatal exposure to polychlorinated biphenyls and attention at school age. J Pediatr. 2003;143(6):780–8.
183. Vreugdenhil HJ, Mulder PG, Emmen HH, Weisglas-Kuperus N. Effects of perinatal exposure to PCBs on neuropsychological functions in the Rotterdam cohort at 9 years of age. Neuropsychology. 2004;18(1):185–93.
184. Guo YL, Lambert GH, Hsu CC, Hsu MM. Yucheng: health effects of prenatal exposure to polychlorinated biphenyls and dibenzofurans. Int Arch Occup Environ Health. 2004;77(3):153–8.
185. Edwards SC, Jedrychowski W, Butscher M, Camann D, Kieltyka A, Mroz E, et al. Prenatal exposure to airborne polycyclic aromatic hydrocarbons and children's intelligence at 5 years of age in a prospective cohort study in Poland. Environ Health Perspect. 2010;118(9):1326–31.
186. Perera FP, Rauh V, Whyatt RM, Tsai WY, Tang D, Diaz D, et al. Effect of prenatal exposure to airborne polycyclic aromatic hydrocarbons on neurodevelopment in the first 3 years of life among inner-city children. Environ Health Perspect. 2006;114(8):1287–92.
187. Takeda K, Tsukue N, Yoshida S. Endocrine-disrupting activity of chemicals in diesel exhaust and diesel exhaust particles. Environ Sci. 2004;11(1):33–45.
188. Yokota S, Mizuo K, Moriya N, Oshio S, Sugawara I, Takeda K. Effect of prenatal exposure to diesel exhaust on dopaminergic system in mice. Neurosci Lett. 2009;449(1):38–41.
189. Perera F, Herbstman J. Prenatal environmental exposures, epigenetics, and disease. Reprod Toxicol. 2011;31(3):363373.
190. Engel SM, Zhu C, Berkowitz GS, Calafat AM, Silva MJ, Miodovnik A. Prenatal phthalate exposure and performance on the Neonatal Behavioral Assessment Scale in a multiethnic birth cohort. Neurotoxicology. 2009;30(4):522–8.
191. Miodovnik A, Engel SM, Zhu C, Ye X, Soorya LV, Silva MJ, et al. Endocrine disruptors and childhood social impairment. Neurotoxicology. 2011;32(2):261–7.
192. Swan SH, Liu F, Hines M, Kruse RL, Wang C, Redmon JB, et al. Prenatal phthalate exposure and reduced masculine play in boys. Int J Androl. 2010;33(2):259–69.
193. Yolton K, Xu Y, Strauss D, Altaye M, Calafat AM, Khoury J. Prenatal exposure to bisphenol A and phthalates and infant neurobehavior. Neurotoxicol Teratol. 2011;33(5):558–66.
194. Perera F, Vishnevetsky J, Herbstman JB, Calafat AM, Xiong W, Rauh V, et al. Prenatal bisphenol A exposure and child behavior in an inner-city cohort. Environ Health Perspect. 2012;120(8):1190–4.
195. Jones DC, Miller GW. The effects of environmental neurotoxicants on the dopaminergic system: a possible role in drug addiction. Biochem Pharmacol. 2008;76(5):569–81.

196. Narita M, Mitagawa K, Mizuo K, Yoshida T, Sizuki T. Prenatal and neonatal exposure to low-dose of bisphenol-A enhance the morphine-induced hyperlocomotion and rewarding effect. Neurosci Lett. 2006;402(3):249–52.
197. Rauh V, Arunajadai S, Horton M, Perera F, Hoepner L, Barr DB, et al. Seven-year neurodevelopmental scores and prenatal exposure to chlorpyrifos, a common agricultural pesticide. Environ Health Perspect. 2011;119(8):1196–201.
198. Rauh VA, Perera FP, Horton MK, Whyatt RM, Bansal R, Hao X, et al. Brain anomalies in children exposed prenatally to a common organophosphate pesticide. Proc Natl Acad Sci U S A. 2012;109(20):7871–6.
199. Bigler ED, Mortensen S, Neeley ES, Ozonoff S, Krasny L, Johnson M, et al. Superior temporal gyrus, language function, and autism. Dev Neuropsychol. 2007;31(2):217–38.
200. Elliott R, Deakin B. Role of the orbitofrontal cortex in reinforcement processing and inhibitory control: evidence from functional magnetic resonance imaging studies in healthy human subjects. Int Rev Neurobiol. 2005;65:89–116.
201. Forbes CE, Grafman J. The role of the human prefrontal cortex in social cognition and moral judgment. Annu Rev Neurosci. 2010;33:299–324.
202. Shelton JF, Hertz-Picciotto I, Pessah IN. Tipping the balance of autism risk: potential mechanisms linking pesticides and autism. Environ Health Perspect. 2012;120(7):944–51.
203. Ohtani-Kaneko R, Tazawa H, Yokosuka M, Yoshida M, Satoh M, Watanabe C. Suppressive effects of cadmium on neurons and affected proteins in cultured developing cortical cells. Toxicology. 2008;253(1-3):110–6.
204. Hines M. Prenatal testosterone and gender-related behaviour. Eur J Endocrinol. 2006;155(1):S115–21.
205. Jirtle RL, Skinner MK. Environmental epigenomics and disease susceptibility. Nat Rev Genet. 2007;8(4):253–62.
206. Klose RJ, Bird AP. Genomic DNA methylation: the mark and its mediators. Trends Biochem Sci. 2006;31:89–97.
207. Li E. Chromatin modification and epigenetic reprogramming in mammalian development. Nat Rev Genet. 2002;3:662–73.
208. Talbert PB, Henikoff S. Spreading of silent chromatin: inaction at a distance. Nat Rev Genet. 2006;7:793–803.
209. Chan TL, Yuen ST, Kong CK, Chan YW, Cahn AS, Ng WF, et al. Heritable germline epimutation of MSH2 in a family with hereditary nonpolyposis colorectal cancer. Nat Genet. 2006;38:1178–83.
210. Suter CM, Martin DI, Ward RL. Germline epimutation of MLH1 in individuals with multiple cancers. Nat Genet. 2004;36:497–501.
211. Anway MD, Cupp AS, Uzumcu M, Skinner MK. Epigenetic transgenerational actions of endocrine disruptors and male fertility. Science. 2005;308:1466–9.
212. Pembrey ME, Bygren LO, Kaati G, Edvinsson S, Northstone K, Sjöström M, Golding J. Sex-specific, male-line transgenerational responses in humans. Eur J Hum Genet. 2006;14:159–66.
213. Rakyan VK, Chong S, Champ ME, Cuthbert PC, Morgan HD, Luu KV, et al. Transgenerational inheritance of epigenetic states at the murine Axin(Fu) allele occurs after maternal and paternal transmission. Proc Natl Acad Sci U S A. 2003;100:2538–43.
214. Andersen HR, Schmidt IM, Grandjean P, Jensen TK, Budtz-Jorgensen E, Kjaerstad MB, et al. Impaired reproductive development in sons of women occupationally exposed to pesticides during pregnancy. Environ Health Perspect. 2008;116(4):566–72.
215. Dolinoy DC, Huang D, Jirtle RL. Maternal nutrient supplementation counteracts bisphenol A- induced DNA hypomethylation in early development. Proc Natl Acad Sci U S A. 2007;104(32):13056–61.
216. Novikova SI, He F, Bai J, Cutrufello NJ, Lidow MS, Undieh AS. Maternal cocaine administration in mice alters DNA methylation and gene expression in hippocampal neurons of neonatal and prepubertal offspring. PLoS One. 2008;3(4):e1919.

217. Perera F, Tang WY, Herbstman J, Tang D, Levin L, Miller R, et al. Relation of DNA methylation of 5'-CpG island of ACSL3 to transplacental exposure to airborne polycyclic aromatic hydrocarbons and childhood asthma. PLoS One. 2009;4(2):e4488.
218. Yaoi T, Itoh K, Nakamura K, Ogi H, Fujiwara Y, Fushiki S. Genome-wide analysis of epigenomic alterations in fetal mouse forebrain after exposure to low doses of bisphenol A. Biochem Biophys Res Commun. 2008;376(3):563–7.
219. Andre SM, Markowski VP. Learning deficits expressed as delayed extinction of a conditioned running response following perinatal exposure to vinclozolin. Neurotoxicol Teratol. 2006;28(4):482–8.
220. Crews D, Gore AC, Hsu TS, Dangleben NL, Spinetta M, Schallert T, et al. Transgenerational epigenetic imprints on mate preference. Proc Natl Acad Sci U S A. 2007;104(14):5942–6.
221. Ottinger MA, Lavoie E, Thompson N, Barton A, Whitehouse K, Barton M, et al. Neuroendocrine and behavioral effects of embryonic exposure to endocrine disrupting chemicals in birds. Brain Res Rev. 2008;57(2):376–85.
222. Skinner MK, Anway MD, Savenkova MI, Gore AC, Crews D. Transgenerational epigenetic programming of the brain transcriptome and anxiety behavior. PLoS One. 2008;3(11):e3745.

Postnatal EDC Exposure and Prevention of Endocrine Disruption

23

Jayshree Annamalai and Vasudevan Namasivayam

23.1 Postnatal Exposure

23.1.1 Lung Cancer

Lung cancer is the most common cancer in the world, and about 87% of lung cancer cases are caused by cigarette smoking. Extensive epidemiologic data clearly suggests cigarette smoking as the major cause of lung cancer. In addition to lung cancer, smoking also causes oral, oropharyngeal, hypopharyngeal, laryngeal, and esophageal cancers as well as pancreas, bladder, and renal pelvis cancer [1, 2].

23.1.1.1 EDCs Involved and Mechanism

About 95% of the smoke is made up of gases, chiefly nitrogen, oxygen, and carbon dioxide. The particulate phase contains at least 3500 compounds and most of the carcinogens [3]. Nickel, chromium, cadmium, and arsenic are all present in tobacco, and their metal carcinogenicity depends on the valence state and anion [4]. Thus, although some metals are effective pulmonary carcinogens, the role of metals in tobacco-induced lung cancer is murky. Substantial levels of cocarcinogens such as catechol, methylcatechols, pyrogallol, decane, undecane, pyrene, benzo[*e*]pyrene, and fluoranthene are also present in cigarette smoke [5]. Among the PAHs, benzo[*a*] pyrene (BaP) is the most extensively studied compound, and its ability to induce lung tumors via acting as ligands of AhR upon local administration or inhalation is well documented [6, 7]).

Numerous studies have shown that the AhR plays an important role in the development of lung cancer. The main mechanism behind lung cancer is the activation of

J. Annamalai (✉) • V. Namasivayam
Centre for Environmental Studies, Anna University,
CEG Campus, Chennai 600 025, Tamil Nadu, India
e-mail: jayshreeannamalai@yahoo.com

F. Capello, A.V. Gaddi (eds.), *Clinical Handbook of Air Pollution-Related Diseases*,
https://doi.org/10.1007/978-3-319-62731-1_23

the AhR that has a variety downstream effects including formation of DNA adducts via CYP1A/1B1-dependent metabolic activation, tumorigenesis, inflammation, cell proliferation, and loss of cell-cell adhesion [8]. Because of its high level of expression in human bronchial epithelial cells, AhR has many physiological consequences in the lung such as its effects on cell proliferation and differentiation, cell-cell adhesion interaction, cytokine expression, mucin production, and xenobiotic metabolism. In a study, urban particulate matter induced lung cancer in wild type ($AhR^{+/+}$) but not in AhR-null mice, suggesting that AhR plays a mechanistic role in the development of lung tumorigenesis by urban particulate matter, and this occurred through CYP1A1 induction [9].

23.1.1.2 Symptoms

Generally, lung cancers do not cause signs and symptoms in its earliest stages. Symptoms of lung cancer in advanced stage may include cough that prolongs, changes in chronic cough, coughing up blood, shortness of breath, chest pain, wheezing, hoarseness, losing weight, bone pain, and headache.

23.1.1.3 Diagnosis

(a) *X-ray and CT scan* – to check for abnormal mass, nodules, and small lesions in the lungs
(b) *Sputum cytology* – to check for the presence of lung cancer cells
(c) *Tissue sampling* – biopsy

Lung Cancer Staging

Spread of cancer beyond lungs is assessed further based on MRI, positron emission tomography (PET), and bone scans. The stages of lung cancer include *stage I*, where cancer is limited to the lung and the tumor is generally smaller than 2 in.; *stage II*, where tumor may have grown larger than 2 in. or may be a smaller tumor involving nearby structures, such as the chest wall, the diaphragm, or the lining around the lungs (pleura) or nearby lymph nodes; *stage III*, where tumor may have grown very large and invaded other organs near the lungs; and *stage IV*, where cancer has spread beyond the affected lung to the other lung or to distant areas of the body.

23.1.1.4 Treatment

(a) *Surgery* – involves removal of nodules or lesions and margin of healthy tissues. Procedures involved include:
   1. *Wedge resection* – removal of small section of lung that contains the tumor along with a margin of healthy tissue
   2. *Segmental resection* – removal of a larger portion of the lung but not an entire lobe
   3. *Lobectomy* – removal of entire lobe of one lung
   4. *Pneumonectomy* – removal of the entire lung
(b) Other treatment methods are chemotherapy, radiation therapy, targeted drug therapies, and additionally palliative care.

23.1.2 Type 2 Diabetes

Occurrence of type 2 diabetes is pronounced after a prolonged period of insulin resistance which is the main factor disrupting blood glucose homeostasis. Insulin resistance is characterized by a decreased action of insulin in peripheral tissue. Insulin is a key hormone involved in glucose homeostasis, synthesis and storage of fat, protein synthesis, and non-metabolic functions such as cell growth and differentiation [10]. Apart from insulin, glucagon, somatostatin, and pancreatic polypeptide are involved in glucose homeostasis; all of these hormones are secreted by islets of Langerhans containing insulin β-cells, glucagon α-cells, somatostatin δ-cells, and pancreatic polypeptide PP-cells.

23.1.2.1 EDCs Involved

Bisphenol A

In vitro and animal studies suggest that low-dose BPA exposure acting through diverse endocrine signaling pathways leads to the impairment of pancreatic islet morphology, insulin resistance, and β-cell dysfunction which are key mechanisms involved in the development of impaired glucose tolerance and diabetes [10–15]. This evidences the association between BPA exposure and the development of type 2 diabetes with increased insulin secretion from pancreatic beta cells, decreased glucose tolerance, increased uptake of glucose by adipocytes, and adiponectin release [10, 14, 16]).

In an experimental study, histological evaluation of pancreas from 12-week-old female mice revealed significantly increased insulitis in mice exposed to 1 mg/l BPA, while the insulitis was less severe at the higher BPA exposure. Serum glucose levels in 1 mg/mL BPA group tended to be hyperglycemic. The high BPA exposure also seemed to counteract the diabetes development in females and also in male nonobese diabetic mice for both BPA concentrations. Prior to insulitis, both BPA concentrations resulted in increased apoptosis and reduced numbers of tissue-resident macrophages in pancreatic islets. Long-term BPA exposure at a dose of three times higher than the tolerable daily intake of 50 mg/kg appeared to accelerate spontaneous insulitis and diabetes development [17].

Phthalate Esters

PEs have been associated with type 2 diabetes and insulin resistance in human populations [18–22]. However, there are limited epidemiologic data in assessing the risk of metabolic dysfunction associated with exposure to EDCs during pregnancy [23, 68].

Arsenic

Among the maternal blood evaluated in first tri-semester for heavy metals, arsenic at elevated level was suggested to have association with impaired glucose tolerance [24].

23.1.2.2 Symptoms

Increased thirst, frequent urination, extreme hunger, unexplained weight loss, presence of ketones in the urine, fatigue, irritability, blurred vision, slow-healing sores, frequent infections, such as gums or skin infections, and vaginal infections are common symptoms of diabetes. Apart from these symptoms, high blood sugar (hyperglycemia), increased ketones in urine (diabetic ketoacidosis), hyperglycemic hyperosmolar nonketotic syndrome, and low blood sugar (hypoglycemia) may accompany.

23.1.2.3 Diagnosis

The risk factors for type 2 diabetes include body mass index of higher than 25 (regardless of age), high blood pressure, sedentary lifestyle, high cholesterol levels, delivered a baby that weighed more than 4 kg (9 lbs), history of diabetes in pregnancy, and with the history of polycystic ovary syndrome, heart disease, or close relative with diabetes. Tests for diabetes include:

(a) *Glycated hemoglobin (A1C) test* – indicates average blood sugar level for the past 2–3 months by measuring the percentage of blood sugar attached to hemoglobin, the oxygen-carrying protein in red blood cells. A1C level of 6.5% or higher indicates diabetes, between 5.7 and 6.4% indicates prediabetes, and below 5.7% is normal.
(b) *Random blood sugar test* – blood sample taken at a random time; a random blood sugar level of 200 milligrams per deciliter (mg dL^{-1}) or higher suggests diabetes.
(c) *Fasting blood sugar test* – taken after an overnight fast, blood sugar level less than 100 mg dL^{-1} is normal, 100–125 mg dL^{-1} is prediabetes, and 126 mg dL^{-1} or higher is diabetes.
(d) *Oral glucose tolerance test* – after fasting blood sugar level test, a sugary liquid is drunk, and blood sugar levels are tested periodically for the next 2 h. A blood sugar level less than 140 mg dL^{-1} is normal, between 140 and 199 mg dL^{-1} is prediabetic, and more than 200 mg dL^{-1} indicates diabetes.

23.1.2.4 Treatment

Eating a healthy diet, maintaining a healthy weight, and participating in regular activity also are important factors in managing diabetes. Treatment of type 2 diabetes primarily involves monitoring of blood sugar along with diabetes medications, insulin, or both.

23.1.3 Metabolic Syndrome

Metabolic syndrome (MetS) is a combination of disorders including impaired glucose tolerance or insulin resistance, dyslipidemia, high blood pressure, and obesity [25]. Insulin resistance is suggested to be the main cause of MetS that in turn increases the risk of cardiovascular diseases, stroke, and type 2 diabetes. The EDCs involved in MetS involve primarily estrogen receptors and other unveiled mechanisms related to the above disorders.

23.1.3.1 Symptoms

MetS is mostly not associated with specific symptoms, though large waist circumference may be a visible sign. Other than this, the symptoms may be related to the disorders as mentioned above.

23.1.3.2 Diagnosis

MetS is diagnosed based on certain criteria that include:

(a) *Large waist circumference* – waistline that measures more than 35 in. (89 cm) for women and 40 in. (102 cm) for men
(b) *High triglyceride level* – 150 mg dL^{-1} or higher in blood
(c) *Reduced high-density lipoprotein (HDL) cholesterol* – >40 mg dL^{-1} in men or >50 mg dL^{-1} in women
(d) *Increased blood pressure* – 130/85 mmHg or higher
(e) *Elevated fasting blood sugar* – 100 mg dL^{-1} or higher

23.1.3.3 Treatment

Healthy lifestyle, exercises, and medications to control blood pressure, cholesterol levels, and blood glucose are recommended.

23.1.4 Cardiovascular Diseases

Cardiovascular disease (CVD) refers to the conditions involving narrowed or blocked blood vessels that can lead to a heart attack, chest pain, or stroke. Other conditions, such as those that affect heart's muscle, valves, or rhythm, are also considered as forms of heart diseases.

23.1.4.1 EDCs Involved

Phthalates

PEs are reported to significantly decrease human embryonic stem cell viability, along with a reduction in cardiac differentiation, following mono-ethylhexyl phthalate (MEHP) exposure [26]. Based on epidemiological studies, higher urinary phthalate levels have been linked to increased blood pressure in adolescent populations and increased coronary risk in elderly populations [27–29]. A significant association between higher MEHP urinary levels and LDL cholesterol levels but not blood pressure was also observed. As the report suggests, a positive association between MEHP urinary levels and the echogenicity of vascular plaques tends to be an indicator of lipid infiltration and a predictor of future cardiovascular death [27].

Bisphenol A

Studies report higher BPA urine concentrations to be associated with an increased risk of coronary artery disease, hypertension, carotid atherosclerosis, angina, myocardial infarction, and decreased heart rate variability [27, 30–32]. Higher BPA urinary levels have also been associated with LDL and HDL cholesterol levels and the echogenicity of vascular plaques [27, 28].

Polychlorinated Biphenyls

The first human evidence to PCBs and cardiovascular toxic effects was focused on the high exposure of PCBs in workers or residents near accidental spills [33–35]. Similarly, epidemiological studies also have reported an elevated incidence of hypertension in populations highly exposed to PCBs [36–39].

23.1.4.2 Symptoms

CVD symptoms depend on the type of heart disease; certain disease and symptoms are as follows:

Atherosclerotic Disease

In this type of CVD, narrowed, blocked, or stiffened blood vessels are observed that prevent heart, brain, or other parts of the body from receiving blood. The symptoms include chest pain (angina); shortness of breath; pain, numbness, weakness, or coldness in legs or arms if the blood vessels in those parts of your body are narrowed; and pain in the neck, jaw, throat, upper abdomen, or back. Among women shortness of breath, nausea, and extreme fatigue are mostly observed.

Heart Arrhythmias

This is characterized by an abnormal heartbeat that may be too fast, too slow, or irregular. The symptoms include fluttering in chest, racing heartbeat (tachycardia), slow heartbeat (bradycardia), chest pain or discomfort, shortness of breath, light-headedness, dizziness, and fainting (syncope).

Heart Defects

Serious congenital heart defects include the following symptoms: pale gray or blue skin color (cyanosis) and swelling in the legs and abdomen or areas around the eyes. In an infant, shortness of breath during feedings leads to poor weight gain.

Dilated Cardiomyopathy

This CVD is characterized by thickening and stiffening of the heart muscle. The symptoms are breathlessness with exertion or at rest; swelling of the legs, ankles, and feet; fatigue; irregular heartbeats that feel rapid, pounding, or fluttering; and dizziness, light-headedness, and fainting.

Heart Infection

There are three types of heart infections:

(a) Pericarditis – affects the tissue surrounding the heart (pericardium)
(b) Myocarditis – affects the muscular middle layer of the walls of the heart (myocardium)
(c) Endocarditis – affects the inner membrane that separates the chambers and valves of the heart (endocardium)
Varying slightly with each type of infection, symptoms include fever, shortness of breath, weakness or fatigue, swelling in your legs or abdomen, changes in heart rhythm, dry or persistent cough, and skin rashes or unusual spots.

Valvular Heart Disease

The four valves of the heart, aortic, mitral, pulmonary, and tricuspid valves, open and close to direct blood flow through the heart. These valves may be damaged by a variety of conditions leading to narrowing (stenosis), leaking (regurgitation or insufficiency), or improper closing (prolapse). Depending on the damage, symptoms include fatigue, shortness of breath, irregular heartbeat, swollen feet or ankles, chest pain, and fainting (syncope).

23.1.4.3 Diagnosis

Physical examination and blood tests are followed by:

(a) Electrocardiogram (ECG) – records electrical signals to detect irregularities in heart's rhythm and structure.
(b) Holter monitoring – also detects heart rhythm irregularities that are not regular during ECG examination.
(c) Echocardiogram – shows detailed images of heart's structure and function.
(d) Cardiac catheterization – shows pressure in heart chamber is measured.
(e) Cardiac CT scan – checks for heart problems.
(f) Cardiac MRI – evaluates heart based on image produced.

23.1.4.4 Treatment

CVD treatment includes changes in lifestyle and food habits, medication, and surgery if needed.

23.1.5 Neurodegenerative Disorder

This is a central nervous system (CNS) disorder characterized by loss of neuronal cell function and atrophy of the affected nervous system structures [40].

The most common chronic neurodegenerative diseases are Alzheimer's disease (AD), Parkinson's disease (PS), and multiple sclerosis (MS). These disorders have complex etiologies with insidious onset, protein aggregates, selected neuronal degeneration, and characteristic clinical syndromes [41].

23.1.5.1 EDCs Involved

Heavy Metals

Metals are crucial for the maintenance of cell homeostasis and preservation of life. They show important structural, regulatory, and catalytic functions in different types of proteins, such as enzymes, receptors, and transporter [42]. Alterations in metal homeostasis have been suggested to cause neurodegeneration via association of metals with proteins and subsequent induction of aggregate formation. In addition, metals can cause neurodegeneration through a vicious cycle by disrupting mitochondrial function, which depletes ATP, induces reactive oxygen species (ROS) production, and ultimately causes cell death by apoptotic and/or necrotic mechanisms [43]. Certain heavy metals that tend to cause neurodegeneration are as follows:

1. *Aluminum* – it induces neuropathy that includes neurofibrillary degeneration, oxidative stress, and inflammatory response [44].
2. *Copper* – long-term exposure showed increased levels of brain amyloid-β protein and in neuro-inflammation, the hallmarks of AD development [45].
3. *Lead* – early exposure is known to impact physiological development that may possibly increase susceptibility to neurodegeneration (AD, PD, MS) in later life by increasing the expression of amyloid precursor protein and increasing the production of amyloid-β protein [46].
4. *Cobalt* – industrial exposure is associated with adverse effects on neuromuscular transmission and neurological status [47, 48].
5. *Cadmium* – occupational exposure has been linked to neurological symptoms and neurobehavioral problems involving loss of attention, psychomotor speed, and memory. *In vitro* studies show self-aggregation of the tau peptide R3, thereby potentially impacting AD pathogenesis by mechanisms including astrocyte and neural cell toxicity [48, 49].
6. *Manganese* – brain biopsy of a single human subject with high manganese (Mn) level revealed multiple neuritic plaques and neurofibrillary tangles, characteristic of AD [49].
7. *Mercury* – a well-known neurotoxin; *in vitro* studies have demonstrated increase in tau protein hyperphosphorylation and the increased formation of amyloid-β protein [50].

Pesticides

Many pesticides target the nervous system of insect pests and similarly are neurotoxic to humans by adversely affecting cell signaling, disturbing neurochemical processes, and causing neurotoxicity [51]. Epidemiological studies illustrate that exposure to organochlorines and organophosphates is associated with an increased risk of AD, PD, and MS later in life [52]. A pilot study in a population of North Indians reported that increased blood levels of β-HCH and the organochlorine compound dieldrin were associated with significant increases in AD risk, independent of the genetic risk factor [53]. Exposure to paraquat and rotenone showed mitochondrial dysfunction in cerebral cortex, which in turn is known to promote impairment of cognition function with elevated levels of Aβ protein associated with AD and PD etiology [54].

Brominated Flame Retardants

Adult mice exposed to PBDEs showed altered spontaneous behavior, impaired learning and memory, and decreased hippocampal cholinergic receptors [55]. *In vitro* studies showed that PBDEs are neurotoxic and amyloidogenic specifically causing Ca2 + -ATPase inhibition, amyloid-β peptide release, and apoptosis a key neurodegenerative pathology observed in AD [56].

Alkylphenol

Estrogenic effects were seen in octylphenol (OP)-exposed turtles together with increased expression of amyloid-like precursor protein-2 and amyloid precursor protein, accumulation of which causes neuronal degeneration in AD brains [57].

Dioxins

The most potent dioxin congener, 2,3,7,8-tetrachlorodibenzo-p-dioxin (TCDD), was observed to increase in neuronal cell calcium levels and tau phosphorylation via upregulation of phospho-glycogen synthase kinase-3 β. These *in vitro* changes are similar to the pathologies of postmortem brain tissues of AD patients [58].

Polychlorinated Biphenyls

Case studies showed that exposure to PCBs produces certain clinical features consistent with AD-type dementia, and cohort studies revealed that women occupationally exposed to PCBs are more susceptible to PD and AD than PCB-exposed men [58, 59]. In postmortem studies, PCBs were higher in striatum of PD cases than controls [60]. In animal models, PCB exposure was associated with relatively selective, persistent, reduction of dopamine, primarily affecting caudate, putamen, substantia nigra, and olfactory tract [61, 62].

Bisphenol A

BPA has been shown to interfere with spine synapse formation in the prefrontal cortex and hippocampus that may have clinical implications resembling the events in AD [63]. BPA also disrupts expression of the Kcc2 gene through epigenetic mechanisms causing neurodevelopmental toxicity [64].

Phthalate Esters

DEHP significantly inhibits acetylcholinesterase activity and upregulates glial fibrillary acidic protein as well as myelin basic protein in zebrafish embryos [65]. It also causes cognitive dysfunction and increased phosphorylation of tau protein in prenatally exposed aged rats [22].

(i) Alzheimer's Disease

AD is characterized by the brain amyloid-β protein that forms dimers and oligomers, leading to protein aggregation visible in the postmortem brains as plaques. Plaques are accompanied by aggregates of phosphorylated tau protein called neurofibrillary tangles. These lesions are thought to cause synaptic loss and neuronal cell death, resulting in cognitive dysfunction [67, 69].

Symptoms

Initially increasing forgetfulness and mild confusion may be the symptoms and over time may lead to loss of memories. People suffering with AD start to lack memory, thinking, and reasoning and are not able to make judgment and decisions or perform familiar tasks. Change in personality and behavior is noticed, characterized by depression, apathy, social withdrawal, mood swings, distrusting others, irritability, aggressiveness, changes in sleeping habits, wandering, and delusions.

Diagnosis

(a) Physical and neurological examination – involves checking reflexes, muscle tone and strength, sense of sight and hearing, coordination, and balance
(b) Blood test – to rule out other potential causes of memory loss and confusion, such as thyroid disorders or vitamin deficiencies
(c) Mental status and neuropsychological test – to asses level of thinking and memory skills
(d) Brain imaging – to check for any abnormalities such as stroke or trauma

Treatment

Though AD is incurable, medications to improve brain cell communication and slow down the disease progression, such as providing safe and supportive environment, doing exercise, and proper nutrition, are recommended.

(ii) Parkinson's Disease

PD is a chronic, progressive neurodegenerative disease of aging characterized clinically by the classical motor signs of parkinsonism (resting tremor, bradykinesia, cogwheel rigidity, postural reflex impairment) and a variable constellation of associated features including autonomic, sensory, cognitive, and psychiatric changes. These clinical features correspond to the anatomic distribution of pathologic changes, consisting of neuronal degeneration and aggregation of the protein alpha-synuclein in specific neuronal populations including the substantia nigra, locus coeruleus, other brainstem and cortical regions, and the peripheral autonomic nervous system [41].

Symptoms

PD symptoms vary from person to person; usually it affects one side of the body and sometimes may affect both sides. The symptoms include tremor (shaking of hands and fingers when they are relaxed), slowed movement (bradykinesia), rigid muscles, impaired posture and balance, loss of automatic movements, and changes in speech and writing.

Diagnosis

No specific test exists to diagnose PD. Sometimes medical history, review of signs and symptoms, and neurological and physical examinations are done.

Treatment

Parkinson's disease can't be cured, but medications can help control your symptoms, often dramatically. In some later cases, surgery for deep brain stimulation may be done where electrodes are implanted in specific part of the brain.

(iii) Multiple Sclerosis

MS is a chronic inflammatory disease, which at least in the early stages is driven by immune-mediated processes. The characteristic hallmark of the disease is primary demyelination in the white as well as in the gray matter. However, neurodegeneration reflected by neuronal and synaptic loss results in permanent neurological disability [66].

Symptoms

MS symptoms vary from person to person based on the location of the affected nerve fiber.

The symptoms include:

(a) Numbness or weakness of one or more limbs that typically occurs on one side of the body at a time or the legs and trunk.
(b) Partial or complete loss of vision usually in one eye at a time, often with pain during eye movement.
(c) Prolonged double vision.
(d) Tingling or pain in parts of your body.
(e) Electric-shock sensations which occur with certain neck movements, especially bending the neck forward (Lhermitte sign).
(f) Tremor, lack of coordination, or unsteady gait.
(g) Slurred speech, fatigue, dizziness, and problems with bowel and bladder function are the other symptoms.

Diagnosis

There are no specific tests for MS. Instead, a diagnosis of multiple sclerosis often relies on ruling out other conditions that might produce similar signs and symptoms, known as a differential diagnosis. The recommended tests include:

(a) *Blood tests* – to help rule out other diseases with symptoms similar to MS.
(b) *Lumbar puncture (spinal tap)* – spinal fluid is checked for abnormalities in antibodies that are associated with MS.
(c) *MRI* – to detect for areas of MS (lesions) in the brain and spinal cord.
(d) *Evoked potential tests* – record electrical signals produced by nervous system in response to stimuli.

Treatment

There is no cure for MS. Treatment typically focuses on slowing the progression of the disease and managing MS symptoms.

23.2 Precautionary Measures to Prevent Endocrine Disruption

On behalf of prenatal and postnatal EDC exposure, there is an urgent need to frame certain steps to prevent exposure and enforce them into action by both government and nongovernment organizations in developing and non-developing countries. The exposure to EDCs begins from maternal exposure to the postnatal—environmental, industrial, and indoor exposures. The complete "zero EDC environment or atmosphere" is unpredictable in this industrialized world; instead the following measures may be followed to reduce the threating health consequences:

(a) Educate about EDCs among students, family members, and public. Initiate awareness about sources, toxic dosage limit, and extent of EDC interference in normal metabolic pathway.
(b) The possibility to use organic food products has to be raised and application of pesticides to be avoided.
(c) Awareness on the lipophilic nature of the EDCs.
(d) The sea animals tend to bioaccumulate more persistent organic pollutants (POPs) that are drained from domestic and industrial wastewater treatment plants and in turn serve as major seafood contaminant. Such contamination needs to be regulated and assured for consumption before proceeding to the market.
(e) Storing and heating of foods in plastic containers have to be avoided.
(f) Incineration of municipal solid waste containing plastics and E-waste has to be regulated as it is the main source for EDC to be released as air and gaseous particulates into the atmosphere.
(g) Raw fruits and vegetables are the direct source of pesticides, thus proper washing must be regularized before consumption.
(h) The closed indoor atmosphere must be dusted and cleaned up regularly to avoid concentrated EDCs in the particular environment.
(i) The traffic regulation and costumed motor service have to be followed to avoid release and exposure to PAHs, dioxin, and air particulates.
(j) The severity of certain chemical usage and its effects on endocrine disruption should be labelled on chemicals to avoid mishandling.
(k) The toddlers and children must be avoided to use plastic teethers and toys.
(l) The use of fragranced cosmetics and other personal care products needs to be avoided.
(m) Since most of the EDC mechanisms are unknown, more research on endocrine disruption by EDCs must be encouraged.
(n) The rules and regulation laid by the government in regard to EDCs preventive measure must be strictly followed.

Conclusion

In this chapter we have discussed about air pollution especially, due to EDCs and their health consequences in the atmosphere and the predictable diseases based on epidemiological evidences. The prenatal exposure insists on neurotoxicity, asthma, changes in epigenome, disease susceptibility, cancer, and infertility in later life, while in addition to these disorders, on postnatal exposure cardiovascular diseases, type 2 diabetes, neurodegeneration, and metabolic syndromes are also accounted. Though these consequences are predicted based on experimental data and continuous research on animal models, a gap of unexplored mechanisms still remains behind EDCs. Daily life depends on chemicals, and there is no way to escape mixture of EDC exposure rather than single EDC. It is difficult to rule out and address the changing patterns on chemical exposure, since it depends on changing needs, chemical production, and their release into the environment.

Based on available data, it may be understood that literature is available on certain EDCs and their consequences; e.g., BPA is best studied in relation to obesity and precocious puberty. Instead there is a need to understand a broad range of chemical-chemical interaction in the atmosphere and antagonistic response of such chemicals toward ligands and receptors of a cell at various species levels. The mechanism of EDCs at low and high doses is not eventually the same and also effect of single EDC varies from other; all the above when the term "mixture of EDCs" emanates, the statement on effect and mechanism is really challenging that has to be surpassed. For this, collaboration among research institutes, universities, government heads, and grant commissions would serve to better understand the exposure variance, population response, transgenerational effects, change in epigenome, and developmental stages. Apart from these, notions on elimination of more potential EDCs replaced with non-EDCs, equal ban of EDCs in both developing and non-developing countries, synthesis, and preferred use of green chemicals would greatly tend to reduce EDC exposure. To the end, present actions will be the mirror images of future generation.

Note: Symptoms, diagnosis, and treatment for the above disorders were referred at www.mayoclinic.org.

Reference

1. Blot WJ, Fraumeni JF Jr. Cancers of the lung and pleura. In: Schottenfeld D, Fraumeni Jr J, editors. Cancer epidemiology and prevention. New York: Oxford University Press; 1996. p. 637–65.
2. Hecht SS. Cigarette smoking and lung cancer: chemical mechanisms and approaches to prevention. Lancet Oncol. 2002;3:461–9.
3. Hoffmann D, Hecht SS. Advances in tobacco carcinogenesis. In: Cooper CS, Grover PL, editors. Handbook of experimental pharmacology, vol. 94/I. Heidelberg: Springer-Verlag; 1990. p. 63–102.

4. Smith TJ, Livingston SD, Doolittle OJ. An international literature survey of "IARC group I carcinogens" reported in mainstream cigarette smoke. Food Chem Toxicol. 1997;35:1107–30.
5. Hecht SS. Carçinogenic effects of cigarette smoke on the respiratory tract. In: Roth RA, editor. Comprehensive toxicology: toxicology of the respiratory system, vol. 8. Oxford: Elsevier; 1997. p. 437–51.
6. Hecht SS. Tobacco smoke carcinogens and lung cancer. J Natl Cancer Inst. 1999;91(14):1194–210.
7. Wolterbeek AP, Schoevers EJ, Rutten AA, Feron VJ. A critical appraisal of intratracheal instillation of benzo[*a*]pyrene to Syrian golden hamsters as a model in respiratory tract carcinogenesis. Cancer Lett. 1995;89:107–16.
8. Tsay JJ, Tchou-Wong KM, Greenberg AK, Pass H, Rom WN. Aryl hydrocarbon receptor and lung cancer. Anticancer Res. 2013;33:1247–56.
9. Matsumotom Y, Ide F, Kishi R, Akutagawa T, Sakai S, Nakamura M, et al. Aryl hydrocarbon receptor plays a significant role in mediating airborne particulate-induced carcinogenesis in mice. Environ Sci Technol. 2007;41:3775–80.
10. Ropero AB, Alonso-Magdalena P, Garcia-Garcia E, Ripoll C, Fuentes E, Nadal A. Bisphenol-A disruption of the endocrine pancreas and blood glucose homeostasis. Int J Androl. 2008;31:194–200.
11. Adachi T, Yasuda K, Mori C, Yoshinaga M, Aoki N, Tsujimoto G, et al. Promoting insulin secretion in pancreatic islets by means of bisphenol A and nonylphenol via intracellular estrogen receptors. Food Chem Toxicol. 2005;43:713–9.
12. Alonso-Magdalena P, Morimoto S, Ripoll C, Fuentes E, Nadal A. The estrogenic effect of bisphenol A disrupts pancreatic beta-cell function in vivo and induces insulin resistance. Environ Health Perspect. 2006;114:106–12.
13. Gerich JE. Contributions of insulin-resistance and insulinsecretory defects to the pathogenesis of type 2 diabetes mellitus. Mayo Clin Proc. 2003;78:447–56.
14. Sakurai K, Kawazuma M, Adachi T, Harigaya T, Saito Y, Hashimoto N, Mori C. Bisphenol A affects glucose transport in mouse 3T3-F442A adipocytes. Br J Pharmacol. 2004;141:209–14.
15. Kawasaki T, Igarashi K, Ogata N, Oka Y, Ichiyanagi K, Yamanouchi T. Markedly increased serum and urinary fructose concentrations in diabetic patients with ketoacidosis or ketosis. Acta Diabetol. 2010;49:119–23.
16. Alonso-Magdalena P, Vieira E, Soriano S, Menes L, Burks D, Quesada I, et al. Bisphenol A exposure during pregnancy disrupts glucose homeostasis in mothers and adult male offspring. Environ Health Perspect. 2010;118:1243–50.
17. Bodin J, Bolling AK, Samuelsen M, Becher R, Lovi M, Nygaard UC. Long-term bisphenol A exposure accelerates insulitis development in diabetes-prone NOD mice. Immunopharmacol Immunotoxicol. 2013;35(3):349–58.
18. James-Todd T, Stahlhut R, Meeker JD, Powell SG, Hauser R, Huang T, et al. Urinary phthalate metabolite concentrations and diabetes among women in the National Health and Nutrition Examination Survey (NHANES) 2001–2008. Environ Health Perspect. 2012;12:130–1313.
19. Lind PM, Zethelius B, Lind L. Circulating levels of phthalate metabolites are associated with prevalent diabetes in the elderly. Diabetes Care. 2012;35:1519–24.
20. Stahlhut RW, van Wijngaarden E, Dye TD, Cook S, Swan SH. Concentrations of urinary phthalate metabolites are associated with increased waist circumference and insulin resistance in adult U.S. males. Environ Health Perspect. 2007;115:876–82.
21. Sun Q, Cornelis MC, Townsend MK, Tobias DK, Eliassen AH, Franke AA, et al. Association of urinary concentrations of bisphenol A and phthalate metabolites with risk of type 2 diabetes: a prospective investigation in the Nurses' Health Study (NHS) and NHSII cohorts. Environ Health Perspect. 2014a;122:616–23.
22. Sun W, Ban J-B, Zhang N, Zu Y-K, Sun W-X. Perinatal exposure to di-(2-ethylhexyl)-phthalate leads to cognitive dysfunction and phospho-tau level increase in aged rats. Environ Toxicol. 2014b;29(5):596–603.

23. Ettinger AS, Zota AR, Amarasiriwardena CJ, Hopkins MR, Schwartz J, Hu H, et al. Maternal arsenic exposure and impaired glucose tolerance during pregnancy. Environ Health Perspect. 2009;117:1059–64.
24. Shapiro GD, Dodds L, Arbuckle TE, Ashley-Martinc J, Fraser W, Fisher M, et al. Exposure to phthalates, bisphenol A and metals in pregnancy and the association with impaired glucose tolerance and gestational diabetes mellitus: the MIREC study. Environ Int. 2015;83:63–71.
25. Casals-Casas C, Deavergne B. Endocrine disruptors: from endocrine to metabolic disruption. Annu Rev Physiol. 2011;73:135–62.
26. Schulpen SHW, Robinson JF, Pennings JLA, van Dartel DAM, Piersma AH. Dose response analysis of monophthalates in the murine embryonic stem cell test assessed by cardiomyocyte differentiation and gene expression. Reprod Toxicol. 2013;35:81–8.
27. Lind PM, Lind L. Circulating levels of bisphenol A and phthalates are related to carotid atherosclerosis in the elderly. Atherosclerosis. 2011;218:207–13.
28. Olsen L, Lind L, Lind PM. Associations between circulating levels of bisphenol A and phthalate metabolites and coronary risk in the elderly. Ecotoxicol Environ Saf. 2012;80:179–83.
29. Trasande L, Sathyanarayana S, Spanier AJ, Trachtman H, Attina TM, Urbina EM. Urinary phthalates are associated with higher blood pressure in childhood. J Pediatr. 2013;163:747–53.e1.
30. Bae S, Kim JH, Lim YH, Park HY, Hong YC. Associations of bisphenol A exposure with heart rate variability and blood pressure. Hypertension. 2012;60:786–93.
31. Lang IA, Galloway TS, Scarlett A, Henley WE, Depledge M, Wallace, Melzer D. Association of urinary bisphenol A concentration with medical disorders and laboratory abnormalities in adults. JAMA. 2008;300:1303–10.
32. Melzer D, Osborne NJ, Henley WE, Cipelli R, Young A, Money C, et al. Urinary bisphenol A concentration and risk of future coronary artery disease in apparently healthy men and women. Circulation. 2012;125:1482–90.
33. Bertazzi PA, Bernucci I, Brambilla G, Consonni D, Pesatori AC. The Seveso studies on early and long-term effects of dioxin exposure: a review. Environ Health Perspect. 1998;106(2):625–33.
34. Gustavsson P, Hogstedt C. A cohort study of Swedish capacitor manufacturing workers exposed to polychlorinated biphenyls (PCBs). Am J Ind Med. 1997;32:234–9.
35. Hay A, Tarrel J. Mortality of power workers exposed to phenoxy herbicides and polychlorinated biphenyls in waste transformer oil. Ann N Y Acad Sci. 1997;837:138–56.
36. Goncharov A, Bloom M, Pavuk M, Birman I, Carpenter DO. Blood pressure and hypertension in relation to levels of serum polychlorinated biphenyls in residents of Anniston, Alabama. J Hypertens. 2010;28:2053–60.
37. Goncharov A, Pavuk M, Foushee HR, Carpenter DO, Anniston Environmental Health Reseach Consortium. Blood pressure in relation to concentrations of PCB congeners and chlorinated pesticides. Environ Health Perspect. 2011;119:319–25.
38. Huang X, Lessner L, Carpenter DO. Exposure to persistent organic pollutants and hypertensive disease. Environ Res. 2006;102:101–6.
39. Valera B, Ayotte P, Poirier P, Dewailly E. Associations between plasma persistent organic pollutant levels and blood pressure in Inuit adults from Nunavik. Environ Int. 2013;59:282–9.
40. Cannon JR, Greenamyre JT. The role of environmental exposures in neurodegeneration and neurodegenerative diseases. Toxicol Sci. 2011;124:225–50.
41. Tanner CM, Goldman SM, Ross GW, Grate SJ. The disease intersection of susceptibility and exposure chemical exposures and neurodegenerative disease risk. Alzhemeirs Dementia. 2014;10:S213–25.
42. Briner W. The role of metal regulatory proteins in brain oxidative stress: a tutorial. Oxidative Med Cell Longev. 2012;2012:981561.
43. Bowman AB, Kwakye GF, Herrero Hernandez E, Aschner M. Role of manganese in neurodegenerative diseases. J Trace Elem Med Biol. 2011;25:191–203.
44. Kawahara M, Kato-Negishi M. Link between aluminum and the pathogenesis of Alzheimer's disease: the integration of the aluminum and amyloid cascade hypotheses. Int J Alzheimers Dis. 2011;2011:276393.

45. Singh I, Sagare AP, Coma M, Perlmutter D, Gelein R, Bell RD, et al. Low levels of copper disrupt brain amyloid-β homeostasis by altering its production and clearance. Proc Natl Acad Sci. 2013;110(36):14771–6.
46. Basha MR, Wei W, Bakheet SA, Benitez N, Siddiqi HK, Ge YW, et al. The fetal basis of amyloidogenesis exposure to lead and latent overexpression of amyloid precursor protein and β- amyloid in the aging brain. J Neurosci. 2005;25(4):823–9.
47. Kim JH, Gibb HJ, Howe PD. WHO concise international chemical assessment document 69. Cobalt and inorganic cobalt compounds. 2006.
48. Mates JM, Segura JA, Alonso FJ, Marquez J. Roles of dioxins and heavy metals in cancer and neurological diseases using ROS- mediated mechanisms. Free Radical BiolMed. 2010;49(9):1328–41.
49. Jiang LF, Yao TM, Zhu ZL, Wang C, Ji LN. Impacts of Cd(II) on the conformation and self-aggregation of Alzheimer's tau fragment corresponding to the third repeat of microtubule-binding domain. Biochim Biophys Acta. 2007;1774(11):1414–21.
50. Mutter J, Curth A, Naumann J, Deth R, Walach H. Does inorganic mercury play a role in Alzheimer's disease? A systematic review and an integrated molecular mechanism. J Alzheimers Dis. 2010;22(2):357–74.
51. Kodavanti P. Cell signaling and neurotoxicity: protein kinase C. In vitro and in vivo. In: Costa LG, Giordano G, Guizzetti M, editors. In vitro neurotoxicology. New York: Humana Press; 2011. p. 307–19.
52. Hayden KM, Norton MC, Darcey D, Ostbye T, Zandi PP, Breitner JCS, Welsh-Bohmer KA, Cache County Study Investigators. Occupational exposure to pesticides increases the risk of incident AD: the Cache County study. Neurology. 2010;74(19):1524–30.
53. Chhillar N, Singh NK, Banerjee BD, Bala K, Sharma D, Mustafa M, et al. β-hexachlorocyclohexane as a risk for Alzheimer's disease: a pilot study in north Indian population. Am J Alzheimers Dis. 2013;1:60–71.
54. Chen L, Yoo S-E, Na R, Liu Y, Ran Q. Cognitive impairment and increased Aβ levels induced by paraquat exposure are attenuated by enhanced removal of mitochondrial H2O2. Neurobiol Aging. 2012;33(2):432.e15–e26 XXX.
55. Viberg H, Fredriksson A, Eriksson P. Neonatal exposure to polybrominated diphenyl ether (PBDE 153) disrupts spontaneous behaviour, impairs learning and memory, and decreases hippocampal cholinergic receptors in adult mice. Toxicol Appl Pharmacol. 2003;192(2):95–106.
56. Al-Mousa F, Michelangeli F. Some commonly used brominated flame retardants cause Ca2+-ATPase inhibition, beta-amyloid peptide release and apoptosis in SH-SY5Y neuronal cells. PLoS One. 2012;7(4):e33059.
57. Trudeau VL, Chiu S, Kennedy SW, Brooks RJ. Octylphenol (OP) alters the expression of members of the amyloid protein family in the hypothalamus of the snapping turtle, Chelydra serpentina serpentina. Environ Health Perspect. 2002;110(3):269–75.
58. Sul D, Kim HS, Cho E-K, Lee M, Kim HS, Jung WW, et al. 2,3,7,8-TCDD neurotoxicity in neuroblastoma cells is caused by increased oxidative stress, intracellular calcium levels, and tau phosphorylation. Toxicology. 2009;255(1–2):65–71.
59. Steenland K, Hein MJ, Cassinelli RTI, Prince MM, Nilsen NB, Whelan EA, et al. Polychlorinated biphenyls and neurodegenerative disease mortality in an occupational cohort. Epidemiology. 2006;17(1):8–13.
60. Corrigan FM, Murray L, Wyatt CL, Shore RF. Diorthosubstituted polychlorinated biphenyls in caudate nucleus in Parkinson's disease. Exp Neurol. 1998;150:339–42.
61. Chu I, Poon R, Yagminas A, Lecavalier O, Hakansson H, Valli VE. Subchronic toxicity of PBC 105 (2, 3, 3′, 4, 4′,-pentachlorobiphenyl) in rats. J Appl Toxicol. 1998;18:285–92.
62. Seegal RF, Bush B, Brosch KO. Decreases in dopamine concentrations in adult, non-human primate brain persist following removal of polychlorinated biphenyls. Toxicology. 1994;86:71–87.
63. Hajszan T, Leranth C. Bisphenol A interferes with synaptic remodeling. Front Neuroendocrinol. 2010;31(4):519–30.

64. Yeo M, Berglund K, Hanna M, Guo JU, Kittur J, Torres MD, et al. Bisphenol A delays the perinatal chloride shift in cortical neurons by epigenetic effects on the Kcc2 promoter. Proc Natl Acad Sci. 2013;110(11):4315–20.
65. Xu H, Shao X, Zhang Z, Zou Y, Chen Y, Han S, et al. Effects of di-n-butyl phthalate and diethyl phthalate on acetylcholinesterase activity and neurotoxicity related gene expression in embryonic zebrafish. Bull Environ Contam Toxicol. 2013;91(6):635–9.
66. Marik C, Felts PA, Bauer J, Lassmann H, Smith KJ. Lesion genesis in a subset of patients with multiple sclerosis: a role for innate immunity? Brain. 2007;130:2800–15.
67. Perl DP. Neuropathology of Alzheimer's disease. Mt Sinai J Med. 2010;77(1):32–42.
68. Robledo C, Peck JD, Stoner JA, Carabin H, Cowan L, Koch HM, et al. Is bisphenol-A exposure during pregnancy associated with blood glucose levels or diagnosis of gestational diabetes? J Toxicol Environ Health A. 2003;76:865–73.
69. Selkoe DJ. The molecular pathology of Alzheimer's disease. Neuron. 1991;6(4):487–98.

Air Pollution and Cancer

24

Andrea A. Martoni

Foreword Air pollution (AP) is associated with an increased risk of lung cancer. This relationship is well established both for outdoor AP and indoor AP. In this chapter we look at the major scientific evidence supporting the existence of this relationship and the current outlook for suitable measures to limit the carcinogenic effects of the AP in humans. The primary and secondary literature on the relationship between AP and cancer is very broad, and, as a result of the "clinical handbook" character of this publication, we will necessarily refer to the major contributions that have been on the subject in recent years. In particular we will have as main reference the monograph of the International Agency for Research on Cancer (IARC) published in 2015 which gathers expert opinions of the IARC Working Group (WG) on the Evaluation of Air Pollution Carcinogenic Risks to Humans, which met in Lyon, 8–15 October 2013 [1].

For more details on the AP characteristics, we refer to other specific chapters of the present book.

24.1 Carcinogens in Outdoor Air Pollution

Outdoor AP is the presence in the air of substances at a concentration or for a duration above their natural levels, with the potential to produce an adverse effect. AP is a mixture of many different substances, and the exact contents vary depending on what sources of pollution are nearby, the location, the time of year, and the weather. The most common air pollutants are represented by vehicle emissions, stationary power generation, other industrial and agricultural emissions, residential heating and cooking, reemission from terrestrial and aquatic surfaces, manufacturing, distribution, and use of chemicals and natural processes.

A.A. Martoni
Medical Oncology Unit, S.Orsola-Malpighi University Hospital, Bologna 40136, Italy
e-mail: andrea.martoni@fastwebnet.it

F. Capello, A.V. Gaddi (eds.), *Clinical Handbook of Air Pollution-Related Diseases*, https://doi.org/10.1007/978-3-319-62731-1_24

Any outdoor AP mixture includes gases and suspended particulated matter (PM) that are constantly interacting. Gaseous pollutants include ozone, nitrogen oxide (NO) and nitrogen dioxide (NO2), sulfur dioxide (SO2), and numerous volatile organic gases. PM is a heterogeneous mixture of liquids and solids with variable and dynamic chemical composition and physical characteristics. PM is characterized by both its size and its chemical composition. Particle sizes range from the nanometer scale up to grains of dust on the order of tens of micrometers, but most measurements capture the mass in specific size ranges; for example, PM2.5 is PM with particles of aerodynamic diameter less than 2.5 μm and PM10 is PM with particles of aerodynamic diameter less than 10 μm. These size classes are relevant to PM dynamics in the atmosphere and uptake in the human body and reflect size ranges used in health studies.

On the whole outdoor air contains many substances that have been classified by IARC as Group 1 (e.g., benzene, benzo[*a*]pyrene, chromium (VI), and dioxin) and Group 2A carcinogens. These are both significant components of urban AP mixtures and include PM and other compounds (Table 24.1). The table includes also diesel engine exhaust in Group 1 substances and gasoline engine exhaust in Group 2B [2]. The category Group 1 is used when there is sufficient evidence of carcinogenicity in humans. Group 2A is applied to the agents with probable carcinogenicity to humans (i.e., limited evidence of carcinogenicity in humans and sufficient evidence in experimental animals), and Group 2B is used for those agents possibly carcinogenic to humans (i.e., limited evidence of carcinogenicity in humans and less than sufficient evidence of carcinogenicity in experimental animals). On these bases the IARC WG concluded that "there is sufficient evidence in humans and in experimental animals for the carcinogenicity of outdoor AP in general and of PM in outdoor AP more specifically."

This statement is supported by strong mechanistic evidence in exposed humans, including studies showing increased frequencies of DNA damage, mutations, and chromosomal damage in individuals occupationally or residentially exposed to polluted air (Table 24.2). The genetic and related effects in humans and experimental systems induced by exposures to outdoor AP and samples derived from outdoor AP are summarized in Table 24.3. A large body of evidence clearly indicates that humans exposed to elevated levels of outdoor AP have increased levels of chromosomal damage, including chromosome breaks and aneuploidy. Similar effects in experimental systems, both in vivo and in vitro, are also well documented. In addition, a variety of other genotoxic effects in humans and experimental animals exposed to elevated levels of outdoor AP or samples derived from outdoor AP are also well documented (e.g., mutations, DNA strand breaks, stable DNA adducts, and oxidized nucleobases). A wide range of other effects related to carcinogenesis, including oxidative stress, inflammation, and epigenetic alterations, have also been observed in exposed humans and animals and in a variety of experimental systems.

Table 24.1 Agents in outdoor air that are established or probable IARC carcinogens

Agent	Carcinogenic group
Metals and fibers	
Arsenic and inorganic arsenic compounds	1
Asbestos	1
Beryllium and beryllium compounds	1
Cadmium and cadmium compounds	1
Chromium (VI)	1
Lead compounds, inorganic/organic	2A/3
Nickel, metallic/compounds	2B/1
Silica dust	1
Organic chemicals	
1,3-Butadiene	1
Benzene	1
Ethylene oxide	1
Formaldehyde	1
Halogenated chemicals	
Ethylene dibromide	2A
2,3,7,8-Tetrachlorodibenzo-*para*-dioxin	1
Tetrachloroethylene	2A
Trichloroethylene	1
1,2,3-Trichloropropane	2A
Vinyl bromide	2A
Vinyl chloride	1
Vinyl fluoride	2A
Polycyclic aromatic hydrocarbons	
Benzo[*a*]pyrene	1
Cyclopenta[*cd*]pyrene	2A
Dibenz[*a,h*]anthracene	2A
6-Nitrochrysene	2A
1-Nitropyrene	2A
2-Nitrotoluene	2A
Mixtures	
Biomass fuel (primarily wood), indoor emissions from household combustion of	2A
Coal, indoor emissions from household combustion of	1
Coal tar pitch	1
Coke production	1
Creosotes	2A
Diesel engine exhaust	1
Frying, emissions from high temperature	2A
Mineral oils, untreated or mildly treated	1
Polychlorinated biphenyls	1
Polybrominated biphenyls	2A
Tobacco smoke, secondhand	1
Wood dust	1

From IARC [1] modified

Table 24.2 Summary of genetic effects reported in molecular epidemiology studies of outdoor air pollution

Exposure scenario	End point reported: DNA adducts	Cytogenetic damage	DNA strand breaks	Changes in gene expression
Traffic police	+	+	+	NA
Taxi drivers	+	+	+	NA
Bus drivers	+	+	NA	NA
Mail carriers	+	+	NA	NA
Urban residents	+	+	+	+
Urban children	+	+	+	+

From IARC [1] modified
+ positive, *NA* not available

Table 24.3 Summary by end point of genetic and related effects induced in humans and experimental systems by exposure to outdoor air pollution or samples derived from outdoor air pollution (from IARC [1])

End point	Humans	Experimental animals	Mammalian cells	Plants	Bacteria	Acellular
Mutations	(+)[a]	+	+	+	+	[b]
Cytogenetic damage (CAs, MN, SCEs)	+	+	+	+	NA	NA
Stable DNA adducts	+	+	+	NE	NE	+
DNA strand breaks	+	+/–[c]	+	NE	NE	+
Oxidatively damaged DNA	+	+/–[d]	+	NE	NE	+
Oxidative stress and inflammation	+	+	+	NE	NE	+
Cell transformation	NA	NA	+	NA	NA	NA
Epigenetic changes	+	+	NE	NE	NE	NA

[a]Limited information available
[b]Not applicable
[c]Few studies, conflicting results
[d]Few studies, conflicting results
+ positive, – negative, *CAs* chromosomal aberrations, *MN* micronuclei, *NA* not available, *NE* not evaluated, *SCEs* sister chromatid exchanges

24.2 Air Pollution and Cancer in Humans

Epidemiological studies of relationships between AP exposure and cancer have required long periods of observation and large populations from as early as the 1970s through 2010. Two large cohort studies—the European ESCAPE study [3] and the CPS II study in the USA [4, 5]—are particularly informative for their large-scale, exposure assessment based on actual measurements, the broad range of exposures considered, very detailed information about potential confounders (e.g., duration and intensity of smoking), and standardization of methods. In particular the ESCAPE study investigated the long-term effects of AP on a broad range of chronic conditions as asthma, allergies in children, adult respiratory and cardiovascular disease, cancer, and life expectancy. Funded by the EU, the project brought together over 20 leading research groups on AP and health from 15 countries to analyze over 30 cohort studies including some 900,000 subjects. Cohort studies follow a population over time, and ESCAPE focused on how different levels of exposure to air pollution affected people's health.

24.3 Lung Cancer

24.3.1 Epidemiologic Notes

Lung cancer has been the most common cancer in the world for several decades. In 2012, 1.8 million new cases (12.9% of the total) have been estimated, 58% of which occurred in the less developed regions. The disease remains as the most common cancer in men worldwide (1.2 million, 16.7% of the total) with the highest estimated age-standardized incidence rates in Central and Eastern Europe (53.5/100,000) and Eastern Asia (50.4/100,000). Low incidence rates are observed in Middle and Western Africa (2.0 and 1.7/100,000, respectively). In women, the incidence rates are generally lower and the geographical pattern is a little different, mainly reflecting different historical exposure to tobacco smoking. Thus the highest estimated rates are in Northern America (33.8) and Northern Europe (23.7) with a relatively high rate in Eastern Asia (19.2) and the lowest rates again in Western and Middle Africa (1.1 and 0.8, respectively) [6]. Lung cancer is the most common cause of death from cancer worldwide, estimated to be responsible for 1.72 million deaths with an increment of 20% from 2005 to 2015 [7].

Lung cancer is a heterogeneous disease characterized by a variety of biomarkers and differing histotypes. Non-small cell lung cancer is the most common histological subtype, occurring in up to 85% of cases. The disease affects both sexes and its incidence is low in people aged <40 years and increases up to age 75–80 years in most populations. Unfortunately, the diagnosis of lung cancer is often made late in the course of the disease, with almost 70% of patients presenting with locally advanced or metastatic disease at initial diagnosis. The 5-year relative survival rate varies markedly depending on the stage at diagnosis, from 49% to 16% to 2% for patients with local, regional, and metastatic disease, respectively. Because of its high fatality (the overall ratio of mortality to incidence is 0.87) and the relative lack of variability in survival in different world regions, the geographical patterns in mortality closely follow those in incidence.

Although the continuous increase of the number of cases and deaths, the age-standardized incidence and mortality rates demonstrate a global declining trend for about two decades. In the period 2005–2015, the drop of the standardized mortality rate was of -8.1% [7]. However, the decline in incidence and mortality involves men, while conversely the global trend in women is increasing. This is considered a direct consequence of the tobacco smoking decrease in man as opposed to the increase in women occurred in most countries.

24.3.2 Risk Factor Notes

The geographic and temporal patterns of lung cancer incidence and mortality on a population level are chiefly determined by tobacco consumption, the main etiological factor in lung carcinogenesis. For smokers, the risk for lung cancer is on average tenfold higher than in lifetime non-smokers. The risk increases with the quantity of cigarettes, duration of smoking, and starting age. Smoking cessation results in a decrease in precancerous lesions and a reduction in the risk of developing lung cancer. Former smokers continue to have an elevated risk for lung cancer for years after quitting [8].

The epidemiological evidence and biological plausibility support a causal association also between secondhand exposure to cigarette smoke and lung cancer risk in non-smokers [9]. The effect of involuntary smoking appears to be present for both household exposures, mainly from spousal and workplace exposure [10, 11] and perhaps from involuntary childhood smoking exposure [12]. Other factors such as genetic susceptibility, poor diet, occupational exposures (asbestos, cadmium, nickel, arsenium, beryllium, silica, polycyclic aromatic hydrocarbons), ionizing radiation, and AP may act independently or in concert with tobacco smoking in shaping the descriptive epidemiology of lung cancer [13].

The actual most important modality of ionizing radiation exposure is represented by radon exposure in the home or workplace. Radon is a naturally occurring radioactive gas that results from the breakdown of uranium in soil and rocks. Indoor radon represents a risk factor for lung cancer. In a pooled analysis of 13 European case–control studies there was strong evidence of an association between the radon concentration at home and lung cancer. The absolute risk to smokers and recent former smokers was much greater than to lifelong non-smokers. Radon in the home accounts for about 9% of deaths from lung cancer and about 2% of all deaths from cancer in Europe [14]. In the USA the Environmental Protection Agency considers radon the second leading cause of lung cancer in this country and the leading cause among non-smokers [15].

24.3.3 Air Pollution and Lung Cancer

The studies considered by the IARC WG to evaluate the carcinogenic effects of the AP in humans are represented primarily by large cohort studies conducted almost exclusively in high-income countries, mostly in North America and Europe. Overall, studies with quantitative measures of outdoor AP showed positive associations with lung cancer in both sexes and in cohort and case–control studies from all regions, with the potential confounding effect modification by tobacco smoking. Some studies provided analyses stratified by smoking or restricted to never-smokers, and associations of lung cancer with outdoor AP were similarly observed. The available studies used a range of quantitative or qualitative estimates of exposure. Quantitative estimates were mainly represented by PM (PM2.5 or PM10) and NO2 or NOx. Qualitative or semiquantitative estimates included traffic density or distance from heavy-traffic roads.

The relative risk (RR) estimates for PM2.5 and PM10 were indicative of positive associations in almost all the studies. When exposure–response relationships were examined, these generally indicated increasing risk of lung cancer with increasing levels of exposure to PM. In the ESCAPE study, with a mean length of follow-up of about 13 years, there were over 2,000 incident lung cancers. The analysis showed a hazard ratio (HR) of 1.22 for each 10 $\mu g/m^3$ increase in estimated PM10 concentration. For PM2.5, the HR was 1.18/5 $\mu g/m^3$. The same changes in PM10 and PM2.5 were associated with HRs for adenocarcinoma histotypes of 1.51 and 1.55, respectively [3]. In addition, in this study the HRs for lung cancer were similar with and without restriction to participants below most of the predefined threshold values, suggesting that exposure of populations to PM AP even at concentrations below the existing European Union air quality limit values for PM10 (40 $\mu g/m^3$) and PM2.5 (25 $\mu g/m^3$) might increase the risk for lung cancer. PM2.5 is generally believed to be most relevant to health effects, including cancer as they penetrate more deeply into the lung and are more likely to be retained. However, it was still not entirely clear what of the PM components are responsible of the carcinogenic effects. Composition of PM and lung cancer was studied within 14 cohort studies in eight European countries. The study indicates that the association between PM in AP and lung cancer can be attributed to various PM components and sources. PM containing S and Ni might be particularly important [16].

After the IARC WG evaluation, a systematic meta-analysis of studies examining the relationship of exposure to PM2.5 and PM10 with lung cancer incidence and mortality was carried out to support the IARC evaluation [17]. The analysis was performed in subgroups defined by geographic region, potential confounders and effect modifiers, and exposure assessment method. The influence of single studies to the overall meta-estimate was also examined. The meta-RR for lung cancer associated with PM2.5 was 1.09 (95% CI, 1.04–1.14). The meta-RR of lung cancer associated with PM10 was similar but less precise, 1.08 (95% CI, 1.00–1.17). Analyses by smoking status showed that lung cancer risk associated with PM2.5 was greatest for former smokers [1.44 (95% CI, 1.04–2.01)], followed by never-smokers [1.18 (95% CI, 1.00–1.39)], and then current smokers [1.06 (95% CI, 0.97–1.15)]. In addition, meta-estimates for adenocarcinoma histology associated with PM2.5 and PM10 were 1.40 (95% CI, 1.07–1.83) and 1.29 (95% CI, 1.02–1.63), respectively. These results support the decision of the IARC WG to classify outdoor AP as a Group 1 carcinogen and further justify efforts to reduce exposures to air pollutants.

After the IARC WG report, PM exposure and incident lung cancer were studied in the Nurses' Health Study cohort [18]. Findings confirmed previous studies indicating increased risk of incident lung cancer associated with ambient PM2.5 exposures, especially among never- and long-term former smokers. Few studies have directly assessed the relationship between ambient PM2.5 and lung cancer among never-smokers. In the AHSMOG-2 Study, an increased risk estimates of lung cancer were observed for each 10 $\mu g/m^3$ increment in ambient PM2.5 concentration (RR,1.43; CI, 1.11–1.84) [19]. The estimate was higher among those with longer residence at enrollment address and those who spent more than 1 hr/day outdoors. A recent Canadian study confirmed the association between lung cancer incidence and long-term exposure to PM2.5, even at lower concentration such as those currently found in Canadian cities and lung cancer [20]. In this study stratified analyses suggested increased PM2.5 risks were limited to those who smoked cigarettes.

Nitrogen oxides (N0x) and N02 are gaseous pollutants used as markers of traffic exposure in epidemiological research. The IARC WG found there was a suggestion of increasing risk with increasing levels of exposure to NO2 and/or NOx, but results were inconsistent. Recently, two meta-analyses evaluated the association between traffic-related AP and lung cancer. The first included 36 studies and found exposure to NO2 (RR, 1.06; 95% CI, 0.99–1.13), NO (RR, 1.04; 95% CI, 1.01–1.07), SO2 (RR,1.03; 95% CI,1.02–1.05), and fine PM (RR, 1.11; 95% CI, 1.00–1.22) were positively associated with a risk of lung cancer. Occupational exposure to AP among professional drivers significantly increased the incidence (RR, 1.27; 95% CI, 1.19–1.36) and mortality of lung cancer (RR, 1.14; 95% CI, 1.04–1.26) [21]. Also the second meta-analysis on studies focused on NOx and NO2 found consistent evidence of a relationship between NO2, as a proxy for traffic-sourced AP exposure, and lung cancer [22]. These two meta-analyses support the IARC classification of outdoor AP and particulate matter as Group 1 carcinogenic drawing particular attention to traffic-sourced AP. Most studies on the association between traffic diesel exhaust exposure and lung cancer suggest a modest, but consistent, increased risk [23]. A very large pooled study on 13,304 lung cancer cases and 16,282 controls from 11 case–control studies conducted in Europe and Canada showed a small, consistent association between occupational exposure to diesel motor exhaust and lung cancer with an odds ratio of 1.31 and a significant exposure–response relationship, after adjusting for potential confounders, such as smoking and other occupational exposures. The effect is similar for non-small cell and small cell lung carcinoma.

A systematic review analyzed the association between professional drivers and lung cancer, taking into consideration the potential confounding effect of cigarette smoking [24]. The search interested all published cohort and case–control studies in English from January 1996 to January 2011. A total of 19 studies were included in the meta-analysis, and a significantly increased risk of lung cancer (pooled smoking-adjusted RR, 1.18; 95% CI, 1.05–1.33) among professional drivers was observed. A higher pooled RR was observed among smoking-adjusted studies reporting 10 years or longer of employment (RR, 1.19; 95% CI, 1.06–1.34) compared with the study reporting shorter duration of employment (6 years; RR 1.00; 95% CI, 0.92–1.09). The IARC WG believes that, while not definitive in themselves because of many limitations, the findings of occupational cohort studies suggesting an association between professional driving and risk of lung cancer are supportive.

Cohort and case–control studies and meta-analyses involving millions of subjects and many thousands of lung cancer cases in different parts of the world consistently showed an association between exposure to outdoor AP and the risk of lung cancer, in both sexes and after adjustment for the main potential confounders. The association was present in almost all studies that specifically investigated the association of lung cancer and outdoor AP among never-smokers. Positive exposure–response relationships were reported in several studies.

It is useful to note also that an epidemiological study involving 352,053 patients with newly diagnosed lung cancer during 1988–2009, ascertained by the California

Cancer Registry, after adjusting for histology and other potential confounders, supports the hypothesis that AP exposures could also shorten survival in patients with lung cancer diagnosis [25].

24.4 Cancer of the Urinary Tract

Compared with studies focusing on lung cancer, the IARC WG found a limited number of studies considering cancers of the urinary bladder as an outcome of exposure to outdoor AP. Seven studies with cohort or case–control designs directly evaluated the association of cancer of the bladder with measurements of exposure to outdoor AP, traffic, or traffic fumes. In several studies, some of which adjusted for tobacco smoking, an increased risk of bladder cancer was associated with these exposure metrics. Several studies also demonstrated a higher risk among people who were occupationally exposed to potentially high levels of outdoor AP after accounting for tobacco smoking. Since these studies involved occupations (specifically taxi, bus, and truck drivers) as surrogate indicators of exposure to outdoor AP, but specific air pollutants were not measured, they were not placed as much weight by the IARC WG.

24.5 Other Tumors

The IARC WG also reviewed in detail studies of breast cancer, leukemia and lymphoma, and several other cancers. The evidence of carcinogenicity for these sites was based on a relatively small number of informative studies, and the observed associations were inconsistent. A recent study carried out on the Danish population found an association between long-term exposure to traffic-related AP (evaluated as long-term NOx and NO2 exposure) and acute myeloid leukemia in the general population, but not for other subtypes of leukemia [26].

As for the association between AP and childhood cancers, the IARC WG could not rule out weak associations with childhood leukemia and in particular with acute lymphoblastic leukemia. Some of these were reported in studies that were informative because they were large, were population-based, used incidence as the end point, used validated exposure assessment methods, had no potential for recall bias, and had no or limited potential for selection bias. Although the associations with childhood leukemia were suggestive, they were inconsistent.

24.6 Effects of Air Pollution Control Measures on Lung Cancer Incidence and Mortality

Because of progressive evidence of health damage by AP exposure, the World Health Organization (WHO) issued over the 1980s the first Air Quality Guidelines based on available scientific evidence. These guidelines are an important resource for governmental authorities as they develop health-based national air quality management strategies, especially in those countries which lack the necessary scientific infrastructure and resources to conduct their own assessments in support of public policy. The WHO has updated its Air Quality Guidelines in 2005 [27] with intention to be relevant and applicable worldwide and taking into consideration large regional inequalities in exposures to AP. The guidelines indicate that by reducing particulate matter (PM10) pollution from 70 to 20 μg/m^3, AP-related deaths can be cut by around 15%.

In North America and Europe, concentrations of major pollutants such as PM, NO2, and SO2 have decreased substantially in the past 30 years [28]. Instead, in many developing countries, concentrations have increased with rapid economic development. The decrease in lung cancer incidence and mortality observed in the last two decades in many countries is attributable to the overall reduction of the tobacco smoking. At present there are few studies that have evaluated the impact of the AP reduction on the lung cancer incidence and mortality. Reasons of this mainly depend on the need to conduct observations over long periods of time and the need to consider the influence of confounding factors, first of all the tobacco smoking. The extended follow-up of the Harvard Six Cities study [29], one of the pioneer studies on AP impact, carried out from the mid-1970s until 1990, reported that an improved overall mortality was associated with decreased mean PM2.5 (10 μg/m^3) between studied periods (RR 0.73; 95% CI, 0.57–0.95). This reduction was observed specifically for deaths due to cardiovascular and respiratory disease and not from lung cancer, a disease with a longer latency period and less reversibility. These findings led the authors to speculate that the mortality effects of long-term AP may be at least partially reversible over periods of a decade. Similar predictions are also spring from the "Ringland project" in the city of Antwerp, Belgium; this project plans to reduce PM 2.5 exposure of the urban population putting the entire urban ring road into a tunnel and thus filtering AP. The model predicts changes between −1.5 and +2 μg/m^3 in PM2.5 within a 1.500 m radius around the ring road, for the "filtered tunneled ring road" scenario as compared to an "open air ring road" [30]. Although the expected change in PM2.5 and NO2 by covering the entire urban ring road is associated with considerable health gains for the approximate 352,000 inhabitants living in a 1.500 m perimeter around the current open air ring road, the estimated decrease in lung cancer mortality and decline between the filtered tunneled ring road versus open air ring road were of borderline relevance.

On the other hand, a possible positive impact on lung cancer incidence has been reported by a quasi-experimental study in Tokyo that evaluated the effect of a diesel emission control ordinance introduced in 2003 on mortality rates in 23 wards of the Tokyo metropolitan area, from October 2000 to September 2012. Comparison was made taking into account change in mortality rates in a reference population with the introduction of such a regulation in 2009. The study reported that a decline in PM2.5 was associated with percent changes in mortality between the first 3-year interval (October 2000 to September 2003) and the last 3-year interval (October 2009 to September 2012) of −6.0% for all causes and −4.9% for lung cancer in particular, suggesting that emission control was associated with improvements in both air quality and health outcomes [31].

The importance to maintain PM under WHO limits is underlined by a recent Italian study that estimated the dose–response relationship between female lung cancer mortality and available long-term outdoor PM10 concentrations for all the Italian province capital city municipalities in the period 2000–2011, considering percentage of smokers and deprivation index as additional explanatory variables. A rough estimate of the impact of PM10 exposure at level above the WHO guideline value of 20 $\mu g/m^{-3}$ is between 2920 and 3449 lung cancer deaths out of 22,162 (13–16%). Maintaining the PM10 concentrations below such WHO recommendation, an overall saving of nearly 300 lung cancer deaths per year in a population of 8,146,520 women has been evaluated [32].

Conclusions

On the basis of large epidemiological, case–control, and cohort studies evaluated by IARC WG as well as the studies on animal models and meta-analyses, there is clear evidence that living in places with high levels of particulate AP confers an increased risk of developing lung cancer. The carcinogenic effect of AP on the respiratory system is independent from the tobacco smoking having manifested also in the exposed population of non-smokers. The Global Burden of Disease collaboration estimated cancers of the trachea, bronchus, or lung represent approximately 7% of total mortality attributable to PM2.5 in 2010 [6]. Various studies demonstrate the existence of a dose–response relationship between AP exposure and lung cancer, and at the same time, there are data demonstrating the lack of any threshold for PM below which there is no risk for lung cancer. These findings support the WHO view that public health benefits will result from any reduction of PM concentrations, whether or not the airborne levels are above or below the limit values. Most sources of outdoor AP are well beyond the control of individuals and demand action by cities, as well as national and international policymakers in sector like transport, energy waste management, buildings, and agriculture.

References

1. IARC. Working group: outdoor air pollution. IARC Monogr Eval Carcinog Risks Hum. 2015;109:1–448.
2. IARC. Diesel and gasoline engine exhausts and some nitroarenes. IARC Monogr Eval Carcinog Risks Hum. 2013a;105:1–704.
3. Raaschou-Nielsen O, Andersen ZJ, Beelen R, Samoli E, Stafoggia M, Weinmayr G, Hoffmann B, Fischer P, Nieuwenhuijsen MJ, Brunekreef B, Xun WW, Katsouyanni K, Dimakopoulou K, Sommar J, Forsberg B, Modig L, Oudin A, Oftedal B, Schwarze PE, Nafstad P, De Faire U, Pedersen NL, Ostenson CG, Fratiglioni L, Penell J, Korek M, Pershagen G, Eriksen KT, Sørensen M, Tjønneland A, Ellermann T, Eeftens M, Peeters PH, Meliefste K, Wang M, Bueno-de-Mesquita B, Key TJ, de Hoogh K, Concin H, Nagel G, Vilier A, Grioni S, Krogh V, Tsai MY, Ricceri F, Sacerdote C, Galassi C, Migliore E, Ranzi A, Cesaroni G, Badaloni C, Forastiere F, Tamayo I, Amiano P, Dorronsoro M, Trichopoulou A, Bamia C, Vineis P, Hoek G. Air pollution and lung cancer incidence in 17 European cohorts: prospective analyses from the European Study of Cohorts for Air Pollution Effects (ESCAPE). Lancet Oncol. 2013;14:813–22.

4. Pope CA 3rd, Burnett RT, Thun MJ, Calle EE, Krewski D, Ito K, Thurston GD. Lung cancer, cardiopulmonary mortality, and long-term exposure to fine particulate air pollution. JAMA. 2002;287:1132–41.
5. Krewski D, Jerrett M, Burnett RT, Ma R, Hughes E, Shi Y, Turner MC, Pope CA 3rd, Thurston G, Calle EE, Thun MJ, Beckerman B, DeLuca P, Finkelstein N, Ito K, Moore DK, Newbold KB, Ramsay T, Ross Z, Shin H, Tempalski B. Extended follow-up and spatial analysis of the American Cancer Society study linking particulate air pollution and mortality. Res Rep Health Eff Inst. 2009;140:5–114.
6. Lim SS, Vos T, Flaxman AD, Danaei G, Shibuya K, Adair-Rohani H, et al. A comparative risk assessment of burden of disease and injury attributable to 67 risk factors and risk factor clusters in 21 regions,1990–2010: a systematic analysis for the Global Burden of Disease Study 2010. Lancet. 2012;380:2224–60.
7. GBD 2015 Mortality and Causes of Death Collaborators. Global, regional, and national life expectancy, all-cause mortality, and cause-specific mortality for 249 causes of death, 1980–2015: a systematic analysis for the Global Burden of Disease Study 2015. Lancet. 2016;388:1459–544.
8. Moolgavkar SH, Holford TR, Levy DT, Kong CY, Foy M, Clarke L, Jeon J, Hazelton WD, Meza R, Schultz F, McCarthy W, Boer R, Gorlova O, Gazelle GS, Kimmel M, McMahon PM, de Koning HJ, Feuer EJ. Impact of reduced tobacco smoking on lung cancer mortality in the United States during 1975-2000. J Natl Cancer Inst. 2012;104:541–8.
9. IARC. Tobacco smoke and involuntary smoking. IARC Monogr Eval Carcinog Risks Hum. 2004;83:1191–413.
10. Boffetta P. Involuntary smoking and lung cancer. Scand J Work Environ Health. 2002;28(Suppl. 2):30–40.
11. Stayner L, Bena J, Sasco AJ, et al. Lung cancer risk and workplace exposure to environmental tobacco smoke. Am J Public Health. 2007;97:545–51.
12. Boffetta P, Trédaniel J, Greco A. Risk of childhood cancer and adult lung cancer after childhood exposure to passive smoke: a meta-analysis. Environ Health Perspect. 2000;108:73–82.
13. Malhotra J, Malvezzi M, Negri E, La Vecchia C, Boffetta P. Risk factors for lung cancer worldwide. Eur Respir J. 2016;48:889–902.
14. Darby S, Hill D, Auvinen A, Barros-Dios JM, Baysson H, Bochicchio F, Deo H, Falk R, Forastiere F, Hakama M, Heid I, Kreienbrock L, Kreuzer M, Lagarde F, Mäkeläinen I, Muirhead C, Oberaigner W, Pershagen G, Ruano-Ravina A, Ruosteenoja E, Rosario AS, Tirmarche M, Tomásek L, Whitley E, Wichmann HE, Doll R. Radon in homes and risk of lung cancer: collaborative analysis of individual data from 13 European case-control studies. BMJ. 2005;330:223–8.
15. US Enviromental Protection Agency: Health Risk of Radon. https://www.epa.gov/radon.
16. Raaschou-Nielsen O, Beelen R, Wang M, Hoek G, Andersen ZJ, Hoffmann B, Stafoggia M, Samoli E, Weinmayr G, Dimakopoulou K, Nieuwenhuijsen M, Xun WW, Fischer P, Eriksen KT, Sørensen M, Tjønneland A, Ricceri F, de Hoogh K, Key T, Eeftens M, Peeters PH, Bueno-de-Mesquita HB, Meliefste K, Oftedal B, Schwarze PE, Nafstad P, Galassi C, Migliore E, Ranzi A, Cesaroni G, Badaloni C, Forastiere F, Penell J, De Faire U, Korek M, Pedersen N, Östenson CG, Pershagen G, Fratiglioni L, Concin H, Nagel G, Jaensch A, Ineichen A, Naccarati A, Katsoulis M, Trichpoulou A, Keuken M, Jedynska A, Kooter IM, Kukkonen J, Brunekreef B, Sokhi RS, Katsouyanni K, Vineis P. Particulate matter air pollution components and risk for lung cancer. Environ Int. 2016;87:66–73.
17. Hamra GB, Guha N, Cohen A, Laden F, Raaschou-Nielsen O, Samet JM, Vineis P, Forastiere F, Saldiva P, Yorifuji T, Loomis D. Outdoor particulate matter exposure and lung cancer: a systematic review and meta-analysis. Environ Health Perspect. 2014;122:906–11.
18. Puett RC, Hart JE, Yanosky JD, Spiegelman D, Wang M, Fisher JA, Hong B, Laden F. Particulate matter air pollution exposure, distance to road, and incident lung cancer in the Nurses' Health Study Cohort. Environ Health Perspect. 2014;122:926–32.

19. Gharibvand L, Shavlik D, Ghamsary M, Beeson WL, Soret S, Knutsen R, Knutsen SF. The association between ambient fine particulate air pollution and lung cancer incidence: results from the AHSMOG-2 Study. Environ Health Perspect. 2016;125:378. doi:10.1289/EHP124.
20. Tomczak A, Miller AB, Weichenthal SA, To T, Wall C, van Donkelaar A, Martin RV, Crouse DL, Villeneuve PJ. Long-term exposure to fine particulate matter air pollution and the risk of lung cancer among participants of the Canadian National Breast Screening Study. Int J Cancer. 2016;139:1958–66.
21. Chen G, Wan X, Yang G, Zou X. Traffic-related air pollution and lung cancer: a meta-analysis. Thorac Cancer. 2015;6:307–18.
22. Hamra GB, Laden F, Cohen AJ, Raaschou-Nielsen O, Brauer M, Loomis D. Lung cancer and exposure to nitrogen dioxide and traffic: a systematic review and meta-analysis. Environ Health Perspect. 2015;123:1107–12.
23. Olsson AC, Gustavsson P, Kromhout H, Peters S, Vermeulen R, Brüske I, Pesch B, Siemiatycki J, Pintos J, Brüning T, Cassidy A, Wichmann HE, Consonni D, Landi MT, Caporaso N, Plato N, Merletti F, Mirabelli D, Richiardi L, Jöckel KH, Ahrens W, Pohlabeln H, Lissowska J, Szeszenia-Dabrowska N, Zaridze D, Stücker I, Benhamou S, Bencko V, Foretova L, Janout V, Rudnai P, Fabianova E, Dumitru RS, Gross IM, Kendzia B, Forastiere F, Bueno-de-Mesquita B, Brennan P, Boffetta P, Straif K. Exposure to diesel motor exhaust and lung cancer risk in a pooled analysis from case-control studies in Europe and Canada. Am J Respir Crit Care Med. 2011;183:941–8.
24. Tsoi CT, Tse LA. Professional drivers and lung cancer: a systematic review and meta-analysis. Occup Environ Med. 2012;69:831–6.
25. Eckel SP, Cockburn M, Shu YH, Deng H, Lurmann FW, Liu L, Gilliland FD. Air pollution affects lung cancer survival. Thorax. 2016;71:891–8.
26. Raaschou-Nielsen O, Ketzel M, Harbo Poulsen A, Sørensen M. Traffic-related air pollution and risk for leukaemia of an adult population. Int J Cancer. 2016;138:1111–7.
27. Krzyzanowski M, Cohen A. Update of WHO air quality guidelines. Air Qual Atmos Health. 2008;1:7–13.
28. WHO. Air quality guidelines. Global updates 2005. WHO; 2006.
29. Laden F, Schwartz J, Speizer FE, Dockery DW. Reduction in fine particulate air pollution and mortality: extended follow-up of the Harvard Six Cities study. Am J Respir Crit Care Med. 2006;173:667–72.
30. Van Brusselen D, Arrazola de Oñate W, Maiheu B, Vranckx S, Lefebvre W, Janssen S, Nawrot TS, Nemery B, Avonts D. Health impact assessment of a predicted air quality change by moving traffic from an urban ring road into a tunnel. The case of Antwerp, Belgium. PLoS One. 2016;11:e0154052.
31. Yorifuji T, Kashima S, Doi H. Fine-particulate air pollution from diesel emission control and mortality rates in Tokyo: A quasi-experimental study. Epidemiology. 2016;27:769–78.
32. Uccelli R, Mastrantonio M, Altavista P, Caiaffa E, Cattani G, Belli S, Comba P. Female lung cancer mortality and long-term exposure to particulate matter in Italy. Eur J Public Health. 2017;27:178–83.

25 Air Pollution and Gynecological Diseases

Lidia La Marca and Giulia Gava

25.1 Introduction

Evidences linking air pollution exposure to woman health are growing. Although respiratory and cardiovascular systems are the most frequently affected by the exposure to air pollution, recent studies showed that common air pollutants could also affect both reproductive health and pregnancy outcomes. In fact, ambient air pollution can be deleterious both to fertility and pregnancy outcomes [1], but it seems also to be associated to higher risk of malignant and benign gynecological diseases.

These results are derived from animal studies but also from several epidemiological investigations that suggested the detrimental effects of air pollution on reproductive health. Anyway, some limitations of these studies should be acknowledged as there are still few evidences of the biological mechanisms and limited available information on personal exposures both for duration and molecules [2, 3]. It should also be considered that these studies often have different study designs and discordant evaluation of confounding variables (e.g., lifestyle, genetic factors, socioeconomic conditions, etc.).

Future toxicological and clinical studies are mandatory to confirm previously hypothesized associations and to clarify the mechanism involved. They should be also useful in identifying which subgroups of population are most prone to air pollution exposure.

L. La Marca (✉)
Delegate of Lega Vita e Salute Foundation, Bologna, Italy
e-mail: lidialamarca@infopro.it

G. Gava
Gynecology and Physiopathology of Human Reproduction, S. Orsola-Malpighi Hospital, University of Bologna, Bologna, Italy

F. Capello, A.V. Gaddi (eds.), *Clinical Handbook of Air Pollution-Related Diseases*,
https://doi.org/10.1007/978-3-319-62731-1_25

In the following chapter, we will analyze the major epidemiological and experimental evidences regarding the association between exposure to common urban air pollution and women reproductive health.

25.2 Breast Cancer

25.2.1 Introduction

Breast cancer (BC) is the most common female cancer worldwide with nearly 1.7 million new cases diagnosed in 2012, and it represents about 12% of all new cancer cases and 25% of all cancers in women [4, 5]. BC rate increased in the last decades in developing countries [6–8], and it represents the second cause of death in several countries [5].

The main risk factors for BC [9] are thought to be genetic mutations, family history of cancer, lifestyle, and reproductive history; anyway, these risk factors can be held responsible only for 30–40% of new cases of BC [10].

In general, BC rates are quite different in developed countries when compared to developing countries [6, 11], and they are greater in white European women and lower in East Asian populations [12, 13]. The geographic location is a strong predictor of the incidence of BC as the highest rates are observed in urbanized areas [14, 15], and it has been associated with the western lifestyle [15, 16]. Many studies suggested that BC rates are greater in urban areas also after adjusting for multiple confounders and for known BC risk factors [17, 18].

Being born, but not living, in an urban area in 1900 was associated with an increased BC risk [19]. Another study, conducted in Australia, did not find any statistically significant difference in BC when comparing rural to urban areas [20]. It should be considered that several studies linking BC to urban or rural settings did not make adjustment for BC individual risk factors such as reproductive and lifestyle factors.

Chen and Bina [21] reported that the long-term trend of female BC in the United States and in some European countries had the same trend of NO2 or NOx emission assuming that the observed differences in BC rates were related to industrialization as the risk was higher in industrialized countries (Fig. 25.1).

25.2.2 Breast Cancer and Air Pollution

Exposure to traffic-related air pollution is ubiquitous and it is linked to a wide range of chronic diseases.

Little is known about the role of air pollution in cancers of reproductive system as few studies have been performed, and they are limited by difficulties in the assessment of long-term exposures and of other confounding factors like exposure to other known carcinogens in food and water.

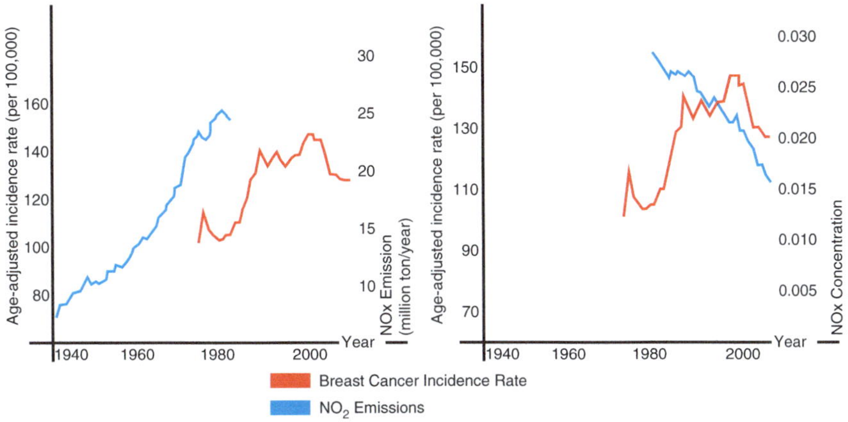

Fig. 25.1. NO_2 emissions and concentration in the United States (SEER data) in relationship to BC incidence rate in white women (from [21])

Recently outdoor pollution (e.g., diesel and gasoline exhaust) has been classified as carcinogenic for humans by the International Agency for Research on Cancer (IARC, [22]).

Rudel et al. [23] has made a report that identified many chemicals, including air pollutants, as associated to an increase in mammary gland tumors in the animal, but few studies have examined the relationship between air pollution exposure and risk of BC in humans. Two case-control studies [24, 25] found some evidence for an association, but they did not evaluate specific compounds.

An ecological study [26] explored the association between some industrial chemical plus metals and BC failing to find an increased personal BC risk.

Though epidemiological data are limited, there is growing evidence that air pollution, both from industrial source and traffic emissions, could be involved in BC risk [27]. Recently many studies have hypothesized that environmental exposure is related to an increased trend in mobility and mortality of BC too [22, 28].

In particular, some studies have shown a positive association between long-term exposure to air pollution and incidence of BC [29–31].

Others several studies have shown an influence of air pollution on BC risk [21, 32] particularly if this occurred in early-life periods [24, 31]. Some authors have stressed the importance of early exposure of traffic-related pollutants in risk of BC. Furthermore, it is known that the time of menarche is associated with an increased risk of premenopausal BC, while the age at the first delivery is associated to an increase in postmenopausal BC risk [31].

In a prospective study, Binachon et al. [19] reported an association between BC and being born in an urban area but no association with the residence in urban area during adult life suggesting the dangerous effects of pollution exposure in early life. Anyway, traffic-related air pollution exposure did not seem to increase breast density as evaluated by mammography, assuming that this is not the mechanism through which it increases BC risk [33].

Long-term air pollution exposure could also play a xenoestrogen role increasing the possibility to have a BC [30] due to traffic-related particles that have significant estrogenic activity [34].

25.2.3 Nitrogen Dioxide (NO2) Exposition and Breast Cancer

It is a marker of traffic-related air pollution.

At molecular level the endogenous nitric oxide seems to play an important role in the biology of cancer. In mouse fibroblasts, NO2 has an important effect on the promotional phase of neoplastic transformation, and, at low levels, it promotes tumor growth, while at higher levels it has a cytotoxic effect on malignant cells [35]. In addition, peroxynitrite and nitric oxide, both in vivo and vitro [36], seem to be able to suppress DNA repairing system. In human BC, NO2 could have a carcinogenic role through DNA damaging.

Crouse et al. [29] reported an association between the incidence of postmenopausal BC and exposure to ambient concentrations of NO2.

Hystad et al. [37] found, in a large case-control study (3310 invasive, primary breast cancer), a positive association between BC incidence and several measures of long-term exposure to ambient NO2 mostly among premenopausal women.

In a systematic review and meta-analysis in 2016, Keramatinia et al. [38] evaluated the association between NO2 and BC concluding that there is a tendency toward a weak association between NO2 exposure and BC at an individual level, while there is a significant association at aggregate level.

25.2.4 Particulate Matter (PM) Pollution and Breast Cancer

PM is a complex mixture of extremely small particles and liquid droplets that get into the air.

Bonner et al., in a population-based case-control study, showed that early-life exposure to PM doubled the risk of BC (OR, 2.42) [24].

In a large prospective study (3416 cases of BC), the relationship between PM concentration and the distance to main roadway with the incidence of BC [39] was evaluated, and the author reported that living near (<50 m) the traffic way was associated with an increase in BC risk.

The risk is increased for long-term exposure (10 years before diagnosis) in women living in China in high-pollution areas [40]. Long-term exposure to PM may contribute to the development of BC through the role of xenoestrogens [40, 41].

Reding et al. [42] after an analysis of 47,591 women from the sister study cohort, suggested that there is no substantial increase risk for BC if PM2.5 and PM10 are considered associated to NO2 overall, but NO2 alone may individually affect BC risk, particularly ER- and PR-positive BC.

Tagliabue et al. [43] in a retrospective study on 2021 BC cases with 7 years of follow-up reported that the mortality from BC increased with PM2.5 exposure, but some limitation of this study should be acknowledged such as possible lifestyle factors and comorbidities and that the exposure was assessed only in the area (10 × 10 km) surrounding the residence of the women making impossible to know the real exposure.

25.2.5 Polycyclic Aromatic Hydrocarbons (PAHs) and Breast Cancer

Sources for PAHs can be both outdoor and indoor, such as active smoking and residential environmental tobacco smoke, diet (grilled/smoked meat), indoor-wood-burning stove and fireplace, and vehicular traffic.

Hung et al. [44] demonstrated that mortality from BC in Taiwan is correlated with PM10 and PM5 levels suggesting that PAHs contained in PM2.5 could have a role in BC evolution.

White et al. [45, 46], in a population-based case-control (1508 BC cases/1556 controls) study, evaluated the relationship between exposure to multiple sources of PHA and BC incidence, concluding that mostly indoor sources were associated to a 30–50% increase in BC incidence. The potential biologic mechanism for the association PAHs and BC [45, 46] could be the abnormal methylation. Furthermore, it should be underlined that these particles can have estrogenic action and when inhaled they can directly enter the systemic circulation bypassing hepatic metabolism [47].

25.2.6 Bisphenol A and Breast Cancer

This compound can be found in the indoor air environment due to its semi-volatility. It is a typical endocrine disruptor, and it can stimulate the epithelial-mesenchymal transition of ER-breast cancer cells with downregulation of several genes including FOXA1 that is a determinant key of endocrine response [48] or inducing expression of HOXC6 that promotes tumor growth factors [49].

Matsumoto et al. in 2005 [50] demonstrated that BPA can be found in the urban outdoor pollution with seasonal variations, increasing from autumn to winter and decreasing from winter to spring. It represents the main contributor to the estrogenic activity in the organic extract from the air particulate.

25.3 Gynecological Cancers

Few data are available regarding the association between air pollution and gynecological cancers.

Globally, ovarian cancer is the sixth most common cancer among women. The annual incidence rate differs among geographic areas. High rates are reported in

Scandinavia (15/100000), intermediate rates (10/100000) in Northern America and Western Europe, and low rates (3/100000) in developing countries and Japan [15]. In high-incidence areas, lifetime risk for a woman to develop ovarian cancer is 1–2%. The prognosis is poor, with 5-years survival rates less than 40% mainly due to advanced stages at the diagnosis. Annually more than 100,000 women are estimated to die from the disease [51].

Up to now, little is known about the etiology of ovarian cancer [52], but some factors such as family history of breast or ovarian cancer (5–10% of ovarian cancers), reproductive factors such as age at menopause, and infertility are involved, whereas pregnancy, tubal ligation, and hysterectomy reduce risk. Oral contraceptive use has clearly been shown to be protective. Lifestyle factors such as obesity and diet may affect ovarian cancer risk.

While cigarette smoking has been clearly documented as a causative factor for other gynecologic cancers, the relationship between smoking and ovarian cancer is not so clear, and evidences from many studies are still lacking. Because of the marked influence of hormones and reproductive factors, it is possible that endocrine disruptors may have an impact on the risk of ovarian cancer. Even if available studies are still limited, there is growing evidence that environmental (e.g., to talc, pesticides, and herbicides) and occupational exposure could increase the risk of ovarian cancer [53].

Iwai et al. [41] in a cross-sectional epidemiological study evaluated the annual statistics of life expectancy in Japan and air pollution, demonstrating a significant increase in mortality rates for breast, ovarian, and endometrial cancer. In this study, breast, uterine, and ovarian cancer correlated significantly with PM levels and in particular with PM2.5. Uterine cancer was further divided into cervical and endometrial cancer, as they differ in pathogenesis. After this subgroup analysis, cervical cancer showed no correlation to air pollution, while endometrial cancer had a significant correlation with PM/PM2.5 levels.

Recently, in Taiwan, Hung et al. [44] showed that women living in areas with high levels of PM2.5 had a significantly increased risk of death from ovarian cancer.

Occupational epidemiologic studies reported, in women working in the printing industry, a strong association with excess risk of ovarian cancer diagnosis [54, 55].

Cervical cancer is considered as the third most common female malignant disease during childbearing years (20–49 years) [56].

The role of human papillomavirus (HPV) infection is well known as a necessary cause in the development of cervical cancer; however, many women infected with HPV do not necessarily develop cervical cancer. For this reason, other factors have been associated to cervical cancer development, such as lifestyle factors, sexual behavior, tobacco smoke, oral contraceptives use, high parity, and co-infections such as human immunodeficiency virus infection. These findings suggest a hypothetic role of both genetic and/or environmental factors [57].

It is difficult to establish the link between environmental factors exposure and cervical cancer mainly because of confounding factors. Some studies have indicated that environmental exposure to chemicals could increase the risk of cervical cancer and that certain groups of women may be especially vulnerable to these environmental risk factors (occupational and domestic settings and general ambient environment). In particular, two chemicals, trichloroethylene and tetrachloroethylene, have been demonstrated to have a potential association to cervical cancer [58].

Ma et al. [59] revealed that nanomolar concentrations of bisphenol A (BPA) are significantly able to promote the in vitro migration and invasion of cervical cancer.

Raaschou-Nielsen et al. [60] performed a study on 54,304 subjects exposed to different levels of air pollution, collecting records of 21 types of cancer over a period of 10 years. People who lived in areas with higher levels of pollution (concentration of NOx molecules) had elevated rates of cervical, brain, and lung cancers. The authors concluded that airways were certainly the primary target organs, but evidence from experiments in animals confirmed that ultrafine particles can migrate to other organs through the bloodstream causing carcinogenesis. Anyway, up to now, the association between high automobile traffic and increased rates of cervical cancer is not explained by any known mechanism. Raaschou-Nielsen pointed out that there may be a possible higher HPV infection prevalence in women living in areas with heavy traffic and air pollution.

The first study in a multiethnic sample of women to evaluate the association between exposure to traffic-related HAPs and cervical dysplasia prevalence [61] suggested that women with high residential exposure to benzene, DPM, or PAHs have an increased prevalence of cervical dysplasia compared to women with relatively low exposure to these pollutants. Anyway, it should be considered that traffic-related HAPs, such as benzene, DPM, and PAHs, are not as well-regulated and monitored as other air pollutants (e.g., ozone), underscoring the need for studies evaluating the potential effects of these molecules on disease risk.

Further research is also needed to determine whether it was the air pollution itself or some related factor that caused higher rates of cervical cancer [61]; this will help us in identifying novel risk factors that may contribute to cervical disease in conjunction with HPV infection. These new findings may also be important in the future for cervical cancer prevention.

25.4 Benign Gynecologic Diseases

25.4.1 Endometriosis

Endometriosis is defined as the presence of endometrial-like tissue outside of the lining of the uterus. This condition can be either asymptomatic or usually characterized by menstrual related pain and pelvic pain of variable severity.

The prevalence of endometriosis varies by population and method of diagnosis from 4% to 50% [62]. The prevalence of undiagnosed endometriosis of all stages in the general population is believed to be 11% while the likely prevalence of severe endometriosis is less than 2% [63].

Several risk factors have been proposed as contributors to endometriosis development and to disease severity, including anatomic, anthropometric, hormonal, immunologic, inflammatory, and genetic factors [64, 65]. Inflammatory markers are elevated in women with endometriosis [66], but this state of inflammation can be considered both as a result or as a promoter of the disease [67].

Traffic-related exhaust (diesel and nondiesel) and particulate matter (PM) through the bloodstream can deposit in peripheral tissues where they promote local and systemic inflammation [68, 69]. In particular, PM of smaller sizes, such as those ≤2.5 μm in diameter (PM2.5), can cross the bloodstream more easily than PM of greater sizes. Diesel exhaust particles have hormonal activity, presenting both estrogenic and androgenic activity [70–72]. Exposure to diesel exhaust in utero and in the postnatal life has been reported to promote the persistence of endometriosis in the rat model [73].

Dioxins, through the aryl-hydrocarbon receptor, and endocrine disruptors found in diesel exhaust may influence the local hormonal levels surrounding the lesion of endometriosis promoting endometriosis progression [74].

In 2014, Mahalingaiah et al. evaluated 84,060 adult women from the Nurses' Health Study II, and of these 2486 had surgically confirmed endometriosis. The author reported no evidence of association between endometriosis risk and distance to main roads or exposure to PM (PM2.5, PM10–2.5, or PM10) averaged over follow-up or during the previous 2- or 4-years period during adulthood. To our best knowledge this is the only available study about the association between air pollution and endometriosis. Although this study did not support an association of adult exposure to PM with endometriosis risk, more research is needed on air pollution exposure during both adolescence and adulthood.

25.4.2 Fibroids

Uterine fibroids, known also as leiomyomata, are hormonally responsive clonal tumors present in several women of reproductive age. They can be or not clinically significant, causing increase in uterine size, heavy menses potentially leading to anemia, pelvic pain, increased urinary frequency, and abdominal pain. Their incidence is approximately 70% in white women and 80% in African-American women by the age of 50 years [75].

Several risk factors for leiomyomata have been proposed including genetic, hormonal, lifestyle, and diet factors. Numerous studies have evaluated the role of exposure to environmental endocrine disruptors, such as diethylstilbestrol, phenols, dioxin, and polychlorinated biphenyls, but no clear relationship has emerged [76, 77].

One study evaluated exposure to polycyclic aromatic hydrocarbons and uterine leiomyomata expression of the aryl-hydrocarbon receptor [78]. It is known that polycyclic aromatic hydrocarbons are typical components of air pollution, and they are well known for their hormonal activity through the aryl-hydrocarbon receptor in the animal model. Bidgoli et al. demonstrated that in women living closer to a polycyclic aromatic hydrocarbon producing company, there was a greater probability to have uterine fibroids with overexpression of the aryl-hydrocarbon receptor [78].

Mahalingaiaha et al., in 2014, described that chronic exposure to PM2.5 may be associated with a modest increase in risk of uterine fibroids (data from 85,251 women aged 25–42 at enrollment in the Nurses' Health Study II) [76, 77].

25.5 Gynecologic Endocrine Disorders

25.5.1 Age at Puberty

Signs of pubertal onset in girls are breast development (thelarche), pubic hair growth (pubarche), and later, menstruation (menarche). Epidemiological available data showed that the age of pubertal maturation has declined over the past 30 years [79, 80]. Girls showing significant signs of puberty and its progression at an age that is 2–2.5 standard deviations (SD) earlier than population norms are considered precocious.

Precocious puberty is associated with adverse psychosocial and health outcomes later in life, including obesity, depression, and higher risk of breast cancer [81–83].

Increases in childhood obesity prevalence can be considered as one of the causes of this decline in age of pubertal onset, though human and animal studies conclude that the trend cannot be explained only by nutritional status [84, 85]. Some studies have reported an association between low birth weight and pollution and early transition to puberty [86].

Growing evidence suggests that exposure to environmental toxicants during critical windows of susceptibility may be associated with altered pubertal timing [79]. Furthermore, children are more vulnerable than adults to the effects of air pollution because they breathe more air per unit of body weight than adults [87], and, sometimes, they spend more time outdoors [88].

Several authors reported a link between endocrine-disrupting chemicals (EDCs) and altered pubertal maturation [89].

Polycyclic aromatic hydrocarbons (PAHs) and heavy metals can be present as a part of particulate matter typical of air pollution from traffic. PAHs are known estrogenic and endocrine-disrupting elements, and, in animal studies, they have been found to interfere with both reproductive and pubertal development [90]. While heavy metal exposure has been associated with delayed pubertal development in girls [91].

A recent study analyzed in 437 girls, the relationship between traffic-related air pollution and pubertal onset, reporting that girls with higher exposure reached one pubertal milestone several months earlier than low-exposed girls [92].

25.5.2 Polycystic Ovary Syndrome

It is estimated that polycystic ovary syndrome (PCOS) affects 5–10% of reproductive-aged females worldwide. The pathogenesis of PCOS is unclear: it has a multifactorial etiology, with both genetic and environmental factors.

Up to now a limited number of studies have explored the connection between pollutants and PCOS [93, 94]. Persistent organic pollutants are ubiquitous and can accumulate in human organism acting as exogenous endocrine-disrupting chemicals that can interfere with hormonal production and function.

In 2015 Yang et al. [95] reported that in 50 women from Northern China, PCOS was significantly associated with elevated serum levels of pollutants, including polychlorinated biphenyls (PCBs), organochlorine pesticides, and polycyclic aromatic hydrocarbons (PAHs).

25.5.3 Age at Menopause

Age at menopause is determined in part by genetic characteristics, but genetic factors have been suggested to explain only up to 63% of the variation in menopausal age. Several factors have been correlated with age at the menopause with equivocal results, such as body weight, alcohol consumption, socioeconomic class, age at menarche, menstrual characteristics, parity, and exposure to pesticides [96].

Apart from genes, the most important influence on age of menopause is smoking, as the relationship between smoking and decreased menopausal age has been observed in several studies. However, a few studies have investigated the effect of passive smoking on menopausal age with conflicting results [97–99]. Smoking is also associated with modifications in hormonal levels during fertile years and in premenopausal years. [100–102].

It is possible that smoke exposure during the intrauterine life may affect the follicle pool and influence menopausal age, suppressing the folliculogenesis or damaging follicles. It has been recently demonstrated that prenatal tobacco exposure was associated with telomere shortening in children, and this can affect biological programming causing premature aging and increased health risks that can be prevented by educating pregnant women to not smoke and motivating smokers to quit in early pregnancy [103].

Lutterodt et al. suggest that prenatal smoke exposure may damage somatic cells in the developing ovary [104].

Few data are available regarding the association between prenatal smoke exposure and age at menopause: a prospective study reported a modest association between prenatal smoke exposure and earlier age at menopause [105] while two other studies did not observe any association [101, 106].

25.6 Fertility and Fecundability

Infertility is a complex situation defined by the incapability to conceive after 1 year of attempts (or after 6 months if 35 years old or more). It involves female factors (tubal, cervical, uterine, ovarian, and hormonal), male factors, and unexplained causes. Worldwide the prevalence of infertility has been estimated to be around 10% in the last 20 years. In Sub-Saharan Africa and South Asia an increase of prevalence has been recently reported [107]. Infertility has been increasing during recent decades, and the most important causes of this increase are changes in lifestyle factors such as a delay in the timing of motherhood, leading to lower ovarian reserve and poorer oocyte quality.

It is known that exposure tobacco smoke is associated with poor reproductive outcomes: a longer time to conception and an increase of IVF cycles needed [108–110]. It is also well established that secondhand smoke causes a reduction in IVF success rate in exposed women [111, 112].

Several studies on animals have highlighted the effects of air pollution on fertility, semen quality, and fertilization success rates in IVF. The relationship between ambient levels of air pollution and ovarian function was well characterized in a mouse study form Veras et al. [113], which described increases in estrus cycle length thereby resulting in fewer estrus cycles and decreased fertility [113].

Up to now the impact of environmental causes on human fertility has not been clearly defined, and few studies have analyzed air pollution impact on female fertility while numerous reports exist for the effects on male fertility.

Air pollution is believed to influence hormonal activity and to be negatively associated with early reproductive outcomes such as fertilization and implantation in human IVF [114, 115].

Interesting results are given by studies of air quality in IVF settings. Slama et al. in 2013 reported a 22% reduction in fecundability associated with an increase of 10 mg/m^3 in PM2.5 [116]. Another study evaluated the effect of air quality on 7403 IVF cycles from 2000–2007 in northeastern USA: the author reported that NO2 concentrations both at the patient's address and at the address of the IVF laboratory and PM2.5 concentrations at the IVF laboratory were negatively associated with odds of pregnancy during the IVF cycle but statistically significantly after embryo transfer [114].

Another IVF cohort study with male infertility factor in Brazil analyzed the city-wide average for PM10, and it reported the association of an increased risk of miscarriage with high preconception air pollution exposure but no difference in pregnancy rates or clinical pregnancy outcomes [115]. Perin et al. also studied the effects of PM10 on embryo quality in women undergoing IVF/ET: the author and his group did not observe any significant differences between patients' subgroups.

In literature, some epidemiological data on the impact of air pollution on fertility are available. Nieuwenhuijsen et al. reported a reduction in fertility rate with the increase of the census tract levels of traffic-related air pollution (particulate matter (PM) and oxides of nitrogen (NOx)) in a cross-sectional study over 1-year period in Barcelona [117]. The relationship was stronger when considering particulate matter with an aerodynamic diameter between 2.5 and 10 μm (PM2.5–10) that was associated with a 13% reduction in fertility.

Mahalingaiah et al. in 2016 [118] evaluated over 213,416 person-years with 2508 reports of infertility. The author observed an association between all size fractions of PM exposure as well as traffic-related air pollution and incidence of infertility. This study reported a small increase in risk of both primary and secondary infertility in those living closer to a major road and that chronic exposure was more important than short-term exposure.

According to Checa Vizcaíno et al., to clarify data present in literature about influence of air pollution on female fertility, it is possible to distinguish four major components of traffic pollution hypothetically impairing human fertility:

1. Particular matter, that is a mixture of solid and liquid particles that remains suspended in the air and vary in size and composition ranging from "respirable" particles (2.5 μm, PM2.5) that can cross the bloodstream to "thoracic" particles (10 μm, PM10) that can penetrate into the lower respiratory system. PM has been reported to be associated with reduced fertility and live birth rates and increased risk of miscarriage in IVF cycles.
2. Nitrogen dioxide is derived from the combustion of fossil fuels in stationary sources (heating and power generation) and motor vehicles. NO2 has been associated with a significant increase in miscarriage rate, but results for fertility rate are inconsistent even if a significant decrease in live birth rates was reported in the IVF setting.
3. SO2 derives from the combustion of fuels containing sulfur. It has been found to induce chromosomal aberrations in in vitro studies. It was also associated with an elevated rate of miscarriages in the general and in the subfertile population.
4. Carbon monoxide (CO) derives mainly from industrial combustion, automobile exhaust, and cigarettes. Its effect is mediated through its binding to hemoglobin forming carbon monoxide–hemoglobin (carboxyhemoglobin), which is more stable than oxyhemoglobin and prevents red blood cells from absorbing oxygen. It is reported to have a statistically significant association with elevated miscarriage rates [119].

Conclusions

Implementation of industry and vehicular traffic led to a great diffusion of air pollution with a major impairment of air quality especially in urban areas over the last decades. Air pollution is an environmental contaminant containing thousands of harmful compounds derived from exhaust emissions containing a mixture of gaseous, liquid, and solid substances. In air pollution, a great number of components can be found, such as CO, NO2, SO2, O3, Pb, polycyclic aromatic hydrocarbons (PAH), and particulate matter (PM) [120].

Epidemiologic and experimental evidences showed that current levels of air pollution have a detrimental role on women's health and reproduction. Exposure to ambient PM2.5 is correlated to increased prevalence of chronic diseases (over 20% increase) in women in a longitudinal color study [121]. Recent data have underlined as common and widespread air pollutants could affect both reproductive health and pregnancy outcomes. Indoor and outdoor air pollution has been associated to higher risk of malignant and benign gynecological diseases. Even if further research is needed, traffic-related air pollution seems to be associated with an increased risk of development of breast cancer especially in premenopausal women. Furthermore, exposure to toxic air pollution particles can have a detrimental role on ovarian function starting from the intrauterine time of life both affecting time of puberty and of menopause but also harming the fertility potential of women. Air pollutants have been also demonstrated to be able to decrease the success rate of IVF techniques.

Actual and previous national choices about energy, industry, and mobility are affecting populations, even those who do not use these facilities and that cannot participate in decision making. Furthermore, levels of air pollution are particularly high in countries where technologies are still old and where laws to preserve the environment are not applied. For these reasons air pollution levels are alarmingly increasing in developing and underdeveloped countries where both population density and natality are high. This situation is leading to the exposure of a growing number of subjects to the deleterious effects of air pollution both in the prenatal period and in childhood and adulthood. Exposures to air pollution could have a profound impact on health leading also to an increased burden on public health.

The best method to reduce health risks associated with air pollution is to implement air quality through national management strategies to meet the air quality standards. Human exposure to air pollution should be controlled and reduced through better traffic management [122], particularly for young women before puberty and for pregnant women [31].

Women living in an environment with high levels of air pollution should be informed about health risks for her reproductive system and for her future pregnancies and children. These women should be advised to reduce also other modifiable risk factors, for example, improving their lifestyle and diet.

Women should be advised about the deleterious effects of tobacco smoke on their reproductive organs both increasing risk of malignant diseases and decreasing their reproductive potential. In fact, both active smoking and secondhand smoke play a detrimental role in fertility potential of women starting their effect during the intrauterine life and continuing through the entire life. Active strategies should be implemented to reduce tobacco smoke exposition, suggesting women to quit smoking and to avoid secondhand smoke in their home, car (even with the windows down), workplaces, and children's schools.

Further educational interventions should be performed to reduce the exposure to indoor pollution sources including changing cooking practices and introducing smoke-free stoves in poor countries. The reported data underline the urge for preventive efforts against the dangerous effects of air pollution through implementation of public health strategies as soon as possible.

References

1. Maisonet M, Correa A, Misra D, Jaakkola JJ. A review of the literature on the effects of ambient air pollution on fetal growth. Environ Res. 2004;95(1):106–15.
2. Chen H, Goldberg MS, Villeneuve PJ. A systematic review of the relation between long-term exposure to ambient air pollution and chronic diseases. Rev Environ Health. 2008;23(4):243–97.
3. Ren C, Tong S. Health effects of ambient air pollution-recent research development and contemporary methodological challenges. Environ Health. 2008;7:56.
4. American Cancer Society. Breast cancer facts and figures 2013–2014. Atlanta, GA: American Cancer Society; 2014.
5. Ferlay J, Soerjomataram I, Ervik M, Eser S, Mathers C, Rebelo M, Parkin DM, Forman D, Bray F. GLOBOCAN 2012 v1.0, Cancer Incidence and Mortality Worldwide: IARC CancerBase No 11. [Internet]. 2012. http://globocan.iarc.fr.
6. Benson JR, Jatoi I. The global breast cancer burden. Future Oncol. 2012;8(6):697–702.
7. Ferlay J, Shin HR, Bray F, Forman D, Mathers C, Parkin DM. Estimates of worldwide burden of cancer in 2008: GLOBOCAN 2008. Int J Cancer. 2010;127:2893–917.
8. Jemal A, Bray F, Center MM, Ferlay J, Ward E, Forman D. Global cancer statistics. CA Cancer J Clin. 2011;61:69.
9. Nelson HD, Zakher B, Cantor A, Fu R, Griffin J, O'Meara ES, Buist DSM, Kerlikowske K, van Ravesteyn NT, Trentham-Dietz A, Mandelblatt JS, Miglioretti D. Risk factors for breast cancer for women age 40 to 49: a systematic review and meta-analysis. Ann Intern Med. 2012;156(9):635–48.
10. Madigan MP, Ziegler RG, Benichou J, Byrne C, Hoover RN. Proportion of breast cancer cases in the United States explained by well-established risk factors. J Natl Cancer Inst. 1995;87:1681–5.
11. Youlden DR, Cramb SM, Dunn NA, Muller JM, Pyke CM, Baade PD. The descriptive epidemiology of female breast cancer: an international comparison of screening, incidence, survival and mortality. Cancer Epidemiol. 2012;36(3):237–48.
12. Ferlay J, Autier P, Boniol M, et al. Estimates of the cancer incidence and mortality in Europe in 2006. Ann Oncol. 2007;18:581–92.
13. Hirabayashi Y, Marugame T. Comparison of time trends in ovary cancer mortality (1990-2006) in the world, from the WHO Mortality Database. Jpn J Clin Oncol. 2009;39:860–1.
14. Dey S, Soliman AS, Hablas A, Seifeldein IA, Ismail K, Ramadan M, et al. Urban–rural differences in breast cancer incidence in Egypt (1999–2006). Breast. 2010;19:417–23.
15. Parkin DM, Fernandez LM. Use of statistics to assess the global burden of breast cancer. Breast J. 2006;12(Suppl 1):S70–80.
16. Prehn AW, West DW. Evaluating local differences in breast cancer incidence rates: a census-based methodology (United States). Cancer Causes Control. 1998;9:511–7.
17. Laden F, Spiegelman D, Neas LM, Colditz GA, Hankinson SE, Manson JE, et al. Geographic variation in breast cancer incidence rates in a cohort of U.S. women. J Natl Cancer Inst. 1997;89:1373–8.
18. Reynolds P, Hurley S, Goldberg DE, Anton-Culver H, Bernstein L, Deapen D, et al. Regional variations in breast cancer among California teachers. Epidemiology. 2004;15:746–54.

19. Binachon B, Laure Dossus L, Danjou AMN, Clavel-Chapelon F, Fervers B. Life in urban areas and breast cancer risk in the French E3N cohort. Eur J Epidemiol. 2014;29:743–51.
20. Wilkinson D, Cameron K. Cancer and cancer risk in South Australia: what evidence for a rural-urban health differential? Aust J Rural Health. 2004;12:61–6.
21. Chen F, Bina WF. Correlation of white female breast cancer incidence trends with nitrogen dioxide emission levels vehicle density patterns. Breast Cancer Res Treat. 2012;132:327–33.
22. Loomis D, Grosse Y, Lauby-Secretan B, El GF, Bouvard V, Benbrahim-Tallaa L, Guha N, Baan R, Mattock H, Straif K, On behalf of the International Agency for Research on Cancer, Monograph Working Group IARC. The carcinogenicity of outdoor air pollution. Lancet Oncol. 2013;14:1262–3.
23. Rudel RA, Attfield KR, Schifano JN, Brody JG. Chemicals causing mammary gland tumors in animals signal new directions for epidemiology, chemicals testing, and risk assessment for breast cancer prevention. Cancer. 2007;109:2635–66.
24. Bonner MR, Han D, Nie J, et al. Breast cancer risk and exposure in early life to polycyclic aromatic hydrocarbons using total suspended particulates as a proxy measure. Cancer Epidemiol Biomark Prev. 2005;14:53–60.
25. Lewis-Michl EL, Melius JM, Kallenbach LR, Ju CL, Talbot TO, Orr MF, et al. Breast cancer risk and residence near industry or traffic in Nassau and Suffolk Counties, Long Island, New York. Arch Environ Health. 1996;51:255–65.
26. Coyle YM, Hynan LS, Euhus DM, Minhajuddin AT. An ecological study of the association of environmental chemicals on breast cancer incidence in Texas. Breast Cancer Res Treat. 2005;92:107–14.
27. IARC Working Group on the Evaluation of Carcinogenic Risks to Humans. Outdoor air pollution. Lyon: International Agency for Research on Cancer; 2015.
28. Brody JG, Moysich KB, Humblet O, Attfield KR, Beehler GP, Rudel RA. Environmental pollutants and breast cancer epidemiologic studies. Cancer. 2007;109(12):2667–711.
29. Crouse DL, Goldberg MS, Ross NA, Chen H, Labreche F. Postmenopausal breast cancer is associated with exposure to traffic-related air pollution in Montreal, Canada: a case-control study. Environ Health Perspect. 2010;118:1578–83.
30. Huo Q, Zhang N, Wang X, Jiang L, Ma T, et al. Effects of ambient particulate matter on human breast cancer: is xenogenesis responsible? PLoS One. 2013;8(10):e76609.
31. Nie J, Beyea J, Bonner MR, Han D, Vena JE, Rogerson P, et al. Exposure to traffic emissions throughout life and risk of breast cancer: the Western New York Exposures and Breast Cancer (WEB) study. Cancer Causes Control. 2007;18:947–55.
32. Wei Y, Davis J, Bina WF. Ambient air pollution is associated with the increased incidence of breast cancer in US. Int J Environ Health Res. 2012;22:12–21.
33. Huynh S, von Euler-Chelpin M, Raaschou-Nielsen O, Hertel O, Tjønneland A, Lynge E, Vejborg I, Andersen ZJ. Long-term exposure to air pollution and mammographic density in the Danish Diet, Cancer and Health cohort. Environ Health. 2015;14:31.
34. Wang J, Ping X, Ying X, Kettrup A, Schramm KW. Differing estrogen activities in the organic phase of air particulate matter collected during sunny and foggy weather in a Chinese city detected by a recombinant yeast bioassay. Atmos Environ. 2004;38(36):6157–66.
35. Xu W, Liu LZ, Loizidou M, Ahmed M, Charles IG. The role of nitric oxide in cancer. Cell Res. 2002;12:311–20.
36. Wink DA, Laval J. The Fpg protein, a DNA repair enzyme, is inhibited by the biomediator nitric oxide in vitro and in vivo. Carcinogenesis. 1994;15:2125–9.
37. Hystad P, Villeneuve PJ, Goldberg MS, Crouse DL, Johnson K, Canadian Cancer Registries Epidemiology Research Group. Exposure to traffic-related air pollution and the risk of developing breast cancer among women in eight Canadian provinces: A case–control study. Environ Int. 2015;74:240–8.
38. Keramatinia A, Hassanipour S, Nazarzadeh M, Wurtz M, Bahadori Monfared A, Khayyamzadeh M, Bidel Z, Mhrvar N, Mosavi-Jarrahi A. Correlation between nitrogen dioxide as an air pollution indicator and breast cancer: a systematic review and meta- analysis. Asian Pac J Cancer Prev. 2016;17(1):419–24.

39. Hart JE, Bertrand KA, DuPre N, Peter James P, Vieira VM, Tamimi RM, Laden F. Long-term particulate matter exposures during adulthood and risk of breast cancer incidence in the Nurses' Health Study II prospective cohort. Cancer Epidemiol Biomark Prev. 2016;25(8):1274.
40. Huo Q, Cai C, Yang Q. Atmospheric particulate matter and breast cancer survival: estrogen receptor triggered? Tumor Biol. 2015;36:3191–3.
41. Iwai K, Mizuno S, Miyasaka Y, Moria T. Correlation between suspended particles in the environmental air and causes of disease among inhabitants: cross-sectional studies using the vital statistics and air pollution data in Japan. Environ Res. 2005;99:106–17.
42. Reding KW, Young MT, Szpiro AA, Han CJ, DeRoo LA, Weinberg C, Kaufman JD, Sandler DP. Breast cancer risk in relation to ambient air pollution exposure at residences in the sister study cohort. Cancer Epidemiol Biomark Prev. 2015;24(12):1907–9.
43. Tagliabue G, Borgini A, Tittarelli A, van Donkelaar A, Martin RV, Bertoldi M, Fabiano S, Maghini A, Codazzi T, Scaburri A, Favia I, Cau A, Bargelletti G, Tessandori R, Contiero P. Atmospheric fine particulate matter and breast cancer mortality: a population-based cohort study. BMJ Open. 2016;6:e012580.
44. Hung LJ, Chan TF, Wu CH, Chiu HF, Yang CY. Traffic air pollution and risk of death from ovarian cancer in Taiwan: fine particulate matter (PM2.5) as a proxy marker. J Toxic Environ Health A. 2012;75(3):174–82.
45. White AJ, Bradshaw PT, Herring AH, Teitelbaum SL, Beyea J, Stellman SD, Steck SE, Mordukhovich I, Eng SM, Engel LS, Conway K, Hatch M, Neugut AI, Santella RM, Gammon MD. Exposure to multiple sources of polycyclic aromatic hydrocarbons and breast cancer incidence. Environ Int. 2016;89-90:185–92.
46. White AJ, JiaChen J, Teitelbaum SL, McCullough LE, Xu X, Cho YH, Conway K, Beyea J, Stellman SD, Steck SE, Mordukhovich I, Eng SM, Beth Terry M, Engel LS, Hatch M, Neugut AI, Hibshoosh H, Santella RM, Gammon MD. Sources of polycyclic aromatic hydrocarbons are associated with gene-specific promotermethylation in women with breast cancer. Environ Res. 2016;145:93–100.
47. Perry NM, Allgood PC, Duffy SW, Mokbel K. Exposure to traffic emissions throughout life and risk of breast cancer. Cancer Causes Control. 2008;19:435.
48. Zhang Y, Yang R, Burwinkel B, Breitling LP, Brenner H. F2RL3 methylation as a biomarker of current and lifetime smoking exposures. Environ Health Perspect. 2014;122:131–7.
49. Hussain I, Bhan A, Ansari KI, Deb P, Bobzean SA, Perrotti LI, Mandal SS.Bisphenol-A induces expression of HOXC6, an estrogen-regulated homeobox-containing gene associated with breast cancer. Biochim Biophys Acta. 2015 Jun;1849(6):697–708.
50. Matsumoto H, Adachi S, Suzuki Y. Bisphenol A in ambient air particulates responsible for the proliferation of MCF-7 human breast cancer cells and its concentration changes over 6 months. Arch Environ Contam Toxicol. 2005;48:459–66.
51. Pisani P, Parkin DM, Ferlay J. Estimates of the worldwide incidence of eighteen major cancers in 1985. Int J Cancer. 1983;54:891–903.
52. Adami HO, Hunter D, Trichopoulos D. Textbook of cancer epidemiology. 2nd ed. New York: Oxford University Press; 2008.
53. Salehi F, Dunfield L, Phillips KP, Krewski D, Vanderhyden BC. Risk factors for ovarian cancer: an overview with emphasis on hormonal factors. J Toxicol Environ Health Part B. 2008;11:301–21.
54. Shen N, Weiderpass E, Antilla A, Goldberg MS, Vasama-Neuvonen KM, Boffetta P, Vainio UH, Partanen TJ. Epidemiology of occupational and environmental risk factors related to ovarian cancer. Scand J Work Environ Health. 1998;24:175–82.
55. Shields T, Gridley G, Moradi T, Adami J, Plato N, Dosemeci M. Occupational exposures and the risk of ovarian cancer in Sweden. Am J Ind Med. 2002;42:200–13.
56. Cancer Research UK. Fertility and cervical cancer. 2010.
57. Couto E, Hemminki K. Heritable and environmental components in cervical tumors. Int J Cancer. 2006;119:2699–701.
58. President's Cancer Panel. 2008–2009 annual report: reducing environmental cancer risk: what we can do now. U.S. Department of Health and Human Services. 2010.

59. Ma XF, Zhang J, Shuai HL, Guan BZ, Luo X, Yan RL. IKKβ/NF-κB mediated the low doses of bisphenol A induced migration of cervical cancer cells. Arch Biochem Biophys. 2015;573:52–8.
60. Raaschou-Nielsen O, Andersen ZJ, Hvidberg M, Jensen SS, Ketzel M, Sørensen M, et al. Air pollution from traffic and cancer incidence: a Danish cohort study. Environ Health. 2011;10:67.
61. Scheurer ME, Danysh HE, Follen M, Lupo PJ. Association of traffic-related hazardous air pollutants and cervical dysplasia in an urban multiethnic population: a cross-sectional study. Environ Health. 2014;13:52.
62. Abbas S, Ihle P, Köster I, Schubert I. Prevalence and incidence of diagnosed endometriosis and risk of endometriosis in patients with endometriosis-related symptoms: findings from a statutory health insurance-based cohort in Germany. Eur J Obstet Gynecol Reprod Biol. 2012;160(1):79–83.
63. Buck Louis GM, Hediger ML, Peterson CM, Croughan M, Sundaram R, Stanford J, Chen Z, Fujimoto VY, Varner MW, Trumble A, Giudice LC, ENDO Study Working Group. Incidence of endometriosis by study population and diagnostic method: the ENDO study. Fertil Steril. 2011;96(2):360–5.
64. Augoulea A, Alexandrou A, Creatsa M, Vrachnis N, Lambrinoudaki I. Pathogenesis of endometriosis: the role of genetics, inflammation and oxidative stress. Arch Gynecol Obstet. 2012;286:99.
65. Nyholt DR, Low SK, Anderson CA, Painter JN, Uno S, Morris AP, MacGregor S, Gordon SD, Henders AK, Martin NG, Attia J, Holliday EG, McEvoy M, Scott RJ, Kennedy SH, Treloar SA, Missmer SA, Adachi S, Tanaka K, Nakamura Y, Zondervan KT, Zembutsu H, Montgomery GW. Genome-wide association meta-analysis identifies new endometriosis risk loci. Nat Genet. 2012;44(12):1355–9.
66. Gentilini D, Perino A, Viganò P, Chiodo I, Cucinella G, Vignali M, Di Blasio AM, Busacca M. Gene expression profiling of peripheral blood mononuclear cells in endometriosis identifies genes altered in non-gynaecologic chronic inflammatory diseases. Hum Reprod. 2011;26(11):3109–17.
67. Agic A, Xu H, Finas D, Banz C, Diedrich K, Hornung D. Is endometriosis associated with systemic subclinical inflammation? Gynecol Obstet Invest. 2006;62(3):139–47.
68. Brook RD, Rajagopalan S, Pope CA 3rd, Brook JR, Bhatnagar A, Diez-Roux AV, Holguin F, Hong Y, Luepker RV, Mittleman MA, Peters A, Siscovick D, Smith SC Jr, Whitsel L, Kaufman JD, American Heart Association Council on Epidemiology and Prevention, Council on the Kidney in Cardiovascular Disease, and Council on Nutrition, Physical Activity and Metabolism. Particulate matter air pollution and cardiovascular disease: an update to the scientific statement from the American Heart Association. Circulation. 2010;121(21):2331–78.
69. Calderón-Garcidueñas L, Engle R, Mora-Tiscareño A, Styner M, Gómez-Garza G, Zhu H, Jewells V, Torres-Jardón R, Romero L, Monroy-Acosta ME, Bryant C, González-González LO, Medina-Cortina H, D'Angiulli A. Exposure to severe urban air pollution influences cognitive outcomes, brain volume and systemic inflammation in clinically healthy children. Brain Cogn. 2011;77(3):345–55.
70. Misaki K, Suzuki M, Nakamura M, Handa H, Iida M, Kato T, et al. Aryl hydrocarbon receptor and estrogen receptor ligand activity of organic extracts from road dust and diesel exhaust particulates. Arch Environ Contam Toxicol. 2008;55:199–209.
71. Oh SM, Ryu BT, Chung KH. Identification of estrogenic and antiestrogenic activities of respirable diesel exhaust particles by bioassay-directed fractionation. Arch Pharm Res. 2008;31:75–82.
72. Wang J, Xie P, Kettrup A, Schramm KW. Inhibition of progesterone receptor activity in recombinant yeast by soot from fossil fuel combustion emissions and air particulate materials. Sci Total Environ. 2005;349:120–8.

73. Umezawa M, Sakata C, Tanaka N, Tabata M, Takeda K, Ihara T, et al. Pathological study for the effects of in utero and postnatal exposure to diesel exhaust on a rat endometriosis model. J Toxicol Sci. 2011;36:493–8.
74. Takeda K, Tsukue N, Yoshida S. Endocrine-disrupting activity of chemicals in diesel exhaust and diesel exhaust particles. Environ Sci. 2004;11:33–45.
75. Baird DD, Dunson DB, Hill MC, Cousins D, Schectman JM. High cumulative incidence of uterine leiomyoma in black and white women: ultrasound evidence. Am J Obstet Gynecol. 2003;188(1):100–7.
76. Mahalingaiah S, Hart JE, Laden F, Aschengrau A, Missmer SA. Air pollution exposures during adulthood and risk of endometriosis in the Nurses' Health Study II. Environ Health Perspect. 2014;122(1):58–64.
77. Mahalingaiah S, Hart JE, Laden F, Terry KL, Boynton-Jarrett R, Aschengrau A, Missmer SA. Air pollution and risk of uterine leiomyomata. Epidemiology. 2014;25(5):682–8.
78. Bidgoli SA, Khorasani H, Keihan H, Sadeghipour A, Mehdizadeh A. Role of endocrine disrupting chemicals in the occurrence of benign uterine leiomyomata: special emphasis on AhR tissue levels. Asian Pac J Cancer Prev. 2012;13(11):5445–50.
79. Biro FM, Greenspan LC, Galvez MP. Puberty in girls of the 21st century. J Pediatr Adolesc Gynecol. 2012;25(5):289–94.
80. Herman-Giddens ME, Slora EJ, Wasserman RC, Bourdony CJ, Bhapkar MV, Koch GG, Hasemeier CM. Secondary sexual characteristics and menses in young girls seen in office practice: a study from the Pediatric Research in Office Settings network. Pediatrics. 1997;99(4):505–12.
81. Biro FM, Deardorff J. Identifying opportunities for cancer prevention during preadolescence and adolescence: puberty as a window of susceptibility. J Adolesc Health. 2013;52(5 Suppl):S15–20. https://doi.org/10.1016/j.jadohealth.2012.09.019. Review. Erratum in: J Adolesc Health. 2013 Jun;52(6);808
82. Laitinen J, Power C, Järvelin MR. Family social class, maternal body mass index, childhood body mass index, and age at menarche as predictors of adult obesity. Am J Clin Nutr. 2001;74(3):287–94.
83. Pierce MB, Leon DA. Age at menarche and adult BMI in the Aberdeen children of the 1950s cohort study. Am J Clin Nutr. 2005;82(4):733–9.
84. Biro FM, Kiess W. Contemporary trends in onset and completion of puberty, gain in height and adiposity. Endocr Dev. 2016;29:122–33.
85. Himes JH, Park K, Styne D. Menarche and assessment of body mass index in adolescent girls. J Pediatr. 2009;155(3):393–7.
86. Yermachenko A, Dvornyk V. Nongenetic determinants of age at menarche: a systematic review. Biomed Res Int. 2014;2014:371583.
87. Arcus-Arth A, Blaisdell RJ. Statistical distributions of daily breathing rates for narrow age groups of infants and children. Risk Anal. 2007;27(1):97–110.
88. Bateson TF, Schwartz J. Children's response to air pollutants. J Toxicol Environ Health A. 2008;71(3):238–43.
89. Özen S, Darcan Ş. Effects of environmental endocrine disruptors on pubertal development. J Clin Res Pediatr Endocrinol. 2011;3(1):1–6.
90. Schug TT, Janesick A, Blumberg B, Heindel JJ. Endocrine disrupting chemicals and disease susceptibility. J Steroid Biochem Mol Biol. 2011;127(3-5):204–15.
91. Gollenberg AL, Hediger ML, Lee PA, Himes JH, Louis GM. Association between lead and cadmium and reproductive hormones in peripubertal U.S. girls. Environ Health Perspect. 2010;118(12):1782–7.
92. McGuinn LA, Voss RW, Laurent CA, Greenspan LC, Kushi LH, Windham GC. Residential proximity to traffic and female pubertal development. Environ Int. 2016;94:635–41.
93. Craig ZR, Wang W, Flaws JA. Endocrine-disrupting chemicals in ovarian function: effects on steroidogenesis, metabolism and nuclear receptor signaling. Reproduction. 2011;142(5):633–46.

94. Gregoraszczuk EL, Ptak A. Endocrine-disrupting chemicals: some actions of POPs on female reproduction. Int J Endocrinol. 2013;2013:828532.
95. Yang Q, Zhao Y, Qiu X, Zhang C, Li R, Qiao J. Association of serum levels of typical organic pollutants with polycystic ovary syndrome (PCOS): a case-control study. Hum Reprod. 2015;30(8):1964–73.
96. Farr SL, Cai J, Savitz DA, Sandler DP, Hoppin JA, Cooper GS. Pesticide exposure and timing of menopause: the Agricultural Health Study. Am J Epidemiol. 2006;163:731–42.
97. Cooper GS, Sandler DP, Bohlig M. Active and passive smoking and the occurrence of natural menopause. Epidemiology. 1999;10:771–3.
98. Fleming LE, Levis S, LeBlanc WG, et al. Earlier age at menopause, work, and tobacco smoke exposure. Menopause. 2008;15:1103–8.
99. Mikkelsen TF, Graff-Iversen S, Sundby J, Bjertness E. Early menopause, association with tobacco smoking, coffee consumption and other lifestyle factors: a cross-sectional study. BMC Public Health. 2007;7:149.
100. Berta L, Fortunati N, Gennari P, Appendino M, Casella A, Frairia R. Influence of cigarette smoking on pituitary and sex hormone balance in healthy premenopausal women. Fertil Steril. 1991;56(4):788–9.
101. Kinney A, Kline J, Kelly A, Reuss ML, Levin B. Smoking, alcohol and caffeine in relation to ovarian age during the reproductive years. Hum Reprod. 2007;22(4):1175–85.
102. Westhoff C, Gentile G, Lee J, Zacur H, Helbig D. Predictors of ovarian steroid secretion in reproductive-age women. Am J Epidemiol. 1996;144(4):381–8.
103. Ip P, Chung BH, Ho FK, Chan GC, Deng W, Wong WH, Lee SL, Chan PY, Ying D, Wong WL, Tung KT, Lau YL. Prenatal tobacco exposure shortens telomere length in children. Nicotine Tob Res. 2017;19(1):111–8.
104. Lutterodt MC, Sorensen KP, Larsen KB, Skouby SO, Andersen CY, Byskov AG. The number of oogonia and somatic cells in the human female embryo and fetus in relation to whether or not exposed to maternal cigarette smoking. Hum Reprod. 2009;24(10):2558–66.
105. Strohsnitter WC, Hatch EE, Hyer M, et al. The association between in utero cigarette smoke exposure and age at menopause. Am J Epidemiol. 2008;167(6):727–33.
106. Steiner AZ, D'Aloisio AA, DeRoo LA, Sandler DP, Baird DD. Association of intrauterine and early-life exposures with age at menopause in the sister study. Am J Epidemiol. 2010;172(2):140–8.
107. Mascarenhas MN, Cheung H, Mathers CD, Stevens GA. Measuring infertility in populations: constructing a standard definition for use with demographic and reproductive health surveys. Popul Health Metr. 2012;10(1):17.
108. Augood C, Duckitt K, Templeton AA. Smoking and female infertility: a systematic review and meta-analysis. Hum Reprod. 1998;13(6):1532–9.
109. Bolumar F, Olsen J, Boldsen J. Smoking reduces fecundity: a European multicenter study on infertility and subfecundity. The European Study Group on Infertility and Subfecundity. Am J Epidemiol. 1996;143(6):578–87.
110. Hull MG, North K, Taylor H, Farrow A, Ford WC. Delayed conception and active and passive smoking. The Avon Longitudinal Study of Pregnancy and Childhood Study Team. Fertil Steril. 2000;74(4):725–33.
111. Benedict MD, Missmer SA, Vahratian A, Berry KF, Vitonis AF, Cramer DW, Meeker JD. Secondhand tobacco smoke exposure is associated with increased risk of failed implantation and reduced IVF success. Hum Reprod. 2011;26(9):2525–31.
112. Meeker JD, Missmer SA, Vitonis AF, Cramer DW, Hauser R. Risk of spontaneous abortion in women with childhood exposure to parental cigarette smoke. Am J Epidemiol. 2007;166(5):571–5.
113. Veras MM, Damaceno-Rodrigues NR, Guimarães Silva RM, Scoriza JN, Saldiva PH, Caldini EG, Dolhnikoff M. Chronic exposure to fine particulate matter emitted by traffic affects reproductive and fetal outcomes in mice. Environ Res. 2009;109(5):536–43.
114. Legro RS, Sauer MV, Mottla GL, Richter KS, Li X, Dodson WC, Liao D. Effect of air quality on assisted human reproduction. Hum Reprod. 2010;25(5):1317–24.

115. Perin PM, Maluf M, Czeresnia CE, Januário DA, Saldiva PH. Impact of short-term preconceptional exposure to particulate air pollution on treatment outcome in couples undergoing in vitro fertilization and embryo transfer (IVF/ET). J Assist Reprod Genet. 2010;27(7):371–82.
116. Slama R, Bottagisi S, Solansky I, Lepeule J, Giorgis-Allemand L, Sram R. Short-term impact of atmospheric pollution on fecundability. Epidemiology. 2013;24(6):871–9.
117. Nieuwenhuijsen MJ, Basagaña X, Dadvand P, Martinez D, Cirach M, Beelen R, Jacquemin B. Air pollution and human fertility rates. Environ Int. 2014;70:9–14.
118. Mahalingaiah S, Hart JE, Laden F, Farland LV, Hewlett MM, Chavarro J, Aschengrau A, Missmer SA. Adult air pollution exposure and risk of infertility in the Nurses' Health Study II. Hum Reprod. 2016;31(3):638–47.
119. Checa Vizcaíno MA, González-Comadran M, Jacquemin B. Outdoor air pollution and human infertility: a systematic review. Fertil Steril. 2016;106(4):897–904.
120. WHO. Air quality guidelines—global update 2005. 2005.
121. To T, Zhu J, PJ V, Simatovic J, Feldman L, Gao C, Williams D, Chen H, Weichenthal S, Wall C, Miller AB. Chronic disease prevalence in women and air pollution: a 30 year longitudinal cohort study. Environ Int. 2015;80:26–32.
122. Rojas-Rueda D, de Nazelle A, Teixidó O, Nieuwenhuijsen MJ. Health impact assessment of increasing public transport and cycling use in Barcelona: a morbidity and burden of disease approach. Prev Med. 2013;57:573–9.

Air Pollution Effects in Pregnancy

26

Lidia La Marca and Giulia Gava

26.1 Introduction

The developing embryo and then fetus is believed to be more susceptible than adults to environmental pollutants, and birth outcomes can be impaired by air pollution exposure affecting gestational length, fetal growth, and possibly causing miscarriage/pregnancy loss and congenital anomalies.

Firsthand smoking in pregnancy has been associated to low birth weight, placental abruption, and sudden infant death syndrome [1]. Anyway, almost 12.3% of pregnant women smokes during pregnancy [2], and globally more than a third of women are exposed to secondhand smoke [3].

Up to now, numerous studies have shown associations between exposure to air pollutants and negative birth outcomes such as low birth weight, preterm birth, and congenital anomalies [4, 5]. Furthermore, it has been hypothesized that air pollution molecules inhaled by the mother during pregnancy may lead to health impairment later in life for the newborn with possible reduction in lung function, increase in respiratory morbidity, and alterations of the immune system [6].

There are growing evidences that common air pollutants can contribute to the risks of hypertension, cardiovascular events, and diabetes in adults. Recently, similar associations have been reported in pregnant women highlighting the increased risk of hypertensive disorders and gestational diabetes after exposure to air pollution.

L. La Marca (✉)
Delegate of Lega Vita e Salute Foundation, Bologna, Italy
e-mail: lidialamarca@infopro.it

G. Gava
Gynecology and Physiopathology of Human Reproduction, S. Orsola-Malpighi Hospital, University of Bologna, Bologna, Italy

F. Capello, A.V. Gaddi (eds.), *Clinical Handbook of Air Pollution-Related Diseases*, https://doi.org/10.1007/978-3-319-62731-1_26

Anyway, though evidences are continuously growing, many questions remain unanswered: from available data it is still not clear whether those effects are due to a specific molecule or to the interactions of different pollutants. Furthermore it is not possible to point out if there is a specific trimester in which the exposure can be more harmful to fetal development. The exact mechanism through which maternal exposure to air pollution can impair gestational outcomes is still partially unknown, even if some authors hypothesized that oxidative stress, inflammation and hemodynamic changes associated with air pollutants can impair oxygen and nutrient transport to the fetus. For these reasons, despite an ever-growing volume of studies, scientific literature regarding this topic is still characterized by enormous heterogeneity in designs and settings, sometimes leading to several inconclusive results.

26.2 Spontaneous Abortions and Stillbirth

Spontaneous abortion (SAB) is believed to be the most common complication of early pregnancy [7] as it occurs in about 17–22% of all pregnancies [8]. Regarding the role of tobacco smoke in SAB risk, several published data are inconsistent, although several studies concluded that smoking was associated to SAB; Pineles et al. [1] recently reported in their review that active and secondhand smoking are associated with increased risk of SAB.

The existence of a possible association between SAB and exposure to outdoor pollutants during pregnancy is still unclear, and few studies with inconclusive and/or conflicting results have been published up to now [9, 10].

Anyway, more recently reports of association between air pollution and SAB are growing. Green et al. analyzed the link between annual average traffic within 50 m and SAB risk in the general population, and the author did not found any association, but he reported a significant negative impact in the African Americans and nonsmoker's subgroups [11]. Further studies underlined that the risk of SAB was increased in the general population in association to exposure to high levels of NO2 and SO2 and coal combustion [12, 13].

The association between spontaneous abortion and seasonal variation of air pollutants in Mongolia has been analyzed, and the author reported a strong correlation between air pollution and spontaneous abortion [14].

Recently Di Ciaula and Bilancia [15] analyzed records of SAB from 514,996 residents, and the authors found a correlation of the mean monthly SAB rate with higher PM10 or ozone concentration values even if below the legal limits (particularly if industrial areas were analyzed).

Regarding studies conducted in women undergoing in vitro fertilization, Perin et al. reported a significant increase in the rate of SAB in women exposed to higher levels of PM10 [16]. Also, Checa Vizcaíno et al. in 2016 reported an increased risk of miscarriage following exposition to PM10 and NO2 in the IVF setting and to SO2 and CO in the in the general and the subfertile population [125].

Stillborn is defined by the Perinatal Mortality Surveillance Report as a baby delivered with no signs of life known to have died after 24 completed weeks of pregnancy. Worldwide, it is believed that 2.6 million children were stillborn at 28 weeks or more in 2015, most of all, in low-income and middle-income countries [17]. About one third of stillbirths are small-for-gestational-age fetuses and half are unexplained [18]. Half of the stillbirths occur during labor, while congenital abnormalities are responsible for less than 10% of stillbirths [17].

Stillbirth is still a poorly understood adverse outcome of pregnancy. Even though further studies are needed and some studies reported insufficient evidences [19], other studies reported that exposure to ambient air pollution can increase the risk of stillbirth. Lacasana et al. reported a positive even if non-consistent relationship between air pollution and stillbirth [20]. More recently Siddika et al. [21] in a systematic review and meta-analysis suggested an elevated risk of stillbirth in relation to air pollution, in particular to mean prenatal exposure to NO2, CO, SO2, PM2.5, and PM10. The authors also reported that the effects of SO2, CO, PM10, and O3 were higher in case of third trimester exposure.

26.3 Preeclampsia and Hypertensive Disorders of Pregnancy

Hemodynamic changes typical of pregnancy lead to a higher stress on the cardiovascular system, and women during pregnancy are more susceptible to hypertensive disorders [22].

Hypertensive disorders of pregnancy (HDP) are hypertensive disorders typically occurring during pregnancy; they include gestational hypertension, preeclampsia (PE), and eclampsia and they usually regress after delivery [23].

HDP are the most common and serious complications of pregnancy affecting 2–10% of women after 20 weeks of gestation [24, 25], and they increase the risk of maternal and perinatal mortality and morbidity [24, 26, 27].

HDP have also a high rate of recurrence in subsequent pregnancies: 30–50% of women who had HDP in their first pregnancy present another HDP in the second pregnancy [28, 29], and these women also have a greater risk of chronic hypertension, heart disease, or stroke later in life [30–32].

In normal pregnancy, blood pressure starts to decrease during the first trimester, reaching its lowest point in midpregnancy, and then gradually returns to nonpregnant levels by the term [33]. Previous research indicates that this pattern is different in women who develop gestational hypertensive disorders, as their blood pressure is stable during the first half of pregnancy and then continuously increases until delivery [33, 34]. Gestational hypertension is defined by the new onset of hypertension (systolic blood pressure $\geq$140 mmHg and/or diastolic blood pressure $\geq$90 mmHg) at $\geq$20 weeks of gestation in the absence of proteinuria or new signs of end-organ dysfunction. Preeclampsia is a multisystem disorder characterized by the new onset of hypertension and proteinuria or end-organ dysfunction or both in the last half of pregnancy. Eclampsia is the occurrence of new-onset, generalized, tonic-clonic seizures or coma in a woman with preeclampsia [25]. The etiology of HDP is complex and remains largely unknown. Obesity, multiple gestation, low parity, abnormal placentation, damage to blood vessels, inflammation, oxidative stress, and immunological changes are believed to increase blood pressure and to contribute to the development of hypertensive outcomes [24, 35].

Environmental exposure to ambient air pollution has been reported to contribute to the risks of hypertension and other cardiovascular events in adults and elderly [36–39].

Recently, reports of association between air pollution exposure during pregnancy and increased risk of HDP are growing [40–42] although the association with PE was not confirmed by some studies [43, 44].

Olsson et al. [45] reported a strong association between high levels of nitrogen monoxide (NO), used as marker of traffic pollution, and HDP in women who did not change their address during pregnancy. van den Hooven et al. [42] pointed out that the exposure to higher PM10 levels but not to nitrogen dioxide (NO2) was associated with an increased systolic blood pressure during pregnancy and with an increased risk of gestational hypertension and early symptoms of PE.

Up to now, several studies have reported positive associations between PE and air pollutants including nitrogen oxides (NOx), NO, NO2, carbon monoxide (CO), ozone (O3), PM2.5, and PM10 [46–48], but others have reported no association with PM2.5 or CO [44], with PM2.5 and PM10 [43], and inconclusive findings for PM10 and NO2 [42].

To live in proximity to roadways is a risk factor [49–51]; moreover, the risk of PE and HDP in pregnancy is increased by the exposure both to ambient air pollution and noise [52].

Regarding the time of exposure, there is a statistically significantly increased risk of HDP in relation to CO, NO2, and SO2 during the entire pregnancy [53]; in relation to NO2, SO2, and CO2 [53, 54]; and in relation to PM2.5 both during the first trimester [54] and the entire pregnancy [53]. Olsson et al. [45], in a large European study, pointed out that the worst period to be exposed to O3 is the first trimester of gestation; for Rudra et al. [44], the evidence is for a weakly association between eclampsia and exposure to CO2 during the first 7 months. Exposure to PM10 and O3 ([42]) and to PM10, PM2.5, and O3 during the first trimester [43] was associated with HPD. Exposure to air pollution 4 h before delivery can also influence blood pressure at the time of admission in the hospital [55].

Traffic-related air pollution significantly increases the risk of late-onset (after 34+0 weeks of gestation) PE, and it has a potential relationship with an increased severity of PE defined by the author as hypertension over 160/110 mmHg and proteinuria above 0.5 g/24 h, neurological symptoms, renal failure, or HELLP syndrome [48].

Moreover, the association between NOx exposure and PE was found statistically significant in a population exposed to air pollution levels below those recommended by the current air quality guidelines [56].

Recently Pedersen et al. [57] published a systematic review and meta-analysis based on 17 papers evaluating the exposure to ambient air pollution (NO2, NOx, PM10, PM2.5, CO, O3) and HDP. Even if few studies evaluated traffic density and the proximity to major roads, which can be used as an indicator of air pollution, this meta-analysis showed an increased risk of HDP in association to exposure to all pollutants. Furthermore, several studies reported that traffic-related air pollution is associated with endothelial dysfunction that is a potential predictor of PE [42].

Not only higher levels of PM10 and PM2.5 were positively associated with HPD but also neighborhood deprivation and social stress [47], lower socioeconomic groups, or previous circulatory or respiratory diseases. The effects were more evident in younger women (<20 years) or older than 40 years and in women with pre-existing or gestational diabetes [58].

It has also been estimated a 10% increased risk in asthmatic women exposed to CO during the second trimester and to NOx and SO2 during the first trimester of pregnancy, suggesting the major vulnerability to air pollution of women with asthma [59].

In conclusion, even if the evidence of the association between air pollution and HDP is growing, it is still limited. Most of the studies rely on crude assessment of exposure and on scarce information regarding potential confounders and the time of exposure. In fact, a large number of epidemiological studies from different part of the world have reported different results with respect to significant pollutants and time widow of susceptibility [46, 51, 56, 58, 60–62]. It should also been taken into account that some pollutants have a large temporal variation and that the approach to estimate pregnant women's exposure was often different.

26.4 Gestational Diabetes Mellitus (GDM)

GDM is a glucose intolerance first diagnosed during pregnancy. It affects 2–6% of pregnant women worldwide and 10–20% of pregnancies in high-risk populations [63]. Pregnancies affected by diabetes are characterized by a greater risk of adverse gestational and perinatal outcomes such as fetal hypoglycemia and birth trauma to both mother and infant. Furthermore, both women with a previous GDM and children born from a diabetic mother are at greater risk of type 2 diabetes development later in life [64]. Known risk factors for GDM are obesity, older age, and family history of type 2 diabetes. Anyway, in about half of women with GDM there cannot be found any known risk factor, suggesting a role for environmental factors.

PM2.5 exposure in the animal model can induce insulin resistance acting on endothelial function and promoting inflammation and oxidative stress [65]. Furthermore, several epidemiologic studies have demonstrated an association between air pollution exposure and increased risk of type 2 diabetes mellitus in the adult [66]. Most of the epidemiologic data regarding the relationship between prenatal air pollution exposure and abnormal glucose tolerance in pregnancy showed an association even if some studies are not concordant [56, 67–69]. Some other studies did not confirm an association [70]; a recent study by Fleisch et al., although reporting no relationship between residential PM2.5 exposure and GFM, evidenced a correlation between greater exposure to PM2.5 during the second trimester and GDM in the youngest [71].

Furthermore, it has been reported that prenatal exposure to endocrine-disrupting chemicals can increase the risk of obesity of childhood [72] and that exposure to air pollutants in the third trimester is associated to more rapid postnatal weight gain in addition to reduced fetal growth [73]. Children born from mothers who lived close to a major roadway at the time of delivery was reported to have an increased prevalence of markers of adverse cardio-metabolic risk, such as higher total fat mass, in early and mid-childhood [71].

26.5 Preterm Birth

Preterm birth (PTB) is defined as birth before 37 weeks of gestation [74]. It represents a leading cause of infant mortality and morbidity, being also associated with adverse effects in child and adult life, such as impaired vision, hearing, and cognitive function, decreased motor function, and behavioral disorders [75].

Many mechanisms have been proposed to explain the increase risk of PTB attributed to air pollution: increase of oxidative stress and inflammation [76], impairment of placentation [70], and increase of maternal susceptibility to infections [77].

Numerous reports of a positive association between PTB and exposure of pregnant women to air pollution have been published, but results vary widely among studies [45, 78, 79]. Many studies have evaluated the link between PTB and traffic-related pollutants or the proximity to traffic sources, and they reported positive associations [80, 46, 49, 51]) with few exceptions [81, 82].

Regarding the association between PM and PTB, some studies reported a positive association [4, 46, 78, 79], while others reported an inverse association [83, 84]. Discrepancies can be due to methodological differences, different population susceptibilities, study settings, and differences in air pollutant molecules considered in the analysis. In fact, air pollution composition varies dramatically through seasons and setting as it is influenced by the nature of pollution sources. In particular, the potential effects of particulate matter (PM) can be mediated by core chemical components of PM, such as the elemental carbon and nitrates, or by organic compounds, such as quinones, polycyclic aromatic hydrocarbons, or metals [85].

Few studies, performed in the United States, have examined the association between PM composition and PTB. The first one was performed in Georgia and reported a positive association between PTB and sulfate and water-soluble metals in PM2.5, despite a lack of association with total PM2.5 mass [86]. In 2011 a second study was conducted in California, and the author found a positive association between PTB and organic carbon (OC), elemental carbon (EC), and ammonium nitrate in fine PM (PM2.5), even if there was an inverse association with total PM2.5 mass [84]. On the counterpart, in 2016 Johnson reported that PM2.5 and NO2 were not significantly associated with PTB [87]. Recently, Laurent et al. analyzed the relationship between PTB and ultrafine PM (PM0.1) in California over 2000–2008 and reported that the exposures to both primary and secondary pollutants were associated with an increase in PTB [88].

A few studies evaluated the association of PTB to other sources of air pollution as oil refineries [89], cement plants [90], or gasoline stations [91] and to exposure to PM from specific sources, such as open-hearth steel mill [92], coal [93], diesel [84], or biomass burning [94].

Analyzing the causes of PTB, preterm premature rupture of membranes (pPROM), that is, the rupture of membranes before 37 weeks before labor onset, is responsible for 1 of every 3 preterm births, and it is associated with significant morbidity for both mothers and newborn, including infection, sepsis, umbilical cord compression, risk of placental abruption, and the short- and long-term adverse consequences of neonatal prematurity. pPROM is believed to be caused by pathological mechanisms associated with infection and inflammation. In 2016, Wallace et al. analyzed the relationship between mean exposures to particulate matter less than 10 μm or less than 2.5 μm, nitrogen oxides, carbon monoxide, sulfur dioxide, and ozone, among 223,375 singleton deliveries. The author reported that the whole-pregnancy exposure to carbon monoxide and sulfur dioxide was associated with an increased risk of PROM but not pPROM [95].

Up to now there is still inconsistence regarding the gestational windows of greatest susceptibility and the specific pollutant most strongly associated with preterm birth. Hao et al. in 2016 analyzed 511,658 births in Georgia, reporting trimester-specific and total pregnancy associations for several pollutants. In particular, all the traffic-related pollutants (carbon monoxide, nitrogen dioxide, PM2.5 elemental carbon) were associated to PTB [96].

26.6 Low Birth Weight

Low birth weight (LBW) neonates are defined by WHO as newborns with a weight at birth of less than 2500 g, and this condition is a major public health problem worldwide especially in low- and middle-income countries [97].

The prevalence is estimated to be 15% worldwide with a range of 3.3–38% [98]. LBW babies are approximately 20 times more likely to die than heavier babies, and this is an indicator of a wild range of future poor health outcomes in terms of morbidity and disability in neonatal, infancy, and childhood resulting in substantial costs to the health systems and imposing a significant burden to the society [98].

For these reasons, it was a goal of the 2012 World Health Assembly to reduce the number of LBW babies by 30% by the year 2025 [99]. LBW can be caused by preterm birth that is the main cause of death, morbidity, and disability, but it can also be due to a restriction of fetal growth that is associated with poor growth in childhood and with a higher incidence of adult diseases, such as type 2 diabetes, hypertension, coronary heart disease, and stroke (also referred to as the "fetal programming" or "Barker hypothesis") [100].

Epidemiological end experimental data regarding effects of air pollution on fetal growth is continuously growing. Particles smaller than 1 μm, corresponding to the size range of anthropogenic air pollution, can penetrate the alveolar wall and enter the maternal bloodstream reaching the placenta and the fetus [101]. Via blood stream, those inhaled particles can affect mother and offspring health at the microvascular level in the human and animal model maybe impairing the mitochondrial function [102, 103]. Air pollution, especially ultrafine particles, has been associated to inflammation through oxidative stress.

Epidemiological studies are pointing out the effects of air pollution on birth weight [77, 104]. Stieb et al. in 2012 reviewed sixty-two studies reporting that the majority of them found a positive association of reduced birth weight and increased odds of low birth weight with exposure to carbon monoxide (CO),nitrogen dioxide (NO2), and particulate matter less than 10 and 2.5 μm (PM10 and PM2.5). Data based on third trimester exposure were precise especially for Co and PM10 while were less consistent for outcomes regarding ozone and sulfur dioxide.

Estarlich et al. evaluated the association between exposure to nitrogen dioxide (NO2) and benzene, metrics for traffic-generated pollution, and anthropometric measures at birth [105]. The author reported that an increase in NO2 exposure was correlated with a decrease in birth length and in birth weight.

Similar results were reported in a different birth cohort study in Spain where the exposure to NO2 above 40 $\mu g/m^3$ during the first trimester was associated with a reduction in birth length and weight and with a significant decrease in head circumference [60].

Also PM exposure has been linked to similar birth outcomes. It has been demonstrated that an increase exposition to PM (PM10) and fine PM (PM2.5) during gestation was associated with an increased incidence of LBW [106, 107]. Another study evaluating a 9-years exposure to PM2.5 in the month and the 3 months before birth reported that an increase PM2.5 was associated with a decrease in birth weight [78].

Similarly, it was recently reported that body weight at birth was inversely associated with PM2.5 exposure and that PM2.5 exposure during the first trimester and entire gestation period showed a stronger association with baby weight than the exposure during the second and third trimesters [108].

Malmqvist et al. recently published the data from the Maternal Air Pollution in Southern Sweden cohort: 48,000 pregnancies from an ultrasound database, birth registry, and exposure data based on residential addresses [109]. The author reported a negative effect of air pollution with a decrease of abdominal diameter and femur length and a reduction of birth weight and of fetal growth late in pregnancy in a geographic area with air pollutant levels below those suggested by the current WHO guidelines.

26.7 Fetal Malformations

Worldwide about 10% of deaths of children under 5 years of age are caused by congenital anomalies. The European Surveillance of Congenital Anomalies (EUROCAT) (a population-based registry for the surveillance of congenital anomalies in Europe) reported a perinatal death rate of 9.3/10,000 births of all congenital anomaly between 2008 and 2012. Among them, 23.7% was due to chromosomal anomalies, 22.6% to congenital heart defects, and 17.2% to nervous system anomalies. About half of all congenital malformations are of unclear etiology and are suggested to have multifactorial causes including environmental exposures.

Numerous studies have investigated the association of congenital anomalies and air pollution, but the evidence of the association between ambient pollutants and birth defects is still limited.

Gianicolo et al. and Vinikoor-Imler et al. have suggested that maternal exposure to air pollution may play an important role in causing birth defects [110, 111]. The potential impact of environmental exposures has been reviewed by Chen et al. in 2012, and the majority of the included studies concluded that exposure to NO2, SO2, and PM2.5 increased significantly the risk of congenital heart diseases [112]. This meta-analysis included 17 articles evaluating the risk for a variety of air pollutants (SO2, NO2, PM10, PM2.5, CO, and O3) and anomaly defect outcomes, and only one significant combination was found: NO2 concentrations were significantly associated with coarctation of the aorta.

Farhi et al. in 2014 reported that increased concentration of PM10 and NOx pollutants during the entire pregnancy were associated with a slight increased risk of congenital malformation; in that cohort, specific malformations were more often evident in the circulatory systems and genital organs [113].

A growing of epidemiologic evidence has suggested a link between ambient particulate matter (PM) exposure and fetal anomalies especially cardiovascular malformations. PM with an aerodynamic diameter less than 10 micrometers (PM10) has been associated to an increased prevalence of fetal cardiovascular malformations, such as ventricular septal defects and pulmonary valve stenosis [114], patent ductus arteriosus [115], and multiple congenital heart defects (Agay-Shay et al. [116]). A meta-analysis of air pollutant-anomaly combinations found that PM10 exposure was related to an increased risk of atrial septal defects [5]. Liu et al. in 2016 reported a positive association between maternal exposure to PM10 during the first 2 months of pregnancy and fetal cardiovascular malformations [117]. Zhu et al. recently evaluated 188,102 live births and reported that exposure to several air pollutants before conception and during early gestation was associated with elevated risk for cleft palate, while cleft lip associated or not with cleft palate was linked only to preconception SO2 exposure [118].

However other studies failed to demonstrate any association between particulate matter and PM10 and cardiovascular malformations [119–121], and no association was found between total suspended particles and congenital heart diseases in Brindisi, Italy [110].

Positive associations between air pollutants and orofacial defects have been reported [122–124], anyway no significant pooled associations were found in meta-analyses [5, 112]. The overall absent associations may be due to the heterogeneity in outcome and exposure evaluation, confounders, and the small number of studies.

Conclusions

Adverse pregnancy outcomes, as preterm birth, low birth weight, gestational hypertensive disorders, and gestational diabetes, are well-recognized public health concern as they can lead to enormous consequences for the children, their families, and the community, with a significant short- and long-term burden to the society. Furthermore, it should be accounted that adverse gestational outcomes are more prevalent and sometimes also more severe in low-income families and in developing countries.

It is well known that children, elderly, and pregnant women are more vulnerable to the effects of air pollution. Embryos and fetuses are particularly subject to damage from air pollutants given the susceptibility of developing organs. Data regarding the potential detrimental effects of air pollution on pregnancy outcomes are continuously growing, providing quite strong evidences of a negative impact. Consistent data from different continents have demonstrated an increased risk of negative gestational outcomes in mothers exposed to air pollution. In particular, air pollutants have been associated with several adverse birth outcomes such as preterm delivery, low birth weight and small for gestational age, and increased risk for gestational diabetes and hypertensive disorders.

Given these data, it is mandatory to provide pregnant women with available knowledge in order to reduce their exposition to known pollutants possibly impairing fetal and mother health. Pregnant women should be counseled about the importance of quitting smoking and to avoid secondhand smoke.

Anyway further studies are required to know the spatial scale and particular windows of vulnerability during pregnancy. Data collection and analysis in this field is particularly difficult as measuring systems may differ and components of air pollution are numerous, and, even if widespread, they can differ by geography depending on local and regional factors such as development, industry, agriculture, and the natural environment. The developing technology in the monitoring systems will help in this field enabling more precise data collection and a more clear understanding of the link between air pollution and pregnancy outcomes. To organize proper prevention measures in the public health system, it is mandatory to know impairing molecules, levels of pollutants, windows of greater susceptibility, and if some pregnant women are more prone than others to the damage caused by air pollution.

References

1. Pineles BL, Park E, Samet JM. Systematic review and meta-analysis of miscarriage and maternal exposure to tobacco smoke during pregnancy. Am J Epidemiol. 2014;179(7):807–23.
2. Tong VT, Dietz PM, Morrow B, D'Angelo DV, Farr SL, Rockhill KM, England LJ, Centers for Disease Control and Prevention (CDC). Trends in smoking before, during, and after pregnancy--pregnancy risk assessment monitoring system, United States, 40 sites, 2000–2010. MMWR Surveill Summ. 2013;62(6):1–19.
3. WHO. Report of the Global Tobacco Epidemic, 2009: implementing smoke-free environments. Geneva: World Health Organization. 2009. http://www.who.int/tobacco/mpower/2009/en/. Accessed 21 Dec 2016.
4. Stieb DM, Chen L, Eshoul M, Judek S. Ambient air pollution, birth weight and preterm birth: a systematic review and meta-analysis. Environ Res. 2012;117:100–11.
5. Vrijheid M, Martinez D, Manzanares S, Dadvand P, Schembari A, Rankin J, Nieuwenhuijsen M. Ambient air pollution and risk of congenital anomalies: a systematic review and meta-analysis. Environ Health Perspect. 2011;119(5):598–606.
6. Proietti E, Röösli M, Frey U, Latzin P. Air pollution during pregnancy and neonatal outcome: a review. J Aerosol Med Pulm Drug Deliv. 2013;26(1):9–23.

7. Regan L, Rai R. Epidemiology and the medical causes of miscarriage. Baillieres Best Pract Res Clin Obstet Gynaecol. 2000;14(5):839–54.
8. García-Enguídanos A, Calle ME, Valero J, Luna S, Domínguez-Rojas V. Risk factors in miscarriage: a review. Eur J Obstet Gynecol Reprod Biol. 2002;102(2):111–9.
9. Hansteen IL, Kjuus H, Fandrem SI. Spontaneous abortions of known karyotype related to occupational and environmental factors: a case-referent study. Int J Occup Environ Health. 1996;2(3):195–203.
10. Hemminki K, Niemi ML. Community study of spontaneous abortions: relation to occupation and air pollution by sulfur dioxide, hydrogen sulfide, and carbon disulfide. Int Arch Occup Environ Health. 1982;51(1):55–63.
11. Green RS, Malig B, Windham GC, Fenster L, Ostro B, Swan S. Residential exposure to traffic and spontaneous abortion. Environ Health Perspect. 2009;117(12):1939–44.
12. Faiz AS, Rhoads GG, Demissie K, Kruse L, Lin Y, Rich DQ. Ambient air pollution and the risk of stillbirth. Am J Epidemiol. 2012;176:308–16.
13. Mohorovic L, Petrovic O, Haller H, Micovic V. Pregnancy loss and maternal methemoglobin levels: an indirect explanation of the association of environmental toxics and their adverse effects on the mother and the fetus. Int J Environ Res Public Health. 2010;7:4203–12.
14. Enkhmaa D, Warburton N, Javzandulam B, Uyanga J, Khishigsuren Y, Lodoysamba S, Enkhtur S, Warburton D. Seasonal ambient air pollution correlates strongly with spontaneous abortion in Mongolia. BMC Pregnancy Childbirth. 2014;14:146.
15. Di Ciaula A, Bilancia M. Relationships between mild PM10 and ozone urban air levels and spontaneous abortion: clues for primary prevention. Int J Environ Health Res. 2015;25(6): 640–55.
16. Perin PM, Maluf M, Czeresnia CE, Januario DANF, Saldiva PHN. Impact of short-term preconceptional exposure to particulate air pollution on treatment outcome in couples undergoing in vitro fertilization and embryo transfer (IVF/ET). J Assist Reprod Genet. 2010;27:371–82.
17. Lawn JE, Blencowe H, Waiswa P, et al. Stillbirths: rates, risk factors, and acceleration towards 2030. Lancet. 2016;387:587–603.
18. Confidential Enquiry into Maternal and Child Health (CEMACH). Perinatal mortality 2007: United Kingdom. London: CEMACH. 2009.
19. Glinianaia SV, Rankin J, Bell R, et al. Particulate air pollution and fetal health: a systematic review of the epidemiologic evidence. Epidemiology. 2004;15:36–45.
20. Lacasaña M, Esplugues A, Ballester F. Exposure to ambient air pollution and prenatal and early childhood health effects. Eur J Epidemiol. 2005;20:183–99.
21. Siddika N, Balogun HA, Amegah AK, Jaakkola JJ. Prenatal ambient air pollution exposure and the risk of stillbirth: systematic review and meta-analysis of the empirical evidence. Occup Environ Med. 2016;73(9):573–81.
22. Yoder SR, Thornburg LL, Bisognano JD. Hypertension in pregnancy and women of childbearing age. Am J Med. 2009;122:890–5.
23. Zhang J, Villar J, Sun W, et al. Blood pressure dynamics during pregnancy and spontaneous preterm labor. Am J Obstet Gynecol. 2007;197:162.e1–6.
24. Steegers EA, von Dadelszen P, Duvekot JJ, Pijnenborg R. Pre-eclampsia. Lancet. 2010;376:631–44.
25. American College of Obstetricians and Gynecologists. Task force on hypertension in pregnancy. Hypertension in pregnancy. Report of the American College of Obstetricians and Gynecologists' Task Force on Hypertension in Pregnancy. Obstet Gynecol. 2013;122:1122–31.
26. Duley L. The global impact of pre-eclampsia and eclampsia. Semin Perinatol. 2009;33:130–7.
27. Lo JO, Mission JF, Caughey AB. Hypertensive disease of pregnancy and maternal mortality. Curr Opin Obstet Gynecol. 2013;25:124–32.
28. Hargood JL, Brown MA. Pregnancy-induced hypertension: recurrence rate in second pregnancies. Med J Aust. 1991;154(6):376–7.
29. Sibai B, Dekker G, Kupferminc M. Pre-eclampsia. Lancet. 2005;365(9461):785–99.
30. Nisell H, Lintu H, Lunnell NO, Mollestrom G, Pettersonn E. Blood pressure and renal function seven years after pregnancy complicated by hypertension. BJOG: An Int J Obstet Gynaecol. 1995;102(11):876–81.

31. Sibai B. Preeclampsia as a cause of preterm and late preterm (near-term) births. Semin Perinatol. 2006;13:16–9.
32. Wang IK, Tsai IJ, Chen PC, Liang CC, Chou CY, Chang CT, Kuo HL, Ting IW, Lin CC, Chuang FR, Huang CC, Sung FC. Hypertensive disorders in pregnancy and subsequent diabetes mellitus: a retrospective cohort study. Am J Med. 2012;125:251–7.
33. Hermida RC, Ayala DE, Mojon A, Fernandez JR, Alonso I, Aguilar MF, Ucieda R, Iglesias M. Differences in circadian blood pressure variabilità during gestation between healthy and complicated pregnancies. Am J Hypertens. 2003;16:200–8.
34. Hermida RC, Ayala DE, Iglesias M. Predictable blood pressure variabilità in healthy and complicated pregnancies. Hypertension. 2001;38:736–41.
35. Trogstad L, Magnus P, Stoltenberg C. Pre-eclampsia: risk factors and causal models. Best Pract Res Clin Obstet Gynaecol. 2011;25:329–42.
36. Basile JN, Bloch MJ. Exposure to air pollution increases the incidence of hypertension and diabetes in black women living in Los Angeles. J Clin Hypertens. 2012;14:819.
37. Coogan PF, White LF, Jerrett M, et al. Air pollution and incidence of hypertension and diabetes mellitus in black women living in Los Angeles. Circulation. 2012;125:767–72.
38. Shah AS, Langrish JP, Nair H, et al. Global association of air pollution and heart failure: a systematic review and meta-analysis. Lancet. 2013;382:1039–48.
39. Sorensen M, Hoffmann B, Hvidberg M, Ketzel M, Jensen SS, Andersen ZJ, Tjonneland A, Overvad K, Raaschou-Nielsen O. Long-term exposure to traffic-related air pollution associated with blood pressure and self-reported hypertension in a Danish cohort. Environ Health Perspect. 2012;120:418–24.
40. Eslami B, Malekafzali H, Rastkari N, Rashidi BH, Djazayeri A, Naddafi K. Association of serum concentrations of persistent organic pollutants (POPs) and risk of pre-eclampsia: a case–control study. J Environ Health Sci Eng. 2016;14:17.
41. Hu H, Ha S, Roth J, Kearney G, Talbott EO, Xu X. Ambient air pollution and hypertensive disorders of pregnancy: a systematic review and meta-analysis. Atmos Environ. 2014;97:336–45.
42. van den Hooven EH, de Kluizenaar Y, Pierik FH, et al. Air pollution, blood pressure, and the risk of hypertensive complications during pregnancy: the generation R study. Hypertension. 2011;57:406–12.
43. Lee PC, Roberts JM, Catov JM, Talbott EO, Ritz B. First trimester exposure to ambient air pollution, pregnancy complications and adverse birth outcomes in Allegheny County, PA. Matern Child Health J. 2013;17:545–55.
44. Rudra CB, Williams MA, Sheppard L, Koenig JQ, Schiff MA. Ambient carbon monoxide and fine particulate matter in relation to preeclampsia and preterm delivery in western Washington State. Environ Health Perspect. 2011;119:886–92.
45. Olsson D, Mogren I, Forsberg B. Air pollution exposure in early pregnancy and adverse pregnancy outcomes: a register-based cohort study. BMJ Open. 2013;3(2):e001955.
46. Wu J, Ren C, Delfino RJ, Chung J, Wilhelm M, Ritz B. Association between local traffic-generated air pollution and preeclampsia and preterm delivery in the south coast air basin of California. Environ Health Perspect. 2009;117(11):1773–9.
47. Vinikoor-Imler LC, Gray SC, Edwards SE, Miranda ML. The effects of exposure to particulate matter and neighbourhood deprivation on gestational hypertension. Paediatr Perinat Epidemiol. 2012;26:91–100.
48. Wu M, Ries JJ, Proietti E, Vogt D, Hahn S, Hoesli I. Development of late-onset preeclampsia in association with road densities as a proxy for traffic-relatetd air pollution. Fetal Diagn Ther. 2016;39:21–7.
49. Miranda ML, Edwards SE, Chang H, Richard L, Auten RL. Proximity to roadways and pregnancy outcomes. J Expo Sci Environ Epidemiol. 2013;23:32–8.
50. Yorifuji T, Naruse H, Kashima S, Murakoshi T, Doi H. Residential proximity to major roads and obstetrical complications. Sci Total Environ. 2014;508:188–92.
51. Yorifuji T, Naruse H, Kashima S, Ohki S, Murakoshi T, Takao S, et al. Residential proximity to major roads and preterm births. Epidemiology. 2011;22(1):74–80.

52. Pedersen M, Halldorsson TI, Olsen SF, Hjortebjerg D, Ketzel M, Grandstrom C, Rashou-Nielsen O, Serensen M. Impact of road traffic pollution on pre-eclampsia and pregnancy-induced hypertensive disorders. Epidemiology. 2017;28:99–106.
53. Xu X, Hu H, Ha S, Roth J. Ambient air pollution and hypertensive disorder of pregnancy. J Epidemiol Community Health. 2014;68:1–19.
54. Mobasher Z, Salam MT, Goodwin TM, Lurmann F, Ingles SA, Wilson ML. Associations between ambient air pollution and hypertensive disorders of pregnancy. Environ Res. 2013;123:9–16.
55. Mannisto T, Mendola P, Liu D, Leishear K, Sherman S, Laughon SK. Acute air pollution exposure and blood pressure at delivery among women with and without hypertension. Am J Hypertens. 2015;28:58–72.
56. Malmqvist E, Jakobsson K, Tinnerberg H, Rignell-Hydbom A, Rylander L. Gestational diabetes and preeclampsia in association with air pollution at levels below current air quality guidelines. Environ Health Perspect. 2013;121(4):488–93.
57. Pedersen M, Stayner L, Slama R, Sorensen M, Figueras F, Nieuwenhuijsen MJ, Raaschou-Nielsen O, Dadvand P. Ambient air pollution and pregnancy-induced hypertensive disorders: a systematic review and metaanalysis. Hypertension. 2014;64:494–500.
58. Pereira G, Haggar F, Shand AW, Bower C, Cook A, Nassa N. Association between pre-eclampsia and locally derived traffic-related air pollution: a retrospective cohort study. J Epidemiol Community Health. 2013;67:147–52.
59. Mendola P, Wallace M, Liu D, Robledo C, Mannisto T, Grantz K. Air pollution exposure and preeclampsia among US women with and without asthma. Environ Res. 2016;148:248–55.
60. Ballester F, Estarlich M, Iñiguez C, Llop S, Ramón R, Esplugues A, Lacasaña M, Rebagliato M. Air pollution exposure during pregnancy and reduced birth size: a prospective birth cohort study in Valencia, Spain. Environ Health. 2010;9:6.
61. Madsen C, Gehring U, Walker SE, et al. Ambient air pollution exposure, residential mobility and term birth weight in Oslo, Norway. Environ Res. 2010;110:363–71.
62. Olsson D, Ekstrom M, Forsberg B. Temporal variation in air pollution concentrations and preterm birth—a population based epidemiological study. Int J Environ Res Public Health. 2012;9:272–85.
63. Galtier F. Definition, epidemiology, risk factors. Diabetes Metab. 2010;36(6 Pt 2):628–51.
64. Buchanan TA, Xiang AH, Page KA. Gestational diabetes mellitus: risks and management during and after pregnancy. Nat Rev Endocrinol. 2012;8(11):639–49.
65. Rajagopalan S, Brook RD. Air pollution and type 2 diabetes: mechanistic insights. Diabetes. 2012;61(12):3037–45.
66. Eze IC, Hemkens LG, Bucher HC, Hoffmann B, Schindler C, Kunzli N, et al. Association between ambient air pollution and diabetes mellitus in Europe and North America: systematic review and meta-analysis. Environ Health Perspect. 2015;123(5):381–9.
67. Fleisch AF, Gold DR, Rifas-Shiman SL, Koutrakis P, Schwartz JD, Kloog I, et al. Air pollution exposure and abnormal glucose tolerance during pregnancy: the project Viva cohort. Environ Health Perspect. 2014;122(4):378–83.
68. Hu H, Ha S, Henderson BH, Warner TD, Roth J, Kan H, et al. Association of atmospheric particulate matter and ozone with gestational diabetes mellitus. Environ Health Perspect. 2015;123(9):853–9.
69. Robledo CA, Mendola P, Yeung E, Mannisto T, Sundaram R, Liu D, et al. Preconception and early pregnancy air pollution exposures and risk of gestational diabetes mellitus. Environ Res. 2015;137:316–22.
70. van den Hooven EH, Jaddoe VW, de Kluizenaar Y, Hofman A, Mackenbach JP, Steegers EA, Miedema HM, Pierik FH. Residential traffic exposure and pregnancy-related outcomes: a prospective birth cohort study. Environ Health. 2009;8:59.
71. Fleisch AF, Kloog I, Luttmann-Gibson H, Gold DR, Oken E, Schwartz JD. Air pollution exposure and gestational diabetes mellitus among pregnant women in Massachusetts: a cohort study. Environ Health. 2016;15:40.

72. Rundle A, Hoepner L, Hassoun A, Oberfield S, Freyer G, Holmes D, Reyes M, Quinn J, Camann D, Perera F, Whyatt R. Association of childhood obesity with maternal exposure to ambient air polycyclic aromatic hydrocarbons during pregnancy. Am J Epidemiol. 2012;175(11):1163–72.
73. Fleisch AF, Rifas-Shiman SL, Koutrakis P, et al. Prenatal exposure to traffic pollution: associations with reduced fetal growth and rapid infant weight gain. Epidemiology. 2015;26: 43–50.
74. Howson CP, Kinney M, Lawn JE, editors. March of Dimes, PMNCH, Save the Children, WHO. Born Too Soon: The global action report on preterm birth. Geneva: World Health Organization; 2012.
75. Saigal S, Doyle LW. An overview of mortality and sequelae of preterm birth from infancy to adulthood. Lancet. 2008;371(9608):261–9.
76. Vadillo-Ortega F, Osornio-Vargas A, Buxton MA, Sánchez BN, Rojas-Bracho L, Viveros-Alcaráz M, Castillo-Castrejón M, Beltrán-Montoya J, Brown DG, O'Neill MS. Air pollution, inflammation and preterm birth: a potential mechanistic link. Med Hypotheses. 2014;82(2): 219–24.
77. Slama R, Darrow L, Parker J, Woodruff TJ, Strickland M, Nieuwenhuijsen M, Glinianaia S, Hoggatt KJ, Kannan S, Hurley F, Kalinka J, Srám R, Brauer M, Wilhelm M, Heinrich J, Ritz B. Meeting report: atmospheric pollution and human reproduction. Environ Health Perspect. 2008;116(6):791–8.
78. Kloog I, Melly SJ, Ridgway WL, Coull BA, Schwartz J. Using new satellite based exposure methods to study the association between pregnancy PM2.5 exposure, premature birth and birth weight in Massachusetts. Environ Health. 2012;11:40.
79. Pereira G, Belanger K, Ebisu K, Bell ML. Fine particulate matter and risk of preterm birth in Connecticut in 2000-2006: a longitudinal study. Am J Epidemiol. 2014;179(1):67–74.
80. Généreux M, Auger N, Goneau M, Daniel M. Neighbourhood socioeconomic status, maternal education and adverse birth outcomes among mothers living near highways. J Epidemiol Community Health. 2008;62(8):695–700.
81. Brauer C, Budtz-Jørgensen E, Mikkelsen S. Structural equation analysis of the causal relationship between health and perceived indoor environment. Int Arch Occup Environ Health. 2008;81(6):769–76.
82. Malmqvist E, Rignell-Hydbom A, Tinnerberg H, Björk J, Stroh E, Jakobsson K, Rittner R, Rylander L. Maternal exposure to air pollution and birth outcomes. Environ Health Perspect. 2011;119(4):553–8.
83. Trasande L, Wong K, Roy A, Savitz DA, Thurston G. Exploring prenatal outdoor air pollution, birth outcomes and neonatal health care utilization in a nationally representative sample. J Expo Sci Environ Epidemiol. 2013;23(3):315–21.
84. Wilhelm M, Ghosh JK, Su J, Cockburn M, Jerrett M, Ritz B. Traffic-related air toxics and preterm birth: a population-based case-control study in Los Angeles County, California. Environ Health. 2011;10:89.
85. Bell ML, Ebisu K, Belanger K. Ambient air pollution and low birth weight in Connecticut and Massachusetts. Environ Health Perspect. 2007;115(7):1118–24.
86. Darrow LA, Klein M, Flanders WD, Waller LA, Correa A, Marcus M, Mulholland JA, Russell AG, Tolbert PE. Ambient air pollution and preterm birth: a time-series analysis. Epidemiology. 2009;20(5):689–98.
87. Johnson S, Bobb JF, Ito K, Savitz DA, Elston B, Shmool JL, Dominici F, Ross Z, Clougherty JE, Matte T. Ambient fine particulate matter, nitrogen dioxide, and preterm birth in New York City. Environ Health Perspect. 2016;124(8):1283–90.
88. Laurent O, Hu J, Li L, Kleeman MJ, Bartell SM, Cockburn M, Escobedo L, Wu J. A statewide nested case-control study of preterm birth and air pollution by source and composition: California, 2001–2008. Environ Health Perspect. 2016;124(9):1479–86.
89. Yang CY, Chang CC, Chuang HY, Ho CK, Wu TN, Chang PY. Increased risk of preterm delivery among people living near the three oil refineries in Taiwan. Environ Int. 2004;30(3):337–42.
90. Yang CY, Chang CC, Tsai SS, Chuang HY, Ho CK, Wu TN, Sung FC. Preterm delivery among people living around Portland cement plants. Environ Res. 2003;92(1):64–8.

91. Huppé V, Kestens Y, Auger N, Daniel M, Smargiassi A. Residential proximity to gasoline service stations and preterm birth. Environ Sci Pollut Res Int. 2013;20(10):7186–93.
92. Parker JD, Mendola P, Woodruff TJ. Preterm birth after the Utah Valley Steel Mill closure: a natural experiment. Epidemiology. 2008;19(6):820–3.
93. Mohorovic L. First two months of pregnancy--critical time for preterm delivery and low birthweight caused by adverse effects of coal combustion toxics. Early Hum Dev. 2004;80(2): 115–23.
94. Wylie BJ, Coull BA, Hamer DH, Singh MP, Jack D, Yeboah-Antwi K, Sabin L, Singh N, MacLeod WB. Impact of biomass fuels on pregnancy outcomes in central East India. Environ Health. 2014;13(1):1.
95. Wallace ME, Grantz KL, Liu D, Zhu Y, Kim SS, Mendola P. Exposure to ambient air pollution and premature rupture of membranes. Am J Epidemiol. 2016;183(12):1114–21.
96. Hao H, Chang HH, Holmes HA, Mulholland JA, Klein M, Darrow LA, Strickland MJ. Air pollution and preterm Birth in the U.S. State of Georgia (2002–2006): associations with concentrations of 11 ambient air pollutants estimated by combining community multiscale air quality model (CMAQ) simulations with stationary monitor measurements. Environ Health Perspect. 2016;124(6):875–80.
97. Martin JA, Hamilton BE, Ventura SJ, Osterman MJ, Kirmeyer S, Mathews TJ, Wilson EC. Births: final data for 2009. Natl Vital Stat Rep. 2011;60(1):1–70.
98. World Health Organization. Regional consultation towards the development of a strategy for optimizing fetal growth and development. Cairo: World Health Organization; 2005.
99. World Health Organisation. Global nutrition targets 2025: low birth weight policy brief (WHO/NMH/NHD/14.5). Geneva: World Health; 2014.
100. Barker DJ, Eriksson JG, Forsén T, Osmond C. Fetal origins of adult disease: strength of effects and biological basis. Int J Epidemiol. 2002;31(6):1235–9.
101. Valavanidis A, Fiotakis K, Vlachogianni T. Airborne particulate matter and human health: toxicological assessment and importance of size and composition of particles for oxidative damage and carcinogenic mechanisms. J Environ Sci Health C Environ Carcinog Ecotoxicol Rev. 2008;26(4):339–62.
102. Janssen BG, Byun HM, Gyselaers W, Lefebvre W, Baccarelli AA, Nawrot TS. Placental mitochondrial methylation and exposure to airborne particulate matter in the early life environment: an ENVIRONAGE birth cohort study. Epigenetics. 2015;10(6):536–44.
103. Stapleton PA, Nichols CE, Yi J, McBride CR, Minarchick VC, Shepherd DL, Hollander JM, Nurkiewicz TR. Microvascular and mitochondrial dysfunction in the female F1 generation after gestational TiO2 nanoparticle exposure. Nanotoxicology. 2015;9(8):941–51.
104. Backes CH, Nelin T, Gorr MW, Wold LE. Early life exposure to air pollution: how bad is it? Toxicol Lett. 2013;216(1):47–53.
105. Estarlich M, Ballester F, Aguilera I, Fernández-Somoano A, Lertxundi A, Llop S, Freire C, Tardón A, Basterrechea M, Sunyer J, Iñiguez C. Residential exposure to outdoor air pollution during pregnancy and anthropometric measures at birth in a multicenter cohort in Spain. Environ Health Perspect. 2011;119(9):1333–8.
106. Dadvand P, Parker J, Bell ML, Bonzini M, Brauer M, Darrow LA, Gehring U, Glinianaia SV, Gouveia N, Ha EH, Leem JH, van den Hooven EH, Jalaludin B, Jesdale BM, Lepeule J, Morello-Frosch R, Morgan GG, Pesatori AC, Pierik FH, Pless-Mulloli T, Rich DQ, Sathyanarayana S, Seo J, Slama R, Strickland M, Tamburic L, Wartenberg D, Nieuwenhuijsen MJ, Woodruff TJ. Maternal exposure to particulate air pollution and term birth weight: a multi-country evaluation of effect and heterogeneity. Environ Health Perspect. 2013a;121(3):267–373.
107. Dadvand P, Figueras F, Basagana X, Beelen R, Martinez D, Cirach M, Schembari A, Hoek G, Brunekreef B, Nieuwenhuijsen MJ. Ambient air pollution and preeclampsia: a spatiotemporal analysis. Environ Health Perspect. 2013b;121:1365.
108. Kumar N. The exposure uncertainty analysis: the association between birth weight and trimester specific exposure to particulate matter (PM2.5 vs. PM10). Int J Environ Res Public Health. 2016;13(9):906.

109. Malmqvist E, Liew Z, Källén K, Rignell-Hydbom A, Rittner R, Rylander L, Ritz B. Fetal growth and air pollution—a study on ultrasound and birth measures. Environ Res. 2017;152: 73–80.
110. Gianicolo EA, Mangia C, Cervino M, Bruni A, Andreassi MG, Latini G. Congenital anomalies among live births in a high environmental risk area--a case-control study in Brindisi (southern Italy). Environ Res. 2014;128:9–14.
111. Vinikoor-Imler LC, Stewart TG, Luben TJ, Davis JA, Langlois PH. An exploratory analysis of the relationship between ambient ozone and particulate matter concentrations during early pregnancy and selected birth defects in Texas. Environ Pollut. 2015;202:1–6.
112. Chen EK, Zmirou-Navier D, Padilla C, Deguen S. Effects of air pollution on the risk of congenital anomalies: a systematic review and meta-analysis. Int J Environ Res Public Health. 2014;11(8):7642–68.
113. Farhi A, Boyko V, Almagor J, Benenson I, Segre E, Rudich Y, Stern E, Lerner-Geva L. The possible association between exposure to air pollution and the risk for congenital malformations. Environ Res. 2014;135:173–80.
114. Padula AM, Tager IB, Carmichael SL, Hammond SK, Yang W, Lurmann F, Shaw GM. Ambient air pollution and traffic exposures and congenital heart defects in the San Joaquin Valley of California. Paediatr Perinat Epidemiol. 2013;27(4):329–39.
115. Strickland MJ, Klein M, Correa A, Reller MD, Mahle WT, Riehle-Colarusso TJ, Botto LD, Flanders WD, Mulholland JA, Siffel C, Marcus M, Tolbert PE. Ambient air pollution and cardiovascular malformations in Atlanta, Georgia, 1986–2003. Am J Epidemiol. 2009;169(8):1004–14.
116. Agay-Shay K, Friger M, Linn S, Peled A, Amitai Y, Peretz C. Air pollution and congenital heart defects. Environ Res. 2013;124:28–34.
117. Liu CB, Hong XR, Shi M, Chen XQ, Huang HJ, Chen JH, Yang K, Chen SQ, Chen HQ, Kan HD, Sun QH. Effects of prenatal PM10 exposure on fetal cardiovascular malformations in Fuzhou, China: a retrospective case-control study. Environ Health Perspect. 2016.; [Epub ahead of print]
118. Zhu Y, Zhang C, Liu D, Grantz KL, Wallace M, Mendola P. Maternal ambient air pollution exposure preconception and during early gestation and offspring congenital orofacial defects. Environ Res. 2015;140:714–20.
119. Ritz B, Yu F, Fruin S, Chapa G, Shaw GM, Harris JA. Ambient air pollution and risk of birth defects in Southern California. Am J Epidemiol. 2002;155(1):17–25.
120. Schembari A, Nieuwenhuijsen MJ, Salvador J, de Nazelle A, Cirach M, Dadvand P, Beelen R, Hoek G, Basagaña X, Vrijheid M. Traffic-related air pollution and congenital anomalies in Barcelona. Environ Health Perspect. 2014;122(3):317–23.
121. Stingone JA, Wing S. Poultry litter incineration as a source of energy: reviewing the potential for impacts on environmental health and justice. New Solut. 2011;21(1):27–42.
122. Gilboa SM, Mendola P, Olshan AF, Langlois PH, Savitz DA, Loomis D, Herring AH, Fixler DE. Relation between ambient air quality and selected birth defects, seven county study, Texas, 1997–2000. Am J Epidemiol. 2005;162(3):238–52.
123. Hwang BF, Jaakkola JJ. Ozone and other air pollutants and the risk of oral clefts. Environ Health Perspect. 2008;116(10):1411–5.
124. Marshall EG, Harris G, Wartenberg D. Oral cleft defects and maternal exposure to ambient air pollutants in New Jersey. Birth Defects Res A Clin Mol Teratol. 2010;88(4):205–15.
125. Checa Vizcaíno MA, González-Comadran M, Jacquemin B. Outdoor air pollution and human infertility: a systematic review. Fertil Steri. 2016;106(4):897–904.

Model for Pollutant and Disease Monitoring

27

Giovanni Rinaldi and Samuele Rinaldi

27.1 Introduction

Epidemiologic studies have demonstrated that traffic-related air pollutants, often indexed by proximity to roadways, have been associated with adverse reproductive outcomes [1], childhood cancers [2], and adult mortality [3]. Growing evidence suggests similar associations with allergies and asthma [4].

All over the world, the air pollutant causing most deaths is the particulate matter with an aerodynamic diameter $\leq$ 10 μm (PM10). Exposure to PM10 in Europe has been associated with increased all-cause, respiratory and cardiovascular mortality, but adverse health effects can occur at lower concentrations [5]. This evidence has been used to develop air quality standards for health protection [6].

In this work is described an approach model to treat complex events in the individual lives, such as complex diseases and the adverse effects caused by pollution, starting from an holistic and comprehensive point of view, integrating the science of different disciplines and taking advance by the recent technological improvements. They affect the way to collect data, to manage it in suitable containers (different from the traditional ones) and to impose new methods for the analysis.

The starting point is the availability of huge amount of structured, unstructured and in many cases geolocated data; the Internet of Things (IoT) technology has allowed the availability (conscious or unconscious, voluntary or involuntary) of data from each type of sensors, making in some cases easier the validation of mathematical models; data from users (appropriately validated and processed) produces useful information, for example, in medicine in which traditional static medical records acquire dynamic functions in order to be compliant with new medical models treating different types of data. Localization on the territory of events (natural, anthropogenic, social, connected to the

G. Rinaldi (✉) • S. Rinaldi
QI International, Ltd., London, UK
e-mail: rinaldi8giovanni@gmail.com

F. Capello, A.V. Gaddi (eds.), *Clinical Handbook of Air Pollution-Related Diseases*,
https://doi.org/10.1007/978-3-319-62731-1_27

health) or models allows to study their modification in time and space adding information for further analysis or for a better understanding and knowledge. These data that in some cases become huge in quantity produce information only if well treated: the usage of mathematical models is fundamental to extract information and produce scenarios not only for future predictions but also for better understanding the current situation. The result of physical events, the overlapping with people's health status or social events, their modification during the time on the territory, the epidemiology analysis and variation in time and the prediction models from mathematical modelling are more effective if represented through maps, if visualisation techniques are used and if innovative approaches to information visualisation are proposed.

To explore and understand the complex relationships between the environment and health, a range of disciplines must be engaged to bring expertise on discrete portions of the project.

Whereas these disciplines are currently handled separately, we propose a coherent and comprehensive production of information in which physicians, policy makers, modellers and physics and engineers could in a co-production work produce real scenarios for improving people's health and lifestyle.

27.2 Technology for Health and Pollution Monitoring

Environmental Protection Agency (EPA) of the USA (https://www.epa.gov/p2) in the official website makes available a set of pamphlets which explain the harmful effects of different pollutants on health. According to the American College of Cardiology, American Heart Association, American Stroke Association and other National Medical Associations, they affirm that pollutants have dangerous effects for the health of people and for people suffering of chronic diseases such as cardiovascular disease, diabetes, pulmonary diseases, blood cholesterol, asthma and chronic bronchitis. Pollutants produce serious effects (https://www3.epa.gov/airtoxics/3_90_022.html) such as birth defects, kidney and liver damage, skin rash and cancer. Pollutants can trigger heart attacks, stroke and irregular heart rhythms especially in people who are already at risk for these conditions.

But these effects reported by EPA are not new to the medical research. Since the early 1990s, there is growing evidence from epidemiological studies of a relationship between air pollution and illness (not only respiratory) and mortality [7].

A substantial number of epidemiological studies report associations between mortality/morbidity and air pollution levels [8–10].

At this stage we wonder what are the instruments used to manage information related to the effects of pollutants on the patient's health.

Among the traditional tools used by clinicians to manage personal health information, the Electronic Health Record (EHR), we haven't found any section regarding the effect of pollution on the illness or how lifestyles are influenced by pollution. This depends also on by a scarce ductility of the traditional EHRs. Despite the results of the medical research, it seems that the enhanced effects of pollution on chronic disease or generally on the health of population are not taken into account in the practice.

We have also to consider that the home monitoring of chronic disease such as cardiovascular disease, diabetes, chronic obstructive pulmonary disease (COPD) and others have reached a high level of maturity [11]. A set of instruments for the home monitoring for each disease has been realized by the industry, and in real cases they are proposed by clinicians and full adopted by patients [12] (in literature we have a long list of projects).

For example, for home monitoring cardiac diseases, the tools generally used are electrocardiograph, scales, pulse oximeter and blood pressure monitor [13].

Data are collected using a specific device at home (e.g. a PDA) and periodically sent to a medical centre where doctors analyse data and continuously keep monitored the health patient state. From the 1990s in the EU, the USA and all over the world, a set of project has been implemented with these logics.

The fundamental concept linked to these projects is not only the home usage of telemedicine devices but the integration of information in the patient EHR, so that the doctor can assess in overall manner the patient's state of health.

To better evaluate the health status of people suffering from chronic disease according to the pollutant effects, we propose to associate to the traditional monitoring tools also methods and tools able to evaluate and monitor the amount of the pollutant in the home air. Instruments for monitoring the most dangerous air pollutant are at the moment expensive (the level of SOx, NOx, COx, PMx, VOC, etc. is traditionally evaluated in many cities through detection control units placed at critical points). So we propose two modalities.

The first is the acquisition of a set of tools monitoring also some pollutants at home: CO and PM (e.g. two of the most devices sold at present). Today the market offers these monitoring instruments at the cost which is around $ 100, but we are confident that when the request will rise, the cost will fall. At the moment the real market of these tools is the Far East (China above all), also if in the West countries, the awareness to know the air condition we breathe is growing. In fact the website of EPA proposes a list of current devices for monitoring the air pollution at home that are on the market, and people can buy.

Of course these monitoring devices must be connected to the health monitoring set in order to send to the medical centre also the data from the air condition.

The second and at this time more realistic method is the indirect analysis performed by models using data from institutional sites.

The Air Quality Index (AQI) is an index for reporting daily air quality. It tells population how clean or unhealthy the air is, and what associated health effects might be a concern. The importance of AQI for the aims here treated is that it focuses on health effects people experience within a few hours or days after breathing unhealthy air. AQI can be calculated in different manner by different countries. Generally AQI is calculated by a set of major pollutants (in the USA, e.g. it is calculated by ground-level ozone, particle pollution, carbon monoxide, sulphur dioxide; see EPA website).

The purpose of the AQI is to help people to understand what local air quality means to people's health.

The AQI levels distinguish the effects for who sufferers of chronic diseases with healthy people; what it is dangerous for sick could still not be for a healthy.

In this option we suggest a connection between the centre that calculates the AQI level and the individual suffering chronic disease so that the clinical centre which evaluates the clinical monitoring data could know in real time the dangerous effect on the health caused by the worsening of the conditions of pollution.

In addition to monitoring the air pollution level at patient's home, we suggest also the devices to clean the home air. They must be switched-on when the air pollutant level exceeds the hazard levels for the patient that undergoes an enhanced pollution effect. Also in this case, we must note that at the moment, the market offers air cleaning devices at relatively high cost; it is around 100/150 $.

This opportunity to know the AQI level can be considered as an alerting system in order to know the air pollution level and in order to act accordingly by activating air cleaning systems.

Moreover, the effect of the pollution can be more dangerous for people who exercise. We know that physical activity is not useful only by healthy people, but it is strongly indicated (moderately and at the level recommended for the degree of disease that you suffer) to people who suffer from chronic cardiovascular disease and elderly.

The same alerting system previously described can be integrated into the exercise bikes and treadmills with the aim to make available the AQI index to people exercising in order to know if it is better exercising outdoors or at home.

At last, we want to emphasize that modern technology can modify the approach to face the dangerous effect of the pollution to the people's health, above all for the people affected by chronic disease, which may undergo worsening.

In addition to traditional monitoring systems of clinical parameters, we propose the acquisition of devices for the measurement of the air quality or alternatively a centralized system that communicates to patient the AQI evaluated in the zone in which the home is located. This monitoring has the aim to alert patients in order to suggest the switching-on devices for cleaning the air. The same type of monitoring can be applied to home device for exercise to suggest physical activities at home or outside.

These new devices of alerting, monitoring and air cleaning are enclosed in the home automation technology.

But it is worth to note that this proposal makes sense within an integrated system in which data can be managed in comprehensive EHR and in which it created a network where data about pollutant level are acquired and managed also through models and directed to those who need it.

This brief analysis led us to consider also a new model of treating health data.

This new type of information we have documented in pollution related to health has created a division between the accumulation of knowledge and the reliable delivery of quality healthcare. Present medical records suffer staticity and are inadequate to treat different types of information ensuring consistency between the information in the capacity of integration and synthesis. Even if EHR managers and developers have tried to improve in timeliness and accessibility of clinical information, most tools do not maximize performance at the point of care as, for example, data coming monitoring devices and telemedicine systems. EHR development has struggled to keep pace with advances in medical knowledge and information technology, but the results are not convincing. Most EHRs use an encounter-oriented ontology similar to the paper chart so replicating many of the paper chart usability issues. It can be difficult for clinicians at the point of care to assemble coherent narrative of illness or therapeutic interventions. In particular the ability to manage narrative (retrieve, analyse, treat) becomes important for complex pathologies and where social, natural or anthropic events affects the individual's health. Moreover there are a lot of usability barriers such as distributed display screens, poor navigational aids, cumbersome input dialogs and confusing on-screen tools that affect the traditional medical records. But in addition to problems in the representation and visualisation of complex data, we want at the moment to focus on the poor tools of data acquisition from monitoring devices or telemedicine tools and the difficulty to organize data in a coherent manner regarding external events (like pollution level, lifestyles, anthropic events). This would allow to produce real health scenarios, personalized for the patients that suffer illnesses (chronic or not) that live in a place with a certain level of pollution, which have a definite genetic predisposition, which have certain lifestyle and so on.

The next generation of EHR, according to the challenge described in this work, must be sensitive to healthcare's complex and culturally textured sociotechnical systems, seamlessly integrating with legacy information technology and workflow process. They must collect, organize and display covering information streams in a coherent manner. So we propose that medical data must be treated in a sort of layers in order to allow completive views or only extracting information on the layers of interests and analysing the correlation between information among the layers.

Consequently, always to meet the challenges that pollution proposes to health, the interface must leverage the best of new technology including multimedia, summative reporting and machine learning.

27.3 Spatial Analysis and GIS: Models in Health Context

What is geographical information system (GIS)? Why can GIS support analysis in health context and in a specific way about pollution related problems? GIS is a powerful tool that can analyse event located on the territory and how they evolve in the time and space. Its fundamental characteristic is the spatial analysis that establishes a relationship between entity spatially defined with its specific information with the purpose of mining and extracting hidden information. Adding the time dimension, it is possible to study and predict the temporal evolution of anthropic, physical, natural or social events.

One of the important and powerful features of GIS is the capability to organize information in logical layers in order to perform analytical and comprehensive analysis.

The breakdown of physical objects in overlapping layers is a facility in order to execute mathematical analysis and physical modelling and at the same time to offer a graphic representation of the selected events during time.

But in addition to the ability to make visible events located on the territory, it is important to note that the GIS tool offers great opportunities to perform modelling of events linking different types of data, typically treated through separate tools.

In the health context, spatial analysis performed by GIS tools is not a new discipline. The type of analysis ranges from the localization of health services in order to evaluate the location of health facilities in relation to population and how to get these services to epidemiological analysis based on the social characteristics of the inhabitants and temporal evolution of local changes.

Air quality is important to our health and environment, but sources of contamination are often difficult to monitor over time and through the changes in the social structure, as well as the effects on the population. GIS technology manages statistical and spatial data to shows the relationship between poor air quality and occurrences of deficient human and environmental health. In this way, a GIS aids to know the state of population health through the monitoring of pollutant emissions and the effects on the population.

GIS has the capability to:

- Examine attributes of individual features or groups of features and their spatial relationships to other features.
- Query the underlying data to analyse the region and export selected data for manipulation in other software.
- Explore the nature of the data through thematic maps.

The last point seems particularly important because it introduces the concept of geospatial visual analytics, a method to facilitate the integration between users and data providing tools for dynamic display and suitable visual interaction opportunities with the data that can support analytical reasoning and the exploration of data from multiple user-customisable aspects, for example, in the area of human health, surveillance, emergency management and epidemiology [14].

The usage of GIS and spatial analysis technique for the analysis of the air quality and the effect of air pollution on the territory is not a new issue [15, 16]. In these works GIS was used as analysis tool in which there were integrated algorithms for the description of the pollution sources (due to transportation, residential and industrial sector), considering also time modification and meteorological conditions and local entities. Emission sources, proximity analysis and environmental sensitivity were located and displayed producing different scenarios, current and as a prediction of the effects. This allowed to produce thematic maps showing the overall emission inventory (e.g. SOx, NOx, CO, VOC, PMx, etc.), to compare directly different zones in the area of concern and to identify the critical ones, regarding each single air pollutant.

The approach to modelling impact is based on the concept of integration. Environmental problems as air pollutant emission analysis are effectively treated through spatial analysis technique, but they also need of modelling, simulator and suitable data preparation methods.

Following this pioneering phase (which unfortunately has not always produced the desired results in terms of usage of the tools), what is new at present regards mainly the increasing of the monitoring station located on the territory.

By leveraging on the availability of accurate georeferenced data, the detection units of pollutants are located at strategic points of the city, the forecast models are more accurate, and instantly it is possible to obtain the map of pollution sources and the dispersion prediction.

Figure 27.1 summarizes in a schematic view the associations between spatial analysis tools distinctive of GIS and mathematical models.

These data must be crossed with the population data and the health of the individual to assess the personal effects of pollution.

As discussed previously, EHRs with the addition of information about genomics and lifestyles contain these personal information.

But again we must note the lack of the use of such tools in practice, integrated with clinical assessment tools of disease.

According to the World Health Organization (WHO) "Environmental health addresses all the physical, chemical, and biological factors external to a person, and all the related factors impacting behaviours. It encompasses the assessment and control of those environmental factors that can potentially affect health. It is targeted towards preventing disease and creating health-supportive environments". [17]. The impact of the environment on human health is substantial. According to WHO, environmental hazards are responsible for approximately 25% of the total burden of disease. WHO also states that as many as 13 million deaths can be prevented every year by making our environments healthier [18].

As we know researchers will continue to explore the complex, multidimensional relationships between pollution and disease. Many of the increasing mentions of GIS in peer-reviewed health literature will make their way into environmental public health practice. A recent Medline search for "geographic information systems" from 1990 to 2009 revealed a total of 3621 results.

Due to the considerable impact of environment on human health, we need tools which aim to:

- Map the emission sources responsible for pollution (in air, water and soil)
- Evaluate the effect on the population located on the territory according to the meteorological conditions and ground structure
- Implement epidemiological analysis and personal health analysis based on the individual
- Formulate plans according to forecasts

Sources of pollutants, population with associated services and facilities, meteorological conditions and segmentation of the diseases localized in the territory are all physical entities which are possible to georeferencing on virtual layers using GIS facilities.

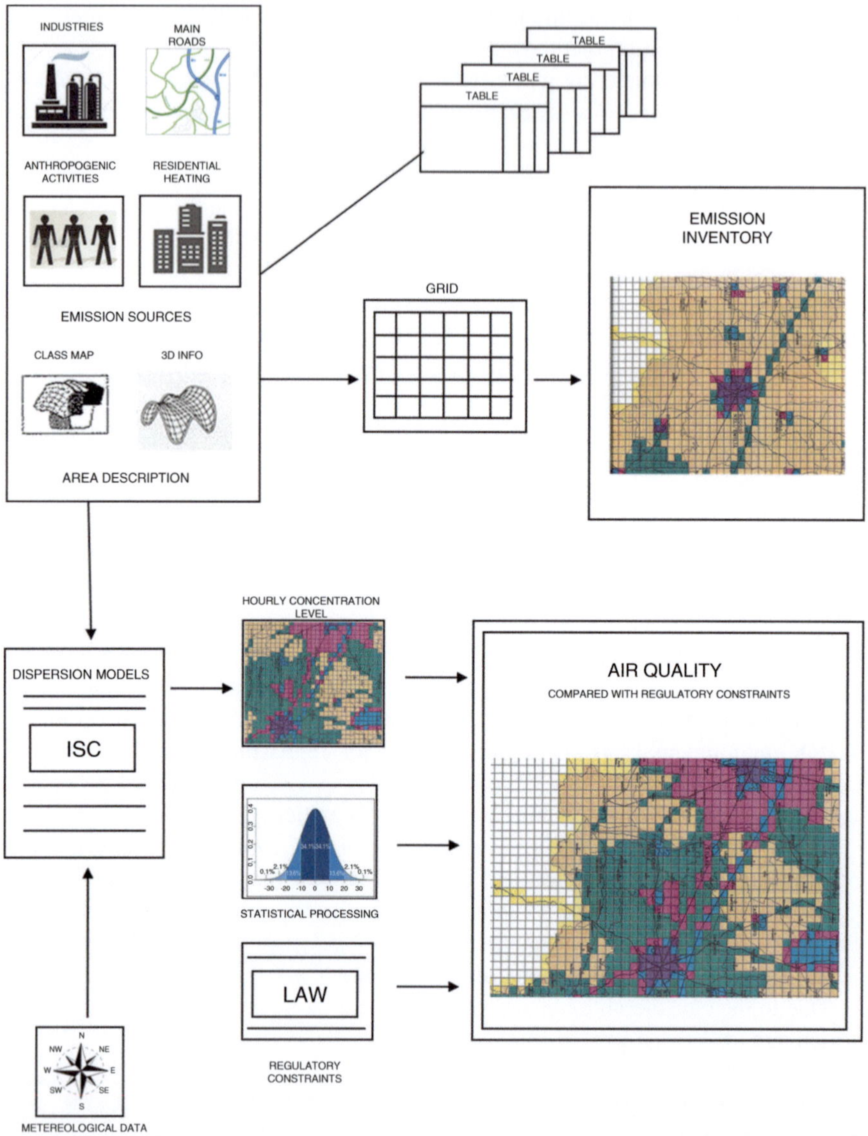

Fig. 27.1 A model for the emission inventory and forecast analysis of pollutant dispersion on the territory

Mathematical models associated to these features enrich the spatial analysis allowing to create models regarding the emission, the forecast of emission and the effect on the population exposition during the time.

These information must be associated to the individual medical records.

They are composed by: data from health events (health services, interventions), data from chronic diseases, information from encounters, prescriptions and medicine taken, genetics data, lifestyles and in some cases data from monitoring home devices.

At the moment additional GIS models include mostly environmental health surveillance and epidemiology, exposure surveillance, laboratory services (e.g. testing for food-borne illnesses, lead), emergency response (e.g. chemical spills) and outdoor air pollution control.

Many researchers and some agencies are focusing on how the built environment influences human health and are taking a more active role in land use planning via public health impact statements and tools.

Despite extensive data collection, we need further consistent analysis of the links between the environmental exposures and health outcomes (such as asthma, birth defects, cancers, lead poisoning and myocardial infarctions). GIS starts out as a stand-alone analytic tool for environmental health investigation, health planning or epidemiologic research, but spatial analysis capability linked to medical records can allow deep analysis about the effect of pollution on the individual health and obviously on the population health.

"Many GIS-based projects have been successful in supporting public and environmental health practice, including those investigating toxic exposure, vector-borne disease, health information access, and the built environment" [19].

GIS during the last years has proved an essential scientific tool for health data processing, analysis of geographical distribution and variation of diseases and mapping, monitoring and management of health epidemics.

In the past, GIS has been applied to estimate the spatial concentrations of air pollutants and many epidemiologic studies have adopted GIS to explore the health impact of air pollutants on major diseases.

GIS provides tools and capabilities for performing a wide array of activities associated with geographic and spatially referenced information. Associating data with location optimizes analysis, visualisation and reporting/communication of information, thus maximizing the value of the data.

The first step of analysis (as we described) was led in recent years about the mapping of pollutants (static or dynamic using dispersion modelling).

In our new model, we suggest some environmental analysis strictly connected to the health of population and to the personal health information.

27.3.1 Outdoor Air Quality

GIS technique allows to improve the accuracy of air pollution health impact assessments according to the health need of the population and according to the health of the single individuals.

Spatial analysis is performed examining residential proximity to heavy-traffic roadways or any harmful emission source and associated adverse health outcomes.

Moreover, estimating at what distances the impact of direct source of emission (e.g. traffic emissions) on ambient particulate matter (PM) or other pollutants concentrations is significant related to the disease evaluation of the single individual.

Models associated to GIS allow the estimation of daily air pollution concentrations and the consequences on the health of the single patient.

In the same way, estimating exposure to pollutants related to the disease evaluation of the patients suffering of chronic disease allows to extract information on the patient's health, through the correlation of data.

27.3.2 Water Quality

Monitoring naturally occurring contaminants in public drinking water (such as arsenic and nitrates) through GIS spatial analysis, allows to identify vulnerable target, growth of diseases and the possible effect on the single patient.

Using maps to display oil spill information, coliform levels for beaches, well water quality data, etc. allows to visualise the cross action between pollution and population, and through integrated models it is possible to build prediction plans which take into account the needs of the population and its health.

Moreover, GIS and spatial analysis can be used to track violations, health advisories and reported illnesses that may be related to drinking water or water contamination.

Examining relationships between pollutants levels in water or air and various cancers is a new method to realize epidemiological analysis based on the analysis of natural and anthropic events located on the territory in which people lives, and knowing the single medical records of the individuals personalized health analysis can be executed.

27.3.3 Toxics and Waste

Controlling and evaluating waste, junkyard and recycling facilities and identifying locations where potentially noxious land uses may be having a disparate adverse impact on minority and low-income populations provide useful information about the health of the people.

Furthermore, conducting surveillance (exposure assessment) for radiation, asbestos particles, radio frequency exposure, etc. using GIS will allow to monitor the population exposure and the effect described on the medical records of the individuals.

27.3.4 Health

Spatial analysis allows to examine environmental conditions (level of pollutants, pollens, mould spores and different types of pesticides) connected to health problems (e.g. asthma, lung disease, cardiovascular disease, the onset of cancers).

Examining the geographic distribution of disease according to exposure to pollutants and its modifications during the time, it is possible to create a map of disease with the connection to the emission sources and to the dispersion of the pollutants during the time.

These tools allow to investigate epidemics detecting regions of higher incidence of diseases according to exposure to pollutants and analysing the spatial distribution of standard morbidity rates per area.

Spatial analysis in different aspect of public health were performed until now to examine disease rates and other health statistics by geographical area to assess the health of population, to observe variation in health and use of health services as comparative approach to need assessments and resources locations and to study time trends in disease at local level.

Analysis of spatial distribution of healthcare facilities and studies of variation in health treatments and outcomes for planning the development of health services have been long used through GIS facilities, so too is study of health and health promotion interventions at community level.

But above all the power of GIS is evident in disease surveillance: analytical epidemiological studies of factor affecting the occurrence, progression or outcome disease; investigation of hazards including industrial sources of pollution, disease cluster investigation and the selection of geographically ordered population samples for surveys are just some examples.

Here the linkage of health data with population characteristics, environmental conditions and healthcare that led us to statistical and epidemiological analysis are stressed. But analysis of correlation among disease factors and environment factors and the lifestyles (as we focused previously) produce new types of analysis that must consider the individual and the personal health.

Now we want to introduce a further element. We have started from the description of the powerful spatial analysis for the analysis of natural and anthropic events both in static (sources emission) and in dynamic (dispersion models) actions. Then we have described the interesting connection between pollution and health, highlighting mainly the epidemiological and statistical studies. Now we want to emphasize as individual and personalized medicine can take a further impulse.

Traditional epidemiology has taken advantage from GIS technology; it posed emphasis on "public health" and on the treatment of "area" of interests. Now this is gradually changing, with spatial analysis methods and GIS software tools also playing an important role in the delivery of analysis focused on individuals, besides their more traditional role in spatial epidemiology which is instead focused on populations.

Our health depends on where we currently live, as well as on where we have lived in the past and for how long in each place (this is "place history") [20]. Place history is particularly relevant in conditions with long latency between exposures and clinical manifestations, as is the case in many types of cancer and chronic conditions.

Patient's place history should regularly be considered by physicians when diagnosing and treating individual patients. It provides useful contextual environmental information (and the corresponding health risks) about the patient and should ideally form (together with genetic/genomic information) an essential and regular part of every patient medical record.

Moreover where a person has lived must be considered as part of the context in which clinical decision making occurs; so “the place” becomes another useful “vital sign”.

Can your address or localization event reveal about your health? It is not only a simple or trivial matter for which we know that people living in poor countries or disadvantaged areas have more and serious health problems.

It is a complex analysis, but as spatial analysis helps market research organizations better understand what is happening around you, the knowledge about your habitat and likely behaviours hidden within your neighbourhood can be revealed when you provide information about your street address.

In this sense, “Exposomics” is the discipline relating to the analysis of exposure to and interaction with all environmental stressors both internal (own body, e.g. inflammation, lipid peroxidation, bacteria, etc.) and external sources, including chemical factors and pollutants. This helps us better understand, manage and prevent chronic disease development. Exposome represents all nongenetic environmental exposure from conception (including lifestyle factors) to complement the “genome” when studying disease aetiology [21].

These analysis techniques support Precision Medicine and System Medicine and target the health and clinical management of individual persons/patients rather than of whole populations.

We want to emphasize that whereas epidemiological analysis using GIS can be a natural extension of geostatistics technique based on spatial analysis crossed with massive health population data, Precision Medicine and System Medicine can get much more effectiveness.

Precision Medicine and System Medicine are two new medical models in which the focus is the posed on the individual health [22]. Personal health data, genetics, lifestyles and the living places, co-morbidity and the influence of various disease, act on individuals in a different way one from the other and allow to build a personalized medicine [23].

At present we need to integrate the environmental space in which individual lives (described by GIS) and the health information (described by personal medical records of the single individuals) with models in which personal health scenario are proposed for the health of the single patient.

The spatial analysis ability added to the capability of the modern medical records to treat and track the individual health and through virtual layers integrate different information (genomics and molecular information, lifestyles, etc.) combined with the power of algorithms to elaborate clinical personalized scenarios is the new frontier that we have to reach to consolidate and improve analysis already partially taken. As part of a comprehensive study on people’s health, full consideration must be given to develop a wider understanding of how both natural and anthropogenic characteristics act and interact to modify the environment and how occupational and environmental exposures, or lifestyle and habits, contribute to disease incidence.

We distinguish the environmental components related to disease in inorganic and organic. For example, dust (particulate matter) can be composed of biologically derived matter from animal or microbial origin and contain fungi/bacteria. Inorganic forms of particulate matter can be minerals, such as asbestos or silica, or a suite of minerals that can be found in a range of geological deposits, such as volcanic ash or sedimentary rock. A great number of information exist in occupational health literature on particulate matter exposure in occupational settings, mainly due to the fact that both the component in question (i.e. the form/type of particulate matter) and the exposure can be well characterised.

We know that the concept of exposure must not only consider the release and distribution of the element within the environment but also the route by which populations are exposed.

The ingestion of elevated arsenic levels in drinking water and diet (mainly through seafood) has been a widely recognised issue, linked with an assortment of health problems, such as cancer, and other noncarcinogenic effects such as skin lesions and neurological and hepatic effects. The health effect can be studied through the analysis of the route of the pollutant during the time and how this path crosses the human actions and population life.

The factor of time (temporal dimension) is fundamental in health studies, where all of the connections that create the way from environment to disease must be taken into account and also the influence of time on the formation of the contaminant in the environment, on the exposure (chronic or acute) and on disease initiation (latency). Spatial analysis techniques are of great help in the spatiotemporal modelling and visualisation of environmental contaminants and their spread over time and space [24]. In the approach supporting epidemiological studies, once a disease has been identified, carefully controlled epidemiological analysis can help constrain the incidence and distribution. Data can be collected as voluntary or incidental data (the latter type describing the use of data originally collected for an alternative purpose). The use of cohort and cross-sectional studies instead of aggregated population data often proves to be valuable in capturing individual exposure. In addition, by fully characterising the nature of the disease, this may allude to the causal agents, as already known and documented associations may exist between elements and metals, for example, and target organs.

One advantage of a more detailed (personal) epidemiological investigation into the disease is the capacity for identifying confounding factors such as disease susceptibility due to physiological and potentially even genetic variations. Defining exposure is difficult and a challenging aspects of any medical investigation, as exposure may occur over long periods of time: days, years and even decades. Further to this, the complication arises from the fact that exposure to the dangerous agent may have been at some time and place in the past. This is the case for asbestos exposure and mesothelioma, where symptoms manifest often several decades after first exposure.

There is a further approach in which, if an environmental survey identifies the presence of a certain element that could potentially be a dangerous to human health, then a wide investigation can be realized to analyse and map the distribution of the component within the domain of interest. Field data and data sourced from other methods (monitoring, remote sensing) are usually used to collect sets of variables that aim to describe the domain of interest and the link with the population. To complete this connection, we need also the information of medical records of the individuals in order to extract correlation that can corroborate the hypothesis.

To study environmental factors contributing to diseases, clinical researchers use the template for genome-wide association studies. In this context GIS will help improve Environmental-Wide Association Study methodologies: by controlling exposition to events, it is possible to reduce the uncertainly and improve the analysis. Type 2 diabetes (T2D) and other chronic diseases are caused by a complex combination of many genetic and environmental factors. Mathematical models associated to GIS features allow to comprehensively associate specific physical environmental factors with disease and analyse in deep all the factors and the correlation between data.

Starting from the theory that most chronic diseases are a result of a complex combination of genetic processes and the environment (e.g. see [25]), different analysis were conducted in which taken into account also are environmental factors in disease. T2D provides a specific example: while genetics play a large role, specific environmental factors are also emerging as risk factors for the disease; for this reason, it is necessary to measure and evaluate both types of factors to better understand complex diseases. In this context GIS tools and spatial analysis associated to clinical data from practice (EHR) and to genetic information provide powerful techniques to analyse complex diseases in deep. About the health risk assessment due to air pollution, GIS allows to support complex analysis.

The results of a health risk assessment can also quantify the mortality [26] and the diseases associated with pollution resulting from transport [27, 28]. The focus on the integration of mathematical models and GIS approaches in the analysis of exposure to pollutant increases the accuracy of the assessment and allows to produce timely and consistent assessment results so that they can support the decision making process on urban planning and contribute to a more sustainable mobility in the urban area [29].

Geolocation of emission sources during the time, associated to the variability of meteorological conditions, is crossed with information about the population and disease collected in EHR; this model produces health risk assessments.

During last decade, GIS-based pollution mapping using interpolation techniques such as inverse distance weighting, kriging and land use regression modelling [30] was explored by many researchers for epidemiological studies.

In general, the kriging method [31] was used as a statistical mapping technique for using data collected at each point location, to predict concentration in each grid cell over a spatial domain.

The study sought to investigate the spatiotemporal association between air pollutants and asthma rate [32]. To understand the interrelationships among predictor variables, both GIS-based correlation analysis (map correlation analysis) and point data correlation analyses were carried out.

We propose accessing and using data that adds intelligence to the medical encounter should be the driving force behind modernizing our health information technology. There are a lot of interesting experiences.

By analysing the spatial pattern of hospitalization and air pollutants, major threat areas can be visualised in the form of maps with the help of GIS. The University of Cincinnati has summarized drinking water quality for US major metropolitan areas, using local community drinking water reports listing EPA-regulated contaminants [33].

Another example of interesting usage of GIS and spatial analysis in medicine is the probable connection between diseases; the project suggests that breast and prostate cancers cluster spatially, this finding corroborates other studies that have found these two cancers share similar risk factors [34].

Using modern geographic mapping techniques, health researchers have concluded that the proximity of roadways to places where children spend a great deal of their day experience higher than acceptable levels of exposure to air pollution.

GIS is an appropriate tool to deduce spatial relationships and to facilitate the proper understanding and resolving of related complex issues. The use of GIS techniques for statistical analysis and modelling (e.g. pollutants and diseases) is rarely done in previous studies. However, the past studies have not been fully explored and applied in analysing the association between exposure of air pollution and human health. In order to better understand the disease of the individuals, we suggest the integration of the outcomes by spatial analysis with the medical records in which also added are the information about genomics, exposomics, lifestyle and patient notes.

Continuing our route now, we would like to present other opportunities that technology has recently made available with the aim to extend the concepts of georeferred data to other types of data.

27.4 Social Networks and Localization of Events: Sharing Information to Understand and Improve

Recent improvements that have invested all the area of technology led us to consideration on the capture and utilisation of geospatial data. Traditionally geospatial data were acquired by the public agencies; authoritative datasets were dominating in topographic domain. But at present new ways of collecting geospatial data have emerged, leading to completely new data sources and data types of geographical nature. We are now facing a situation where geospatial data acquisition is a commodity implemented in everyday devices used by many people. The devices are capable of acquiring environmental geospatial information at an unprecedented level, with respect to geometric and temporal resolution and thematic granularity. They are small, easy to handle and able to acquire data even unconsciously.

Of course this poses problems of accuracy of data, so that it becomes information which can be used profitably and becomes important according to the situation to treat. For example, in some situations such as disaster context, it is important to get any information even if it is not very accurate; social media and services are considered as new information sources, for example, for early response and crisis management [35].

But it is necessary to emphasize that in the area of georeferenced data, there is a rich of new sensors and data sources which lead to large collections of miscellaneous and "dirty" data (inaccurate, incomplete or erroneous data). They acquire relevance by intelligent integration and fusion with complementary data; this is the specific work of big data technique.

Participatory sensing (PS) [36] combines the use of everyday mobile devices, such as cellular phones, GPS technology and location-based services and sensors to form interactive, bidirectional mobile sensing information systems. In PS, each mobile phone becomes a mobile sensing device capable of capturing, processing, transmitting, and receiving information about a particular variable or variables of interest. The collective participation of many cellular users in a common sensing application brings the possibility to acquire large amounts of data in a simple and cost-effective manner with the potential to address large-scale societal or community-oriented problems. Data collected in this way are natively georeferenced and can be treated by GIS.

Participatory data acquisition is conducted in a conscious process by a user, who explicitly selects objects and their features and with this information contributes for the comprehensive knowledge. Opportunistic data capture occurs unconsciously, mostly with no specific purpose or even a completely different purpose. This produces unconscious knowledge; information collected for certain scope is used also for different one, currently unknown.

So as to carry out air pollution monitoring over an extensive area, a combination of ground measurements through inexpensive sensors and wireless GIS will be used for this purpose.

At present there are different experimental projects in which portable device, comprising solid state gas sensors (e.g.) are integrated to a personal digital assistant (PDA) linked through Bluetooth communication tools and Global Positioning System (GPS). These integrated devices will allow rapid dissemination of information on pollution levels at multiple sites simultaneously. The report on air quality generated can be then published using Internet GIS to provide a real-time information service. The aim in this case is an increased public awareness and enhanced public participation. Of course data collected must be treated. Local deterministic and geostatistical interpolation methods can be used for spatial prediction and to find out the most suitable method for studying air pollution, based on observations at each monitoring site.

We have to consider that the cost of establishing and implementing ordinary monitoring systems in each strategic point of the city is extremely high; the use of analytical instruments is time-consuming, is expensive, and can seldom be applied for real-time monitoring in the field, even though these can give a precise analysis.

Hence, we are conscious that a new generation of detectors, solid state gas sensors, offers an excellent alternative for environmental monitoring due to low cost, light weight, extremely small size and also due to the reason that they can be deployed anywhere so as to receive data that can eventually be transmitted through a wireless GIS network system as a rapid monitoring tool to the general public. But at the same time, we have to do necessary consideration about data quality and data precision issues.

Data crowdsourced can be useful for growing awareness, but it brings significant problems with the accuracy and quality of information.

The vision of pervasive sensing through "citizens as sensors" poses considerable challenges in terms of interoperability involving data formats, service interfaces, semantics and measurement uniformity. Thus, one key prerequisite to achieve this vision is the broad usage of open sensor standards, we think.

For this reason, we propose to integrate data from authoritative detection control units (we assume to be accurate and maintained over time) with data crowdsourced by portable miniaturized devices. These information must be treated according to algorithms for describing the map of pollutants and the dispersion models.

These projects aim at providing fine grained air quality data to allow citizens and urban decision makers to assess environmental conditions instantaneously and intuitively. The goal is to provide real-time information on urban processes to support short-term decisions at multiple levels, from personal to governmental.

Data collection (voluntary or unconscious), processing and information provision architecture must be designed and implemented, covering the whole process chain from sensor network development via measurement integration to data analysis and information visualisation using GIS facilities; hence, this infrastructure can potentially serve as the architectural bridge between domain independent sensor network developments and use case specific requirements for end user tailored information output for public health monitoring.

A non-exhaustive list of participatory sensing deployments is provided in [37].

We find collection of CO, NOx, noise, air temperature and relative humidity, together with the geographic position and time via GPS. Both CO and CO2 emission and distribution depend on very high temporal and spatial fluctuations which are induced by a variety of factors including temperature variability, urban topography, time during the day, the urban tunnelling effect, traffic emergence or plant natural action (we know plants release major amounts of CO2 overnight).

My Place History app, produced by ESRI (Redlands, CA, USA), despite being active only in the USA and without connection to any medical records, offers a first demonstration of the concept of place history associated to crowdsourced information. It is a way to link public health information with personal environmental experience bringing everyday information into the physician/patient relationship to achieve a greater understanding of how local environments affect health [38]. Geolocated information from people becomes important for all; but as we will analyse further, we must take into account the possible mismatching and senseless considerations that can be provided. The proximity to an industrial site can be problematic depending on the distance and on the weather conditions. The same is true for the concentration and distribution of various natural contaminants. Whereas living in a zone with high concentration of people who have undergone heart attack, it is not sufficient claim to be at high risk. These data are poor and without significance if not linked to others in order to became information. That's why in this work, we want to stress the importance of the interconnection of the various sources of information (personal medical records composed also by images, lifestyle and narrative, geolocated data, anthropic data, IoT data, sensors, crowdsourcing data, meteorological data, epidemiological data, etc.), especially a correct analysis made by robust algorithms ensuring holistic and comprehensive approach.

At this stage it is interesting to introduce the implications that can have in emergency management the crowdsourced data management. EM involves both spatial analysis, environmental structure of the area of interest and obviously the medical records of the population. In many cases if well analysed also crowdsourced geolocalized information can help the organizational structure of the rescue. The immediate and far reaching impacts of the disasters (natural or not) highlight the need for rapid and effective emergency management. This issue emphasizes the necessity for effective regional preparation, response, recovery, and restoration. For emergency management personnel to make accurate and timely decisions, they must have situational awareness, an accurate perception of the situation that they are facing and its complex data to be treated. Essential to good situational awareness is the ability to provide relevant and timely information to decision makers and the public. Effective information collection and sharing has long been recognised as a challenge in emergency management.

With the addition of social media as another input, there is often more information available than can be understood at the rate needed (information overload).

Precision Information Environments (or PIEs) provide tailored access to information from disparate sources augmented with decision support capabilities in a system that is designed around the multiple user roles, contexts and phases of emergency management, planning and response [39].

In those cases, profile and source assessment is just as important as the situation. Is the source trustworthy?

The global profile system inside an information environment can be applied to emergency management professionals and PIE content sources such as government professionals, nongovernmental organizations, private industry and citizens. Each type of source is assigned a reputation score (can vary during the time) that is adapted over time based on the content they create (and how it is used by others) and the people they interact with.

To conclude this short part about the participatory sensing paradigm, we have to introduce the implications that can have in the people rising awareness about pollution and its effect on health.

We have started an experience in China in which WeChat (http://www.wechat.com/) crowdsourced platform is used in order to publish short message to the people about the effects of pollution on the health of individuals.

The platform is conceived in a bidirectional way; a centre provides periodically information to people about pollution, whereas people can meet the demands in different ways. Growing the awareness about the effect of the pollution and increasing the incidence of messages involving other viral threads. Being source of new information triggering data from personal sensors, data about traffic jam, the increasing of pollution, the personal effect on own pathology, alert about pollution trend and so on.

The social and physical environment provides the context that can enable healthy behaviours. Social networks have the ability to propose the change in lifestyles; obesity within a person's social network becomes a predictor of their own body mass index, for example. The possibility of walking can impact the amount of exercise that people get, and access to supermarkets may affect their ability to buy fresh fruits and vegetables. Furthermore, the physical environment (e.g. air quality, pollution, noise, public transportation access) has direct impacts on health that need to be better understood at both population and individual levels. The real power and uniqueness of crowdsourcing lies in the active participation of intelligent humans in a task assigned to them.

But at last, data from personal sensors, voluntary or unconscious geolocalized data, even if well managed can be used both in the ordinary management of the effects on the population, tracing the events in health people's lives, and finally in emergency management, they impose the treatment of a huge amount of data.

27.5 Big Data Technique and GIS for Pollution Analysis and Health

So far we have stressed as the main problem in order to face the new challenges brought by new medical models is the inability to adapt the old data structure of the medical records to new source of information and the inability of treating complex data in order to extract useful information to support medical practice. Traditional EHRs are statics based on method to save and storage data rather than provide clinical scenarios.

This requires integration of information (in type, structure and meaning), capability of treating different types of data (not only numerical but also images, unstructured information, etc.) and above all the ability to correlate and extract hidden information.

We have also explained as to address studies about the effect of pollution on the health we need spatial analysis tools, performed by GIS, and that these data will provide useful information if appropriately integrated with personal EHRs.

At last we assist to the huge growth of data about pollution, health and environment voluntarily or involuntarily taken from sensors or simply managed by people. Linking these communicating sensors to computing intelligence is the challenge called the Internet of Things or the Industrial Internet. We try to improve the paradigm from real-time cities to real-time pollution and health.

We are conscious about the richness of data we can have, and we must to search methods for using it.

Big data consists of huge, dynamic, varied, comprehensive, detailed, interconnected datasets that can be linked and utilised in different ways. These characteristics offer the opportunity of studies shifting from data-scarce to data-rich static snapshots to dynamic unfolding, coarse aggregation to high resolution and relatively simple hypotheses and models to more complex, sophisticated simulations and theories [40].

Big data features in brief are the ability to treat huge data in volume; high in velocity, being created in or near real time; diverse in variety, being structured and unstructured in nature, and often temporally and spatially referenced; exhaustive in scope, striving to capture entire populations or systems ($n = \text{all}$), or at least much larger sample sizes than would be employed in traditional, small data studies and detailed as possible fine-grinned in resolution [41].

In general, big data is considered as structured and unstructured datasets with massive data volumes that cannot be easily captured, stored, manipulated, analysed, managed and presented by traditional hardware, software and database technologies. This is the environment that we have described so far about the connection between pollution and people's health.

Big data platform for health has these logical features about information: it is shared horizontally; more timely and more complete; higher in quality (in terms of sensitivity), as local populations know what is the baseline behaviour of their own communities; higher in quality (in terms of specificity), as being local allows to quickly validate, verify information and dismiss false positives earlier and geo-aware, noting that local citizens have a high authority on what is nearby to them; this piece of information is essential for the contextualisation and interpretation of data.

It is common opinion among researchers in health data that big data can offer interesting way capable of supporting data processing models useful for the clinical practice and obviously for the medical research.

The main idea is to consider the health of patients as "individual" matter strictly connected to different information, not only included in a cluster of epidemiological status; in this sense other data depending on the peculiarity of the individuals make sense: genetic heritage, lifestyles, way of life and environmental constraints also geolocalized.

Big data analytics is proving a useful technique that can facilitate action on the modifiable risk factors that contribute to a large fraction of the chronic disease burden, such as physical activity, diet, tobacco use and exposure to pollution. It can do so by facilitating the discovery of risk factors for disease at population, subpopulation and individual levels and by improving the effectiveness of interventions to help people achieve healthier behaviours in healthier environments.

At the moment we address the usage of big data in healthcare to four directions: the analysis of compliance of health services assigned to the patients as a function of the disease; hidden relationships between the information about the pathology suffered by the patient and co-morbidity; increasing the capacity to understand the behavioural, social, and environmental determinants of health in populations and enabling disease prevention efforts. As previously described geocalized data about natural and anthropic events, exposomic and other complex information directly related to the medical records of the individual (genomics, molecular data, lifestyle, physical activity, etc.) complete and integrate the information base necessarily.

We must identify modifiable risk factors for disease, such as environmental pollution, diet, exercise, smoke, alcohol, lifestyles, etc., in order to propose intervention to mitigate the risks factor. We want to emphasize that as described previously, we need geolocalized information and data from sensors devices (passive – physical activity, sleep, movement, etc. – or active, telemedicine). Correlations between risk factors and disease can be identified by applying analytics on these datasets.

Identifying personalized risk factors meets the need (from System Medicine and Precision Medicine) of giving single patients more effective information about how to prevent their own disease and doing so in a way that is more convincing for them to act upon because it is targeted to them specifically as opposed to the average person typical of epidemiological studies. Specific diet, physical activity and so on depend on the individual and may be beneficial to some people but harmful to others.

We have already focused on the fact that the current medical records are inadequate to face the modern medicine: a lot of stored information is unstructured such as images, narrative and texts, and is not provided with tools for using these information.

At this we add a crowdsourcing information on chronic disease that introduces the personal expectation during the illness and personal notes. Whereas we know that this crowdsourcing information is used for health surveillance and epidemics control, these are a source of information for personalized medicine also.

The first step to improving health behaviours is to monitor and measure them, and recent technological advances have provided many new ways of doing this. Devices or smartphone apps can be used to monitor health behaviours such as physical activity, sleep quality, medication adherence [42] and diet [43].

A large amount of environmental data is regularly collected in non-health sectors and could be an important input into health-related big data. A few examples of available data on the physical environment include weather patterns, pollution levels, allergens, land use change, forest fires, particulate matter, traffic patterns, pesticide applications or water quality. Growing capacity for ambient environmental sensing, citizen science and the use of drones will expand access to remotely and passively captured environmental data.

Genomics is an important risk factor for disease, both as a direct cause of disease and as a marker of susceptibility or resistance to other risk factors. High throughput genetic information is becoming increasingly easy and inexpensive to collect.

Once research has identified the appropriate risk factor targets for a given person, the next step in the disease prevention process is to help that person achieve these goals. Big data offers the potential for this important advice to reach each person outside of the clinic in a personalized manner, which increases the likelihood of its impact improving the traditional advice of the physician during the annual checkup visit.

Moreover monitoring health behaviour and providing real-time feedback on performance in comparison to personalized targets can help people to reach their behaviour goals. If possible, behavioural data should be continuously collected without additional effort from the patient.

But the core of the system, we know, is a set of shared algorithms able to treat all types of data and in order to analyse deep and hidden correlation among data with the aim to produce different personalized health scenarios.

27.6 The Comprehensive GIS Model

With these considerations, we can complete our GIS model.

In our model, GIS is used to estimate the association between air pollution (e.g. fine particulate matter (PM2.5), sulphur dioxide (SO2) and ozone (O3), (the most important pollutants generally used in the AQI index) and human health (e.g. asthma emergency department visit rate and asthma discharge rate for children, elderly or people suffering of respiratory or cardiovascular diseases or T2D).

First, we estimated the pollutant level by constructing a spatial model representing a geographical area using daily average pollutant concentration data. We provide a static model (map of sources), a dynamic model and a dispersion model. We consider three entities: point, describing the point source of emission; line, representing the emission on routes; and area, representing the area of dispersion.

Second, we linked air pollutant concentration to disease rate (cardiovascular, asthma, etc.) within the defined study area. All the pollutants are among the most studied of environmental hazards and are at significant levels in air to adversely affect human health. Spatial analysis includes the following: proximity analysis in which calculated is the distance of the patients by the emission sources (point, areas, lines) and inclusion analysis in which we estimate the presence of the patients in the area of interest and the persistence in the time.

Third, we extract individual EHRs and update information related to environment and lifestyles. Data about hospitalization and adverse events (suspected) caused by pollution are related to the spatial analysis. Exposomic, genomic, lifestyles and data from practice are integrated, and suitable algorithms are used to mine information and extract correlation in order to provide personalized solutions to the patients.

Fourth, we add to the dataset information from sensors. We insert different typology of sensors: IoT (devices monitoring source of emissions, environmental sensors, anthropic events, environmental events, etc.), personal sensors (monitoring health status, physical activities, diet, drug assumption, etc.), voluntary sensors (images, personal notes, place history, etc.). These are all georeferenced and provide local information. Data must be cleaned according to typical big data homogenization treatments, before using it. At this stage the personal EHR is completed. In addition to the previous information, we collect also information from users, voluntary notes on disease and voluntary information from devices but also involuntary data.

Fifth, algorithms are used in order to extract correlation among information and mine hidden data. The aim is to provide health scenarios according to Precision Medicine model, but also traditionally epidemiological study is possible.

Sixth, provision analysis is provided producing maps and visual outcomes for proposing to policy makers new solutions for better living.

Conclusions

Individual lives own life on a specific territory and moves on it that is permeated by anthropogenic or natural events (during the time) which affect the health. Moreover he/she has an own genetic heritage that responds in different manner (one from the other) to the stresses produced by the external events. During the life in this permeated space he is subject to health events. In the modern age the data of health events is stored in specific databases regarding exclusively the modification of the individual's health (Electronic Health Records – EHR). Lifestyles, social heritage, educational level and social relationships have a reason and a meaning in the heath life of the individuals. Not only individual notes by the patients regarding the reaction to therapy but also the own feelings regarding the disease is an important matter to take into account during the course of the disease.

At present all this information are treated in separated manner by professionals in order to describe local events, individual's health, genetic map and social events.

But each of these information may be related to the individual through a comprehensive model [44]. We think that these issues led us to consider a personalized medicine that takes into account not only data coming from health events but also genetic data, lifestyles information, social relationships, anthropic and natural events that permeate the living space of the individuals. This allows to understand correlation between different type of information, through mining data.

In recent years Precision Medicine [45] and System Medicine [46, 47] have been presented as possible innovative models in healthcare; these solutions have their foundation as well as medical research including on the wide usage of innovative technical solutions according to the new technical solution in health context.

The concepts underlying these approaches concern the systemic perspective in order to consider the holistic and composite characteristics of a health problem and evaluate the problem with the use of computational and mathematical tools.

They require a coordinated approach across disciplines and across research, industry and all the relevant stakeholders.

The creation of a network among the specialists and professionals skilled in managing, storing, analysing and mining different sources and types of information, according to systems biology projects, research projects and practice information, will support new technical methodologies, in order to share information/resources, and will support suitable algorithms in order to analyse in detail and extract correlation by huge amount of data with the aim to offer to patients individual health services.

This is the natural result of the adoption of the basic ideas of System Medicine and Precision Medicine.

As described, we think that spatial analysis is an important methodology for medical research and practice and pollution. GIS technology's ability to author, publish and share critical information about the spatial dynamics of disease makes it, without exception, the technology of choice for accelerating the detection and identification of disease clusters or personal events. GIS technology's capacity to reach beyond geopolitical boundaries makes it generally usable in public health emergencies and responses, but it makes highly desirable in Precision Medicine.

We know that this not only requires a modification of the organizational aspects or a different approach to care but is also a technical challenge.

We have already stressed that the existing information systems are deeply unable to create relationships with data of other types, losing a wealth of hidden information to existing systems. This is the aim of the big data analytics technique.

Therefore there seems increasingly essential the professional figure of the data scientist in the field of practical medicine. He/she must analyse different types of data and extract information in order to support clinicians in the development of personalized medicine scenarios for individuals.

The goals of prediction, prevention and personalized treatment in the case of complex, rare, chronic diseases can be effectively realized through an intensive multiple co-production approach, where the various relevant experts and professionals collaborate together and complement each other in interactive and holistic manners.

References

1. Wilhelm M, Ritz B. Residential proximity to traffic and adverse birth outcomes in Los Angeles County, California, 1994–1996. Environ Health Perspect. 2003;111(2):207–16.
2. Raaschou-Nielsen O, Reynolds P. Air pollution and childhood cancer: a review of the epidemiological literature. Int J Cancer. 2006;118(12):2920–9.
3. Brunekreef B, Beelen R, Hoek G, et al. Effects of long-term exposure to traffic-related air pollution on respiratory and cardiovascular mortality in the Netherlands: the NLCS-AIR study. Res Rep Health Eff Inst. 2009;139:5–71.
4. Pershagen G, Rylander E, Norberg S, et al. Air pollution involving nitrogen dioxide exposure and wheezing bronchitis in children. Int J Epidemiol. 1995;24(6):1147–53.
5. Anderson HR, Atkinson RW, Peacock JL, Marston L, Konstantinou K. Metaanalysis of time-series studies and panel studies of particulate matter (PM) and ozone (O3). Report of a WHO task group. Geneva: World Health Organization; 2004.
6. World Health Organization. Air quality guidelines: global update 2005. Geneva: World Health Organization; 2005.
7. Schwartz J. Air pollution and daily mortality: a review and meta-analysis. Environ Res. 1994;64:36–52.
8. Analitis A, Katsouyanni K, Dimakopoulou K, Samoli E, Nikoloulopoulos AK, Petasakis Y, Touloumi G, Schwartz J, Anderson HR, Cambra K, et al. Short-term effects of ambient particles on cardiovascular and respiratory mortality. Epidemiology. 2006;17:230–3.
9. Yang C-Y. Air pollution and hospital admissions for congestive heart failure in a subtropical city: Taipei, Taiwan. J Toxic Environ Health A. 2008;71:1085–90.
10. Halonen JI, Lanki T, Yli-Tuomi T, Tiittanen P, Kulmala M, Pekkanen J. Particulate air pollution and acute cardio-respiratory hospital admissions and mortality among the elderly. Epidemiology. 2009;20:143–53.
11. Field MJ, Grigsby J. Telemedicine and remote patient monitoring. JAMA. 2002;288(4):423–5. https://doi.org/10.1001/jama.288.4.423.
12. Paré G, Jaana M, Sicotte C. Systematic review of home telemonitoring for chronic diseases: the evidence base. J Am Med Inform Assoc. 2007;14:269–77. https://doi.org/10.1197/jamia.M2270269-277.
13. Koehler F, Winkler S, Schieber M, Sechtem U, Stangl K, Böhm M, Boll H, Kim SS, Koehler K, Lücke S, Honold M, Heinze P, Schweizer T, Braecklein M, Kirwan B-A, Gelbrich G, Anker SD, TIM-HF Investigators. Telemedical interventional monitoring in heart failure (TIM-HF), a randomized, controlled intervention trial investigating the impact of telemedicine on mortality in ambulatory patients with heart failure: study design. Eur J Heart Fail. 2010;12:1354–62. https://doi.org/10.1093/eurjhf/hfq199.
14. Kamel Boulos MN, Viangteeravat T, Anyanwu MN, Ra Nagisetty V, Kuscu E. Web GIS in practice IX: a demonstration of geospatial visual analytics using Microsoft live labs pivot technology and WHO mortality data. Int J Health Geogr. 2011;10:19. https://doi.org/10.1186/1476-072X-10-19.
15. Cavalone G, Rinaldi G, Gagno F, Vestrucci P. Air pollutant emission analysis: an approach using geographical information system. Annual conference AICA (Associazione Italiana Calcolo Automatico), vol. 2. Siena, Italy. 25–27 Sept 1991.
16. Rinaldi G, Cavallone G, Stanghellini S, Vestrucci P. Air quality assessment using geographical information system. EGIS '93 conference. Fourth European Conference and Exhibition on Geographical Information System. Genoa, Italy. Mar 29–Apr 1 1993.

17. World Health Organization. Environmental health. Retrieved 29 Aug 2016. http://www.who.int/topics/environmental_health/en/.
18. World Health Organization. 10 facts on preventing disease through healthy environments. Retrieved 29 Aug 2016. http://www.who.int/features/factfiles/environmental_health/en/index.html.
19. Miranda ML, et al. Building geographic information systems capacity in local health departments: lessons learned from a North Carolina project. Am J Public Health. 2005;95:2180–5.
20. Davenhall B. Your health depends on where you live. In: TEDMED 2009 (filmed Oct 2009). New York: TED Conferences LLC. http://www.ted.com/talks/bill_davenhall_your_health_depends_on_where_you_live?language=en. Accessed 16 July 2016.
21. Wild CP. Complementing the genome with an "exposome": the outstanding challenge of environmental exposure measurement in molecular epidemiology. Cancer Epidemiol Biomark Prev. 2005;14(8):1847–50.
22. Lawler M, Sullivan R. Personalised and precision medicine in cancer clinical trials: panacea for progress or Pandora's box? Public Health Genomics. 2015;18:329. https://doi.org/10.1159/000441555.
23. Nishi A, Milner DA Jr, Giovannucci EL, Nishihara R, Tan AS, Kawachi I, Ogino S. Integration of molecular pathology, epidemiology and social science for global precision medicine. Expert Rev Mol Diagn. 2015;4:1–13. https://doi.org/10.1586/14737159.2016.1115346.
24. Matejicek L, Janour Z, Benes L, Bodnar T, Gulikova E. Spatio-temporal modelling of dust transport over surface mining areas and neighbouring residential zones. Sensors. 2008;8:3830–47. https://doi.org/10.3390/s8063830.
25. Patel CJ, Bhattacharya J, Butte AJ. An environment-wide association study (EWAS) on type 2 diabetes mellitus. PLoS One. 2010;5(5):e10746. https://doi.org/10.1371/journal.pone.0010746.
26. Dockery DW, Pope CA III, Xu X, Spengler JD, Ware JH. An association between air pollution and mortality in six US cities. N Engl J Med. 1993;329(24):1753–9.
27. Dab W, Segala C, Dor F, Festy B, Lameloise P. Air pollution and health: correlation or causality? The case of the relationship between exposure to particles and cardiopulmonary mortality. J Air Waste Manag Assoc. 2001;51:220–35.
28. Dockery DW, Pope CA. Acute respiratory effects of particulate air pollution. Annu Rev Public Health. 1994;15:107–32.
29. Davenhall B. The missing component: adding place when evaluating health risks (genetics + lifestyle + environment = risks). ArcUser. ESRI, Redlands; 2010. Winter issue: 10–11.
30. Jerrett M, Burnett RT, Ma R, Pope CA 3rd, Krewski D, Newbold KB, Thurston G, Shi Y, Finkelstein N, Calle EE, Thun MJ. Spatial analysis of air pollution and mortality in Los Angeles. Epidemiology. 2005;16:727–36.
31. Wong DW, Yuan L, Perlin SA. Comparison of spatial interpolation methods for the estimation of air quality data. J Expo Sci Environ Epidemiol. 2004;14:404–15.
32. Croner CM, Sperling J, Broome FR. Geographic information systems (GIS): new perspectives in understanding human health and environmental relationships. Stat Med. 1996;15:1961–77.
33. http://www.uc.edu/gissa/projects/drinkingwater/. Accessed 1 Aug 2016.
34. Mandal R, St-Hilaire S, Kie JG, Derryberry D. Spatial trends of breast and prostate cancers in the United States between 2000 and 2005. Int J Health Geogr. 2009;8:53. https://doi.org/10.1186/1476-072X-8-53.
35. Goodchild MF, Glennon A. Crowdsourcing geographic information for disaster response: a research frontier. Int J Digital Earth. 2010;3(3):231–41.
36. Mendez D, Perez A, Labrador M, Marron J. P-sense: a participatory sensing system for air pollution monitoring and control. Proceedings of IEEE international conference on pervasive computing and communications. WiP Workshop. 2011. P. 344–347.
37. Tilak S. Real-world deployments of participatory sensing applications: current trends and future directions. ISRN Sens Netw. 2013:583165, 8 pages. https://doi.org/10.1155/2013/583165.
38. My Place History by ESRI (iOS app). USA. https://itunes.apple.com/us/app/my-place-history/id396902848?mt=8. Accessed 24 July 2016.

39. Bengtsson L, Lu X, Thorson A, Garfield R, von Schreeb J. Improved response to disasters and outbreaks by tracking population movements with mobile phone network data: a post-earthquake geospatial study in Haiti. PLoS Med. 2011;8(8):e1001083. https://doi.org/10.1371/journal.pmed.1001083.
40. Kitchin R. Big data and human geography: opportunities, challenges and risks. Dialog Hum Geogr. 2013;3:262.
41. Boyd D, Crawford K. Critical questions for big data. Inf Commun Soc. 2012;15(5):662–79.
42. Dayer L, Heldenbrand S, Anderson P, et al. Smartphone medication adherence apps: potential benefits to patients and providers. J Am Pharm Assoc. 2013;53:172–81.
43. Carter MC, Burley VJ, Nykjaer C, et al. Adherence to a smartphone application for weight loss compared to website and paper diary: pilot randomized controlled trial. J Med Internet Res. 2013;15:e32.
44. Rinaldi G, Gaddi A, Capello F. Medical data, information economy and federative networks: the concepts underlying the comprehensive electronic clinical record framework, vol. 1. Hauppauge, NY: Nova Science. p. 1–396. isbn:978-1-62257-845-0.
45. National Academy of Science. Toward precision medicine: building a knowledge network for biomedical research and a new taxonomy of disease. Washington, DC: The National Academies Press; 2011.
46. Ahn AC, Tewari M, Poon CS, Phillips RS. The limits of reductionism in medicine: could systems biology offer an alternative? PLoS Med. 2006;3(6):e208. https://doi.org/10.1371/journal.pmed.0030208.
47. Ahn AC, Tewari M, Poon CS, Phillips RS. The clinical applications of a systems approach. PLoS Med. 2006;3(7):e209. https://doi.org/10.1371/journal.pmed.0030209.

Basic Principles for Treatment of Air Pollution-Related Diseases

28

Antonio Vittorino Gaddi, Benini Enrica, and Fredrick O. Odoo

28.1 Overview

At present, in the field of air, water and food pollution, it is almost impossible to plan and conduct clinical trials with proper audit systems or to follow the methodological criteria adopted in other areas of biomedical research. However, it is paramount to reach a sufficient level of knowledge on the efficacy, effectiveness and safety of different methods and tools for individual protection and for prevention of pollution-related diseases (PRD).

Scientific authorities, governments and institutions worldwide agree with the general idea that a decrement of the polluting emissions related to health benefits and an increment of the level of wellness in the people [1–8].

In the past and in other fields of research, the general approach was and is different when risk factor (RF) for human health is considered. The idea that risk factors are reversible (viz. the idea that decreasing the level of risk, the incidence of a given disease decreases) is not an obvious characteristic also for factors with very high predictivity, consistency, coherence, strength of association, specificity or temporality [9, 10]. However, in other fields of preventive medicine, the "reversibility of the RF[1]" must be demonstrated with ad hoc clinical trials before starting "active" interventions on a population or to apply counselling and education strategies [10].

[1] The risk factor must be reversible. However, the reversibility alone has no value, if it is not proven that reducing the risk factor level also reduces mortality (or other measures of effectiveness). It is also essential to know the function which correlates the variation of factor with the variation of the expected events.

A.V. Gaddi (✉)
Scientific Director of Health Lab of GTechnology Foundation, Modena, Italy
e-mail: profgaddi@gmail.com

B. Enrica
Caravelli Lab—Technobios EuroGenLab, Bologna, Italy

F.O. Odoo
Henrico Doctors' Hospitals, Richmond, VA 23229, USA

F. Capello, A.V. Gaddi (eds.), *Clinical Handbook of Air Pollution-Related Diseases*,
https://doi.org/10.1007/978-3-319-62731-1_28

The overall burden associated with pollution is a dramatic emergence globally. Therefore, we must accept higher degree of approximation in the decision-making process. These aspects are discussed also in the chapter "Complexity in pollution-diseases management" (see Chap. 14).

The theatrical benefits coming from the reduction of some pollutants, as the PM2.5, have been also calculated [6, 11], whilst models that consider many factors were created; those include especially those related to climate and the circulation of air masses in the atmosphere [12].

Moreover, we have some convincing evidences on the health benefits secondary to the reduction of pollution coming from studies that analyzed air pollution in specific work places, from ecological studies [13] and from medium- [14, 15] and long-term [16] observations.

There are also relevant studies that have observed the positive effects of the implementation of some public health policies, like the one promoted in China to prevent or reduce infant mortality, that resulted in a reduction of the 20% in those areas subjected to a better pollutant emission control [17].

The risks associated to pollution and the decisions needed to fight this modern plague have been widely debated. A relevant publication is the reference book by Daniel Vallero [9]. A further analysis in our book would be, therefore, redundant. However, we can claim that we are facing today a global emergency; because life, health, ability and quality of life of billions of people and of the whole civilization are at risk, we "must" move forward, looking for more reasonable preventative measures and/or adopting already those supported by favourable preliminary experimentation.

Furthermore, the adoption of political strategies aimed to the containment of the emissions, that is, a precise duty of governments and industries, is not sufficient to give guarantees to the population for the following reasons:

(a) The measures adapted to fight the emissions are not efficient or efficacious.
(b) The prediction models—when available—aimed to establish the usefulness/utility of such measures are not perfect or incomplete.
(c) Even if the criticalities indicated in (a) and (b) were resolved, the time needed to reverse the course and to restore the minimal acceptable levels of pollution in given areas is very long (decades) although this process can hopefully bring benefits to the future generations (optimistically from the third generation forward). In the meantime, millions of people will die or will get sick, being the contribution of pollution to total mortality estimated between 8% and 12% (depending on to the cause of death) [18].

We therefore agree with the conclusions of a recent meta-analysis that suggests that "projecting air pollution-related mortality requires a systematic consideration of assumptions and uncertainties, which will significantly aid policymakers in efforts to manage potential impacts of PM2.5 and O3 on mortality in the context of climate change" [12].

Thus, equipment and devices aimed to protect from air pollution (prevent and/or reduce exposure in populations and in clusters of or in single individuals) must have some essential requisites. Those (a) are based on logical considerations, (b) must be extremely effective (best odds of positive outcomes) and (c) must be easy to understand, easy to use and widely available, highly versatile so that anyone could benefit from them.

It is also necessary to define "what people must know" about air pollution so to defend themselves from its harms. This does not differ from what was recently proposed for pulmonary diseases [19], but it extends such indications to the entire population and should be applied for all the diseases.

Two main actions are consequently needed: the implementation of policies aimed to reduce the emission (reducing the chance of exposure) and the widespread use of protective equipment and devices (reducing the entity of exposure).

Anyhow, those strategies are sufficient, even when they are properly planned. Education, counselling and empowerment of the people must be fostered, so to create an extensive awareness of all the problem related to air pollution, promoting proactive behaviours and positive attitudes aimed to protect from toxic substances individuals and communities (*do not pollute, defend yourself, defend others*).

Figure 28.1 summarizes this idea.

It is immediately evident that the three points are interdependent on one another; particularly the second and the third, in order to be effective, imply an active involvement of people promoting as we have seen active and proactive behaviours.

From these considerations come the essential requisites that we mentioned above. The main features are synthetized in Table 28.1.

Those features represent a *conditio sine qua non*; if those conditions are not met, there is no chance of favourable results in the middle and long term. No model of prevention and of health protection, however, flawless from a theoretical point of view, can be effective without the active involvement of people. It means that everyone or at least the most prevalent part of the population should take part.

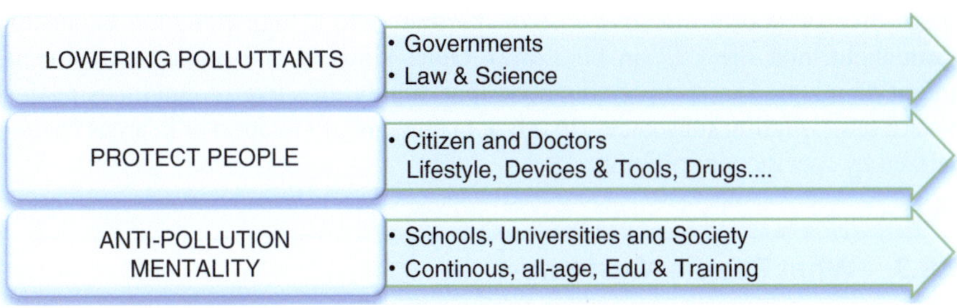

Fig. 28.1 Optimal treatments against pollution diseases should be highly effective against several pollutants and are effective in all individuals and also in presence of other risk factors (i.e. hypertension, age, etc.). C, little stars shown in Fig. 28.2

Table 28.1 The essential requisites of equipment, devices and strategies aimed to reduce the exposure to air pollution and to protect from its effects on health

Feature	Aim
Comprehensibility	Easy to understand and to use also for less educated people
Applicability	Easy to implement also in limited resource settings, in complex scenarios or in non-standard conditions
Economic	Cost-effectiveness of the devices that should be also available for anyone in spite of their social and economic status
Community acceptance	Interventions that are accepted and do not disrupt or minimally affect the normal routines of people or do not create discomforts that may displease people
Long-term compliance	Because of the previous features, people do not cease the utilization of equipment and devices, nurture positive behaviours for all their lives, ask for anti-pollution policies
Capacity to motivate and to move to action	Because of the previous features, people do not accept passively the effect of pollution and actively seek for solutions
Lifelong effect	Secondary to the above features, the use of devices and equipment and the diffusion proactive policies and individual behaviours promote a long-lasting effect in people and communities

28.2 Basic Principles of Treatment: People's Viewpoint

To better understand the implications of this issue, we need to better define the term "treatment" as it is used in this and in the following paragraphs. We consider treatment of the people at risk to develop air pollution-related disease or that already have a disease or a medical condition caused by pollution, all the measure aimed to prevent or cure the disease itself. It includes also the concepts of primary, secondary and tertiary prevention. The pre-primary prevention (or the prevention of those conditions that are responsible to the exposure to a risk factor) is not included in this definition as it is a responsibility of government and industries to prevent/reduce the polluting emissions (Fig. 28.1).

Some traditional preventive interventions against individual risk factors might be ineffective or poorly effective in patients that have a high pollution exposition throughout their life ("E" in Fig. 28.2). Other kinds of anti-pollution treatments might be adopted in specific situations (e.g. highly effective programmes against one or few polluting substances (like B)), in emergency situation or in areas characterized by specific air pollution.

28.3 What People Must Know

28.3.1 Statement One: Pollutant Exposure, Time and Life

From a clinical point of view, we organize adverse or toxic effects of pollutants in function of time. Timing in fact is crucial because the when and how of exposure can determine the outcome. The main classification relates to:

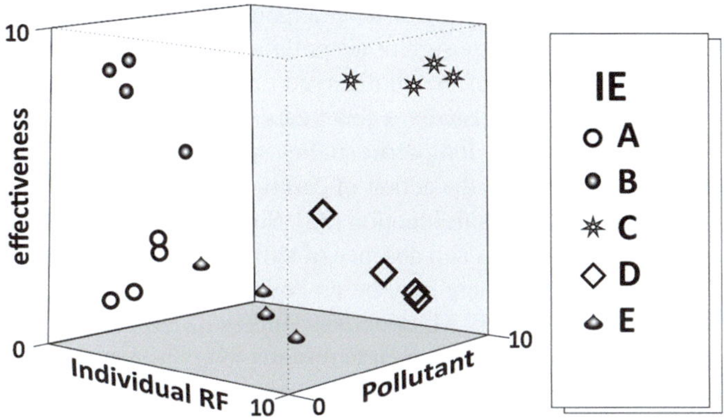

Fig. 28.2 How to decide on which preventive treatments to implement. Individual RF, individual risk factors, like genetic or personal RF (hypertension, hypercholesterolemia, age, family history, etc.). Pollutants, number of pollutant substances removed/concentration reduced. Effectiveness, real effectiveness (including the cost/benefit ratio and the feasibility). See text for details

1. Time of exposition of a subject to a given pollutant (or time*dose, or AUC of time and dose, and other indexes). In this category, we can divide different cases as:
   (a) Toxic action of a pollutant (or intoxication)—effect's timescale: very short times (acute exposure to high doses) that mainly show immediate effects that may or may not result in permanent damages
   (b) Acute action that reveals or unleashes damages resulting in chronic process and in a lag phase—effect's timescale: mainly medium or long term, but some effects may show up after the exposure
   (c) Chronic action (inflammatory, carcinogenic, degenerative, etc.)—effect's timescale: mainly long term, but these actions should be distinguished on the basis of the duration of the exposure (months, years, decades, whole life)
2. Time of life in which the exposure began (e.g. foetal life, childhood, elderly, etc.)
3. Age of the subject especially for acute and short-term exposure, as age corresponds to specific physiopathological characteristics and specific behaviour and environmental factors
4. Time of exposure in correlation to the presence/absence of other individual risk factors that can contribute to damage primarily or secondarily caused by pollution

To these different "times", we could add the time considered as absolute value (expressed according to a known unit of measurement), because the probability of any pathological phenomenon is always a function of the time.

We find in literature some experimental data on how pollutants act over the time: it has been described acute, chronic, relapsing, variable or constant toxic actions. For example, acute exposures exacerbate chronic pollution diseases, like the increased number of hospitalizations or emergency room visits in patient with COPD [20–22]; the chronic exposure of air pollution may have negative effects on obesity and on the incidence of metabolic syndrome [23]. There are many cohort studies on long-time effects (usually a few years, rarely decades) particularly on PM2.5 and on ozone [24]; the long-term studies suggest, in addition, new effects, initially not hypothesized, like the action of carbon particles and, in a broad sense, of PM2.5 on the autonomic health function [25]. Some authors have tried to analyse the bias caused by the measures and absence of temporal series in the different studies and have concluded that there may be an "overestimation of the impact of air pollution on health effects" [26]. Most studies concern different diseases or respiratory symptoms and measurable pollutant exposures (e.g. the particle) [27–30].

In recent years, also because of the high increment of the levels of toxicity of some polluting substances in some large cities around the world, it has been recorded an increase in the number of studies on the acute effects of air pollution (from hypertension and cardiovascular to the number of suicides) [31–37].

Almost all the cited studies and many others do not examine specifically the problem of timing of the exposure, which is essential when providing correct advices to people. Literature data are more consistent when the identification of threshold doses in a very short time is considered; in those cases, generally the dose range is well defined (e.g. it is known that level of PM2.5 between 35 μg/mc and 115 μg/mc can cause, in the following days, acute bronchial illnesses that include also respiratory tract infection) [38]. On the other hand, we have few information on long periods and the effects of the interactions between the polluting substances [39].

When the problem is considered from another point of view, namely, the need to offer proper timely protection throughout the entire life, the current index of pollution and consequently of possible exposure is defective. The Air Quality Index (AQI), widely adopted worldwide; the Air Quality Health Index (AQHI), used in Canada; and the Common Air Quality Index (CAQI), used in Europe, together with numerous other specific indices for single pollutants, do not give the right information to the population. In fact, they are used as suggestions (an example of which is reported in Table 28.1) that underline the usefulness of avoiding some types of activity or exposures for short periods; they on the contrary do not give any information in relation to the continuous exposure, to the long-term effects and to the delayed effect of air pollution or on the second- or third-hand effect of high concentration of pollutants. People therefore do not know when and how to defend themselves when the whole day is considered, let alone when greater units of time are considered. Such types of information offer an illusory idea in contrast that adopting specific behaviours in a limited range of time (e.g. limiting outdoor physical activity for half an hour a day) can be effective in defending people from pollution.

These indexes may be useful to reduce the chance of acute intoxications in case of very high AQI levels, but do not clarify the pollution problem, the cumulative effects of exposure to pollutants and its delayed or long-term consequences. Moreover, some preventive measures try to offer information and protection especially for the subjects at risk; yet, it is very difficult to understand who and what really is at risk.

Table 28.2 Examples of key messages used in different countries were addressed to the people in case of "moderate" air pollution

At-risk individual: health message	General population: health message
Adults and children with lung problems, and adults with heart problems, who experience symptoms, should consider reducing strenuous physical activity, particularly outdoors	Enjoy your usual outdoor activities
Consider reducing or rescheduling strenuous activities outdoors if you are experiencing symptoms	No need to modify your usual outdoor activities unless you experience symptoms such as coughing and throat irritation
None reported	Reduce unnecessary outdoor activity and strenuous outdoor exercise
None reported	Air quality is acceptable; however, for some polluting substances, there may be a moderate health for a very small number of people who are unusually sensitive to air pollution
Usually sensitive people should consider limiting prolonged outdoor activity	None reported
Adults and children with lung problems, and adults with heart problems, should reduce strenuous physical exertion, particularly outdoors and particularly if they experience symptoms. People with asthma may find they need to use their reliever inhaler more often. Older people should also reduce physical exertion	Slight irritations may occur; individuals with breathing or heart problems should reduce outdoor exercise
Adults and children with lung problems, adults with heart problems and older people should avoid strenuous physical activity. People with asthma may find they need to use their reliever inhaler more often	Reduce physical exertion, particularly outdoors, especially if you experience symptoms such as cough or sore throat

From Table 28.2, we can also observe that the definition of "moderated alteration of air quality" is not identical in the guidelines of different countries, even if those are usually very similar. In any case, for most of the people, the meaning of "moderated" is positive, or not negative, and does not raise any alarm. The messages used around the world to give information on the level of air pollution and summarized in Table 28.2 give suggestions diverse enough to be contradictive; from these messages, people cannot understand that even a chronic "moderated" exposure can cause very serious diseases and can highly impact on people's health in the long term. The whole message that people must defend themselves 24 h a day is missing.

In some cases, the simplicity with which people are told that a moderate AQI is not harmful or alarming is worrisome. For example, levels of PM2.5 and PM10 that can cause acute infectious and noninfectious pulmonary diseases increasing the risk of lung cancer are frequently classified as "mild or moderate".

Therefore, there would be no treatment, also intended as prevention, as we have seen, if the main criteria are not met. That is, we must avoid exposition to pollution 24 h a day, every day of the year and during the various phases of life.

Men, women and children should always, almost instinctively, automatically, adopt behaviours, attitudes and choices that reduce exposure. Cigarette smoke can be used as a paradigm in this sense, as most people know that tobacco smoking can lead to serious diseases; nevertheless, smokers do not easily give up smoking even when they are aware of the risks. Knowledge in this case is not enough. The same idea can be applied when discussing strategies to avoid the personal exposition to pollution. The key points are:

(a) Knowledge of what is important to do (see also Chaps. 29 and 34)
(b) Transform this knowledge into "automatic" and efficient actions that become real habits or routines (like washing the hands after going to the toilet or brushing the teeth after eating)

However, to avoid alarmism and to maintain a good and serene quality of life, common sense must be used. From this perspective, it is reasonable to consider individual threshold of exposure that can be considered acceptable, even when it means rising the tolerance. People must have an active defensive attitude from toxic substances: it implies the dissemination of useful and trustworthy information, avoiding catastrophism and asking people to slowly adopt healthy lifestyles, without disrupting their quality of life, provided that the level risk-benefit ratio remains acceptable.

In other words, an ideal educational campaign should contain among the others (see Chap. 34) the ideas exposed in the following paragraphs.

28.3.1.1 Timing of Exposure and Interventions

The risk of disease/symptoms depends on the time and the dose of exposure (Table 28.3). If people are aware of that, public health interventions can be implemented more easily, even when they may appear extreme in case of major emergencies or long-lasting high level of pollution (potentially lethal or not). On the other hand, aware people may effortlessly accept solutions that may sound unpopular like limiting the access to cars to some area of a city or reducing the hours of heating in private or public places. The optimal condition would be a right balance between a watchful and constant request of defence and the diffusion of information on the real level of risk, avoiding false alarms, plus the capacity to adopt drastic measures when the alarm is real and present.

There are analogies with attitudes and forms of education and empowerment suggested in disaster medicine and in major emergency and in mass casualty

Table 28.3 Scales of exposures and consequent appropriate action and reactions

Dose	Acute toxicity	Exposure time	Alert reaction	Scare	Consciousness
Low	Low	Short	Not justified	Not justified	Yes
Low	Middle	Short	Monitor	Not justified	Yes
Low	Middle	Lifelong	Defend	Not justified	Yes
High	High	Short	Defend	Justified	Yes
High	Low	Decades	Defend	Not justified	Yes

incidents. The main difference remains the fact that disaster medicine mainly applies to acute crisis, whilst air pollution maintains a background chronic activity whose effects are still not perfectly known.

In Table 28.3 are summarized the basic actions and reactions to different level and timing of pollution. The actual possible combinations are more than the ones exposed, but it can offer an overview of the possible and most appropriate response to different concentrations of pollutants in the air. It also highlights the differences in terms of risk when the timing of exposure changes even when the level of pollution is similar.

From that summary, the statement "always defend and protect" from air pollution should follow these rules:

(a) Do not induce fear, except when there are conditions of grave danger (but in accordance with the basic rules of risk communication).
(b) Induce protective reactions and prompt useful behaviours aimed to defend and protect from pollution.
(c) Increase awareness, also empowering the people.

28.3.1.2 Chronic Exposure and Interventions

Any diseases caused by pollution depend on the chronic action (in years or decades) of one or more (additive effect or synergic effect) toxic substances. That risk is frequently underestimated, and people only concentrate on the acute outcomes of acute exposures. In the Middle Ages, they used to say "*gutta cavat lapidem non vi sed saepe cadendo*" or, better, as an antique Greek doctor used to say, "with the constancy, a dripping of water also pierces a cliff" (Κοιλαίνει πέτραν ῥανὶς ὕδατος ἐνδελεχείῃ). In other and modern words, a constant background activity, even when apparently harmless, can cause disruptive damages in the long term.

In conclusion, the key message must be "always defend yourself, through your entire life". The difference from the prior point (Sect. 28.3.1.1) is that in case of chronic exposure, it is very difficult to educate people and periodically update them, give the right messages and suggest new defensive strategies as they emerge.

We believe that the key element to obtain these results depends on making people aware of the problem and of the real risks for individual, communities and societies.

28.3.2 Statement Two: Protect Yourself, Your Family and Your Friends

This remains a crucial element for the implementation of effective health policies for prevention. The empowerment of the people is a first step to reduce the risks and consequently the effect of pollution; as we have seen, to reach this goal, people must become aware of the problem and of the risks for their health and their quality of life, so to incept healthy behaviours that could help to reduce the exposure. Hence, it is paramount to create a new consciousness in people and at a social level. In these terms, the general picture is maybe positively changing.

It has been observed that among the youth, there is a growing understanding of the relationships between health and environmental issues [40]; in other cases, as in farmers, three factors (age, income, level of education) influence the degree of cognizance of the pollution problem [41, 42]; other studies underline the importance of social determinants that influence the environmental risk perception [43]. Many other factors have been studied. Unfortunately, families, among which elderly and children, who are for a number of reasons the most exposed and the frailest ones (see Chaps. 10 and 11), have not been sufficiently studied.

The advantage of involving the family, in addition to those ethical terms that largely justify that approach, serves to empower the degree of responsibility of adults and correctly educate the young.

It serves also to optimize the behaviours, or the necessary cost of health protection, and also amplify the final result in terms of health gained in the future (in the hypothesis that the whole family adopts effective and efficient measures).

From this point of view, it seems useful to give unified and equally valid messages to all the family members, other than just identify the ones at high risk, or merely the most exposed ones. The message mentioned above (to protect the frailest) is still valid, for obvious reasons; however the suggestions on what to do (save exceptional cases) must be adopted or be adoptable by the whole family.

An example derived from a real case can be useful to better explain this concept: two parents have much concern for their 12-year-old child with mild intermittent asthma. They live in an industrial city and are exposed to both indoor and outdoor pollution. Measure to protect the child (integrators, medications, anti-pollution masks and anti-allergens) are taken only when he has symptoms. The entire focus of the family is for the child to avoid contact with pollen and, in days of visible smog, to impose the use of a mask. There is no defensive policy adopted to the rest of the family. The father, exposed to his environment's pollution (indoor and outdoor, at the workplace), dies prematurely of lung cancer. A fatality that was probably avoidable, with protective measures extended to the whole family.

A recent study conducted in Shanghai points out that the majority of the ones interviewed "agreed that improving air quality is the responsibility of every citizen, and the joint action of governments and all the citizens should be utilized for enhanced control" [42]. The level of education and family income results to be the best predictors of knowledge and awareness. This indicates a strong social conscience, at least in the analysed sample; however, it is not sufficient to guarantee that the degree of education is sufficient to protect the family as a whole.

For this reason, individual measures must get along with the implementation of actions aimed to safeguard clusters of people (families) that share common aims and, at a higher level, to protect the communities where these same people live.

28.4 What You Must Know to Defend Yourself

As we have seen and as we will see in several parts of this book, knowing does not necessarily mean adopting active and useful habits. However, knowledge is the foundation of any life-changing decisions, as the acceptance of behaviours and measures that may impact normal routines as the one needed to have an active protection again pollution. It is also the base to claim and require actions from governments and industries aimed to reduce the level of exposure in the population. An informed consent is also needed from the ethical and juridical point of view.

In fact, to "educate" is not enough. Concepts can be explained in different ways, and some are more effective than others depending also on the target, as reported in detail in the case of education and new technologies [44].

To confront air pollution, we must be informed on what the pollution is, the risks encountered, etc. Information means, technically, the acquisition of previously unknown knowledge (data, ideas). This alone is not enough: we must educate, to facilitate the acquisition of theoretical knowledge to develop a know-how and enhance personal faculties and vocations.

However, the most important task is to prompt active anti-pollution behaviours in people and patients. It means not only to train but rather to give a deeper background, to improve sensibility and to mould the behaviour, "affecting all the intellectual abilities of the human being, including the emotional one".

Again, "knowing" does not mean "to endorse" and knowing and endorsing do not necessarily mean "adopting" the knowledge to improve the personal state of health.

You may know that air pollution can cause damages, but you cannot endorse the idea that air pollution should be stopped; you can be perfectly aware of the air pollution damages, and you may also share the idea that an action should be taken to prevent and fight tobacco addiction (as reported in the Shanghai Study quoted above), but nevertheless you do not defend yourself against pollution.

To obtain consistent results, we must choose the main purpose of our education; we must define the population targets to which the messages can be sent; we must define when and with what the messages must be proposed (Table 28.4).

Table 28.4 Anti-pollution education strategy (see text) upper part shows some theoretical examples of possible strategies for antipollution education

eHEdu main aim	Who	For which purpose	When	With what	Expected outcome
To educate	All people	Rise personal knowledge on pollution (in food, air and water)	Always, long life	Internet and mass media	To be defined
To inform	Pregnant women	Avoid newborn brain damage	Expectant mother during the pregnancy	Ad hoc courses with obstetrician	Reduction of AP exposition
To form	Students	Preventive measures against water pollution	Every year (10-h course) during the primary school	By teacher and books	Verify personal skill and capacity in avoiding polluted water

The examples might be near to infinite. The table underlines the need of correct Edu-Strategy. AP, air pollution

It is clear that a general director is needed to orchestrate the educative action against air pollution; in this process, all the actors and possible stakeholders, from the family to school, places of work, advertisers and the media (Internet, WhatsApp, WeChat and others), have to take part and have their say. Those industries that work in the field of air pollution should also take part in the discussion, giving their contribution and sharing their experience; it can result in a good educational model, as long as industries have a strong know-how and are particularly involved.

It is similarly evident that the educational campaigns must not be done at random, expressing hearsay rather than scientific knowledge. Those must be coherent, convincing, usable and reusable. They must be clear to understand considering the different audiences (that potentially come from any social and cultural background and any group of age).

There must be procedures adopted to measure the outcomes of these educational campaigns.

Carbon monoxide can offer a valid example of that. CO is a toxic and lethal substance whose dangerousness is very underestimated, due to ignorance (lack of knowledge, mainly in rural areas, where people do not think they may be at risk) and due to excessive confidence (mainly in urban areas, where people believe that in cities technology is safe enough); the risk associated with CO in fact is undervalued also by some so-called experts. On the contrary, CO is extremely dangerous not only for its known acute and possibly lethal effects for high-dose exposition but also at lower concentrations. One recent study for underlines how CO must be studied in depth: even a mere increase of 1 mg/mc of CO determines the increase of cardiovascular risk of 4.2% [45]. In addition, it has been recently underlined how some toxic and still not perfectly studied CO effects could affect also the central nervous system and should lead to public health concerns [46]. The same study suggests that these findings must be widely available and that people must be aware of the risks related to CO. In terms of education control, the example of CO is particularly helpful, because it involves those who smoke [29], those who have a home source of chronic production or acute production of CO and those who are exposed to CO coming from traffic-related pollution or other particular environments.

Table 28.5 gives an overview of the possible actions aimed to reach this goal. For each proposed action (those in the table are just a few examples), there shall be done a valuation of opportunity, of cost and of feasibility in order to choose which campaign to activate. Another parameter is the urgency of the single interventions. Anyhow, as described above, the idea of imposing only or principally educational campaigns as a reaction to emergency situations is not convincing. Air pollution is a global problem and complex, which will still last decades. It however deserves well programmed strategies and not extemporaneous reactions dictated by emergency.

The basis of all the future campaigns against pollution is still to make people aware that "to prevent you must know". If this basic motivation is missed, all the actions of information, learning and education will be more difficult and less effective.

A last recommendation is needed: all the teaching acts aimed to reach the population that are done by free citizens, students and professionals must be clear, simple and scalable; they must solicit the natural curiosity of man. The fight against

Table 28.5 An example of different Edu-Goal for lowering carbon monoxide deaths and incidents

Aim	Who/purpose	Verify
To inform	All people: CO is highly toxic, lethal gas. Smokers: Smoking produces an increase of CO in your blood and in expired air. Adults: CO is produced by traffic, by warming systems (etc)	Per cent of population informed—what to understand (by questionnaires in population sample)
To educate	Adults, university students, secondary school students (other): On CO effects on human health and on means to avoid CO production (omissis) Families, workers, etc.: As measure of CO concentration in air (in function of actual risk)	Learning verification
To form	All people: Automatic, instinctual, recognition of places at risk of high concentration of CO and immediate adoption of appropriate defensive measures (omissis)	Individual skill and behaviour Ad hoc questionnaires
To promote anti-pollution actions	Technicians are more likely to control and correct CO production. Citizens avoid highly and congested traffic. Citizens are aware that normal masks (including nanofibre masks) are not useful against CO. Those who buy, sell or install air conditioning equipment, gas ovens, etc. adopt preventive measures against the production of CO. Have in home CO sensors and air vents to avoid accumulation of CO, etc	Per cent of adhesion to correct action by each category. Specific test and examine for technicians and person designed at security. Certifications. Law fulfilment

The table is simplified (real table must be designed as the 28.4 with a row for each correspondence, row by row, of aim, what, when, etc.). Complete anti-CO table is a matrix of about 25 row × 10 columns. Each educative action should be coherent and linked to each other

pollution must not become a long list of laws, norms and precepts; their use must be limited to selected cases (discipline on security, on equipment, etc.). But the existence of rules, even when appropriate, cannot substitute the education of the population.

28.5 Where and How to Defend Themselves

Talking about complexity, there is an exact idea of how much information would be sufficient to answer to the question on how to defend (from what) and where. The question is still more complex if we think of the timespan we have mentioned in Sect. 28.1.

From an educational point of view, we retain that it is not plausible to give all the possible answers, in teaching programmes, media, pamphlets and ads or through Apps, social media or specifically designed software. The risk of inducing confusion to the population can have highly counterproductive results.

We therefore propose a simplified response, hopefully not too different from the most complex solution, dividing the substances that make up air pollution according

Table 28.6 Some anti-pollution defence tools: matrix between type of toxic substance and living place

Pollutant	∝ ($)	INDOOR		OUTDOOR		JOB/SCHOOL	
		Individual	Home	Individual	Street	Individual	Workplace
Gas	0.0001	Professional AG Mask (°)	Difficult ! Measure and open the	Professional AG Mask (°)	(^)	Professional AG Mask	Centralized Professional Air Systems
Heavy Metals and Toxic Inorganic Molecules	0.001	Professional AG Mask (°)	Difficult ! Measure and open the	Professional AG Mask (°)	(^)	Professional AG Mask	Centralized Professional Cleaner
Toxic Organic Compounds (*)	0.01	Perhaps Specific Pro Mask	Air Cleaner (usually poor effective/ efficient)	Perhaps Specific Pro Mask	(^)	Specific Pro Mask	Centralized Professional Cleaner
Virus	0.5	Perhaps Specific Pro Mask (often used surgical or N95 mask) and	Only High Efficient Semi-pro Cleaner (like HEPA 14/ULPA filters)	Perhaps Specific Pro Mask (often used surgical or N95 masks)	No	Perhaps Specific Pro Mask (often used surgical or N95 masks)	Public Health Measures
Ultrafine Particulate Matter	0.5	Specific Pro Mask	HEPA14/U LPA filters	Specific Pro Mask	(^)	Specific Pro Mask	Centralized Professional Cleaner
Humidity/ Flugge Drops	1	No	Dehumidifiers/ humidifiers	none	No	No	Centralized Professional Air Systems
Fine Particulate Matter	5	Quality masks and quality	HEPA14/U LPA like filters	Quality masks and quality nose filter	(^)	Quality masks and quality	Centralized Professional Air Systems
Bacteria, small molds	5	Quality masks and Hygiene	HEPA14/U LPA like filters	Quality masks and perhaps quality	(^)	Quality masks	Public Health Measures
Almost visible dust (+)	50	Mask and Nose Filters	EPA >11 and other filtration	Mask and Nose Filters	(^)	Mask and Nose filters	Centralized Professional Air Systems
Pollens, Allergens, Molds	50	Quality masks and quality nose filter	EPA >11/12 and other filtration systems	Quality masks and quality nose filter	No	Quality masks and quality nose filter	Centralized Professional Air Systems

Grey box, preventive measures suggested and/or reasonably effective and/or extent of the citizen. Light grey box, possible usefulness. Professional AG masks are specific anti-gas masks, professional and costly, usually specific for one or few gas types. These professional masks are equipped with highly specific filters (usually external); perhaps the new nanofiber masks could be used in the future (see the following chapters). The table is applicable only to AIR contaminants (pollutant or microbes dispersed in the air). The table refers to systemic protection (not included eye, skin, hair protection)

to a different classification, based on the dimension of the pollutant, and that depend on the location of exposure (Table 28.6).

The table serves to create a problem, rather than find or propose solutions. The contents of each box are realistic, the information given are reliable, but the table is

Table 28.7 Simplified matrix of basic "minimal" information about pollutants

Pollutant	mm	Visible	Example	Key Word
Very small molecules	0.000001	No, not even with microscope	Carbon monoxide (gas), alcohol or gasoline vapours	Measure, escape, avoid, prevent
Very fine particulate matters	0.001	No, only with microscopes	PM2.5, other fine dusts, aggregates of chemicals	Defend, measure, avoid
Microbes	0.001	No, only with microscopes	Virus, bacteria, moulds	Defend, hygiene measures
"Dust"	0.1	Very little: In few cases visible by lenses	Pollens, house dust, others	Defend, avoid

clearly difficult to understand and use by people with no specific competences in this field.

Meanwhile, in imposing educational campaigns, we must carefully consider the intersection of the matrix presented in Table 28.6, but we have to propose a simpler and clearer version to the general population (Table 28.7). We think that this type of diagram is simple enough to be understood by all and can contain enough of the basic information needed to understand how to protect oneself.

This distinction is based on the general idea that some physical barriers against pollutants (e.g., masks, air purifier) offer a valid protection, while in other cases (e.g., toxic gases) there are no adequate defensive means, or they are not available to the general population. The microbes represent a separated class because there can be valid and different measures: like filtration, similar to that used for fine particles, to pest control, hygiene and public health strategies to avoid contact or reduce the microbial load.

28.6 Contrasting Wrong Information and Cliché on Air Pollution

In the last decade, the way people access to information has largely changed. Especially the development and the easy accessibility of the World Wide Web and lately of the social media have made an incredible mass of information available. These same contents are often not filtered or come from unreliable sources that may (or may not) have interest in diffusing false information. When it comes to air pollution, a number of sites offer a number of solutions that mainly refers to defensive measures (medications, diet, equipment). Some of these sites are reliable, some are untrustworthy, some have also political claims or endorse ideology linked also with ecological issues, some are promoted by retailers that offer useless solutions, and so on. Most of those websites give contrasting information, and unfortunately the less reliable ones are the ones that often become viral in the socials. The risk of confusion, or of doubt and indecision, is very high. In our opinion at this moment, some of the most dangerous cliché to knock down are:

Wrong:

1. Air pollution causes harms to the lungs and respiratory tract, and consequently:
   (a) Air pollution causes harm when the subject manifests symptoms (cough, dyspnoea, etc.).

Correct: Air pollution cause harm to the entire body (due to systemic diseases) even when there are no symptoms. In many cases, when symptoms are manifested, it is often too late to prevent.

Wrong:

1. The air is polluted when we see smog or clouds or when there is bad weather.
   (a) When the sky is clear, the air is clean.

Correct: The air in some cities is almost always polluted (e.g. Air Quality Index >100 for 200 days a year), even when the sky is clear. Smog may be a sign of a high concentration of pollutants; just as in some cases, we only have clouds and the concentration is not very high.

Wrong:

1. You can see pollution or smell its bad odour.

Correct: Many toxic gases are odourless, colourless and invisible. Even fine and very fine particles cannot be seen under many lighting conditions. In some cases, they can simulate a romantic and beautiful sky.

Wrong:

1. I can decide when to defend myself from pollution in relation to what I see and feel.

Correct: We defend against pollution following the values of Air Quality Index (or other similar indices), available in the whole world. It is useful to measure the pollution in your own living space. Measuring pollution is complex, but doable even in the non-professional level.

Wrong:

1. Air pollution is the principal enemy to health.
   (a) Therefore, we must defend solely or mainly from air pollution.

Correct: The harms of air pollution are summed to those of water, food and soil. Pollutants often carry infectious diseases. The harms of pollution are summed to those caused by other risk factors (e.g. hypertension, diabetes, hypercholesterolemia, hypercysteinaemia, age, etc.)

Correct.2: Health is unique; we must defend from all that causes harm. Air pollution causes a lot of harm.

Arguments against commonplaces, prejudices and beliefs on health (and on pollution) have no boundaries and exceed the objectives of this book. The above-mentioned examples serve to reiterate the principal facts that (a) we must understand what people think and believe; (b) then a study will be done on coherent and comprehensible messages; (c) when possible, there will be a measurement of the learning outcomes and then of the real outcome on health; and (d) when necessary, there will be an improvement and update of the messages but always with respect to the basics of "educational continuity", with "interdisciplinary and collaborative governance" as proposed for academic education in clinical settings [47]:

1. The principle of "coherence" among the teachings and consequently of the sources of information and knowledge. It implies the convergence of the teachings (intended as educational actions) toward shared educational targets
2. The principle of "non-contradiction", that is secondary to the previous one. It is more focused on the idea that those information and notions that are due to become practical actions do not have to be in contradiction among each other. It is referred to all those actions related to prevention, prophylaxis, treatment or rehabilitation that the patient has to go though.
3. The principle of "selection" of the teaching contents and of the educational targets: a rational scheduling has to be planned according to priority scales, secondary- to general-shared criteria (i.e. ethical, political or scientific criteria). This education plan can be also secondary to the vocational profile of a single student, to personal agreements and to arranged choices that do not follow those same shared criteria, but that are more focused to other less known or explored issues. Is on the contrary to be banned a casual choice of the educational pathways and target, or a planning secondary to biases choices, or to transitory trends. This selection is strongly needed also because of the exceptional number of available information. The arguments to be discussed and taught as well have to be selected according to the previous criteria: educational continuity, coherence and non-contradiction.
4. The principle of "evaluation" of the outcomes that come from the educational action. In some cases, this can be done only through the use of surrogate end-points (e.g. with the use of questionnaires that evaluate the level of understanding of a given information), in a proper approach that should require a continuous feedback that could set a starting point, followed by short- and long-term evaluations. Whenever it is possible, this kind of evaluation has to be done on the real outcomes or at least on close indicators of the outcomes. Nevertheless, this is a difficult approach that is uneasy to put into practice

28.7 Do We Need the Ten Commandments?

Governments, institutions responsible for education, mass media and managers of information and the citizens themselves must face a complex problem of information, education and learning, for the reasons explained in the previous paragraphs

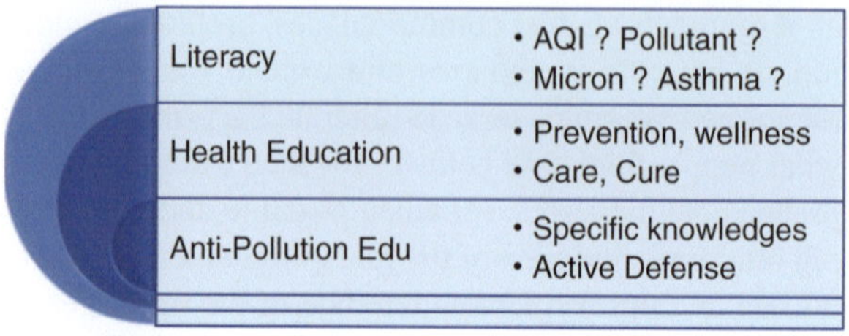

Fig. 28.3 Education on pollution defence, health education and knowledge and culture are coexistent (see text)

but also because part of an unique cultural and learning process, as explained in Fig. 28.3; in fact, the field of pollution protection is intertwined with many others.

To achieve a good education and pollution defence, it is necessary to know the specific terms regarding chemistry, physics, medicine, etc., and one must know how to picture the pollution problem in general terms of health protection and well-being. Many studies show factors in socioeconomic characteristics [43, 48–51], together with the cultural level or scholastic title with the capacity to defend oneself from pollution and to adhere to preventive programmes [41, 42, 52, 53] though with some exception, in the case of certain areas in Latin America [54].

In particular, the social disease is a factor that becomes very fragile with respect to environmental diseases and, together, is often combined with the level of "literacy" and of the very low knowledge of one's own health state.

This calls for wise indications, shareable with everyone, men, women and children, easy to understand and valid through time (at least in the reasonable limits we can hypothesize today) that are at the base of specific learning acts.

These indications must be useable by all, starting from the weakest people, with minimal resources, and must be extended, deepened and made to perfection in function of the level of basic culture of the single population group.

These "laws" (suggestions) must therefore constitute "common knowledge" of the population on health and its protection, on which to build a solid education; with reference to the emergency of pollution, some Chinese author has recently defined it the "*airpocalypse* of this century" and perhaps of this millennium, but still applicable in another sense to the health sector.

28.8 Some Key Points to Ponder

1. To protect health, you must know the enemy (what it is, what it causes and how to confront it). Inform yourself choosing your sources. Ask for advice from doctors and experts.
2. Protect yourself and your family and your friends. Defend with more attention to the more fragile (children, aged, pregnant women, the sick) and the more exposed. Inform those who do not know of the risk and means of defence.
3. Protect your entire body. In particular, the nose, mouth and respiratory tracts are entry ways of pollution, but also the eyes, ears, skin, face and hair are always exposed. Pollution creates systemic diseases that affect the entire body (heart, brain, lungs, reproductive organs, sense organs, etc.).

4. Protection every day, every week, and every year. The battle against pollution is a resistance match to fight for the entire life.
5. Pay attention to chronic harm, tied to polluting substances (e.g., lung cancer, atherosclerosis and heart attacks or stroke), but also defend from acute causes from gases or highly lethal substances or high smog levels (e.g. AQI > 300).
6. We almost never see pollution and it has no odour. If you want to defend, you must know what it is and what it has. Consult the air quality indices (e.g. Air Quality Index, AQI) to know, day-to-day, the pollution of your local area or the measure of principal pollutants present in your home or garden.
7. Consider that there are toxic substances in the air but also in food and soil and water and have more harmful effects on persons predisposed to diseases of risk factors (e.g. hypertension and diabetes).
8. Choose with maximum attention the instruments with which to defend (e.g. masks, intranasal spirals, air purifiers, protective creams, goggles). Use them at the useful time, even day and night when appropriate. Choose eco-friendly products.
9. The defence against pollution requires an active behaviour from the citizens. Maintain an active role against pollution and also avoid the act of pollution. Follow attentively the government and health authority advice.
10. Do not let fear and irrational behaviour prevail. We can battle pollution and the risk of disease that it brings can be reduced.

Conclusions

The strategies needed to protect people from air pollution are several, but most of them still need to a better forge. Air pollution remains a complex matter and most of its effects on human health are unknown.

People have a crucial role, as changes in behaviours can represent the only escape to the devastating consequences of this new global catastrophe. They also have to accept some measure, although unpopular, provided that those do not impact on the quality of life.

Education remains the basic; people must become aware of the problem and understand that prompt actions are needed now. But awareness may come only from a proper dissemination of knowledge and of messages that everyone should be able to understand, in spite of the origin, the cultural, social or economic background or of the age group.

The goal is the implementation of measures able to constantly protect, every hour and every day, people and particularly those more exposed to the effects of air pollution as elderly or children.

References

1. Thanh BD, Lefevre T. Assessing health benefits of controlling air pollution from power generation: the case of a lignite-fired power plant in Thailand. Environ Manag. 2001;27(2):303–17.
2. Li J, Guttikunda SK, Carmichael GR, Streets DG, Chang YS, Fung V. Quantifying the human health benefits of curbing air pollution in Shanghai. J Environ Manag. 2004;70(1):49–62.

3. Mestl HE, Aunan K, Seip HM. Health benefits from reducing indoor air pollution from household solid fuel use in China--three abatement scenarios. Environ Int. 2007;33(6):831–40.
4. Cesaroni G, Boogaard H, Jonkers S, Porta D, Badaloni C, Cattani G, et al. Health benefits of traffic-related air pollution reduction in different socioeconomic groups: the effect of low-emission zoning in Rome. Occup Environ Med. 2012;69(2):133–9.
5. Voorhees AS, Wang J, Wang C, Zhao B, Wang S, Kan H. Public health benefits of reducing air pollution in Shanghai: a proof-of-concept methodology with application to BenMAP. Sci Total Environ. 2014;485-486:396–405.
6. Pope CA 3rd, Cropper M, Coggins J, Cohen A. Health benefits of air pollution abatement policy: role of the shape of the concentration-response function. J Air Waste Manage Assoc. 2015;65(5):516–22.
7. Li S, Williams G, Guo Y. Health benefits from improved outdoor air quality and intervention in China. Environ Pollut. 2016;214:17–25.
8. Meng J, Liu J, Fan S, Kang C, Yi K, Cheng Y, et al. Potential health benefits of controlling dust emissions in Beijing. Environ Pollut. 2016;213:850–9.
9. Vallero DA. Fundamentals of air pollution. Amsterdam: Academic/Elsevier; 2014.
10. Fuster V, Ross R, Topol EJ. Atherosclerosis and coronary artery disease. Philadelphia: Lippincott-Raven; 1996.
11. West J, Cohen A, Dentener F, Brunekreef B, Zhu T, Armstrong B, et al. What we breathe impacts our health: improving understanding of the link between air pollution and health. Environ Sci Technol. 2016;50:4895.
12. Madaniyazi L, Guo Y, Yu W, Tong S. Projecting future air pollution-related mortality under a changing climate: progress, uncertainties and research needs. Environ Int. 2015;75:21–32.
13. Pope CA 3rd, Rodermund DL, Gee MM. Mortality effects of a copper smelter strike and reduced ambient sulfate particulate matter air pollution. Environ Health Perspect. 2007;115(5):679–83.
14. Clancy L, Goodman P, Sinclair H, Dockery DW. Effect of air-pollution control on death rates in Dublin, Ireland: an intervention study. Lancet. 2002;360(9341):1210–4.
15. Clancy L. Reducing lung cancer and other tobacco-related cancers in Europe: smoking cessation is the key. Oncologist. 2014;19(1):16–20.
16. Laden F, Schwartz J, Speizer FE, Dockery DW. Reduction in fine particulate air pollution and mortality: extended follow-up of the Harvard six cities study. Am J Respir Crit Care Med. 2006;173(6):667–72.
17. Tanaka S. Environmental regulations on air pollution in China and their impact on infant mortality. J Health Econ. 2015;42:90–103.
18. Evans J, van Donkelaar A, Martin RV, Burnett R, Rainham DG, Birkett NJ, et al. Estimates of global mortality attributable to particulate air pollution using satellite imagery. Environ Res. 2013;120:33–42.
19. Jiang XQ, Mei XD, Feng D. Air pollution and chronic airway diseases: what should people know and do? J Thorac Dis. 2016;8(1):E31–40.
20. Martins LC, Latorre Mdo R, Saldiva PH, Braga AL. Air pollution and emergency room visits due to chronic lower respiratory diseases in the elderly: an ecological time-series study in Sao Paulo, Brazil. J Occup Environ Med. 2002;44(7):622–7.
21. Tsai SS, Chang CC, Yang CY. Fine particulate air pollution and hospital admissions for chronic obstructive pulmonary disease: a case-crossover study in Taipei. Int J Environ Res Public Health. 2013;10(11):6015–26.
22. Ko FW, Chan KP, Hui DS, Goddard JR, Shaw JG, Reid DW, et al. Acute exacerbation of COPD. Respirology. 2016;21:1152.
23. Wei Y, Zhang JJ, Li Z, Gow A, Chung KF, Hu M, et al. Chronic exposure to air pollution particles increases the risk of obesity and metabolic syndrome: findings from a natural experiment in Beijing. FASEB J. 2016;30:2115.
24. To T, Zhu J, Villeneuve PJ, Simatovic J, Feldman L, Gao C, et al. Chronic disease prevalence in women and air pollution—a 30-year longitudinal cohort study. Environ Int. 2015;80:26–32.

25. Mordukhovich I, Coull B, Kloog I, Koutrakis P, Vokonas P, Schwartz J. Exposure to subchronic and long-term particulate air pollution and heart rate variability in an elderly cohort: the Normative Aging Study. Environ Health. 2015;14:87.
26. Boffetta P, Vecchia C, Moolgavkar S. Chronic effects of air pollution are probably overestimated. Risk Anal. 2015;35:766.
27. Assad NA, Balmes J, Mehta S, Cheema U, Sood A. Chronic obstructive pulmonary disease secondary to household air pollution. Semin Respir Crit Care Med. 2015;36(3):408–21.
28. Gao Y, Chan EY, Li L, Lau PW, Wong TW. Chronic effects of ambient air pollution on respiratory morbidities among Chinese children: a cross-sectional study in Hong Kong. BMC Public Health. 2014;14:105.
29. Zhang Q, Li L, Smith M, Guo Y, Whitlock G, Bian Z, et al. Exhaled carbon monoxide and its associations with smoking, indoor household air pollution and chronic respiratory diseases among 512,000 Chinese adults. Int J Epidemiol. 2013;42(5):1464–75.
30. Mao W, Xia W, Chen J. Air pollution and chronic cough in China. Chest. 2013;144(1):362–3.
31. Zhang Q, Qi W, Yao W, Wang M, Chen Y, Zhou Y. Ambient particulate matter (PM/PM) exposure and emergency department visits for acute myocardial infarction in Chaoyang District, Beijing, China during 2014: a case-crossover study. J Epidemiol. 2016;26:538.
32. Ye X, Peng L, Kan H, Wang W, Geng F, Mu Z, et al. Acute effects of particulate air pollution on the incidence of coronary heart disease in Shanghai, China. PLoS One. 2016;11(3):e0151119.
33. Dominguez-Rodriguez A, Abreu-Gonzalez P, Rodriguez S, Avanzas P, Juarez-Prera RA. Short-term effects of air pollution, markers of endothelial activation, and coagulation to predict major adverse cardiovascular events in patients with acute coronary syndrome: insights from AIRACOS study. Biomarkers. 2016;22:389–93.
34. Yang Y, Cao Y, Li W, Li R, Wang M, Wu Z, et al. Multi-site time series analysis of acute effects of multiple air pollutants on respiratory mortality: a population-based study in Beijing, China. Sci Total Environ. 2015;508:178–87.
35. Wilson WE. The relationship between daily cardiovascular mortality and daily ambient concentrations of particulate pollutants (sulfur, arsenic, selenium, and mercury) and daily source contributions from coal power plants and smelters (individually, combined, and with interaction) in phoenix, AZ, 1995-1998: a multipollutant approach to acute, time-series air pollution epidemiology: I. J Air Waste Manage Assoc. 2015;65(5):599–610.
36. Mannisto T, Mendola P, Liu D, Leishear K, Sherman S, Laughon SK. Acute air pollution exposure and blood pressure at delivery among women with and without hypertension. Am J Hypertens. 2015;28(1):58–72.
37. Bakian AV, Huber RS, Coon H, Gray D, Wilson P, McMahon WM, et al. Acute air pollution exposure and risk of suicide completion. Am J Epidemiol. 2015;181(5):295–303.
38. Zhang A, Zhou F, Jiang L, Qi Q, Wang J. Spatiotemporal analysis of ambient air pollution exposure and respiratory infections cases in Beijing. Cent Eur J Public Health. 2015;23(1):73–6.
39. Yu HL, Lin YC, Kuo YM. A time series analysis of multiple ambient pollutants to investigate the underlying air pollution dynamics and interactions. Chemosphere. 2015;134:571.
40. Pluhar ZF, Piko BF, Kovacs S, Uzzoli A. "Air pollution is bad for my health": Hungarian children's knowledge of the role of environment in health and disease. Health Place. 2009;15(1):239–46.
41. Chen M, Chen M, Lu Y, Wang L, Huang Y. The farmers' perceptions of ANPS pollution and its influencing factors in Poyang Lake region, China. Water Sci Technol. 2016;73(7):1591–8.
42. Wang R, Yang Y, Chen R, Kan H, Wu J, Wang K, et al. Knowledge, attitudes, and practices (KAP) of the relationship between air pollution and children's respiratory health in Shanghai, China. Int J Environ Res Public Health. 2015;12(2):1834–48.
43. Vandermoere F. Hazard perception, risk perception, and the need for decontamination by residents exposed to soil pollution: the role of sustainability and the limits of expert knowledge. Risk Anal. 2008;28(2):387–98.
44. Gaddi AV, Capello F, Manca M. eHealth, care and quality of life. Milan: Springer; 2014.

45. Chen R, Pan G, Zhang Y, Xu Q, Zeng G, Xu X, et al. Ambient carbon monoxide and daily mortality in three Chinese cities: the China air pollution and health effects study (CAPES). Sci Total Environ. 2011;409(23):4923–8.
46. Levy RJ. Carbon monoxide pollution and neurodevelopment: a public health concern. Neurotoxicol Teratol. 2015;49:31–40.
47. Hirsh DA OB, Thibault GE, Cox M. "Continuity" as an organizing principle for clinical education reform. N Engl J Med. 2007;356(8):858–66.
48. Schultz AS, Nowatzki J, Ronson G. Effects of household socialization on youth susceptibility to smoke: differences between youth age groups and trends over time. Am J Public Health. 2013;103(7):e39–42.
49. Hicken MT, Adar SD, Diez Roux AV, O'Neill MS, Magzamen S, Auchincloss AH, et al. Do psychosocial stress and social disadvantage modify the association between air pollution and blood pressure?: the multi-ethnic study of atherosclerosis. Am J Epidemiol. 2013;178(10):1550–62.
50. Yadama GN, Peipert J, Sahu M, Biswas P, Dyda V. Social, economic, and resource predictors of variability in household air pollution from cookstove emissions. PLoS One. 2012;7(10):e46381.
51. Naess O, Piro FN, Nafstad P, Smith GD, Leyland AH. Air pollution, social deprivation, and mortality: a multilevel cohort study. Epidemiology. 2007;18(6):686–94.
52. Kan H, London SJ, Chen G, Zhang Y, Song G, Zhao N, et al. Season, sex, age, and education as modifiers of the effects of outdoor air pollution on daily mortality in Shanghai, China: the public health and air pollution in Asia (PAPA) study. Environ Health Perspect. 2008;116(9):1183–8.
53. Cakmak S, Hebbern C, Cakmak JD, Vanos J. The modifying effect of socioeconomic status on the relationship between traffic, air pollution and respiratory health in elementary schoolchildren. J Environ Manag. 2016;177:1–8.
54. O'Neill MS, Bell ML, Ranjit N, Cifuentes LA, Loomis D, Gouveia N, et al. Air pollution and mortality in Latin America: the role of education. Epidemiology. 2008;19(6):810–9.

Basic Characteristics of Antipollution Devices and PPE

29

Claudia Bandini, Ibar Sabatini, Michela Dimilta, and Giulia Baldessari

29.1 Risks and Effects of Air Pollution

> There's so much pollution in the air now that if it weren't for our lungs there'd be no place to put it all. (Robert Orben)

Usually jokes are used to cause healthy and abundant laughter, but, unfortunately, the mentioned quote sums up in a few words a sad and dangerous reality which affects everyone. There are 5.5 millions of people who are victims of indoor and outdoor air pollution every year, as reported by a recent research from Canadian researchers of British Columbia University and Health Effects Institute [1].

This data is upsetting. To make a comparison, consider that in 2001 Twin Towers disaster in New York, there have been 2974 victims. Due to exposition to toxic chemicals, death rate is equivalent to the victims caused by five 9/11 attacks a day, every day. Chemical and toxic exposure mortality has been widely investigated from researchers in the last 50 years: hundreds of publications are available in the scientific literature, proposing statistical analysis, panel studies, and reviews on air pollution outcomes. In particular, it has been verified the association between air pollution chronic exposure and cardiovascular diseases and mortality [1–11].

C. Bandini (✉)
University of Bologna, Bologna, Italy
e-mail: cbandini3@gmail.com

I. Sabatini
Proeon, San Benedetto del Tronto, Italy

M. Dimilta
University of Bologna, Bologna, Italy

G. Baldessari
BLS Group, Milan, Italy

F. Capello, A.V. Gaddi (eds.), *Clinical Handbook of Air Pollution-Related Diseases*,
https://doi.org/10.1007/978-3-319-62731-1_29

The WHO database shows that in the first decades of the twentieth century [12, 13], cardiovascular diseases (ischemic heart disease, stroke, lower respiratory infections, and chronic obstructive lung disease) are the principal causes of mortality worldwide.

The majority of the studies [4, 8] are concerned on the effects of short- and long-term exposure to the particulate matter (PM) which is mainly produced by automotive vehicles, by domestic heating, and by industrial plants: PM10, PM2.5, and ultrafine particles (UFPs).

In particular, fine and ultrafine particulates are considered more dangerous because their small size allows the particles to reach easily the alveoli in the lung [14–16].

Studies reveal a certain association between dysfunction of the autonomic nervous system [2, 17, 18]and air pollution, especially the occurrence of myocardial infarction [19–21] linked to short exposure, the incident of stroke [22, 23], and lung cancer.

Not only particulate matter causes bad effects on human health, but several studies show that also ozone (O_3) [24–30], nitrogen oxides (NOx) [25, 31, 32], sulfur oxides (SOx) [25, 33], carbon oxides (COx), and volatile organic compounds (VOC) impact negatively on cardiovascular system.

Other dangerous sources are diseases related to microscopic size pathogens or submicroscopic ones which are carried by indoor and outdoor air both. Among them, we can mention the H1N1 virus, the avian flu, and many other diseases. They are carried by animals or by men through continuous immigration flows.

Several pathogens are easily carried by fine particles of dust, droplets, or aerosol. They become optimal vehicles toward lower airways until alveolar level. Recent studies showed that most of the flu viruses are carried by dust or aerosol particles with diameters between 1 and 4 μm. These viruses are carried by PM with diameter sizes lower than 1 μm (till 200–600 nm) too. These sizes allow them to reach the alveoli and to promote the diffusion and the pathogenic capacity of viruses.

In Fig. 29.1, sizes of particle diameters are reported in order to clarify the differences between them and their capacity to reach human respiratory system.

We will explain the basic characteristics of antipollution devices and PPE and their clinic benefits for civilians:

1. Abatement of infectivity and pathogenicity indices for both bacterial diseases (with a drastic reduction or cancelation of pathogenic bacteria) and in the majority of aero-transmitted viral diseases.

   This feature translates into:

   (a) Containment of the number of infected/infectious subjects and sick individuals, in the course of epidemic or pandemic diseases (influenza, avian influenza, etc.)
   (b) Individual protection of the patient's family in the case of contagious diseases at home (highly contagious measles up to 98%; paramyxovirus with 150–300 nm), communities, or restricted environments
   (c) Effective protection of the support staff and host staff working for immigrants from other countries

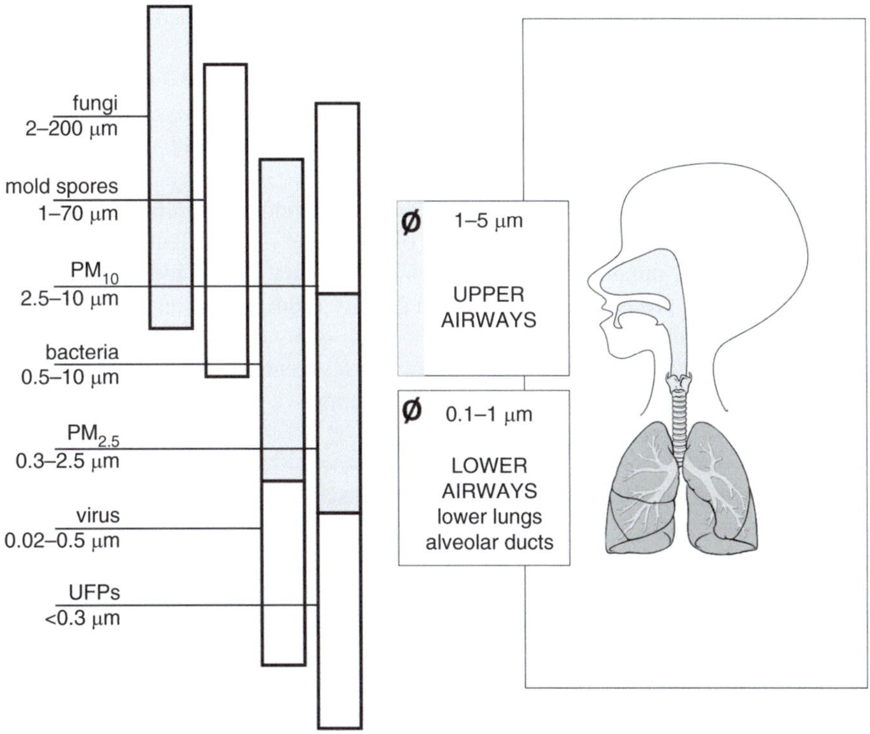

Fig. 29.1 Sediments of dangerous particles for human health based on types and size

2. Drastic reduction of the risk of importing new strains, bacteria, and fungi through uncontrolled immigration flows and to facilitate its dissemination. Personal protective equipment against fine and ultrafine toxic dust. All citizens of all ages exposed to outdoor pollution are considered (in some cases indoor too).
3. Protection of travelers in areas with high concentration of particulates and toxic aerosols.
4. Highly effective protection for patients already suffering from cardiovascular impairment (e.g., chronic obstructive pulmonary disease stage > IIa or pulmonary hypertension) from dust. We consider outdoor and indoor dust which causes or encourages the progression of disease. Reduction in the same patients of the risk of bacterial or viral superinfection.
5. Reduction of the incidence of nosocomial infections.
6. Protection for patient and staff (of surgery, medication, etc.) from the risk of contamination due to open skin lesions, operating rooms, and inhaling pathogens.
7. Protection of health workers, patients, and their families in all at-risk departments (e.g., infectious diseases, intensive care units, long hospitalizations of immobilized or frail patients, etc.).

We will list the main commercial areas based on clinic and medical prevention view. They are:

(a) People and individuals: personal defense from air pollutants, fine dust, and ultrafine dust particularly (not retained by the masks sold on the market currently)
(b) Health and non-health border staff. Bilateral and mutual protection from infectivological risks. Reduction of the risk of importing new virulent strains
(c) Medical staff, nursing staff, and patients in hospitalization and care environments (better protection for patients in surgery room, reduction of nosocomial infections, etc.)
(d) General population, population clusters and communities, and individual citizens: protection against epidemics and pandemics

29.2 Selection Parameters and Evaluation of Antipollution Devices

> Instead of trying to invent the methods to eliminate the microbes in the wounds, wouldn't it be smarter not to introduce them in the first place? (Louis Pasteur)

Surely Louis Pasteur in the nineteenth century had a brilliant idea, but unfortunately actual commercial masks can't still provide a perfect barrier. Aware of high risks for human health established in several epidemiological studies on the effects of air pollution, we are going to collect the information described in international standards, scientific results present in literature, and some medical and technical assessments in order to produce solid guidelines which allow the consumer to really evaluate the parameters described above.

It is important not to confuse the devices designed for an industrial application from those which have an "urban" use (also said home and personal devices). In this chapter, we will examine precisely the latter, in particular home detectors, respirator masks, and nasal filters.

We talked about how the risk of disease due to pollution is related to how much the chemical and physical properties of a substance are dangerous for human health and how much time a person is exposed to that substance ($R = H \times E$) [34]. It's easier intervening in the exposure factor for common people. They can move to another place or protect themselves in the same location. Antipollution devices have this goal: reducing the exposure in order to reduce the risk of disease.

We will focus on describing the logic behind these devices and how producers rate them. But now we start considering consumer's point of view, listing the main characteristics to take into account when buying an antipollution device (outlined also in Fig.29.2):

1. Work efficiency: is the product compliant to the most important world standards (USA and Europe)? Does the device perform as intended?
2. Possibility to detect the efficiency: does the device show an immediate/evident validation of its filtering/monitoring efficiency?

Fig. 29.2 Scheme of evaluation parameters of antipollution devices

3. Health effectiveness: is it theoretically possible that the device could have some advantageous clinical effect if used for an adequate time? Does any trial exist, proving that there are favorable effects on people or epidemiologic observations on the outcomes?
4. Dangerousness: does it expose to risks? Could it produce unknown or toxic subproducts, stop working, or become dangerous in certain conditions?
5. Comfort/portability: does the device fit well/is inconvenient to carry?
6. Ease of use: is it a "mistake-proofing" device? Is it easy to use even by inexpert people as kid and ancients?
7. Cleanliness/maintenance and reusability: does the object easy to clean and to fix in case of break or malfunction? Can it be used more than once? Is it clearly stated in instruction handbook how many times the device can be reused effectively?
8. Longevity: is the device resistant? Is there a risk that it would break down when used?
9. Versatility: could it be used in different conditions, environment, or applications?

10. Appearance: does it appear beauty or ugly and awkward?
11. Cost: has it a fair cost/benefit ratio?
12. Commercial availability: is the technology commercially available?
13. Ecology: is the device useful to protect from pollution without affecting negatively the environment? For instance, is it recyclable?

This list represents the major aspects that consumer will have to take into account. It's obvious that the producers are aware of these needs, but they can't cover them all with a single product. We may find, for example, masks with a high grade of comfort but with a poor level of reusability, or we may notice other masks with a beautiful appearance but that are extremely uncomfortable during operation or a detector that is able to reveal very low concentration but just for a restricted selection of substances. The final choice is up to the consumer.

29.3 Monitoring and Data Collecting

In the previous paragraph, we have indicated the parameters that help to quantify the risk of getting disease due to that substance, i.e., hazard and exposure. In this section, we want to deepen the characteristics of devices used in monitoring parameters to consider in order to use them effectively and technical considerations to improve performances.

Let's start considering the Air Quality Index (AQI), which is an index used by worldwide government agencies to inform the population on the rate of air pollution and consequently on the degree of risk for human health. Each country has developed own standards and procedures for the calculation of this index; usually only the main pollutants are considered (ground-level ozone, particle pollution, carbon monoxide, sulfur dioxide, and nitrogen dioxide), and the final data is evaluated based on a scale that shows all the possible range of risks (from low to very high).

The substances are not measured all with the same frequency; in fact some are monitored every hour, others only once or few times a day.

Although the AQI is a useful information tool, it has got some limits. In fact, this index describes a general condition and an average estimate of air quality, and it does not allow the population to know if there are specific risks related to certain substances, and especially it doesn't allow to restrict the valuation to a limited area, because it describes a condition of large outdoor areas. This means that the data is not always truthful and trustworthy for population.

The necessity of local-scale monitoring is dramatically increasing, especially in developing countries where indoor pollution, due to solid fuels used for domestic uses as cooking and heating, causes millions of deaths. The monitoring of PM and other toxic substances has not been implemented only by the governmental agencies but by every citizen who wants to protect his own health and that of the people with whom he lives.

Large-scale monitoring is therefore not sufficient, so mobile and personal air quality monitoring and pollutant detectors were developed to reveal in real time both the indoor and outdoor air quality.

Sampling should be carried out by the device both qualitatively (identifying different types of particles) and quantitatively (measuring the concentration).

Capturing data can also be differentiated according to the extent of the inspected area: in addition to monitoring on large scale, there are devices that can monitor specific indoor areas or make detections around the person that wears it; in this last case, the portable device should measure according to the characteristics of respiration; in this way, it could be very useful for those doctors who have to identify the causes of given pathologies [34].

Table 29.1 contains the main gas hazardous to human health.

Table 29.1 Main harmful gases with sources and effects on human health

Chemical formulation	Name	Source	Main effects
CO	Carbon monoxide	Combustion processes (hydrocarbon engines in vehicles, power plants, and heating systems), tobacco smoke	Headache, fatigue, and breathing problems (it prevents tissue oxygenation) Death
CO_2	Carbon dioxide	Chemical and industrial combustion processes, humans and animals respiration	It is the greenhouse gas most dangerous for the increase of planet temperature
NO_x	Nitrogen oxides	Combustion processes	Irritation to eyes and respiratory tract, pulmonary edema and methemoglobin formation risks, photochemical smog, soil acidification, and eutrophication
SO_2	Sulfur dioxide	Sulfur-containing fuel combustion in industrial plants	Acid rains, others
O_3	Ozone		Irritation of the respiratory system
VOC (including C_6H_6)	Volatile organic compounds	Paints, solvents, cleaning products, gasoline, and natural gas	Benzene is carcinogenic; others contribute to the greenhouse effect
$C_{20}H_{12}$	Benzo[*a*]pyrene (polycyclic aromatic hydrocarbons, PAHs)	Petrol and wood combustion	Carcinogenic effects
PM	Particulate matter	Combustion in internal combustion engines, in domestic heating, asphalt and tire wear, emissions from agricultural and mechanical manufacturing, tobacco smoke	Cardiovascular diseases

The first patents related to air pollution detectors date back to the 1970s, and they have been deposed in the USA.

The first forerunner (1972) of currently available devices was covered by a data recording surface that allowed to measure the presence of pollutants by evaluating the deterioration of the coating that could have been composed by an oily or vaporous chemical substance that reacts to different substances floating in the air in different ways, such as change in color or erosion. A couple of years later in 1974 in the USA, a new environmental monitoring device patent was deposed, which was innovative for the time but almost rudimentary compared to modern devices; it was unable to give an immediate reading of the substance quantity in the air, as the air had to pass and react with several layers of dry reagents (each one sensible to a different substance) and the chemical reaction products could have been analyzed just at a later time by opportunely equipped technicians, therefore just by sector specialists.

Thanks to advancements in electronics and communications, as well as the development of nanotechnologies and new materials, nowadays different types of professional mobile monitoring devices were developed to measure the quantity, the size, the type, and the mass of particles and nanoparticles [35]. Also for personal use, nowadays low-cost sensors are available in commerce (semiconductors, infrared, oxidation, catalytic, electrochemical, etc.). With respect to the professional device, they have got both advantages and disadvantages.

Among positive aspects, there are small dimensions, low power consumption, low cost, easy application in portable devices, and large distribution for personal use; the major disadvantage is in general the low accuracy and sensitivity to particle mass concentration with respect to professional instruments.

Nevertheless, in literature research reveal that also dust detector (Sharp GP2Y1010) can effectively monitor PM if calibrated appropriately [35]. The low-cost detector SYhitech DSM501 was deeply studied: data show the same monitoring accuracy of fixed stations and the capacity of detect the concentration of particles with size bigger than 2.5 μm.

An ideal mobile particle monitoring device should permit also the characterization of particles based on chemical composition and morphology, and it should identify the emission sources [36].

Although it is desirable to develop devices that can detect even small concentrations of all dangerous substances, nowadays available technology does not yet allow this, so it is wise to choose the device that best suits the needs of the consumer.

Now we try to evaluate in Table 29.2 the parameters previously discussed in reference to home pollutant detectors.

29.4 Respiratory Protective Devices

"A respirator is a device that protects you from inhaling dangerous substances, such as chemicals and infectious particles. Respirators are among the most important pieces of protective equipment for working in hazardous environments. Selecting

Table 29.2 In deep evaluation of basic characteristics of home pollutant detectors

General parameter	Description of parameters	Guidelines
Work efficiency	Is the detector able to monitor low concentration of dangerous substances? Does the device recognize the chemical and morphology composition of the particles?	As a cure, you should buy a device able to filter the substance which causes the disease. As prevention, you should buy a device able to filter all the substances which cause diseases you fear the most
Possibility to detect the efficiency	Does the detector show an immediate/evident validation of its monitoring efficiency? Can consumers understand if device have a good efficiency?	You should check product features. If the product is an electric device, you should check its consumption and then if cost-effect ratio is worth. You can find these information in product instructions or in producer's website
Ease of use	How much easy is to use the detector even by inexpert people as kids and ancients? It is easy to understand the recorded concentration and if this could be dangerous for health? Does the device produce an alarm sound in case of high concentrations of pollutants? Is available a detailed instruction handbook? Is it necessary to read it carefully?	You have to check product instructions always. You shall know all its features before buying it. You can find them in producer's website, sell points, and consumer's forums. In the last case, you have to check if the website contains valid information put by reliable people
Cleanliness, maintenance, reusability	Does the detector need to be cleaned often? Is it easy to fix in case of break or malfunction?	Every device has an expiring time. You should use it under the rules written in the instructions. This behavior will help the device to work longer so to delay the time to buy a new one
Health effectiveness	Does the device provide some advantageous clinical effect if used in certain circumstances with an adequate attention? Does any trial exist, proving that there are favorable effects on people, or epidemiologic observations on the outcomes?	These devices protect from pollutants. Your health benefit is based on the controlled pollutant. Producers show which medical standards their device follow usually. This is a competitive asset for producers and a guarantee of effectiveness for buyers
Dangerousness	Does the device produce toxic substances?	Producers must show this information but they are not displayed adequately. You should check product instructions and producer's website wisely

(continued)

Table 29.2 (continued)

General parameter	Description of parameters	Guidelines
Comfort/ portability	Is the detector handy and light or bulky, heavy, and difficult to move?	The device management is based on consumer's possibilities. You should choose your product valuating weight and measures. This recommendation should help you for choosing a product which fits your habits and lifestyle
Versability	Can the device monitor a high number of different pollutants efficiently? Can it be used both for indoor and outdoor monitoring?	You buy a device for two reasons mainly: cure or prevention. You should choose your product based on its purpose
Appearance	Does it appear beautiful/ well-looking or is it ugly and awkward?	Product appearance is another competitive asset for producers. If you consider the aesthetic factor, you should buy a device which fits better the environment
Longevity	Is the device resistant? Is there a risk that it would break down when used? How much time does it work well?	You know how much time you must use a device. You should check information about devices expiring time. Choose the most convenient device based on this information
Cost	Has it a fair cost/benefit ratio for consumer?	Choose how much you would pay to cure your disease or to prevent one. Then compare your idea with market prices
Commercial availability	Is it easy to buy? Is it also available in the shops or just online? How much is shipping time?	In the worst scenario, choose if you prefer to wait to obtain a particular product or to buy a similar one immediately
Ecology	Is the detector useful to monitor pollution without affecting negatively the environment, generating additional pollution?	You should buy an antipollution device which doesn't create pollution. This would be a collateral damage for yourself and for other people

the right respirator requires an assessment of all the workplace operations, processes or environments that may create a respiratory hazard" [37]; this is the definition published in General Respiratory Protection Guidance for Employers and Workers by OSHA (Occupational Safety and Health Administration-USA).

The prevention from the spreading of airborne epidemics and personal protection from contagion and from exposure to toxic particulates can be contained through the use of suitable facial filter equipment: wearing a face mask also for short-term expositions can remarkably reduce the negative effects of pollution on cardiovascular (arterial pressure and heart rate) and breathing apparatus [15]. The quality of a facial filtering device is not measurable only by its filtering efficacy, but it depends on several aspects, such as configuration, material, way of use, or cost. Typologies are obviously numerous, and the consumers have to pay attention in evaluating the best for their needs.

European and US standards classify these instruments as PPE (personal protective equipment) if their function is protection of who will wear them and surgical devices, in case if they have to limit the transmission of infective agents between patients and staff during the surgical operations [38].

Since there are no specific regulations for "urban" antipollution devices, we will consider and analyze both types trying to underline advantages and negative features, limits, and applications of both although the direct comparison is not so easy.

29.5 Classification, Legislation, and Standards

As mentioned above, the European (EN) and the American Standards are the most detailed regarding the occupational respirators and surgical masks, as they classify and describe accurately the functional and technological characteristics of such devices, as well as the methodology to evaluate them. In order to consider a device in compliance, all its components must satisfy the technical requirements and positively pass all the practical tests required by each relevant standard. Among the Chinese Standards, there is only one standard regarding respiratory protective devices, (YY/T 0969-2013), which specifies use, requirements, test methodologies, packing, transport, and storage of single-use medical masks, used for generic medical (not surgical) applications.

On Table 29.3 are reported the main standards which regulate breathing apparatus protection devices.

In the subsequent paragraphs, we will study deeply the technical characteristics and peculiarities of particulate and chemical cartridge/gas mask respirators that are those products able to filter air pollutants.

In the USA, NIOSH (National Institute for Occupational Safety and Health) "develops new knowledge in the field of occupational safety and health and to transfer that knowledge into practice" as reported on its website. One of its many tasks is approving Particulate Filtering Facepiece Respirators that comply the standard; all the approved devices are reported online, divided according to the type. There are seven classes of filters that differ according to the filtering capacity; they are summarized in table GGG.2 (http://www.cdc.gov/niosh/).

Table 29.3 Principal International Standards about respiratory protective devices in the workplace

Issuing organization	Standards/ regulamentations	Title
Occupational Safety and Health Administration (USA)	OSHA 29 CFR 1910.134	Respiratory protection standard
The National Institute for Occupational Safety and Health (USA)	NIOSH 42 CFR Part 84	Respiratory Protective Devices
American National Standards Institute (USA)	ANSI Z88.2-1992	American National Standard for Respiratory Protection
	ANSI Z88.10-2001	Respirator fit test methods
National Fire Protection Association (USA)	NFPA 1404-2006	Standard for Fire Service Respiratory Protection Training
	NFPA 1500-2007	Standard on Fire Department Occupational Safety and Health Program 2007 Edition
Food and Drug Administration (USA)	ASTM F2100-11,ASTM F2299, ASTM F1862	Standard specification for performance of material used in medical face masks
Council of the European Union	89/686/CEE	Council Directive of 21 December 1989 on the approximation of the laws of the Member States relating to personal protective equipment
European Parliament and Council of the European Union	(UE) 2016/425[a]	Regulation(EU) 2016/425 of the European Parliament and of the Council of 9 march 2016 on personal protective equipment and repealing Council Directive 89/686/EEC
European Standards—European Committee for Standardization (EU)	EN 136:1998	Respiratory protective devices—full face masks—requirements, testing, marking
	EN 140/AC:2000	Respiratory protective devices—half masks and quarter masks—requirements, testing, marking
	EN 143/AC:2006	Respiratory protective devices—particle filters—requirements, testing, marking
	EN 149+A1:2010	Respiratory protective devices—filtering half masks to protect against particles—requirements, testing, marking
	EN 405+A1:2010	Respiratory protective devices—valved filtering half masks to protect against gases or gases and particles—requirements, testing, marking
	EN 529:2006	Respiratory protective devices—recommendations for selection, use, care and maintenance—guidance document
	EN 14683:2014	Medical face masks—requirements and test methods
	EN 14387+A1:2008	Respiratory protective devices—gas filter(s) and combined filter(s)—requirements, testing, marking
Health and Safety Executive (UK)	HSE 282/28	Fit testing of respiratory protective equipment used for controlling exposure to asbestos fibres
China National Standards and Industry Standards (CN)	YY/T 0969-2013	Single-use medical face mask
Canadian Standards Association (CA)	CSA Z94.4-02	Selection, use, and care of respirators
Australian Standard (AU)	AS/NZS 1715:1994	Selection, use and maintenance of respiratory protective devices

[a]New regulation on DPI repealing the 89/686/CEE from 21 April 2018

Table 29.4 Classes of filters for NIOSH-approved filtering facepiece respirators

Type	Particle filtering efficiency (penetrating particle size is considered equal to 0.3 μm)	Resistance to oil
N95 (most common)	95%	NO
N99	99%	NO
N100	99.97%	NO
R95	99.97%	Partial
P95	95%	Strong
P99	99%	Strong
P100	99.97%	Strong

Table 29.5 Classes of filters according to EN Standards

EN 143	Tax of protection	Comments	EN 149	Particle filtering efficiency (at 95 L/min airflow)	Breathing resistance Inhal. 30 L/min	Inhal. 95 L/min	Exhal. 160 L/min
P1	Low protection: max filtering capacity =4 × TLV (threshold limit value)	All devices are designed to protect against both solid and liquid aerosols	FPP1	80	0.6	2.1	3
P2	Medium protection: max filtering capacity = 10 × TLV		FPP2	94	0.7	2.4	3
P3	High protection: max filtering capacity = 30 × TLV	Both P3 and FPP3 are indicated in case of pandemic influenza or against biological agents [40, 41]	FPP3	99	1	3	3

In Table 29.4 the classification according to European Standards is reported: P1, P2, and P3 in EN 143, where P means "particle," and FPP1, FPP2, and FPP3 in EN 149, where FFP means "filtering facepiece particle." From the first class to the third one, the filtering efficiency increases, and the penetration of the particles, tested with sodium chloride and paraffin oil, decreases as shown in the table (particles with diameter 0.6 μm for sodium chloride and 0.4 μm for paraffin oil [39]). European Standards specify both the percentage of airborne particles blocked from the filter and the threshold limit value (TLV): environmental concentrations of airborne chemicals below which workers are able to operate during working time without any adverse effect (according to actual medical knowledge) (Table 29.5).

Filters that are adopted in PPE devices employed in working environments must be selected as a function of:

- The typology of toxic substances which the device should prevent the contact
- The job that has to be performed
- The execution modality and the length of the intervention

These simple rules can be also considered true for urban use; it is fundamental that the person wearing the device is efficiently protected by toxic substances in the air for the whole exposure time for every daily activity (for instance, antipollution urban face mask should guarantee filtration efficiency, without affecting negatively the breathing during sport activities).

Concerning surgical mask, the European Standard EN 14683, in add to determine construction, performance requirements, and test methods for the surgical masks, classifies them according to bacterial filtration efficiency (tested using particles of mean size of 3.0 ± 0.3 μm), differential pressure, and splash resistance (IIR).

In this standard, surgical masks are considered device of class 1, and there are no references to filter capacity against inert particles.

In Table 29.6, the classification according EN 14683 is reported.

In the USA, the organism that regulates medical devices is the FDA (Food and Drug Administration) (http://www.centexbel.be/surgical-masks-in-europe-and-the-usa), which provides in ASTM F2100-11, to test the pressure drop, the bacterial filtration efficiency and splash resistance as EN standard does, and in addition it determines also the filtration efficiency related to particulate matter, the fire resistance, and the biocompatibility.

Table 29.7 shows classification criteria of surgical masks in base of US norm.

Table 29.6 Classification of surgical mask types according to EN 14683:2014

Test	Type I	Type II	Type IIR[a]
Bacterial filtration efficiency (BFE) (%)	≥95	≥98	≥98
Differential pressure (Pa)	<29.4	< 29.4	<49.0
Splash resistance pressure (mmHg)	Not required	Not required	≥120

[a]Splash-resistant models

Table 29.7 Classification of surgical mask types according to ASTM F2100-11, ASTM F2299, and ASTM F1862 (USA)

Test	Level 1	Level 2	Level 3
Bacterial filtration efficiency (BFE) (%)	≥95	≥98	≥98
Differential pressure (mmH_20/cm^2)	<39.23	<49.1	<49.1
Filtration efficiency of particles of 0.1 μm	≥95	≥98	≥98
Min splash resistance pressure(mmHg)	80	120	160
Flame spread	Class 1	Class 1	Class 1

Table 29.8 Requirements that must be printed directly on the filtering device of EN-approved and NIOSH-approved respirators (67)

EN-approved device	NIOSH-approved device
1. CE mark 2. The classification of the mask (e.g., FFP2) 3. The manufacturer's name	1. NIOSH logo 2. NIOSH Testing and Certification approval number (e.g., TC-84A-XXXX) 3. NIOSH filter series and filter efficiency level (e.g., N95, N99, etc.) 4. Model number or part number: The approval holder's respirator model number or part number, represented by a series of numbers or alphanumeric markings (e.g., 8577 or 8577A) 5. The manufacturer's name

Table 29.9 Requirements that must be printed on the packaging of EN-approved and NIOSH-approved respirators (67)

EN-approved device	NIOSH-approved device
1. Identification number of notified body 2. The manufacturer's name 3. Type-identifying marking 4. Classification (class and reusability) 5. The number and year of publication of the European Standard 6. At least the year of end of shelf life (usually mm/yyyy) 7. The sentence "see information supplied by the manufacturer" or the related pictogram  8. The manufacturer's recommended conditions of storage (temperature, humidity) or the related pictograms 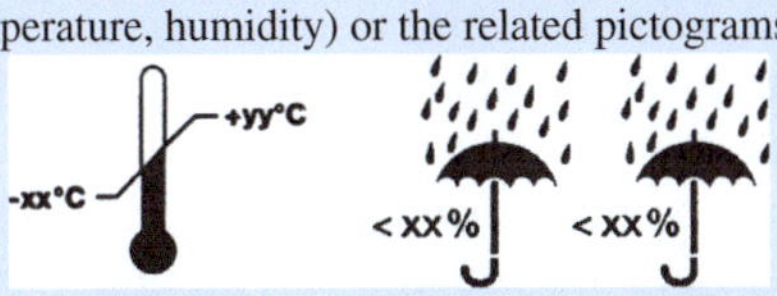 9. If the filtering device pass the dolomite test, it shall be additionally marked with the letter "D"	1. NIOSH logo 2. NIOSH Testing and Certification approval number (e.g., TC-84A-XXXX) 3. Seal of the Department of Health and Human Services 4. Applicant's name and address

It is recommended to purchase face masks tested according to the latest European or American Standard); both standards provide different recommendations for the marking of the device itself and the information printed on the packaging (which is intended as the smallest unity commercially available). In Tables 29.8 and 29.9, the requirements for EN-approved and NIOSH-approved respirators are reported (http://www.cdc.gov/niosh/npptl/topics/respirators/disp_part/) [42, 43].

29.6 Technical Evaluation of Characteristics and Features of Facial Masks

As previously written, active standards rule on the professional and surgical uses. There is no legislation or standard which imposes pressing technical characteristics for urban use devices. The absence of quality standards for face masks for personal use has caused an overflow on the market of devices absolutely inefficient, almost a scam, especially in China, where the demand of antismog goods like face masks is highly elevated.

According to World Health Organization, N95 products, certified by NIOSH, can be considered devices that provide good protection from air pollution, but we think that, due to the high complexity of the topic, a deeper evaluation of most important characteristics is necessary.

For instance, surgical mask, although is not considered in International Standard a PPE, may similarly be a great effective protection device because its quality has to be evaluated according to different points of views. In Table 29.10, each general parameter abovementioned was investigated specifically for face masks.

29.7 Nasal Filters

This kind of device is similar to masks. It's meant to be used by common people and domestic use. Its purpose is to protect the respiratory tract filtering air pollutants and pollen. It's composed by two "wings" (filters) which are connected by a plastic arc as shown in Fig. 29.3.

The wings are built in order to filter pollen, dust, bacteria with size of 1–3 μm, and pollution such as PM10 and PM2.5. We know masks, their use, and their features. Why should people prefer nasal filters instead of masks? There are some differences. Masks cover both mouth and nose but nasal filters work in the nose only. Consumers can do actions such as eating, properly speaking, etc. without the typical problem of a mask. If you were a mask, you may risk air pollution infiltration from the sides. Nasal filters work for every direction because they are disposed in the nose. You have to pay attention how you breathe. You may do physical exercises or suffer for some form of sinusitis. These (and other) reasons may force you to breath with the mouth. This fact makes nasal filter ineffective. You should prefer this kind of device for aesthetic reasons. A mask covers the face and it may be not well-looking, depending on the model. Nasal filters are more difficult to detect because the only visible part is the plastic arc. The Table 29.11 summarizes the aspects.

Table 29.10 In deep evaluation of basic characteristics of facial masks

General parameter	Description of parameters	Guidelines	Practical example
	How well and how long does the mask have to filter particulate and bacteria in the air? Does the mask block a great amount of PM, bacteria, and virus? What is the expiring time of the filter?	We can consider a good filtering efficiency if the facial mask is able to filter 100% of particles with diameters up to 300 nm, 99,X% with size between 250 nm and 300 nm and 80% with size up to 50 nm	
		Cotton or nonwoven tissue filters are to be avoided because they are inefficient against particulate and more easily perishable	
		Filtering particles with size of 300 nm at least with absolute-deterministic valuation (100% of filtered particles) and particles with diameter bigger than 50 nm with probabilistic valuation using sigmoidal function with a minimum of 80% of filtered particles for diameters smaller than 99,X% and between 250 nm and 300 nm	
	Does the mask show an immediate/evident validation of its filtering efficiency? Can consumers understand if device have a good filtering efficiency?	It is very important that the filtering efficiency has been successfully tested by the manufacturer and results have to be available to the consumers	
	How much easy is to use the mask? Is it easy to use even by inexpert people as kids and ancients?	The mask should be easily usable by all (also kids, elders, population of low-income countries), possibly it should not require a specific training, and in case of wrong positioning, it should guarantee a certain filtering capacity loss less than 5–10%. The mask should be "mistake proofing"	Imagine you need to buy a mask because you live in a city with a high grade of pollution. You don't want to pay too much for a mask so you decide to buy a cheap one. The product you have chosen is cheap but you have to put a new one after few days. Of course, the producers wrote in the instructions all the rules for a correct use of the product. What happens if you don't read those instructions properly? You may wear the mask for too much time. The device fails its purpose because it has already reached its expiring time. So you feel safe wearing an item which is not protecting you at all
	Is available a detailed instruction handbook? Is it necessary to read it carefully?		

(continued)

Table 29.10 (continued)

General parameter	Description of parameters	Guidelines	Practical example
	Does the mask need to be cleaned often? Is it easy to fix in case of break or malfunction? Can it be used more than once? If yes, is it easy to clean? Is it clearly stated in instruction handbook how many times the device can be reused effectively?	Cleaning with water and natural soap for reusable devised in rubber or silicon. For devices made of nonwoven materials, a cotton wool with disinfectant should be used for cleaning the part of the device that touches the face	After you used and properly cleaned the device, you can use a plastic bag for storage. In this way, the device will not be exposed to possible contamination present in the air
		Keep the devices away from dust and toxic substances when they are not worn	
		While the devices are not worn, respect the manufacturer's recommendations for conditions of storage (temperature, humidity)	
	Does the device provide some advantageous clinical effect if used for an adequate time and with the correct modality? Does any trial exist, proving that there are favorable effects on people or epidemiologic observations on the outcomes?	The fake sense of security brings the individual to expose for longer time to pollutant agents; clearly the use of ineffective devices causes a worsening of the health of the user	Chinese people wear masks which can filter dusts with sizes of 10–20 μm barely. They breath PM10 and PM2.5 pollutants and/or all kind of fine and ultrafine dusts (<0.3 μm)
		To increase health effectiveness scrupulous respect of suitable hygiene standards is advisable: washing of the hands before and after the removal of the mask from the face, as well as a proper disposal of the device if not reusable to avoid the contact with other human beings. This precaution is essential if the subject is affected by certain pathologies that weaken the immunity system and becomes useful to allow protection on virus and bacteria	

Dangerousness	Does the mask cover both mouth and nose? There are air leaking between mask and face?	Face mask has to protect both the nose and the mouth although a certain percentage of inhaled air passes through the noses, based on the type of particles	
	Does it cause health problems related to skin or allergic reactions or inhalation/ingestion of toxic substances?	It has to stick perfectly to the face to avoid dangerous air leaking	
		Materials must be nontoxic and hypoallergenic and they must maintain wholeness	
		Replace immediately the device if it gets damaged or if there are any issue with respiration or other negative effects	
Comfort/ portability	Does the mask fit comfortable on face? Does it permit a good breathability? Does it avoid smelling unpleasant odors? Can the mask cause fog on glasses?	When you choose a device, it is important to take account of the breathability, as not every mask allows for comfortable ventilation	
		The mask should not be heavy, irritating, and uncomfortable	
		It should provide systems to not smell unpleasant odors and shouldn't cause fog on glasses	
		Silicone is often preferred for its comfort	
Versatility	Does the mask work as protection in the consumer's usual routine without compromising it? Does it able to perform correctly his function also while person is doing exercises such as jogging or biking?	A valid device is designed to be equally effective in both inspiratory and expiratory flow side, in order to protect both people who inhale dust or pathogens (viral and antipollution protection) and people who are close to the bearer of the mask, from exhaled air (e.g., protection of the surgical field during surgery)	
	Does the mask protect both from air pollution and bacteria and virus?		
Appearance	Does the mask appear beautiful/ well-looking or is it ugly and awkward? Are people embarrassed to wear it in public?	It's advisable to choose a mask that can be worn without discomfort. If the wearer feels its appearance pleasant, he will wear it more often	

(continued)

Table 29.10 (continued)

General parameter	Description of parameters	Guidelines	Practical example
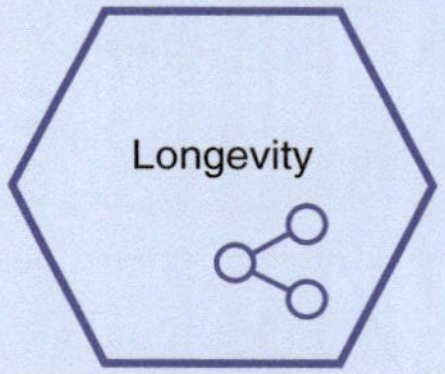	Is the mask resistant? Is there a risk that it would break down when used? How much time does the mask work well?	It's advisable that the mask is made with wear- and humidity-resistant materials If it is not a reusable device, it has to guarantee equally wholeness and protection for all the time indicated as its durability Typically rubber and neoprene are more rigid and consequently more durable than other materials	
	Has it a fair cost/benefit ratio for consumer?	The cost of the mask has to be adequate to its performance and to the reduction of the risk of disease. In any way, the price has to be affordable for almost a certain number of habitants of medium-low-income countries	
	Is it the mask easy to buy? Is it also available in the shops or just online? How much is shipping time?	Consumers should be able to buy quickly and easily masks both online and directly into the store; the products should be available also for people living in more isolated zone as countryside	
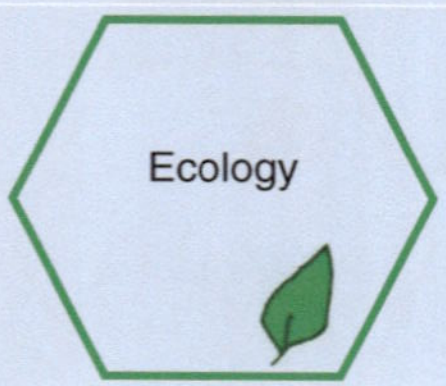	Is the mask useful to protect from pollution without affecting negatively the environment, generating additional pollution?	It is important to take into account if the mask is eco-friendly, that is, if the product has a good impact on environment (for instance, recyclable materials, nonpolluting manufacturing and packaging)	

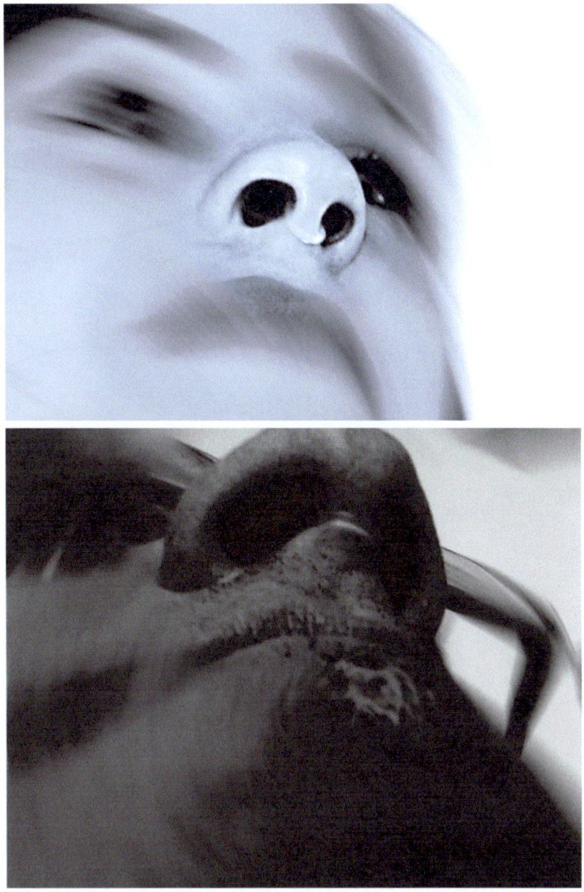

Fig. 29.3 Application of nasal filters

Table 29.11 Evaluation of basic characteristics of nasal filters

Parameters	Nasal Filter	Masks
Protection	Nose	Mouth and nose
Infiltration	Breathing by mouth	Sides of the mask
Actions not permitted	Smoking	Speaking clearly, eating, etc.
Portability	Small size	Dedicated package
Aesthetic	Hard to detect	Visible

Conclusions

We shall require a substantially new manner of thinking if mankind is to survive. - Albert Einstein

The educational aspect is fundamental; people don't need just to receive information and data, but they need also to be adequately motivated so the need of action will be perceived as a common necessity. Only if all individuals consider

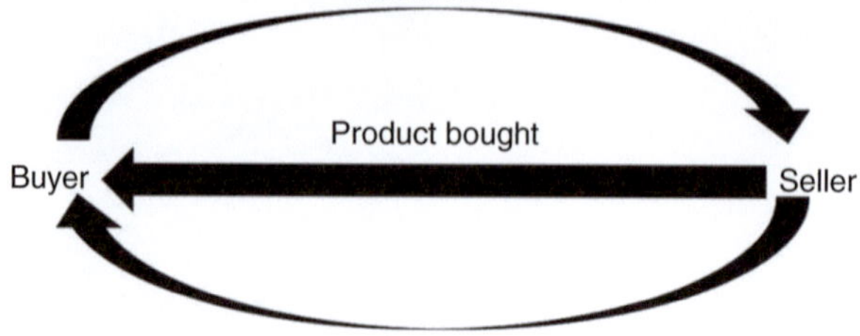

Fig. 29.4 Information exchange during the purchase process

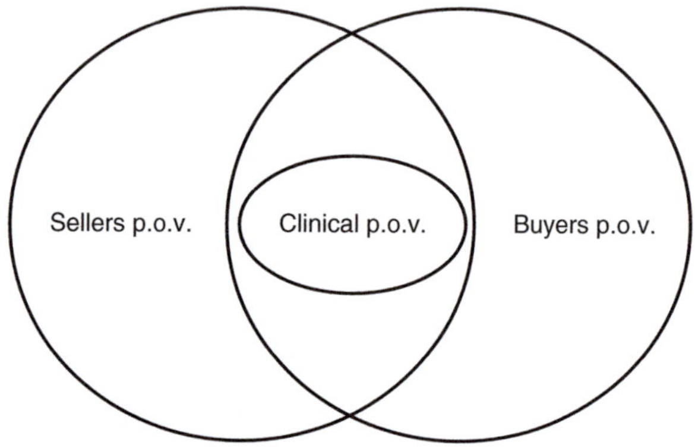

Fig. 29.5 Sellers, buyers, and clinical point of view

the health of the whole population worthy of protection as its own, they will be able to participate actively to the improvement of the condition of their community, by limiting pollutant emissions, by protecting themselves and their families, by installing monitoring devices in their own house, or by informing other people who are still ignoring the risks involved in air pollution.

Another aspect to improve is the quality of information. We know sellers inform about the features of their product and buyers gather information based on their needs. At the end of the process, buyers purchase the product. Fig. 29.4 shows a schematization of the interaction between sellers and buyers.

Buyers of these devices look for a remedy, if they are already sick, or prevention, if they are sane. Sellers should communicate the clinical aspects of the devices clearly, and buyers should understand those aspects correctly. We know buyer's point of view (p.o.v.) and seller's point of view. How should we consider clinical information between the two p.o.v.? (Fig. 29.5).

The clinical point of view is one of those aspects that regards both parts. We think that improving this aspect may be one important step to lower risk factor of disease.

References

1. Dockery DW, Pope CA, Xu X, Spengler JD, Ware JH, Fay ME, Ferris BG, Speizer FE Jr. An association between air pollution and mortality in six U.S. cities. N Engl J Med. 1993;329:1753–9.
2. Brook RD. Cardiovascular effects of air pollution. Clin Sci. 2008;115(6):175–87.
3. Committee on the Medical Effects of Air Pollutants (COMEAP). Final Report: Longterm exposure to air pollution: effect on mortality. 2009.
4. Martinelli N, Olivieri O, Girelli D. Air particulate matter and cardiovascular disease. A narrative review. Eur J Intern Med. 2013;24(4):295–302.
5. Franklin BA, Brook R, Pope CA III. Air pollution and cardiovascular disease. Curr Probl Cardiol. 2015;40:207–38.
6. Franchini M, Mannucci PM. Air pollution and cardiovascular disease. Thromb Res. 2012;129(3):230–4.
7. Mills NL, Donaldson K, Hadoke PW, Boon NA, Mac Nee W, Cassee FR, Sandström T, Blomberg A, Newby DE. Adverse cardiovascular effects of air pollution. Nat Clin Pract Cardiovasc Med. 2009;6(1):36–44.
8. Franchini M, Mannucci PM. Particulate air pollution and cardiovascular risk: short-term and long-term effects. Semin Thromb Hemost. 2009;35(7):665–70.
9. Chen H, Goldberg MS, Villeneuve PJ. A systematic review of the relation between long-term exposure to ambient air pollution and chronic diseases. Rev Environ Health. 2008;23(4):243–97.
10. Pope CA 3rd, Hansen JC, Kuprov R, Sanders MD, Anderson MN, Eatough DJ. Vascular function and short-term exposure to fine particulate air pollution, 1995. J Air Waste Manag Assoc. 2011;61(8):858–63.
11. Buteau S, Goldberg M. A structured review of panel studies used to investigate associations between ambient air pollution and heart rate variability. Environ Res. 2016;148:207–47.
12. World Health Organization (WHO). Ambient air pollution attributable deaths, by region, 2012, Global Health Observatory visualizations. 2012. http://apps.who.int/gho/data/node.wrapper.ENVHEALTH3?lang=en&menu=hide.
13. World Health Organization (WHO). The top 10 causes of death. Fact sheet N°310. 2014. http://www.who.int/mediacentre/factsheets/fs310/en/.
14. Dominici F, Peng RD, Bell ML, Pham L, McDermott A, Zeger SL, Samet JM. Fine particulate air pollution and hospital admission for cardiovascular and respiratory diseases. JAMA. 2006;295:1127–34.
15. Langrish JP, Mills NL, Chan JKK, Leseman DLAC, Aitken RJ, Fokkens PHB, Cassee FR, Li J, Donaldson K, Newby DE, Jiang L. Beneficial cardiovascular effects of reducing exposure to particulate air pollution with a simple facemask. Part Fibre Toxicol. 2009;6:8.
16. Janssen NAH, Fischer P, Marra M, Ameling C, Cassee FR. Short-term effects of PM2.5, PM10 and PM2.5–10 on daily mortality in the Netherlands. Sci Total Environ. 2013;463-464:20–6.
17. Nelin TD, Joseph AM, Gorr MW, Wold LE. Direct and indirect effects of particulate matter on the cardiovascular system. Toxicology. 2012;208(3):293–9.
18. Vanderlei LC, Pastre CM, Hoshi RA, Carvalho TD, Godoy MF. Basic notions of heart rate variability and its clinical applicability. Rev Bras Cir Cardiovasc. 2009;24(2):205–17.
19. Brook RD, Rajagopalan S, Pope CA III, Brook JR, Bhatnagar A, Diez-Roux AV, et al. Particulate matter air pollution and cardiovascular disease. An update to the scientific statement from the American Heart Association. Circulation. 2010;121(21):2331–78.
20. Bhaskaran K, Hajat S, Haines A, Herrett E, Wilkinson P, Smeeth L. Effects of air pollution on the incidence of myocardial infarction. Heart. 2009;95(21):1746–59.
21. Mustafić H, Jabre P, Caussin C, Murad MH, Escolano S, Tafflet M, Périer MC, Marijon E, Vernerey D, Empana J, Jouven X. Main air pollutants and myocardial infarction: a systematic

review and meta-analysis. JAMA. 2012;307(7):713–21. http://jama.jamanetwork.com/article.aspx?articleid=1104975
22. Shah ASV, Lee KK, DA MA, Hunter A, Nair H, Langrish JP, Newby DE, Whiteley W, Mills NL. Short term exposure to air pollution and stroke: systematic review and meta-analysis. BMJ. 2015;350:h1295.
23. Maheswaran R. Air pollution and stroke—an overview of the evidence base. Spat Spatiotemporal Epidemiol. 2016;18:74–81.
24. Medina-Ramon M, Zanobetti A, Schwartz J. The effect of ozone and PM10 on hospital admissions for pneumonia and chronic obstructive pulmonary disease: a national multicity study. Am J Epidemiol. 2006;163:579–88.
25. World Health Organization (WHO). Air quality guidelines-global. Update 2005. 2006.
26. Chen C, Zhao B, Weschler CJ. Assessing the influence of indoor exposure to "outdoor ozone" on the relationship between ozone and short-term mortality in U.S. communities. Environ Health Perspect. 2012;120:367–72.
27. Bell ML, Dominici F. Effect modification by community characteristics on the short-term effects of ozone exposure and mortality in 98 US communities. Am J Epidemiol. 2008;167:986–97.
28. Bell ML, Dominici F, Samet JM. A meta-analysis of time-series studies of ozone and mortality with comparison to the national morbidity, mortality, and air pollution study. Epidemiology. 2005;16:436–45.
29. Ito K, De Leon SF, Lippmann M. Associations between ozone and daily mortality: analysis and meta-analysis. Epidemiology. 2005;16:446–57.
30. Bell ML, Peng RD, Dominici F. The exposure–response curve for ozone and risk of mortality and the adequacy of current ozone regulations. Environ Health Perspect. 2006;114:532–6.
31. Mills NL, Törnqvist H, Gonzalez MC, Vink E, Robinson SD, Söderberg S, Boon NA, Donaldson K, Sandström T, Blomberg A, Newby DE. Ischemic and thrombotic effects of dilute diesel-exhaust inhalation in men with coronary heart disease. N Engl J Med. 2007;357:1075–82.
32. Giles LV, Barn P, Künzli N, Romieu I, Mittleman MA, van Eeden S, Allen R, Carlsten C, Stieb D, Noonan C, Smargiassi A, Kaufman JD, Hajat S, Kosatsky T, Brauer M. A temporal, multicity model to estimate the effects of short-term exposure to ambient air pollution on health. Environ Health Perspect. 2008;116:1147–53.
33. Ware JH, Ferris BG Jr, Dockery DW, Spengler JD, Stram DO, Speizer FE. Effects of ambient sulfur oxides and suspended particles on respiratory health of preadolescent children. Am Rev Respir Dis. 1986;133(5):834–42.
34. Fiorina A, Bonifazi F. Aerobiologia ed Allergeni Stagionali. ECIG. 2006.
35. Budde M, Busse M, Beigl M. Investigating the use of commodity dust sensors for the embedded measurement of particulate matter, networked sensing systems (INSS). 2012 Ninth International Conference on IEEE; 2012. p. 1–4.
36. Flagan R. Probing the chemical dynamics of aerosols. In: Newman L, editor. Measurement challenges in atmospheric chemistry. Washington, DC: American Chemical Society; 1993.
37. OSHA. OSHA bulletin: general respiratory protection guidance for employers and workers. 2011.
38. EN 14683:2014. Medical face masks—Requirements and test methods.
39. EN 13274-7:2008. Respiratory protective devices. Methods of test. Determination of particle filter penetration.
40. Department of Health, Health protection agency: Guidance for pandemic influenza: infection control in hospitals and primary care settings. 2005.
41. HSE-Health and Safety Executive (UK), HSG53 "Respiratory protective equipment at work, A practical guide". 2013.
42. EN 149:2001+2009, chapter 9 Marking and Packaging.
43. C89/686/CEE: Council Directive of 21 December 1989 on the approximation of the laws of the Member States relating to personal protective equipment.

30 Exploring the Technologies Behind Air-Cleaning Systems and Indoor Applications

Patrizio Gagliardi

Despite the fact that it's not perceived at the same level in all the countries, it's quite assessed that indoor air pollution is among the top five environmental health risks [1]. It's also known that (according to the United States Environmental Protection Agency) pollution levels in an urban contest are five times higher inside buildings than outside. If you think that it's quite common that a man working in an urban environment spends up to 90% of his time inside a building [2], it's very simple to understand why there are growing interests around indoor pollution problem.

Talking about indoor air pollution often drives to a very simple answer: the best way to address this risk is to ventilate a home (or an office) with clean outdoor air, together with the elimination or the control of the sources of pollutants. This simple answer is not so simple to execute: the opportunities for ventilation may be limited by weather conditions, by the building design, or simply by contaminants in the outdoor air, a significant problem on the modern cities.

Given that, there is a growing market of air filters and other air-cleaning devices that are designed to remove pollutants from indoor air, helping in those contexts where it's not possible to ventilate and/or control pollution sources with success.

Indoor air filters and cleaners can be found in a big variety: some are installed in the ductwork of a home's central system (HVAC) to clean the air in the entire house or office (they are quite common in medical facilities and services, often defined by law requirements). Some others are portable and can be used to clean the air in a single room or in specific areas, clearly not intended to filter the air in the whole house. Air-cleaning devices are categorized by the type of pollutants that the device is designed to remove or destroy: particulate and gaseous.

P. Gagliardi
Sipra Engineering, Bologna, Italy

Member of Information Technology commission of Association of Engineers of Bologna, Bologna, Italy
e-mail: patrizio.gagliardi@sipraengineering.it

F. Capello, A.V. Gaddi (eds.), *Clinical Handbook of Air Pollution-Related Diseases*,
https://doi.org/10.1007/978-3-319-62731-1_30

There are two types of air-cleaning devices that can remove particles from the air: mechanical air filters and electronic air cleaners, while in order to remove gases and odors by either physical or chemical processes, a gas-phase filter has been designed. Finally, if the aim is to destroy or at least deactivate indoor air pollutants, the air-cleaning devices feature ultraviolet germicidal irradiation (UVGI), photocatalytic oxidation (PCO), and ozone generation. Each technology has some pros and cons, and a comprehensive analysis must consider factors like effectiveness against pollutants, installation cost, running costs, and collateral health effects. The filtration system itself can be analyzed in order to identify strengths and weaknesses of a typical indoor air-cleaning machine.

30.1 Technology Analysis: The Different Filter Media

30.1.1 Mechanical Air Filters

The purpose of a mechanical air filter is quite self-explanatory: they remove airborne particles as they pass through it. This is quite effective in removing bigger particles such as dust, pollen, dust mite and cockroach allergens, some kinds of mold, and animal dander, and in superior equipments, they can remove smaller particles like humidifier dust, lead dust, milled flour, and auto-emission particles. Moreover, in professional buildings featuring real HEPA filters, they can stop viruses, carbon dust, sea salt, and all combustion smoke.

This kind of technology is quite diffused, especially in HVAC systems, so there is a standard efficiency measurement method: the minimum efficiency reporting value (MERV) [3], developed by the American Society of Heating, Refrigerating and Air-Conditioning Engineers (ASHRAE). Thanks to this measurement method, we can make comparisons of air filters made by different companies (Fig. 30.1).

A first consideration must be done: the cost of a centralized solution could be very different, depending on the filters implemented and the destination. The impact over a single house could be significant, where the same architecture in a multi-apartment building can be well accepted because it's divided in multiple quotes. Given that and focusing on mechanical filters, usually a better performance involves a higher overall costs: top-performance filters (like HEPA filters, despite they are not commonly used in common house environments because of the costs) are expensive and require a proper design, together with a suitably sized power source—high efficiency requires high horsepower and higher electricity costs. A central solution, however, is the only way to assure an effective pollution control over multiple rooms and if correctly sized and developed can reach very high cleaning performances with very low noise.

It may be obvious but a mechanical filter decreases its efficiency with the usage: the more pollution it catches, the less pollutants it will be able to catch. Maintenance is required, so during the design process a sufficient access for inspection during use must be considered. With portable mechanical air purifiers, the maintenance is simpler but with two main disadvantages: the local effect, often limited to one room (or possibly a part of it), and the fan noise (fanless units typically have low effectiveness).

Fig. 30.1 A mechanical air filter

Another problem is that these filters are effective against particles floating on the air. Smaller particles float longer than bigger ones, so these filters are ineffective against big particles. Finally, because the particles that are in the range of air filtration (10 μm–0.3 μm) settle rather quickly, mechanical air filters are not very good at removing them completely from indoor areas. Of course, normal human activities (moving, vacuuming, etc.) can stir up particles, but most of the larger particles will resettle before an air filter can remove them.

Finally, mechanical air cleaners don't remove gases from air (some of them can remove some smoke particles but not the gas part of smoke).

30.1.2 Electronic Air Cleaners

Electronic air cleaners use a physical process called electrostatic attraction to trap charged particles. They divide in two types: electrostatic precipitators and ion generators.

Electrostatic precipitators must feature an external power source in order to power an ionization section and a collecting plate section. The air cleaner draws air through the ionization section (featuring a fan and ducts in some models), in order to give particles an electrical charge. The charged particles accumulate on the collector that is oppositely charged: keeping the collector plates clean is necessary in maintaining adequate performance (Fig. 30.2).

Ion generators (also called ionizers) work similarly to the electrostatic precipitator, but without collecting plates: they disperse charged ions into the air. These ions attach to particles and give them a charge, making them adhere to any nearby surface or combine with other particles and settle on room surfaces. They come in different forms and are the simplest form of electronic air cleaner.

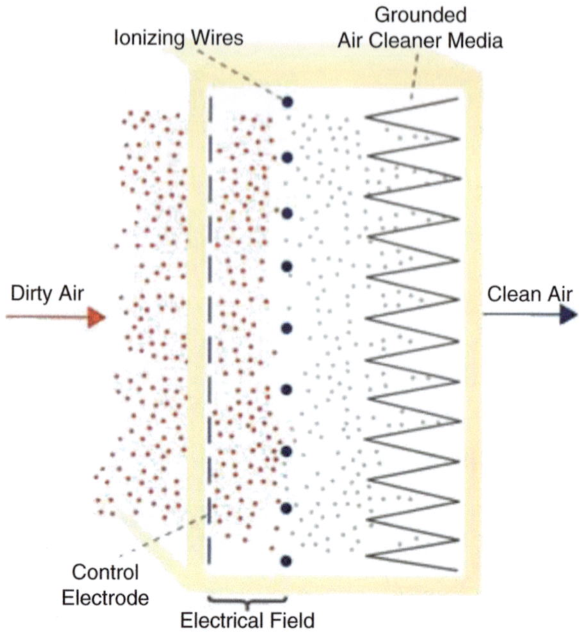

Fig. 30.2 Electronic air cleaner functioning principle

Electronic air cleaners can come in a stand-alone form or can be integrated inside a HVAC system. Although they are usually effective in filtering small particles and are typically not expensive, they have a major disadvantage: they produce ozone, someone by design [4]. Ozone is a risk for human health, because it can react with a high variety of domestic substances and materials: these chemical reactions produce harmful by-products that may be associated with adverse health effects in some sensitive populations and can produce ultrafine particles (diameter smaller than 0.1 μm), formaldehyde, ketones, and organic acids [5].

Another problem comes from the maintenance: the collection plates need frequent cleaning, so there's more maintenance required than with filter-based units and it's possibly more complicated, due to the fact that the mechanical filter is designed to be properly disposed, where the collection plate must be manually cleaned and the cleaning procedure can be tricky and can pollute water.

As the most electronic air cleaner producers offer them in a portable form, the positioning inside the room can affect the performance of the filter itself.

Finally, like mechanical air filters, the electronic air cleaners cannot remove gases from air.

30.1.3 Gas-Phase Air-Filtration Devices

Gas-phase air filters are designed to remove gases and odors by using a so-called sorbent, a material (such as activated carbon) which adsorbs the pollutants using either physical or chemical processes. These filters are typically intended to remove one or more specific gaseous pollutants from the airstream that passes through

Fig. 30.3 Activated carbon filter

them, being useless against those pollutants for which they were not designed. A premium HVAC system can feature one or more gas-phase air filters, but at present no system is expected to remove all of the gaseous pollutants present in the air of a typical indoor environment.

For example, one of the most dangerous gases that you can find in a typical home is the carbon monoxide, produced from any kind of fuel combustion: many people are killed every year in closed areas (offices, plants, at home, etc.) by this gaseous pollutant, but despite that no carbon monoxide effective filters can be found in a common HVAC system (Fig. 30.3).

Commonly, only activated carbon is implemented inside HVAC and portable systems. Less commonly, permanganate alumina or zeolite (incorporated into fibrous filter media) can be found. Integrating such filters inside an existing plant is usually not complicated.

This specificity is one of the disadvantages of these filters: in order to design a HVAC system capable of gas filtering, several different sorbents should be implemented adding costs both in realization and in maintenance (ignoring the fact that each layer could introduce a power-related issue over the motors).

Another issue related to these filters is the fact that they usually wear-out very fast and some kind of sorbents not only become ineffective but can release in the are some of the pollutants they catch before. Moreover, the right maintenance is complicated by the fact that there are no reliable measurement systems available to be integrated inside indoor air filters, so the user must schedule the correct maintenance by himself, without a feedback. The lack of reliable indoor gas-measuring systems is so critical that there is no standard to measure and classify the performance of the gas-phase air filters [6–8].

30.1.4 Ultraviolet Germicidal Irradiation Cleaners

The purpose of integrating an UVGI (ultraviolet germicidal irradiation cleaners) inside a HVAC or a portable air filter is to improve residential IAQ by deactivating (killing) indoor biological pollutants that are airborne or growing inside HVAC surfaces (e.g., cooling coils, drain pans, or ductwork, where moist can stack up).

Again, there is no standard test method to evaluate and compare the effectiveness of UVGI cleaners used in an indoor environment. Considering that the effective destruction of some viruses and most mold and bacteria spores usually requires much higher UV exposures than a standard home unit provides [9], the UVGI cleaners used in domestic contexts have very limited effectiveness in killing bacteria and molds [10].

UVGI systems must be used in conjunction to other filtration systems (like the mechanical ones previously mentioned); whether the UVGI is designed to keep the ducts of the air cleaner free from mold and bacteria or is intended to kill pollutants from the airflow, the maintenance is absolutely critical: the UV lamps must be kept working and clean in order to have a result [11, 12].

The positive thing of the UVGI is that, according to the available literature [13], no dangerous generated products originate from the usage of these filtering systems.

30.1.5 Photocatalytic Oxidation Cleaners

In indoor environments, PCO cleaners are intended to destroy gaseous pollutants and their odors by converting them into harmless products, but they are not designed to remove particulate pollutants [14].

PCO cleaners feature a UV lamp and a photocatalyst (usually titanium dioxide), in order to create oxidants that destroy gaseous contaminants. A quite complex reaction, called photocatalytic oxidation, converts organic pollutants into the carbon dioxide and water. This is a young technology, still under development, where some manufacturers claim PCO devices can remove tobacco smoke, microorganisms, and other indoor particulate pollutants (even though the devices are not meant to remove particles), but the market penetration is still low and the effectiveness is limited, because of the limited capability of available photocatalysts of destroying gaseous pollutants in indoor air [15].

Moreover, PCO of certain VOCs may create by-products that are indoor pollutants if the system's design parameters and catalyst metal composition do not match the compound targeted for decomposition, particularly in the presence of multiple reactive compounds commonly found in residential settings [16, 17].

From an engineering point of view, these systems draw a lot of energy to work properly, so at now there is no evident benefit in featuring a PCO inside a HVAC system (Fig. 30.4).

30.1.6 Ozone Generators

Although we must cite this "technology" talking about air-filtration technologies and media, we will not examine in detail the ozone generators because it's quite established that they are wrongly sold as air purifiers and they can cause serious adverse health effects (ozone is lung irritant).

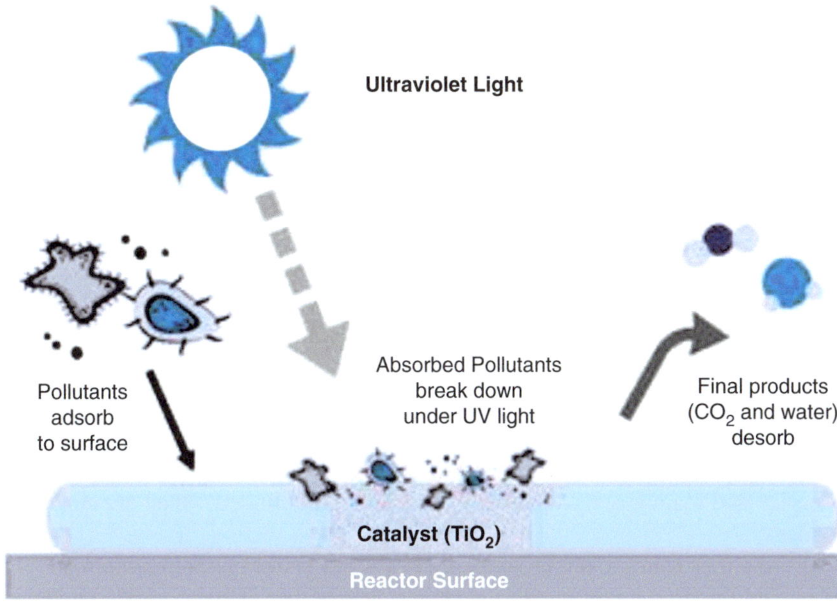

Fig. 30.4 Photocatalytic cleaning reaction

30.2 Analyzing an Indoor Air Filter as System

30.2.1 Power Sources

The choice of a power source depends on many factors: the number of filtering stages, the dimensions of the whole plant, the acceptable noise, the costs, etc. Nowadays, the smaller systems feature quite commonly high-efficiency brushless motors, while the HVAC systems often rely in older (but reliable) technology. Switching to constant torque motors (high-efficiency, brushless DC motors that are based on the same ECM technology described in ECM 2.3 variable-speed motors) can improve the cost-effectiveness, reduce the utility bill, and be sustainable.

A fan that has sufficient capacity (pressure and airflow ratings) to move air through the filter media must be included and the power source must be capable of driving that fan with the performance required: this is quite simple to achieve in small devices but not so obvious in the design of a HVAC system, especially if an existing installation is upgraded over time (Fig. 30.5).

30.2.2 Design and Choice

An air-cleaning system must be chosen according to the destination: covering a building is different than serving a little house, and professional activities (such as medical facilities) have totally different requirements from domestic environments.

Fig. 30.5 An electronically commutated motor

A portable air cleaner is the most affordable way to purify domestic air: they are quite cheap and they can be moved from one room to another without big issues. Of course, being small and cheap is a disadvantage in terms of filtering capabilities (limited filtering stages and media can be packed), and typically it cannot treat the air of a big room in a satisfactory percentage. Many of them are also not so simple to service and maintain, having also a life cycle quite limited—they are consumer products.

On the other hand, a HVAC centralized system is always preferable when the budget and the destination justify the choice. The plenty of space inside the ducts, the choice of the power source, and the capability of delivering the airflow where it is needed more make them tremendously efficient compared to the portable systems. In case of buildings to be covered, the centralized solution is proportionally more convenient than the portable one. The design process is very important, because many factors like the facility of maintenance and operation depend on a good initial design.

A centralized system is the only way to feature filtering stages like advanced UVGI or gas-phase filters or complex HEPA filters.

30.2.3 Costs and Maintenance

As previously said, costs can vary according to the complexity of the whole air-filtering system, and it's quite obvious that a portable small device will be cheaper than a centralized system. Things change if you take a more scientific approach and start to consider a parametrized cost, reporting that cost to efficiency against particles, total cleaning capacity, unitary power consumption, and so on. In this case, the parametrized cost can tell another truth.

During the life of the cleaner, filters and sorbents must be replaced, and the plates or charged media of electronic air cleaners must be also kept clean. Electronic air cleaner efficiency decreases as the collecting plates become loaded with particles, so the plates must be cleaned, sometimes frequently, as required by the manufacturer. The cleanings should be scheduled to keep the unit operating at peak efficiency; some filters (like the ones commonly adopted over laser printers) feature an indicator that helps in remembering when to replace the filter (http://www.cleanoffice-feinstaubfilter.de/en/clean-office/clean-office.html), but there is a general lack of devices capable of autochecking for normal maintenance.

During cleaning or replacement of filters, it would be important to ensure that pollutants are not reemitted into the air. Excessive movement or air drafts should be avoided when filters are removed, and used filters should be placed in plastic bags or other containers for disposal. Bigger systems can be serviced by professionals who can ensure a high quality at a cost that, divided by the users of the system, could be not so significative.

30.2.4 Noise

One could easily argue that the smaller portable devices are the most silent; this is not always true, because it depends from the effectiveness of the design and it's possible to find small devices that at maximum fan speed are noisier than bigger ones [18].

Large HVAC can feature big motors (and auxiliary ones) that are really noisy, but they can be placed out of human activities by design reducing (or eliminating at all) the disturbance of the noise itself.

30.2.5 Feedback and Closed-Loop Control

Finally, a big issue with common air-cleaning devices is a lack of effective air pollution measurement system on board. To be honest, the industry of consumer-grade pollution measurement devices is growing and has reached contradictory results, mainly due to the lack of a scientific approach to the problem.

Giving this, air-cleaning devices feature poor measurement systems, and often they have no onboard system at all. This has two big consequences: the user cannot rapidly check the effectiveness of the cleaning process, making it difficult to discover faulty situations and obliging to execute maintenance on a scheduled basis and not according to the need, and the system cannot feature a reliable closed-loop control that could permit a more efficient functioning and a cost reduction (Fig. 30.6).

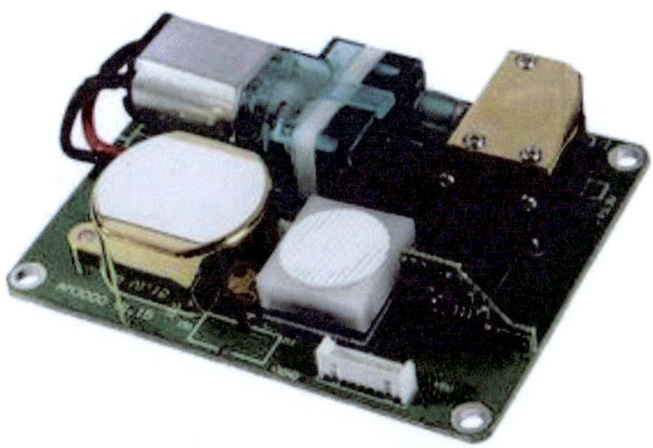

Fig. 30.6 An air quality monitoring device

References

1. EPA. Residential air cleaners (second edition): a summary of available information. Washington DC: EPA; 2009. EPA 402-F-09-002, Revised August 2009.
2. Centers for Disease Control and Prevention, U.S. Department of Housing and Urban Development. Healthy housing reference manual. Atlanta: US Department of Health and Human Services; 2006.
3. Wilkinson, Ron. "MC² market & competitive convergence air filters: new facilities, New Standard." E. L. Foust Co. April 2001.
4. Ozone generators that are sold as air cleaners: an assessment of effectiveness and health consequences. U.S. Environmental Protection Agency. https://www.epa.gov/indoor-air-quality-iaq/ozone-generators-are-sold-air-cleaners-assessment-effectiveness-and-health
5. Shaughnessy RJ, Sextro RG. What is an effective portable air-cleaning device? A review. J Occup Environ Hyg. 2006;3:169–81.
6. Khazraei Vizhemehr A, Haghighat F, Lee CS, Kholafaei H, Lakdawala N. Evaluation of gas-phase filter performance for a gas mixture. Clean Soil Air Water. 2015;43(4):469–78.
7. Liu RT. Field methods to measure contaminant removal effectiveness of gas-phase air filtration equipment–phase 1: search of literature and prior art. Washington, DC., EPA/600/R-97/092 (NTIS PB98-111677): U.S. Environmental Protection Agency; 1997.
8. Vizhemehr AK, Haghighat F, McKay G. Modelling comparison of relative performance of gas-phase filter at high and low challenge concentration. In: Indoor air 2014–13th international conference on indoor air quality and climate; 2014 (p. 474).
9. Phillips Lighting Division. Disinfection by UV–radiation: Phillips Lighting Division; 1992.
10. Jelden K. Limitations of ultraviolet germicidal irradiation (UVGI) in hospital room disinfection. In: APHA annual meeting & expo (Oct. 31–Nov. 4, 2015). APHA; 2015.
11. Kowalski WJ, Bahnfleth W. Airborne respiratory diseases and mechanical systems for control of microbes. Heat Pip Air Condition. 1998;70(7):34–48.
12. Phillips Lighting Division. Germicidal lamps and applications: Phillips Lighting Division; 1985.
13. VanOsdell D., Foarde K. Defining the effectiveness of UV lamps installed in circulating air ductwork. Prepared for the air-conditioning and refrigeration technology institute, RTI international. ARTI-21CR/610–40,030–01; 2002.
14. Chen W, Zhang J, Zhang Z. Performance of air cleaners for removing multiple volatile organic compounds in indoor air. ASHRAE Transac. 2005;111(2005):1101–14.
15. Tompkins DT, Lawnicki BJ, Zeltner WA, Anderson MA. Evaluation of photocatalysis for gas-phase air cleaning–Part 2: economics and utilization. In: ASHRAE Research Project 1134-RP. Conducted: December 1999–November 2002, University of Wisconsin; 2003.
16. Jardim WF, Alberci RM. Photocatalytic destruction of VOC in the gas-phase using titanium dioxide. Appl Catal B Environ. 1997;14(1):55–68.
17. Lake DM, Jacoby WA, Nimlos M. Identification of by-products and intermediates in the photocatalytic oxidation of gas-phase trichloroethylene. In: Proceedings. 6th international symposium on solar thermal concentrating technologies, September 28–October 2, 1992; 1992.
18. Clearing the air: A guide to reducing indoor pollution. Consumer Reports Magazine. p. 41–49. 2002.

Basic Characteristics of PPE (II): Industrial Applications

31

Giulia Baldessari

31.1 Introduction

Throughout history, people, citizens and workers have had to deal with air pollution in both the workplace and their homes. For centuries coal was heavily used for domestic heat, industrial processes and transportations, causing indoor and outdoor air pollution leading to serious health issues including respiratory and cardiovascular diseases and even death [1]. Since the thirteenth century, London has been afflicted by "pea soup fog", a thick smog caused by air pollution that contains particulates and sulphur dioxide gas; in China, a study carried out on 3010 miners exposed to silica dust and employed for at least 1 year during the 1960s in tin mines pointed out that 33.7% of the subjects were identified with silicosis, and most of them developed silicosis years after exposure ended [2]; in the 1930s in America, over 470 out of 5000 workers died from silicosis while building the Hawks Nest Tunnel in West Virginia [3].

An inscription at Hawks Nest State Park reads as follows: "Tragedy brought recognition of acute silicosis as occupational lung disease and compensation legislation to protect workers".

In the early 1900s, one of the first legislations regarding the use of respiratory protection devices in the workplace comes from the United States Bureau of Mines following a decade in which the number of coal mine fatalities exceeded 2000 annually.

Facts like the ones described above pointed out a worldwide lack of knowledge on work-related disease and contributed to create international congresses and associations focused on discussing this theme.

Nowadays, many institutes regulate workers' health and safety through insertion of the knowledge coming from occupational health and safety, whose aim is to

G. Baldessari
BLS Group, Milan, Italy
e-mail: g.baldessari@blsgroup.it

F. Capello, A.V. Gaddi (eds.), *Clinical Handbook of Air Pollution-Related Diseases*,
https://doi.org/10.1007/978-3-319-62731-1_31

maintain and promote workers' health and working capacity together with the improvement of the working environment and work to become conducive to health and safety [4].

Among all institutes regarding occupational health and safety, international acknowledgments go to the International Labour Organization (187 member states), European Agency for Safety and Health at Work (Europe), Health and Safety Executive (UK) and Occupational Safety and Health Administration (US).

As a result of the international effort to improve workers' health, it is important to mention that between 1981 and 2004, silicosis mortality rates in America declined by 70.6% [5].

31.2 Risk Assessment and Respirator Selection

In the industrial and production sector, employers have a duty to assess the health and safety risks faced by their employees. This procedure is known as "risk assessment", and it is a systematic exam of all aspects of a specific task that can contribute to harm the worker.

The first step is the identification of the hazard that could cause injury or harm; once the hazard is recognized, the advisable approach is to work on systematic solutions starting with technical and procedural strategies:

- Eliminate the hazardous substance.
- Substitute the hazardous substance.
- Reduce the time of exposure to contaminants with a procedural approach.
- Separate the working environment and install collective protective equipment (e.g. ventilation).

However, even a correct approach can leave a residual risk. The use of a personal protective equipment (PPE) should be limited to the residual risk, after the preliminary approaches have been taken into account.

In selecting the appropriate respirator, one must consider the nature and extent of the hazard; the following categories of information must be taken into account [6]:

- Nature of the hazard: the physical and chemical properties of the air contaminant will affect the respirator selection. Physical properties include factors such as particle size for dusts and vapour pressure for gases. Chemical properties of the air contaminant can affect breakthrough times and the ability of the filter material to remove or adsorb the contaminant.
- Concentrations of contaminants: sampling and analysis of the air in the workplace determine what degree of exposure is occurring and thus which is the required degree of protection. The result becomes a point of comparison with the occupational exposure level (TLV) and will determine how much the concentration must be lowered by the respirator in order to bring it back to a safe level for employees.

- Occupational exposure limit: the main purpose for a respirator is the protection against overexposure, by reducing and maintaining exposure to or below the threshold limit (TLV) for the related contaminants.
- Nature of the work operation or process: during the selection of the respirator, the motions and the movements that the job requires alongside the equipment used should all be taken into account.
- Time period in which the respirator is worn: selection of the respirator should consider the period of time, during the entire shift, in which the employee will use the device. For example, it is possible that one single respirator cannot guarantee the same level of protection during the shift for all the contaminants present in the air because the duration of the respirator depends strongly on the concentration in the workplace air. As another example, the use of a respirator for a long period may impose the choice of a device with the minimum possible physical weight.
- Physical/psychological stress coming from work activities: heavy works can be physically draining; thus certain types of respirators must be excluded from the selection in order to ensure that better conditions are also provided when working under extreme temperature and humidity, which are an important factor contributing to the overall risk.
- Physical characteristics of the respirator: the protection device can lead to physical limitations due to its intrinsic design; the selected equipment should not impair the worker's vision, hearing, communication and physical movement necessary to perform jobs safely.

31.2.1 What are the Risks?

Selection of the appropriate filter will depend firstly on the physical nature of the harmful contaminant in the air; other parameters contribute to the respirator selection, including extreme climatic conditions (temperature and humidity) (Table 31.1).

31.2.2 Occupational Exposure Limits

ACGIH (American Conference of Governmental Industrial Hygienists) developed some guidelines to assist in the control of health hazard, introducing the TLV (threshold limit value) for over 600 chemical substances and physical agents (www.acgih.org). ACGIH defines the TLV "as the level to which is believed a worker can be exposed day after day for a working lifetime without adverse effects" and refers to the concentration of the substance in the air usually expressed in ppm (part per million) or mg/m^3.

Definition of limit values is based on data coming from scientific literature and epidemiological studies in the industrial sector, together with experimental research on human, animals and cell cultures [8].

Table 31.1 List of risk and their definitions

Risk	Definition [7]	Examples
Dusts	Solid and fine particles that are generated from solid organic or inorganic materials usually by reducing their size through mechanical processes	Silica, coal dust, wood dust, asbestos fibres
Fumes	When a volatilized solid, such as a metal, condenses in cool air, it forms airborne particles known as fumes	Formation of fumes in welding process, magnesium burning
Mists	Mists are formed when tiny liquid droplets are suspended in the air. Droplets can be generated by a change from gaseous to liquid state (condensation) or by breaking up a liquid into a dispersed state (atomization)	Acid mists from electroplating, oil mist produced during cutting and grinding operations
Gases	Gases are formless fluids formed by individual molecules present in the air at room temperature and pressure	Welding gases, carbon monoxide from internal engine combustion, hydrogen sulphide
Vapour	Vapours are the gaseous form of substances that are normally in the solid or liquid state at room temperature and pressure. They are formed by evaporation from a liquid or solid	Solvents in paintings, gasoline
Oxygen deficiency	An oxygen-deficient atmosphere has an oxygen content below 17–19% by volume	Oxygen deficiency may occur in confined spaces, such as storage tanks, process vessels, towers, drums, bins, sewers, underground utility tunnels and pits

The TLV is just a guideline and the limit can be crossed during a working day if other values are taken into consideration:

- Time-weighted average (TLV-TWA) is the average exposure during 8 h/day, 40 h/week work schedule.
- Short-term exposure limit (TLV-STEL) is the spot exposure for 15 min that cannot be repeated more than four times per day with at least 60 min between exposure periods.
- Ceiling limit (TLV-C) is the absolute exposure limit that should not be exceeded at any time.

Even though limits specified by ACGIH are the most commonly accepted and the most complete in terms of number of substances included, similar lists are compiled, among others, by OSHA (PEL, permissible exposure limits) [9] and CE Commission (IOELV) [10].

Limit values represent a guideline that can only be applied to the workplace and cannot be transferred on the everyday life. In this case, in fact the risk of exposure

Table 31.2 Example of occupational exposure limit for H_2S (in gaseous form) according to different organizations

Authority/country	Name	Time-weighted average	Short-term exposure limit
NIOSH	REL	10 ppm (TWA)	10 ppm (for 10 min)
OSHA	PEL	20 ppm (ceiling)	50 ppm (for 10 min)
ACGIH	TLV	1 ppm (TWA)	5 ppm (STEL)
UK	WEL	5 ppm (TWA)	10 ppm (STEL)
Australia	OEL	10 ppm (TWA)	15 ppm (STEL)

has to be applied to 24 h instead of an 8-h work shift, and it is important to consider that the entire population can be exposed, including children, pregnant women, the elderly and the infirm (Table 31.2).

31.2.3 Filtration Methods

Once the hazardous substance is identified, the next step is to match the right type of respirator and filtration technology that will be different depending on the hazardous substance and its form.

Nowadays, the two most widely used filtering technologies in respirator devices available on the market are particle filters and gas/vapour filters. All substances in the liquid form (fine spray and mists), dusts, fibres, smoke and fume are intended to be filtrated by a particle filter: these filters do not protect from gas or vapour, for which is used a carbon-based filter (Fig. 31.1; Table 31.3).

Both particle and gas/vapour filters are not fitted to protect against oxygen deficit in the air: in this case, an air supplying breathing apparatus has to be considered.

It is common to find more than one risk and more than one chemical in a single workplace; for this reason manufacturers develop combined devices that can be used to protect against various hazards.

31.2.4 Deciding the Protection Factor

APF (assigned protection factor) is defined by OSHA as "the workplace level of respiratory protection that a respirator or class of respirators is expected to provide to employees when the employer implements a continuing and effective respiratory protection program". The introduction of APF is necessary in order to ensure the correct selection of the respirator based on the concentration of the hazardous agent in the air. The protection factor needed is based on the ratio of two variables:

$$\text{assigned protection factor}\left(\text{APF}\right) = \frac{C_I\left(\text{concentration of the contaminant inside the respirator}\right)}{C_O\left(\text{maximum allowable concentration inside the respirator}\right)}$$

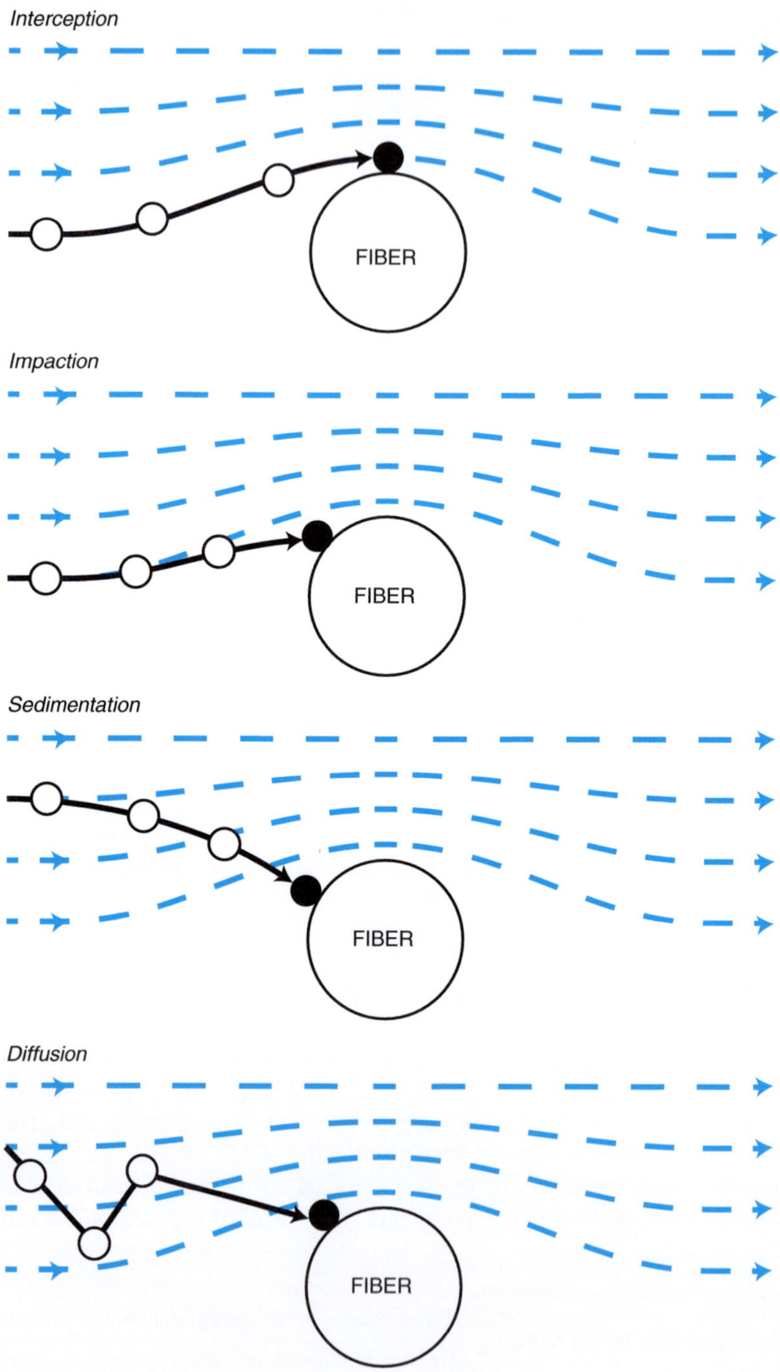

Fig. 31.1 Mechanical filtering mechanisms [11]

Table 31.3 Filtering mechanisms

Particle Filters		Gas/vapour filters
Mechanical filtration	Electrostatic filtration	Adsorption
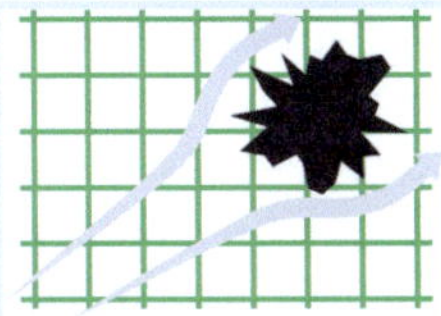	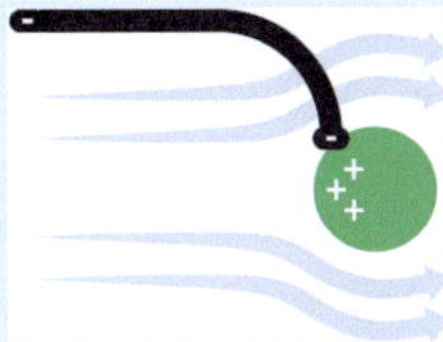	
Filtering material has a small structure net of fibres that stop particles bigger than the mesh	Smaller particles are attracted to the fibres and held by the material electrostatic charge	Adsorption is a process in which atoms and molecules present in a gas or liquid adhere to a solid surface, in this case made of activated carbon grains
Also particles smaller than the mesh are stopped, thanks to the combination of other mechanical capture mechanisms: interception, sedimentation, inertial impact, diffusion	This capture mechanism helps other capture mechanisms, especially interception and diffusion	Each grain is formed in a way to have small pores that increase the surface area, where the adsorption mechanism takes place

C_O value is determined by measurements in the workplace; C_I factor is the maximum breathable concentration for a worker during a day shift, which is the TLV-TWA.

Each respirator is categorized by an assigned protection factor, which indicates the performance of protection of the respirator and the level of protection that can provide. For example, a respirator with an APF = 10 (if used properly) will reduce the wearer's exposure by a factor of 10. The user will therefore breathe 1/10 of the amount of the chemical present in the workplace.

Practical example [12]:

- Substance: Toluene (solvent).
- Measured airborne toluene concentration: 350 ppm within an 8-h time-weighted average (TWA).
- Toluene TWA: 20 ppm (from ACGIH recommendation)
- Required APF to reduce to TWA = 350/20 = 18
- Select RPE device with an APF above the required protection factor. In this case an APF of 20 will be required (it's important to choose a protection factor above the calculated value).

Lists of APD for each respirator are available in OSHA 29 CFR 1910.134 and EN 529.

31.3 Respirators Description and Their Characteristics

The European Directive 89/686/EEC (replaced by UE 2016/425 from April 2018) defines essential requirement which PPE must satisfy at the time of manufacture and before it is placed on the market. The Directive divides the PPE in three different categories according to the degree of risk: increasing the category level, the risk the PPE protects against is higher, and the certification procedure becomes more stringent. All respiratory protection devices are Category III PPE (Table 31.4).

The purpose of any respirator is to protect the human respiratory system from inhalation of hazardous elements; protection can be provided by removing contaminants from the inhaled air using filtering technology (air-purifying respirator) or supplying an independent source of respirable air (atmosphere-supplying respirator) [11]. Both types can be further subclassified by the mode of operation and their basic characteristics.

31.3.1 Filtering Disposable Facepiece

These filtering devices are designed for particle filtration by covering the nose and mouth. The mask consists entirely of one or more layers of nonwoven filtering material eventually coupled with neutral non-filtering material for structural support. Facepieces usually present a flexible clip-band in the nose area and a facial gasket, both used to improve fitting on the user's face. Facial gasket is a thick layer made of a soft and flexible material that can be attached to the whole mask profile covering the nose, cheeks and chin (total gasket) or covering only the nose area (partial gasket) (Fig. 31.2).

The mask is maintained into position by the head harness, composed by elastic tapes fixed directly on the facepiece body.

Contaminated air enters the filtering material during inhalation, where it is purified, while the exhaled air passes through the filtering material and the exhalation valve (if fitted, both valved or non-valved facepieces are available on the market): exhalation valve remains closed during inhalation and it is opened during exhalation, improving the air exchange inside the mask.

Disposable masks can be equipped with an additional carbon layer coupled with the filtering material, composed of a non-filtering base material impregnated with

Table 31.4 Risk degree for protective equipment

Category	Description	Examples
I—low risk	PPE protecting against minimal risk, where user himself can assess the protection needed	Gardening gloves, sunglasses
II—medium risk	PPE protecting against normal risk	Life jackets, hearing protection, eye protection
III—high risk	PPE protecting against mortal danger or can cause irreversible damage to health or where the effects cannot be identified in sufficient time	Fall arrest harnesses, respiratory products, head protection

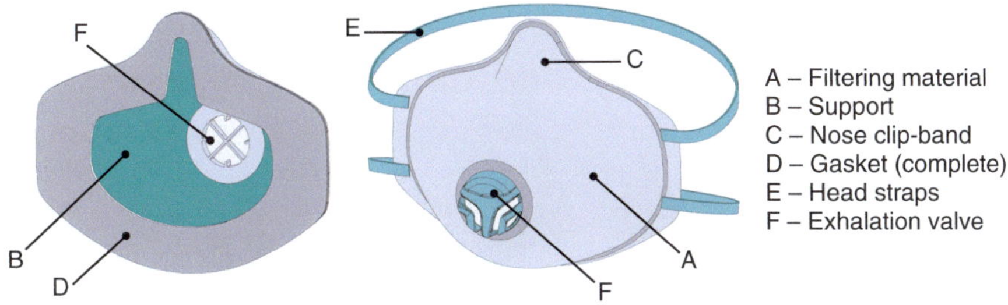

Fig. 31.2 Disposable facepiece

Fig. 31.3 Layers in disposable facepieces with additional carbon material

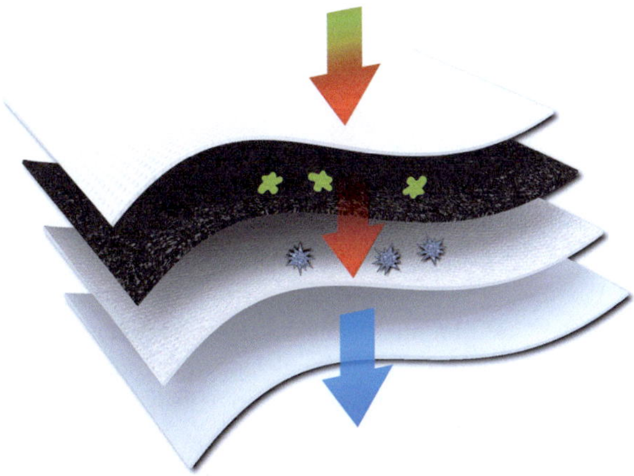

carbon grains or dust. This layer works as a filter for nuisance odour and vapours in small concentration (lower than TLV level), and it is not intended for considerable concentration of gas (in fact no requirements of adsorption for carbon layers are included in the standards) (Fig. 31.3).

Disposable masks are divided into nonreusable ("NR", intended for one single shift) and reusable ("R", intended for more than one shift); reusable devices use to have complete gasket in order to allow better cleaning after daily usage.

An additional test can be done on all particle filters: the clogging test consists in a simulation of the level of blockage that the particles may procure to the device after a long-time usage. If the particle filtering device passes the requirement, the letter D will be added to the marking.

31.3.2 Half-Face Mask

Half masks are also devices covering both mouth and nose. The main body, also known as orinasal, is made of flexible and shaped rubber or silicon that adheres to the user's face; it is kept in the right position through a head harness that buckles together at the back of the head and of the neck and can be adjusted to gain the perfect adherence of the orinasal to the skin.

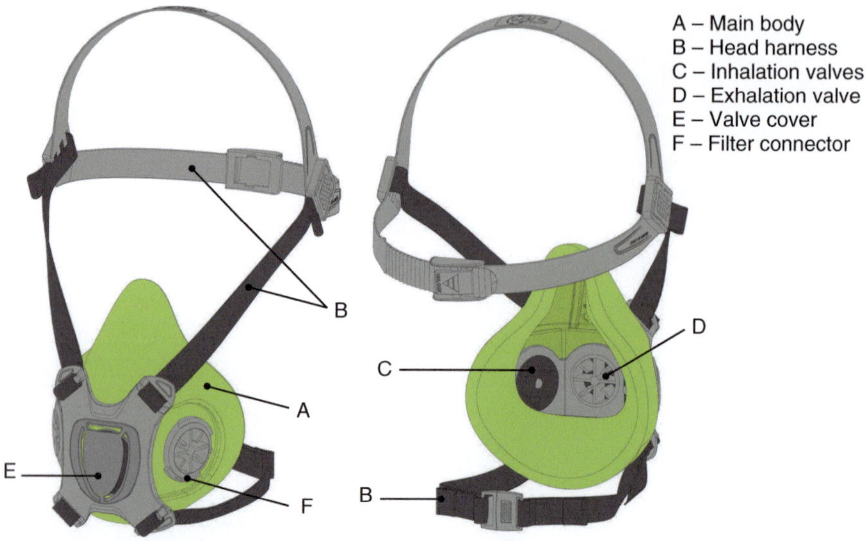

Fig. 31.4 Half-face mask

Filters are mounted through a specific connector on half masks for air purification; filter connections can be designed according to specific standards with regulatory dimensions or can have a custom design specifically made by the manufacturer. In this second case, the manufacturer will also provide the correct filters that fit on the connection (Fig. 31.4).

A half-face mask can be intended for one single filter or for two twin filters depending on the design. For twin filters, the weight is divided in two allowing a better load distribution on the head.

Coming through the filters, breathable air then enters in the inhalation valve/valves and comes out from the exhalation valve. Exhaled air will not pass again through the filters because the inhalation valves are closed during exhalation; likewise during inhalation the exhalation valve is sealed and there is no possibility for contaminated air to enter into the mask. The housing of the exhalation valve is covered to avoid accidental damage of the valve during usage or storage.

31.3.3 Full-Face Mask

Full-face mask is a device that covers the mouth, nose and eyes. It is usually composed of a visor made of glass or polycarbonate attached to a facial gasket that permits a safe fitting on the user's face. To the visor the harness is connected usually made from five to six adjustable rubber straps. Air circulation works in the same way than in half-face masks: when the air (purified by filters) is dragged into the lungs, the inhalation valve opens while the exhalation valve closes; exhaled air is then discharged through the exhalation valve.

The orinasal inside the mask, equipped with inhalation valves, directs the flux of exhaled air directly through the exhalation valve avoiding steam formation on the visor (Figs. 31.5 and 31.6).

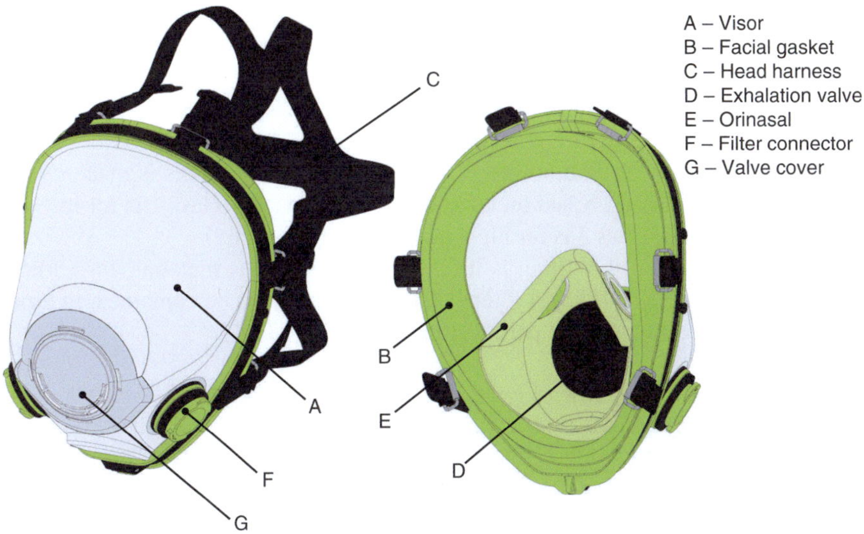

Fig. 31.5 Full-face mask

Fig. 31.6 Inhaled and exhaled air flow in a full-face mask

31.3.4 Filters

Gas masks and half masks are effective only if used with the correct cartridge, canister or filter (these terms are often used interchangeably) for a particular biological or chemical substance. Selecting the proper filter can be a delicate process. There are cartridges available that protect against more than one hazard, but there is no "all-in-one" cartridge that protects against all known gases. It is important to know the exact substance present in the workplace and its concentration before choosing the right filter.

Filters are colour-coded to help select the right one. European standard and NIOSH standard use two different codes for colours (Tables 31.5 and 31.6).

In the European standard, a further classification for gas filters is made by the division in three classes, based on their adsorption capacity (class 1 is for low gas adsorption capacity, class 3 is for high gas adsorption capacity).

Inside the canister, the filling is usually made of fibrous material (for particle filters) or carbon grains (for gas filters); filter types can also be combined, in order to achieve a protection to both particles and gases (Fig. 31.7).

Table 31.5 EN filter classification according to EN 143 and EN 14387

Contaminant	Filter type	Colour code	
Particles	P	White	
Organic gases and vapours (boiling point < 65°C)	A	Brown	
Inorganic gases and vapours	B	Grey	
Acid gases	E	Yellow	
Ammonia and its organic derivatives	K	Green	
Mercury	Hg (incorporates P3 filter)	Red-white	
Oxides of nitrogen	NO (for single use only, incorporates P3 filter)	Blue-white	
Specific organic gas and vapours specified by the manufacturer	AX (for single use only)	Brown	
Specific substances specified by the manufacturer	SX Marked with the name of the chemical	Violet	

Table 31.6 NIOSH filter classification according to 42 CFR 84

Contaminant	Filter type	Colour code	
Particle	P100	Magenta	
Organic vapour	OV	Black	
Acid Gas	AG	White	
Organic vapours / Acid Gas	OG/AV	Yellow	
Ammonia and basic gas	AM/MA	Green	
Mercury vapour	MV	Orange	
Multi-Gas and vapour		Olive green	

Fig. 31.7 Section for particle, gas and combined filters

Carbon grains are activated to filter a specific group of chemicals: for example, the filters marked with letter A contain carbon specifically activated for adsorption of all organic gas and vapours; this carbon will not be able to filter other chemical groups, as acid gases or ammonia. Carbon can also be activated for more than one chemical group with the capacity of adsorption, for example, not only of organic vapours but also of ammonia and acid gases.

In activated carbon filters, adsorption capacity is measured by a parameter called "breakthrough time". It defines how long the carbon grains are able to purify the contaminated air containing a certain concentration of gas/vapour. Standards regulate breakthrough time for different filter type and classes.

Filter designation must be marked on the canister, together with the related colour for gas protection. An example of a canister that protects from particles and multiple gases according to European standard is A1B1E1K1 P3, where A, B, E,

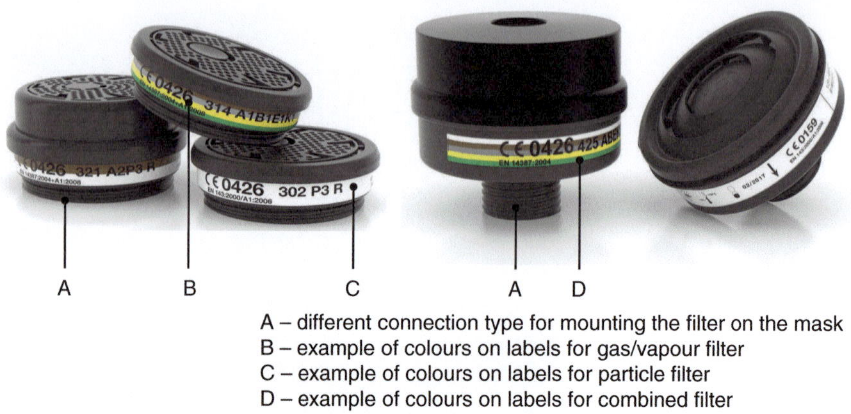

Fig. 31.8 Filter examples: labelling and connections

and K are the letters for the gas protection type, 1 is the class protection for gas adsorption capacity and P3 is the maximum protection for particles (99.95% for filters according to EN 14387) (Fig. 31.8).

31.3.5 Filtering Half Mask Incorporating Combined Filters

Filtering half mask incorporating combined filters is a particular half mask protecting against both particle and gas/vapour, where all the filters form an inseparable part of the device. Both inhalation and exhalation valves are provided, together with adjustable harness and buckles. It is a reusable mask, with a shorter life compared to a standard half-face mask equipped with filters (Fig. 31.9).

According to the design specification and protection capacity, according to European standards, the filtering type is designated with prefix FF followed by the gas filter type and class and the particle class protection (e.g. FFA2P3, where A is the gas type, 2 is the class protection for gas adsorption capacity and P3 is the protection for particles—99% for this device, according to EN 405).

31.3.6 Power-Assisted Filtering Device

There are two types of assisted filtering devices: power-filtering devices equipped with a full-face mask (negative pressure) and equipped with a hood (positive pressure).

The device consists of at least one turbo unit, a battery, one or more particle filters or combined filters, a corrugated flexible hose and the full-face mask or the hood. The powered turbo unit usually is worn fixed to the waist and/or shoulders and pulls the ambient air through the filters; then the purified air goes through the hose to the mask or the hood.

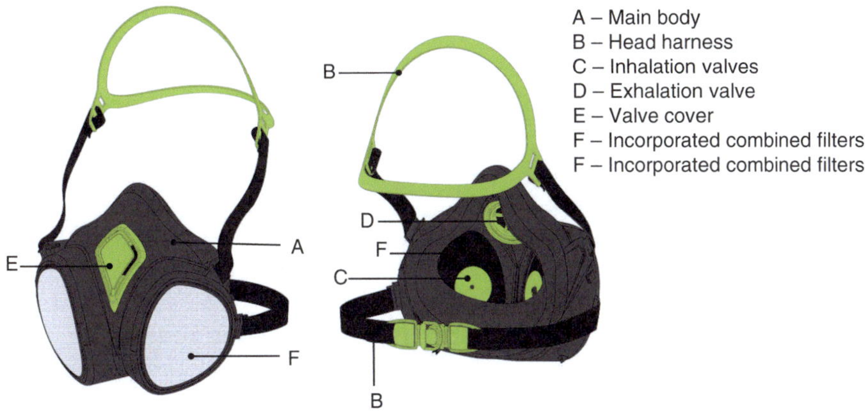

Fig. 31.9 Filtering half mask incorporating combined filters

Depending on the protection level and the filters used (particle, gas or combined), the devices are designated in different classes both for full-face mask and hood devices.

31.3.7 Atmosphere-Supplying Respirator

Those respirators that provide a respirable air to the wearer are called atmosphere-supplying respirators: this category includes supplied air respirators and self-contained breathing apparatus.

Supplied air respirators provide breathing air through air hoses from an independent source, while self-contained breathing apparatuses use a high-pressure air tank connected to the mask.

These types of equipment are used when an air purifier is not allowed, for example, not only when oxygen concentration in ambient air is low or the characteristics of the contaminants are unknown but also if the contaminant is evaluated as extremely dangerous.

31.3.8 Fit Testing: A Measure of a Mask's Correct Fitting

Fit testing is recommended for all tight respirators. The correct placement of the device to have a tight seal on the face at all times is essential to prevent exposure. The purposes of the test are to check whether the subject is able to wear the respirator correctly and check that the mask ensures a good fit on to the subject's face.

It can be carried on the whole range of tight-fitting respirators: disposable face-pieces, half-face mask and full-face mask.

The most widely used method is known as "quantitative fit test" and involves the use of a machine that gives a numerical measure of the fit: the concentration of dust in the air is measured inside the tight-fitting respirator (purified air) and outside (ambient air), and the ratio between the two values gives the fit factor (FF), which is the parameter that indicates whether the test is passed or not.

A second method is increasingly widespread due to its practicality and simplicity, the "qualitative fit test", which is suitable exclusively for disposable facepieces. It is based on the wearer detecting leakage through the face seal region by nebulizing a sweet/bitter aerosol towards the mask inside a hood: if the subject perceives the sweet/bitter taste, this means that the mask does not guarantee the perfect sealing.

31.3.9 Importance of Comfort and User's Subjective Feeling

Apart from a technical performance point of view, comfort may represent the most important parameter to take into consideration during the development and the selection of a respirator. In some cases, the perception of the wearer during the use of the respirator is fundamental. Sense of comfort arises from a subjective feeling and is the combination of many factors:

- Ease to breathe through the respirator: this parameter is expressed by means of the breathing resistance, defined as the pressure drop before and after the respirator subjected to an air flow. It is measured during both inhalation and exhalation, and the lower this parameter is, the more comfortable the respirator will result.
- Accumulation of moisture and heat: especially during activities that require physical effort, the use of a respirator leads to condensation of the water contained in the exhaled air. Hot and humid air together with water droplets accumulated in the mask can produce a feeling of discomfort. The capability to eliminate those elements from the mask is determined by the material transpirability for disposable facepieces and by the effectiveness of the exhalation valve for all kind of respirators.
- Effectiveness of air recycling: if exhaled air is not expelled completely from the mask, the content of carbon dioxide in the mask will increase, and the wearer will again inhale CO_2 causing breathing difficulties.
- Pleasantness of the materials in contact with the skin: materials, orinasals, head harness and buckles must be smooth and flexible in order to avoid itching sensation; moreover, materials must be hypoallergenic to avoid skin irritation.
- Respirator encumbrance: unnecessary weights can cause a premature fatigue on the wearer. Additionally if the mask is too large or cumbersome, it may reduce the wearer's field of view.

31.4 Overview on Occupational Diseases and Industrial Applications

"Occupational diseases cause huge suffering and loss in the world of work. Yet, occupational or work-related diseases remain largely invisible in comparison to industrial accident, even though they kill six time as many people each year". With this sentence, the International Labour Organization (ILO) opened its report on prevention of occupational diseases in 2013. The statement wants to focus on those occupational and work-related diseases, which may manifest in the long period, after years of work in a harmful environment and after exposure even to small quantities of noxious elements. Extent of those exposures may seem less dangerous and the importance of prevention and protection does not always seem so obvious.

In the same paper, ILO estimated that two million people die each year from work-related diseases, which is equivalent to more than 5000 death every day. Moreover, 160 million cases of nonfatal work-related diseases occur annually [13].

Similar studies are carried out worldwide, investigating the incidence of occupation and work-related diseases on the global death rate and marking a separation of occupational and nonoccupational causes. An example of the delicate distinction between occupational and nonoccupational disease is COPD (chronic obstructive pulmonary disease) which is one of the leading causes of morbidity and mortality both in industrialized and developing countries. Although one of the most important causes is cigarette smoke, studies demonstrate that almost 15% of the COPD cases in the world are work related (with no significant connection with workers' smoking habits) [14, 15].

Work-Related Lung Disease Surveillance System is a surveillance carried out by NIOSH institute presenting tables and figures of occupational-relevant respiratory diseases, investigating mortality rate and morbidity rate in the last decades (1968–2004) in the US working population. The surveillance reveals some remarkable facts about the trend changing for some work-related diseases in the USA. The rate of mortality for major pneumoconiosis (including silicosis, byssinosis and coal workers' pneumoconiosis) has decreased between the 1960s and the mid-2000s, but from this picture, death for asbestosis must be excluded: death cases grow from 78 in 1968 to 1470 in 2004 [16].

The decreasing of mortality rate for diseases as silicosis in countries as the USA (while exposure to silica dust on the workplace has not decreased) shows the importance of prevention and protection on the workplace: silicosis is in fact one of the first diseases of public interest and one of the first recognized causes of work-related death.

Despite silicosis had become less of a problem in developed countries, in China this is still a major concern, and in the last part of the decade 2000–2010, pneumoconiosis represented the 70–80% of the total number of cases reported as occupational diseases [17].

In the last 20 years, as a result of China's rapid economic development, regulations regarding health and safety on the workplace have become necessary, and only in 1995 the first Labour Law was promulgated, and lately many institutes of occupational diseases and prevention and occupational service agencies have been founded, each accredited by Chinese health administrative department. The Chinese Center for Disease Control and Prevention (China CDC) is an important network registering and reporting each verified case of occupational disease and poisoning: between 2005 and 2013, the number of occupational diseases reported every year has increased by more than 100% (from 12,212 to 26,393 cases per year) [18].

Various other occupational diseases originated from work exposure to chemical agents are still a pressing topic. Occupational asthma is a major concern for workers that are exposed to miscellaneous chemicals, mineral and organic dust, cleaning materials, pyrolysis products and solvents (HSE states that during 2012 and 2014, vehicle paint technicians and bakers and flour confectioners had the highest rate of new cases per year [19]). Lung cancer is a growing disease caused by occupational exposure: in 2013 more than 2000 death were caused by asbestos-related lung cancer and more than 2800 cases of death for lung cancer due to other agents; the statistic branch of HSE sows that the rate of death for mesothelioma continues to increase from 1990 to 2010 [20].

The data reported show that although worldwide there is a growing interest to the topic of occupational safety and health, a conspicuous number of death cases and nonfatal disease are registered every year, both in industrialized and developing countries. In this scenario, the importance of prevention and information towards workers is undeniable, coming from institutions and employers; indeed the cooperation between these three subjects (workers, institutions, employers) is fundamental to assure the enhancement of the current situation on occupational diseases. Knowledge of the risk that workers are facing must be clearly transmitted in order to familiarize them with the real problems that can originate from lack of protection on the workplace. The use of protections for breathing apparatuses is still perceived today as an obligation imposed by the employer and not a real need for the health of individual workers.

The following section will give an overview of some significant industry sectors, focusing on the main characteristics of the hazard and how it is possible to reduce the risk.

31.4.1 Welding Industry: A Whole Range of Risks

Welding is a fabrication process that joins materials, usually metal, by causing fusion of the surfaces of two workpieces to form one.

Occupational risks associated with welding operations originate mainly from welding fumes, a mixture of a large number of organic and inorganic chemicals, generated in the heating and melting of the metallic parts. Surveys in multiple production sites in different parts of the world revealed the presence of many chemical fumes [21–23]: iron, manganese, nickel, chromium and zinc are between the most frequent chemicals present in welding fumes.

Many other agents involved in the welding processes represent a potential hazard for inhalation exposure and may lead to acute or chronic respiratory disease: acetylene gas, oxides of nitrogen, phosgene, tungsten, arsenic, beryllium, cadmium, cobalt, copper, lead, silver and tin. Additionally, the presence of dust and ozone gas is observed in almost all welding production sites and aggravates the hazard on the workplace. This high variability of chemical components contained in the air during exposure (combined with other factors such as heat, burns, exposure to radiation, noise) allows us to understand why welding is considered one of the most dangerous occupation [24].

Presence of nickel and chromium fumes (especially in welding process of stainless steel) raises alarms because of their high level of toxicity [25, 26].

Presence of manganese is detected with the highest concentration: manganese fumes are a big concern for welding industries (and, as a consequence, for all those jobs where manganese fumes are detected, as ferroalloy smelters and mines) since these are toxic if inhaled in high concentration [9].

HSE states that the principal effects caused by welding fumes on breathing apparatus are pneumonia, asthma, cancer, metal fume fever, irritation of throat and lungs and temporary reduced lung function. The development of chronic bronchitis is more likely after exposure to a combination of dust and irritant gases in the work place, as demonstrated in several case studies [27–29].

In the welding industry, the use of facepieces, half mask or full-face mask is determined based on the concentration of the welding fumes. When welding fumes are close to the exposure limit value, e.g. if a good ventilation system is implemented and the working area is adequately big, a disposable facepiece is recommended. The filtration efficiency of the facepiece depends on the concentration of the fumes, and usually it used a facepiece with a carbon layer to filter ozone and has relief from its nuisance odour (in average welding activities, ozone gas is beneath allowable exposure limit). If fume concentration is highest, a device like a half-face mask with particle filter should be considered.

31.4.2 Coating and Painting Industry

In the world of painting and coating, various processes exist and for each one the worker is exposed to a different class of chemical substances, therefore to a different risk.

Powder coating is a process involving the use of dry powder applied electrostatically to the item surface that is placed after into an oven to melt and become a uniform layer. The advantage of this procedure is the absence of solvents in the painting; however, the presence of dust becomes the bigger concern, and a device for protection against dust and particles must be used. The choice of the device (disposable face mask, half- and of full-face mask with particle filters) depends on the dust concentration on the workplace.

Some powder coating (as polyurethane power coating) contains isocyanates, which are internationally recognized as the leading attributable chemical cause of occupational asthma. Specific measures must be taken in this case because isocyanates has a low odour threshold (the concentration at which the substance can be smelled by the human olfactory system) that is higher than the allowed exposure limit. Therefore a gas/vapour filter is not recommended because when the filter is saturated and the worker begins to sense the odour, the exposure limit has already been exceeded and the worker is receiving no protection. For this reason, if any concentration of isocyanate close to the exposure limit is measured, a supplied air respirator should be considered to ensure the correct protection.

Electrophoretic painting process consists in the immersion of metal part in a water-based solution containing paint: an electric voltage is applied to the part (which is metallic, therefore conductive) causing the paint to condense on it.

Risk in this process originates from the galvanic coating bath and depends on the galvanic bath type: mists of different elements can be present (chromium, nickel, cadmium, zinc, copper or silver and many of their compounds). Usually acid vapours evaporate from the liquid bath, causing the presence of H_2S and its compound.

Depending on the chemical element/compound, a different particle/gas/vapour filter will be selected and mounted on a half-face mask (or full-face mask, if a higher protection factor is required). If H_2S vapours are detected, a gas filter for acid gas should be used, combined with a particle filter for mist filtration.

The technique of spray painting consists of covering the item with paint sprayed from a device (air gun spray) through the air on the item surface. Paints used in this process contain solvents, which are the main concern for this employment. To protect against solvents, the best solution is to choose a device (half mask usually for this application) with filters for organic gases; since during the spraying a considerable amount of mist can be created, a combined filter (particle plus vapour/gas) should be used.

However, elements like epoxy resins and isocyanates can be contained in the varnish used in this process: this leads to the selection of a supplying air respirator.

31.4.3 Asbestos Removal: Danger Caused by Asbestos Fibres

Asbestos is composed by silicate minerals with physical properties that make it largely used in addition to building construction materials (e.g. cement) for thermal insulation, acoustic insulation and fireproofing. During operation on materials that contain asbestos (as drilling, sawing and sanding), a dust made of tiny and respirable asbestos fibres is released in the air. These fibres have an elongated shape that can be easily inhaled and carried in the lower region of the lung causing asbestosis, reduced respiratory function, pleural mesothelioma and lung cancer and can lead to death.

ACGHI set the exposure limit value (TWA) to 0.1 f/cm^3 and recognize asbestosis fibres as carcinogen (carcinogenicity designation A1) based on the weight of evidence from epidemiologic studies.

Due to its high-risk nature, manufacturers do not recommend the use of disposable facepieces for protection against asbestos; however, a particle filtration system must be considered. For lower concentration (1 f/cm^3), a half-face mask shall be used, and exposure up to 5 f/cm^3 implicates the use of a full-face mask. For higher concentrations, it becomes necessary to use a device with high protection factor: full-face powered air-purifying respirator or full-face supplied air respirator.

31.4.4 Foundries: High Exposure to Respirable Silica Dust

The iron and steel industry is very diverse in materials and processes, resulting in occupational exposures to a wide variety of substances.

Exposure to respirable quartz (a form of respirable crystalline silica, RCS) continues to be a major concern in all iron foundry industry; it is used in a metal casting process (also known as sand mould casting) characterized by using sand as the mould material. Silica is the most commonly used sand base, and for this particular process, a large amount of silica dust has to be taken into account (for high melting point casting, such as steels, a base sand with a minimum of 98% pure silica sand is used [30]). Foundry dust consists predominantly of particles below 2.5 μm [31].

A Swedish study [32] tried to determine the occupational exposure to respirable dust and respirable quartz in 11 Swedish iron foundries, representing different sizes of industrial operation and different manufacturing techniques. Results show that for respirable quartz, 23% of all the measurements exceeded the EU-OEL and 56% exceeded the ACGIH-TLV. The overall geometric mean for the quartz levels was 0.028 mg/m^3, ranging from 0.003 to 2.1 mg/m^3. Similar values were confirmed by an assessment in Taiwan foundries [33].

Together with dust, in foundries the presence of metal fumes has a great importance in choosing the correct protection device. Using X-ray diffraction, different compounds can be identified inside foundries: iron oxides, manganese oxide, zinc oxide and calcium oxide [34].

In foundries, the presence of dust and fumes requires the use of a particle filtering device: disposable facepieces should be selected for lower concentration; filters (usually a particle filter mounted on half mask) are necessary for highest concentrations.

Effects on health coming from occupational exposure to silica dust have been investigated for many years and in many different ways, starting from animal experimentation and epidemiological studies. By breathing RCS there is the possibility to develop lung diseases as silicosis, chronic obstructive pulmonary disease (COPD) and lung cancer [35].

31.5 Conclusions: Are Industrial Respirators Suitable for Air Pollution?

As we have gradually deduced in this chapter, respiratory protection is a delicate issue, and the selection of the right respirator must be submitted to the knowledge of the type and the extent of the chemicals contained in the air. A transition from the methods used in the industrial sector and the evaluation of protection against air pollutants present in our everyday life is not straightforward.

In fact, the introduction of the threshold limit values (TLV) is a key factor in the device selection for industrial application, but those same values were not developed for evaluation of the potential harm deriving from environmental air pollution.

In the last few years, institutions have drafted several lists of exposure limit values that shall not be exceeded during a period of 1 h, 8 h or 24 h and the number of permitted exceedances over the limit values each year. Limits and target values are suggested for specific air pollutants (some pollutants are not considered in all standards): sulphur dioxide, nitrogen dioxide and oxides of nitrogen, particulate matter (PM10 and PM2.5), lead, carbon monoxide, benzene, ozone, arsenic, cadmium, nickel, benzo(a)pyrene and many others (Table 31.7).

By analysing this list of substances, it is easy to understand that a combination of different elements constitutes the air pollution, each with different physical properties. If we consider the seven pollutants present in the highest percentage, we can divide them in two categories: particle pollutants and gas/vapour pollutants (Table 31.8).

For each category a different filtering technology should be considered. For particle pollutants (PM2.5 and PM10), disposable facepieces are commonly preferred, but many of the serious health concerns regarding air pollution are not filtered by this type of respirator. Particulate respirator does not reduce the exposure to gases and vapours (such as sulphur dioxide, nitrogen dioxide, carbon monoxide and ozone) but is effective uniquely against dusts: in this case a particulate filter respirator should be combined with a carbon filtering device.

Theoretically, the best solution for filtration of air pollutants seems to be a half-face mask equipped with cartridges containing carbon suitably activated for adsorption of those chemicals present in the polluted air combined with a particle filter. Nevertheless this kind of device proves to be bulky and too heavy for the daily use (the weight for a common half-face mask with combined filter can easily reach 300–400 g), apart from the fact that the protection given by this ensemble is more than what is needed.

Disposable facepieces with an integrated carbon filtering layer/material are an appropriate choice for protection against air pollution as long as concentration of

Table 31.7 Comparison for SO_2 maximum allowed values in industrial regulation by ACGIH and guideline for outdoor air quality by WHO

SO_2 industrial TLV (by ACGIH)	SO_2 guideline values for outdoor air quality (by WHO)
5 mg/m^3 (TWA) (equal to 5,000 μg/m^3)	20 μg/m^3 (24-h mean)
13 mg/m^3 (STEL) (equal to 13,000 μg/m^3)	500 μg/m^3 (max 10 min)

Table 31.8 Most common air pollutants divided on the basis of their physical nature

Particle pollutants	Gas/vapour pollutants
PM2.5	Sulphur dioxide (SO_2)
PM10	Nitrogen dioxide (NO_2)
Lead (Pb)	Carbon monoxide (CO)
	Ozone (O_3)

gases remains low. Carbon layers are usually activated for adsorption of minimum quantities of one or both organic and acid substances: specific activated carbon should be considered for the adsorption of SO_x gases (sulphur oxides) and NO_x gases (nitrogen oxides) in addition to carbon monoxide. As a matter of fact, the majority of industrial disposable masks with carbon layer that can be found on the market are intended to filter low concentration of organic (VOC compounds) and acid gases. Therefore, the research for a specific activated carbon-based material is essential while choosing an antipollution mask and together with the technical filtering performance should be explicitly explained which gases can be filtered by the mask.

However, specific regulations regarding adsorption of the carbon layer in disposable facepieces, both for industrial exposure and environmental pollution, do not exist. Along with the already existing guidelines on the maximum concentrations allowed, the definition of test methods and specific requirement for filtration against air pollutants (of all physical nature) should be taken into account by institutions and world organizations, in order to ensure the best protection for the entire world population and give some parameters to correctly choose the appropriate antipollution protection device.

Beside the composition of the mask, which is a fundamental aspect to take into consideration, the importance of donning the mask correctly should not be underestimated. Gaining the perfect sealing of the mask on the face means that the unfiltered air will not enter in the mask (therefore will not be breathed by the wearer). In fact, even if a small quantity of unpurified air goes through the small interspaces between the body of the mask and the user's face, that air will bring noxious chemicals into the lungs and the mask becomes useless.

Manufacturers' instructions explain how to don the respirator correctly and may explain an easy and immediate test to check if the mask is fitted right. As an example, to check the fit of a disposable facepiece, it is recommended to cover the front of the respirator with both hands being careful not to disturb the facepiece position. If the mask position is correct, by inhaling and exhaling deeply, no air should be perceived escaping from the mask boundaries, from the eyebrows, chin and cheeks. This is a rough test and sometimes seems to be not enough. In fact, the use of a respirator device should go through a more detailed training, even with audio-video support if a training face-to-face with a specialized expert is not possible.

References

1. World Health Organization. Residential heating with wood and coal: health impacts and policy options in Europe and North America.
2. Chen W, Zhuang Z, Attfield M, Chen B, Gao P, Harrison J, Fu C, Chen J, Wallace W. Exposure to silica and silicosis among tin miners in China: exposure-response analyses and risk assessment. Occup Environ Med. 2001;58:31–7.
3. West Virginia Department of Culture and History. Hawk's nest tunnel disaster. Retrieved 2008-11-25.

4. Rantanen J, Fedotov IA. Standards, principles and approaches in occupational health services. Article, July 2002. http://www.ilo.org/safework/info/publications/WCMS_110439/lang--en/index.htm
5. Bang KM, Attfield MD, Wood JM, Syamlal G. National trends in silicosis mortality in the United States: 1981–2004. Am J Ind Med. 2008;51(9):633–9.
6. OSHA. Technical manual (OTM), TED-01-00-015. 1999. https://www.osha.gov/dts/osta/otm/otm_toc.html
7. Informational Booklet on Industrial Hygiene Industrial Hygiene. U.S. Department of Labor and Occupational Safety and Health Administration. OSHA 3143. 1998.
8. Campurra G. Manuale medicina del lavoro. Ipsoa; 2013.
9. 29 CFR 1910.1000 toxic and hazardous substances.
10. Directives 91/322/EEC, 2000/39/EC, 2006/15/EC, 2009/161/EU.
11. Bollinger NJ, Schutz RH. A guide to industrial respiratory protection. Washington, D.C.: US Department of Health and Human Services; 1987.
12. Health and Safety Executive. Respiratory protective equipment at work. A practical guide. 2013.
13. The prevention of occupational diseases, International Labour Organization. 2013.
14. Boschetto P, Quintavalle S, Miotto D, Lo Cascio N, Zeni E, Mapp CE. Chronic obstructive pulmonary disease (COPD) and occupational exposures. J Occup Med Toxicol. 2006;1:11.
15. Hnizdo E, Sullivan PA, Bang KM, Wagner G. Association between chronic obstructive pulmonary disease and employment by industry and occupation in the US population: a study of data from the Third National Health and Nutrition Examination Survey. Am J Epidemiol. 2002;156(8):738–46.
16. National Institute for Occupational Safety and Health. Work-related lung disease surveillance report 2007. 2008.
17. Xueyan Z, Zhongxu W, Tao L. The current status of occupational health in China. Environ Health Prev Med. 2010;15(5):263–70.
18. Fenghong W, Yan C. Regulatory system reform of occupational health and safety in China. Ind Health. 2015;53(3):300–6.
19. Health and Safety Executive. Work-related respiratory disease in Great Britain 2014: an overview of the current burden of disease in Great Britain. 2015.
20. Health and Safety Laboratory. The Great Britain Asbestos Survey 1971-2005, Mortality of workers listed on the Great Britain Asbestosis or Mesothelioma Registers. HSE Books, 2010. http://www.hse.gov.uk/research/rrpdf/rr805.pdf.
21. Korczynski RE. Occupational health concerns in the welding industry. Appl Occup Environ Hyg. 2000;15(12):936–45. https://www.ncbi.nlm.nih.gov/pubmed/11141606.
22. Fethke NB, Peters TM, Leonard S, Metwali M, Mudunkotuwa IA. Reduction of biomechanical and welding fume exposures in stud welding. Ann Occup Hyg. 2016;60(3):387–401.
23. Stanisławska M, Janasik B, Trzcinka-Ochocka M. Assessment of occupational exposure of welders based on determination of fumes and their components produced during stainless steel welding. Med Pr. 2011;62(4):359–68.
24. Tierney MP. Analysis of mine injuries associated with maintenance and repair in metal and non-metal mines. Washington, D.C.: U. S. Dept. of the Interior, Mining Enforcement and Safety Administration; 1977.
25. Agency for Toxic Substances and Disease Registry
26. International Agency for Research on Cancer. IARC monographs on the evaluation of carcinogenic risks to humans. 1990. ©International Agency for Research on Cancer, 1999. https://monographs.iarc.fr/ENG/Monographs/vol71/mono71.pdf
27. Meo SA, Al-Khlaiwi T. Health hazards of welding fumes. Saudi Med J. 2003;24(11):1176–82. https://www.ncbi.nlm.nih.gov/pubmed/14647549.
28. Ozdemir O, Numanoğlu N, Gönüllü U, Savaş I, Alper D, Gürses H. Chronic effects of welding exposure on pulmonary function tests and respiratory symptoms. Occup Environ Med. 1995;52(12):800–3.

29. Sobaszek A, Edme JL, Boulenguez C, Shirali P, Mereau M, Robin H, Haguenoer JM. Respiratory symptoms and pulmonary function among stainless steel welders. J Occup Environ Med. 1998;40(3):223–9.
30. Rao TV. Metal casting: principles and practice. New Delhi: New Age International; 2003.
31. Beeley P. Foundry technology. Oxford: Butterworth-Heinemann; 2001.
32. Andersson L, Bryngelsson I-L, Ohlson C-G, et al. Quartz and dust exposure in Swedish iron foundries. J Occup Environ Hyg. 2009;6:9–18.
33. Chen PC, Doyle PE, Wang JD. Respirable dust exposure and respiratory health in male Taiwanese steelworkers. Ind Health. 2006;44:190–9.
34. Tossavainen A. Metal fumes in foundries. Scand J Work Environ Health. 1976;1:42–9.
35. Health and Safety Executive, Control of exposure to silica dust.

32 Sociological Theories on Air Pollution: Between Environmental Justice and the Risk Society Approach

Antonio Maturo and Veronica Moretti

32.1 Introduction

Substantially, sociology has addressed the problem of pollution from three general perspectives: environmental sociology, medical sociology, and theoretical sociology. Environmental sociology has focused in large part on social movements and deliberative democratic processes originating in opposition to technologies considered to be pollutant or dangerous, such as incinerators or nuclear reactors. In correlation with these topics, environmental sociology has also studied the nexus between social inequalities and exposure to pollution, an issue which is also central to medical sociology's important investigation of the relationship between socioeconomic inequalities and health [1]. Indeed, pollution is given a fundamental role in the social determinants of health model [2]. Both environmental and medical sociology have provided solid scholarship on the consequences of pollution in the job place. Theoretical sociology has analyzed pollution through the concept of risk. According to Giddens' [3] well-known observation, at a certain point, we stopped worrying so much about what nature could do to us, and we started worrying more about what we have done to nature, marking a shift from a society dominated by dangers to one dominated by risks. The German sociologist Ulrich Beck [4] has proposed what is probably the most consistent theory of risk. In his "risk society" theory, Beck emphasizes the globalized character of risks along with their "democratizing" nature: nowadays both the poor and those who are better-off are exposed to risks – a position challenged by many scholars doing empirical research on pollution.

A. Maturo (✉) • V. Moretti
Department of Sociology and Business Law,
University of Bologna, Bologna, Italy
e-mail: antonio.maturo2@unibo.it

F. Capello, A.V. Gaddi (eds.), *Clinical Handbook of Air Pollution-Related Diseases*,
https://doi.org/10.1007/978-3-319-62731-1_32

Air pollution can be analyzed according to different categories. Firstly, we can distinguish between outdoor and indoor pollution: "Some of the major sources of outdoor pollution include exhaust fumes from vehicles, emissions from manufacturing facilities (i.e. factories) and power generation (i.e. smoke stacks of coal fired power plants)," while "the main cause of indoor air pollution is inefficient fuel combustion from rudimentary technologies used for cooking, heating and lighting. There are also natural indoor air pollutants, like radon, and chemical pollutants from building materials and cleaning products that also impact health" [5]. Undoubtedly, outdoor air pollution can be an important factor affecting the quality of the air of indoor spaces as well, in particular when houses have ventilation systems or are located next to sources of pollution. Conversely, indoor air pollution can worsen the quality of outdoor and urban air, especially in cities where houses use biomasses or coal for heating and cooking. The combination of indoor and outdoor pollution is particularly harmful in poor areas, especially considering that in many developing countries, indoor pollution is linked to the combustion of solid fuels used for cooking and heating [6]. Industry and automobiles are the first and the second causes of air pollution even if there are substantial differences among different countries: "In the United States, there is one car for every two people, in Mexico one for every eight, and in China one for every 100"; moreover: "The average number of miles travelled/car/year in the United States has more than doubled, from 4570 in 1965 to 11,400 in 1999" [7, p. 574]. Yet, car owners are not always the category of people most exposed to pollution. In fact, the WHO estimates that more than 600,000 people die each year as a result of outdoor air pollution in India; most of them are poor people who live in the most polluted areas and who do not own a car [8].

Another important distinction involves visible and invisible forms of air pollution. Several studies have shown that the visibility of pollution plays a crucial role in shaping the way it is perceived [9–11]. Individuals are more aware of bad air conditions when dust and fallout are visible and can, above all, be smelled: "it is very likely that the 'invisibility' aspect of air pollution discourages people from drawing strong links between air pollution and health" [11, p. 12]. "Annoyance" is a factor that strongly influences the perception of air pollution. Individuals seem to be more worried by the short-term consequences of pollution, rather than by the long-term ones: "Annoyance involves individual perceptions and attitudes towards the exposure influenced by factors such as sex, age and prior exposure to the pollutant" [6, p. 4853].

Working within this general framework, this chapter will investigate the role that sociology can play in preventing and containing air pollution. In particular, we believe that sociology has an important contribution to make from the environmental justice perspective [12]. Such an approach demonstrates how the most deprived people are the most exposed to health risks from pollution and as such should be at the focal point of research and intervention. In addition, sociology can provide new data and solid scholarship to institutional programs based on environmental determinants and health promotion demonstrating the most appropriate communication strategies for stimulating responsible and "greener" lifestyles and behaviors.

More specifically, in this chapter, we describe and analyze the main sociological theories on the environment and pollution—with a special focus on recent contributions to risk sociology—and discuss environmental justice and its main pillars: popular epidemiology, contested illnesses model, and community-based participatory research. As can be inferred from its terminology, environmental justice is closely connected to a social justice perspective and to a governance approach based on social participation and prevention.

32.2 Sociological Theories and Pollution

In sociological scholarship, the interaction between individuals and environment has been studied from three theoretical perspectives: functionalism, conflictualism, and social construction theory.

Functionalism has been the dominant theory for several decades in sociology. Its main proponents were Emile Durkheim, Talcott Parsons, Robert K. Merton, and Niklas Luhmann. Functionalism defines society as a biological organism. Like in biology, society is seen as a system made up of different interrelated parts. Every part has a specific function that is necessary to the survival of whole. As such, this holistic paradigm emphasizes the conservation and the equilibrium of the system. In other words, the social system has a series of needs that must be satisfied through reciprocal activities—i.e., functions—carried out by the various components of the system. Logically, functionalists consider society as a whole that has specific features, this whole being greater than the sum of the individuals comprising it.

For example, according to Durkheim [13], social phenomena should be explained by taking into consideration the functions they carry out for the "social organism" and not for their essence or for their social meanings. According to Durkheim, a certain amount of crime and deviance is "normal" or a society. Making an extreme example from the present, we could state that the violence of the hooligans during soccer matches is like a relief valve that functions to lessen the frequency of worse crimes during the rest of the week. According to Niklas Luhmann's [14] "cynical" functionalism, pollution emerges as a social problem only when it is "perceived" by the political system. In other words, policies against pollution will be implemented only when the political system realizes that pollution can cause a loss of consensus among the voters. Before that, pollution has no social relevance.

Functionalism has been criticized for its static and conservative view of society, its emphasis on equilibrium in the system, and its predilection for the status quo. The functionalist approach to environmental sociology sees pollution as a necessary evil, which, however, may become dysfunctional for society beyond a certain limit; at which point, it must be reduced.

The conflictualistic theory—also called "radical" perspective or critical theory—is in many respects the contrary of functionalism. Though it has several versions, it has its foundation in Marxism. According to this theoretical approach, society should be considered an arena in which different social classes struggle to get control of a scarce quantity of resources. The economic structure or material basis of society—that is, the ownership of the means of production—heavily influences the "superstructure" (the culture, laws, and religion and the political institutions) of a given society. Therefore, the elements of the superstructure become a tool for legitimizing and maintaining the dominant class. And consequently, according to the proponents of conflictualism, not even the State functions as a neutral power but rather supports the capitalistic system. According to conflictualism, capitalism and neoliberalism are the cause of environmental deterioration, but the State and political institutions cannot do much as they are themselves an instrument of the economic power. Pollution, then, can only be reduced following a profound reorganization of society.

Critical consumption and degrowth theory are two "byproducts" of conflictualism which are closely connected to environmental issues. Indeed, critical consumption practices are based on the assumption that individuals should choose the goods—or services, like transportation—they consume according to anti-neoliberalistic values, such as environmental sustainability, fair trade, labor conditions, health and safety conditions, animal welfare, and human rights. Critical consumption can also be seen as the practical expression of degrowth theory. Serge Latouche is the principle thinker and proponent behind degrowth theory, which proposes a paradigm shift away from neoliberalism and therefore from the greedy and irresponsible use of natural resources [15]. According to this view, society should be remodeled in order to foster an ecologically sustainable form of development that achieves a state of harmony between humans and nature. Practical examples of critical consumption include the sharing economy, local markets, energy saving, fair trade groups, and other practices that resist neoliberalism. Critics to the conflictualistic perspective cite its utopian aspects and economic determinism. Finally, another frequently adopted critical approach is the environmental justice perspective, the assumption (also based on empirical studies) that the most deprived individuals in society are also more exposed to pollution in comparison with members of the upper class.

The third sociological theory that is central to research on the environment and pollution is the social construction perspective. This theory holds that the real world, regardless of its material basis, is also represented as a series of socially and culturally legitimated ideas, practices, and things. Just as there are several social representations of cancer, beauty, success—just to make few examples—there are also social constructions of environmental problems. As such, it is fundamental to investigate: "the process through which environmental claims-makers influence those who hold the reins of power to recognize definitions of environmental problems, to implement them and to accept responsibility for their solution" [16, p. 185]. According to some scholars that adhere to this theory, the "successful" construction of an environmental problem requires six components: "the scientific authority for and validation of claims; the existence of "popularisers" (activists, scientists) who can frame and package the "problem" to journalists, political leaders, and other opinion makers; media attention that frames the problem as novel and important (such as the problems of rainforest destruction or ozone depletion); the dramatization of the problem in symbolic or visual terms; visible economic incentives for taking positive action;

and the emergence of an institutional sponsor who can ensure legitimacy and continuity of the problem" [16, p. 55].

Mayer [17] carried out an exhaustive study of the print media's coverage of the relationship between asthma and air pollution, focusing on the way in which the three largest US newspapers depicted a causal link between exposure to air pollution and the disease's occurrence over a twenty-year span. Although the print media's coverage of asthma and its environmental correlates had increased over time, Mayer found that there was little consistency in how the relationship was depicted; at times air pollution was portrayed as the direct cause of asthma, and at others it was simply represented as a triggering agent. On this basis, Mayer concluded that: "This lack of coherence and specificity in the portrayal of asthma as an environmental disease may weaken regulators' ability to act in passing air pollution reform by lowering the public interest and concern" [17, p. 892].

The social construction perspective, which is less "ambitious" in proposing a theoretical framework for understand the relationship between individuals and their environment, is strategically different from functionalism and critical theory. Rather than offering a comprehensive sociological view of pollution, the social construction perspective provides a solid methodological basis for empirical investigations. This is both its strength and its weakness.

Of the perspectives discussed above, the critical approach is the most prominent in the field of pollution studies. Indeed, while it contains some utopic and unrealistic elements, there is a lot of empirical evidence that demonstrates how uncontrolled consumerism and neoliberal deregulation are at the root of the larger part of environmental deterioration and pollution. Also, there is evidence that the negative externalities of capitalism—e.g. air pollution—are more burdensome for the most vulnerable members of the population.

32.3 Risk Approaches to Air Pollution

The concept of risk has only recently entered the field of sociology, but it has rapidly become a pillar of sociological theory and empirical studies. As we already mentioned, according to Anthony Giddens, when we stopped worrying so much about what nature could do to us, and we started worrying more about what we have done to nature, we moved from a society dominated by dangers to a society dominated by risks [3]. Risks are a product of human ability and are concerned with catastrophes and harmful phenomena occurring in our natural and social environments. They arise from technological and scientific development and cause unanticipated consequences. Our society is dominated by risk because it is future-oriented, and therefore decisions taken today will produce effects tomorrow. Yet, only rarely can these effects be anticipated and foreseen; the world of today is being too complex and interdependent to be studied with a linear model of forecasting. For example, financial speculation on mortgages in Florida could lead to pension reform in Italy. Moreover, risks are global: a nuclear disaster in the Ukraine can cause an increase of cancer in Lithuania. Also, nowadays, (perceived) risks have increased as a result of the infinite set of scientific tools that we have to analyze reality. In medicine, for example, sophisticated diagnostic tools make it impossible for anyone to be considered completely healthy: we are all "at risk" of something.

Presently, several issues related to the environment (air pollution, climate change, water pollution, toxic garbage) can be framed by the sociological concept of risk because of the strong impact they have on human life.

There are three sociological approaches that address pollution—mainly air pollution—through the category of risk. The first is the *risk society* proposed by Ulrick Beck and based on the concepts of the globalization and the democratization of risk. The second is made up of various groups of scholars who focus their analysis on inequalities, demonstrating that the health risks derived from pollution are inversely correlated with the socioeconomic position. These scholars emphasize issues of *social justice*. The approach is more of an empirical research orientation than a full-fledged theory: through adopting a combined perspective of *risk society* and environmental justice while carrying out empirical investigations.

32.3.1 Risk Society and Air Pollution

The risk society approach stems from the sociological theory of Ulrich Beck [4]. Beck has dedicated several books [4, 18, 19] to the analysis of the new risks of modernity and the manners by which society organizes itself in order to respond to them. According to Beck, accelerated technological development together with an irresponsible form of capitalism is responsible for the huge increase in the number of risks to which we are exposed. Indeed, the growth of wealth has been parallel to the growth in risks—capitalism does not only create "goods," it also creates "bads" [4]. Citing the Chernobyl disaster and climate change, Beck claims that the upsurge in technological development has in some cases had devastating effects on a global level, with unexpected consequences [20]. Moreover, and this is one of most original but also most controversial parts of his theory, Beck states that environmental threats these days have a "democratizing" element to them. Climate change affects everyone—even the wealthy. Therefore, it reduces social differences. Although some individuals are "well equipped" to face the negative consequences of climate change, some ubiquitous rather than hierarchical risks (air pollution and climate change) affect everyone equally: "hunger is hierarchical, smog is democratic" [4, p. 48. Here, Beck mentions the *boomerang effect*, a process that subverts class structure, exposing the wealthy to the risks they contributed to create. Hence, the protagonists of modernization become victims of their own actions: "air pollution, climate change and other "bads" that cannot be circumscribed by human boundaries (…) have an equalizing effect" [21, p. 111]. Consider smog, for example. It is easy to observe that its harmful effects also spread to the most affluent areas of cities and to highly modern societies as well. Environmental risks—according to the risk society theory—due to their mobility and pervasiveness, cannot be delimited [21].

According to Beck, political policies have also changed in their approach to dealing with air pollution, adapting to new forms of pollution. Indeed, while in the past the governments focused more on visible problems (like smog coming from chimney or vehicles), nowadays there is a particular attention given to hidden and invisible threats (like toxic substances in food). According to Beck, with a nod to social constructionism theory, risks become "visible"—and therefore gains social existence—when socially defined within knowledge or knowledge-processing systems such as science, the legal system, and the mass media: "The hazards, which are not merely projected onto the world stage, but really threaten, are illuminated under the mass media spotlight" [22, p. 101]. Following this same line of reflection, the results of an investigation conducted in three cities in Latin America—Bogota, Mexico City, and Santiago—confirm that "association between air pollution levels and socioeconomic vulnerabilities did not always correlate within the study cities. Furthermore, the spatial differences between socioeconomic vulnerabilities within cities do not necessarily correspond with the spatial distribution of health impacts" [21, p. 110]. Indeed, the researchers, analyzing local temperature, air pollution, mortality, and socioeconomic vulnerability data concluded that health risks stemming from atmospheric conditions and pollutants have no boundaries and do not discriminate between social classes, as proposed in Beck's boomerang effect concept.

Yet, at the same time, it is hard to deny—and this is the focus of environmental justice perspective—the relationship between exposure to air pollution, vulnerability, and socioeconomic position. Therefore, Beck's proposal about the "democratization" of risks is true only for some aspects.

32.3.2 Environmental (In)Justice

Bauman's book *Liquid Fear* [23] provides a good introduction to the environmental justice perspective. Everyone was aware that hurricane Katrina was going to come. People had been promptly alerted and had more than sufficient time to find a safe place; however, many people were unable to get away, by flying elsewhere or booking a room in a motel. As such, the poor people, as well as those who could not afford the cost of leaving their houses, were the most affected. This is not only because they lost their property or were impacted more than others; they could not prepare themselves in a fair way, compared to the wealthy part of the population. To use the example of the Titanic, most of the victims of Katrina were people living in the economic class of society, rather than in the business class. The relationship between poverty and air pollution has led several scholars and social activists to highlight the idea of an environmental injustice: "almost two thirds of the most polluting industrial facilities are found in areas of below average income" ([24]; quoted in [25, p. 214]. Indeed, the concept of environmental injustice originates from the fact that some communities are much more exposed to environmental risks than others.

Typically, these minorities—who can be very numerous—have a higher risk of becoming sick from air pollution. As the sociology of health has demonstrated, there is a correspondence between lower socioeconomic status and an increase in vulnerability to illnesses. For example, Clark et al. [26, p. 1] found that "low-income nonwhite young children and elderly people are disproportionately exposed to residential outdoor NO_2." On average in the United States, minorities are exposed to 38% higher levels of outdoor NO2 than whites. As such, the expression "environmental racism" is particularly appropriate [27]. A study carried out in Montreal found a significant relationship between the concentration of NO_2 and neighborhood-level indicators of material deprivation, such as median household income, and with indicators of social deprivation, such as the proportion of people living alone [28]. In France, a group of epidemiologists found that the most advantaged social classes are less exposed to air pollution from traffic in comparison to deprived people [29]. Indeed, several studies have shown that "spatial variability of level of ambient air pollution can be greater between different neighborhoods within cities than those between cities" [28, p. 971]; often the most deprived neighborhoods are also severely polluted.

Often, the inequality pattern with relation to exposure to air pollution runs parallel to access to health services. An epidemiological study carried out in California over the period 2005–2007 investigated "racial and ethnic disparities in hospital admission and emergency room visit rates resulting from exposure to ozone and fine particulate matter levels in excess of federal standards ('excess attributable risk')" (p. 1163). This study found that: "Black residents experienced the greatest EAR, roughly 2.5 times that of white residents. Differences also existed in exposure to ambient ozone and PM2 5 pollution, with Hispanics experiencing the highest levels of both on average. Baseline event rates differed substantially across groups, and in many cases it was these differences that drove the overall disparity" [30, p. 1167].

Yet, such inequalities and discriminations do not only occur with outdoor pollution but are also present in the case of indoor pollution. Burning solid fuels is one of the most serious environmental problems in developing countries [31, 32], and "it appears that young children in households that use solid biomass fuels are 2–3 times more likely to suffer ARI than children in households that use other fuels" [33, p. 572]. Using data from four poor provinces in China (Gansu, Guizhou, Inner Mongolia, and Shaanxi), Jin et al. [34] examined the relationship between technology, user knowledge and behavior, and access and infrastructure in exposure to IAP from household energy use. Also here, there was correlation between socioeconomic status and exposure to pollution. Poor people and minorities were the most affected by indoor air pollution.

Moreover, "the role of heating as a source of IAP exposure has also created unique epidemiological patterns in China, in which both female and male Chinese non-smokers have substantially higher rates of mortality from lung cancer and chronic respiratory disease than other populations, with those in northern, and colder, parts of the country having even higher rates [35]" [34, p. 3169].

In the United States, the abundance of evidence on inequalities and discrimination in the exposure to air pollution led to the foundation of the *The United States*

Environmental Protection Agency—Office of Environmental Justice. This Agency defines environmental justice as:

> the fair treatment and meaningful involvement of all people regardless of race, color, national origin, or income, with respect to the development, implementation, and enforcement of environmental laws, regulations, and policies. It will be achieved when everyone enjoys the same degree of protection from environmental and health hazards, and equal access to the decision-making process to have a healthy environment in which to live, learn, and work.[1]

Actually, in several cases, the environmental justice approach does not just "certify" the nexus between socioeconomic inequalities and exposure to pollution and poor health conditions but rather is a vociferous lobbyist of institutions. In several instances, this critical perspective results in the direct participation of researchers in political actions carried out to modify pathogenic environmental conditions. Moreover, as we will see with the CIRG example, this kind of social activism is often implemented together with the affected individuals of the polluted areas. In this type of so-called "action-research" approach, lay persons sometimes also contribute to the construction of scientific data.

32.3.3 Air Pollution and "Urban Vulnerability As Impact"

The "urban vulnerability as impact" approach should be considered more of a research hypothesis than a theoretical framework. In fact, it should be seen as an intermediate perspective between Beck's theory and the critical theory on which the environmental justice perspective is based.

It is important to delineate the two concepts that characterize this perspective: "urban" and "vulnerability." The last decades have witnessed a steady growth in the number of urban residents, as "(increasingly) urban centers are home to a large proportion of the world's population, economic activity, and physical infrastructures" [36, p. 142]. Indeed, between 2007 and 2050, the world population should increase by 2.5 billion shifting from 6.7 billion to 9.2 billion, and this growth will take place mostly in the urban areas of the world. As a consequence, cities and megacities—most of them in developing countries—will face what is called the "urbanization of poverty" [37].

The second concept is vulnerability. Downing and Patwardhan [38] indicate four components of vulnerability: a social system (population group, natural system, or an economic sector), the attribute of concern (the level of the threat or the exposure to a hazard), the hazard, and the temporal reference (time period of interest). Therefore, it is possible to talk about either natural vulnerability—harm caused by climate or pollution—or socioeconomic vulnerability—the capacity (or incapacity) to recover from negative events and/or to face unhealthy living conditions [39]. The researchers who adopt this third approach point out that scholarship that has

[1] https://www.epa.gov/environmentaljustice.

investigated socioeconomic factors of adaptive capacity, such as income, poverty, and ethnicity, have given mixed results. Some cases have confirmed the environmental justice hypothesis, while others have demonstrated the opposite, the boomerang effect, proposed by Beck: "If the health risks are indeed non hierarchical, as proposed by Beck, the differences in vulnerability will not be correlated with these risks; but if the health risks are socio-economically differentiated, the differences in vulnerability will mirror differences in risks" [21, p. 112]. The "urban vulnerability as impact" approach emerges to untangle this—at first sight—etiological dilemma. Consequently: "To test this hypothesis, we conceive of health risks as a function of exposure and socioeconomic vulnerability" [21, p. 112]. Exposure is a concept that is relatively easy to operationalize by indicators and indexes since there are several tools that measure the levels of air pollution.

As already mentioned, the research conducted by three proponents of the human vulnerability as impact approach—Patricia Romero-Lankaob, Hua Qina, and Mercy Borbor-Cordovac—is focused on the effects of air pollution and climate change in three South American cities: Bogotà, Mexico City, and Santiago. More specifically, they investigated whether the health risks related to air pollution and temperature extremes are spatially and socioeconomically differentiated. Their main finding confirms Beck's theory: "health risks from atmospheric conditions and pollutants act without boundaries or social distinctions within urban areas." However, at the same time, other results resonated with the environmental justice perspective, leading the researchers to affirm that: "the influence of socioeconomic status should not be underestimated as it plays a complex role in driving and explaining health risks, and interacts in intricate ways with the other dimensions of health risks" [21, p. 117].

32.4 Environmental Justice and Air Pollution

The burden of air pollution on health is significant, if not tragic: "airborne particulates have been estimated to account for more than 100,000 deaths annually in the US from pulmonary and cardiac disease—more than breast cancer, prostate cancer and AIDS combined" [40, p. 113], while indoor pollution from biomass and coal is a leading environmental risk in many developing countries [34]. To reduce these frightening numbers, institutions have implemented several programs of health promotion and prevention. However, also thanks to the web, there are a growing number of bottom-up initiatives driven by lay people that act as social activists fighting for a healthier place to live.

32.4.1 Air Pollution: Individual Responsibility and the Double Burden of Deprivation

In 2009, an impressive group of experts from different disciplines—researchers, public policy makers, air quality managers, and NGO representatives – gathered in Vancouver in order to brainstorm and plan effective actions to reduce the health impact of air pollution both at the community and individual-level [41]. Community-level interventions mainly focused on emissions reductions. In general, the reduction of the emissions of air pollutants can be pursued—according to the experts—through three primary actions: source substitution (like woodstove replacements and bans on the sale of coal), urban/transportation planning (e.g., the Congestion Charge Scheme in London and the separation between sources of air pollution and residential locations and schools), and technology upgrades (e.g., zero-emission vehicles like electric cars). These actions require legal restrictions and regulations in order to be concretely implemented. Moreover, on a national level, generous incentives should be introduced to decrease commercial traffic on the roads and foster the use of railways—the global economy relies heavily on the transport of goods over long distances, and these often affect local communities.

With regard to interventions directed at individuals, the experts proposed four areas of action: exercise, nutrition, medication, and health information about air quality.

There is a large amount of research that proves that physical activity reduces the risk of cardiovascular diseases and diabetes; moreover air pollution is a significant risk factor for these kinds of conditions [42]. Hence, physical activity is a proactive factor for some health conditions that can be co-caused by air pollution. Tragically, outdoor exercise in polluted areas is associated with several illnesses, first of all asthma [43] as it leads to an increased inhaled PM doses [44]. In order to avoid adverse effects on health, it is therefore necessary to practice physical activity away from traffic and pollution in general. City councils and political institutions should limit community dependence on vehicles and in so doing promote physical activity and reduce traffic pollution. Examples include "active transportation 'green' corridors that are separated from major traffic arteries, design of neighborhoods that are both walkable and high density (…) and incentives to reduce emissions in urban centers" [41, p. 32].

In addition to physical activity, changes in nutrition can play a significant role in decreasing "the risk of disease development and therefore reduce susceptibility to the effects of air pollution" [41, p. 32]. Indeed, cardiovascular health can be protected by long-term supplementation with omega-3. According to current evidence of the relationship between oxidative stress and air pollution effects, "it is logical to consider efforts to increase the body's anti-oxidant defenses as potential interventions to ameliorate the negative health impacts of pollution exposure" [41, p. 32]. These positive effects—according to the experts of Vancouver workshop—can be achieved, for some degrees, also by an optimal therapeutic management.

Last, but not least, there is the need to increase health information on air quality, about the timing, duration, and location of outdoor activities in polluted areas. It is not only long-term exposure to air pollution that carries with it an increased risk of cardiovascular mortality; also short-term exposure has negative effects on health. Therefore, as little time as possible should be spent in traffic jams, at bus stops, and/or in indoor parking garages. Moreover, since ozone levels have temporal patterns, outdoor human activities should be done outside of the peak hours of the day. Also, the indoor/outdoor distinction should be reframed. Indoor pollution should not be under evaluated, especially because outdoor pollution can easily penetrate indoors. In addition, "indoor sources such as cooking can generate high concentrations of PM indoors (…) Not surprisingly, following advice to stay indoors can reduce exposure to some pollutants while increasing exposure to others" [41, p. 33].

It cannot be denied that modifying individual behaviors and lifestyles to cope with air pollution and its consequences should be part of a health promotion strategy, yet if considered separately from the structural factors connected to environment, they could be seen as the outcome of neoliberal and functionalist analysis on air pollution. This perspective—bracketing socioeconomic disparities—emphasizes the aspect of individual responsibility in the exposure and vulnerability to pollution.

Moreover, also mainstream media present, as demonstrated by Mayer [17], a "normalized" idea of air pollution risk. Such a portrayal of risk is exemplified in this quote from the *Los Angeles Times*: "It is a an average summer day in Southland (…) Smog levels exceed state standards (…) The air quality for Simi Valley is expected to approach levels considerable unhealthful for sensitive groups, such as asthma sufferers" [45, p. 1]. The readers of the *Los Angeles Times* are therefore invited to think about "poor outdoor air quality as part and parcel of living in the region without much reflection on the anthropogenic sources of the dirty air" [17, p. 899]. In other words, pollution and its health risks are "normal," and individuals should just find the best way to adapt to this situation. In this framework, people affected by asthma, for example, are seen as partially responsible for their condition and implicitly invited to change their lifestyles: "This lifestyle critique implies that good care and the daily practices supporting it are within the reach of any individual or family" [17, p. 899]. Yet, as already seen, more deprived people tend to live in areas characterized by higher levels of environmental pollution, and "because lower-SEP [Socio-Economic Position] groups already experience compromised health status due to material deprivation and psychosocial stress, they may be more susceptible to the health effects of air pollution" [46, p. 1861. Thus, the combination of greater exposure and higher susceptibility puts disadvantaged individuals in situations in which they are likely to suffer more serious health effects from their exposure to air pollution. Their deprivation makes them subject to a double burden [28].

32.4.2 Popular Epidemiology and Contested Illnesses

The Contested Illness Research Group (CIRG) represents a particularly innovative sociological approach. The CIRG is composed of an interdisciplinary group founded by the sociologist Phil Brown in 1999 at Brown University. The peculiarity of this center is that it does not limit itself to doing pure research but also operates as a political actor, engaging in forms of activism and advocacy. Specifically, the CIRG has defined its activities as three "challenges to the boundaries of traditional research" [12, p. 145]:

1. The first boundary is the disciplinary separation between medical and environmental sociology; the means by which this boundary is challenged is the concept and the practice of *popular epidemiology*.

2. In order to deal with what Brown termed *contested illnesses*, "diseases or conditions in which there is a dispute over environmental causation" [12, p. 150], the second boundary is overcome through the linkage between social science and environmental health science.
3. The third boundary is the distinction between scholarship, on one side, and advocacy or activism, on the other: the practice that integrates these two activities is the sociological research called *community-based participatory research.*

In order to better grasp the three challenges, it can be useful to reproduce a piece taken from the description of a study conducted by Phil Brown and his group. This part will shed light on the work of the CIRG "in action."

> Both breast cancer and environmental justice advocates have sought to reshape scientific approaches in ways that highlight potential causes of disease and suggest opportunities to reduce chemical exposures and safeguard community health. Our group chose to conduct an exposure study, because an epidemiologic breast cancer study within Richmond would not have been informative given the community's small size and the lack of data on relevant historical chemical exposures. (…) An exposure study of potentially hazardous compounds can assess the extent of a pollution problem and inform action to reduce exposures. Moreover, exposure studies are not as resource-intensive and are more conducive to an academic and community partnership incorporating collaborative design of study protocols, data interpretation, and dissemination of results. To decide how to collect data that would advance the shared goals of the three project partners, we conducted a year-long deliberative process. This entailed gathering information on community health concerns through community meetings, building on CBE's relationships with public officials in Richmond to gain support for our work, and convening an advisory council of neighborhood activists, breast cancer, and biomonitoring activists, a state health official, an environmental justice advocate, and an outside academic researcher. Using the results of this input, we designed research to assess the cumulative exposures and specific sources of indoor pollution originating from outdoor industrial emissions, transportation sources, and consumer products. CBE staff were equipped and trained to conduct interviews and collect air and dust samples that were subsequently analyzed for industrial and traffic-related pollutants, such as particulates, metals, polycyclic aromatic hydrocarbons (PAHs), ammonia, and sulfates, and also for compounds found in consumer products, such as pesticides, flame retardants, phthalates, and phenols. (…) The scientific rigor of the study was ensured through collective discussion and negotiation of study design issues such as choosing study sites, recruiting study participants, finalizing the list of chemicals for analysis, and developing protocols for reporting study results. For example, CBE encouraged the study team to collect a subset of air and dust samples from a community that did not have significant outdoor industrial and transportation source emissions, so that these results could be compared to those from Richmond. As a result, we added Bolinas as a comparison site. Similarly, the relevance of the study was bolstered through the development of bilingual (Spanish and English) graphic displays to communicate aggregate and individual sampling results for dissemination to individuals and at community meetings. This effort helped to ensure that: study results were transparent and scientifically valid and conveyed uncertainties and elucidated strategies for exposure reduction. Finally, CBE's engagement in the study helped us to reach broader audiences and thus apply scientific results to making better decisions about land use. For example, scientists and CBE trained community residents to effectively present scientific data at community meetings and to testify before regulatory and policy forums (p. 8–10).

With regard to the first challenge, the concept of popular epidemiology was conceived by Brown during a study in Woburn, near Boston, that he carried out from 1984 to 1986. There had been a very high number of cases of infant leukemia in the area. A group of mothers thought the cause was the drinking water, but the association with municipal water supply was denied by state and federal health officials. The Woburn families recruited experts from the Harvard School of Public Health. In a joint effort, which also relied on the legwork of dozens of Woburn residents, the alliance between lay people and researchers discovered that the municipal water was contaminated by a solvent (trichloroethylene) that had leaked from a factory located nearby. Therefore, popular epidemiology can be considered a form of bottom-up, community-driven epidemiology which sees lay people from a "contaminated community" engaging in science. As described by Brown [12], the practices of popular epidemiology typically develop along the following stages:

- Lay observations of health effects and pollutants
- Hypothesis of a connection between them
- Creation of a community group
- Social mobilization in order to seek answers from institutions and science
- Non-confirmation of the connection from official studies
- Creation of an alliance between lay people of the contaminated community and researchers.
- Corroboration of positive findings that result from lay-professional collaboration in popular epidemiology

Clearly, in many ways, popular epidemiology is in opposition to the "dominant epidemiological paradigm," which has often functioned "to limit scientific, medical and political attention to environmental causes of disease, limiting the ability of environmental health advocates to communicate to the general public" [17, p. 893]. The dispute over the connection between polluted water and childhood leukemia led to the formation of the contested illnesses model. This model emphasizes the health competence of patients and their capacity to act as experts thanks to their lived experience of a disease. In particular, at times with the help of scientists and social activists, community groups of patients identify connections between illnesses and environmental contests, challenging the dominant epidemiological paradigm. As such they put pressure on the institutions and request the implementation of urgent public health measures and legal action with regard to the source of pollution. Through this process: "people with a disease or a condition developed a common identity around their shared problem, and when they added political organizing to that, they joined ranks to form a politicized collective illness experience" [12, p. 152]. Hence, a fundamental element of the contested illness model is the presence of an "embodied" health movement formed by scientists and patients—both experts of that illness—that contest the official view of clinical medicine and its etiology. These *boundary movements* are opposed to the dominant scientific and political institutions founded on an idea of "science and policy as distinct and separate worlds depicting science as the world of neutral and independent facts and policy making as the world of values" [47, p. 255]. On the contrary, Jasanoff—with her concept of *boundary work*—underlines just how much the borders of these boundaries between science and policy are blurred, intertwined and negotiable [48]. It is not hard to notice that this kind of environmental justice is based on a health paradigm that emphasizes the central role of social determinants of health (SDH) in shaping health trajectories: a perspective that is distant from the institutional version of biomedicine [49]. The SDH model assumes that health conditions are mainly affected by social and economic factors, like education, income, housing, neighborhood, social capital, and, of course, environment [8]. Therefore, health is strictly connected to social inequalities and social justice [50].

At this point, it should be clear that this kind of sociological perspective can be characterized by a "triple openness": it is open to other disciplines, it is open to lay people, and it is aimed at changing and intervening with institutions. In other words, from an epidemiological stance, CIRG "does" popular epidemiology; from a theoretical stance (as object of study), CIRG activities focus on contested illnesses; and from a methodological point of view, CIRG investigates/acts on reality through community-based participatory research:

> Community-based participatory research projects strive to increase community engagement in research in order to generate scientific knowledge, improve public trust and understanding of science, inform culturally and socially appropriate intervention methods, improve public health decisions, and stimulate action. [12, p. 154]

The long piece we reproduced at the beginning of this section illustrates the high level of involvement—if not of "fusion"—between researchers and lay persons. The great reliability of the methodology used in the research should also be stressed, a quality which signals methodological accuracy but also makes the group a trustworthy counterpart of political institutions. Indeed, today we are witnessing the "scientization of decision making" [51, 52], i.e., the predominance of scientific or pseudoscientific frames in political and ethical contests with the consequence that approaches that are not measurable or cannot be verified are excluded as a basis for decision-making. As a consequence, it is necessary to demonstrate a mastery of scientific methodology in order to deconstruct this technocratic approach, which hides behind a façade of rational argumentation.

The most innovative results of CIRG's research have probably come from their biomonitoring studies (described in the long citation reproduced at the beginning of this section). These studies have matched individual clinical tests (sampling human tissues, such us blood, urine, breast milk, and hair, for environmental contaminants) with household air and dust or with outdoor pollution. The researchers not only analyzed the level of the individual's exposure indoor/outdoor but also, anthropologically, investigated the ways in which they interpreted the results of the clinical tests. The studies also had practical consequences: "The results of our exposure study were successfully used to prevent the expansion of a major oil refinery literally next door to two minorities communities" [12, p. 155].

In other cases, these studies were challenged by institutional review boards that underlined what they saw as a lack of evidence connecting the illnesses and the environmental contexts. Indeed, according to some experts, it was irresponsible to inform people of their individual or household exposure as this could create social alarm. According to Brown, however, the "precautionary principle" should be followed, i.e., it is advantageous to avoid actions whose harmfulness is theoretically present, even if not yet scientifically proven. Nowadays research based on biomonitoring is a standard practice at the *Centers for Disease Control and Prevention* and one commonly carried out by academic researchers and advocacy organizations [12].

Conclusions

The peculiarities of sociological research on air pollution can be delineated as follows:

- There is solid evidence that the most deprived people face a double burden of inequality; they are more exposed and (due to their health status) more vulnerable to air pollution.
- In some cases, health risks from pollution are globalized and affect all the social classes.
- Individual interventions based on lifestyles changes are desirable but cannot solve the problem of health threats caused by air pollution.
- Environmental justice can be pursued using popular epidemiology practices, contested illnesses model, and community-based participatory research.
- Several preventive measures can arise thanks to social participation and activism.
- The institutional prevention of air pollution has a lot to learn from observing social activism.
- Sociology can provide a theoretical framework, evidence from empirical research, and data for both institutional forms of prevention and social mobilization against sources of pollution.

Acknowledgements While this article is the result of several discussions between the authors, Antonio Maturo has written the sections "Introduction"; "Social justice and air pollution" and its sub-sections; "Conclusions"; Veronica Moretti has written "Sociological theories and pollution" and "Risk approaches to air pollution", enclosed the sub-sections.

References

1. Link B, Phelan J. Social conditions as fundamental causes of health inequalities. In: Bird CE, Conrad P, Fremont AM, Timmermans S, editors. Handbook of medical sociology. 6th ed. Nashville: Vanderbilt University Press; 2010.
2. WHO. Governance for health equity. http://www.euro.who.int/__data/assets/pdf_file/0020/235712/e96954.pdf?ua=1. 2010.
3. Giddens A. The consequences of modernity. Cambridge: Cambridge University Press; 1990.
4. Beck U. Risk society. London: Sage; 1992.
5. WHO. Public health, environmental and social determinants of health (PHE). http://www.who.int/phe/health_topics/outdoorair/en/. 2016.
6. Egondi T, Kyobutungi C, Ng N, Muindi K, Oti S, van de Vijver S, Ettarh R, Rocklöv J. Community perceptions of air pollution and related health risks in Nairobi slums. Int J Environ Res Public Health. 2013;10:4851–68.

7. Donohoe M. Causes and health consequences of environmental degradation and social injustice. Soc Sci Med. 2003;56:573–87.
8. WHO. The cost of air pollution health impacts of road transport. http://www.oecd.org/env/the-cost-of-air-pollution-9789264210448-en.htm. 2014.
9. Wakefield SE, Elliott SJ, Cole DC, Eyles JD. Environmental risk and (re) action: air quality, health, and civic involvement in an urban industrial neighbourhood. Health Place. 2001;7(3):163–77.
10. Saksena S, Prasad RK, Shankar VR. Daily exposure to air pollutants in indoor, outdoor and in-vehicle micro-environments: a pilot study in Delhi. Indoor Built Environ. 2007;16(1):39–46.
11. Oltra C, Sala R. A review of the social research on public perception and engagement practices in urban air pollution. Informes Técnicos Ciemat. 2014;1317
12. Brown P. Integrating medical and environmental sociology with environmental health: crossing boundaries and building connections through advocacy. J Health Soc Behav. 2013;54(2):145–62.
13. Durkheim E. Les Règles de la Méthode Sociologique. Paris: Alcan; 1985.
14. Luhmann N. Ecological communication. Chicago: University of Chicago Press; 1989.
15. Latouche S. Le pari de la décroissance. Paris: Fayard; 2006.
16. Hannigan JA. Environmental sociology: a social constructionist perspective. Milton Park: Psychology Press; 1995.
17. Mayer B. 'Relax and take a deep breath': print media coverage of asthma and air pollution in the United States. Soc Sci Med. 2012;75(5):892–900.
18. Beck U. World at risk. Cambridge: Polity Press; 2009.
19. Beck U. World risk society. Cambridge: Polity Press; 1998.
20. Jarvis DSL. Risk, globalisation and the state: a critical appraisal of Ulrich beck and the world risk society thesis. Glob Soc. 2007;21(1):23–46.
21. Romero-Lankao P, Qin H, Borbor-Cordova M. Exploration of health risks related to air pollution and temperature in three Latin American cities. Soc Sci Med. 2013;83:110–8.
22. Beck U. Ecological politics in the age of risk. Cambridge: Polity Press; 1995.
23. Bauman Z. Liquid fear. Cambridge: Polity; 2006.
24. McLaren D, Cottray O, Taylor M, Pipes S, Bullock S. Pollution injustice: the geographic relation between household income and polluting factories. In: Report for friends of the Earth. 1999. http://www.foe.org.uk/pollution-injustice/poll-inj.html.
25. Burningham K, Thrush D. Pollution concerns in context: a comparison of local perceptions of the risks associated with living close to a road and a chemical factory. J Risk Res. 2004;7(2):213–32.
26. Clark LP, Millet DB, Marshall JD. National patterns in environmental injustice and inequality: outdoor NO2 air pollution in the United States. PLoS One. 2014;9(4):1–50.
27. Skelton R, Miller V. The environmental justice movement. New York: Natural Resources Defense Council; 2006.
28. Crouse DL, Ross NA, Goldberg MS. Double burden of deprivation and high concentrations of ambient air pollution at the neighborhood scale in Montreal, Canada. Soc Sci Med. 2009;69:971–81.
29. Havard S, Deguen S, Zmirou-Navier D, Schillinger C, Bard D. Traffic- related air pollution and socioeconomic status: a spatial autocorrelation study to assess environmental equity on a small-area scale. Epidemiology. 2009;20(2):223–30.
30. Hackbarth AD, Romley JA, Goldman DP. Racial and ethnic disparities in hospital care resulting from air pollution in excess of federal standards. Soc Sci Med. 2011;73(8):1163–8.
31. Smith K, Mehta S. The burden of disease from indoor air pollution in developing countries: comparison of estimates. USAID/WHO Global consultation on the health impact of indoor air pollution and household energy in developing countries. In: Setting the agenda for action. Washington, DC. 2000.
32. World Bank. World development report 1992. New York: Oxford University Press; 1992.
33. Larson BA, Rosen S. Understanding household demand for indoor air pollution control in developing countries. Soc Sci Med. 2002;55:571–84.

34. Jin Y, Ma X, Chen X, Cheng Y, Baris E, Ezzat M. Exposure to indoor air pollution from household energy use in rural China: the interactions of technology, behavior, and knowledge in health risk management. Soc Sci Med. 2006;62:3161–76.
35. Liu BQ, Peto R, Chen ZM, Boreham J, Wu YP, Li JY. Emerging tobacco hazards in China: 1. Retrospective proportional mortality study of one million deaths. Br Med J. 1998;317:1411–22.
36. Romero-Lankao P, Qin H. Conceptualizing urban vulnerability to global climate and environmental change. Curr Opin Environ Sustain. 2011;3:142–9.
37. Tipple G. Housing, urban vulnerability and sustainability in rapidly-developing cities. Built Environ. 2006;32(4):387–99.
38. Downing TE, Patwardhan A. Assessing vulnerability for climate adaptation. In: Lim B, Spanger-Siegfried E, editors. Adaptation policy frameworks for climate change: developing strategies, policies, and measures. Cambridge: Cambridge University; 2004.
39. Klein RJT, Nicholls RJ. Assessment of coastal vulnerability to climate change. Ambio. 1999;28:182–7.
40. Brown P, Mayer B, Zavestoski S, Luebke T, Mandelbaum J, McCormick. The health politics of asthma: environmental justice and collective illness experience in the United States. Soc Sci Med. 2003;57(3):453–64.
41. Giles LV, Barn P, Künzli N, Rom I, Mittleman MA, van Eeden S, Allen R, Carlsten C, Stieb D, Noonan C, Smargiassi A, Kaufman JD, Haiat S, Kosatsky T, Brauer M. From good intentions to proven interventions: effectiveness of actions to reduce the health impacts of air pollution. Environ Health Perspect. 2011;119(1):29–36.
42. Newby DE, Mannucci PM, Tell GS, Baccarelli AA, Brook RD, Donaldson K, Forastiere F, Franchini M, Franco OH, Graham I, Hoek G, Hoffmann B, Hoylaerts MF, Kunzli N, Mills N, Pekkanen J, Peters A, Piepoli MF, Rajagopalan S, Storey RF. Expert position paper on air pollution and cardiovascular disease. Eur Heart J. 2015;36(2):83–93.
43. Daigle CC, Chalupa DC, Gibb FR, Morrow PE, Oberdorster G, Utell MJ. Ultrafine particle deposition in humans during rest and exercise. Inhal Toxicol. 2003;5(6):539–52.
44. Panis L, Geus B, Vandenbulcke G, Willems H, Degraeuwe B, Bleux NM. Exposure to particulate matter in traffic: a comparison of f cyclists and car passengers. Atmos Environ. 2010;44(19):2263–70.
45. Surman M. Highs in 90s give taste of summer; weather: Simi Valley and Thousand Oaks are the hot spots. Temperatures and smog levels are expected to fall by the end of the week. Los Angeles, CA: Los Angeles Times; 2001.
46. O'Neill MS, Jerrett M, Kawachi I, Levy JI, Cohen AJ. Health, wealth, and air pollution: advancing theory and methods. Environ Health Perspect. 2003;111:1861–70.
47. Jungcurt S. Taking boundary work seriously: towards a systemic approach to the analysis of interactions between knowledge production and decision-making on sustainable development. In: Meuleman S, editor. Transgovernance advancing sustainability governance. New York: Springer; 2013.
48. Jasanoff S. Contested boundaries in policy-relevant science. Soc Stud Sci. 1987;17(2):195–230.
49. WHO. Closing the gap in a generation: health equity through action on the social determinants of health. 2006. http://www.who.int/social_determinants/thecommission/finalreport/en/.
50. Venkatapuram S. Health justice: an argument from the capabilities approach. Boston: Polity Press; 2011.
51. Morello-Frosh R, Zavestoski S, Brown P, editors. Environmental justice and collective illness experience contested illnesses: citizens, science, and health social movements. Berkeley: University of California Press; 2011.
52. Morello-Frosh R, Brown P, Zavestoski S. Introduction. Environmental justice and contested illnesses. In: Brown P, Morello-Frosch R, Zavestoski S, editors. Environmental justice and collective illness experience contested illnesses citizens, science, and health social movements. Berkeley: University of California Press; 2011.

The Contribution of Digital Sociology to the Investigation of Air Pollution

33

Linda Lombi

33.1 Introduction

External environmental pollution has serious implications for human quality of life and health in the developed world. Current evidence appears to support the suggestion that air pollution is associated with eye, skin and throat irritations [1], lung cancer [2–4], cardiovascular and pulmonary system ailments [5] and cognitive decline [6]. Problems generated by airborne pollutants, which obviously cannot be contained at national borders, are phenomena with the potential to transform from local to global.

The many implications of air pollution mean that the monitoring of air quality is a priority for national governments of states with high rates of industrialisation and urbanisation. To this end, nations focus on problem areas, conducting constant studies to remain informed regarding a range of the relevant parameters. An example of these is the Air Quality Index (AQI) for five major air pollutants. Government agencies inform the public of the calculated index in order to communicate possible health warnings. However, monitoring systems require onerous economic investments, and the relevant control systems are not sufficiently widespread to be capillary. For this reason, scholars investigating air pollution have undertaken to analyse alternative methods for timely measurement which may be applied at a territorial level and be less expensive to monitor.

In recent years, innovative studies on air pollution have focused on several objectives in particular, including the support of cheaper new sensors, the promotion of network data sharing and the development of new personal sensors linked to social media [7].

L. Lombi
Università Cattolica del Sacro Cuore, Milan, Italy
e-mail: linda.lombi@unicatt.it

F. Capello, A.V. Gaddi (eds.), *Clinical Handbook of Air Pollution-Related Diseases*,
https://doi.org/10.1007/978-3-319-62731-1_33

The paper will discuss new approaches and examples of air quality monitoring which take advantage of digital technologies and, in particular, the tools provided by the development of Internet 2.0.

33.2 Towards New Approaches to Data Collection: People As Sensors, Citizen Science and Collective Sensing

The emergence of the digital society has enabled scientists to develop new approaches and research methods which use the Internet as a channel for data collection in fields such as public safety, road traffic, environmental monitoring and public health. In the area of health studies, for example, several scholars have researched web sources to predict the spread of influenza [8–10], to analyse and predict behaviours linked to lifestyles that impact on health [11, 12] and to assess the satisfaction levels of patients for medical care received [13].

Among the factors that have contributed to the implementation of new data collection methods based on digital technologies, two phenomena should be pointed out for mention: the spread of mobile devices and the emergence of Internet-based social media networks. As seen in the discussion to follow, mobile devices allow two types of data to be recorded as information: "objective" data (collected through sensors applied to the device) and "subjective data" (linked to users' opinions, impressions and perceptions).

The various social media may be regarded as social sensors capable of providing real-time information on social activities and collective human behaviours [14]. The widespread diffusion of social media has contributed to the development of the Internet 2.0 era (the participatory or dynamic web), during which users acquire the ability to edit user-generated content (UGC) becoming at once the producers and consumers of information, the *prosumers* [15].

The popular use of mobile devices and social media has led to a significant expansion of the availability of data on the opinions, perceptions, behaviours, facts and parameters that affect people and their life environments. Scholars have even referred to a data deluge [16].

Big data warrants particular attention within this context. The term refers to the sum of data in digital format which is collected, stored and managed through large datasets and which cannot be processed through the software and hardware systems traditionally used within social research. Attention has been focused on three main characteristics: *volume* (date connotes huge quantities), *velocity* (data is generated and analysed rapidly) and *variety* (heterogeneousness, with the presence of structured and unstructured data). In addition to these features, Kshetri [17] stresses *variability* (data undergoes fluctuations related to trending periods) and *complexity* (data originates from various sources which need to be linked, matched, "cleaned" and transformed), while Kitchin [18] emphasises further characteristics, defining data as *exhaustive* (entire populations or systems may be captured), fine-grained in *resolution* and uniquely *indexical* in identification, *relational* in nature, *flexible* and characterised by *extensionality* and *scalability*.

As mentioned above, smartphones and tablets (and their applications) have increased the ability to record objective data related to everyday experiences thanks to sensors able, for example, to record motions, vibration, pressure, sound, temperature and images through hardware sensors (i.e. the camera, microphone, GPS, digital compass or other connection tools such as gyroscopes, accelerometers, light intensity and proximity detectors).

In effect, mobile devices are potential "sensing instruments" which permit the real-time recording, viewing and sharing of information, complete with geolocation. A sensor society has been created, one marked by "the increasing deployment of interactive, networked devices as sensors; the resulting explosion in the volume of sensor-generated data; the consequent development and application of data mining and machine learning techniques to handle the huge amounts of data; and the ongoing development of collection, storage, and analytical infrastructures devoted to putting to use the sensor-derived data" [19, pp. 2–3].

However, the sensors installed on mobile devices are not the only possible mode of data registration. A user's subjective observations on the phenomena of everyday life may also be uploaded through many of the apps currently available. The process has led scholars to speak of *people as sensors* [20], the citizen as sensor [21] or humans as sensors [22]. As Resch states, "People as Sensors defines a measurement model, in which measurements are not only taken by calibrated hardware sensors, but in which also humans can contribute their subjective 'measurements' such as their individual sensations, current perceptions or personal observations. These human sensors can thus complement—or in some cases even replace—specialised and expensive sensor networks" (2013, p. 393).

The emergence of these monitoring methods has brought several advantages: cost-effectiveness due to a reduced need for technical sensors, timeliness because the real-time information is immediately available from the data warehouses without processing delays and geographical, as the mere presence of individuals equipped with connected mobile devices is sufficient to ensure diffusion. Examples of apps which take advantage of citizen sensor logic include those allowing the recording and sharing of information regarding traffic incidents, accidents endangering public safety or, even more simply, weather conditions.

The people as sensor approach should not be confused with another data collection template, citizen science or participatory sensing. The latter is a large body of related projects which enable citizens to act as agents of change through the use of mobile technologies by promoting their active participation in the collection and dissemination of information regarding their living environments and their activities. In other words, citizen science represents a form of public participation in scientific research where amateur or non-professional scientists contribute by conducting or informing a research project. It is also called crowd-sourced science, volunteer monitoring or networked science.

Projects based on this approach collect data detected in part by hardware sensors and in part by citizens (defined as "virtual sensors"; see Boulos et al. [23] whose participation in a formal project is in response to an invitation addressed to volunteers, amateurs and enthusiasts.

For Paulos et al. [24], citizen science forms the basis of participatory urbanism, a movement "more directly focused on the potential for emerging ubiquitous urban and personal mobile technologies to enable citizen action by allowing open measuring, sharing and remixing of elements of urban living marked by, requiring, or involving participation, especially affording the opportunity for individual citizen participation, sharing, and voice" [24, p. 2]. Other scholars have discussed participatory environment sensing (PES) [25], referring to the tools that "enable participatory or citizen-initiated sensing and transform the role of the citizenship from a passive receiver of information to an active agent participating in the elaboration of that information" [25, p. 125].

Citizen science is on the increase in particular in urban centres which aspire to become smart cities, cities which "on the one hand, are increasingly composed of and monitored by pervasive and ubiquitous computing and, on the other, whose economy and governance is being driven by innovation, creativity and entrepreneurship, enacted by smart people" [26, p. 1].

The third innovative approach using digital technologies for the analysis of living environments is collective sensing, "an infrastructure-based approach, which tries to leverage existing ICT networks to generate contextual information. Unlike smartphone-based or specialised web apps, which examine single input data sets, collective sensing holistically analyses events and processes in a network" [20, p. 395]. Collective sensing, therefore, analyses the UGC published on social media such as Facebook, Twitter, Flickr, Foursquare or Sina Weibo. Here, the aggregated data is collected through the geolocation of the input, without disclosing the identity of the individual user. The aim here is to capture the data stored by users on the net, either in real time or near real time, providing a snap-shot of the object of study and aiding investigations into collective representations and the resulting sentiments of the users. The level of reliability is usually lower than that of the other two approaches and the level of knowledge required is low or zero.

Table 33.1, from Resch [20], summarises the three approaches.

To follow are examples of tools and projects relative to these three innovative air pollution research approaches.

Table 33.1 Comparison of concepts: people as sensors, citizen science and collective sensing

	People as sensors	Citizen science	Collective sensing
Voluntary/involuntary	Voluntary	Voluntary	Involuntary
Content	Layman observations	Semi-professional observations	Raw geodata (images, tags, etc.)
A priori knowledge	Medium	High	Low/none
Contextual data	Yes	Yes	Yes
Reliability	Medium	Good	Mediocre
Analysed datasets	Individual	Individual	Aggregated
Specific infrastructure	No	No	No

Source: Resch [20, p. 397]

33.3 People As Sensors: The New Frontier of Mobile Devices

People as sensors is a new approach expanding the possibility to gather, learn, visualise and share information about ourselves and the world we live in. As noted above, the most important factor in the development of people as sensors has been the proliferation of smartphone use. Two main tools permit citizens to collect, summarise, represent and share information: applications (apps) and the Global Positioning System (GPS). An app is a downloadable software designed for a particular purpose on a mobile device, and GPS allows individuals to register the physical location of the data collection.

The mobile device apps on which the approach is based fall in the main two categories: people-centric (documenting the activities and behaviours of individuals, such as sporting activities) and environment-centric, where environmental parameters such as air pollution, noise, road traffic, etc. are collected [27]. This paper will focus on this second group of apps, with specific reference to those which monitor air pollution.

On the individual level, people carrying sensors are equipped to make informed decisions about how they may move through the city. According to Burke et al., "As sensors, network-connected mobile handsets will be embedded near the ultimate elusive subjects: people and their built environments. Instead of being in the hands of a central observer, these sensors are always on and under their owners' control" [46, p. 1].

The benefits of such an approach can be seen not only in the ability to collect and store data but also in the fact that this information may be summarised as graphics and shared with other device users. This latter factor renders the approach particularly significant for its potential in timely event or accident warnings and emergency management based on the information uploaded [21].

It is important to note that not all apps offer the same functions. One variation is informative, allowing only the visualisation of officially released risk data and information, while others offer interactive and participatory functions which permit the user to record personal observations, a people as sensor function. It should however be noted that informative apps may also allow a certain degree of user engagement. The information may be shared via social media, thus contributing to its dissemination.

An example from the informative group is the Global Air Quality Monitoring and Pollution Forecast PM 2.5 AQI app available for iOS. It offers a panoramic view of real-time global pollution indexes and tracks real-time concentrations of PM 2.5, PM10, ozone, nitrogen dioxide, sulphur dioxide and carbon monoxide. Users may receive personalised health recommendations, advice on how to lower health risks and minimise their exposure to pollutants as well as up-to-date information regarding air pollution, current events, medical findings and breaking news. The Plume Air Report app offers similar functions but includes information on "safe" activities such as the advisability of conducting outdoor activities or allowing children to remain in the open air. The notices are based on the amount of smog recorded and expected.

Although the software may increase user awareness of air pollution issues, and therefore are particularly useful for vulnerable persons such as asthmatics or those

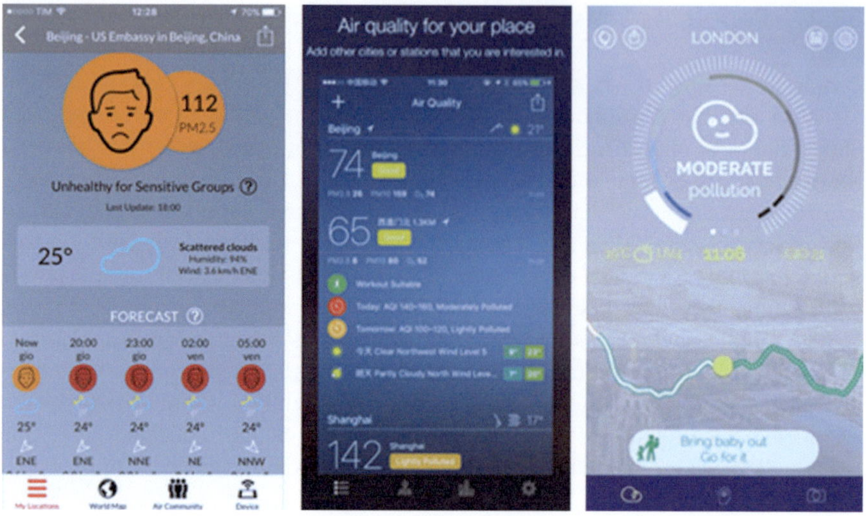

Fig. 33.1 Screenshots from apps for monitoring air quality

suffering from other respiratory system disorders, the apps in the informative group take a different approach to those in the human as sensor group. These are inspired by the AirForU app, an American software available for both iPhone and Android mobile devices. It provides real-time updates about air quality but also permits the user to publish alerts when pollution levels could threaten the health of sensitive groups. Users are encouraged to participate by communicating whether they remained indoors due to the poor air quality and suffered a recent asthma attack or whether they had installed indoor air filters.

Figure 33.1 includes screenshots of several of the apps described.

33.4 Citizen Science: People As Agents of Change

As mentioned earlier, citizen science is driven by a desire to empower the public, through their smartphones, in the collection and sharing of sensed data from their environments [27, 28]. Citizen science is a form of crowdsourcing whereby companies use social media and Internet forums to invite the participation of volunteers in specific projects [29]. Participation is often motivated and stimulated by a gamification of the project, a process where non-playful activities are represented in a game-like form. This may take the form of, for example, rewards such as point distribution, level leaps, badges, etc. [30, 31].

A well-known citizen science air pollution project is CITI-SENSE-MOB (http://www.citi-sense.eu/), initiated in Norway and today involving 29 institutions from 14 nations. This European Commission-funded project, launched in 2012, installed portable sensors on bicycles and other vehicles in the participating cities. The sensors record air quality parameters. The information registered is made available to

Fig. 33.2 CITI-SENSE project: the web interface

local citizens, allowing them to better understand the conditions present in their environments. The project additionally aims to use the data provided both automatically and manually through users to develop citizens' observatories, thereby empowering citizen participation in environmental governance [32]. However, the platform is not based on the registration of records through external sensors alone. It also allows the user to integrate the information by inputting data into a web interface or a mobile phone application, the CityAir app, that allows users to express their perception of the outdoor air quality at their location, to indicate the assumed source of the air pollution opinions and to write a comment (Fig. 33.2).

Once collected, the data is processed and returned to citizens and interest groups in user-friendly, clearly defined graphs to ensure that the information is easily understood by members of the public. According to a study by Nikzad et al. [33], the CITI-SENSE-MOB project allows the collection of information in considerably more detail than that from conventional stationary systems.

The Haze Watch project, an initiative by a group of students from the School of Electrical Engineering at the University of New South Wales, is another good example of citizen science as applied to air pollution. The project aims to measure the concentration of pollutants by integrating information collected by the mobile sensors, smart phones and citizens involved in the study. Haze Watch has also implemented an app to record the movement of individual users via the GPS sensor built into their iPhones. This then allows the cross referencing of an individual's journey time with the in-house pollution database server, an estimation of the air pollution exposure during the journey, the display of a graph illustrating an individual's exposure over time and the identification of regions of heavy pollution on their daily commute (http://www.hazewatch.unsw.edu.au).

The PEIR (Personal Environmental Impact Report) project has also gained much attention in the scientific community [34]. Promoted by the University of California in Los Angeles and Nokia Corporation, it is based on a mobile to web app that uses GPS installed on smartphones to explore, measure and share citizen environmental footprints with the purpose of providing participants with information on their environmental impact and exposure. Moreover, the platform is capable of producing a map amalgamating information such as location, time, air quality and weather data to calculate a personal carbon footprint and exposure to airborne fine particle matter (PM 2.5).

Finally, another example of a project inspired by the citizen science approach is the Air Quality Egg, a community-led air quality sensing network that offers the public a way to participate in the air quality conversation (http://airqualityegg.com/). In this case, the data is not collected via a smartphone but by a small electronic sensing system which plugs into a USB wall wart or AC/DC adapter. This transmits the data over WiFi to the cloud at OpenSensors.io, an open data service which both stores and provides free access to the data.

From the point of view of the citizen, involvement in citizen science projects delivers the twin benefits of an increase in the knowledge of the phenomenon being studied and an increased sense of participation. However, the benefits of these initiatives are much broader: current research appears to validate the view that citizen science will have a strong importance in the future as a discipline for decision support, education, outreach and improved scientific literacy [35, 36].

33.5 The Collective Sensing Approach: How to Use Social Media to Investigate Air Pollution

The emergence of user-generated content has provided an important contribution to computational social science that may be considered "an approach to social inquiry defined by (1) the use of large, complex datasets, often - though not always - measured in terabytes or petabytes; (2) the frequent involvement of" naturally occurring "social and digital media sources and other electronic databases; (3) the use of computational or algorithmic solutions to generate patterns and inferences from these data; and (4) the applicability to social theory in a variety of domains" [37, p. 7]. The attractiveness of these sources lies not only in their ability to grasp the flow of real-life time through continuous observation but also the availability of data in a format that allows cross-search operations, analysis and graphical representation [38].

The Internet thus becomes not only a window to explore the online world but also one to aid the understanding of offline phenomena. Rogers [39, 40] argues that these new research techniques, included in the more general group of digital methods, mean that a groundedness approach should be adopted, the application of a search path that follows the medium, one which adapts to the specific needs (regarding nature and affordances) of the study platforms and their evolution over time.

A wide range of message types are posted on social media, including the relation of first-hand experiences, a sharing in the experiences of others, and the republication of news and data from official sources as retweets, for example. There are apps which allow the media user to monitor the pollution levels detected and disseminated by agencies responsible for data collection, the parameters of which can be shared on social networks. This data, while not reporting first-hand experience, is of importance because the levels of audience engagement may be measured through the levels of impact observed and interest generated.

A multiple of techniques may be used to analyse the material collected through social media. The simplest involves a scan of message contents (data mining) to highlight the topics which generate the greatest degree of discussion. A word cloud is the most common graphic representation employed to synthesise data mining results. The higher the occurrence of a word, the larger its dimensions (Fig. 33.3).

Co-occurrence analysis and network analysis techniques may also be used to illustrate the correlations between the words in order to, for example, underline topic associations. Social network analysis, a method of inquiry involving the mapping and measuring of relationships between the subjects, is also useful when investigating the identities of those who contribute most to the debate (so-called influencers) and their mutual relations. Figure 33.4, representing a data extraction for the tag #airpollution on Twitter on 20 July 2016, is a clear example of the technique's use. The nodes correspond to users (the larger the node, the greater the number of tweets), while the arcs denote relationships and correspond to exchange activities, for example, retweets and replies. Arc width is proportional to the number of activities. Users who belong to the same cluster or subjects involved in a mutual exchange are depicted in the same colour.

Fig. 33.3 A word cloud representing a social media debate on air pollution

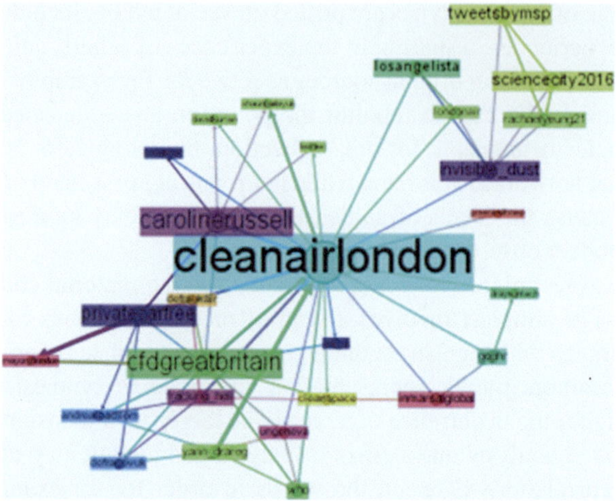

Fig. 33.4 An example of social network analysis depicting the most influential users and their relationships on Twitter for the tag #airpollution (20 July 2016)

Sentiment analysis is another interesting technique available for the analysis of data collected from social media. It is a data mining technique used to identify and extract the emotional tone behind a series of messages posted on a social media platform in order to understand the attitudes, opinions and emotions expressed by users. Sentiment analysis can, for example, recognise the increase in fear or anger expressed in response to changes in air pollution levels or to specific related situations.

China, where the problem of air quality is at the centre of scientific and public debate, is the main producer of the well-established literature investigating the validity of applying computational social science studies to the topic under discussion.

In particular, numerous studies have concentrated on Sina Weibo, a microblogging platform which may be considered the Chinese equivalent of Twitter (see Mei et al. 2013 and [14, 41–43]). Figures released in March 2016 show Sina Weibo as having 261 million active users and 120 million daily users [44].

For example, Wang et al. [43] analysed 93 million messages posted on Sina Weibo using web crawling procedures based on keyword matching and topic models.

Keyword matching refers to a search for posts via the identification of keywords considered by scholars to be relevant to the topic investigated. A topic model "is a probabilistic model of text data, which has two sets of parameters: each document has a discrete distribution over topics", "and each topic has a discrete distribution over words" [43, p. 2]. This means that an analysis of a text through topic modelling allows the detection of the words having the highest probability of existing within the text. This analysis, exploratory in nature, makes a mapping of the discussion possible.

The study by Wang et al. also established a further qualitative detail through the manual analysis of the content of a sample of 170 messages selected on the basis of relevance and pertinence. The object was to detect the presence of reactive behaviours or health concerns (e.g. the wearing of a mask or limits on outdoor activities). 67.1% of messages were relevant to air quality and of those 79% were a first-hand report. Of the latter 28.4% indicated a reactive behaviour and 18.9% expressed a health concern.

To assess the reliability of the methodology design, the results were compared with data relating to the volume of air particle pollution rate detected in 74 cities. The results showed a strong correction with air pollution levels recorded through monitoring tools (Pearson correlation values up to .718, $P < 0.001$). The authors conclude by stating that social media may be considered useful and inexpensive air quality monitoring instruments and allow the inference (though not measurement) of health concerns, including assumptions regarding the associations between certain diseases and air pollution, such as cough or sore throat, and possible corrective behaviour implemented by citizens.

A study with similar methodology, but with elements of originality, was undertaken by Jiang et al. [14]. As in the previous study, posts published on Sina Weibo were examined in detail in an attempt to identify a correlation between the number of posts and the AQI. The messages were divided into three types: retweeted messages, mobile app messages and original individual messages. The authors established that the latter group presented the highest rates of correlation (and thus reliability) with air pollution levels recorded through instrumental monitoring systems.

Although almost all of the studies applying the collective sensing approach have been based on text examinations, several attempts have been made to analyse other data formats such as images. Li et al. [45] studied the presence of haze in photos acquired from online social media in an attempt to establish a correlation between the Air Quality Index registered by hardware sensors installed in the cities. Their work found significant associations between images and data recorded by survey instruments installed in the city.

Studies conducted applying collective sensing therefore suggest interesting and economic prospects for future research in so far as that they indicate a sufficient or even high reliability of results, in particular when the data sources and analytical techniques are integrated. However, as we shall see in the concluding paragraph, limitations do exist.

33.6 Conclusions: New Perspectives and Their Pros and Cons

Research undertaken confirms that the innovative digital tools related to the emergence of the digital society offer interesting prospects for the collection of information related to air quality monitoring. In addition, the use of these tools may potentially assist real-time decision-making in environmental monitoring and the protection of public health.

The advantages of the approaches discussed derive primarily from the ability to use existing tools and infrastructures (mobile devices connected to the Internet) which allow a particularly economic monitoring of air quality and pollution. Furthermore, the movement and mobility of the devices guarantee a greater spatial and temporal coverage and the ability to capture unexpected events. The data collected allows the collection of information considered significant in understanding the phenomenon, offering advantages to the public interest and a means of raising awareness regarding possible personal risks. In the words of Deborah Lupton, "When people are able to collect 'small data', as in the case of people using digital self-tracking devices or engaging in citizen science or citizen journalism practices, they are contributing to 'home-made big data' that may be used for broader research or political as well as personal purposes" [49, p. 151].

We have seen how the approaches based on the "people as sensor" and "citizen science" models represent ways in which people knowingly collect information on air quality by providing an active contribution to the investigation of the phenomenon. However, "[p]articipation means more than data collection (...). Citizens have intimate knowledge of patterns and anomalies in their communities and enabling them to respond is both empowering and valuable to long-term research. If such knowledge can be effectively gathered, it can also impact professional research and planning" [46, p. 2].

By employing collective sensing approaches, social media may in fact become surveillance instruments and social sensors capable of integrating the data obtained through physical monitoring sensors and, where these are absent, of replacing the sensors themselves. The data when shared spontaneously by users is supplied in real time and on a massive scale. As Lupton states, "the vast bulk of digital data is generated unobtrusively, as a part of other activities" [49, p. 44]. Furthermore, the extraction and analysis of specific data enable not only fluctuations in air pollution levels to be predicted but also the subjective air quality perceptions of citizens and, of at least equal importance, the presence of reactive behaviours to counteract the negative effects of pollution.

Despite these advantages, concerns have been raised regarding several relevant limits. The literature underlines criticalities where the approach is applied to social media.

A first point of criticism is that digital data is often raw data and requires an important initial cleaning phase, in particular in the presence of unsupervised analysis. When conducting a study based on computational analysis, it is important to remove extraneous elements, the so-called rumours. In addition, human languages possess a complexity that renders fully automatic investigations unreliable. The precise interventions required by researchers consume time and financial resources, undermining several of the principal advantages associated with the methods, the timeliness and the lower cost of the information. A significant factor is that user-generated content often exploits folksonomy, emoticons and ironic expressions which are difficult to encode through text mining software.

Boccia Artieri also raises a "contingency problem", referring to the tendency of many studies to crush the analysis down to a photograph of the collected data, analysing it according to a cumulative logic that disregards the gradualist nature of the development of online conversations [47, p. 33].

The most significant difficulties are found in the analysis of reactive behaviours which are dependent on cultural components and variations in the social acceptance of measures against smog. For example, the use of masks as a protection against pollution differs from country to country. The practice is considered socially accepted in some major Asian cities, while it may be stigmatised in some urban centres of European cities.

This point also relates to another question of criticality, the sampling procedure. Big data, as seen above, contains data from a widely spread source universe. However, it is clear that a high sample size is not sufficient (nor necessary) to build representative samples of a population. Representativity derives, instead, from the adoption of a suitable sampling design based on probabilistic procedures that legitimise the statistical inference or allow the generalisation of results to be applied to the universe from which the sample is drawn. The shrinking of the digital divide certainly means the inclusion and participation of many individuals who had initially been excluded from the digitalisation process. However, this does not permit the generalisation of the results and their application to the wider population. In addition, the challenge of representativeness must come to terms with the fact that accessibility and participation do not always go hand in hand. Scholars have estimated that, out of a hundred Internet users, only 1% actively participates in discussions on social networks, while on average 9% participate only occasionally and 90% remain as so-called lurkers, users who read posts from others but remain passive participants, leaving no comment [48]. Representativity becomes decisive above all when investigating reactive behaviours. As Kanhere states, "samples collected from mobile phones are typically randomly distributed in space and time, and are incomplete. The challenge thus is to recover the original spatiotemporal profile of the phenomenon being monitored from random and incomplete samples" [27, pp. 23–24].

Another critical point when applied to big data studies concerns the assumed objectivity of the information. In fact, the algorithms used to extract information do not merely describe the data. They also help to create it, as Lupton claims, because they represent social and cultural artefacts of the global digital society, the production and use of which are affected by political, social and cultural processes. She states that "numbers can play a part in constituting phenomena, bringing them into being, making them as well as making sense of them. Numbers are not neutral and objective, (…). The digital objects that are rendered into numbers by digital technologies are both the products of sociotechnical devices and such devices themselves, possessing their own agency and power" [49, p. 101].

In addition, the fact that the data is uploaded by a user introduces the problem of the reliability and veracity of the information. While a technical sensor offers an objectivity related to its instrumentation, human sensors may mistake or even falsify information.

A last point is the concern expressed over the protection of user privacy. The boom in the dispersion and use of devices capable of collecting huge data quantities have given enormous impulse to dataveillance, the systematic monitoring of persons or groups through the accumulation of personal data in order to control their behaviours. Without adequate controls and protections, cell phones risk becoming "miniature spies" [27].

Notwithstanding the limitations of some of the digital techniques implemented to collect information on air pollution, the sources that take advantage of web-based technologies described in this paper represent modes, in part alternative and in part integrative, capable of supplying real-time air quality data and information regarding reactive measures undertaken by the relevant citizen groups. Moreover, given the importance of air quality in public health, it is important to ensure free, easy and direct citizen access to information on the presence of air pollutants. For this reason, future research will be influenced by the importance of using the Internet not only to collect data on air quality but also to disseminate reliable and timely results in public information released through early warning systems.

We conclude therefore that the innovativeness of these approaches stems above all from the fact that the described techniques not only allow citizens to receive more information but that they also contribute to an increase in active citizen participation in air quality monitoring, completing in full the role of the citizen as prosumer.

References

1. Kampa M, Castanas E. Human health effects of air pollution. Environ Pollut. 2008;151(2):362–7.
2. IARC. Outdoor air pollution a leading environmental cause of cancer deaths. International Agency for Research on Cancer, WHO, Press release 221. 2013. Available at: http://www.iarc.fr/en/media-centre/iarcnews/pdf/pr221_E.pdf. Last Access: 15 July 2016.
3. Raaschou-Nielsen O, Andersen ZJ, Beelen R, Samoli E, Stafoggia M, Weinmayr G, et al. Air pollution and lung cancer incidence in 17 European cohorts: prospective analyses from the European study of cohorts for air pollution effects (ESCAPE). Lancet Oncol. 2013;14(9):813–22. doi:10.1016/S1470-2045(13)70279-1.
4. Yorifuji T, Kashima S. Air pollution: another cause of lung cancer. Lancet Oncol. 2013;14(9):788–9. doi:10.1016/S1470-2045(13)70302-4.
5. Kan H, Heiss G, Rose KM, Whitsel EA, Lurmann F, London SJ. Prospective analysis of traffic exposure as a risk factor for incident coronary heart disease: the atherosclerosis risk in communities (ARIC) study. Environ Health Perspect. 2008;116:1463–8.
6. Weuve J, Puett RC, Schwartz J, Yanosky JD, Laden F, Grodstein F. Exposure to particulate air pollution and cognitive decline in older women. Arch Intern Med. 2012;172(3):219–27.
7. Engel-Cox J, Kim Oanh NT, van Donkelaar A, Martin RV, Zell E. Toward the next generation of air quality monitoring: particulate Matter. Atmos Environ. 2013;80:584–90. doi:10.1016/j.atmosenv.2013.08.016.
8. Achrekar H, Gandhe A, Lazarus R, Ssu-Hsin Y, Benyuan L. Predicting flu trends using Twitter data. In: Proceedings of the 2011 IEEE computer communications workshops. Shanghai; 2011. p. 702–7. doi: 10.1109/INFCOMW.2011.59289038.
9. Broniatowski DA, Paul MJ, Dredze M. National and local influenza surveillance through Twitter: an analysis of the 2012–2013 influenza epidemic. PLoS One. 2013;8(12):e83672.
10. Nagel AC, Tsou M-H, Spitzberg BH, An L, Gawron JM, Gupta DK, et al. The complex relationship of realspace events and messages in cyberspace: case study of influenza and pertussis using tweets. J Med Internet Res. 2013;15(10):e237. doi:10.2196/jmir.2705 PMC38413596.
11. Moreno MA, Parks M, Richardson LP. What are adolescents showing the world about their health risk behaviors on MySpace? MedGenMed. 2007;9(4):9.
12. Myslín M, Zhu SH, Chapman W, Conway M. Using Twitter to examine smoking behavior and perceptions of emerging tobacco products. J Med Internet Res. 2013;15(8):e174.
13. Nakhasi A, Passarella RJ, Bell SG, Paul MJ, Dredze M, Pronovost PJ. Malpractice and malcontent: analyzing medical complaints in Twitter. In: Proceedings symposium on information retrieval and knowledge discovery in biomedical text; 2012.

14. Jiang W, Wang Y, Tsou M-H, Fu X. Using social media to detect outdoor air pollution and monitor air quality index (AQI): a geo-targeted spatiotemporal analysis framework with Sina Weibo (Chinese Twitter). PLoS One. 2015;10(10):e0141185. doi:10.1371/journal.pone.0141185.
15. Ritzer G, Jurgenson N. Production, consumption, prosumption: the nature of capitalism in the age of the digital 'prosumer'. J Consum Cult. 2010;10(1):13–36.
16. Kitchin R. Big data and human geography Opportunities, challenges and risks. Dialogues Hum Geogr. 2013;3(3):262–7.
17. Kshetri N. The emerging role of Big Data in key development issues: opportunities, challenges, and concerns. Big Data Soc. 2014;1(2):1–20.
18. Kitchin R. Big Data, new epistemologies and paradigm shifts. Big Data Soc. 2014; 1(April–June):1–12.
19. Andrejevic M, Burdon M. Defining the sensor society. Telev New Media. 2014;16(1):19–36. doi:10.1177/1527476414541552.
20. Resch B, Resch B. People as sensors and collective sensing- contextual observations complementing geo-sensor network measurements. In: Krisp JM, editor. Progress in location-based services. Berlin: Springer-Verlag; 2013. p. 391–406.
21. Goodchild MF. Citizens as sensors: the world of volunteered geography. GeoJournal. 2007;69(4):211–21.
22. Wang D, Amin MT, Li S, Abdelzaher T, Kaplan L, Gu S, et al. Using humans as sensors: an estimation-theoretic perspective. In: Proceedings of the 13th international symposium on Information processing in sensor networks. IEEE Press; 2014.
23. Boulos MNK, Resch B, Crowley DN, Breslin JG, Sohn G, Burtner R, Pike WA, Jezierski E, Chuang KYS. Crowdsourcing, citizen sensing and sensor web technologies for public and environmental health surveillance and crisis management: trends, OGC standards and application examples. Int J Health Geogr. 2011;10:67.
24. Paulos E, Honicky R, Hooker B. Citizen science: enabling participatory urbanism. In: Foth M, editor. Urban informatics: community integration and implementation. Hershey, PA: IGI Global; 2008.
25. Karatzas KD. Participatory environmental sensing for quality of life information services. In: Golinska P, et al., editors. Information technologies in environmental engineering. Berlin: Springer-Verlag; 2011. doi:10.1007/978-3-642-19536-5.
26. Kitchin R. The real-time city? Big data and smart urbanism. GeoJournal. 2014;79:1–14. doi:10.1007/s10708-013-9516-8.
27. Kanhere S.S.. Participatory sensing: crowdsourcing data from mobile smartphones in urban spaces. In: International conference on distributed computing and internet technology. Berlin: Springer; 2013.
28. Hand E. Citizen science: people power. Nature. 2010;466(7307):685–7. doi:10.1038/466685a.
29. Muller CL, Chapman L, Johnston S, Kidd C, Illingworth S, Foody G, Overeem A, Leigh RR. Crowdsourcing for climate and atmospheric sciences: current status and future potential. Int J Climatol. 2015;35:3185–203. doi:10.1002/joc.4210.
30. Kuntz K, Shukla R. Choices. How many points for that? A game-based approach to environmental sustainability. In: ACEE proceedings; 2012. Available at: http://www.aceee.org/files/proceedings/2012/data/papers/0193-000221.pdf. Last Access: 17 July 2016.
31. Maturo A, Setiffi F. The gamification of risk: how health apps foster self-confidence and why this is not enough. Health Risk Soc. 2016;17:7–8.
32. Castel N, Kobernus M, Liu HY, Schneider P, Lahoz W, Berre JN. Mobile technologies and services for environmental monitoring: the Citi-Sense-MOB approach. Urban Clim. 2015;14:370–82.
33. Nikzad N, Verma N, Ziftci C, Bales E, Quick N, Zappi P, et al. CitiSense: improving geospatial environmental assessment of air quality using a wireless personal exposure monitoring system. In: Proceedings of the wireless health, October 23–25, 2012, San Diego, CA, USA; 2012.
34. Mei S, Li H, Fan J, Zhu X, Dyer CR. Inferring air pollution by sniffing social media. In: Proceeding of IEEE/ACM International conference on Advances in social networks analysis and mining, 2014. Doi: 10.1109/ASONAM.2014.6921638.

35. Mun M, Reddy S, Shilton K, Yau N, Burke J, Estrin D, Hansen M, Howard E, West R, Boda P. PEIR, the personal environmental impact report, as a platform for participatory sensing systems research. In: Proceedings of the 7th international conference on Mobile systems, applications, and services (MobiSys '09). New York, NY: ACM; 2009. p. 55–68. DOI: 10.1145/1555816.1555823.
36. Newman G, Wiggins A, Crall A, Graham E, Newman S, Crowston K. The future of Citizen science: emerging technologies and shifting paradigms. Front Ecol Environ. 2012;10(6):298–304. doi:10.1890/110294.
37. Shah VD, Cappella JN, Russel NW. Big data, digital media, and computational social science: possibilities and perils. Ann Am Acad Pol Soc Sci. 2015;659:6–13.
38. Boyd D, Crawford K. Critical questions for big data, information. Commun Soc. 2012;15(5):662–79.
39. Rogers R. The end of the virtual: digital methods. Amsterdam: Amsterdam University Press; 2009.
40. Rogers R. Digital methods. Cambridge: The MIT Press; 2013.
41. Mei S, Li H, Fan J, Zhu X, Dyer CR. Inferring air pollution by sniffing social media. In: IEEE/ACM international conference on Advances in social networks analysis and mining (ASONAM). IEEE; 2014. p. 534–9.
42. Sukosd M, Fu K. How Chinese Netizens discuss environmental conflicts? Framing and functions on Sina Weibo, 11. 2013. Available at SSRN: http://ssrn.com/abstract=2277395. Last Access: 20 July 2016.
43. Wang S, Paul MJ, Dredze M. Social media as a sensor of air quality and public response in China. J Med Internet Res. 2015;17(3):e22.
44. DMR. Millenial statistics report 2016. 2016. Available at http://expandedramblings.com/index.php/downloads/dmr-millennial-statistic-report/. Last Access: 15 July 2016.
45. Li Y, Huang J, Luo J. Using user generated online photos to estimate and monitor air pollution in major cities. In: Proceeding ICIMCS '15, August 19–21, 2015, Zhangjiajie, Hunan, China; 2015. p. 1–5. Doi: 10.1145/2808492.2808564.
46. Burke JA, Estrin D, Hansen M, Parker A, Ramanathan N, Reddy S, & Srivastava MB. Participatory sensing. In: Proceedings of the WSW'06 at SenSys '06, October 31, Boulder, Colorado, USA; 2006.
47. Boccia AG. The direction of the network. Le scienze sociali computazionali e la sfida dei Big Data, Salute e Società, XIII. 2014;3:25–38.
48. Van Mierlo T. The 1% rule in four digital health social networks. J Med Internet Res. 2014;16(2):e33.
49. Lupton D. Digital sociology. London: Routledge; 2015.

Basic Principles of Risk Communication in Air Pollution 34

Fabio Capello

34.1 Background

Risk communication is part of the response to any emergency or to the rise of any hazard able to affect human health [1]. It consists in transferring knowledge between experts of similar or different fields, between experts and decision makers, and in its proper sense between experts and/or decision-makers and the public. The idea of data and information sharing among the different parts involved in the process may appear a natural action in case of catastrophes, humanitarian crisis, natural disaster, epidemics, or pandemics. As a matter of fact, it is a revolutionary concept, as it implies that all the actors are actively involved in their own care.

The basic concept is to inform people that a problem exists meaning that the same people's lives, lifestyles and personal and general well-being can be at risk for a specific reason. The aim is to allow people to take action or to accept actions coming from decision-makers based on advices coming from experts to protect themselves, prevent or reduce exposition and prevent or reduce damages. It also means that in case of emergencies or hazards that can massively and/or abruptly affect the whole population, decision-makers are called to find a balance so to protect the health of individuals, the level of health of the whole population and national and international interests and security, the latter defined also as health security (whose key points are "achieving widespread access to essential health services, and protection from environmental and behavioural risks to global public health" and "systems and agreements to prevent the cross-border spread of communicable diseases and other threats to public health" [2]).

F. Capello
Alder Hey Children's Hospital, Liverpool, UK
e-mail: info@fabiocapello.net

F. Capello, A.V. Gaddi (eds.), *Clinical Handbook of Air Pollution-Related Diseases*,
https://doi.org/10.1007/978-3-319-62731-1_34

However simple in its principles, risk communication is not trivial; the simple sharing of information, no matter how accurate, is not communication. Communication in risk management and in health education can be summarized in three phases:

1. Knowing that a problem exists: people have no previous knowledge of the existence of a problem.
2. Understanding the implication of the problem: people already know about a problem, and understand the implication. This knowledge is not sufficient to trigger actions.
3. Awareness of the problem: people become aware that a problem actively affects their lives. This awareness is sufficient to trigger actions.

In a simplify scheme, problems that have short-term consequences and where the relationship between cause and effect is evident go rapidly through the three phases; the major consequences are that people are prompt to take immediate action often overreacting (i.e. the "mad cows" meat causing the Creutzfeldt-Jakob disease and the reduction of beef consumption). Problems that have long-term consequences can be ignored or can trigger actions only in a minority of the population even when the consequences are severe; people may not take action at all or take it when it is too late, often underreacting (i.e. tobacco smoking causing lung cancer and people quitting smoking). The scheme can be summarized in the Table 34.1.

Because of that, communication has a central role, spreading knowledge among the population in relation to a risk for people's health. This has to be a conscious process. In fact, an improper dissemination of information can lead to disastrous results for a number of reasons, among which:

- The information can be unreliable.
- The contents of the information can be accurate, but the message is unclear.
- The information can be accurate, but the source is perceived as untrustworthy.
- The information can be distorted because it is oversimplified or because the same broadcaster does not understand it.
- The information can be voluntarily distorted to create false reassurance.
- The risk can be statistically irrelevant, but presents immediate and clear consequences for the ones affected.
- People consequently can get scared and act irrationally.
- The information can be accurate and reliable and broadcasted in an understandable form, but offer no solutions, no reassurances if any are due, no hope.
- The information can be accurate and reliable, but other sources are broadcasting contrasting information.

Table 34.1 Type of consequences related to a problem and the likelihood of people taking action (expressed in – and +) to avoid or reduce exposure to the hazard

	Short term	Medium term	Long term
Severe	++++	++	+
Moderate	+++	+	–
Mild	+	–	–

- The official sources (mainly experts and decision-makers) are perceived as untrustworthy because they give information that contrast with the other coming from unofficial sources.
- The latter can be true or false, especially when the official source voluntarily decides to conceal the problem, distort, or minimize it in the attempt of avoiding scaremongering.

In addition, the improper diffusion of information produces:

(a) *Misinformation* (intended as unintentional)
(b) *Disinformation* (intended as intentional)
(c) *Prejudices* (intended as fragments of incorrect knowledge that prevent the transmission and the acquisition of correct information) [3]

Not surprisingly, people reactions to the communication of a risk overlap with the so-called NIMBY (Not In My Back-Yard) syndrome, where people understand that an action to a problem is due, and agree that it may be the only possible solution to the same problem, but do not want to be personally affected by that decision. As a consequence, the syndrome in its simpler definition depicts people as "overly emotional, uninformed, and unscientific in their opposition; motivated by narrow and selfish interests; and obstructing policies that would provide for collective good" [4].

However, especially when it comes to risk communication, this is known to be partially true: people may react rationally or irrationally to an emergency or a danger as part of their human nature and in response to a physiological survival strategy; risk communication is asked to offer information so to modulate the natural and spontaneous response to a threat in the attempt to produce the better solution for the community and its individuals.

In the last years, with the rise of the Internet and lately of the social media, this classic scenario changed [5]. While official sources are becoming perceived as less reliable than in the past, people seek actively and 24/7 information on a Web that never close and where news spread instantly from every corner to every other corner of the world in a matter of seconds, becoming potentially accessible to everyone on the planet as soon as someone publish them.

For these reason, communicating the risk cannot be an improvised action taken light heartily or by people with no experience and formal education in communication and risk management.

34.2 Risk Communication in Air Pollution

Air pollution is a global threat and the risk for human health is becoming to be known to scientists and to the public. Some of these information are already part of the cultural baggage of many, especially in high-developed countries where these problems have been treated for years now. As field research shows (see Chap. 35), people living in highly polluted areas are aware of the problem and of some of the consequences on health. Yet, only a few take proper action in the attempt to protect themselves or to reduce exposure for themselves or for the one they care for (as elderly or children).

Differently to other health hazards as major pandemics and epidemics (e.g. Ebola, Cholera, Zika, SARS), humanitarian crisis or catastrophes (e.g. Chernobyl, Fukushima), air pollution is a silent threat, mainly invisible, and that is constantly part of our life, surrounding our environment and the places where we normally spend our life. Most of the consequences are still unknown, and the ones that are acknowledged also by the majority of the public are not so scary, although very serious, because the cause-effect relationship is not evident or because the consequences show up only in a limited number of people and often after years or decades. Because of the many outcomes and the multifactorial aetiology of some of the diseases related to air pollution, people have nothing to focus on, unless a clear cluster in a given population appears (like several cases of the same cancer in a population known to be particularly exposed to a particular environmental pollutant as in the case of the asbestos).

This is possibly the most difficult task for science communicators, health education promoters, or experts in risk management.

For most of the effects, people are still in the first phase of health education: most of those are prominently known to researchers and some medical doctors. For others, like lung diseases, people understand that air pollution is a treat and action needs to be taken; most are not aware of the real impact, of the sources, of the long-term consequences on different organs; a few understand that individual actions and change in behaviours can reduce the risk, and even fewer are eager to change their lifestyles in order to protect themselves or reduce the active or passive exposure.

On the other hand, air pollution is a universal problem, and policy-makers, government, and enterprises play a major role in the regulation of the emissions and the reduction of the sources of pollution. At the same time, public health officers are in need of policies able to actively change the way people live so to promote primary, secondary and tertiary prevention. These same policies, though, may create annoyances or major concerns in the population. This is particularly true when air pollution is related to technological development (both in high- and low-income settings) that from one hand is an achievement of the human potentiality and on a simpler level a way to massively improve people's lives and from the other hand is a major source of pollutants.

For this reason, people must be actively involved in the discussion, in a clear and transparent way, where the main issues are honestly presented as well as the consequences of action or inaction. As a general rule, the following principles should guide a good communication [6] and should result in some practical actions as summarized in Table 34.2:

1. Openness
2. Transparency
3. Independence
4. Responsiveness/timeliness

Table 34.2 Practical consequences of some of the principles that should guide risk communication

Action to be fostered	Main principles	Risk to be avoided
Publish/make available all key documents	Openness Transparency	Reveal complex information that the public or other stakeholder may misunderstand
Understandable and usable communications	Openness Transparency	Simplify excessively the problem, missing key information
Timely communications	Responsiveness/timeliness	Create unnecessary alarms
Sharing between who assess the risk and the decision-makers	Openness Transparency Independence	Inadequate responses or overreaction that may lead to useless policies or policies that compromise excessively other aspects of life
Sharing between stakeholders and the public	Openness Transparency Independence	Inadequate responses or overreaction that may lead to useless policies or policies that compromise excessively other aspects of life
Acknowledging and communicating uncertainty	Openness Transparency	Showing unjustified weakness that can undermine the trust between the different actors involved

The main aim in risk communication in fact is to establish a useful and productive relationship among the different parties involved, so that a constructive dialogue can be started. This is crucial in every field of risk communication, but is paramount in those areas where the hazard is only partially known, the outcome lost in the future, and the effective measures to reduce the problem still unclear.

Because air pollution is a global concern, a global effort is necessary: scientists need to present their data in understandable way and be aware that people's opinions can affect the policies as well as the knowledge produced by research, even when those are based on inaccurate and incorrect information; government and organization have the duty to inform people and to listen to their concerns; people have to understand that changes may come from above but can come also from everyday decisions like taking public transports or avoiding children being exposed to second- and thirdhand smoke.

Conclusions

Risk communication remains a key part in the fight against air pollution and environmental health. As airborne pollutants are everywhere and are able to deeply affect people's life, there are no aspects of everyday life that is untouched. In other words, it means that deep changes are needed provided that people are able to understand why. If telling the truth is the first rule of risk communication, making people understand is the final goal. Otherwise there would be no difference between risk communication and a gossip from the evening news.

Therefore, science communicator and experts in risk management are needed and are called to action, in order to spread correct information, with the appropriate media, and to the correct audience.

Risk communication as a part of the medical encounter and the doctor-patient relationship can also have practical consequences in the clinical practise, becoming part of the medical education process that eventually can lead to prevention.

No matter how accurate this job will be, the accessibility to unreliable information and the dissemination of reasonable worries in the population will be part of the game. Thus, a right balance should be found in giving and receiving information, taking into account the opinion of the people that must always be involved in the discussion, the need of trustworthiness that comes from the science and the fears of policy-makers upon whose shoulder lay the weight of decisions.

References

1. WHO. Risk communication 2017 [13/01/2017]. Available from: http://www.who.int/risk-communication/en/.
2. ESSACHESS. 21st century challenges and opportunities for risk communications. J Commun Stud. 2012; 5(1). Available from: http://ssrn.com/abstract=2161601. Accessed on 13 Jan 2017.
3. Capello F, Luini MGG, eHealth Policy. eHealth, care and quality of life. Milan: Springer; 2014. p. 129–55.
4. McAvoy GE. Controlling technocracy: citizen rationality and the NIMBY syndrome. Washington D.C.: Georgetown University Press; 1999.
5. Gamhewage G. An Introduction to risk communication. 2014. Available from: http://www.who.int/entity/risk-communication/introduction-to-risk-communication.pdf?ua=1. Accessed on 13 Jan 2017.
6. EFSA. When food is cooking up a storm – proven recipes for risk communications [Corporate Publication]. 2015. Available from: https://www.efsa.europa.eu/sites/default/files/corporate_publications/files/20120712_EFSA_RCG_EN_WEB.pdf.

Marketing Issues in Air Pollution Prevention

35

Erio Malagoli

35.1 Overview

Pollution marketing is an extremely new topic, mostly unexplored and extremely open. Nowadays, it is very difficult even to define exactly what the phenomenon of air pollution is from a marketing point of view; from this standpoint, is pollution only a threat or, on the contrary, an opportunity as well?

Air pollution has been largely discussed during the last 10 years. However, it remains strange that a mass awareness has not been formed yet; this task probably shall be vested on marketing.

Our studies and analyses highlight scarce or nil awareness in the population of the real impact of this phenomenon, even in those regions where air pollution manifests itself more bluntly. The role of training and education has been reported by many specialists [1–3], and there are some studies showing that education, also increasing awareness of the pollution issues, may entail favourable results [4–7]. A very recent study has shown that about 90% of the respondents agree that the interest in becoming acquainted with air pollution-related issues shall be raised, to a greater extent, in every and all residents [8]. All the same, even when the general level of knowledge is maintained [9], an increment of awareness of this phenomenon can be observed, particularly within groups of residents, which already have higher education.

Air pollution depends on different pollutants, including PM_{10}, $PM_{2.5}$, $PM_{0.3}$, NO_2, SO_2, CO and CO_2 that are the best known and measured ones, whose concentration changes during different seasons and within different geographical regions of the world. For this reason, different marketing approaches are needed. Those should be tailored on different individual situations: the recorded air quality index (AQI) of air

E. Malagoli
Qi International Ltd., 1st Floor 87/89 High Street, Hoddesdon, EN118TL London, UK
e-mail: eriom@usa.net

F. Capello, A.V. Gaddi (eds.), *Clinical Handbook of Air Pollution-Related Diseases*, https://doi.org/10.1007/978-3-319-62731-1_35

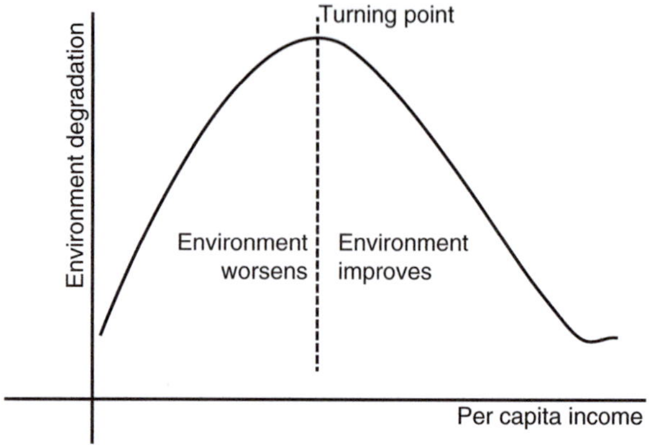

Fig. 35.1 Approximate Kuznets curve of increase of pollution and economic growth

pollution in London may be of 65, whereas in Beijing AQI reaches a value of 480, namely, making it 7.5 times higher.

Air pollution does enormous damages to the people; so all sources of pollution should be theoretically removed; and this is a world-known issue. Governments and organisations make efforts to reduce and limit emissions of pollutants into the atmosphere. The timing to achieve results is too long so that the results cannot be always appreciated; therefore, there we need to find prompt solutions and different alternatives. The typical Kuznets curve (Fig. 35.1) leaves a door open and hope that in increasing the economic resources of a country, raising jointly people's awareness and improving the laws and procedures regulating environmental discharges, we will come to a moment when the economic benefits continue to increase, whilst pollution decreases (instead of growing as expected).

However, this is a theoretical hypothesis. In reality, protection against air pollution needs a structured plan to be carefully drafted and then widely spread within the population, this being a prevention plan to be applied using both public and private forces and resources at the same time.

Large territories containing a significant number of people are concerned about the air pollution phenomenon, which can be settled on a large scale only joining the efforts of private and public companies.

We have noticed in many geographical areas that public information shared by private companies clash too often with medical data, due to 'political' reasons. In other cases, such data are complex and not always completely clear even for scientists.

The governments tend to diminish negative effects related to air pollution, and many private companies do not have adequate knowledge and skills in expressing appropriate and clear messages. Medical doctors are still launching alarm messages about disasters people cannot avoid. Such a mix of information generates a significant ambiguity in messages and doubts that undermine the right information; so often the wrong one is spread creating a low level of awareness of what are the prevailing challenges and damages for the human body.

It is clear that an enormous investment is needed in order to raise the population's awareness worldwide; yet, nobody seems to be interested now. Media campaigns are very expensive and hardly acknowledged by governments, which see in such campaigns the recall of their guilt and responsibility, entailing negative political repercussions.

On the other hand, if such advertising campaigns are carried on by the companies selling antipollution products, and by security or protection companies, a final consumer is entitled to ask whether such information is real and trustworthy or just exaggerated 'in order to sell more'.

Only few medical doctors write articles, books or publications, using the scientific resources they possess; however, from a marketing point of view, when it comes to pollution, such works are hardly effective and not able to raise awareness in people; those, on the contrary, may scare people without giving that basic knowledge required to accept those everyday changes needed to achieve an adequate personal protection.

The impact of marketing researches and publications is very high now in the field of medical preparations; however, it is declining slowly, in the field of nutrition and dietetics, of foodstuffs and diets and of water and air quality.

35.2 Marketing and Communication and Perception of Air Pollution

The experience of these years of work in the field of air pollution shows that there still are objective difficulties within the population when it comes to acceptance of preventive measures.

We are still very far from the achievement of a widespread awareness about the problems related to air pollution. People are not ready to accept those actions needed to restrain the problem; policy-makers do not have or do not offer solutions, or are unwilling to take decisions that may not be accepted by population.

If, on the one hand, everybody is a bit aware of this subject, particularly the young generations, few of them are ready to make changes in their everyday life in order to protect themselves in a better and more comprehensive way. This is a well-known problem.

Marketing teaches us that since there is a need, such a need shall be fulfilled, regardless of objective or subjective demands.

Now when it comes to air pollution, there are objective medical data, which show a challenge and a scale thereof, as well as solutions deemed to be effective for protection and reduction of exposure, and, consequently, the impact on health. They are presented in more details in further chapters of this book. Then, there arise subjective needs, such as hypothesis ('I do not believe in such messages'), economic factors ('How much will it cost me to protect myself'), aesthetic factors (the majority of masks sold are, for instance, a remade version of those ones designed and conceived for the factory workers but not for everyday use in the cities) and deep knowledge of this problem so that protective equipment and devices, as air filters, can be properly used.

Quite the other thing is to buy an air-conditioning unit that will promptly reduce the temperature and then will immediately repay the investments made. But when buying air filters, the way of action is not so evident: it is not clear oftentimes when to switch it on or off; it is also not clear beforehand whether it is suitable for a certain type of polluted air; and, especially, the specific prompt effects are not seen, since the function of an air filter is, as a matter of fact, prevention or a preventive medicine device.

As a consequence, marketing shall reveal the information very precisely, and such information shall be disseminated in a professional but simple way. Otherwise, the opposite effect may be achieved, i.e. people not able to cope with sophisticated information, that therefore renounce to know more.

We have presently identified different challenges and various remedies, in order to make air pollution marketing a truly effective tool able to serve people. We believe that it is necessary to:

1. Understand and let people know what air pollution is.
2. Understand and let people know where it is and how it occurs and why develops.
3. Understand and let people know what are the health problems caused by air pollution.
4. Design some mask-resembling products that would be less invasive, for everyday use in the cities.
5. Define quality and standards for the masks: manufacturers need to know, follow and certify simple but clear safety and quality standards, and those standards should be easy to understand for the final consumers (this is not always evident throughout the world).
6. Promote a more careful production and certification of air cleaning tools having the simple functions for everyday use and simple maintenance.

We are sure that if all those engaged in the activities related to the problems generated by air pollution follow the example presented in those six points, the final communication with clients will lead to better understanding and thus to better results than what we have today.

Using air pollution marketing, we would like to describe a range of actions to be simultaneously undertaken by several persons concerned, in order to understand and to satisfy the needs and demands coming from many sides, particularly medical doctors and people; the aim is to reach a greater level of awareness so to increase the use of preventive tools drastically reducing at the same time the negative consequences caused by air pollution to the human body.

Our information and our analyses, carried on in different areas, show that those people most aware of this problem are the young ones aged from 18 to 40; this factor is a doubtless advantage when it comes to communication, because this population can be reached with all the tools offered by the new technologies including the social media, which are certainly used by such cadre of people on a large scale all over the world.

If the information travelling on the Internet are carefully managed, the benefits from the economic point of view are clear as the web offers a safe and widespread way to spread knowledge.

It will be necessary to inform about what air pollution is, and it will be also important to do it using the phrases, arguments and examples that can be easily understood by the majority of residents. The use of too technical images, words, trends and schemes as those ones often used by medical doctors entails the opposite effect—negligence instead of interest.

The problem is well-known in this case too. Simplifying contents leads to reduction of information, based on the following equation:

$$\frac{C}{S} = K$$

where '*C*' means scientific contents, '*S*' is the level of simplification of the scientific content and '*K*' is a constant value. The more the contents are simplified, the more information is lost; the more technical information is kept, the more complex the narration becomes; and, as a consequence, the more difficult the comprehension thereof is for the wide public.

The majority of people do not know what are the pollutants that form the so-called air pollution. They confuse them and do not understand in full their meaning and their dangerousness. Hence, there is a need for replacing the current information strategy with a simpler, more interesting and more emotional one. It will be used to raise awareness, to familiarise, to apply simple methods of communication with high emotional contents and to inform people all over the world. As a matter of fact, warnings and medical bulletins are not enough.

The reason is that the word of a familiar and trustworthy witness is sometimes more effective than the word of a scientist. For instance, we have seen in China, how a person from the Chinese show business gathered a great number of followers in his blog after writing about air pollution.

People have only general information about the fact that air pollution is dangerous; however, those same people do not possess any specific knowledge of the different types of pollutants, of the way they appear, where, when and in what state, to what extent they are dangerous and what to use and how to use them to limit damages to the human body.

This is why an appraisal of the available products and devices against air pollution is needed.

Relying on my experience in the field of air pollution during these years, I have seen that there are a great number of certifications and standards for antipollution equipment and devices; however, all of them refer to the workplace products.

Perhaps, the differences between the work environment and air pollution shall be clearly exposed, since these are two resembling but still different realities.

If a person is exposed to specific forms of pollution, a mask protection can be used at a workplace; the choice of mask depends on the activities conducted, and such a mask is worn even for many consecutive hours in case of very high levels of pollution in the workplace.

Quite different is the case of general exposure to air pollution, when a person needs different products at different moments of the day.

There is a need for a mask to be used only when a person is outside, subject to considerable pollution, differing from that one used in a working environment. Such pollution may be even lower, equal or higher; however, duration thereof is short or very short whilst commuting and little longer afterwards when such a person goes to office or school where he or she spends relatively few hours; then he or she comes home where spends a great number of hours. This is when full protection is also needed.

This is where things get complicated.

The majority of the masks sold are mainly designed for the industry, and then those are used and re-proposed for urban application with all the consequences thereof.

Their fitness, colours, styles, design and technical features are seldom perfect for such a use, and this results in great discomfort for people that wear these products. Then, there comes the tendency not to use them or to use them as seldom as possible.

It would be quite different, if there were models more suitable for this final use. As long as we know, it was only in 2016 when the products cited above—able to protect with a more effective filtering capacity and with at the same time the other characteristics indicated—appeared on the market. However, there are not enough as there is still plenty to do, to invent and to develop in this field.

I believe that the use of nanotechnology, nanomaterials, innovative materials, 'second skin' tissues and filtering materials of another type and kind, like those that can be fitted in the nasal or oral cavity, would deliver protective devices easier to use and tolerate, leading consequently to positive outcomes in terms of pollution prevention.

A reference market offering new technologically advanced products, new ideas and inventions would be very huge, giving at the same time new alternative to the masks that we see on the market today.

The regulations shall be revised, simplified and standardised worldwide. Nowadays, we see a great number of products for sale both online and offline; however, it is difficult to find out their level of certification that may be more or less appropriate. Too many reference certifications of such products hamper their appreciation by people and arouse mistrust. A final customer does not understand in full the meaning of individual certifications and feels disoriented. Finally, such a customer feels unsure when it comes to product quality, real filtering capacity thereof and, as consequence, real preventive capacities. Hence, it results in disloyalty to that product.

A correct information sharing and the simplification of the number and quality of informational messages will, on the contrary, help consumers to know their way around the market of consumption.

The results from our marketing studies and analyses show that with this approach the use of the masks and therefore of prevention would grow exponentially.

This is true also for the air cleaning products currently on the market. It is clear from a detailed study we made that those various ways of air cleaning relate, as a matter of facts, to a brand and/or product sales; when it comes to their quality, the results remain doubtful.

Some of these products for air cleaning claim combined technological solutions, which do not seem reliable; many questions thus remain, as well as suspicions regarding their real filtering capacity.

Another known problem is that there is no air cleaning device able to filter all the possible pollutants (in the form of particles, volatile organic compounds and gases). So, people may be confused in selecting, buying and using such a product; otherwise, they need to buy tens of various devices, some of which are extremely sophisticated, very professional and very expensive.

People also question about the real usability of such products, as well as about the selling price. As a consequence, poorer people cannot access to these solutions for air cleaning that remain a status symbol for the richer ones.

People living in most polluted areas are worried. They are aware of existing problem and realise that its consequences may be severe. However, the majority of them consider all those issues and express their concerns about the real help offered by the current preventive products; they have a lower level of trust in such products; all those issues lead to a low level of trust in those solutions so that people are not motivated to buy products to prevent the effect of exposure to pollution. People then will rather choose to be exposed to air pollution and to its consequences renouncing in facts to any kind of prevention.

Air pollution marketing shall take into account all such elements and shall become a meeting point for all the actors involved as medical doctors, manufacturers and consumers.

This will give an impulse and will define a common direction for further orientation.

Medical doctors need to increase their knowledge and to improve people's awareness on air pollution prevention in order to reduce deaths and casualties; they shall use a more popular language and easier forms of dissemination. Manufacturers, to a great extent, shall be stimulated to perform better studies in order to find better solution to the specific problem of air pollution. People, seeing such improvements, would be more apt to get informed, to pass on the words and to use such technologies for reduction, prevention and protection against different forms of air pollution.

This shall be the way to follow in air pollution marketing, and this is a message to send: a constructive feedback for the actions and policies of the governments and of the environmental agencies. This is something useful for the promotion of macroeconomic manoeuvres aimed to reduce the emissions, as well as to promote prevention. Medical doctors and researchers give an imminent warning; this cannot be ignored any further. We rather need real and prompt actions aimed to achieve immediate results. On the contrary, nothing essential had occurred during from 2013 to 2016.

In order to achieve results in the field of air pollution, even using different marketing approaches, the nearest future objectives shall be the following:

- Motivate the manufacturers to produce products properly designed for everyday use.
- Motivate governments to make more standardised quality certification and clear to understand both for the manufacturers and for the final users.
- Motivate the medical doctors to give clear and adequate messages, interesting and easy to understand, thus promoting the viral spreading of the information.
- Motivate people to use more widely preventive devices.

Doing this, marketing will help not only to sell or to earn incomes. It will revolutionise itself increasing the chances to change the way of thinking and acting of all the involved actors.

Testimonial mondiali di Marketing della Air Pollution che parlino di questi punti e ne chiedano impego ai vari attori per la sua realizzazione sarebbe un grande successo su scala mondiale.

The involvement of worldwide-known testimonials promoting the marketing of air pollution, asking for concrete actions to all the stakeholder implicated, would be a great success at the global level.

New technologically advanced products more suitable for correct daily prevention, simplifications and clear services for quality certification, as well as simple and efficient medical messages to be disseminated to everyone, as well as an opinion of air pollution marketing—these are the things we regard as necessary to be followed more during the years to come, in order to get an appropriate and well-balanced global air pollution market and to ensure that the degree of people's awareness and prevention grows, whilst deaths and diseases related to air pollution decrease.

This is the objective we have set in order to promote the marketing of air pollution in the years to come.

Pollution is one of the subjects most people are concerned about. The sales of any and all products related to prevention like masks, air cleaning and skin protection creams are growing. We would be pleased to see such a marketing promoted by manufacturing companies and retailers: seeing entrepreneurs deal not only with sales but also with all the implications of air pollution, fostering marketing and web marketing actions aimed to inform people, in order to improve and raise their level of awareness, creating at the same time roots deeper than those presently existing.

35.3 Field Research

35.3.1 China's Experience

The analysis and the observation we have gathered during the last 2 years in Beijing help us to make some conclusions.

Those living in Beijing know very well about the phenomenon of air pollution, since very high values of pollution are always recorded throughout the year (three to four times higher than the maximum limit equal to 50). It is interesting that such values are not in any way perceived as worrisome. Both the Chinese and foreigners regard the challenge of air pollution with an extreme serenity.

Nevertheless, it results from the interviews carried on during that period that a major concern is the fear that the government and public healthcare professionals are not making enough efforts (see Table 35.1).

Younger people aged from 18 to 40 are more aware of this challenge, whilst elder people are less alert. Hence, there is gap between younger and more aware

Table 35.1 Marketing research as per interview on air pollution perception and prevention (Beijing, Summer 2015 and Spring 2016)

Main issues reported
1. Most of the interviewees agree that air pollution is a major concern in Beijing, China
2. A few of them know what air pollution properly is or have a clear perception of the health risk connected to air pollution
3. Most of them are interested in receiving more information on air pollution
4. Only a few of them use personal protective equipment or devices whilst travelling/commuting, at home or in the place of work
5. A few pregnant women and elder people at risk know the risk related to high level of air pollution exposure
6. A very limited number of children use face mask as a protection for air pollution
7. A very few of the interviewees know what an air cleaner is and how it works
8. A very few of them know how to differentiate between an effective, an ineffective and a useless face mask
9. Many think that in any case there are no alternatives, that however air pollution is not so dangerous and does not practically impact health or quality of life, that it is better to live the life with no further and unneeded complications and that things will settle by themselves in any case
10. Although we know the scientific and medical basis of the negative effect of air pollution on human health, people's perception of the problem is not the same of medical doctors, public health specialists and researchers. People in fact consider those claims coming from the science community overestimated and overrated
11. As a conclusion, health education and information are needed in order to explain the actual consequences of exposition to air pollution, what equipment and devices can be used to reduce personal exposure, who have to use those and when

generations that are closer to global scientific thinking that affirms that air pollution is very dangerous to human body and elder generations that seem to reject this concept and that are more resistant to the use of protective devices.

In this case, young people shall insist and teach elder ones, but not otherwise.

The Chinese culture has a great respect for the old people whose families revere and praise. They are keepers of life and of the secrets of wisdom; however, it seems that such respect is not applicable in this particular case.

It results from the interviews made in Beijing that it is also important that the population should have shops or news websites where problems related to air pollution are found and discussed. A possible application could be a network dedicated to air pollution where professional information and recommendations of products to be used in different conditions and for different people could be available; another may be a forum where specialised medical doctors would answer to specific questions submitted by single individuals that could in this way express their concern. This would be an application of preventive telemedicine networks, able to spread information to promote prevention also in the neighbouring polluted areas.

Certainly, the Central Beijing Government has to undertake serious actions in order to solve drastically the problem of air pollution in metropolis (such actions were already massively initiated by the Beijing Government since 2015, and the first results are coming; according to the 2016 statistics, a tendency for reduction of air pollution in the Chinese capital has been observed), closing or moving the polluting factories and solving the problems in the villages where agricultural residues are burnt with the help of used coal coming from the power plants. Such actions are not prompt, for sure, and they may give negative outcomes in all other areas (labour force employment, profitability, life quality and style of the people living in Beijing and so on).

However, tens of years shall pass before the first results of such direct antipollution actions of the policies will be observed. Such spread of time shall be used to create and to raise, to a greater extent, awareness of the air pollution problems and the ways of prevention and protection against it.

If the government is directly responsible for such policies, then marketing shall be responsible too for promoting the messages that may improve the perception of this problem and endorse, at the same time, the adoption of correct behaviours in the population.

Here is where marketing and medicine overlap. Sometimes, some undeliberate penetrations into one territory or another are even allowed. However, if we take these data and what we know to be true according to medicine, and we compare them with the habits (monitored throughout 2016) of the Beijing population (marketing area), it will be clear that here is a huge gap and a great discrepancy that shall be eliminated as soon as possible (please, see Table 35.2).

Many nations use different parameters to check the air quality index; however, as a result, the final messages about marketing contain severe errors; plus, a great deal depends on different pollutants: an example in this regard may be the increased contents of particles (for instance, particles of PM_{10}, against which we protect ourselves using a good mask of surgical type, or the very same high contents of NO_2 and CO, against which we protect ourselves only using real gas masks with selective filters).

Table 35.2 AQI and behaviour

AQI	Health hazards (–, none; ++++, potential permanent/ fatal damages)	Recommended behaviour	Behaviour of the Beijing residents
0–50	–	No protection	Normal life
50–200	+	Hazard (protection is recommended)	No warnings
200–300	++	Danger (protection is mandatory)	Approximately 1/1000 uses masks
300–400	+++	Protection is mandatory	Approximately 1/100 uses masks
>400	++++	Avoid exposure Avoid leaving the home	The government closes schools and reduces traffic; 1/10 uses masks; however, few people change their life habits

35.3.2 Lessons Learnt from China's Experience

The Beijing example serves to explain the misunderstanding between what shall be done for prevention and public health and what is going on in reality when air pollution is regarded as something normal. This is just one possible example but on another scale may be applied to other realities both in the high-income countries and in the developing ones.

Marketing actions shall be aimed to cover the difference between what must be done and what the local policies and individual residents do. The delivery of this message shall be a chore for professionals from the international health area, for science communicators and for marketing experts, and it shall be addressed to all parties involved: government, certifiers, antipollution product manufacturers, medical doctors and people of any age and from any social class.

It is clear that changing routine behaviours means changing the mindset and the way millions of people think; this is not a simple thing to do, and it is not something that may be done by a single individual using scarce financial resources.

International media only announce that Beijing is covered with the smoke and fog, without promoting any type of expert information or giving any training and/or appropriate advice. By contrast, western websites make flourishing advertising at the local level, but they still remain in the shadow, since they speak about air pollution only if AQI exceeds 400. Local entrepreneurs and transnational companies are engaged only in sales of more or less efficient products (not very efficient in the majority of cases), without spending or reinvesting money coming from the selling of these products in communication; they do not offer information about behaviours in case of air pollution, trainings or warnings to the population and do not give any clear explanations about their products and their quality. Then, when products of doubtful origin and doubtful efficacy emerge in the market, everybody seems concerned and everything seems to change afterwards, but no one does the most important thing – informing and training for pollution prevention.

It seems, as a consequence, that informational marketing is very needful at this stage at a global level and above all where this problem is greater, in order to inform and teach the population about air pollution. Again, there seems to be a great need for information on prevention in this regard.

We also like to signal how in Beijing different lines and forms of masks have been produced and put on the market. The youngest people have begun sharing (particularly since winter in 2016) favourable behaviours able to transform the masks into element of fashion. This process will help marketing through specific actions. And this will help to decrease the current difficulties when it comes to mask wearing, thus giving a natural impulse aimed to increase the preventive protection.

The strongest motivation for enterprises in designing, producing and offering innovative products to be used in case of air pollution is the fact that the market includes at least 100 million pieces sold per year.

Conclusions

Air pollution marketing shall settle many issues and bring together people engaged therein but that do not know each other.

There seems to be too many things that need to be done; however, it seems clear that all the parties involved shall work together both medicine and marketing, in this sense, are strongly tied. Besides cooperation could help to find better solutions and to better inform and help companies manufacturing antipollution equipment and devices. This approach would improve the quality and the efficacy of the products needed to massively protect people from air pollution. This shall be done in the governmental sphere in order to standardise the quality.

The medical sector cannot just say that people must wear masks and use clean air, but has to highlight the problem that using the products actually existing on the market is almost impossible in the everyday life. As a consequence, the medical sector shall contribute to correctly inform and to try to engage other responsible entities, such as the government and manufacturers, so to design useful policies and strategies, as well as appropriate devices.

Our experience in China also shows that there is still a discrete confusion: local and international manufacturing standards use a number of codes that only few persons understand. Consumers and retailers, in fact, do not know the meaning of symbols and numbers reported on the labels, and are unable to understand the differences, in terms of quality and safety, of the same products they are buying or selling.

The product quality is not a secondary issue; a great clarity of the product and manufacturing standards will contribute not only to a better sharing of information about the same products and their use but, as a matter of practice, to create in people and in the final consumers a greater trust in such products.

The quality will mean searching for and improving both masks and air cleaning. A standard regulation regarding production, time of use and kinds of protection offered by the masks will be another step to satisfy the requirements, as results from the interviews made, during which people requested only more information about this issue.

Thus, the in-depth study of air pollution in a real environment, such as Beijing, has confirmed the initial thesis of this work in the sphere of air pollution marketing. Such experience shall be, as a consequence, extended to the international level and shall promote those policies and actions that will contribute to protect communities and individuals.

References

1. Finkelstein MM. Pollution-related mortality and educational level. JAMA. 2002;288(7):830; author reply.
2. O'Neill MS, Bell ML, Ranjit N, Cifuentes LA, Loomis D, Gouveia N, et al. Air pollution and mortality in Latin America: the role of education. Epidemiology. 2008;19(6):810–9.
3. Branis M, Linhartova M. Association between unemployment, income, education level, population size and air pollution in Czech cities: evidence for environmental inequality? A pilot national scale analysis. Health Place. 2012;18(5):1110–4.
4. Dorevitch S, Karandikar A, Washington GF, Walton GP, Anderson R, Nickels L. Efficacy of an outdoor air pollution education program in a community at risk for asthma morbidity. J Asthma. 2008;45(9):839–44.
5. Kan H, London SJ, Chen G, Zhang Y, Song G, Zhao N, et al. Season, sex, age, and education as modifiers of the effects of outdoor air pollution on daily mortality in Shanghai, China: the public health and air pollution in Asia (PAPA) study. Environ Health Perspect. 2008;116(9):1183–8.
6. Ramirez DM, Ramirez-Andreotta MD, Vea L, Estrella-Sanchez R, Wolf AM, Kilungo A, et al. Pollution prevention through peer education: a community health worker and small and home-based business initiative on the Arizona-Sonora border. Int J Environ Res Public Health. 2015;12(9):11209–26.
7. Cakmak S, Dales RE, Judek S. Do gender, education, and income modify the effect of air pollution gases on cardiac disease? J Occup Environ Med. 2006;48(1):89–94.
8. Wang R, Yang Y, Chen R, Kan H, Wu J, Wang K, et al. Knowledge, attitudes, and practices (KAP) of the relationship between air pollution and children's respiratory health in Shanghai, China. Int J Environ Res Public Health. 2015;12(2):1834–48.
9. Pluhar ZF, Piko BF, Kovacs S, Uzzoli A. "Air pollution is bad for my health": Hungarian children's knowledge of the role of environment in health and disease. Health Place. 2009;15(1):239–46.

Air Pollution: Actions to be taken

36

Antonio Vittorino Gaddi and Fabio Capello

The implementation of health policies worldwide nowadays must take into account the role of air pollution and its possible effects on humans. The burden associated with air pollution has such a devastating impact on people and countries that immediate actions by government, health systems, and organizations is needed. A rising body of evidence shows links with respiratory and nonrespiratory diseases and life-changing consequences of antenatal, early-life, short-term, and long-term exposure to air pollutants in children and adults.

Individual actions and changes in behavior can help people to detect, monitor, and reduce the production of indoor and outdoor air pollution, and protect themselves against its effects. Technological solutions that are already available can help in the cleaning up of indoor areas and in protecting children and adults from toxic emissions.

That implies that people need to become aware of the problem and understand that air pollution is an invisible hazard that constantly surrounds us. Besides, single and personal initiatives may not be enough, because people are exposed to pollution even in environments that they cannot control, such as schools, workplaces, or meeting places. These assumptions mean that decision makers and policy makers need to act at different levels to achieve real administration of the environment, aimed at reducing the entity of air pollution and consequently the impact that it has on the population.

In addition, developing countries are a rising concern, as they are changing from rural low-tech societies to advanced and highly polluting societies. In these same settings, poor awareness and scarce regulation secondary to weak or nonexistent

A.V. Gaddi (✉)
Scientific Director of Health Lab of GTechnology Foundation,
Modena, Italy
e-mail: profgaddi@gmail.com

F. Capello
Alder Hey Children's Hospital, Liverpool, UK

F. Capello, A.V. Gaddi (eds.), *Clinical Handbook of Air Pollution-Related Diseases*,
https://doi.org/10.1007/978-3-319-62731-1_36

environmental policies lead to exponential growth of the sources of pollution with scarce limitations on the emission of toxins and inadequate implementation of measures to protect the population.

A possible solution requires multilevel action and application of rules and policies that could act on different sides of the problem to reduce, on the one hand, the level of emissions, and, on the other hand, the exposure of individuals and populations. A coordinated approach is therefore needed to act on the different strategies that could help to achieve the goal:

(a) Health education and health communication
(b) Institution at local, national, and international levels of coordinated and competent bodies
(c) Global and national research networks
(d) Implementation of worldwide antipollution policies
(e) Implementation of worldwide policies for prevention
(f) Reduction of the pollution divide, leveling the inequalities among different countries and among different areas within countries

36.1 Health Education and Communication

Although the health risks secondary to air pollution are known, the level of medical literacy is still poor and uneven. The environmental question is surrounded by myths and urban legends, while knowledge of the actual facts is inadequate in the general population. This brings about the development of unrealistic theories and beliefs, which may overestimate some unimportant aspects of the problem, while, at the same time, minimizing the real concerns.

The advent of social networking, while simplifying the spread of scientific knowledge, has amplified untruthful and untrustworthy sources of information because of the very low threshold that allows any user to publish and disseminate content and unreliable data, potentially reaching a global population in real time.

This leads to the spread of dangerous information, which could undermine the whole process of prevention and implementation of adequate policies for the environment.

Consequently, it is imperative that health education campaigns, aimed at increasing awareness in the population, focus on the real aspects of the problem and the possible solutions. Government agencies and institutions have to take into account people's opinions, but they have a duty to provide correct education that can allow people to understand the problem and demand appropriate actions.

For these reasons, a global approach is needed, starting from schools upward and including the involvement of media, industries, and institutions; health education programs should be implemented starting from school age through to higher education.

The role of the school in this sense is crucial and is the fundamental key. Schools must educate children in complexity to avoid the spread of take-away information. Universities should promote individual research and in-depth theoretical knowledge together with the practical aspect of their teaching. Scientific education, in fact, cannot be formed by lists of facts or ready-to-take-home messages. On the other side, media and institutions need to promote diffusion of knowledge and communication of information in adequate terms; people should be able to understand complex facts of science and use their own minds and judgment to make reasoned decisions.

Governments need to define competencies and assign expert to key positions, avoiding their choices being guided by the interests of lobbies and political parties.

Scientific institutions must learn how to communicate through mediation by professionals to reach out to populations, industries, institutions, governments, and, finally, schools and universities.

The same scientific knowledge has to become part of human heritage and, for this reason, it must become accessible and open to the whole world population, promoting innovation and original research, and opposing the logic of dogmatic science and the mainstream.

At the very least, properly educated medical doctors and health workers should inform the people they care for, helping them to take proper action and make decisions aimed at reducing their risk of exposure and protecting themselves and the ones they care for.

36.2 Coordinated Actions at the Local, National, and International Levels

Coordinated and continuous action is needed to keep the guard high on the matter of air pollution. Air pollution is a complex matter because of the numerous sources of emissions and the role that each one of them has in the economy and in society (e.g., energy production, industry, husbandry, transportation, and public and private heating). Any measure aimed at reducing production of pollutants can have consequences for other aspects of social organization, which may ultimately lead to an increase in pollution in other sectors.

Environmental administration requires knowledge and coordination.

A global effort is needed to reduce the risk of implementing wrong policies forced by the pressure of badly informed public opinion, the interests of lobbies or companies, and the policies of single countries to the detriment of others.

A logical and methodological scale of priorities needs to be defined, as well as scales of ethical, health, economic, political, demographic, and geographic values.

In terms of health and clinical policies, the idea of prevention is paramount: it implies reduction of exposure and implementation of behavior aimed at protecting individuals and communities, and improving the quality of indoor and outdoor air in urban and rural settings.

Therefore, specifically designed and coordinated bodies and institutions must be created, working in strict collaboration with all stakeholders involved, including universities and consumer organizations. They would have the role to gather information, create priorities and strategies, and communicate with each other at the national and international levels, promoting research and the spread of knowledge.

36.3 Global and National Research Networks

Research has a key role in the definition of a coordinated strategy to fight air pollution and, consequently, its effects on human health. An international network should be created so that researchers can communicate and work in collaboration, filling the gaps and improving the quality of the scientific evidence that is produced. A whole new approach may be needed, in contrast to the idea of the mainstream, which could push research institutions and universities to amplify known notions, killing original research and therefore the production of innovative knowledge.

A global network should also be able to overcome the pressure that comes from lobbies and political parties, which may ask for specific results in order to promote policies based on personal or economic interests rather than on scientific models.

The knowledge that is produced must be accessible to the whole population, and research institutions need to communicate with the public, disseminating knowledge and promoting outreach activities.

Because of the complexity of this matter, and the possible economic and ethical implications, universities and organizations must be independent, promoting free research and free information exchange and sharing.

36.4 Implementation of Worldwide Antipollution Policies

Reduction of emissions into the atmosphere is a key point to reduce or eliminate the risk of air pollution. As we have seen, it is not possible to contain the production of pollutants that act only on single elements of the chain of production without affecting the others. Environmental policies therefore need to be coordinated so the whole system can be improved. A global approach need to be fostered, and people must become aware of the risk of pollution and the measures that should be taken even when they affect everyday activities such as using a car or buying an item whose production create pollution.

Governments may make unpopular decisions or implement policies that act against the interests of groups of people or organizations. This is a burden that single countries cannot bear by themselves, and actions have to be planned in agreement with all parties to reduce the adverse effects of necessary decisions.

World and permanent conferences involving governments and nongovernmental bodies and agencies may be a starting point, provided that all aspects of the problem are considered. International agencies should coordinate the work of institutions and private stakeholders, also helping decision makers to implement appropriate policies at the international, national, and local levels.

Industries must have their say but need to comply with regulations, and that may imply increasing the costs for the population; on the other hand, people must become aware that health and well-being have a cost themselves, and that anti–air pollution policies start from the decisions that people make in their everyday activities.

At the very least, free thinking must be always fostered, particularly when the outcomes of innovative solutions are not immediately clear. This also means that all parties can have their say and that policy makers must consider all aspects, including the long-term consequences of their choices. Popular decisions may reduce the production of air pollution in the short term but could create obstacles to research and industry in the production of knowledge that may subsequently yield less polluting solutions. For this reason, fundamental research should always be prioritized even if its practical applications may not be readily evident or immediately usable.

36.5 Implementation of Worldwide Prevention Policies

Although, in the long term, air pollution affects the human economy and societal development, the most evident effects are on human health. We have seen the many consequences and the harmful effect on organs and systems. For most diseases related to air pollution, there is no treatment, and chronic changes result in lifelong disorders or continuous relapses of acute conditions, which negatively impact the quality of life of people and their families.

The role of prevention consequently becomes clear, aimed at minimizing exposure to pollutants for populations and individuals, consequently reducing the risk of developing diseases. To achieve this goal, international coordination is needed, defining plans for intervention at the transnational and local levels, and also considering the peculiarities of fast-developing countries and of low-income and rural settings.

The crucial steps can be defined as follows:

1. *Definition of strategies for pre-primary prevention,* reducing the production of emissions and air pollution (see also the previous paragraph): This implies definition of reliable thresholds that should never be exceeded and a set of regulations that industries need to comply with; it also means that the population must become aware of what the consequences of these policies are, how they may impact the production and distribution of energy and goods, and what the risks are if—for various reasons—they cannot be achieved.
2. *Definition of strategies for primary prevention,* reducing or preventing the risk of exposure: This means implementation of policies aimed at reducing the chance of exposure for communities and individuals, promotion of behavioral changes aimed at avoiding and protecting against exposure, and reducing the concentration of pollutants in the environment where people live or spend most of their time. It also requires proper communication of health information and properly designed programs of health education to avoid dissemination of confounding messages or promotion of behaviors suggesting that even reducing exposure to air pollutants could be otherwise unhealthy.
3. *Definition of strategies for secondary prevention,* aimed at attaining early diagnosis of all diseases that are directly or indirectly related to, triggered, promoted, or caused by air pollution.
4. *Definition of strategies for tertiary prevention* to reduce the burden for those individuals who are already affected by an air pollution–related disease and for

those who have developed a chronic condition secondary to early-life or long-term exposure to pollutants, even when the effect is only partially related to or caused by air pollution.

5. *Definition of strategies to reduce the global burden* related to air pollution–related diseases for individuals, communities, and societies.
6. *Definition of strategies to reduce mortality* related to air pollution–related diseases in developed and developing countries; in high-, middle-, and low-income settings; and in urban, semirural, rural, and extreme rural settings.

36.6 Reducing the Pollution Divide and Leveling Inequalities Among and Within Countries

Sociodemographic factors can lead to different levels of exposure for selected groups of people. This is clear when the differences between high- and low-income counties are considered, and also because in these different scenarios, people are exposed to different types of indoor and outdoor air pollution. On the other hand, social and demographic differences are present also at the national level, especially—but not limited to—those countries and regions with a large territorial extension, a high level of immigration, or large metropolitan areas in contrast to large rural extensions (e.g., the USA, the European Union, and China).

In addition, several studies have underlined differences in terms of health outcomes within the same urban area, highlighting the fact that in those places where clusters of disadvantaged people and families live, the effects of air pollution on health are greater [1–4].

Age, race, and the level of education also affect exposure to air pollutants and health outcomes [5], underlining the fact that inequities represent a risk factor for development of air pollution–related diseases.

Implementation of policies aimed at reducing the effects on human health and promoting well-being to improve the quality of life of communities and individuals must take into account the pollution divide; everyone, in fact, should be able to access information aimed at reducing the risk, and take actions that can minimize exposure to airborne toxins. At the same time, people have the right to live in a clean environment regardless of their age, social status, ethnicity, or level of education.

Conclusions

Air pollution is a clear and present danger, and proper actions need to be taken promptly to avoid a worldwide catastrophe, which is reaching a point of no return. The effects on human health are partially known, but the actual burden is probably underestimated. Joint action by governments, international organizations, and research institutions is needed to better understand the complexity of air pollution, the difficulties involved in establishing its real impact on human societies, the peculiarities of the different components and how they are produced and spread in the atmosphere, and how they interact with the environment, as well as with human tissues, organs, and systems.

The real key to overcoming this hazard is knowledge, and building and accessing that knowledge is a priority. This implies that schools and universities must educate people about complexity and that all ways of research must remain open.

People need to be empowered so they will become able to understand the fine implications of this invisible enemy, which has the capacity to negatively shape human existence. That is a due process; no policy or decision can change the course of events if people are not directly involved. Besides, changes in behavior at the community and individual levels may be enough to reduce exposure to air pollutants and, consequently, their acute and chronic effects on health.

Finally, cooperation among all stakeholders at the international, national, and local levels is essential to address the issue, so that everyone can become able to access information, prevention, and decisions to protect themselves from pollution, regardless of their origins, status, education, or age.

References

1. Pinault L, Crouse D, Jerrett M, Brauer M, Tjepkema M. Socioeconomic differences in nitrogen dioxide ambient air pollution exposure among children in the three largest Canadian cities. Health Rep. 2016;27(7):3–9.
2. O'Lenick CR, Winquist A, Mulholland JA, Friberg MD, Chang HH, Kramer MR, et al. Assessment of neighbourhood-level socioeconomic status as a modifier of air pollution–asthma associations among children in Atlanta. J Epidemiol Community Health. 2016;71:129.
3. Lee JT, Son JY, Kim H, Kim SY. Effect of air pollution on asthma-related hospital admissions for children by socioeconomic status associated with area of residence. Arch Environ Occup. 2006;61(3):123–30.
4. Qian Z, Chapman RS, Hu W, Wei F, Korn LR, Zhang JJ. Using air pollution based community clusters to explore air pollution health effects in children. Environ. 2004;30(5):611–20.
5. Zou B, Peng F, Wan N, Mamady K, Wilson GJ. Spatial cluster detection of air pollution exposure inequities across the United States. PLoS. 2014;9(3):e91917.

Conclusions

37

Fabio Capello and Antonio Vittorino Gaddi

Air pollution is a global concern. As far as we know, the planet Earth is the only celestial body where life has developed and evolved. This is due to its unique peculiarities in our system among others: the right temperature, the presence of liquid water, the presence of a magnetic field, and, above all, the presence of an atmosphere composed of gases that allow life, as we know it, to survive. Other possible environments and life-forms may exist in the universe, whose physiology and biochemistry may be completely alien and irrational from our perspective. Nonetheless, the air that we breathe is the environment where the human race has bred, and it is what keeps us alive.

Over the eras, climate changes and different compositions of the atmosphere have selected, nurtured, and extinguished different species and forms of life; through a process of natural selection, those organisms able to adapt to the changes in the environmental conditions survived and evolved. It takes eons to create and destroy ecosystems and the fine balance that regulates them. Also, when catastrophes have happened—such as the one that likely caused the extinction of the dinosaurs on our planet—life has survived and found its way.

This is where we come from.

Fine changes in the composition of the air we respire cannot be free of consequences. The changes in society and the development of a technologically based civilization, with the rapid changes this brings with it, are affecting the environment at different scales. Although the actual impact of human emissions—and how those

F. Capello (✉)
Alder Hey Children's Hospital, Liverpool, UK
e-mail: info@fabiocapello.net

A.V. Gaddi
Scientific Director of Health Lab of GTechnology Foundation, Modena, Italy

F. Capello, A.V. Gaddi (eds.), *Clinical Handbook of Air Pollution-Related Diseases*,
https://doi.org/10.1007/978-3-319-62731-1_37

affect the composition of the atmosphere as a whole—remain unclear and controversial, it is clear and undisputable that in urban and high-density areas, the spread of airborne pollutants is a fact.

We are therefore facing the rise of a new era, or a new epochal change; in the past, there have been similar occurrences when environmental, technological, ecological, and epidemiological conditions changed: famines, wars, pandemics and plagues, and, recently—especially after the development of new models of husbandry or the discovery of antibiotics—the epidemics of cardiovascular disease, noncommunicable diseases, and, generally speaking, diseases of affluence.

Yet, unlike climate change—which is another major factor for evolution or extinction of races and species—air pollution works on a different timescale and is likely to speed up its pace, accordingly to geographical and demographical changes.

For this reason, humans may not have time to adapt themselves to the new conditions; certainly, there will not be space for a Darwinian selection. Or, otherwise, the interaction with the many components of air pollution may cause epigenetic changes or adaptive processes, which, however, are difficult to accept according to what we know of evolution, development, and selection.

These may appear to be mere speculations, but the rise and fall of ecosystems and, later, of civilizations is a leitmotiv in the history of our planet.

37.1 Definition of Air Pollution–Related Diseases

The effects of air pollution on human health have been studied for decades, but we have only a partial and limited picture of the real extension of the problem. Air pollution is a compound of airborne chemicals, particulates of different sizes, organic molecules, and biologically active or inert matter; every single component can lead to consequences, alone or in synergy with the others. Although inhalation is the main form of assimilation, toxins can enter the human body through other ways such as ingestion or skin absorption. Different components can have direct or indirect effects, alone or in combination with other toxins or microorganisms, interacting with humans cells, tissues, organs, and systems; impeding normal physiological or biochemical pathways; interfering with passive or active protections of the organism; triggering abnormal responses; physically damaging structures at the macro, micro, and molecular levels; facilitating the action of other pathological triggers or expression of genetic conditions; and preventing normal development, multiplication, and maturation of cells, tissues, organs, and structures. Different sources, concentrations, times, and modes of exposure are linked to different effects; people subject to air pollution have unique susceptibility and vulnerability. The effects can show themselves immediately with a clear cause-and-effect relationship or after days, weeks, months, or years, so that a clear causal connection cannot be established.

This noncomprehensive picture illustrates the complexity of this subject and why we are still far from a complete mapping of the damage to human health that comes from air pollution.

Most of the studies have focused on direct effects of air pollution on the respiratory tract; nevertheless, it is becoming clearer that the mixture of components in air pollution has effects on different organs and systems, causing acute, chronic, and permanent damages at all ages.

For these reasons, it is crucial to identify a new branch of medicine, introducing the idea of air pollution–related diseases, defined as all diseases and acute, subacute, chronic, and permanent medical conditions that are directly or indirectly related to, triggered, promoted, facilitated, or caused by air pollution at the macro, micro, genetic, and molecular levels, regardless of the type of exposure or the timing of the exposure.

This definition clearly includes a wide range of pathological manifestations, which cross the boundaries of different disciplines in medicine, biology, and research. Yet the demarcation of boundaries is a primary bias and one of the strongest limitations in research. For the same reason, the concept of different pathologies, grossly clustered under the same blurred dome but linked by a common root, is essential.

This view opens new road to research, aimed at gathering knowledge from the different fields of fundamental science and human science.

The clinical implications may be not so clear, as the clinical presentation of diseases linked to air pollution may not differ from what we already know about similar conditions. Yet, the definition of individual risk and of tailored plans for prevention for individual patients are some of the practical consequences of this new field of medicine. As this is a relatively young discipline, the treatment options aimed at reducing or directly curing those diseases caused by air pollution are still limited; however, knowledge of the fine physiopathological mechanisms that regulate the development of pathologies secondary to exposure to air pollution, and elaboration of models that could predict future damage to tissues and organs, may open new roads. The final goal is discovery of new medicines, new nonpharmaceutical therapies, and new strategies for prevention.

At the same time, it is critical to develop realizable diagnostic tests that could help to detect those subjects who are at higher risk or who have already been exposed to doses and types of pollutants that are known to cause, trigger, or facilitate disease.

Then again, inclusion of air pollution–related diseases in the medical classification of the diagnosis is paramount in order to create plans for intervention in public health settings. Primary, secondary, and tertiary prevention are crucial to reduce the extent of the problem and to promote health and well-being for communities and individuals.

This is particularly true for those categories of people who may be more affected, such as children and elderly people. Children, in particular, are more vulnerable than adults because of nonmodifiable physiological and behavioral characteristics; because the effects of air pollution have time to show themselves years after the exposure; and because chronically exposed children are subject to the effects of air pollutants for more time than adults and elderly people are.

On the other hand, people from different backgrounds can be subjected to damage or air pollution regardless of their origin, race, social status, education, or age.

Consequently, medical doctors must become familiar with the idea of air pollution–related diseases, and this discipline should be included in the core curricula of medical schools worldwide, as well as in specialist training courses and in postgraduate education.

37.2 The Extent of the Problem

Air pollution is not a plague of our time. It has been known since the time of the first industrial revolution, and its effect on human health has been suspected since the earliest stages. However, the demographic expansion and global technological development; the fast urbanization in developing countries or in those countries with strong discrepancies between rural and urban or metropolitan areas; and the use of unregulated and outdated polluting technologies and means of transportation are contributing today to the explosion of the problem and its consequences.

China, in this sense, can be considered a reliable model, with a very high rate of industrialization and urbanization, and a level of pollution that—at particular times of the year and in selected areas—reaches a level barely compatible with life. This does not mean that the awareness of the people who are exposed grows at the same pace, although the morbidity and mortality in those regions secondary to the effects of air pollution are rising.

Pollution, in fact, is often an invisible enemy, which creates concern only when—as on very smoggy days—its presence is unequivocally evident or when it has immediate consequences for human health. This is an old lesson that should have already been learned: first- and second-hand tobacco smoke—by chance, a major source of indoor pollution—shows its deadly effects after years of active or passive smoking.

The critical point is that people may be constantly exposed to air pollution even when they think they are safe; that can happen at home and in residential areas, while traveling or commuting, at school or in the workplace, or in all the other locations where people go for leisure, for prayer, or for any other reason.

The burden for individuals and societies is becoming too heavy to bear, and immediate actions are needed to reduce the amount of pollutants emitted into the atmosphere locally and globally, and to protect the environment—and, ultimately, people—from their effects.

37.3 Solutions

Improving the quality of the air that people breathe every day is vital; it requires coordinated action at the local, national, and international levels, with implementation of shared policies of environmental administration. Because any solution can cause unknown consequences, the role of research is critical. A network of universities, organizations, governments, and nongovernmental agencies and institutions should be promoted, together with sharing of information and ideas.

Education remain a key factor, in terms of improvement of scientific and medical literacy, and development of curricula for doctors, health workers, scientists, engineers, researchers, and policy makers. It also means that all the public and private stakeholders must be involved and should have their say.

Medical doctors must become aware of the problem and should always investigate their patients to rule out or detect possible air pollution–related diseases.

Plans for primary, secondary, and tertiary prevention at the community and individual levels have to be implemented, while individual actions aimed at reduction of emissions, detection and monitoring of exposure, protection against air pollution, and cleaning up of the environment should be encouraged.

The most vulnerable groups should be protected—particularly children and people from disadvantaged backgrounds—to limit the pollution divide and globally improve the quality of life.

The cost effectiveness of similar interventions has to be wisely evaluated, but measures aimed at reducing exposure to air pollution and its effects on the environment, on people's health, and on their quality of life should always be prioritized, even when they lead to implementation of unpopular measures or social sacrifices.

At the very least, a global effort and organization is needed to improve the effects of action against pollution. This implies implementation of coordinated and nonchaotic measures by everyone involved, so they can act in synergy and not in opposition. The ultimate risk, in fact, is the creation of a Tower of Babel, where every actor metaphorically speaks a different language. This remains a major challenge also because of the different interests at stake.

Even so, humanity has the force and the means to overcome this deadly threat that—like the mythical Sword of Damocles—hangs over the heads of us all. Division and confusion could yet sign our fate, whereas sharing and coordination could lead to salvation.

The final goals are improvement of the level of education, health promotion, data sharing, research, coordination, and planning; implementation of environmentally friendly policies, monitoring and data analysis processes, and community-based and individual protection measures; and, ultimately, enhancement of our quality of life and the quality of the air we breathe.

Glossary

Absorption A physical or chemical phenomenon or a process in which atoms, molecules, or ions enter some bulk phase—gas, liquid, or solid material.

Adaptive Immune System A subsystem of the overall immune system that is composed of highly specialized, systemic cells and processes that eliminate or prevent pathogen growth.

ADME Absorption, distribution, metabolism, and excretion.

AhR "Aryl hydrocarbon receptor." The aryl hydrocarbon receptor, a protein encoded in humans by the AHR gene, is a ligand-activated transcription factor implicated in the regulation of biological responses to aromatic hydrocarbons.

AIDS "Acquired immune deficiency syndrome." The acquired immune deficiency syndrome is a multiplicity of conditions caused by contagion with human immunodeficiency virus (HIV).

Air Filter A device composed of fibrous materials which removes solid particulates such as dust, pollen, mould, and bacteria from the air.

Air Pollutants Particulates or biological molecules that cause diseases, allergies, death to humans, and damage to other living organisms or the natural or built environment.

Airborne Carcinogens Potentially carcinogenic chemicals are sources involving combustion.

Allergy A number of conditions caused by hypersensitivity of the immune system to something in the environment that usually causes little or no problem in most people.

Alveoli Grouped together like a lot of interlinked caves, and this allows the gas exchange into lungs.

Ambient Nanoparticles Derive from gas to particle reactions in the atmosphere or are produced unintentionally in high temperature processes.

Apoptosis A process of programmed cell death that occurs in multicellular organisms.

APP "Amyloid precursor protein." It is a single-pass transmembrane protein expressed at high levels in the brain that have major roles in the regulation of several important cellular functions like synaptogenesis and synaptic plasticity.

Arachidonic Acid A polyunsaturated omega-6 fatty acid 20:4(ω-6).

F. Capello, A.V. Gaddi (eds.), *Clinical Handbook of Air Pollution-Related Diseases*,
https://doi.org/10.1007/978-3-319-62731-1

Arteriosclerosis The thickening, hardening, and loss of elasticity of the walls of arteries.

Asthma A common long-term inflammatory disease of the airways of the lungs characterized by variable and recurring symptoms, reversible airflow obstruction, and bronchospasm.

Atherosclerosis Specific form of arteriosclerosis in which an artery-wall thickens as a result of invasion and accumulation of white blood cells (WBCs) (foam cell) and proliferation of intimal-smooth-muscle cell creating a fibrofatty plaque.

Atmosphere Residence Time Is defined as the average time that an aerosol spends in the atmosphere as a mutual result of the processes of formation/production against those of removal.

Atmosphere A layer of gases rounding a planet or other material body, that is held in place by the gravity of that body.

ATSDR "Agency for Toxic Substances and Disease Registry." The Agency for Toxic Substances and Disease Registry (ATSDR), in Atlanta, is a federal public health agency of the U.S. Department of Health and Human Services.

Autophagy The natural, destructive mechanism that disassembles, through a regulated process, unnecessary or dysfunctional cellular components.

AZT "Azidothymidine." Azidothymidine is an antiretroviral drug for the treatment and prevention of HIV/AIDS.

Aβ "Amyloid-β peptide." About 36-43 amino acids that are crucially involved in Alzheimer's disease as the main component of the amyloid plaques found in the brain of Alzheimer patients.

B Cells A type of white blood cell of the lymphocyte subtype that secreting antibodies play function in the humoral immunity component of the adaptive immune system.

B(e)P "Benzo(e)pyrene." It is a toxic element contained in cigarette smoke.

Bacteria A large domain of prokaryotic microorganisms.

Benzene Organic chemical compound with the chemical formula C_6H_6.

Benzo[*a*]pyrene Isomer of benzopyrene, a polycyclic aromatic hydrocarbons.

Biomarkers Or biological marker, generally refers to a measurable indicator of some biological state or condition.

BPA "Bisphenol A." It is an organic synthetic compound, belonging to the group of diphenylmethane derivatives and bisphenols, characterized by two hydroxyphenyl groups.

Bradikynsia Slowness of movement. It is one of the cardinal manifestations of Parkinson's disease often associated with an impaired ability to adjust the body's position.

Brushfield Spots A small white or grayish/brown slightly elevated spots on the periphery of the iris in the human eye due to aggregation of connective tissue. These spots occur in normal children but are far more frequent in Down's syndrome (trisomy 21).

BUT "Break-up time." It is the most important test for diagnosing evaporative dry eye.

BVOC "Biogenic Volatile Organic Compounds." It is emitted by mainly by vegetation, plays a significant role in the generation of SOA.

Cancer A group of diseases involving abnormal cell growth with the potential to invade or spread to other parts of the body.

Car "Constitutive androstane receptor." Constitutive androstane receptor or nuclear receptor subfamily 1, acts as a sensor of endobiotic and xenobiotic substances. It is a human protein encoded by the NR1I3 gene.

Carcinogen Any substance, radionuclide, or radiation that is an agent directly involved in causing cancer.

Cardiac Hipetrophy A response of myocardium to various physiologic and pathologic stimuli including mechanical and hormonal. The common causes are high blood pressure (hypertension) and heart valve stenosis resulting in the thickening of myocardium, in turn decreasing the size of chambers and ventricles of heart.

Carotid Atherosclerosis A condition of building-up of plaque inside carotid artery walls resulting in narrowing and reducing blood flow to brain.

Cataracts A clouding of the lens in the eye leading to a decrease in vision. It can affect one or both eyes.

Cellular Pathways Part of a complex system of communication that governs basic activities of cells and coordinates cell actions.

Chromosomal Aberrations A missing, extra, or irregular portion of chromosomal DNA.

Chromosome A packaged and organized structure containing most of the DNA of a living organism.

Chronic Condition A human health condition or disease that is persistent or otherwise long-lasting in its effects or a disease that comes with time.

Climate Change A change in the statistical distribution of weather patterns when that change lasts for an extended period of time.

CNS "Clinical nurse specialist." It is a healthcare worker who can provide expert advice related to specific conditions or treatment pathways.

CO "Carbon monoxide." It is a colorless, odorless, and tasteless gas that is slightly less dense than air.

Coarse Particles Derive from mechanical processes such as solid abrasion, wind erosion of rocks and soil, organic debris, volcanic emission, and sea-salt aerosol.

Coarse Particulate Matter Particulate matter present in air is divided into different categories depending on the size of the particles.

Concentration The abundance of a constituent divided by the total volume of a mixture.

COPD "Chronic Obstructive Pulmonary Disease." It is progressive lung diseases including emphysema, chronic bronchitis, refractory (non-reversible) asthma, and some forms of bronchiectasis.

CP "Cerebral palsy." It is a group of neurological disorders that appear in infancy or early childhood which permanently affect body movement, muscle coordination, and balance. CP affects the part of the brain that controls muscle movements.

CS2 "Carbon disulfide." It is a colorless, toxic, highly volatile and flammable liquid chemical compound.

CYP "Cytochromes P450." Cytochromes P450 are a family of hemoproteins characterized by the heme as a cofactor, well-known as terminal oxidase enzymes in electron transfer chains.

DED "Dry Eye Disease." It is the most frequent disorder in ophthalmological practice.

Dendritic Cells Antigen-presenting cells of the mammalian immune system.

Determinants of Health Factors which influence health status and determine health differentials or health inequalities.

DEWS "Dry Eye WorkShop." It is the most detailed scientific report about dry eye disease.

Diabetes A group of metabolic diseases in which there are high blood sugar levels over a prolonged period.

Distribution The movement of a drug from one location to another within the body.

DNA A molecule that carries the genetic instructions used in the growth, development, functioning and reproduction of all known living organisms and many viruses.

DNA "Deoxyribonucleic acid." Deoxyribonucleic acid is a molecule that contains the genetic heritage used in the growth, development, functioning, and reproducing of living organisms.

DOHaD "Developmental origins of health and disease." A new scientific branch exploring the influence of adverse events that occur during early phases of human development (particularly exposure to environmental pollution), on the health of the newborn and throughout life (including the development of chronic/not communicable diseases).

Drugs Any substance other than food, that when inhaled, injected, smoked, consumed, absorbed via a patch on the skin or dissolved under the tongue causes a physiological change in the body.

Eco-Friendly Marketing and sustainability terms referring to goods and services, laws, guidelines, and policies that inflict reduced, minimal, or no harm upon ecosystems or the environment.

Ecosystem A community of living organisms in conjunction with the nonliving components of their environment (things like air, water, and mineral soil), interacting as a system.

EDC "Endocrine-disrupting chemical." Endocrine disruptors are chemicals that at particular doses can interact with endocrine systems causing cancer, birth defects, and other growing disorders.

EDCs "Endocrine Disrupting Chemicals." Chemical substances of various origin mainly introduced by ingestion, inhalation, or dermal contact and able to interfere at different levels with hormonal systems. During fetal life EDCs, may act as (pseudo)morphogens, interfering with fetal programming.

EDED "Environmental Dry Eye Disease." It is a subtype of dry eye disease caused by environmental factors.

Epidemiology The study and analysis of the patterns, causes, and effects of health and disease conditions in defined populations.

Epigenetics The study of cellular and physiological phenotypic trait variations that result from external or environmental factors that switch genes on and off and affect how cells express genes.

Epithelial Cells Cells of epithelium, one of the four basic types of animal tissue.

ETS "Environmental tobacco smoke."

Evidence Base The evidence base refers to a body of information, drawn from routine statistical analyses, published studies and "grey" literature, which tells us something about what is already known about factors affecting health.

Excretion The process by which metabolic wastes and other non-useful materials are eliminated from an organism. In vertebrates this is primarily carried out by the lungs, kidneys, and skin.

Exposure Assessments A branch of environmental science and occupational hygiene that focuses on the processes that take place at the interface between the environment containing the contaminant(s) of interest and the organism(s) being considered.

Extracellular Matrix A collection of extracellular molecules secreted by cells that provides structural and biochemical support to the surrounding cells.

Fetal Programming The fetal environment is influenced by several factors including the intake of nutrients and toxic compounds by the mother. The genome as such is habitually not directly changed by these factors, which could induce the "epigenetic software" related to the single genes to change in an adaptive and predictive way. This could have lasting effects on the programming of metabolic functions in the fetus and even in the next generations.

FMO "Flavin-containing monooxygenase." The flavin-containing monooxygenase is a protein family specialized in the oxidation of xenosubstrates with the aim of helping their excretion from living organisms.

Formaldehyde A naturally occurring organic compound with the formula CH_2O.

Gas A state of matter distinguished by relatively low density and viscosity, relatively great expansion and contraction with changes in pressure and temperature.

Glycemia The presence, or the level, of glucose in one's blood.

GSH "Glutathione." The tripeptide glutathione is able to prevent damage to cellular components produced by reactive oxygen species. It is a significant antioxidant in animals, plants, fungi, and some bacteria.

Health The level of functional or metabolic efficiency of a living organism.

HSA Human serum albumin. Human serum albumin, produced in the liver, is the most abundant protein in human blood plasma.

Hyperglycemia A condition in which an excessive amount of glucose circulates in the blood plasma.

Hypoglycemia When blood sugar decreases to below normal levels.

Immune System A host defense system comprising many biological structures and processes within an organism that protects against disease.

Immunity The balanced state of having adequate biological defenses to fight infection, disease, or other unwanted biological invasion, while having adequate tolerance to avoid allergy, and autoimmune diseases.

Indoor Air Constitutes an increasing matter of concern covering both the field of residential environments together with hospitals and other working environment.

Inflammation Part of the complex biological response of body tissues to harmful stimuli, such as pathogens, damaged cells, or irritants, and is a protective response involving immune cells, blood vessels, and molecular mediators.

Innate Immune System An important subsystem of the overall immune system that comprises the cells and mechanisms that defend the host from infection by other organisms

Insulin A peptide hormone produced by beta cells of the pancreatic islets.

Integrator Measurement and control applications is an element whose output signal is the time integral of its input signal.

In-Transit Is a term which refers to the particles characterizing the transportation condition for the considered individual.

Led "Light-emitting diode." It is a semiconductor device that emits visible light when an electric current passes through it.

Lipid A group of naturally occurring molecules that include fats, waxes, sterols, fat-soluble vitamins (such as vitamins A, D, E, and K), monoglycerides, diglycerides, triglycerides, phospholipids, and others.

Lymphocytes One of the subtypes of white blood cell in a vertebrate's immune system. Natural killer cells, T cells and B cells are included.

Macrophages A type of white blood cell that engulfs and digests cellular debris, foreign substances, microbes, cancer cells, and anything else that does not have the types of proteins specific of healthy body cells on its surface in a process called phagocytosis.

Mask An object normally worn on the face, typically for protection.

MDR1 Multidrug resistance protein. Multidrug resistance protein 1 is an essential ATP-dependent efflux pump protein of the cell membrane it pumps substances out of cells, mediating the efflux of drugs from cells.

Mediators The natural, destructive mechanism that disassembles, through a regulated process, unnecessary or dysfunctional cellular components.

Metabolism The set of life-sustaining chemical transformations within the cells of living organisms.

Metabolites The intermediates and products of metabolism.

Metal Poisoning The toxic effect of certain metals in certain forms and doses on life.

Nervous System The part of an animal's body that coordinates its voluntary and involuntary actions and transmits signals to and from different parts of its body.

Neutrophils The most abundant type of granulocytes and the most abundant (40% to 75%) type of white blood cells that form an essential part of the innate immune system.

Nitrogen Dioxide the chemical compound with the formula NO2 that is an intermediate in the industrial synthesis of nitric acid.

NK Cells A type of cytotoxic lymphocyte critical to the innate immune system.

NO_2 "Nitrogen dioxide." It is the chemical compound with the formula NO2.

NO_x "Nitrogen oxide." It is a molecular, chemical compound produced during combustion, especially at high temperature.

O_2 "Oxygen." It is the third most abundant element in the Universe after hydrogen and helium.

O_3 "Ozone." Ozone or trioxygen is an inorganic molecule with the chemical formula O3.

Oat "Organic anion transporter." Organic anion transporter, a member of the organic anion transporter family, is a human protein coded by the SLC22A6 gene.

Occupational Environments Is the place in which an individual is exposed to specific PM sources.

OR "Odds ratio." It is a measure of association between an exposure and an outcome.

OSDI Ocular Surface Disease Index is the validated questionnaire for scoring dry eye symptoms.

Outdoor Air Refers to the basic environmental compartment in which air pollution including PM is formed and emitted.

Oxidative Potential A measure of the tendency of a chemical species to acquire electrons and thereby be reduced.

Oxidative Stress Reflects an imbalance between the systemic manifestation of reactive oxygen species and a biological system's ability to readily detoxify the reactive intermediates or to repair the resulting damage.

PAHs "Polycyclic aromatic Hydrocarbons." Hydrocarbons—organic compounds containing only carbon and hydrogen—that are composed of multiple aromatic rings.

PAPS "3'-phosphoadenosine-5'-phosphosulfate." 3'-Phosphoadenosine-5'-phosphosulfate is the most common coenzyme in sulfotransferase reactions; it derives from 3-phosphorylated adenosine monophosphate with a sulfate group attached to the 5' phosphate.

P-gp "P-glycoprotein." P-glycoprotein is a transporter protein localized to the cellular membrane of drug-resistant cells.

PM "Particulate matter." Inhalable particles of different composition and size able to induce adverse health effects and produced by combustion processes (primary PM) or in the ambient air (secondary PM) following photo-chemical reactions between precursors.

PM10 Particulate matter with a diameter of particles <10 μm.

PM2.5 Particulate matter with a diameter of particles <2.5 μm.

Pneumonia An inflammatory condition of the lung affecting primarily the microscopic air sacs known as alveoli.

Pollutants A substance or energy introduced into the environment that has undesired effects, or adversely affects the usefulness of a resource.

POPs Persistent organic pollutants. Persistent organic pollutants are organic compounds resistant to environmental degradation through chemical, biological, and photolytic processes. They can bioaccumulate in the different species of the food chain.

PPAR Alpha "Peroxisomal proliferator activating receptor alpha." Peroxisome proliferator-activated receptor alpha is a nuclear receptor protein encoded in humans by the *PPARA* gene and involved in hepatic lipid metabolism.

Preeclampsia A disorder of pregnancy characterized by high blood pressure and a large amount of protein in the urine.

Prevention Activities that reduce the amount of pollution generated by a process.

Primary Aerosol Is constituted by particles which are emitted into the atmosphere as soon as they formed.

PXR "Pregnane X receptor." The pregnane X receptor is an important component of the body's adaptive defense mechanism against toxic substances including foreign chemicals.

Radiations The emission or transmission of energy in the form of waves or particles through space or through a material medium.

RB1 "Retinoblastoma 1 gene." It is a protein coding gene.

Respiratory System A biological system consisting of specific organs and structures used for the process of respiration in an organism.

Reumathic Disease An umbrella term for conditions causing chronic, often intermittent pain affecting the joints and/or connective tissue.

Risk The potential of gaining or losing something of value.

RNA "Ribonucleic acid." Ribonucleic acid is a chain of nucleotides often found in nature as a single-strand folded onto itself. It has the function of coding, decoding, regulating and expressing of genes.

ROS Reactive oxygen species. Reactive oxygen species are chemically extreme reactive organic species containing oxygen. Usually they derive from the physiological by-product of the metabolism of oxygen. They can create significant damage to cell structures.

SAM "S-adenosylmethionine." S-Adenosylmethionine is a compound found in almost every tissue and fluid in the body involved in the immune system, cell membranes' maintenance, and production and breakdown of chemicals, such as serotonin, melatonin, and dopamine.

Secondary Aerosol Are those produced following the emission of their gaseous precursors subjected to irreversible oxidation processes whose polar products are contextually condensed.

Smoke A collection of airborne solid and liquid particulates and gases emitted when a material undergoes combustion or pyrolysis, together with the quantity of air that is entrained or otherwise mixed into the mass.

SO_2 "Sulfur dioxide." It is a chemical compound.

Species One of the basic units of biological classification and a taxonomic rank.

Sulphur Dioxide (SO_2) Chemical compound. In a standard atmosphere, it is a toxic gas with a pungent, irritating smell.

Syndrome Metabolic A clustering of at least three of the five following medical conditions which are: abdominal (central) obesity (cf. TOFI), elevated blood pressure, elevated fasting plasma glucose, high serum triglycerides, and low high-density lipoprotein (HDL) level.

T Cells A type of lymphocyte (a subtype of white blood cell) that plays a central role in cell-mediated immunity.

TCE "Trichloroethylene." The halocarbon trichloroethylene is a common industrial solvent. In the past, it was also used as a volatile anesthetic and as an inhaled obstetrical analgesic. It is a non-flammable liquid characterized by a sweet smell.

Terpene A large and diverse class of organic compounds, produced by a variety of plants

TLRs "Toll Like Receptors." They are a class of proteins that play a key role in the innate immune system. They are single, membrane-spanning, that recognize structurally conserved molecules derived from microbes.

TNT "Trinitrotoluene." It is an organic solvent.

Transboundary Pollution Is related to the aim of quantitatively detecting the amount of atmospheric pollutants produced in a country with respect to aliquots produced by neighbor countries, adding to local sources through atmospheric transport.

Tubercolosis An infectious disease caused by the bacterium Mycobacterium tuberculosis (MTB).

UDP "Uridine diphosphate." Uridine diphosphate is an ester of pyrophosphoric acid with the nucleoside uridine. It is a key factor in glycogenesis.

UDPGA "UDP activated glucuronic acid." UDP-glucuronic acid is formed in the liver and composed of glucuronic acid linked via a glycosidic bond to uridine diphosphate. It is often involved in the phase II metabolism of lipophilic xeno- and endobiotics to facilitate the elimination of toxic species from the organism.

UFP "Ultrafine particles." They are particulate matter with a diameter of particles $<0.1 \mu m$.

UV "Ultraviolet light." It has shorter wavelengths than visible light.

VAT "Visceral Adipose Tissue." It is a hormonally active component of total body **fat**, which possesses unique biochemical characteristics that influence several normal and pathological processes in the human body.

Virus A small infectious agent that replicates only inside the living cells of other organisms.

Vitamin D Refers to a group of fat-soluble secosteroids responsible for increasing intestinal absorption of calcium, iron, magnesium, phosphate, and zinc.

WHO "World Health Organization." It is a specialized agency of the United Nations that is concerned with international public health.

ZDV "Zidovudine." Zidovudine is an antiretroviral drug also known as azidothymidine (AZT).

MIX
Papier aus verantwortungsvollen Quellen
Paper from responsible sources
FSC® C105338

If you have any concerns about our products,
you can contact us on
ProductSafety@springernature.com

In case Publisher is established outside the EU,
the EU authorized representative is:
Springer Nature Customer Service Center GmbH
Europaplatz 3, 69115 Heidelberg, Germany

Printed by Libri Plureos GmbH
in Hamburg, Germany